SEVENTEENTH TEXAS SYMPOSIUM ON RELATIVISTIC ASTROPHYSICS AND COSMOLOGY

ANNALS OF THE NEW YORK ACADEMY OF SCIENCES

Volume 759

SEVENTEENTH TEXAS SYMPOSIUM ON RELATIVISTIC ASTROPHYSICS AND COSMOLOGY

Edited by Hans Böhringer, Gregor E. Morfill, and Joachim E. Trümper

The New York Academy of Sciences
New York, New York
1995

∞ *The paper used in this publication meets the minimum requirements of American National Standard for Information Sciences-Permanence of Paper for Printed Library Materials, ANSI Z39.48-1984.*

Cover Art: *The X-ray sky over Munich according to the ROSAT all-sky survey. The field is roughly 75° x 50° wide and contains many well-known X-ray sources. The diameter of the "points" is proportional to the logarithm of the source flux; the colors contain information about the source spectra (blue: hard spectra; yellow: soft spectra). The image was produced for the conference poster by K. Dennerl and W. Voges.* Courtesy of Max-Planck-Institut für Extraterrestrische Physik.

Library of Congress Cataloging-in-Publication Data

Texas Symposium on Relativistic Astrophysics (17th: 1994: Munich, Germany).
Seventeenth Texas Symposium on Relativistic Astrophysics and Cosmology/edited by Hans Böhringer, Gregor E. Morfill, and Joachim E. Trümper.
p. cm. — (Annals of the New York Academy of Sciences, ISSN 0077-8923; v. 759).
Includes bibliographical references and index.
ISBN 0-89766-941-X (cloth: alk. paper). — ISBN 0-89766-942-8 (pbk.: alk. paper).
1. Relativistic astrophysics—Congresses. 2. Cosmology—Congresses. I. Bohringer, Hans. II. Morfill, G. E. III. Truemper, J. (Joachim), 1933-. IV. Title. V. Series.
Q11.N5 vol. 759
[QB462.65]
500 s—dc20
[523.01]

95-34165
CIP

JOS
Printed in the United States of America
ISBN 0-89766-941-X (cloth)
ISBN 0-89766-942-8 (paper)
ISSN 0077-8923

ANNALS OF THE NEW YORK ACADEMY OF SCIENCES

Volume 759
September 30, 1995

SEVENTEENTH TEXAS SYMPOSIUM ON RELATIVISTIC ASTROPHYSICS AND COSMOLOGY[a]

Editors
HANS BÖHRINGER, GREGOR E. MORFILL, and JOACHIM E. TRÜMPER

Local Organizing Committee
H. BÖHRINGER, W. BRINKMANN, G. DIRNBERGER, H. HIPPMANN, M. IHLE, R. LANGE, G. E. MORFILL, E. RIEGER, H. SCHEINGRABER, V. SCHÖNFELDER, J. E. TRÜMPER, W. VOGES, and G. WIEDENMANN

Scientific Organizing Committee
G. E. MORFILL, J. E. TRÜMPER, I. APPENZELLER, J. AUDOUZE, J. N. BAHCALL, J. EHLERS, G. G. FAZIO, C. E. FICHTEL, R. GIACCONI, W. HILLEBRANDT, J. R. JOKIPII, T. KIRSTEN, R. P. KUDRITZKI, R. NARAYAN, K. NOMOTO, F. PACINI, M. J. REES, B. SADOULET, B. SCHUTZ, and J. SILK

Texas Symposium International Organizing Committee
J. AUDOUZE, J. D. BARROW, P. BERGMANN, A. CAMERON, J. EHLERS, L. Z. FANG, E. J. FENYVES, R. GIACCONI, J. C. JONES, M. LIVIO, L. MESTEL, L. MOTZ, Y. NEEMAN, I. OZSVATH, R. RAMATY, I. ROBINSON, R. RUFFINI, B. SADOULET, A. SALAM, D. SCHRAMM, E. L. SCHUCKING, G. SETTI, M. M. SHAPIRO, G. SHAVIV, L. C. SHEPLEY, J. J. STACHEL, A. TRAUTMAN, V. TRIMBLE, S. WEINBERG, J. A. WHEELER, and J. C. WHEELER

CONTENTS

Preface. *By* JOACHIM E. TRÜMPER and GREGOR E. MORFILL xv

[a] The papers in this volume were presented at the Seventeenth Texas Symposium on Relativistic Astrophysics, which was held on December 11-17, 1994, in Munich, Germany.

Part I. Plenary Talks

Solar Neutrino Problems. *By* Till A. Kirsten 1
Superluminal Motions in Our Galaxy. *By* I. F. Mirabel and L. F. Rodriguez ..21
The Galactic Center. *By* R. Genzel, A. Eckart, and A. Krabbe 38
MACHOS and Dark Matter. *By* Roger Ferlet 56
Clusters of Galaxies. *By* Hans Böhringer 67
High Redshift Quasars. *By* P. A. Shaver 87
The Extragalactic X- and γ-Ray Backgrounds. *By* Giancarlo Setti 110
Gravitational Radiation. *By* Kip S. Thorne 127
Inflation: From Theory to Observation and Back. *By* Michael S. Turner 153
Particle Physics and Cosmology. *By* John Ellis 170

Part II. Highlights—Space-based Astronomy

HST Highlights: New Views of the Universe. *By* H. S. Stockman 188
HST Highlights: The Extragalactic Distance Scale. *By* Wendy L. Freedman ..192
Discovery of Explosion Fragments of the Vela Supernova. *By* Bernd Aschenbach 196
The X-Ray Background. *By* Günther Hasinger 200
An Overview of the ASCA Mission. *By* Yasuo Tanaka 206
ASCA Observations of Supernova Remnants. *By* Robert Petre and the *ASCA* Team 208
ASCA Results on Active Galactic Nuclei. *By* K. Mitsuda 213
ASCA Results on Clusters of Galaxies. *By* Takaya Ohashi 217
Recent Highlights of the Compton Gamma Ray Observatory. *By* Carl E. Fichtel 221
Highlights of the Energetic Gamma Ray Experiment Telescope. *By* Carl E. Fichtel 222
Highlights from the COMPTEL 1 to 30 MeV Sky Survey. *By* V. Schonfelder, K. Bennett, H. Bloemen, W. Collmar, R. Diehl, W. Hermsen, L. Kuiper, G. Lichti, M. McConnell, J. Ryan, A. Strong, and C. Winkler 226
BATSE Highlights. *By* Gerald J. Fishman 232
Highlights from OSSE on the Compton Observatory. *By* J. D. Kurfess 236

Part III. Pulsars and Neutron Stars

Detections of Neutron Stars in the Extreme Ultraviolet. *By* Stuart Bowyer ..241
Isolated Neutron Stars and Their Emission throughout the Electromagnetic Spectrum. *By* Patrizia A. Caraveo 246
ROSAT Observations of Neutron Stars. *By* Werner Becker 250

Plasma Configurations around Rotating, Magnetised Objects—Particle Trapping by the Force-free Surface. *By* K. O. Thielheim 257
Boundary Effects in the Pulsar Magnetosphere. *By* Estelle Asseo 261
Supernova Remnants vs. Pulsar Nebulae. *By* Wolfgang Kundt 265
Neutron Star Kicks and Multi-Dimensional Supernova Models. *By* H.-Thomas Janka and Ewald Müller 269
A Long Term Study of High Energy Gamma Ray Emission from the Vela, Geminga, and Crab Pulsars. *By* P. V. Ramanamurthy, D. L. Bertsch, C. E. Fichtel, G. Kanbach, D. A. Kniffen, H. A. Mayer-Hasselwander, P. L. Nolan, P. Sreekumar, and D. J. Thompson 275
CGRO/OSSE Observations of Pulsars: An Update. *By* M. P. Ulmer, P. C. Schroeder, J. M. Cordes, Matthew Bailes, and Jonathon Bell 279
Gamma-Ray Bursts and the Structure of the Galactic Halo. *By* Philipp Podsiadlowski, Martin J. Rees, and Malvin Ruderman 283
Accretion and Magnetic Field Decay in Neutron Stars Entering Binaries. *By* U. Geppert and V. Urpin 287
Anisotropic Cooling and Atmospheric Radiation of Neutron Stars with Strong Magnetic Field. *By* Yu. A. Shibanov, G. G. Pavlov, V. E. Zavlin, L. Qin, and S. Tsuruta 291
An Internal Friction Model for an Unusually Large Second Derivative of the Pulsar Rotation Rate. *By* N. Shibazaki and S. Hirano 295
Effects of Magnetic Fields on Neutron Star Thermal Evolution. *By* S. Tsuruta and L. Qin 299
Fast Pulsars, Strange Stars, and Strange Dwarfs. *By* Fridolin Weber and Norman K. Glendenning 303

Part IV. Binaries and Transients

BATSE Observations of GRS 1915 + 105 and GRO J1655 - 40. *By* W. S. Paciesas, K. J. Deal, B. A. Harmon, C. A. Wilson, S. N. Zhang, and G. J. Fishman 308
SIGMA Survey of the Galactic Bulge Region. *By* J. Paul, A. Goldwurm, M. Vargas, J. Ballet, J.-P. Roques, L. Bouchet, G. Vedrenne, P. Mandrou, R. Sunyaev, E. Churazov, M. Gilfanov, A. Finogenov, A. Vikhlinin, A. Dyachkov, N. Khavenson, and V. Kovtunenko 312
ASCA Observations of SS 433. *By* Nobuyuki Kawai 316
Globular Cluster X-Ray Sources in M31. *By* Rodrigo Supper 320
Observed Characteristics of Supersoft ROSAT Sources in the LMC and Other Galaxies. *By* Peter Kahabka 324

On the Nature, Population, and Implications of Luminous Supersoft X-Ray Sources. *By* R. Di Stefano and S. Rappaport 328
Evidence for Particle Acceleration and Nuclear Reactions around Compact Relativistic Objects. *By* Eduardo L. Martín 332
Evaporation of Companions in VLMXBs and Binary Millisecond Pulsars. *By* Jacob Shaham 336
Accretion Discs around Nonmagnetized Stars. *By* G. S. Bisnovatyi-Kogan 340
Inclination Effects in Z Sources? *By* Erik Kuulkers and Michiel van der Klis 344

Part V. Supernovae

SN Ia: Light Curves, Spectra, and H_0. *By* P. Höflich, C. Dominik, A. Khokhlov, E. Muller, and J. C. Wheeler 348
Chandrasekhar Mass Models for Type Ia Supernovae. *By* S. E. Woosley, D. Garcia-Senz, J. Niemeyer, W. Hillebrandt, S. Blinnikov, and P. Sasorov 352
Type II Supernovae: The Oldest New Kid on the Block. *By* Jason Spyromilio and Bruno Leibundgut 356
Type Ib/Ic/IIb/II-L Supernovae and Binary Star Evolution. *By* K. Nomoto, K. Iwamoto, H. Yamaoka, T. Suzuki, O. R. Pols, E. P. J. van den Heuvel, and P. Höflich 360
Convection in Type-II Supernovae: The First Second. *By* Ewald Müller and H.-Thomas Janka 368
The Physics of Core-Collapse Supernova Explosions. *By* Adam Burrows and John Hayes 375

Part VI. Gamma-Ray Lines

Introductory Remarks to the Gamma-Ray Line Astrophysics Symposium. *By* V. Schönfelder 382
Gamma-Ray Line Observations with CGRO-COMPTEL. *By* R. Diehl 384
Radioactivities Made in Supernovae. *By* S. E. Woosley, F. X. Timmes, R. D. Hoffman, D. H. Hartmann, P. A. Pinto, and T. A. Weaver 388
Analysis and Implications of the Nuclear Line Emission from the Orion Complex. *By* R. Ramaty, B. Kozlovsky, and R. E. Lingenfelter 392
Evidence for ^{56}Co Line Emission from the Type Ia Supernova 1991T Using COMPTEL. *By* D. J. Morris, K. Bennett, H. Bloemen, W. Hermsen, G. G. Lichti, M. L. McConnell, J. M. Ryan, and V. Schönfelder 397
Gamma-Ray Line Spectroscopy with INTEGRAL. *By* Peter von Ballmoos, Tony Dean, and Christoph Winkler 401
Particle Acceleration in Orion-like Objects. *By* Andrei M. Bykov 406

Part VII. Gamma-Ray Bursts: Observation and Theory

Gamma-Ray Burst Mini-Symposium—Introductory Remarks. *By* GERALD J. FISHMAN 410
Temporal Properties of Gamma-Ray Bursts. *By* CHRYSSA KOUVELIOTOU 411
BATSE Observations of the Spectra of Gamma-Ray Bursts. *By* MICHAEL S. BRIGGS 416
EGRET Observations of Gamma-Ray Bursts. *By* E. J. SCHNEID, D. L. BERTSCH, C. E. FICHTEL, R. C. HARTMAN, S. D. HUNTER, D. J. THOMPSON, G. KANBACH, H. A. MAYER-HASSELWANDER, Y. C. LIN, P. F. MICHELSON, P. L. NOLAN, B. L. DINGUS, C. VON MONTIGNY, P. SREEKUMAR, and D. A. KNIFFEN 421
COMPTEL Observations of Gamma-Ray Bursts. *By* R. M. KIPPEN, J. RYAN, A. CONNORS, M. MCCONNELL, V. SCHÖNFELDER, J. GREINER, M. VARENDORFF, W. COLLMAR, W. HERMSEN, L. KUIPER, C. WINKLER, L. O. HANLON, and K. S. O'FLAHERTY 425
Search for GRB Counterparts. *By* J. GREINER 429
Global Distribution Constraints on GRB Models. *By* D. H. HARTMANN, M. S. BRIGGS, and G. N. PENDLETON 434
Gamma-Ray Burst Models: General Requirements and Predictions. *By* P. MÉSZÁROS 440
Gamma-Ray Bursts—What Are They? *By* S. E. WOOSLEY 446

Part VIII. Ultra-High-Energy Cosmic Rays

The Ultra-High-Energy Cosmic Rays: An Astrophysical Puzzle. *By* MURAT BORATAV 450
The Cosmic Ray Composition towards the Knee: Experimental Status and Implications. *By* DIETRICH MÜLLER and SIMON SWORDY 454
The Highest Energy Cosmic Rays. *By* BRUCE R. DAWSON 460
Propagation and Acceleration of Ultra-High Energy Cosmic Rays. *By* VLADIMIR S. PTUSKIN 464
Possible Extragalactic Sources of the Highest Energy Cosmic Rays. *By* JORG P. RACHEN 468
Search for Cosmic Gamma-Rays above 24 TeV with the HEGRA Detector. *By* E. LORENZ 472
Necessity and Reality of Experimental Investigation of Ultrahigh Energy Cosmic Rays (10^{19}-10^{21} eV). *By* YU. A. FOMIN, G. B. KHRISTIANSEN, and G. V. KULIKOV 477

Part IX. Gravitational Radiation

LISA—Laser Interferometer Space Antenna for Gravitational Wave Measurements. *By* KARSTEN DANZMANN 481

Wide-Band Spherical Gravitational Wave Detector. *By* M. KARIM, M. BOCKO, L. E. MARCHESE, and G. ZHANG 485
Binary Neutron Star Inspiral, LIGO, and Cosmology. *By* LEE SAMUEL FINN 489
On the Detectability of Post-Newtonian Effects in Gravitational-Wave Emission of a Coalescing Binary. *By* ANDRZEJ KRÓLAK, KOSTAS D. KOKKOTAS, and GERHARD SCHÄFER 493
Gravitational Wave Signals from Collapsing Rotating Polytropes. *By* EWALD MÜLLER and THOMAS ZWERGER 498
Gravitational Waves from Coalescing Neutron Stars. *By* XING ZHUGE, JOAN CENTRELLA, and STEPHEN MCMILLAN 503
Coalescing Compact Binaries to Second-Post-Newtonian Order. *By* LUC BLANCHET 507
Gravitational Waves from a Particle Orbiting around a Rotating Black Hole: Post-Newtonian Expansion. *By* MASARU SHIBATA, MISAO SASAKI, HIDEYUKI TAGOSHI, and TAKAHIRO TANAKA 512

Part X. Active Galactic Nuclei

Active Galactic Nuclei across the Electro-magnetic Spectrum. *By* THIERRY J.-L. COURVOISIER 517
Recent X-Ray Spectral Results from Seyfert 1 Galaxies. *By* A. C. FABIAN, R. F. MUSHOTZKY, K. NANDRA, and C. S. REYNOLDS 521
Correlated Optical and Gamma-Ray Variability in Blazars. *By* STEFAN WAGNER 526
Decelerating Relativistic Jets, BL-LAC Objects, and the Fanaroff-Riley Classification. *By* GEOFFREY V. BICKNELL 530
Pair Cascade Models of Gamma-Ray Blazars. *By* AMIR LEVINSON and ROGER BLANDFORD 534
Soft X-Ray Excesses as a Probe of the Conditions at the Innermost Part of Accretion Flow in AGN. *By* B. CZERNY, P. T. ŻYCKI, and Z. LOSKA 538
Activity of Rotating Magnetospheres in AGNs: Collimated Propagation of MHD Waves near a Black Hole. *By* KOUICHI HIROTANI and AKIRA TOMIMATSU 542
Accretion Disks in Active Galaxies: The Sub-Keplerian Paradigm. *By* SANDIP K. CHAKRABARTI 546
On the Hot Spot near a Kerr Black Hole. *By* ALEXANDER ZAKHAROV 550
Non-stationary Accretion with Ordered Magnetic Fields and Outflows in AGNs and Quasars. *By* M. M. ROMANOVA and R. V. E. LOVELACE 554
Cosmological Origin of Quasars. *By* ABRAHAM LOEB 558

Part XI. Gravitational Lensing

Testing Cosmogonic Models with Gravitational Lensing. *By* Joachim Wambsganss, Renyue Cen, Jeremiah P. Ostriker, and Edwin L. Turner 563

Giant Luminous Arcs. *By* François Hammer, Isabella Gioia, Olivier Le Fèvre, and Gerry Luppino 568

The Investigation of Cluster Mass Distribution by Gravitational Lenses. *By* G. Soucail 573

Determining the Mass Distribution of Clusters from Gravitational Lensed Background Galaxies. *By* Carolin Seitz 578

Ray Propagation in the Cheese Slice Universe. *By* Sylvie Landry and Charles C. Dyer 583

Gravitational Lensing by Cosmic Strings. *By* F. Marleau, C. C. Dyer, and J. H. Palmer 587

The Correlation of 1-Jansky Sources to Zwicky Clusters. *By* Stella Seitz 591

Dark Matter from Quasar Microlensing. *By* M. R. S. Hawkins 596

Cosmological Parameters and Gravitational Lensing Statistics. *By* Phillip Helbig 600

Gravitational Lensing with Polarization to Determine Galaxy Masses. *By* C. C. Dyer, P. P. Kronberg, R. A. Perley, and H.-J. Röser 604

AGAPE, an Experiment to Detect MACHO's in the Direction of the Andromeda Galaxy. *By* R. Ansari, M. Aurière, P. Baillon, A. Bouquet, G. Coupinot, C. Coutures, C. Ghesquière, Y. Giraud-Héraud, P. Gondolo, J. Hecquet, J. Kaplan, A. L. Melchior, M. Moniez, J. P. Picat, and G. Soucail .608

Part XII. Galaxy Formation

The Canada-France Redshift Survey. *By* O. Le Fèvre, S. J. Lilly, D. Crampton, F. Hammer, and L. Tresse 612

Emission-Line Galaxies at $z \leq 0.3$ in the Canada-France Redshift Survey. *By* Cláudia Rola, Laurence Tresse, Grazyna Stasinska, and François Hammer ..616

Redshift Distribution & Nature of μ-JY Radio Sources. *By* François Hammer, David Crampton, Olivier Le Fèvre, and Simon Lilly 620

From X-Ray Observations to Galaxy Formation. *By* G. Fabbiano 624

Photoionization Effects during Galaxy Formation. *By* Matthias Steinmetz ...628

On the Origin of Halo Globular Clusters and Spheroid Stars. *By* Mario Vietri and Enrico Pesce 632

Part XIII. Large-Scale Structure

Clusters and Large-Scale Structure. *By* Neta A. Bahcall 636

The Primordial Perturbation Spectrum and Large Scale Structure. *By* Stefan Gottlöber 650
The Density of the Universe from the Peculiar Velocities of Sc Galaxies. *By* Wolfram Freudling, Luiz N. Da Costa, Riccardo Giovanelli, Martha P. Haynes, John J. Salzer, and Gary Wegner 654
HST Cepheid Distance to Leo Group Galaxy M96. *By* T. Shanks, N. R. Tanvir, H. C. Ferguson, and D. R. T. Robinson 658
E.R.O.S. Search for Microlensing of Stars by Low Mass Objects in the Galactic Halo. *By* Christophe Magneville 664

Part XIV. Microwave Background

CMB Anisotropies: An Overview. *By* Douglas Scott 668
Mapping with the Jodrell Bank-Tenerife Radiometers. *By* R. D. Davies, C. M. Gutiérrez, R. Rebolo, R. A. Watson, A. N. Lasenby, and S. Hancock 672
Comments on the Comparison of the COBE DMR and Tenerife Data. *By* Charles H. Lineweaver 676
The Slope of Matter Density Perturbations from Tenerife and COBE/DMR. *By* F. Atrio-Barandela, L. Cayón, and J. Silk 680
Estimating Microwave Power Spectra. *By* Max Tegmark 684
Structure Formation with Global Texture. *By* Ruth Durrer and Zhi-Hong Zhou 688
Chasing CMB Photons through the Nonlinear Universe. *By* Robin Tuluie and Pablo Laguna 692
CMB Anisotropies Numerically vs. Analytically. *By* Naoshi Sugiyama and Wayne Hu 697
The Grishchuk-Zeldovich Effect in the Open Universe. *By* David H. Lyth 701
How Anisotropic Can a Universe Be? *By* John D. Barrow 706
The Signatures of Voids and the CMBR. *By* Sharon L. Vadas 710
Ω from the *COBE*-DMR Anisotropy Maps. *By* J. L. Sanz, L. Cayón, E. Martinez-González, N. Sugiyama, and S. Torres 714
Topology of the Microwave Background on Scales 1°-2° and Reionization. *By* P. D. Naselsky and D. I. Novikov 718
CMB Anisotropy due to Compton Scattering in Clusters of Galaxies. *By* S. Colafrancesco, P. Mazzotta, Y. Rephaeli, and N. Vittorio 722

Index of Contributors 726

Financial assistance was received from:

- DAIMLER BENZ AG
- DEUTSCHE AEROSPACE (DASA)
- DEUTSCHE FORSCHUNGSGEMEINSCHAFT (DFG)
- DEUTSCHE RAUMFAHRT AGENTUR (DARA)
- EUROPEAN SOUTHERN OBSERVATORY (ESO)
- INTERNATIONAL SCIENCE FOUNDATION (ISF)
- MAX-PLANCK-GESELLSCHAFT (MPG)
- MAX-PLANCK-INSTITUT FÜR ASTROPHYSIK (MPA)
- MAX-PLANCK-INSTITUT FÜR EXTRATERRESTRISCHE PHYSIK (MPE)
- NATIONAL SCIENCE FOUNDATION (NSF)
- SIEMENS AG

Preface

JOACHIM E. TRÜMPER AND GREGOR E. MORFILL

Max-Planck-Institut für Extraterrestrische Physik (MPE)
85740 Garching, Germany

At the 1992 Texas/PASCOS Symposium at Berkeley, one of us (J. E. T.) was asked by Ivor Robinson and Engelbert Schücking whether we could organize the 1994 Texas Symposium at Munich because the original European candidate could not make it. After some time for reflection and a discussion at MPE, we agreed and thus Munich became the first city outside Texas that has hosted two meetings of this illustrious series of Texas Symposia. Actually, after Munich 1978, Jerusalem 1984, and Brighton 1990, this was the fourth Texas Symposium held outside the United States.

The meeting took place in the Park Hilton Hotel in Munich, which is located close to the city center and the English Garden. Altogether 580 scientists participated. On Sunday evening, December 11, before the meeting, most of the participants gathered for first discussions, debates, and exchanges of news during the welcome reception at the invitation of the Local Organizing Committee. In the opening ceremony, the Symposium participants were welcomed by Hans F. Zacher, the President of the Max-Planck-Society, and by Gerhard Merkl, Secretary of State, representing the Bavarian government. On the evening of Tuesday, December 13, a reception was held in the Kaisersaal of the Residence of the Bavarian Kings, where Secretary of State Rudolf Klinger addressed the Symposium participants. The Symposium Dinner was held in the Ballroom of the Park Hilton on the evening of Thursday, December 15, with Maarten Schmidt as the after-dinner speaker, who was introduced by Engelbert Schücking.

The five mornings were devoted to plenary lectures and each of the four afternoons to three parallel mini-symposia. We are indebted to the organizers of these mini-symposia for their excellent work:

H. Ögelman: Pulsars and Neutron Stars
V. Schönfelder: Gamma-Ray Line Astrophysics
P. Schneider: Gravitational Lensing
E. J. P. van den Heuvel: Binary and Transient X-Ray Sources
C. M. Will: Gravitational Radiation
J. I. Silk: Microwave Background
M. J. Rees: Active Galactic Nuclei
K. I. Nomoto: Supernovae
N. A. Bahcall: Large-Scale Structure and Dark Matter
G. J. Fishman: Gamma-Ray Bursts—Observation and Theory
J. R. Jokipii: Ultra-High-Energy Cosmic Rays
S. D. M. White: Galaxy Formation and Evolution

In addition to these almost traditional program elements, we included four short plenary sessions, in which the latest results from current space missions (HST, ROSAT, GRO, and ASCA) were presented and discussed. In total, 26 invited and highlight talks, 146 contributed papers, and about 190 poster papers were presented at the meeting. Almost all the plenary lectures and the contributions to the mini-symposia are collected in this volume. We wish to thank all contributors for the careful preparation of their manuscripts. The extended two-page abstracts of the poster papers shall be published separately as an MPE report [No. 261 (1995)].

The members of the Local Organizing Committee were supported by scientists, students, and staff of the Max-Planck-Institut für extraterrestrische Physik. We thank all of them for their efficient and valuable help, without which the meeting could not have been conducted.

The financial support from the International Science Foundation (ISF), from the Deutsche Forschungsgemeinschaft (DFG), and from the Deutsche Agentur für Raumfahrtangelegenheiten (DARA) made it possible for a few dozen scientists from Eastern Europe to attend. Further important financial support came from the National Science Foundation, the European Southern Observatory, the Daimler Benz AG, the Siemens AG, the Max-Planck-Institut für Astrophysik, the Deutsche Aerospace, the Max-Planck-Gesellschaft, and the Max-Planck-Institut für extraterrestrische Physik. We wish to express our thanks to these agencies for their support.

Solar Neutrino Problems

TILL A. KIRSTEN

Max-Planck Institut für Kernphysik, Heidelberg, Germany D-69029

INTRODUCTION

There is an astrophysical and a particle physics motivation to be interested in the detection of solar neutrinos. For the astrophysicist, the neutrinos produced in the solar interior allow a unique real time look into the stellar center and provide a test of the theory of stellar structure and evolution. The neutrino spectrum

$N_\nu(E)dE = f\,(T(r);\ \rho(r);\ X,Y,Z(r);\ t)$ reflects uniquely the present stage of stellar evolution where T(r) and ρ(r) are the radial temperature and density distributions and X,Y,Z are the chemical abundances. All these quantities follow from the mass, the age, t, and the initial chemical abundances X_0, Y_0, Z_0 of the Sun.

From the particle physics point of view, the Sun is a very intense low-energy neutrino source which is ideal to test neutrino properties through propagation phenomena both, in vacuo and in very dense matter. Before the neutrinos arrive at the detector about 8 minutes after their birth they must travel 700 000 km in solar matter of changing density and 150 million km in vacuo. Non-zero neutrino mass and non-zero flavor mixing would lead to vacuum flavor oscillations according to

$$\nu_e = \nu_1 \cos\theta_{e\mu} + \nu_2 \sin\theta_{e\mu} \quad ; \quad \nu_\mu = -\nu_1 \sin\theta_{e\mu} + \nu_2 \cos\theta_{e\mu}$$

and for the mass eigenstates not all being equal to zero we get the oscillation length

$$L \propto E / \Delta m^2 \quad \text{where } \Delta m^2 = |m_{\nu_1}{}^2 - m_{\nu_2}{}^2|$$ [1].

For the Sun - Earth distance of 1 Astronomical Unit this is sensitive down to

$$\Delta m^2 \approx 10^{-11}\ (eV/c^2)^2\ !$$

However, the mixing amplitude and hence, observable effects might be rather small. This is completely different inside the Sun with large electron densities ρ_e. Coherent neutrino-electron forward scattering experiences the flavor asymmetry due to the absence of muons and tauons. Mikheyev and Smirnov [2] discovered that this leads to a resonance behavior $tg\theta_{eff} = \sin^2 2\theta_{vac} / \{\cos 2\theta_{vac} - \text{const } L\, \rho_e\}$

such that a maximal flavor conversion can occur even for small vacuum mixing angles θ_{vac} in a region where ρ_e and $L(E_\nu)$ fit the resonance condition. Because of the large variability of ρ_e within the Sun, resonance can occur for large parts of the neutrino spectrum. This 'MSW'-flavor flip is not oscillatory like vacuum oscillations, eventual back flips depend on the nature of the density profile traversed later on within the earth. If only the disappearance of electron neutrinos is observable, the depression factor R becomes a function of the neutrino energy E and of the mixing angle θ, both in matter and in vacuo. Then, observations can define the allowed parameter space in the $(\Delta m^2,\theta)$ plan.

The Standard Solar Model (SSM) deduces the interior structure and the evolutionary stage of the Sun from first principles such as hydrostatic and thermal equilibrium, equation of state, and radiation dominated energy transport in the interior[3)]. Solar input parameters are the solar mass, the present luminosity, the age, and the initial chemical composition. Energy production is assumed to be ultimately due to fusion of hydrogen into helium irrespective of the particularities of the actual fusion chains. Neutrino emission is an immediate consequence, since weak interactions are involved in the reaction network. To describe the energy balance, the nuclear cross sections (S-factors) and the opacities are also entering the solar model. It predicts now, for the 4.6 b.y. old Sun, a central temperature of 15.6 million centigrade, a central density of 148 g/cm^3 and a central pressure of 2.3 x 10^{11} Bar [3)4)]. The solar luminosity is 3.83 x 10^{23} kW. Under these conditions, it turns out that the pp-cycle dominates the energy production in the Sun. The expected neutrino fluxes arriving on earth are still tremendous: 60 billion/cm^2,s from the reaction $p+p \rightarrow d+e^++\nu_e$ ('pp'-neutrinos); 4.6 billion/cm^2,s from the electron capture of ^{7}Be and (only) ≈5 million/cm^2,s from the positron decay of ^{8}B. Figure 1 shows the expected solar neutrino spectrum [4)].

The origin of ^{7}Be is from ^{3}He[from d+p]+^{4}He; that of ^{8}B from ^{7}Be+p. In each cycle: $4H \rightarrow {}^4He + 2e^+ + 2\nu_e$, 26.73 MeV of energy and 2 neutrinos are produced. The latter take away .59 MeV. The chain starts with the pp-reaction and the pp-neutrino flux is consequently strongly coupled to the energy production. In the steady state, this means that the pp-neutrino flux follows from the solar luminosity. For that part, for the particle physicist the solar neutrino beam intensity at its origin is equally well known as the beams from accelerators he is normally used to deal with, and any measured deficits would have to be related to neutrino properties rather than to the Sun. The opposite applies to the higher energy neutrinos. Their fluxes are of little (^{7}Be) or no (^{8}B) influence on the solar luminosity. They depend sensitively on delicate branching ratios such as ^{3}He+^{3}He vs. ^{3}He+^{4}He and ^{7}Be+e$^-$ vs. ^{7}Be+p. These in turn depend on details of the solar structure and are much harder to predict. This is drastically reflected in the fact that the ^{8}B neutrino flux scales with the central temperature to the 18th power.

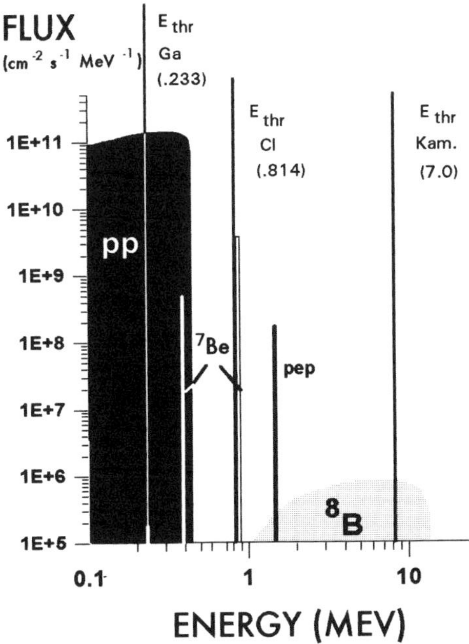

FIGURE 1. Solar neutrino energy spectrum expected from the SSM[4)]. Energy thresholds for solar neutrino detectors are indicated.
The shading code: * black = pp (and pep)-ν
* white lines = ^{7}Be -ν
* gray = ^{8}B -ν
applies also to Figure 2 (below).

SOLAR NEUTRINO DETECTION

The detection of low energy neutrinos is painstaking because of the incredibly low cross sections of MeV (and even sub-MeV) neutrinos for interaction with anything. The consequence are multi-ton sized detectors in which only a few events occur in days, weeks, or months. At such low rates, the task to avoid side reactions which mimic the neutrino interaction is enormous and constitutes one of the two major experimental challenges in solar neutrino experiments, the other one being the detection of the signal in which the few events must be clearly discriminated from the detector background. Both (very different) tasks demand extreme radiochemical purity of the equipment (target, detector, environment) and effective shielding from cosmic rays. This is why solar neutrino experiments are always low-level underground experiments, a technology quite distinct from that of accelerator physics. Two principle reaction types have been

successfully applied: Inverse Beta decay and neutrino-electron scattering. Inverse Beta decay $\nu_e + A(Z) \rightarrow A(Z+1) + e^-$
is based on neutrino capture in some suitable target isotope A(Z) to form the product nucleus A(Z+1) which must be chemically so distinct from the element Z that it is possible to quantitatively remove it from the target, with separation factors typically being of order 10^{30} ! Subsequently, the electron capture back-decay of the product A(Z+1) into A(Z) must be recorded to determine the neutrino induced production rate. Key considerations in the choice of Z to form the basis of such a 'radiochemical' solar neutrino experiment are the energy threshold, the neutrino capture cross sections and their predictability, a suitable life time, the existence of a suitable counting technique for A(Z+1), the isotopic [A(Z)] and the natural elemental [Z] abundances, the existence of a suitable chemical separation scheme, and last but not least, affordability.

The two successfully applied cases are:

$\nu_e + {}^{37}Cl \rightarrow {}^{37}Ar + e^-$, $T_{1/2} = 34$ d, E_{thr}=814 keV

in the HOMESTAKE chlorine experiment in South Dakota; and

$\nu_e + {}^{71}Ga \rightarrow {}^{71}Ge + e^-$, $T_{1/2} = 11.4$ d, E_{thr}=233keV

in the GALLEX gallium experiment at Gran Sasso, Abbruzzi, Italy and in the SAGE gallium experiment at Baksan, Caucasus.

Real time neutrino-electron scattering is detected in a large light water Cerenkov detector at a threshold of ≈7 MeV neutrino energy in the KAMIOKANDE experiment at Kamioka mine in Central Japan[5]. The scattering cross section for e^- (ν_f , ν_f) e^- is about 7 times larger for flavor f=e than for f=μ, hence the signal is mainly from electron neutrinos, not too different in this respect from radiochemical experiments which are sensitive exclusively to electron neutrinos. The big advantages are time resolution and directionality. The potential of these properties must compensate the big disadvantages of a very many orders of magnitude higher background (trigger rate) and, dictated by this, the higher energy threshold.

The neutrino fluxes predicted from solar models must be folded with the detector response, that is, the respective cross sections and excitation functions. In general, such cross sections for inverse β-decay can be deduced rather reliably from the ft-values resulting from the β-decay characteristics of the product nucleus. However, this applies to ground state transitions only and excited state contributions to the production rate may cause uncertainties. They may be significantly reduced if the Gamov-Teller strength of such transitions is inferred from (p,n) forward angle scattering experiments[6)7)]; however, there remain both, principal and experimental problems. In the case of ^{37}Cl, the capture rate is (fortunately) dominated by the transitions to the isobaric analogue state near 5 MeV. In this case, the transition strength is deduced from the ^{37}Ca-β-decay properties. Overall, the errors of the capture (or scattering-) rates for the various detectors are included in the errors quoted for the SSM-rate predictions.

HOMESTAKE

The HOMESTAKE experiment[8] is collecting data since 1970. Without this pioneering effort of R.Davis jr., experimental solar neutrino astrophysics might still not exist. Davis demonstrated to the skeptic world that the above conception can be made to work in practice if rigorous care is applied at all critical steps but also that, without such responsible experimenting, the probability to produce meaningless results in such difficult experiments at the edge of practicability is close to 100%.

The Homestake target is 615 tons of perchlorethylene in a single tank installed in a gold mine below 4100 m.w.e. of shielding. Every few months, the radioactive ^{37}Ar is purged with helium from the target, collected on a charcoal trap, purified, and admixed to the counting gas of small gas-proportional counters. Measured are the Auger electrons resulting from the electron capture decay of ^{37}Ar.

Since the energy threshold for neutrino capture on ^{37}Cl is 814 keV, the Cl-detector is blind for the most abundant pp-neutrinos (maximum energy 420 keV), but it can detect ^{8}B- and ^{7}Be-neutrinos from the PPIII and PPII branches. The dominant contribution is expected from ^{8}B (see Figure 2).

The result of 25 years of data collection at Homestake is (2.55 $\pm$ .25) SNU (1σ)[10] (SNU=Solar Neutrino Unit, 1 capture per second and 10^{36} target atoms). The difference between this figure and the Standard Solar Model ('SSM') prediction of Bahcall [4], (8 $\pm$ 3) SNU (3σ), has long been referred to as "the Solar Neutrino Problem" (SNP). The quoted average result assumes a production rate constant in time (secular solar equilibrium), but the data of certain time periods are barely consistent with this assumption[10][11] (Figure 3). This need not be unreasonable if one considers e.g. the 'VVO'- mechanism[12]. Here one envisions the interaction of time variable magnetic fields in the outer convective zone with a hypothetical magnetic moment of the neutrino. However, the results of the Kamiokande experiment (see below) do not exhibit a temporary variation of the ^{8}B-flux since 1987, the year when this detector was tuned for solar neutrino observation (Figure 4). The discrepancy is unresolved.

KAMIOKANDE

The achievement of the Kamiokande experimenters has been their ability to reduce the intrinsic detector contamination (U,Th-series) to a level at which it became feasible to observe neutrino-induced recoil electrons at energies as low as 7 MeV, that is, within the ^{8}B-neutrino energy range (Figures 1 and 2). Nevertheless, the task to single out solar neutrino induced events is formidable and requires extremely good track recognition criteria and timing. Note that with an expected signal of order 1/d in a fiducial volume of 680 tons the total trigger rate of the detector is ca 150 000/day! The Cerenkov light cones are observed with a dense photo multiplier network. From the number, orientation, and

pp 7Be N+O 8B

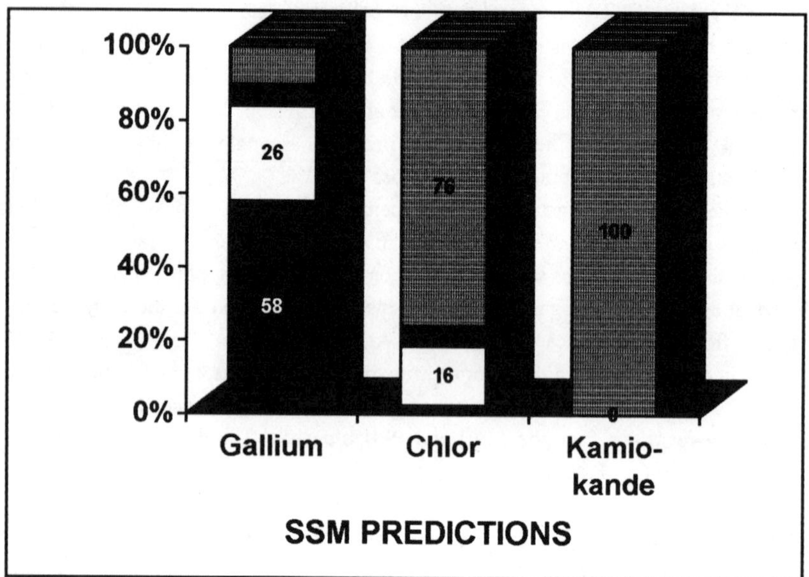

FIGURE 2. Percentage contributions from the various solar neutrino sources to the signals expected in the presently operating solar neutrino experiments. Mean of Ref.[4] and Ref.[9]. The shading code is the same as in Figure 1.

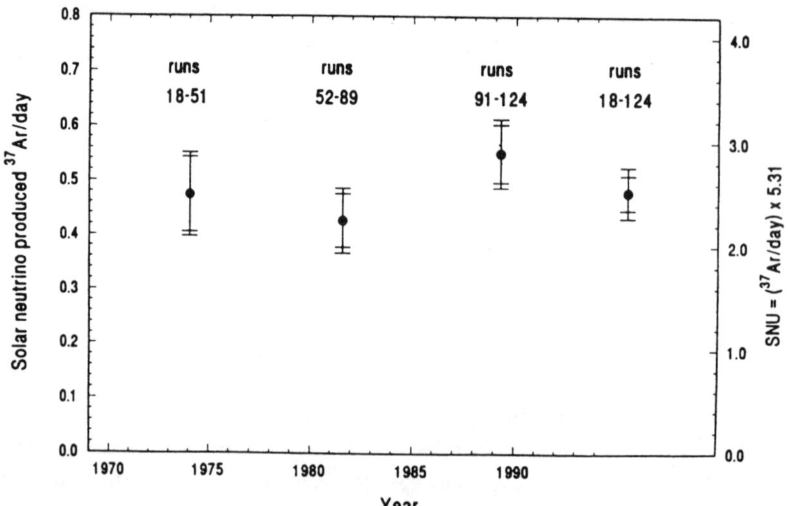

FIGURE 3. From Ref.[10]. The time averaged ^{37}Ar production rates in the Homestake chlorine detector, divided into three subsequent time periods. The last entry is for the grand total mean value. Errors shown are $1\sigma_{statist.}$ (inner bars) and $1\sigma_{total}$ (outer bars) where σ_{total} is from adding the statistical and systematical errors in quadrature.

relative time of the PM-cells hit, the energy and the vertex of an event are reconstructed. The spectrum of the recoil electrons reflects the initial neutrino spectrum in a well understood fashion.

Kamiokande definitely observes solar (^{8}B) neutrinos, a clear signal is seen in the direction pointing towards the Sun (Figures 5,6). But also here, the SSM-expectation is not met. In essence, the result is a confirmation of the SNP, yet the reduction factor is ca. 1/2 rather than 1/3 in the Homestake experiment. In numbers[13)14)15)]: the ratio Kamiokande data/SSM-prediction = .51 ± .07 for the model of [4)] or = .65 ± 0.09 for the model of [9)]; 1σ-errors, theoretical uncertainties *not* included.

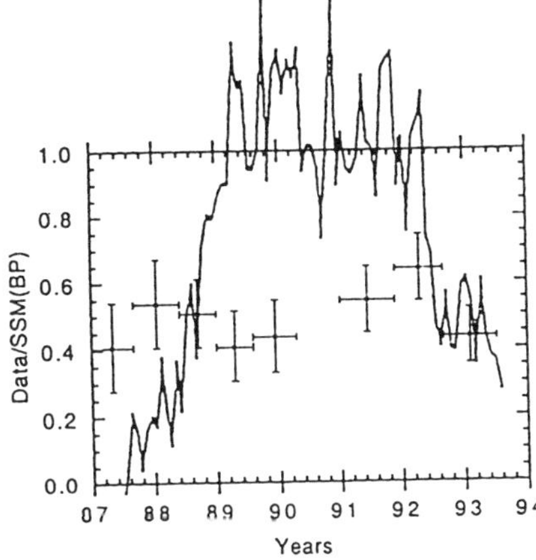

FIGURE 4. (from Ref.[14)]). Kamiokande results for the ^{8}B solar neutrino flux grouped in 200 day bins, expressed relative to the flux expected from a Standard Solar Model [4)]. Superimposed is the relative sunspot number for this time period (linear scale). No respective correlation is obvious.

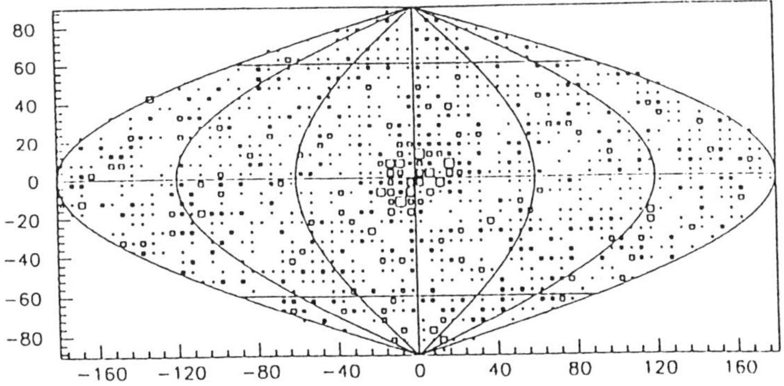

FIGURE 5. (from Ref.[14)]). The angular distribution of the Kamiokande signal plotted in a "heliocentric" coordinate system. Bins are 4x4 degree, each box size represents the signal size in that bin. The neutrino image of the Sun is clearly visible: "neutrino heliograph".

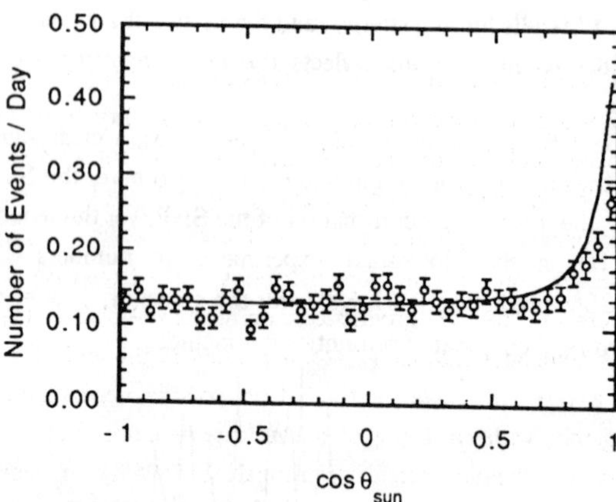

FIGURE 6. (from Ref.[14]). Angular distribution for 5.4 years Kamiokande data after all other cuts are done. The solar signal is seen in the Sun's direction. It is however smaller than predicted from the SSM (solid line).

The SSM-prediction for the tenuous ^{8}B-neutrino flux is extremely difficult and surely has a large error. From the SNP alone it is therefore not save to conclude neutrino oscillations (hence massive neutrinos and "new" physics). Negligence of details describing the solar interior within the SSM (e.g.,collective plasma effects, screening), inaccurate input parameters for cross sections and opacities, or non-standard astrophysics (e.g., rotating inner core, heterogeneous accretion of the protosun resulting in a radial heavy element abundance gradient, ...) do all add to an uncertainty which prohibits to exclude "astrophysical solutions" of the SNP. What one would like to know for a decision is the flux of pp-neutrinos. Contrary to ^{8}B-neutrinos, this is tightly coupled to the energy output of the Sun. As mentioned before, 2 pp-neutrinos must be produced for each completed fusion cycle $4H \rightarrow {}^4He + 26.7$ MeV, no matter of the network branching. Here, a measured shortage would enforce the conclusion of neutrino transformations since the solar luminosity must be accounted for. The inner logic of our curiosity about solar neutrinos as probes of the solar core would surely have called first for the measurement of the bulk pp-neutrinos and only later one would have nursed the more demanding desire to study the fine-tuning of the solar interior by means of ^{8}B neutrino detection. However, this order was reversed for a very practical reason, the non-availability of multi-ton quantities of gallium. Ga is practically the only reasonable choice for an experiment with threshold low enough to be sensitive to pp-neutrinos, but in the Seventieth when the pioneering Chlorine experiment started there was simply not a single ton of gallium available. Only later turned this problem from a principal one into a financial one such that finally, GALLEX became feasible.

GALLEX

The GALLEX collaboration [16,17,18,19] aims primarily at pp-neutrino detection. To this end for many years to come the only realistic possibility is to use the ^{71}Ga-^{71}Ge scheme in a radiochemical experiment. GALLEX uses 30 tons of Ga in 101 tons of aqueous gallium chloride solution within a single target tank. The SSM- predicted capture rate is comparably high and rather well known: (123 [9] - 132 [4]) SNU. 58 % (74 SNU) are expected from pp (and pep) neutrinos, 26% from ^{7}Be neutrinos and only 10% from ^{8}B (Figure 2). The ^{71}Ge production rate in the GALLEX target is 0.89 atoms per 100 SNU. This translates into ca 12 atoms present at the end of a 4-week exposure period (per 100 SNU). They are extracted as volatile germanium tetrachloride by nitrogen purge together with ca 1 mg of a stable germanium isotope which serves as carrier and allows the chemical yield determination (>98%) in each individual run. Subsequently the $GeCl_4$ is converted into germanium hydride which, after 70% xenon admixture, is a perfect counting gas for proportional counting. For a detailed description of the chemical extraction and preparation techniques see [20]. Big efforts went into the realization of a counting system with unprecedented low background rates. That such backgrounds are crucial can be seen from the fact that, in an individual run, typically only about 5 decay events of ^{71}Ge are observed over a period of about one month. Detected are the Auger electrons and X-rays from the electron capture decay of ^{71}Ge to ^{71}Ga at nominal energies of 10.4 keV (K) and 1.2 keV (L). Extremely low background rates (order of 1 count per month) have been achieved[21]. After >180 days of counting for each run a maximum likelihood analysis is applied to split the measured events into an exponentially decaying signal and a time constant background.

The GALLEX solar measurements began on 14 May 1991 and the first ever observation of pp-neutrinos was published in June 1992[16] (15 runs, see Table 1). The subsequently published data remained stable while the errors decreased in a consistent manner (Table 1 and Figure 9 [below]). In so far altogether 30 runs, 148 ^{71}Ge atoms were seen to decay[18]. Side reactions producing ^{71}Ge in the target are well studied, only 12 of the 148 events are due to them (residual cosmic ray muons, fast neutrons, radon and others). The results of the individual runs are consistent with statistical fluctuations, multiple consistency checks are applied to the data and confirmed by Monte Carlo simulations. After each solar run, an instrumental 'Blank' run is performed. It copies the solar runs in all aspects except that the exposure time is kept as close to zero as possible. This checks for instrumental artifacts as opposed to time related production processes. The combined result of all blank runs is (-1 ± 7)SNU[18], contrasted by the result of all solar runs. Only GALLEX performs such blanks.

The GALLEX data do not exhibit any statistically significant time variation. A respective analysis for an eventual correlation of the production rate P(t) with the variation of the sunspot number SS(t) relative to the mean, $\langle SS \rangle$, for the GALLEX data taking period according to $P(t) = a + b\,\{SS(t) - \langle SS \rangle / \langle SS \rangle\}$ yields $a = 82 \pm 11$ SNU for the constant part and b = +10 ± 27 SNU for the variable amplitude[23)]. Note that b is not only <1σ but also >0, that is, 'correlated' rather than 'anticorrelated', as one might extract from the data of the Homestake detector.

For the Sun, the overall result for our later discussion is: (79 ± 12) SNU (1σ)[18)]. This is ≈ 62 % of the SSM expectation or 107% of what is expected for pp and pep neutrinos alone. It constitutes the first experimental observation of hydrogen fusion in the solar interior. This is fundamental for the theory of stellar structure and evolution, it transfers stellar models from the realm of theory into the sphere of observational facts, at least as far as the basics are concerned. At the same time, the deviation from 100% indicates a deficit of higher energy neutrinos from ^{7}Be and ^{8}B, confirming what has been found in the Homestake and Kamiokande experiments.

TABLE 1. (from Ref.[18)]).Results from GALLEX solar exposure periods. Errors quoted[22)] are 1σ.

	GALLEX I	GALLEX II[1)]	combined
time period	14.5.91-29.4.92	19.8.92-13.10.93	14.5.91-13.10.93
exposure days	324	406	730
runs	15	15	30
L-result, [SNU]	105 ± 28	66 ± 19	80 ± 16
K-result,[SNU]	64 ± 21	87 ± 17	78 ± 13
(L + K) main result [SNU]	81 ± 17 ± 9	78 ± 13 ± 5	79 ± 10 ± 6 79 ± 12 [2)]
October-March[3)]			82 ± 15
April - Sept.[3)]			78 ± 14

[1)] data taking continued beyond 13 October 1993.
[2)] statistical and systematical errors combined in quadrature.
[3)] The expected rate due to the Sun-Earth distance variation in the winter half-year is 4% above that for the summer half-year. Evidently, statistics are insufficient for this.

The GALLEX Chromium Source Experiment

GALLEX solar neutrino runs continue, but they were interrupted from 23 June through 10 October 1994. During this period, a calibrated manmade ^{51}Cr neutrino source was inserted into the GALLEX tank[19)] in order to demonstrate the reliability of the experiment in the concerted interplay of all its components and to exclude any unidentified sources of systematical errors such as e.g. unspecified withholding mechanisms or hot atom chemistry. Such a straightforward general demonstration of the reliability of the radiochemical rare event detection technique has long been demanded but this is the first time that it has actually been done.

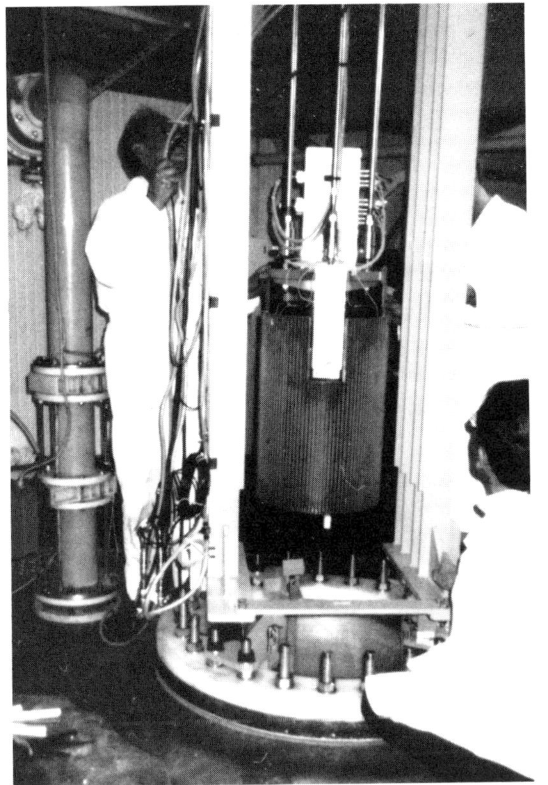

FIGURE 7. Insertion of the ^{51}Chromium neutrino source into the GALLEX target tank at the Gran Sasso Underground Laboratory. The source was activated at the Siloe reactor in Grenoble/France. It remained inside the gallium tank from 23 Juni - 10 October 1994. The activity declined with a 27.7 d halflife. At time of removal, the source neutrino induced production of ^{71}Ge from gallium had dropped below the production rate due to the Sun.

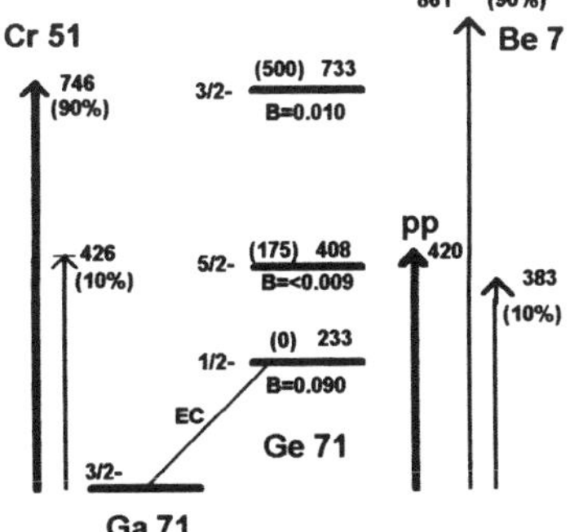

FIGURE 8. Level scheme for the production of ^{71}Ge from ^{71}Ga through (i)solar pp-neutrinos; (ii) solar ^{7}Be-neutrinos and (iii) man-made ^{51}Cr-neutrinos. Apart from the dominant ground states transition, the 500 keV and 175 keV excited states are inclusive in the ^{51}Cr source experiment. (Gamov-Teller strengths (B) are from [7]).

The required low energy neutrinos are emitted in the electron capture decay of ^{51}Cr whereby the latter is produced by neutron activation of ≈36 kg high-purity metallic chromium granulate enriched to 38% of ^{50}Cr (natural abundance 4.35%) in the 35 MW Siloe reactor at Grenoble: $^{50}Cr\ (n, \gamma)\ ^{51}Cr\ ; + e^- \rightarrow\ ^{51}V + \nu$; (EC, $T_{½}$ = 27.7 d).
The strength of the source has been determined independently by calorimetry, gamma counting of the ≈ 320 keV γ-branch (9.86%, $^{51}V^* \rightarrow\ ^{51}V + \gamma$) and by neutron and γ-monitoring during the irradiation. The results are consistent and give a source strength of 1.67 ± 0.03 MCi, the strongest neutrino source ever made.

With the source suspended within a reentrance tube (Figure 7) an initial ^{71}Ge production rate of ≥15 times that measured for the Sun was expected. The exposure lasted for ≈ 3½ months, until the rate had decreased to about the solar rate (the ^{51}Cr-half-life is 27.7 d). During these 3½ months, 11 extractions were performed, aiming for a precision of ≈10%. The results from the first 7 extractions are published[19]. The outcome is very satisfactory: the ^{71}Ge production due to the source is 104 ± 12 % of what was expected. This indicates that "there are no significant experimental artifacts or unknown errors at the 10% level that are comparable to the ≈40% deficit in observed solar neutrino signal."[19] In this respect it is particularly relevant that the ^{51}Cr neutrinos very much resemble ^7Be-neutrinos (Figure 8). Apart from the contribution from the ground state the measured rate does also include the 175 and 500 keV excited state transitions and thus it provides a check on these cross sections.

SAGE

Another gallium-germanium experiment is carried out by the Russian-American SAGE collaboration in the INR underground laboratory below 4700 m.w.e. of shielding in the Baksan Valley, Caucasus. They use metallic gallium which is liquid below 30 centigrade. Batches of 7 tons are contained in special vessels equipped with a stirring device. For Ge-extraction, a HCl/H_2O_2 mixture is added and emulsified with the gallium by mixing. This is a heterogeneous two phase system in which the germanium diffuses out of the gallium droplets into the liquid phase. After about 30 minutes, the emulsion is separated into metal and solution by adding excess HCl. The germanium is now contained in an aqueous solution as $GeCl_4$ and further processed similar as in GALLEX, already described.

The counting procedures are also similar in principle to what has been sketched for GALLEX, even though particularities and details are quite different. The target size applied has gone up with time from ≈ 30 to ≈ 57 tons, but the integral number of solar ^{71}Ge atoms detected (amount of data) is lower than for the 30 t - GALLEX experiment in spite of SAGE's earlier start of data taking (in January 1990). This is because of a smaller duty factor, a lower extraction efficiency, and because of a lower counting efficiency due to the fact that the L-peak of ^{71}Ge decay could not be put to use because of excessive background in the L-energy region (≈ 1.2 keV).

SAGE was first in announcing data in 1990[24)], when they reported a zero signal and reflected on the dramatic consequence that this implied a non-zero neutrino rest mass. However, this result did not stand up (Figure 9). GALLEX announced its first result, (83 ± 19(stat.) ± 8(syst.)) SNU (1σ) in May 1992[16)],allowing the full presence of the pp-neutrinos expected from the solar luminosity. As shown in Figure 9 the SAGE results have meanwhile approached the GALLEX level and the results are now in good agreement: (73 ± 18(stat.) ± 7(syst.)) SNU (1σ) (SAGE)[29)]. In the following discussion we use the GALLEX result, (79 ± 10(stat.) ± 6(syst.)) SNU (1σ)[18)].

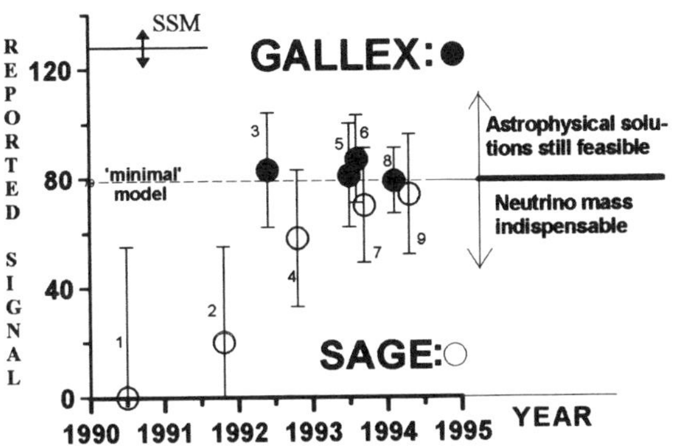

Figure 9.(from Ref.[25)]) Chronology of GALLEX and SAGE results. The labeled points have been reported in conference proceedings and in open-literature publications. Errors shown are statistical and systematic errors added quadratically. In the upper left the predictions of the SSM are displayed as a line with their 1σ-error extension (arrows). The horizontal line below is drawn at 78-80 SNU, which is the value for a 'minimal' solar model, in which the total luminosity is (artificially) assigned to the PPI chain. In the Figure, label 1 = [24)], 2 = [26)], 3 = [16)], 4 = [27)], 5 = [17)](GALLEX I)), 6= [17)], 7 = [28)], 8 = [18)], 9 = [29)].

DISCUSSION

Data Summary

To support the joint discussion of the available evidence from solar neutrinos we summarize the theoretical predictions in Table 2 and the experimental results in Table 3. The essence of these findings is further condensed in Table 4 and illustrated in Figure 10. This Figure visualizes at a glance the present status of the solar neutrino problem. All experiments yield signals below expectation, but a judgement of the significance of the measured deficits δ (defined in Table 4) requires discussion of the confidence ranges for both, predictions and experimental results. This issue has been addressed by many authors (see e.g., [3)4)9)30)31)32)33)]) and this is not the place to repeat the (sometimes heated) argu-

TABLE 2: **Theoretical** predictions (1σ-error) for the major solar neutrino fluxes ϕ_x and for the respective response of the solar neutrino experiments (for radiochemical experiments: in SNU). For the spectral partitioning of the signal in the various neutrino detectors see Figure 2.

Φ_{pp}	$6.00 \pm 0.04 \times 10^{10}$ /cm²,s 6.02 x 10^{10} /cm²,s 6.07 x 10^{10}/cm²,s	BP TCL DS
Φ_{7Be}	4.89 ± 0.3 x 10^{9}/cm²,s 4.33 x 10^{9}/cm²,s 4.10 x 10^{9}/cm²,s	BP TCL DS
Kamio-kande Φ_{8B}	5.7 ± 0.8 x 10^{6}/cm²,s **4.43 + 0.8 -1.2**x 10^{6}/cm²,s 2.77 ± 0.55 x 10^{6}/cm²,s ***2.8 - 5.7*** x 10^{6}/cm²,s	BP TCL DS **nominal range**
Cl-experiment	8.0 ± 1.0 SNU **6.4 ± 1.4 SNU** 4.2 ± 0.6 SNU ***4.2 - 8.0 SNU***	BP TCL DS **nominal range**
Ga-experiment	131.5 ± 7 SNU **122.5 ± 7 SNU** 113 SNU ***113 - 132 SNU***	BP TCL DS **nominal range**

BP = Bahcall and Pinsonneault, Ref. [4]).TCL=Turck-Chieze and Lopes. Being intermediate, this data set is used for illustration in Fig.10 (bold), Ref.[9]). DS = Dar and Shaviv, 1994. Ref.[33])

TABLE 3. Recent update of **experimental** results for the running solar neutrino detectors. Results are also expressed in percent of SSM-expectations.

Experiment	**Data time**	**Result** (SNU or Φ) Errors: 1σ	**Percent of Expectation** *) 2σ- Errors	**Mass** of Target Element	**Sample and References**
Homestake **Cl**	1970.8- 1992.4	2.55 ± 0.25 SNU (±0.17 st ± 0.18 sys)	BP 32 ± 10 TCL 40 ± 20 DS 61 ± 21	610 t	Runs 18-124 Ref.[10])
Kamio-kande Cerenkov	1987.0- 1993.6	2.89± 0.41 x10^{6}/cm²,s (±0.22 st ± 0.35 sys)	BP 51 ± 20 TCL 65+40-30 DS 104 ± 51	680 t fiducial	Kam.II+III Ref.[14])
GALLEX **Ga** (chlor-ide solution)	1991.4 - 1993.8	79 ± 12 SNU (±10 st ± 6 sys)	BP 60 ± 20 TCL 65 ± 21 DS 70	30.3 t	GX I + GX II (SR1-SR30) PLB, Ref.[18])
SAGE **Ga** (metal)	1990.0- 1993.0 (interr.)	73 ± 18 SNU (± 17 st ± 6 sys)	BP 56 ± 28 TCL 60 ± 30 DS 65	27-57 tons (variable)	SAGE I + II PLB, Ref.[29])

*)The reference code for the SSM predictions used is as in Table 2. See text for meaning of '2σ'-errors.

TABLE 4: Solar neutrino deficits δ (in percent)

Experiment	Deficit $\delta = (1 - \frac{\text{signal}}{\text{prediction}}) \cdot 100\%$	R
Homestake	60 ± 20 $(2\sigma)^{*)}$	2.5
Kamiokande	$35\,^{+30}_{-40}$ $(2\sigma)^{*)}$	1.5
GALLEX	35 ± 21 $(2\sigma)^{*)}$	1.5

$\sigma_{theor.}$ and $\sigma_{experim.}$ are added in quadrature.

The reduction factor $R = \frac{\text{prediction}}{\text{signal}} = \frac{100}{100-\delta}$.

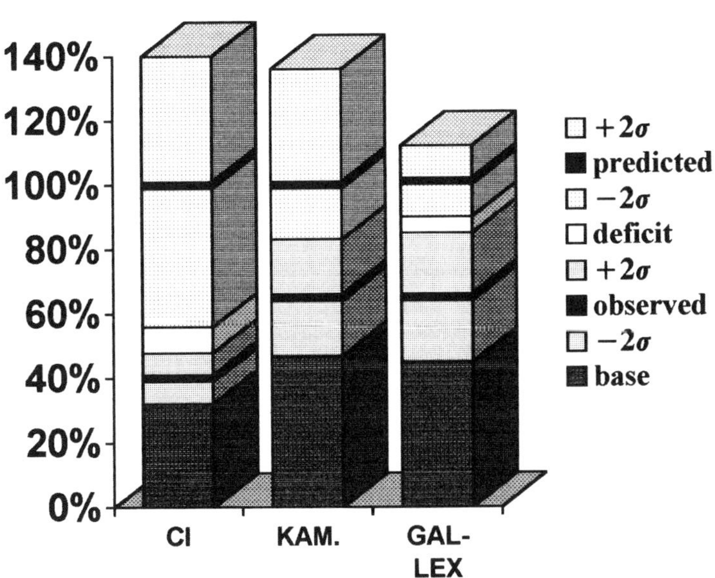

Cl		KAMIOKA		GALLEX
(40 ± 8)%	|	(65 ± 18)%	|	(65 ± 20)%

of SSM EXPECTATION [9] [2σ, experiment only]

FIGURE 10. Major results from solar neutrino experiments in percent of the SSM-expectation with their respective ±2σ errors. The ±'2σ' uncertainty of the theoretical prediction (set=100%) is also shown. For the spectral sensitivity of the experiments, see Figure 2. Data and theoretical predictions as quoted in the text. For Kamiokande, the theoretical error extends all the way down to ['observed' -2σ]. The overlapping shadings could not be displayed in the Figure.

ments in any detail. Instead I will focus only on the main key lines of argumentation and on some new aspects due to shrinking statistical errors and to the increased systematical confidence as a consequence of the Cr-source experiment in GALLEX.

MODELS: To stand in place for all, we consider here the range of predictions of 3 prominent up-to date SSM- calculations: Bahcall and Pinsonneault[4)] (BP), Turck-Chieze and Lopes[9)] (TCL), and Dar and Shaviv[33)]. This choice covers the range of values under discussion. We believe that the degree of confidence for predicted quantities is more or less reflected in the *range* of the proposed values. The pp-flux is well fixed from the solar luminosity (Table 2), predictions for ϕ_{7Be} differ only within 20%, yet the ϕ_{8B} predictions differ by up to a factor of 2! The boron neutrino flux ϕ_{8B} determines the signal of the Chlorine experiment dominantly and that of Kamiokande exclusively (Fig.2). Consequently, the predicted signals for these two experiments differ also by a factor of ≈2. This is mainly caused by new data for the ^{7}Be(p,γ)^{8}B cross section[34)35)] and by different judgment of some other input parameters. Contrary, for gallium the signal is dominated by the well known pp-flux, hence even the extreme range of rate predictions varies only by ≈20% (Tab.2).

EXPERIMENTS: All experiments are low rate, hence the statistical errors shrink significantly as measurements go on. Systematical errors are more subjectively judged from outside observers. Suspicion arises if data change significantly with time or if subsequently released error statements are inconsistent. As mentioned, there was also a certain skepticism with respect to the radiochemical method in general, but the GALLEX source experiment has convincingly demonstrated the credibility of this technique. Hence we take the data and their errors at face values (Table 3).

ERROR PROPAGATION: Error treatment is crucial in the evaluation of the implications from the observed deficits δ and the reduction factors R (Table 4). We have chosen to add the experimental and theoretical 2σ- errors in quadrature in order to deduce the errors given in Table 4 (and in the fourth column of Table 3). This may not be justified. With linear treatment, deficits shrink substantially. This is illustrated in Fig. 10 where the 2σ- errors of predictions and experiments are displayed individually. As evident, not much is left between (signal + $2\sigma_{exp.}$) and (prediction - $2\sigma_{theor.}$) for any of the experiments.

Three Solar Neutrino Problems

The most significant deficit is for the chlorine detector, it constitutes the 'original' solar neutrino problem, a reduced ^{8}Boron neutrino flux Φ_8 . This is
⇒ the first solar neutrino problem , **"SNP #1": $\Phi_8(Cl_{exp}) < \Phi_8(SSM)$**
This problem could in principle be fixed by non-standard solar models tailored to reduce the central core temperature as needed (by about 5%, from 15.6 to 14.8 million °C)[36)] or even in standard models[33)] by changing the ^{7}Be(p,γ)^{8}B cross section, σ_{17}, and others (σ_{33}, σ_{34}), as well as by changing coupled input parameters such as opacities or initial abundances, X_0,Y_0,Z_0. In short, a reliable and accurate prediction of the tenuous

(10^{-4} branch) ^{8}B neutrino flux is not yet possible. However, the problem is aggravated in that the Kamiokande deficit δ_{Ka} is less than δ_{Cl}, hence $\Phi_8(Ka_{exp})$ measured in Kamiokande alone should produce more signal as is seen in the chlorine detector, leave alone that the ^{7}Be neutrino flux Φ_7 would further add to the Chlorine detector signal (Figure 2). This defines

⇒ the second solar neutrino problem, **"SNP #2" :** $\mathbf{\Phi_{8+7}(Cl_{exp}) < \Phi_8(Ka_{exp})}$

Logically, this inversion is not solvable within astrophysical scenarios[37)38)39)40)41)42)43)] while it could be explained if lower energy neutrinos were preferentially suppressed through neutrino flavor oscillations in the energy range

$E_{threshold}(\text{Chlorine}) < E_\nu < E_{threshold}(\text{Kamiokande})$; [0.8 - 7 MeV , see Fig.1]].

Such a solution can quite well have the property to leave the pp-neutrinos (E_{pp}<0.42 MeV) untouched such that all observations are consistently explained. With shrinking errors in GALLEX, the 3rd solar neutrino experiment created also the 3rd solar neutrino problem. It consists in the apparent absence of most or all ^{7}Be neutrinos, the second largest expected contributor to the Ga-signal (Figure 2). The measured rate in GALLEX is exactly what is expected from the PPI cycle alone ('minimal model', 78-80 SNU[3)]), hence there is little space for the sizable contribution expected from ^{7}Be. ^{8}B neutrinos are not distinct in the Ga signal, their direct contribution is small. However, ^{8}B neutrinos are made from ^{7}Be, hence their (partial) presence (see heliograph, Figure 5) requires the precursor ^{7}Be to also be present at least in that proportion. However, even for the neutrinos associated with this reduced fraction there is no place in the Ga-signal. This leads to [25)41)42)43)44)]

⇒ the third solar neutrino problem, **"SNP #3":** $\langle\Phi_7\sigma_{Ga}\rangle_{exp} < \langle\Phi_7\sigma_{Ga}\rangle_{theor.}$

In a more stringent formulation, the 3rd SNP is described by

$$\langle\Phi_7\sigma_{Ga}\rangle_{exp} \approx \langle\Phi_{total}\sigma_{Ga}\rangle_{exp} - \langle\Phi_{pp}\sigma_{Ga}\rangle_{min.mod.} - (\approx 10\text{SNU}) < \langle\Phi_7\sigma_{Ga}\rangle_{theor.} / R_{Ka}$$

where R_{Ka} is the reduction factor (for ^{8}B neutrinos) in the Kamiokande experiment (Table 4) and the 10 SNU correction is to account for small (Figure 2) contributions from ^{8}B and CNO neutrinos, reduced by the factor R. In numbers at 2σ level, this reads $\text{SNU}(^7\text{Be}) \approx (79 \pm 24) - 79 - 10 = -10 \pm 24 < 34/1.5 = 23$. This numerical illustration reveals that the 3rd SNP has only gradually evolved as a result of error reduction in GALLEX. Contrary to PPIII (^{8}B), the 14% PPII (^{7}Be) branch is much more robust and there are not many possibilities to significantly affect it. One of the few possibilities would be a revision of the $^3\text{He}+^4\text{He} \rightarrow {}^7\text{Be}$ cross section σ_{34}. As errors shrink further, it becomes increasingly difficult and eventually at some point even inescapable to conclude ν_e-disappearance for reasons other than those related to the Sun. In view of the severe consequences of such a conclusion and recalling the very many past and present 'neutrino mass discoveries', the plea is that such a conclusion must be forced by the data against the critical resistance of the observers, rather than to accelerate it guided by wishful thinking. The statistical significance of the 3 solar neutrino problems decreases from SNP#1 to SNP#2 to SNP#3 , but the difficulties to find classical escape routes increase in that order. SNP#1 (Cl) is a > 6σ problem but potentially solvable, while SNP#3 (Ga) is < 3σ but may become very severe indeed.

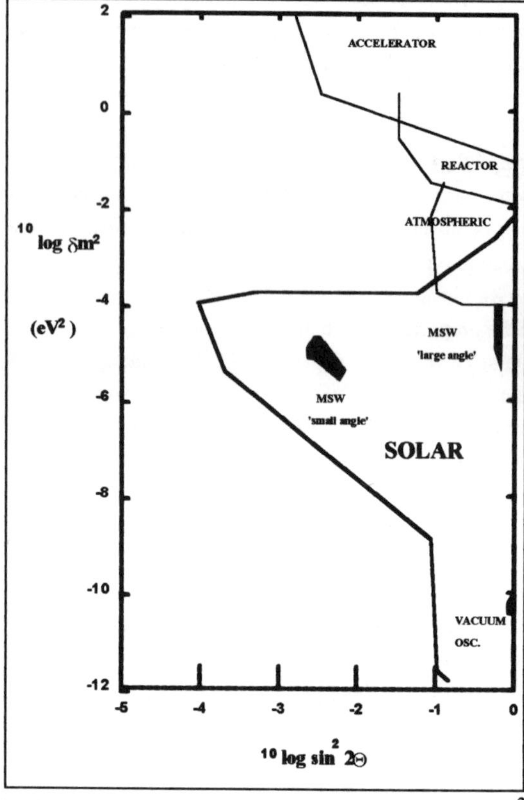

FIGURE 11. Neutrino mixing parameters and domains of sensitivity for neutrinos (antineutrinos) from accelerators, reactors, cosmic-ray-atmosphere interactions, and from the Sun.

Conditional Neutrino Masses

In this last section, we describe what one could conclude about neutrino masses if matter enhanced neutrino oscillations, as described in the introduction, are responsible for the observed reduction factors. After stressing the conditional character of this scenario, we note that there is a consistent solution for all solar neutrino experiments centered at $\Delta m^2 \approx 7 \cdot 10^{-6}$ and $\sin^2 2\theta \approx 5 \cdot 10^{-3}$ or, $\theta \approx 2°$.[30)31)39)42)43)] This is the so called 'small angle solution' in the (Δm^2, θ) plot shown in Figure 11. The second ('large angle') solution became disfavored with shrinking errors in GALLEX [36)]. In any case, Δm^2 is about the same for both solutions.

It is natural (but not certain) to assign the dominant mixing to $\nu_e \leftrightarrow \nu_\mu$ and to assume $m_{\nu_e} \ll m_{\nu_\mu}$. This leads to $m_{\nu_\mu} \approx 2.5$ meV. Intriguingly, the mixing angle $\theta_{sm.angl.sol.}$ interpreted as $\theta_{e\mu}$ comes close to an estimate expected from lepton-quark symmetry considerations,

$\theta_{e\mu} \approx (m_u/m_c)^{1/2} \cdot \theta_{max} \approx (5.6 \text{ MeV}/ 1350 \text{ MeV})^{1/2} \cdot 45° \approx 2.9°$ (m_u, m_c are quark masses). In the same spirit, if lepton-quark family mass scaling according to

$m_u^2 : m_c^2 : m_t^2 \approx m_{\nu_e} : m_{\nu_\mu} : m_{\nu_\tau}$ is relevant, one obtains

$m_{\nu_e} \approx 40$ neV ; $m_{\nu_\mu} \approx 2.5$ meV and $m_{\nu_\tau} \approx 40$ eV. This would leave the τ-neutrino intact as a dark matter candidate but qualify all attempts to measure the electron neutrino mass directly as a hopeless task. Future must tell whether this intriguingly consistent solution is accidental or whether it properly reflects the properties of leptons and quarks.

REFERENCES

1. Wolfenstein,L.1978. Phys.Rev. **D17;** 2369.
2. Mikheyev,S.P. and Smirnov,A.Y.1986. Nuovo Cimento **9C;**17.
3. Bahcall,J.N.1989.Neutrino Astrophysics. Cambridge Univ.Press.
4. Bahcall,J.N. and Pinsonneault,M.H. 1992. Rev.Mod.Phys. **64;** 885.
5. Hirata,K. et al.1989 and 1990. Phys.Rev.Lett. **63;**16 and **65;** 1297 and 1301.
6. Rapaport,J. et al.1985. Phys.Rev.Lett.**54**; 2325.
7. Krofcheck,D. et al.1985 and 1987. Phys.Rev.Lett. **55**;1051 and Phys.Lett. **B189**;299.
8. Davis,R.,Harmer,D.S. and Hoffmann,K.C.1968. Phys.Rev.Lett **20;** 1205.
9. Turck-Chièze,S. and Lopes,I.1993.Ap.J. **408;** 347.
10. Cleveland,B.1995. Nucl.Phys.B (Proc.Suppl.)**38;** 47.
11. Davis,R.1993. Proc. Int.Symp.on Neutrino Astrophysics "Frontiers of Neutrino Astrophysics", editors Y.Suzuki and N.Nakamura, Univ.Acad.Press, Tokyo p.47.
12. Voloshin,M.B.,Vysotsky,M.I. and Okun,L.B.1986. Sov.Phys. JETP **64;** 446.
13. Suzuki,Y.1994. Nucl.Phys. B (Proc.Suppl.) **35;** 407.
14. Suzuki,Y. 1995. Nucl.Phys.B (Proc.Suppl.) **38;** 54.
15. Suzuki,Y.1994. 6th'Neutrino Telescopes',Venice,1994. Proceed. (M.Baldo-Ceolin, ed.) 191.
16. GALLEX Collaboration, P.Anselmann et. al.1992. Phys. Lett. **B285;** 376.
17 GALLEX Collaboration, P.Anselmann et. al.1993. Phys. Lett. **B314;** 445.
18. GALLEX Collaboration, P.Anselmann et. al.1994. Phys. Lett. **B327;** 377.
19. GALLEX Collaboration, P.Anselmann et. al.1995. Phys. Lett. **B342;** 440.
20. Henrich,E. and Ebert,K.H.1992. Angew.Chem.,Int.Ed. (Engl.) **31**; 1283.
21. Wink,R. et al.1993. Nucl.Instr.Methods **A 329;** 541.
22. Kirsten,T.,Hartmann,F.X.,Wink,R.and Anselmann,P.1994.Nucl.Phys.B (Proc.Suppl.)**35;**418.
23. Hampel,W.1994. 3rd'Nuclei in the Cosmos',Gran Sasso, July 1994, to appear in AIP Proc.
24. Abazov,A.I. et al.1991. Nucl.Phys.B (Proc.Suppl.) **19;** 84. Talk given by V.Gavrin at the Int.Conf. "Neutrino 90", Geneva , June 1990.
25. Kirsten,T.,et al.1995. Nucl.Phys.B (Proc.Suppl.) **38;** 68.
26. Abazov,A.I. et al., SAGE collaboration.1991. Phys. Rev. Lett. **67;** 3332.
27. Gavrin,V.1993. Proc.XXVI.Int.Conf.High Energy Physics, Dallas, August 1992, J.Stanford (ed.) 1101.
28. Gavrin,V.N. et al.1994. Nucl.Phys.B (Proc.Suppl.) **35;** 412.
29. Abdurashitov, J.N.et al., SAGE collaboration.1994. Phys.Lett. **B328;** 234.
30. Hata, N., Bludman,S. and Langacker,P.1994.Phys Rev. **D 49**,3622.
31. Krastev,P. and Petcov.S.1993. Phys.Lett. **B299**; 99.
32. Berezinsky,V.1994. Nucl.Phys.B (Proc.Suppl.) **35;** 484.
33. Dar,A. and Shaviv,G.1994. Preprint Technion Ph-94-5 and: Shaviv,G.1995. Nucl.Phys.B (Proc.Suppl.) **38;** 81.
34. Motobayashi,T. et al.1994.Preprint Yale 40609-1141, subm. to Phys.Rev.Lett.,Oct.1994.
35. Gai,M. 1995. Nucl.Phys.B (Proc.Suppl.) **38;** 77.
36. GALLEX Collaboration, P.Anselmann et. al.1992. Phys. Lett. **B285;** 390.

37.Hampel,W.1990. Physics World **3**,20 and 2nd.'Nuclei in the Cosmos',Karlsruhe. Käppeler,F. and Wisshak,K. (ed.). 1992. Inst.Phys.Bristol Publ. 1993. p.629.

38.Bahcall,J. and Bethe,H. 1990. Phys.Rev.Lett. **65**;2233.

39. Castellani,V.,del'Innocenti,S. and Fiorentini,G.1993. Astron.Astrophys. **271**;601.

40. Castellani,V. et al.1994. Phys.Lett. **B324**;425.

41.Bahcall,J. 1994. Phys.Lett.**B338**;276

42.Kwong,W. and Rosen,S. 1994. Phys.Rev.Lett. **73**; 369.

43.Berezinsky,V.,Fiorentini,G. and Lissia,M.1994.Phys.Lett **B341**,38.

44.Kirsten,T.1994. Internat.Workshop 'Solar Neutrino Problem: Astrophysics or Oscillations? Gran Sasso , February 1994. Berezinsky,V. and Fiorini,E. (ed.) **1** ,34.

Superluminal Motions in our Galaxy

I.F. MIRABEL[a] AND L.F. RODRIGUEZ[b]

[a]CEA/DSM/DAPNIA/Service d'Astrophysique
Centre d'Etudes de Sacaly. 91191 Gif-sur-Yvette, France

[b]Instituto de Astronomía, UNAM
Apdo Postal 70-264, México, DF, 04510, Mexico

INTRODUCTION

Astronomical observations in the two extremes of the electromagnetic spectrum, in the domain of the hard X-rays on one hand, and in the domain of radio wavelengths on the other hand, reveal the existence of a new class of objects that we call *microquasars*. They have the particularity of combining two relevant aspects of relativistic astrophysics: black holes (of stellar origin) which are identified by the hard X-rays and gamma-rays, and relativistic jets of particles which are observed by means of their synchrotron radio emission. Because of their relative proximity, the apparent superluminal motions observed in microquasars in our own Galaxy offer the best opportunity to gain a general understanding of relativistic ejections seen elsewhere in the Universe.

The multi-wavelength approach to the gamma-ray sources observed by the telescope SIGMA abord the satellite GRANAT[1] that have spectral characteristics of black hole candidates, lead in the year 1992 to the discovery of "microquasars" in the galactic center region[2,3]. In these miniature versions are found the three basic ingredients of quasars; a black hole (in this case of stellar mass), an accretion disk with temperatures of a few tens of keV, and collimated jets of high energy particules that extend over a few light years. We have proposed that the analogy between extragalactic quasars and these scaled down versions in our own galaxy is more that morphological, and that the same physical mechanisms apply. Strictly speaking the concept of *quasar* ("quasi-stellar-radio-source") would have suited better the stellar mass versions rather than the super-massive analogs at the centers of galaxies.

The discovery of the first superluminal galactic source[4] was the result of a combination of theoretical expectations and a variety of observational signatures. In the comparative picture of *microquasars* there was something missing: although from a theoretical point of view microquasars should produce relativistic ejections with apparent superluminal motions, the latter had been observed only in AGN's and quasars. On the other hand, the radio observations of microquasars had shown time variations in the brightness of the radio lobes[5], and sometimes, also intriguing time variations in the position of the radio components[6]. Until the end of 1993 we did not know if the observed changes in the position of radio components were due to radio flashes from jets in the interstellar medium or, as confirmed later, due to apparent superluminal motions of particles streaming away from the compact objects at relativistic velocities.

SUPERLUMINAL MOTIONS IN GRS 1915+105

GRS 1915+105 (also known as the hard X-ray transient in the constellation of Aquila) was discovered by the WATCH experiment aboard GRANAT[7] on August 15, 1992 and further located at arcmin accuracy by SIGMA[12]. Although there is no optical counterpart brighter than 21 mag (the source is on the galactic plane at $l = 45.4°$, $b = -0.3°$), a time variable infrared counterpart of K $\sim$ 12-14 mag was identified[8]. The broad band infrared spectrum suggests that GRS 1915+105 is beyond 30 magnitudes of visual absorption, which is consistent with a column density along the line of sight of $\sim 4\ 10^{22}$ H cm^{-2} that is derived from radio observations of atomic[6] and molecular gas[9]. A column density of $\sim 5\ 10^{22}$ H cm^{-2} has been estimated from the ROSAT X-ray spectrum of the source[10]. Such large interstellar column densities and visual obscuration imply that the source is at a distance $\geq$ 10 kpc. From HI absorption at λ21 cm a kinematic distance of 12.5 $\pm$ 1.5 kpc is derived[6].

GRS 1915+105 is likely to be a black hole because: 1) as most black hole candidates the X-ray spectrum shows a hard X-ray tail at $\geq$ 150 keV[11,12], and 2) it has often been observed with integrated X-ray luminosities $L_x \geq 4\ 10^{38}$ erg s^{-1}, which is larger than the Eddington luminosity of an object of mass $\sim 3\ M_\odot$, the theoretical upper limit for neutron stars.

This candidate black hole draw our attention since 1992 because among the GRANAT hard X-ray sources of our VLA monitoring program it was the one that had exhibited the most striking changes in the position of compact radio components. On December 1993 GRS 1915+105 produced a strong radio outburst, and by February 1994 we already realized the existence of jets that appeared to change in a few days time. At a distance of $\geq$ 10 kpc the observed displacements appeared superluminal. To understand the nature of such displacements, we remained on the look-out for another strong radio outburst. Such outburst occured in March-April 1994[6], when we were already prepared to follow the evolution of the source by regular observations with the VLA in the configuration that provides the highest angular resolution.

Figure 1 shows a pair of bright radio condensations emerging in opposite directions from a compact, variable core[4]. We find that both before and after

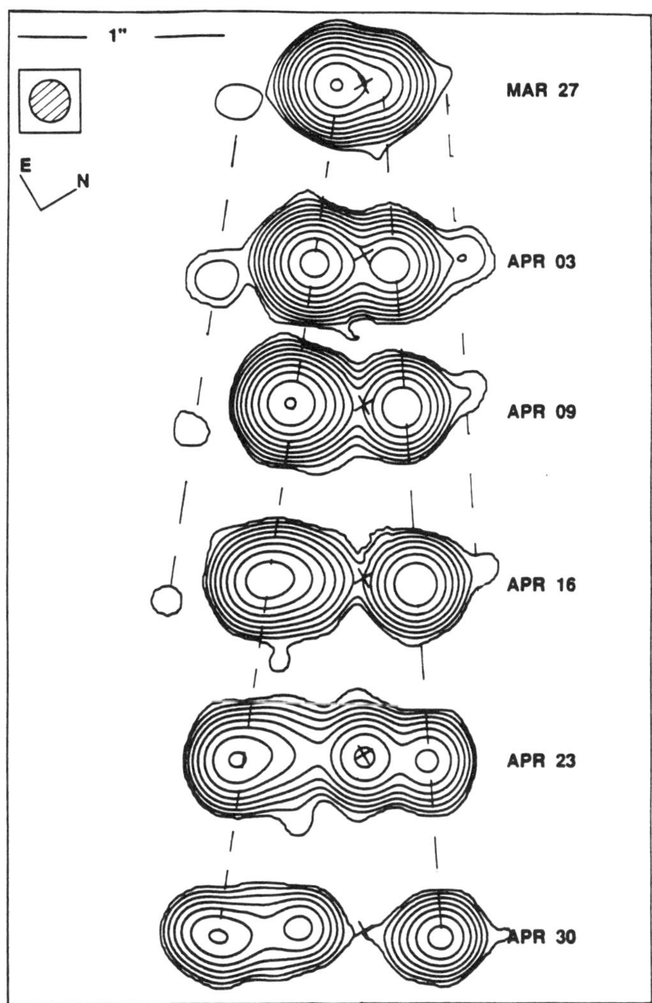

Figure 1: Pair of radio condensations moving away from the hard X-ray source GRS 1915+105. These uniform-weight VLA maps were made at λ3.5-cm for the 1994 epochs on the right side of the figure. Contours are 1,2,4,8,32,64,128,256 and 512 times 0.2 mJy/beam for all epochs except for March 27 where the contour levels are in units of 0.6 mJy/beam. The half power beam width of the observations, 0.2 arc sec, is shown in the top left corner. The position of the stationary core is indicated with a small cross. The maps have been rotated 60° clockwise for easier display. Note in the first four epochs the presence of a fainter pair of condensations moving ahead the bright ones and in the last epoch the presence of a new southern component.

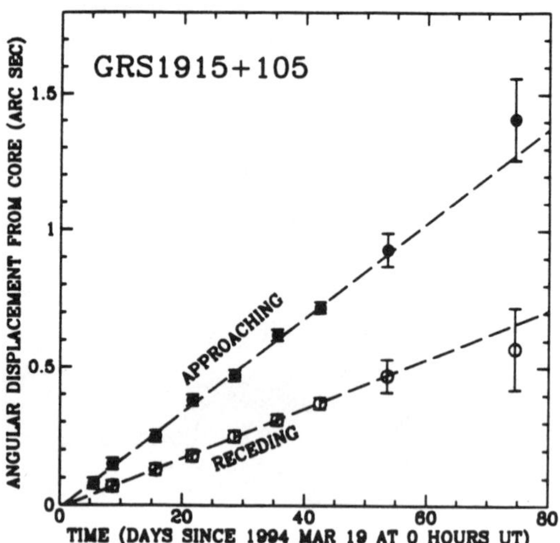

Figure 2: Angular displacements of the bright radio-emitting components as they move from the stationary core as a function of time. The regression lines converge toward the same point on the time axis. This implies that the bright pair of plasma clouds were ejected on March 19, 1994 at 20 ± 5 UT. The proper motions are consistent with ballistic motion for the time interval of 75 days over which we could follow the condensations.

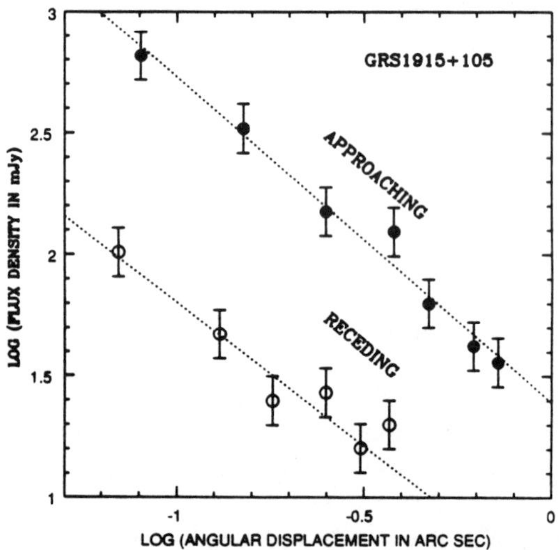

Figure 3: Flux densities at $\lambda 3.5$ cm as a function of the angular displacements from the core. The flux ratios of the condensations at a given angular displacement are consistent with Doopler-boosting of radiation from a twin pair of condensations moving apart with relativistic bulk velocities.

the remarkable ejection event shown in Figure 1, GRS 1915+105 ejected other pairs of condensations but with flux densities one to two orders of magnitude weaker. One of these weaker pairs can be seen in the first four maps of Figure 1, where one can see a fainter pair of condensations moving ahead of the bright ones at about the same speed and direction. So far we have followed the motions of three pairs of plasma clouds in aproximately the same direction on the sky with similar proper motions. The time separation between ejections suggests a quasiperiodicity with intervals in the range of 20 to 30 days.

In Figure 2 we plot the angular displacements from the stationary core of the bright pair ejected on March 19, 1994. The angular displacements are consistent with ballistic (that is, unaccelerated) proper motions over an interval of 75 days while they remained detectable. At a distance of 12.5 kpc the proper motions of the approaching (17.6 $\pm$ 0.4 mas d^{-1}) and receding (9.0 $\pm$ 0.1 mas d^{-1}) condensations imply apparent velocities on the plane of the sky of 1.25c and 0.65c, respectively. The analysis of relativistic distorsion effects shows that the ejecta move with a true speed of 0.92c at an angle θ = 70° to the line of sight.

RELATIVISTIC EFFECTS IN GRS 1915+105

Figures 1 and 2 show two asymmetries: one in apparent transverse motions, another in brightness. The cloud moving faster appears brighter. Both asymmetries in proper motions and in brightness are consistent with the hypothesis of an anti-parallel ejection of twin clouds moving at relativistic velocities. The asymmetry in apparent proper motions is well known and has already been explained in publications quoted by Mirabel and Rodríguez[4]. Here we concentrate on the effects of relativistic boosting.

Due to relativistic aberration the brightness ratio of the approaching and receding condensations (measured at equal distances from the core) are given by

$$\frac{S_a}{S_r} = \left(\frac{1 + \beta\ cos\theta}{1 - \beta\ cos\theta}\right)^{k-\alpha}, \tag{1}$$

Since β = 0.92, θ = 70°, and α = - 0.8, the ratio of the apparent surface brightnesses for a given angular distance from the ejection center is predicted[4] to be in between 6 (k = 3 for discrete clouds) and 12 (k = 2 for continuous jets). Figure 3 shows that for a given angular separation, the observed flux ratio between the approaching and receding condensations is 8 $\pm$ 1, and that during the first few weeks since ejection the two condensations fade out as they move apart with an exponential law $S_\nu \propto \phi^{-1.3\pm0.2}$. The flux ratio of 8 is consistent with Doopler-boosting due to relativistic bulk motions of the matter that emits the radio waves. Because GRS 1915+105 is the first superluminal source of anti-symmetric ejections in which the motions of both the approaching and receding ejecta have been followed, it is the first case where one can resolve the ambiguities that so far have dominated the interpretation of superluminal sources. From our observations we can conclude that *the proper motions and brightness ratios of the clouds indicate that the superluminal motions in GRS 1915+105 correspond to true bulk motions of matter, rather than to the propagation of shocks, or radio echoes.*

The ratio of the observed flux densities S_a for the approaching and S_r for the receding ejecta relative to the emitted (in the frame of reference of the blob) flux density S_o is

$$\frac{S_{a,r}}{S_o} = \delta_{a,r}^{k-\alpha}, \tag{2}$$

where $\delta_{a,r}$ is the Doppler factor $\frac{\nu_{a,r}}{\nu_o}$ for the approaching and receeding condensations

$$\delta_{a,r} = \gamma^{-1}(1 \mp \beta \; cos\theta)^{-1}, \tag{3}$$

In Figure 4 are shown the observed brightness relative to the intrinsic brightness as a function of the angle between the direction of the ejection and the line of sight, computed with eq. (2) and using the parameters found in the case of GRS 1915+105, namely, α = - 0.8, $\beta = 0.92$, and an intermediate case for the structure of the ejecta (k = 2.5). This figure shows that for $\theta \geq 48°$ *the approaching ejecta are also de-boosted. This is a consequence of the competition between Doppler boosting, relativistic beaming, and the stretching of the time scales in an object moving at a relativistic velocity of 0.92c.*

Looking at Figure 4 it is easy to understand why the motions of receding ejecta in extragalactic superluminal sources have never been followed, whereas this was easy in the galactic source GRS 1915+105. Superluminal radio components in distant quasars can be observed when the brighness is greatly enhanced by Doppler boosting, and this occurs when the approaching component is viewed almost "head on", namely, if θ is small. Figure 4 shows that the brightness ratio for an anti-symmetric jet with $\gamma = 2.6$ seen "head-on" ($\theta \leq 10°$) would be of the order of 10^4. But in quasars usually $\gamma \gg 1$ and the boosting and de-boosting factors are $\sim 8\gamma^3$ and $1/8\gamma^3$, respectively, which for $\gamma \sim 10$ imply flux ratios $\geq 10^7$. Therefore, it is not surprising that *the uncertainties in the interpretation of extragalactic superluminal motions can be better resolved by the study of superluminal sources*

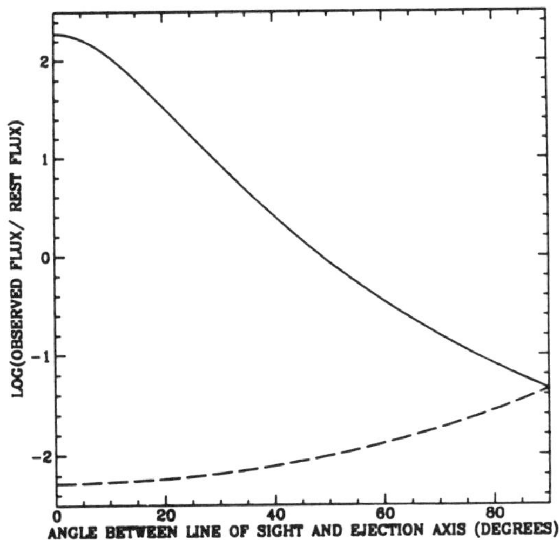

Figure 4: Observed brightness relative to the intrinsic brightness for the approaching (continuous line) and receding (dashed line) clouds as a function of the angle of the direction of motion to the line of sight θ, for $\beta = 0.92$c, α = - 0.8, and k = 2.5. Note that for $\theta \leq 10°$ the flux ratio is $\geq 10^4$, and that for $\theta \geq 48°$ the flux of the approaching ejecta are de-boosted.

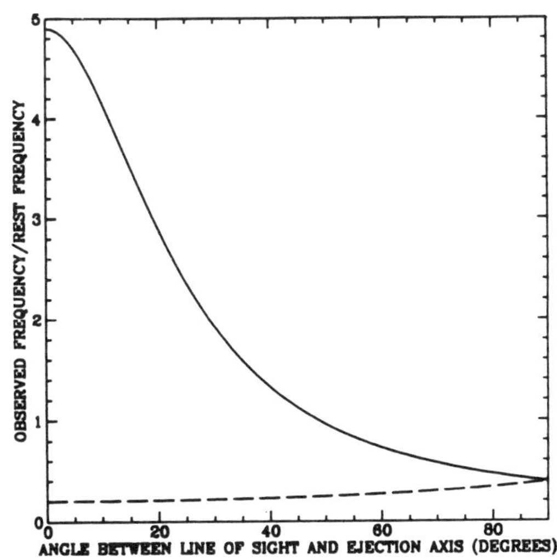

Figure 5: Doppler factors ν_{obs}/ν_o for the approaching cloud (continuous line) and receding cloud (dashed line) as function of θ. Note that for $\theta \geq 48°$ the lines from the approaching ejecta should also be redshifted.

within our own Galaxy, where the proper motions of both the approaching and receding components can be measured.

In Figure 5 are plotted the Doppler factors for the approaching and receding condensations as a function of θ for $\beta = 0.92$, computed with equation (3). Due to the stretching of the time scales eventual emission lines from the approaching jets at angles $\geq 48°$ should also be redshifted. In the case of GRS 1915+105 ($\theta = 70°$) the redshifts of the receding and approaching condensations should be 2.36 and 0.75. Since the source is behind 30 mag of visual absorption the best atom lines for a search of entrained ions in the jets would be H_α redshifted by a factor of 2.36 into the infrared K band, and/or iron lines in the X-rays[4].

In Figure 6 are shown the evolution of the fluxes of the receding and approaching condensations as a function of the time Δt since the ejection. The approaching condensation appears to fade out according to the equation $(S_a/\mathrm{mJy}) = 7.8\ 10^3\ (\Delta t/\mathrm{days})^{-1.4}$, whereas the receding condensation seems to fade somewhat more slowly. This may be due to the interval of time required for the radiation to travel accross the source. As the source expands the crossing time increases and in the sequence of instantaneous pictures of the source taken from Earth the approaching cloud is progressively older than the receding one. However, we note that in the time evolution of the radio fluxes must be involved other factors, such as changes in the direction of motion and/or interactions with the interstellar medium. An example of this may be seen in the map of April 30 (Figure 1), when a brightening of the receding component was observed.

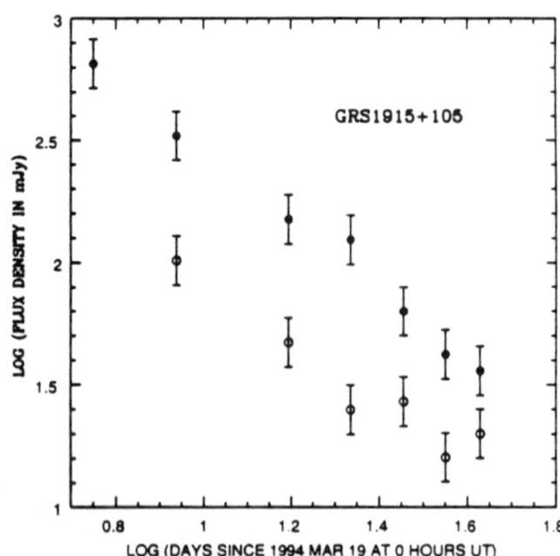

Figure 6: Flux densities at λ3.5 cm as a function of time since the ejection. Fluxes for the approaching and receding condensation are represented by filled and open symbols, respectively. Note that the receding cloud seems to fade out somewhat more slowly than the approaching cloud.

MASS AND ENERGY OF THE EJECTA IN GRS 1915+105

To estimate the mass and energy of the clouds ejected by GRS 1915+105 on the major event of March 19, 1994[4] we first calculate the magnetic fields. For the approaching condensation on 1994 March 24 the angular dimensions were of 60 $\times$ 20 mas (see ref. 4), with the major axis approximately aligned along the outflow axis. The flux density was 655 mJy at 3.5-cm. The spectral index between 1.4 and 15 GHz was about -0.4. These parameters have to be corrected for relativistic effects. Since $\beta = 0.92$ and $\theta = 70°$, we get $\gamma = 2.6$, and $\delta_a = 0.56$. Using equation (2) with k = 3, the estimated flux density S_o (in the frame of reference of the blob) is about 4.7 Jy (about 7 times larger than the observed flux density).

The real major axis, L_o, is related to the observed (in projection) major axis, L_a, by

$$\frac{L_a}{L_o} = sin\theta \; \delta_a, \tag{4}$$

That is, $L_o = 1.9\ L_a$. Then, the true dimensions of the jet are 110 $\times$ 20 mas. We then take as characteristic dimension a geometric mean of 35 mas (or 7 10^{15} cm at a distance of 12.5 kpc).

Using the formulation of Pacholczyk[12] for minimum energy and integrating from 1.4 GHz to 15 GHz (the range over which we measured the flux densities on that date[6]), *and assuming only relativistic electrons*, a magnetic field of about 50 mGauss and an energy of about 1.5 10^{46} ergs in the relativistic electrons is obtained. Multiplying by two to account also for the receding condensation one obtains $\sim$ 3 10^{46} ergs in relativistic electrons for that epoch. This is, of course, the *internal* relativistic energy in the radio blobs. In addition, the plasma clouds have a kinetic energy due to their bulk motion at 0.92c.

To calculate the mass of the blobs one can use the usual formula for critical frequency for synchrotron radiation

$$\nu_c(GHz) = 4.2\ (\gamma_i/1000)^2\ B(mGauss), \tag{5}$$

and since we have observed strong radiation up to 240 GHz (IRAM observations made on 1993 Dec 6, ref. 6) with a magnetic field of tens of mGauss, we adopt $\gamma_i = 1000$ as a rough average value for the (internal) motion of the relativistic electrons. Then, dividing the total energy over $\gamma_i \times m_e \times c^2$, where m_e is the electron rest mass, one can obtain the number n of (relativistic) electrons in the blobs.

The mass of the plasma clouds can be obtained in two different ways. 1) Assuming that there is one (non-relativistic) proton per (relativistic) electron one gets a proton mass estimate in the order of 10^{25} gr. 2) Assuming that there are no protons in the plasma clouds (with positrons being the positive ions) one obtains a similar mass because the electrons are moving so fast that their relativistic mass (as opposed to their rest mass) is comparable to that of the non-relativistic protons. In other words, the energy estimate in relativistic electrons equals n $\times$ $\gamma_i \times m_e \times c^2$. On the other hand, the relativistic mass of the blob (from electrons alone) is n $\times$ $\gamma_i \times m_e$. So, taking the relativistic energy and dividing over c^2 gives the relativistic mass in electrons, which also is about 10^{25} grams. This is the mass one has to use to estimate the relativistic kinetic energy of the blobs (due to their bulk motion at 0.92c or $\gamma_b = 2.6$). Therefore, *the presence or absence of non-relativistic protons affects the mass estimate by a factor of 2 only.*

It is interesting that the blobs have comparable amounts of internal and kinetic bulk energy. The internal energy is $E_i = n \times \gamma_i \times m_e \times c^2$ while the kinetic bulk energy is of order $E_b = (\gamma_b - 1)E_i$. Since $\gamma_b = 2.6$, both energies are comparable.

The estimated total mass of the condensations of $\sim 2\ 10^{25}$ g (~ 3 % the mass of the Earth) must be a large fraction of the mass of the accretion disk, since it is equivalent to the total mass that would be accumulated during one year at the typical rate of 10^{-8} $M_\odot$ yr^{-1} that is required to account for the X-ray light curve of GRS 1915+105[11].

The kinetic energy of the plasma clouds suggests an acceleration mechanism with very large power. From the radio monitoring with the Nançay radiotelescope[6] we know that the ejection event of 19 March 1994 lasted $\leq$ 12 hours, requiring a *minimum* power of 7.2 10^{41} erg s^{-1}. Our study of a more recent ejection event indicate that these phenomena may last only a few minutes, which applied to the 19 March 1994 event would imply a power of $\sim 10^{44}$ erg s^{-1}. This is $\sim 10^6$ times the maximum steady photon luminosity of GRS 1915+105, suggesting that a radiation acceleration mechanism is improbable.

GRO J1655-40: A SECOND SUPERLUMINAL SOURCE

It was not by accident that soon after the superluminal motions in GRS 1915+105 had been reported, similar phenomena were searched and immediately found by two different groups in the first hard X-ray nova (GRO J1655-40; X-ray nova Scorpii) that appeared during August-September 1994. By that time the radio observations of hard X-ray microquasars[2,3,4] had already shown the emergence of jets as a relatively common phenomenon in energetic galactic sources.

In GRO J1655-40 a hard X-ray tail was detected up to 600 keV and as GRS 1915+105 it is believed to be a black hole in a binary system[14]. The fast rise of the X-ray flux and the appearance of an optical nova with a characteristic low mass X-ray binary spectrum[15] suggests that GRO J1655-40 is an X-ray nova. In contrast, GRS 1915+105 cannot be classified as such since in the hard X-rays it had a rise time of a few months, it has been active from 1992 until present, usually undergoing long duration flares interspersed with quiescent periods[14].

GRO J1655-40 is at $\sim$ 3.2 kpc from the Sun and has an optical counterpart similar to other accreting-black-hole candidates in low mass binaries. In the optical it showed a blue spectrum with prominent broad hydrogen emission lines during outburt[15], but during its quiescent state six months later it showed a redder spectrum with narrower emission lines[16]. In contrast, GRS 1915+105 is obscured by intervening dust and no *prominent* emission lines have been detected in the J, H, and K band spectra of the variable infrared counterpart[9]. The compact objects are likely to have binary companions since the optical light curve of GRSO J1655-40[15], and the 0.9 mag variation at 2.2 μm in 24 hours observed in GRS 1915+105[8], hint for eclipses. GRO J1655-40 is likely to be a low mass binary[15] whereas in GRS 1915+105 there are indications for the presence of a red supergiant[9].

The relativistic ejections observed in the radio in GRO J1655-40 have striking similarities as well as differences with those in GRS 1915+105. Bright components moving apart with proper motions in the range of 40 to 65 mas d^{-1} were independently observed by the Southern Hemisphere VLBI Experiment array[17], the VLA and VLB[18]. At a distance of 3.2 kpc the motions of the ejecta have been fit[18] by a kinematic model with a velocity of 0.92, and a jet axis inclined 85° to the line of sight at a position angle of 47°, about which the jets rotate every three days at an angle of 2°.

In contrast to the main ejection observed in GRS 1915+105[4] the flux ratios of the blobs on either side of GRO J1655-40 cannot be ascribed to relativistic Doppler boosting. In GRO J1655-40 the asymmetry in brightness flips from side to side[18]. Not only the jets appear to be intrisically asymmetric, but also the sense of that asymmetry changes from event to event[18]. Therefore, although similar spatial velocities of 0.92c are found in both superluminal sources, *due to intrinsic asymmetries the ultimate physical interpretation of the superluminal expansions in GRO J1655-40 remains ambiguous.*

GALACTIC AND EXTRAGALACTIC SUPERLUMINAL SOURCES

It is noteworthy that, while both superluminal sources in the Galaxy have jet axes close to the plane of the sky, the axes of superluminal ejections seen in extragalactic sources are close to the line of sight. *A priori*, and without taking other effects into account, one expects the axis of an ejection to be closer to the plane of the sky, since the probability of a source having an angle θ between the axis of the ejection and the line of sight is proportional to $\sin(\theta)$. Then, the galactic sources appear to be exhibiting a behavior more in accord with a simple probabilistic argument than the extragalactic sources. If no other effects are important, one should detect preferentially sources with $\theta \sim 90°$. What is probably happening is that the galactic sources appear, given their proximity, much brighter in the sky, and the large Doppler boostings required to bring extragalactic ejecta to detectable flux densities with VLBI techniques are not indispensable.

We then expect that this difference in θ between galactic and extragalactic superluminal sources will be characteristic.

Of course, the detection of sources that depart from this tendency would be of great astrophysical interest. An extragalactic source with $\theta \sim 90°$ would be very weak and difficult to detect, but if this is achieved, it is likely that both the receding and approaching ejecta could be detected and important restrictions to the distance could be set. On the other hand, a galactic source with $\theta \sim 0°$, would be short-lived and difficult to monitor, but could have very large flux densities and exhibit extraordinary proper motions and apparent velocities far in excess of c. One of the characteristics of sources with $\theta \sim 90°$, such as the two detected in the Galaxy, is that the apparent velocities exceed c only modestly.

JETS FROM GALACTIC BLACK HOLES AND NEUTRON STARS

In Table 1 we list the galactic sources with relativistic ejections. Synchrotron radio jets have been reported in four neutron star candidates and in four black hole candidates. It is still unclear if Cygnus X-1 has or has not associated radio jets[19]. References on the jets in Circinus X-1, Cygnus X-3, SS 433, 1E 1740.7-2942 and GRS 1758-258 were given in a previous review (ref. 20) on radio jets.

Proper motions of the ejecta have been measured for 5 out of the 8 sources listed in Table 1. All measurements were in the radio, except for the fast-moving optical features in the Crab nebula[21]. The more steady hard X-ray sources 1E1740.7-2942 and GRS 1758-258 have, as Cygnus X-1, radio counterparts of only a few mJy. Although from year to year slight changes in the brightness of the radio lobes of 1E1740.7-2942 are detected[22], in these relatively steady and weak radio sources it is difficult to measure proper motions. By contrast, proper motions can be measured in high energy sources with strong outbursts in the X-rays and radio.

The sources in Table 1 with soft X-ray spectra produce ejecta with velocities in the range of 0.2c to 0.3c, whereas sources with hard X-ray tails produce ejecta with velocities $\geq 0.90c$. The two superluminal microquasars appear to be scaled-up versions of the stellar sources of radio jets SS433 and Cygnus X-3. Although GRS 1915+105 and GRO J1655-40 may be reminiscent of SS433 and Cygnus X-3, the actual velocity of the ejecta and the X-ray luminosity are much larger in the first two sources. In particular, the kinetic energy per unit mass of the ejecta, that is proportional to (γ-1), is $\sim$40 times larger in the black hole candidates. However, similar large ratios of kinetic to photon luminosity are found for these four sources. In all four sources the kinetic energy of the electrons due to the bulk motion of the ejecta is comparable to the high energy cut-off in the X-ray spectra, suggesting *strong coupling via Compton scattering.*

We note that well before the discovery of pulsars, Oort and Walraven[21] proposed that remarkable fast-moving optical-ripples that had been observed by Baade in the Crab nebula in 1942-45 were large scale injections of relativistic particles ejected by the stellar remnant. In contrast to the other seven sources in Table 1, the X-ray spectrum of the Crab is synchrotron and the acceleration mechanism of the ejecta is likely to be different, since the ultimate source of energy for the injections into the Crab nebula is the rotational energy of the single neutron star.

Table 1 shows that the velocity of the ejecta from black hole candidates is $\geq$ 0.90c whereas it is $\leq$ 0.30c in neutron star candidates. We wonder if this could be related to the fact that the escape velocity (of order of the circular velocity in the inner orbital radius) for a neutron star is $\sim$ 0.3c, whereas it is $\geq$ 0.9c near the Schwarzchild radius of a black hole. Although it is clear that five case studies are not sufficient to derive general conclusions, and that at present there is no firm theoretical basis, perhaps this correlation indicates *a profound relation between the properties of the ejecta and the nature of the stellar remnants.*

TABLE 1: RELATIVISTIC JETS IN OUR GALAXY

Source	X-ray emission	Collapsed Object	v/c	Other Information
Circinus X-1	persistent;soft	neutron star	-	
Cygnus X-3	persistent;soft	neutron star	0.3	
SS 433	persistent;soft	neutron star ?	0.26	
1E 1740.7-2942	persistent;hard	black hole	-	changes in lobes
GRS 1758-258	persistent;hard	black hole	-	
GRS 1915+105	transient;hard	black hole	0.92	
GRO J1655-40	nova;hard	black hole	0.92	
Crab	Synchrotron	neutron star	0.2	

LIGHT CURVES IN THE RADIO AND HARD X-RAYS

In GRO J1655-40 the radio outbursts at centimeter wavelengths generally follow bursts of hard X-ray emission (20-400 keV) with a delay of a few days[14]. In general, the feeding of material into the relativistic jets seems to be coincident with a decline in the hard X-rays, and the radio spectrum which first exhibits an optically thick flat spectrum, evolves to an optically thin synchrotron spectrum.

In GRS 1915+105 the detailed comparison between the hard X-rays measured by BATSE and the radio monitoring[6] for a period of 6 months is still in progress. However, we know that the major ejection event of March 19, 1994 occured at the time of a maximum in the 8-60 keV photon counts[23], and the rise in the radio flux was coincident with a decline in the X-ray flux. During the radio outbursts of GRS 1915+105 rapid changes in the spectral index from flat to optically thin synchrotron have been observed[6]. Therefore, in GRS 1915+105 the ejection of the radio emitting clouds seem to have occured at the time of a drop in the hard X-rays.

We propose that a fraction of the mass in the immediate environment of the black hole is ejected in collimated jets perpendicular to the accretion disk. As the plasma moves away it expands and cools down, moving its radiation peak from the X-rays into the radio waves. In the context of this hypothesis one also expects during ejections a rapid evolution of the X-rays spectral index, namely, the drop in hard X-rays should preceed a maximum in the softer X-rays.

As mentioned above, the clouds ejected from GRS 1915+105 on March 19, 1994 must be a large fraction of the accretion disk mass, since they have the mass that would have accumulated in a year at a rate of 10^{-8} $M_\odot$ yr^{-1}. In this context one would expect that after that major ejection event the source has come back into a quiescent state in the hard X-rays.

There are indications that the intrinsic expansion of the radio-emitting clouds is relativistic. Rodríguez and Mirabel[22] found in GRS 1915+105 an increase of the radio flux by a factor of 2 in 4 minutes. At a distance of 12.5 kpc this implies that the radio emitting clouds initially have sizes smaller than an astronomical unit and that they have internal expansions at velocities $\geq 0.8c$.

GAMMA-RAY BURSTERS AS MICROQUASARS

Relativistic bulk motion and beaming of the radiation appear as essential ingredients of γ-ray bursts[24] and Paczyński[25] has proposed that the γ-ray bursts come from microquasars in galaxies at cosmological distances. This extragalactic scenario calls for more extreme microquasars formed by a black hole with an extremely dense, possibly neutron star torus spinning around it and producing super-relativistic ejections with bulk Lorentz factors of hundreds. In this context, the study of the few less extreme microquasars in our own Galaxy may provide clues for the understanding of one of the most intriguing current mysteries in astrophysics.

We have addressed the question on the possibility that GRS 1915+105 is a repeating source of soft gamma-ray bursts[26]. The ejection events observed in GRS 1915+105 require a power that far exceeds the typical luminosity of the soft-gamma-ray bursts (SGRs). Whereas the luminosity of the three short ($\leq$ 1 sec) soft γ-ray bursts that were observed in 1992 from the same region of the sky[27] was $\sim 10^{40}$ erg s^{-1}, we have shown above that the power of the major ejection event observed in GRS 1915+105 was of 10^{42-44} erg s^{-1}. Although coincidences in position and in time suggest that GRS 1915+105 may be a SGR, until more accurate positions for the SGRs are obtained, the question on the association of the transient source GRS 1915+105 with the SGRs observed by BATSE must remain open[26].

A RELATIVISTIC METHOD TO DETERMINE DISTANCES

We have proposed a new method to determine the distance to sources of relativistic ejections using special relativity constraints[4]. If one can measure the proper motions of the approaching and receding ejecta μ_a and μ_r, and the Doppler factor $\delta_{a,r}$ of spectral lines arising in either the approaching or receding jets, using the three equations

$$\beta\ cos\theta = \frac{\mu_a - \mu_r}{\mu_a + \mu_r}, \tag{6}$$

$$D = \frac{c\ tan\theta}{2} \frac{(\mu_a - \mu_r)}{\mu_a \mu_r}. \tag{7}$$

$$\delta_{a,r} = \gamma^{-1}(1 \mp \beta\ cos\theta)^{-1}, \tag{8}$$

one can resolve the system and find β = v/c, θ, and the distance D to the source.

Since in GRS 1915+105 no redshifted lines have been detected so far, we could only derive a relativistic upper limit for the distance[4]. Obviously, to apply this method one has to observe from the jets Doppler-shifted lines, as in the relativistic jets of the famous stellar source SS 433[28]. At present it is not clear if the rapidly moving ions in the jets of SS 433 are intrinsic to the jets or if they have been entrained from the stellar wind of a binary companion by a pure relativistic pair jet. If the latter is true, this method could be applied mostly to high mass binaries, and in the decades to come to extragalactic AGNs.

ASTROPHYSICAL IMPLICATIONS

The discovery of superluminal motions within our own Galaxy has several general implications in different fields of astrophysics.

1) Because real faster-than-light motion is impossible according to relativity, the observation of superluminal motions had been used as an "argument" for a non-cosmological nature of the redshifts in quasars, and hence, against big-bang cosmologies. The observation of apparent superluminal motions in galactic microquasars at known distances eliminate this argument against an expanding Universe.

2) The discovery of superluminal motions in galactic high energy sources suggests an underlying unity in the physics of relativistic jets over an enormous scale range, from the galaxy-size ejecta seen in active galactic nuclei and quasars, to their comparatively smaller analogues in the Milky Way.

3) Because of their relative proximity, microquasars in our Galaxy offer the best opportunities to gain a general understanding of the relativistic ejections seen elsewhere in the Universe, and to solve the uncertainties that have so far dominated the interpretation of the superluminal motions. The double-sided motions observed in GRS 1915+105 are consistent with true relativistic bulk motions of the radiating matter rather than with the propagation of pulses and/or echos in the surrounding medium.

4) The study of microquasars in our own Galaxy may provide clues for the understanding of one of the most intriguing mysteries in astrophysics: that of the nature of the gamma-ray bursts.

5) The study of GRS 1915+105 revealed the possibility of applying a new method to determine distances in astronomy that is based on relativity constraints. If one can measure the proper motions and Doppler factors of the jets, the distance to the source of relativistic ejections can be determined. Although this may seem technologically difficult at present, perhaps this method will be used in the years to come.

ACKNOWLEDGEMENTS

We thank A. Poveda for calling our attention to the Crab data and J. Paul for his comments on Table 1.

REFERENCES

1. Goldwurm, A. et al. 1994, Nature 371, 589
2. Mirabel, I.F., Rodríguez, L.F. Cordier, B., Paul, J., Lebrun, F. 1992, Nature 358, 215
3. Rodríguez, L.F., Mirabel, I.F., & Martí, J. 1994, ApJ 401, L15
4. Mirabel, I.F. & Rodríguez, L.F. 1994, Nature 371, 46
5. Mirabel, I.F. & Rodríguez, L.F. 1994, AIP Conference Proceedings 304, 413
6. Rodríguez, L.F., Gerard, E., Mirabel, I.F., Gómez, Y., Velázquez, A. 1995, ApJS in press
7. Castro-Tirado, A. et al. 1994, ApJS 92, 469
8. Mirabel, I.F. et al. 1994 A&A 282, L17
9. Chaty, S., Mirabel, I.F., Duc, P.A., Wink, J., Rodríguez, L.F. 1995, in preparation
10. Greiner et al. 1994, AIP Conference Proceedings 304, New York, p 260
11. Harmon, A. et al. 1994, AIP Conference Proceedings 304, New York, p 210
12. Finoguenov, F. et al. 1994, ApJ 424, 940
13. Pacholczyk, A.G. 1970, Radio Astrophysics, Freeman, San Francisco
14. Harmon, B.A., et al. 1995, Nature 374, 703
15. Bailyn, C.D., et al. 1995, Nature 374, 701
16. Mirabel I.F. et al. 1995, in preparation
17. Tingay, S.J. et al. 1995, Nature, 374, 141
18. Hjellming, R.M. and Rupen, M.P. 1975, preprint
19. Martí, J, Rodríguez, L.F., Mirabel, I.F. & Paredes, J.M. 1995, submitted to A&A
20. Mirabel, I.F. 1994, ApJS 92, 369
21. Oort, J.H. & Walraven, Th. 1956, BAN 462, 285
22. Rodríguez, L.F. & Mirabel, I.F., 1995 in preparation
23. Sazonov, S. et al. 1995, in press
24. Baring, M.G. Proc. of the Erice NATO ASI (May 7-17, 1994), Kluwer Academic, in press
25. Pacyński, B. 1993, Annals of the New York Academy of Sciences, 688, 321
26. Mirabel, I.F. & Rodríguez, L.F. 1995, Proceeding of the 29th ESLAB Symposium, in press
27. Kouveliotou, C. et al. 1993, Nature, 362, 728
28. Margon, B.A. 1988, Rev. Astr. Astrophys. 424, 940

THE GALACTIC CENTER

R.Genzel, A.Eckart and A.Krabbe
Max-Planck Institut für extraterrestrische Physik
Garching, FRG

ABSTRACT

The Galactic center is a superb laboratory of modern astrophysics where astronomers can study at unprecedented spatial resolution and across the entire electromagnetic spectrum physical processes that may also happen at the cores of other galaxies. Following a brief review of the phenomena observed in the central few hundred parsecs, we discuss two main issues in more detail. The source(s) of luminosity in the central parsec and the nature of the compact radio source ***SgrA****, the most likely candidate for a massive central black hole.

I. INTRODUCING THE PHENOMENA

Until about 20 years ago, investigations of the Galactic nucleus were severely impeded by the large amount of interstellar dust and gas situated in the Galactic plane between the Sun and the nucleus, and preventing observations in the visible, ultraviolet and soft X-ray bands. In the past two decades, however, our knowledge about the Galactic nucleus has improved dramatically, as sensitive high resolution observations in the radio, infrared, hard X-ray and γ-ray bands have become available. In the following, a brief ***summary of these phenomena*** as revealed by a number of observations is given, followed by a more detailed account of recent results on the ***exploration of the central parsec***. For more extensive discussions we refer to recent reviews (1,2,3).

The nuclear mass is dominated by stars, except perhaps in the innermost parsec. Infrared observations on a scale of 100 to 1000 pc show that these stars appear to be distributed in a ***rotating bar*** (4). Binney et al.(5) have pointed out that the gravitational torque of this bar may account for the ***non-circular motions*** of interstellar gas clouds found by radio spectroscopy (6,7). The non-circular motions in turn may trigger ***gas infall*** into the nucleus (5). X-ray spectroscopy of 6.7keV Fe K-shell emission (and recently of other K- and L-shell lines) has revealed a component of ~10^8 K intercloud, ***coronal gas*** that permeates the central bulge/disk on a scale of about 100 pc (8,9). The coronal medium cannot be gravitationally bound to the nuclear bulge and probably escapes as a ***Galactic wind***. There is increasing evidence from γ-ray spectroscopy of the 1.8 MeV ^{26}Al line (10) and from infrared stellar spectrophotometry (11,12, see below) that ***(massive) star formation*** has occured throughout the Galactic center region no longer than 10 million years ago. Hence the energy input responsible for the coronal gas could be from ***supernovae*** that have been exploding in the central bulge (3). Also on a scale of 100 pc several variable, spectacular hard X-ray and γ-ray sources have been found (13). They may represent ***stellar black holes or neutron stars*** accreting gas from a companion or from nearby dense gas clouds. A broad emission bump at ~500 keV and a twin radio jet seen in one of these sources (the „***Great Annihilator***" 1E1740.7-2942) may signify the presence of a relativistic electron-positron jet annihilating in the environment of the compact source (14,15,16). Elsewhere in these proceedings, Mirabel (see also 17) gives a detailed account of the recent detection by radio interferometry of ***superluminal motions*** of radio knots in another hard X-ray/γ-ray source near the center, providing direct evidence for the above interpretation of relativistic radio jets in the nuclear region of our Galaxy.

Throughout the central few hundred parsecs ***giant molecular clouds*** ($10^{5.5}$-10^7 $M_\odot$, $N(H_2)$~$10^{23.5}$ to 10^{24} cm^{-2}) are found whose gas density ($n(H_2)$~10^4 to 10^6 cm^{-3}) and temperature (40 to 200 K) are significantly greater than those of the clouds in the Galactic disk (18). The dust temperature of these clouds, on the other hand, is fairly low indicating that most of them are presently not heated internally by active star formation. The dynamics of this central molecular cloud layer is characterized by large internal random motions and unusual streaming velocities that can be partially explained by the presence of the central bar potential (18, 19, 20). Extended 6.4 keV line emission from 'neutral' iron (9, 21) and 8.5-22 keV hard X-ray continuum emission (22) appears to be correlated with these very dense molecular clouds. This surprising finding suggests that the molecular clouds act as ***dense reflectors*** of hard X-ray emission originating somewhere in the central 100 pc, perhaps analoguous to what is observed in some active galactic nuclei (23). The origin of the scattered X-ray emission is yet unkown. It could be identical with the known compact sources discussed above (22), or perhaps with an active source at the very center (9).

Magnetic fields as large as ~1 mGauss appear to permeate the central 50pc and are aligned approximately perpendicular to the Galactic plane (24). Where they interact with neutral gas clouds spectacular filaments of nonthermal radio synchrotron emission are seen. The most spectacular such set of filaments, the ***Radio Arc***, located about 13' (34 pc) north of the dynamic center and the central radio source ***SgrA*** also contains a set of more than 20pc long filaments of thermal radio emission that are associated with ionized plasma of temperature about 8000 K (25). The origin of the widespread ionization, whether caused by collisions of fast moving clouds, by magneto-hydrodynamic effects or by photoionization has been the subject of a lively debate for some time (26-29), with the odds recently turning in favor of photoionization by a number of OB associations (11, 3). At radii between 1.5 and 5pc from the dynamic center there is a system of orbiting molecular filaments approximately arranged in form of a circum-nuclear 'ring' or 'disk' (30). This ***circum-nuclear disk (CND)*** is probably fed by gas infall from dense molecular clouds at ≥10pc (31) and appears to drop filaments or streamers into the central 1.5pc which is comparatively devoid of interstellar matter (the „***central cavity***", 32). The ultraviolet (UV) radiation in this cavity is sufficiently high to ionize and heat to about 6000 K a significant fraction of these streamers. The central ionized streamers orbit the dynamic center (33, 34) and are arranged in form of a „***mini-spiral***" that comprises the brightest part of the central thermal radio source ***SgrA (West)***. While the streamer velocities are largely dominated by the gravitational field (34, 35), an ***intense nuclear wind*** (mostly from massive stars discussed below) probably also affects their motions within about 1 pc from the center (36,37, 39). Infrared polarimetry and radio observations of Zeeman splitting indicate that the streamers are permeated by mG-magnetic fields that are dragged along their orbits (40-43). The magnetic fields may account for macroscopic viscosity and angular momentum transport in the clumpy, turbulent circum-nuclear gas streamers. The eastern part of SgrA is a synchrotron shell source that appears to be expanding into and accelerating dense molecular gas (44, 29, 45-46). ***SgrA (East)*** thus may be direct evidence for a ***recent explosion*** ($t < 10^5$ years) in the central 10 parsecs, with an energy of $10^{52\pm1}$ ergs (29,45-46).

The density of the nuclear star cluster increases with decreasing radius R approximately as R^{-2} outside of its core radius of less than a parsec in size (47,48). Inside that core radius, the stellar density is certainly a few 10^6 (47,48) and possibly several 10^7 times greater (49) than in the solar neighborhood. Direct collisions and mergers between stars and collisional destruction of stellar envelopes may thus play a role there (50,3). Infrared spectroscopy has identified individual ***blue and red supergiants*** in the core (51-53). These supergiants are very likely massive stars that have formed in the core within the last few million years. They probably provide a significant and perhaps dominant fraction of the total luminosity of the central few parsecs. A discussion of the present state of our knowledge about this nuclear cluster, its distribution and evolution will be given in the next section. Also inside that radius, the stellar and gas velocities are observed to increase, which may signal the presence of a ***black hole*** with mass 1 to 3×10^6 $M_\odot$ at the dynamic center (34,1). A compact radio source, ***SgrA****, lies approximately at that position and is close to, but not

FIGURE 1.

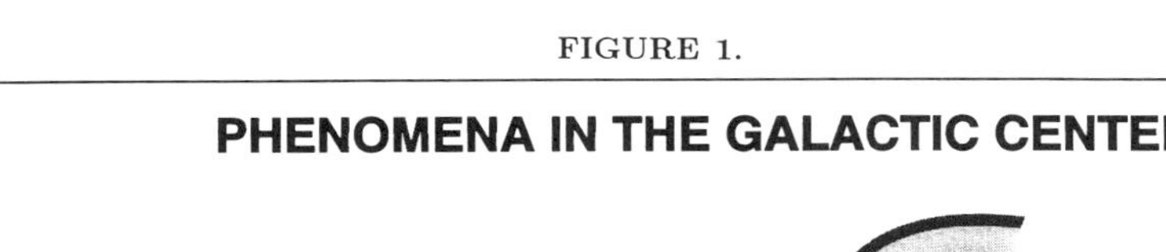

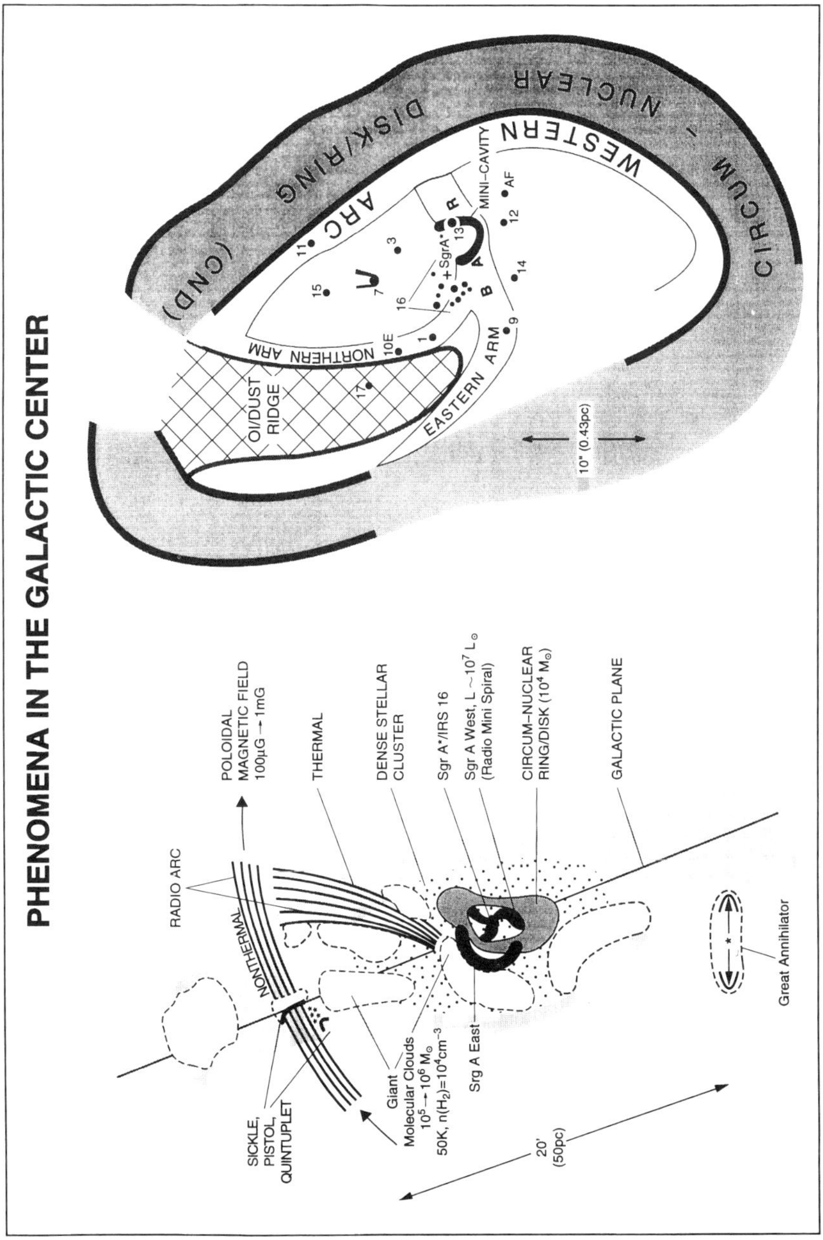
PHENOMENA IN THE GALACTIC CENTER
RADIO ARC
NONTHERMAL
POLOIDAL MAGNETIC FIELD 100μG → 1mG
THERMAL
DENSE STELLAR CLUSTER
Sgr A*/IRS 16
Sgr A West, L ~10^7 L⊙ (Radio Mini Spiral)
CIRCUM-NUCLEAR RING/DISK (10^4 M⊙)
GALACTIC PLANE
SICKLE, PISTOL, QUINTUPLET
Giant Molecular Clouds 10^5 → 10^6 M⊙ 50K, n(H2)=10^4cm^-3
Srg A East
20' (50pc)
Great Annihilator
(CND) ARC DISK/RING
CIRCUM - NUCLEAR
WESTERN
MINI-CAVITY
NORTHERN ARM
EASTERN ARM
OI/DUST RIDGE
+SgrA*
A
B
R
AF
1
3
7
9
10E
11
12
13
14
15
16
17
10" (0.43pc)

coincident with a bright near-infrared source (***IRS16***) of blue color. ***SgrA**** has stellar dimensions (size a few AU, 54), but is presently not very conspicuous in any wavelength range other than the radio range. This is somewhat surprising if ***SgrA**** is in fact a massive black hole surrounded by stars and gas clouds.

Clearly the modern multi-wavelength observations of the Galactic nucleus tell a fascinating story and show that a broad range of phenomena involving a number of physical processes are at work. To help the reader find the way in the jargon of names Fig.1 gives a „road map" of the central 100pc.

II. WHAT POWERS THE CENTRAL PARSEC ?

Fig.2 shows a composite spectral energy distribution of the central few parsecs (from 3). Strong 20 to 300μm continuum emission from 50 to 100 K dust grains originates in the circum-nuclear disk and in a cloud ridge associated with the northern arm of the mini-spiral (55, 32). The far-infrared emission can be used as a ***calorimeter*** for estimating the total short-wavelength (visible and UV) luminosity of the central parsec that cannot be accounted for by the main sequence and late type stars in the nuclear stellar cluster. Taking into account the (~50%) fraction of the nuclear radiation not intercepted by the CND and mini-spiral the total UV and visible luminosity of the central parsec has been estimated to lie between 1 and 2×10^7 L (55, 56). The Lyman continuum flux is about 2 to 3×10^{50} s^{-1} (~1.3 to 1.9×10^6 L_o) as determined from the thermal radio continuum (44).

Infrared spectroscopy of the ionized gas demonstrates that ***SgrA (West)*** is a ***low excitation HII region*** with electron temperature ~6000 K and ***a range of electron densities*** (10^3 to 10^5 cm^{-3}, 57-59). Combining the line ratios of the different fine structure lines, the flux density of the radio continuum and the fluxes of the hydrogen and helium radio recombination lines in an excitation analysis, a ***relatively low effective temperature*** (~35,000 K) is derived for the ionizing radiation field (57, 58, 60, Fig.2). Effective temperatures significantly in excess of 40,000 K are only possible if the strong 12.8μm [NeII] line and most of the radio continuum come from very dense gas (10^6 to 10^7 cm^{-3}, 58), a solution that is not entirely excluded but not supported by the data. More detailed spectroscopy of species with higher ionization potentials and of density sensitive line ratios, planned with the spectrometers onboard the Infrared Space Observatory (***ISO***), should give a clearcut answer. The line ratios also suggest a ***metal abundance of about twice solar*** (57, 61, 62, 58). What powers this low excitation HII region with a bolometric luminosity of $10^{7\pm0.3}$ L_o ?

At this point we need to turn our attention to a number of recent high resolution ***near-infrared*** observations that pertain to the distribution and

FIGURE 2.

FREQUENCY [Hz]
10^10 10^12 10^14 10^16 10^18 10^20
L(H→H+)
L=10^7 L⊙
Teff=3→3.5x10^4 K
IRS16
L(He→He+)
SgrA*
T~10^7K
L~30L⊙
log (νLν [erg s^-1])
log (L_LINE [erg s^-1])
NeII
ArII
Bα
OI
SIII SIII
HeI
OIII
ArIII
Bγ
NIII
FeII
SiII
CII
OIII
SIV
FeIII
OI
CO
H2
SgrA West
(R=0.5pc)
1keV 10keV 100keV 1MeV
10cm 1mm 10μm 1000Å 10Å 0.1Å
Wavelength

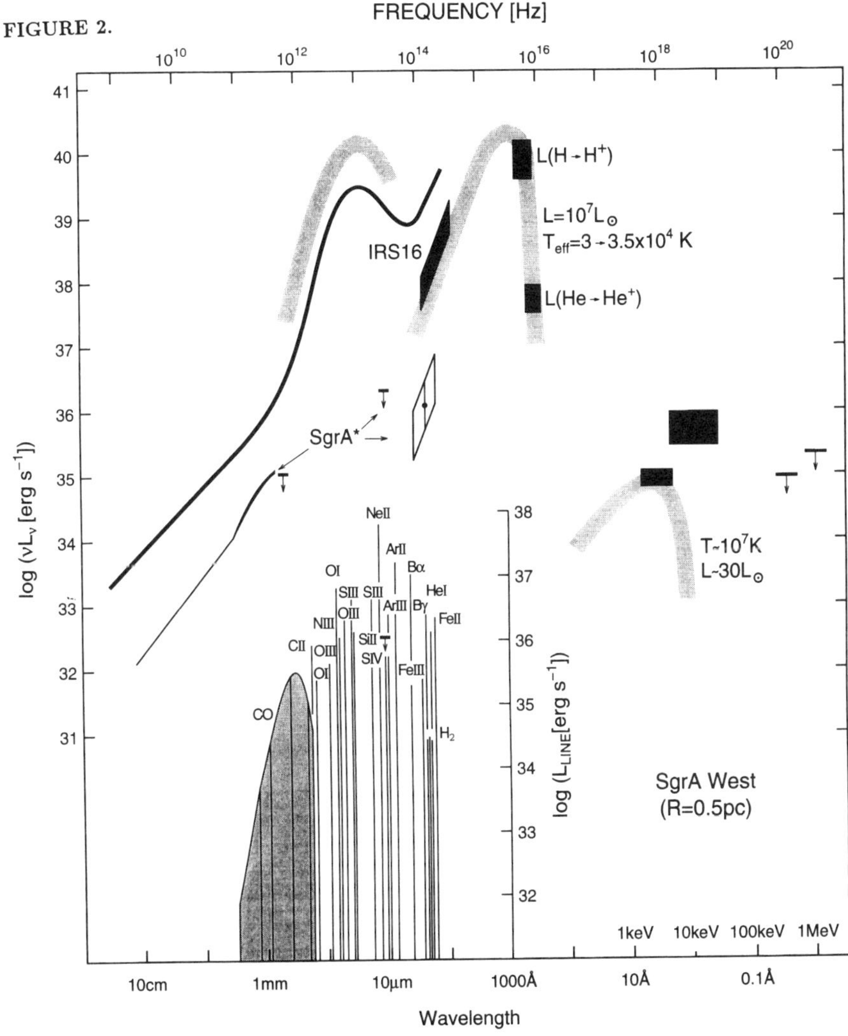

characteristics of the nuclear stellar cluster. These measurements have become possible through the advent in the last half a dozen years of sensitive, large format detector arrays. In only a few years time it has become possible to first record high quality seeing limited images (~1", e.g. 63) and then further to improve through speckle imaging the resolution to the diffraction limit of 4m class telescopes (~0.15" at 2μm, 49, 64). The presently best such 2μm image of the central parsec is shown in Fig.3 (adapted from 65) and was taken with the MPE ***SHARP*** camera on the 3.5m ESO NTT during superb (~0.3") seeing and resolves the near-infrared emission of the central parsec (23") into about 700 stars with K-band (2.2μm) magnitudes ≤16. The central ***IRS16*** complex located within 1" of the compact radio source ***SgrA**** consists of about two dozen single (or perhaps multiple) stars. From the number distribution of the near-infrared sources Eckart et al.(49, 65) conclude that the centroid of the stellar cluster is more likely on ***SgrA**** than on the ***IRS16*** complex and that the core radius of the K≤14 2μm stellar number density distribution is about 0.2pc. If the stars with K≤14 are representative of the overall mass distribution of the cluster (an assumption that still needs to be proven by spectroscopic identification of the stars) such a small core radius would imply that the stellar density in the core is in excess of $10^{7.5}$ $M_{\odot}$ pc^{-3}. At these densities stellar mergers and disruption of giant star atmospheres by direct stellar collisions become very important, with a number of interesting consequences for stellar and cluster evolution (50, 66, 67, 3).

Another important ingredient of the near-infrared story has been the discovery by Allen et al.(52) and Forrest et al.(68) of a 2.06μm HeI/2.16μm Brγ near-infrared ***emission line star*** (the ***AF-star***), followed by the discovery of an entire cluster of about 15 such stars in the central parsec and centered on the ***IRS16*** complex (53, 69). In fact recent subarcsecond HeI line imaging (53, 69, 65) and imaging spectroscopy with the new MPE ***3D***-spectrometer (Fig.4, 69) now unambiguously show that several of the brightest members of the ***IRS16*** complex are HeI-stars, as is the nearby bright source ***IRS 13*** (see also 70). The HeI „broad line region" discovered a decade ago by Hall et al. (71, see also 72) is now identified as a group of mass losing, luminous He-rich stars. Non-LTE stellar atmosphere modeling of the observed emission characteristics of the ***AF-star*** (73) confirms and quantifies the earlier conclusion (52) that the ***AF-star*** is similar to WN9/Ofpe stars, a rare class of evolved, luminous blue supergiants related to Luminous Blue Variables and Wolf-Rayet stars. According to the analysis in (73), the ***AF-star*** has a luminosity of about $10^{5.5}$ $L_{\odot}$, effective temperature near 20,000 K and ZAMS mass between 25 and 40 $M_{\odot}$. The ***AF-star*** has a surface He/H abundance ratio near unity and loses mass at a velocity of 700 km/s and rate of 6×10^{-5} $M_{\odot}$ yr^{-1}. Based on the most recent results from ***3D*** (Fig.4) a preliminary analysis of the brightest HeI stars (***IRS16NE,C,SW, IRS 13***) suggests that these stars may in most respects be similar to the AF-star,

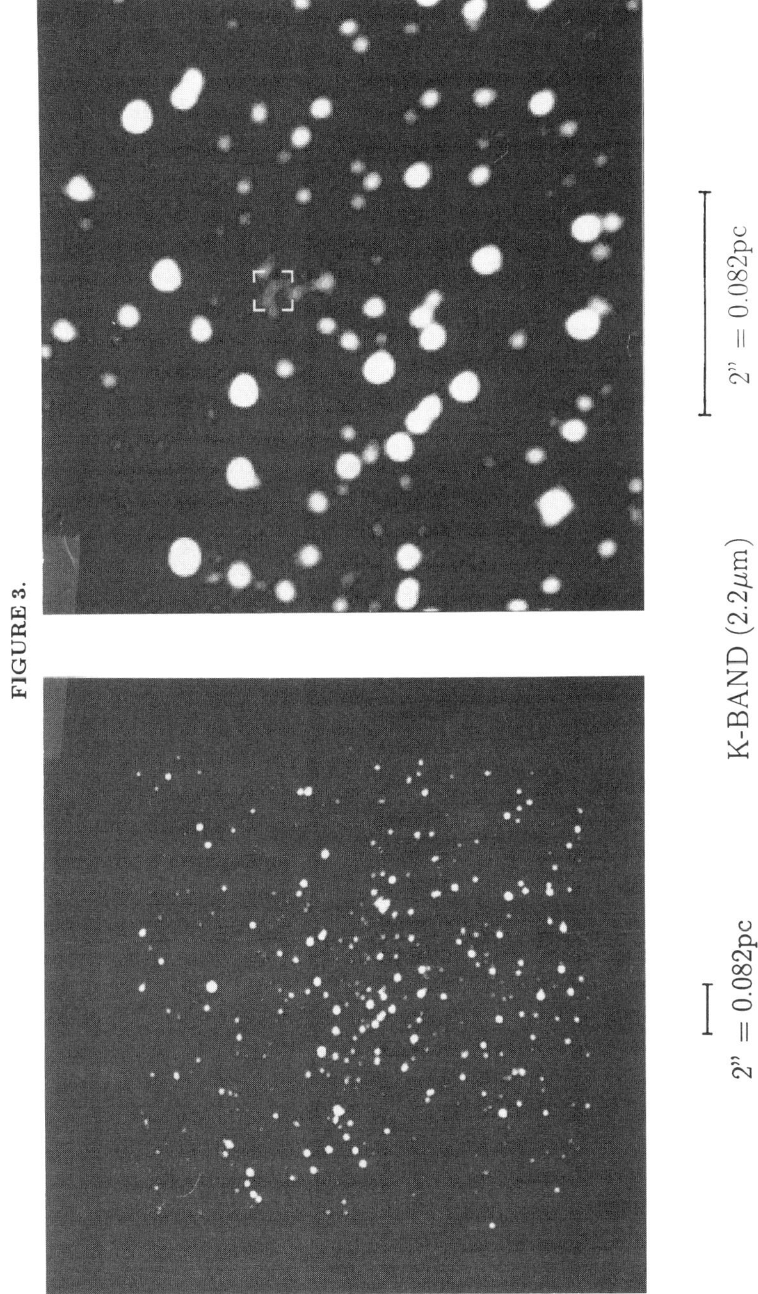

FIGURE 3.

FIGURE 4.

Galactic Center: K-Band Spectra of Central Cluster Stars
3D
IRS 16NE
IRS 16C
IRS 3
MPE-01.35-02.80
IRS 16SE
IRS 34
IRS 21
IRS 6
IRS 33
MPE-00.70-08.80
IRS 13
MPE-04.35-07.50
2.0″
Wavelength [μm]

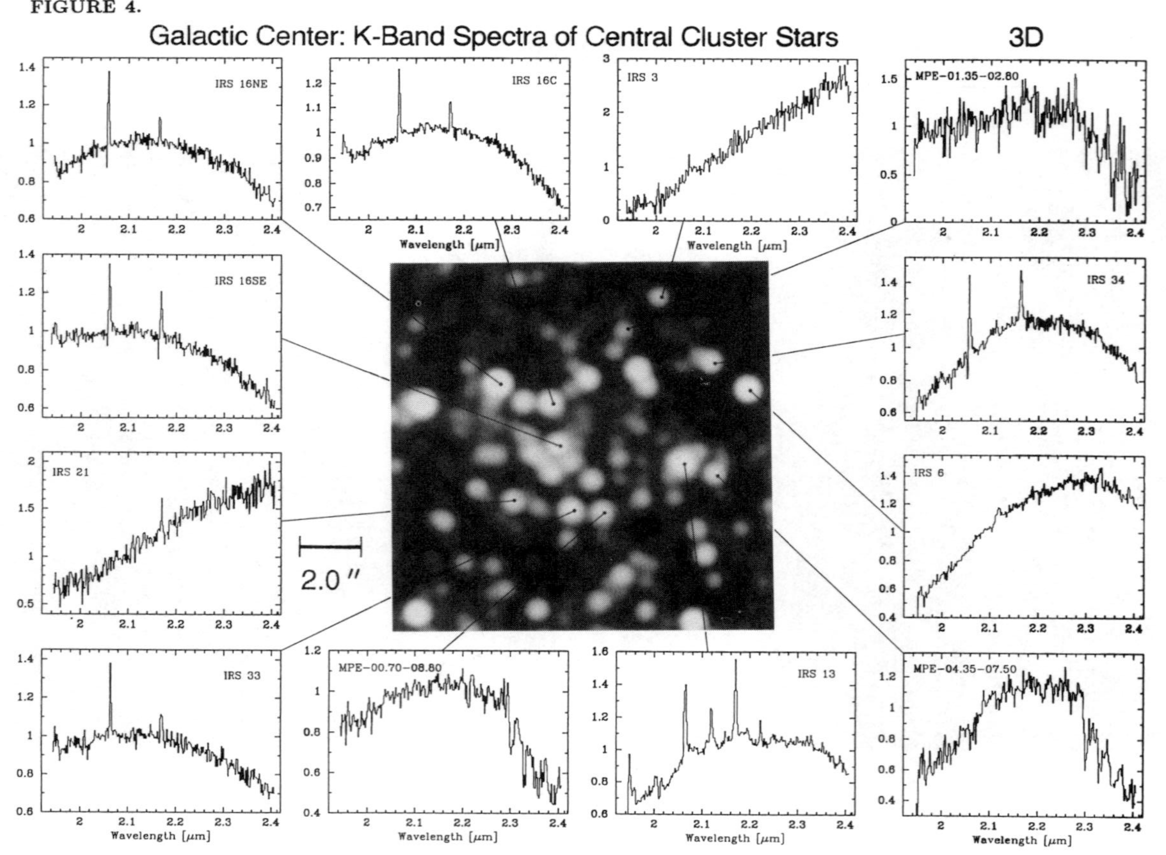

but about 5 to 10 times more luminous (Najarro, priv.comm.). Of particular interest is the spectrum of ***IRS 13*** that shows 2.19μm HeII emission in addition to the very bright and broad wing HeI lines (Fig.4, see also 70). This finding implies that ***IRS 13*** is significantly hotter than most of the other He I stars (~30,000 K) and very luminous (several 10^6 L_o, Najarro, priv.comm.). ***IRS 13*** and the ***IRS16*** HeI stars appear to have a very large He abundance ({He}/{H}>>1, 74) and thus are probably very late type Wolf-Rayet stars. They probably each emit several 10^{49} Lyman continuum photons per second. Finally Blum et al. (75) and Krabbe et al. (69) have also found the first evidence for classical WC (9) Wolf-Rayet stars, further strengthening the interpretation of the HeI stars as evolved massive stars. Combining the contributions from all its members, ***the HeI-star cluster can plausibly account for most of the bolometric luminosity (~1 to 2x10^7 L_o) and a significant fraction of the Lyman-continuum luminosity (~$10^{6.3}$ L_o) of the central parsec***. The HeI-star cluster also provides in excess of 10^{38} erg/s in mechanical wind luminosity which may have a significant impact on the gas dynamics in the central parsec. In fact the outward acceleration of the winds may explain why a possible central massive black hole is not likely to accrete very much interstellar gas at the present time (3).

In agreement with earlier proposals by Rieke and Lebofsky (76) and Allen and Sanders (77), the recent stellar infrared spectroscopy in the central parsec perhaps is best explained by a ***star formation burst*** 5 to 7x10^6 ago in which a few hundred OB stars and perhaps a total of a few thousand stars were formed (3, 78, 69). The HeI stars may be the most massive cluster members that in the mean time have evolved off the main sequence. In this scenario the central parsec is now in the late, ***wind-dominated*** (30 Doradus in LMC!) phase of the burst. This model accounts naturally for the low excitation of the ***SgrA (West)*** HII region. There are several difficulties, however. First the present gas density in the central parsec is too low for conventional gravitational collapse of gas clouds to stars in the presence of the very strong tidal forces (79). Perhaps the burst was triggered by a large influx of dense gas less than 10 million years ago. Second if the HeI stars are in fact closely related to LBVs, conventional wisdom would give an estimate of the duration of that phase of less than 10^5 years. Consequently, either there would have to be $10^{6.7}/(\leq 10^5)$x15 $\geq 10^{2.7}$ additional OB stars lurking in the cluster (if star formation has been semi-continuous) that have not been identified yet. Such a large number of OB stars, however, is inconsistent with the properties of the bright near-infrared sources (69, 65) and the far-infrared luminosity constraint. Alternatively, the burst would have to have been very sharp in time. In that case we would happen to observe this last nuclear burst in the LBV phase (with an a priori probability of less than 1%). The latter explanation, however, cannot explain the simultaneous presence of stars in the HeI-phase that are different by a large factor in luminosity (***AF*** and ***IRS 13, IRS16 NE,C,SW***), and hence in mass.

Based on earlier theoretical work by Lee (80) Eckart et al.(49) have proposed ***sequential merging*** as an alternative to the starburst scenario, a possibility whose likelihood strongly depends on a very high density of the nuclear cluster (3). In a recent Fokker-Planck calculation of an evolving Galactic center type, dense cluster with merging Lee (81), however, has found that there are even not enough 20 M_o stars and no $\geq$30 M_o stars as a core of density $10^{7.5}$ M_o pc^{-3} or greater cannot be maintained for a long enough time. Morris (79) has suggested that the HeI stars are not classical blue supergiants at all but objects that have been created in ***collisions between (~10 M_o) stellar black holes and red giants.*** Both accounts of the HeI stars just cited are very specific to the high density environment of the central parsec. But a number of stars that look just like the ***SgrA*** HeI stars have now been found in several clusters 2 to 13' away from the central, high density ***SgrA*** region (11). In the case of the Morris (79) scenario one would also probably expect a much larger X-ray emission as a result of the collisions than is observed (81).

In conclusion then of this chapter we think that a ***moderate star formation burst 5 to 7x10^6 years ago*** is in fact the most likely interpretation when coupled with the additional proposal that the '***HeI phase'*** in the evolution of a metal rich cluster is much longer than the traditional estimates of the LBV phase. It may comprise a significant fraction of the main sequence lifetime of massive stars. Such a proposal is consistent with recent stellar evolution work. Meynet et al. (82) and Langer et al. (83) propose that in high metallicity stars core nuclear synthesis products are dredged up earlier and mass loss rates are higher than previously thought. We finally note that there is other evidence that star formation in the Galactic center has been time variable. The fact that there are only a few red supergiants (L$\geq$$10^{4.5}$ L_O: ***IRS7***, ***IRS12 N***, ***IRS 14N***) when compared to the number of blue supergiants suggests that there was relatively little star formation before 10 million years ago. Yet there a number of late type stars with luminosities 10^3 to 10^4 L_O , both inside (65, 69) and outside (84, 12) the central parsec. These moderate luminosity stars are likely asymptotic giant branch stars of moderate mass (2 to 8 M_O) that may signify another star burst episode ~10^8 years ago. We are convinced that the high quality near-infrared imaging spectroscopy will now permit within the next year a much better understanding of the evolution of the nuclear cluster and the massive stars it contains.

III. Is SgrA* a Massive Black Hole?

The next key issue that we want to discuss is the evidence for a central massive black hole. Ever since the original discovery of the nonthermal compact radio source ***SgrA**** at the core of the nuclear star cluster (85-87) that source has been the primary black hole candidate, in analogy to compact nuclear radio sources in other nearby normal galaxies (88, 89).

In fact ever more detailed radio observations have confirmed the unique nature of ***SgrA**** in the Galaxy. Recent very long baseline interferometry (VLBI) observations at 7mm show its size to be less than a few AU (54, 90). Its proper motion relative to a background quasar is now known to be less than about 38 km/s (54), almost 5 times smaller than the (2d-) velocity dispersion of the stars. It is therefore almost inevitable that ***SgrA* is a massive object*** with a mass in excess of about 100 M_o. The radio observations (54) have also provided compelling evidence for intrinsic flux density variability on time scales of months/years. The source shows a mm/submm excess above the flat cm-spectral energy distribution (91, Fig.2) probably indicative of the presence of a very compact ($\sim 10^{12}$ cm) radio core of stellar dimensions.

Yet observations at shorter wavelengths show nothing particularly impressive toward the position of ***SgrA****. The high resolution maps of Eckart et al. (49, 64) for the first time did show that there is near-infrared emission toward ***SgrA****, within the present ±0.2“ relative positional uncertainty of the infrared/radio frames. However, the most recent data suggest that the near-infrared emission is most likely due to a local concentration of bright stars near ***SgrA**** (65). The 2μm maps displayed in Fig.3 clearly indicate that SgrA*(IR) is resolved into about half a dozen compact sources. 2μm speckle polarimetry shows that the polarizations of these sources, with the exception of one knot ~1“ south-east of the nominal position of ***SgrA****, are small and consistent with the dichroic absorption of magnetically aligned dust grains in the interstellar medium along the line of sight to the Galactic center (65). SgrA*(IR) also does not show intrinsic variability on scales of minutes or years, or significant line emission (65). Any one of the knots in Fig.3 that might be the true infrared counterpart of ***SgrA**** has an absolute K-magnitude of about -3, similar to an early B main sequence star ($\sim 10^{4.2}$ L_o) or an early K giant ($L \sim 10^{2.2}$ L_o) .

(Variable) X-ray emission is usually considered a key signature of black holes. However, in contrast to the fairly bright infrared emission the X-ray luminosity of ***SgrA (West)*** is fairly low (Fig.2). The ≤2.5 keV X-ray luminosity (corrected for interstellar extinction) is only a few L_o (92, 93) and the hard X-ray luminosity (2.5 to 100 keV) is less than a few hundred L_o (94, 14, 95). The upper limit originates from the recent finding by ***ASCA*** that there is a compact hard X-ray source about 1' south-west of ***SgrA**** that may have contributed significantly to some of the lower resolution hard X-ray observations just mentioned (9). The inevitable conclusion is that any central active source (***SgrA**** ?) presently contributes only a small fraction of the bolometric luminosity of the central parsec, with a conservative upper limit of perhaps a few 10^5 L_o (3, 96). There are currently three possible constraints on any past activity of ***SgrA****. The first two are based on the scattering of hard X-ray emission from ***SgrA****. Considering the extended 8.5 to 22 keV emission measured with the ART-P telescope on ***GRANAT***, Sunyaev, Markevitch and

Pavlinsky (22) conclude that 'if there is a supermassive black hole in the Galactic nucleus, it has not emitted at the level of its Eddington luminosity for even a day over the past 400 yr'. A somewhat more conservative calculation (3) shows that the average 8-22 keV luminosity of ***SgrA**** was not larger than $10^{5.2}$ L_o during the past 400 years (the Eddington luminosity of a 10^6 Mo black hole is ~10^{10} L_o). Extrapolating from the extended 6.4 keV Fe-line emission, Koyama (9) estimates that the hard X-ray luminosity of a central source (***SgrA**** ?) is on average larger than $10^{5.4}$ L_o. The third possible constraint comes from the thermal energy of the 10^8 K gas that ***ASCA*** observes to be associated with ***SgrA(East)*** which may be associated with an energetic recent explosion ($\geq 10^{52\pm1}$ ergs, see above) from the nucleus.

The evidence for a central mass concentration in the Galactic center, perhaps in form of a massive black hole, thus is based entirely on observations of the gas and stellar dynamics. The presently available measurements, most recently reviewed by Genzel, Hollenbach and Townes (3), are summarized in Fig.5. It shows the enclosed mass as function of radius. The various mass estimates derived from the gas/stellar dynamics are compared to the mass distribution derived from the stellar light and a constant mass to light ratio (M/L~1 M_o/L_o, dashed in Fig.5). Compared to the situation more than a decade ago when the first velocity measurements of ionized gas clouds became available (57, 60) and also 6 years ago when the case was last reviewed at a Texas symposium (97) the evidence for a central dark mass of 1 to 3×10^6 M_o has become substantially stronger. Numerous independent gas and stellar dynamics measurements within the central parsec are now available, agree reasonably well and suggest an excess mass within ~0.5 pc above what accounted for by the stellar light. The excess indicated by any single one of the data points, however, is only at the 2 to 3.5 σ level. Only when all measurements are taken together is the evidence for a central mass distribution quite compelling. Of greatest importance for this conclusion are the three data points at the innermost radius (0.2pc or 5") sampled so far. The large cross indicates the mass derived from the velocity field of the central part of the mini-spiral (35, 98, 99). The open circle denotes the mass derived from the stellar velocity dispersion of late type stars sampled in their 2.3μm CO overtone absorption feature (~125 km/s: 100, 101). The filled circle shows the mass derived from the velocity dispersion of the HeI stars (~135 km/s: 102). Taken at face value the density of the dark mass indicated by the data in Fig.5 is at least 10^8 M_o pc^{-3} with a mass to luminosity ratio of at least 10 M_o/L_o.

While the available dynamical measurements strongly point toward the existence of a 1 to 3×10^6 M_o black hole they are clearly not sufficient as a proof. The density of the dark mass referred to above is only a factor of two or so greater than the largest value for the core density of the stellar cluster (49). It is already clear, however, that the dark mass cannot be due

FIGURE 5.

stellar 2 µm light (M/L ~ 1.2, ρ ~ R^-1 8)
OH/IR-stars + individual giants
Boltzmann equation for OH/IR and giants
Boltzmann equation integrated stellar light
He I-stars
HI/CO rotation curve
Circum nuclear gas disk
mini-spiral
Enclosed Mass [M☉]
R[pc]
R0=8.5kpc
Rc~1pc
Rc=0.17pc
10^9
10^8
10^7
10^6
0.1
1
10
100

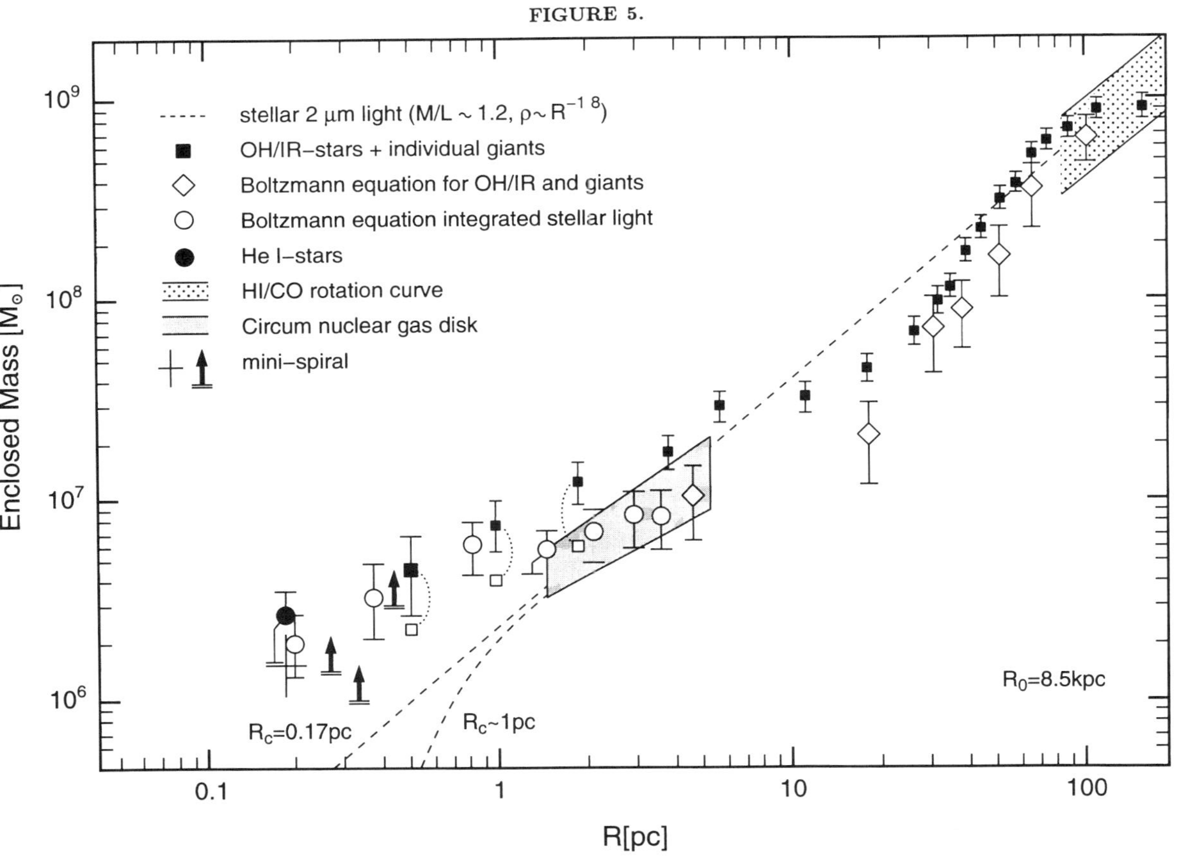

to a concentration of old, low luminosity and low mass stellar remnants (e.g. neutron stars). Such low mass remnants could not have a core radius smaller than the massive HeI-stars or the intermediate mass late type stars sampled with the CO absorption feature. A possible configuration without a central massive black hole may be a central cluster of $\geq$10 M_o stellar black holes (79) of which ***SgrA**** perhaps may be one member. Clearly the key for distinguishing between the case of a single massive hole and a concentration of massive stellar remnants will be to determine the dynamic mass significantly inside of 0.2 pc. Two experiments are now well underway in our group at MPE to a) determine with ***3D*** the radial motions of emission/absorption line stars between 1" and 5" from ***SgrA**** and b), measure the proper motions of all brighter stars within $\leq$2" (0.08 pc) of ***SgrA**** from repeated high resolution near-infrared imaging with the ***SHARP*** camera on the ESO NTT. First results from a ***3D*** data set taken in summer 1994 and from the first four epochs of the proper experiment (covering ~2.5 years) are very encouraging and promise to confirm (or disprove) a 1 to $3x10^6$ M_o black hole in an about 6 year measurement period.

Adopting now the notion that SgrA* is indeed a million solar mass black hole, the riddle remains why it is presently so inactive. It is very interesting that the Galactic center shares this 'luminosity deficiency' or 'blackness' problem with essentially all nearby nuclei for which there is substantial evidence for dark central masses (103), including the perhaps best case, the 'super' H_2O maser source NGC4258 (104). It is possible that the tidal disruption and accretion of stars by the hole (happening in the Galactic center at a rate of ~10^{-4} yr^{-1}) occurs very efficiently albeit at low duty cycle (105). Accretion of interstellar gas streamers by the hole may be prevented by the need to overcome the angular momentum problem, coupled with the outward force of the stellar winds as discussed above. Finally, the wind gas itself may be accreted largely spherically, with very low radiation efficiency (106). ***Nevertheless current models of black hole accretion have to be stretched to be comensurate with SgrA* being an underfed million solar mass black hole*** (107).

Acknowledgement

We are grateful to our MPE colleagues M.Cameron, S.Drapatz, R.Hofmann, H.Kroker, D.Lutz, B.Sams, L.Tacconi-Garman, N.Thatte and L.Weitzel , as well as F.Najarro for letting us report ***3D, SHARP*** and stellar modelling results prior to publication. We also thank K.Koyama and Y.Tanaka for access to unpublished ***ASCA*** results.

REFERENCES

1. Genzel, R. and Townes, C.H. 1987, Ann.Rev.Astr.Ap. 25, 377
2. Blitz, L., Binney, J., Lo, K.Y., Bally, J. and Ho, P.T.P. 1993, NATURE 361, 417

3. Genzel, R., Hollenbach, D. and Townes, C.H. 1994, Rep.Progr.Phys. 57, 417
4. Blitz, L. and Spergel, D.N. 1991, Ap.J. 379, 631
5. Binney, J.J., Gerhard, O.E., Stark, A.A., Bally, J. and Uchida, K.A. 1991, MNRAS 252, 210
6. Liszt, H.S. and Burton, W.B. 1980, Ap.J. 236, 779
7. Bally, J., Stark, A.A., Wilson, R.W. and Henkel, C. 1987, Ap.J.(Suppl.) 65, 13
8. Koyama,K. et al. 1989, NATURE 339, 603
9. Koyama, K. 1994, preprint
10. Diehl, R. et al. 1993, Astr.Ap.(Suppl.) 97, 181
11. Cotera, A.S. et al. 1994, in (38),217
12. Glass, I. 1994, in (38), 209
13. Skinner, G.K. 1993, Astr.Ap.(Suppl.) 97, 149
14. Sunyaev, R. et al. 1991, Ap.J. 383, L49
15. Mirabel, I.F., Rodriguez, L.F., Cordier, B., Paul, J. and Lebrun, F. 1992, NATURE 358, 215
16. Ramaty, R., Leventhal, M, Chan, K.W. and Lingenfelter, R.E. 1992, Ap.J. 392, L63
17. Mirabel, I.F. and Rodriguez. 1994, NATURE 371,46
18. Güsten 1989, in (112), 89
19. Bally, J., Stark, A.A., Wilson, R.W. and Henkel, C. 1988, Ap.J. 324, 223
20. Binney, J.J. 1994, in (38), 75
21. Tanaka, Y. et al. 1994, in prep.
22. Sunyaev, R.A., Markevitch, M. and Pavlinsky, M. 1993, Ap.J. 407, 606
23. Awaki, H., Koyama, K., Inoue, H. and Halpern, J.P. 1991, PASJ 43, 195
24. Morris, M. 1990,in ***Galactic and Extragalactic Magnetic Fields***, eds. R.Beck, P.Kronberg and R.Wielebinski (Dordrecht:Kluwer), 361
25. Sofue, Y. 1994, in (38), 43
26. Serabyn, E. and Güsten, R. 1987, Astr.Ap. 184, 133
27. Heyvaerts, J., Norman, C. and Pudritz, R. 1988, Ap.J. 330, 718
28. Morris, M. and Yusef-Zadeh, F. 1989, Ap.J. 343, 703
29. Genzel, R. et al. 1990, Ap.J. 356, 160
30. Genzel, R. 1989, in (112), 393
31. Ho, P.T.P. et al. 1991, NATURE 350, 309
32. Jackson, J. et al. 1993, Ap.J. 402, 173
33. Lo, K.Y. and Claussen, M, J. 1983, NATURE 306, 647
34. Serabyn, E. and Lacy, J. 1985, Ap.J. 293, 445
35. Serabyn, E., Lacy, J., Townes, C.H. and Bharat,R. 1988, Ap.J. 326, 171
36. Geballe, T.R. et al. 1984, Ap.J. 284, 118
37. Yusef-Zadeh, F. 1994, in (38), 355
38, Genzel, R. and Harris, A.I. 1994, ***Nuclei of Normal Galaxies: Lessons from the Galactic Center***, (Dordrecht:Kluwer)
39. Lutz, D., Krabbe, A. and Genzel, R. 1993, Ap.J. 418, 214
40. Aitken, D.K. et al. 1991, Ap.J. 380, 419
41. Hildebrand, R. H. et al. 1993, Ap.J. 417, 565
42. Schwarz, U. and Lasenby, J. 1990, in ***Galactic and Extragalactic Magnetic Fields***, eds. R.Beck, P.Kronberg and R.Wielebinski (Dordrecht:Kluwer), 383
43. Killeen, N.E.B., Lo, K.Y. and Crutcher,R. 1992, Ap.J. 385, 585
44. Ekers, R.D., van Gorkom, J.H., Schwarz, U. and Goss, W.M. 1983, Astr.Ap. 122, 143
45. Mezger, P.G.et al. 1989, Astr.Ap. 209, 337
46. Zylka,R., Mezger, P.G. and Wink, J. 1990, Astr.Ap. 234, 133

47. Allen, D.A. 1994, in (38), 293
48. Rieke, G. and Rieke, M. 1994, in (38), 283
49. Eckart, A., Genzel, R., Hofmann, R., Sams, B.J. and Tacconi-Garman, L.E. 1993, Ap.J. 407, L77
50. Phinney, E.S. 1989, in (112), 543
51. Lebofsky, M.J., Rieke, G.H. and Tokunaga, A. 1982, Ap.J. 263, 736
52. Allen, D.A., Hyland, A.R. and Hillier, D.J. 1990, MNRAS 244, 706
53. Krabbe, A., Genzel, R., Drapatz, S. and Rotaciuc, V. 1991, Ap.J. 382, L19
54. Backer, D. 1994, in (38), 403
55. Davidson, J.A., Werner, M.W., Wu, X., Lester, D.F., Harvey, P.M., Joy, M. and Morris, M. 1992, Ap.J. 387, 189
56. Becklin, E.E., Gatley, I. and Werner, M.W. 1982, Ap.J. 258, 134
57. Lacy, J.H., Townes, C.H., Geballe, T.R. and Hollenbach, D. 1980, Ap.J. 241, 132
58. Shields, J.C. and Ferland, G.J. 1994, Ap.J.430,236
59. Erickson, E.F., Colgan, W.J., Simpson, J.P., Rubin, R.H. and Haas, M.R. 1994, in (38), 249
60. Lacy, J.H., Townes, C.H. and Hollenbach, D.J. 1982, Ap.J. 262, 120
61. Herter, T., Houck, J.R., Shure, M., Gull, G.E. and Graf, P. 1984, Ap.J. 287, L15
62. Lester, D.F., Dinerstein, H.L., Werner, M.W., Watson, D.M., Genzel, R. and Storey, J.W.V. 1987, Ap.J. 320, 573
63. Forrest, W.J., Pipher, J.L. and Stein, W.A. 1986, Ap.J. 301, L49
64. Eckart, A., Genzel, R., Krabbe, A., Hofmann, R., van der Werf, P.P. and Drapatz, S. 1992, NATURE 355, 526
65. Eckart, A., Genzel, R., Hofmann, R., Sams, B. and Tacconi-Garman, L. 1995, Ap.J. in press
66. Gerhard, O. 1994, in (38),267
67. Lee, H.M. 1994, in (38), 335
68. Forrest, W.J., Shure, M.A., Pipher, J.L. and Woodward, C.A.1987 in ***„The Galactic Center"***, ed. D.Backer, AIP Conference Proceedings 155, 153
69. Krabbe, A. Genzel, R. et al. 1995, in prep.
70. Blum, R.D., dePoy, D.L. and Sellgren, K. 1995a, Ap.J. in press
71. Hall, D.N.B., Kleinmann, S.G. and Scoville, N.Z. 1982, Ap.J. 262, L53
72. Geballe, T.R. et al. 1987, Ap.J. 320, 562
73. Najarro, F. et al. 1994, Astr.Ap. 285, 573
75. Blum, R.D., Sellgren, K. and DePoy, D.L. 1995b, Ap.J. submitted
76. Rieke, G.H. and Lebofsky, M.J. 1982, in ***„The Galactic Center"***, eds. G.R.Riegler and R.D.Blandford, AIP conference series 83 (New York), 194
77. Allen, D.A. and Sanders, R.H. 1986, NATURE, 319, 191
78. Tamblyn, P. and Rieke, G.H. 1993, Ap.J. 414, 573
79. Morris, M. 1993, Ap.J . 408, 496
80. Lee, H.M. 1987, Ap.J. 319, 801
81. Lee, H.M. 1994, in ***„The Galaxy",*** ed.L.Blitz (Dordrecht:Kluwer), in press
82. Meynet, G., Maeder, A., Schaller, G., Schaerer, D. and Charbonnel, C.1994, Astr.Ap.(Suppl.) 103, 97
83. Langer, N. et al. 1994, Astr.Ap. in press
84. Haller, J.W. and Rieke, M.J. 1989, in (112), 487
85. Ekers, R.D. and Lynden-Bell, D. 1971, Ap.Lett. 9, 189
86. Downes, D. and Martin, A. 1971, NATURE 233, 112
87. Balick, B. and Brown, R.L. 1974, Ap.J. 194, 265

88. Lynden-Bell, D. and Rees, M. 1971, MNRAS 152, 461
89. Lo, K.Y. 1989 , in (112),527
90. Krichbaum, T.P., Schalinski, C.J., Witzel, A., Standke, K.J., Graham, D.A. and Zensus, J.A. 1994, in (38), 411
91. Zylka, R., Mezger, P.G. and Lesch, J. 1992, Astr.Ap.261, 119
92. Watson, M.G., Willingale, R., Grindlay, J.E. and Hertz, P. 1981, Ap.J. 250, 142
93. Predehl, P. and Trümper, J. 1994, Astr.Ap. in press
94. Skinner, G.K. et al. 1987, NATURE 330, 544
95. Goldwurm et al., 1994, NATURE 371, 589
96. Mezger, P.G. 1994, in (38), 415
97. Genzel, R. 1987, in Proceedings of 13th Texas Symposium, ed. M.P.Ulmer (World Scientific, Singapore), 388
98. Lacy, J.H., Achtermann, J.M. and Serabyn, E. 1991, Ap.J. 380, L71
99. Herbst, T.M., Beckwith, S.V.W., Forrest, W.J. and Pipher, J.L. 1993, A.J. 105, 956
100. Sellgren, K., McGinn, M.T., Becklin, E. and Hall, D.N.B. 1990, AP.J. 359, 112
101. Haller, J.W., Rieke, M.J. and Rieke, G.H. 1992, B.A.A.S. 24, 1178
102. Krabbe, A. et al. 1994, in ***„Infrared Astronomy with Arrays (3)“,*** ed. I.McLean (Dordrecht:Kluwer), 484
103. Kormendy, J. 1994, in (38), 379
104. Miyoshi, M., Moran, J.M., Hernstein, J., Greenhill, L., Naakai, N., Diamond, P. and Inoue, M. 1995, NATURE 373, 127
105. Rees, M. 1988, NATURE 333, 523
106. Melia, F. 1992, Ap.J. 387, L25
107. Ozernoy, L. and Genzel, R. 1995, in ***„The Galaxy“,*** ed.L.Blitz (Kluwer: Dordrecht),in press
108. Lindqvist, M., Habing, H. and Winnberg, A. 1992, Astr.Ap. 259, 118
109. Rieke, G.H. and Rieke, M.J. 1988, Ap.J. 330, L33
110. Güsten, R. et al. 1987, Ap.J. 318, 124
111. Schwarz, U.J., Bregman, J.D. and van Gorkom, J.H. 1989, Astr.Ap.215, 33
112. Morris, M. (ed.) 1989, ***„The Center of the Galaxy“*** (Dordrecht:Kluwer)

MACHOS and Dark Matter

ROGER FERLET
Institut d'Astrophysique de Paris, CNRS
98bis Bd. Arago, 75014 Paris, France

INTRODUCTION

The Universe itself will more and more serve as the ultimate laboratory for particle physics. The cosmology is thus becoming the meeting point for physicists and astronomers, as it is revealed by one of the key issues presently unsolved: the dark matter.

In the late twenties, Hubble and Humason recognized the expansion of the Universe, whose evolution is governed by its density ρ. In the framework of the General Relativity and assuming the Universe to be homogeneous and isotrope and the cosmological constant zero, one can define a critical density $\rho_c \equiv 3H_o^2/8\pi G$, in which H_o is the Hubble constant, still poorly known within a factor of ~ 2, and G the gravitation constant; ρ_c is of the order of few protons per m^3. If $\rho < \rho_c$, then the Universe will expand forever; if $\rho > \rho_c$, it will recollapse; if $\rho = \rho_c$, the Universe is flat and asymptotically expanding. This latter case corresponds to the closure density: $\Omega = \rho/\rho_c = 1$, which translates into a mass to luminosity ratio M/L required to close the Universe of the order of $800h_{50}$, depending on the Hubble parameter in units of 50 km/s/Mpc. A major problem is therefore to derive the density parameter Ω.

In a first section, we will very briefly summarize evaluations of Ω performed at different scales, and compare them to the luminous matter in the Universe. The presence of dark matter being thus established, we will discuss in a second section the amount of baryonic matter and highlight very recent data concerning the primordial deuterium abundance. We will then concentrate in a third section on the Milky Way, the existence of a massive baryonic halo and its possible nature. The following section will introduce the effect of microlensing to detect massive compact objects in the Galactic halo. In section 5 are describe the different observational programmes and their first results. Finally, preliminary conclusions are drawn in the last section.

THE DENSITY PARAMETER

At the largest scales (> 10 Mpc), bulk flows and peculiar velocities are used (see e.g. *Cosmic Velocity Fields* edited by Bouchet and Lachièze–Rey[1]), as well as IRAS data (see e.g. a recent study by Dekel et al.[2]). Perhaps it is the extremely precise measurements of the cosmic background radiation at 2.7 K which will provide more new insights (see these proceedings). Although very delicate, one can nevertheless note that these approaches seem to agree on high values of Ω, compatible with $\Omega = 1$.

At scales of the galaxy clusters (about 1–10 Mpc), a dark matter problem was first mentionned by Zwicky[3] through dynamical studies. More recently, from X-ray measurements and assuming gravitational equilibrium[4] (see also these proceedings), one seems to converge to $M/L \approx (80\text{–}160)h_{50}$, which corresponds roughly to the range 0.08–0.3 for Ω. Observations of gravitational arcs have provided independent confirmations of these values (see these proceedings, and a recent review by Refsdal and Surdej[5]). It is worth noting that departures from virialization have been pointed out in poor groups of galaxies[6], leading to even higher values.

At scales of $\sim$ 10–100 kpc, those of haloes of galaxies, the main method for estimating the mass density uses the rotation curves, mostly constructed from 21 cm observations of neutral H I gas since the pionneering work of Rubin and others in the seventies (see e.g. the review by Faber and Gallagher[7]). For spirals, it is well documented that the measured velocity distributions as a function of the distance from the center of the galaxy clearly disagree with the velocity distributions calculated by assuming that the mass distribution is the same as the surface brightness distribution. According to the simple newtonian law $V^2 = GM/R$, the flat shape of the rotation curves implies that the total mass M within the radius R still increases with R far beyond the observed stellar population (but this has been tentatively questionned[8] through introduction of magnetic fields). M/L ratios around $(\text{few–}20)h_{50}$ are deduced, and even higher for dwarf spirals and irregulars, corresponding to 0.01–0.06 for Ω. For ellipticals, X-ray data seem to indicate M/L about 100, or $\Omega \sim 0.1$.

What about the luminous matter? Following Persic and Salucci[9], Peebles[10] and Vangioni-Flam and Cassé[11], one has to add the contributions from E and S0 galaxies [0.0015], from spirals [0.0007], from hot gas in clusters $[0.0015h_{50}^{-1.3}]$ and from H I in galaxies $[(3.8 \pm 0.9)10^{-4}h_{50}^{-1}]$. Ω_L appears quite firmly established to lie in the range $(3\text{–}7)10^{-3}$. Therefore, at any scale, the existence of dark matter in the Universe seems inevitable.

BARYONIC OR NON-BARYONIC DARK MATTER

The Big Bang theory predicts in which quantity the light elements deuterium, helium 3 and 4 and lithium 7 were synthesized when the Universe was at a temperature around 0.1 Mev (the first hundreds seconds or so). As a proof of the maturity of this theory, it is worthwhile to recall for instance that the predictions for ^{4}He depend on the number of neutrino flavors. The primordial abundance of ^{4}He is infered from the extrapolation to zero metallicity of the abundances measured through emission lines in extragalactic very metal poor H II regions, and the comparison with the predictions implied only three ν flavors. It was a great success when particle physics experiments with LEP at CERN beautifully confirmed in 1989 that indeed only three flavors do exist[12]. Thus, in the framework of the Big Bang theory, the density of baryons is $\Omega_B \sim 1.5\ 10^8\ \eta\ h_{50}^{-2}$, in which η is the baryons to photons ratio, the one astrophysical parameter of the primordial nucleosynthesis[13,14]. The comparison between computed and observed primordial abundances seems to indicate an overall agreement for a present baryonic density of the Universe well below the closure: $0.01 \leq \Omega_B \leq 0.2$. On these grounds, lot of the dark matter if not all of it could be baryonic.

However, while the primordial ^{7}Li abundance seems to be fairly well evaluated in very old, very metal poor stars, the deuterium one is not directly observable but only derived through quite poorly known galactic chemical evolution models from its present day interstellar abundance (assuming we know it very well; see e.g. Ferlet[15]) and Solar System abundances. Note that the ^{3}He primordial abundance is even more poorly determined, mostly because of unsolved stellar production problems. Moreover, the D/H ratio shows a steep dependance on η, and is therefore a more crucial issue. This is why one tries to look for deuterium in astrophysical sites closer to the early Universe. In 1994, two independant groups reported the first detection of a deuterium Lyman–α line in an absorption system toward the same quasar, in a so-called "primordial" cloud at high redshift (z_{abs}=3.32). One group with the 10 m Keck Telescope at Hawaii[16] and the other with a 4 m telescope at Kitt Peak[17], both used about the same resolving power and agreed on a D/H ratio (1.9–2.5)10^{-4}. Compared with the Big Bang predictions, this unexpectedly very high D abundance implies $\Omega_B \sim 0.01$, if we take a high value for the Hubble constant (~80 km/s/Mpc) as it is favored by recent observations of Cepheids in the Virgo cluster[18]. If this is confirmed, two drastic consequences follow: i) there is almost no more room for baryonic dark matter, and ii) standard models of galactic chemical evolution are very wrong, as they predict a deuterium astration factor in the range 2 to about 7, while the above QSO D/H ratio would on the contrary suggest at least a factor 12 of D destruction to reach the present local interstellar value[10]. However, if the Hubble constant is indeed low, then baryonic dark matter is again viable.

This topic is in fact so hot that another group led by Tytler announced at an ESO Conference[19] held at the end of 1994 a new D/H determination toward a second quasar. Also using the HIRES echelle spectrograph at the Keck Telescope, they claimed D/H $\sim 2\ 10^{-5}$, with thus no more "dramatical" consequences. As a matter of fact, when searching for a deuterium absorption line toward a QSO, there is always a not negligeable probability that an hydrogen line blue shifted by the right amount is mimicing the deuterium one and wrongly increasing the derived D/H ratio. Moreover, another key observational point is the instrumental spectral resolution. It has to be sufficient to resolve eventual multi-component absorptions, as in the case of the Tytler's observations, otherwise leading to perhaps very large error bars on the derived column-densities. These two first deuterium abundances toward QSOs are in complete disagreement; either one (or both) does not reflect the real primordial D/H ratio, or primordial nucleosynthesis was dramatically inhomogeneous. Many more observations are badly needed.

THE MILKY WAY

Historically, dark matter is not at all a new problem. Orbital anomalies in the Solar System led Le Verrier to the discovery of Neptune in 1846; but note that Vulcan has not been found. Shortly before, Bessel similarly derived the existence of the Sirius and Procyon companions. Observations of indirect effects are an efficient way to reveal invisible matter. In the solar neighborhood, as early as in 1932 Oort[20] analyzed stellar velocities and oscillations perpendicular to the galactic plane and determined a dynamical mass density. The most recent related studies[21] tend to indicate that about half of the mass in the solar vicinity must be unobserved matter, with a corresponding volume density of about 0.08–0.12 $M_\odot$ pc^{-3} and an exponential scale height of less than 0.7 kpc.

Although it is difficult to measure from the inside the rotation curve of our Galaxy, the compilation of Fich and Tremaine[22] from many different indicators clearly shows that the curve is also flat out to $\sim$ 20 kpc, with a rotation velocity around 220 km/s (according to Robin et al.[23], the edge of the stellar galactic disk is at $\sim$ 14 kpc from the center). Within $R_\odot$, the M/L ratio is $\sim$ 10 and the mass $\sim$ 1.5 10^{11} $M_\odot$. Beyond $R_\odot$, and as for other spirals, M/L is increasing, together with the mass. In order to explain a constant velocity of $\sim$ 220 km/s, a total mass 5 to 10 times larger than the luminous mass is needed.

One solution is to assume the existence of this dark matter in the form of an isothermal spherical halo with a mass distribution in R^{-2}. Some earlier indirect evidences have already been suggested in favor of such a halo, as for instance the H I flaring of the disk (widening at large radii), the detailed deconvolution of the photometry and stability arguments[24]. The total mass of the Galaxy is currently estimated[25] to be in the range (3–10)10^{11} $M_\odot$ beyond $R_\odot$, perhaps out to 100 kpc. Except if $0.01 \leq \Omega_B \leq 0.2$ is proven to be wrong, it is thus reasonable to search for the galactic dark matter in a baryonic form. But which form?

Plausible forms include massive astrophysical compact halo objects (MACHOS), namely brown dwarves[26,27], with masses below the thermonuclear threshold $\sim 0.08\, M_\odot$ but above the evaporation limit infered to be $\sim 10^{-7}\, M_\odot$. It seems in effect that the initial mass function is still increasing toward the low mass end, but in any case the extrapolation toward brown dwarves is poorly known[28]. It would not be impossible to detect brown dwarves in the infrared, especially the most massive ones; but one will have to wait for sensitive enough future generations of IR space missions. In the last two years, another possibility has been raised, namely molecular hydrogen gas clouds with a fractal mass distribution[29], not emitting because at extremely cold temperature (~3 K). In this latter case, the invisible mass should be more confined toward the galactic plane[30]. If these kinds of objects do really exist, how to detect them?

THE MICROLENSING EFFECT

The genesis of the gravitational lens concept goes all the way back to Newton (bending of light rays), then Soldner in Munich in 1804 (angular deflection by the Sun), the General Relativity of Einstein, Lodge and Eddington in 1919 (multiple images by the Sun) – also the year of a solar eclipse which spectacularly confirmed the Einstein theory – Chwolson in 1924 (ring of light from a star) and finally Einstein[31] in 1936 who calculated the microlensing effect by a distant massive object in the vicinity of the line of sight toward a background light source. It is interesting to note that Einstein was very pessimistic about the possibility of detecting such an effect.

It is Paczyński[32] who proposed to apply this effect in searching for MACHOS by using stars in the Magellanic Clouds. In the special case of perfect alignment between the observer, an LMC star (at 50 kpc) and a 1 $M_\odot$ deflector located 10 kpc from the Sun, the observer should see the so-called Einstein ring with a radius $R_o \sim 1.4\ 10^9$ km. If a similar deflector is lying at a distance R_o from the sight line in the plane perpendicular to it (or equivalently if the impact parameter equals 1), the angular deflection is then 5 mas, that is out of reach from present instrumental capabilities. However, there is also an apparent amplification of light by a factor 1.34, or ~ 0.3 mag, an effect which is detectable, even for the faint stars of the Magellanic Clouds! As the deflector is moving, the microlensing effect will show up as a light curve, with specific characteristics which should allow to distinguish it from intrinsically variable stars: symetry in time, achromaticity, unicity and same magnitude before and after the event.

A critical parameter needed before undertaking such a search is the probability of deflection at a given time. It is the solid angle fraction occupied by the Einstein circles of the deflectors. For a standard halo of ~ $5\ 10^{11}\ M_\odot$, the "optical depth" τ for having a microlensing event with a magnification factor larger than 1.34 is of the order of $0.5\ 10^{-6}$. Since the transverse velocities of the deflectors are unknown as are the impact parameters, the evaluation of the expected number of events requires Monte-Carlo simulations with input distributions for the different parameters. It follows that for larger masses deflectors, events are longer but very few, implying to survey millions of stars, while for smaller masses, events are shorter but more numerous, implying to survey less stars but at higher frequencies.

OBSERVATIONAL PROGRAMMES

Two groups are presently monitoring LMC stars. The first one to start was the french EROS (Expérience de Recherche d'Objets Sombres) collaboration between astrophysicists and particle physicists. Since 1990, they adopted a double observational approach from ESO at La Silla in Chile: i) to explore larger masses, $5^o \times 5^o$ Schmidt plates in two colors, every two nights or so during the observing seasons, which provide about 8 million usable stars, and ii) to explore smaller masses, a special 16 CCDs camera giving more than 10^5 stars in a $1^o \times 0.4^o$ field in the LMC bar, at the focus of a 0.4 m telescope, again in two colors, and providing as many as about 50 images per night[33,34] (for more details, see Magneville et al., in these proceedings). In September 1993, the EROS team isolated from their Schmidt data two light curves presenting all the characteristics, within the error bars, of microlensing events[35]. Photometric and spectroscopic follow-up of both microlensed candidate stars did not infirm this result[36]. However, at the time of writing (February 1995), it seems that one of this star is showing signs of variability which are of course actively studied.

The second group to look at the LMC is the Australian-US MACHO collaboration who started to observe in 1992 from Mount Stromlo in Australia. They use a 1.3 m telescope with an 8 CCDs camera covering a $0.5^o \times 0.5^o$ field in two colors simultaneously and giving about 9 million stars through multi-field operation. Because of this last choice, they are more sensitive to larger masses than to smaller ones (for more details, see Bennett et al., in these proceedings). In September 1993 too, the MACHO team detected one light curve compatible with microlensing[37]. It has to be noted that, while both EROS events occured when MACHO was not yet in operation, the first MACHO event has also been seen by EROS but was not kept as a positive detection because of very poor signal in the blue, avoiding thus the achromaticity criterion to be tested. Since then, MACHO has identified three more events; thus, at the time of writing, a total of six are claimed toward the LMC.

The duration of a microlensing event is a function of the mass of the deflector, its transverse velocity, and its position with respect to the observer. Using the standard halo model of Griest[38], the estimated deflectors mass range for the six LMC events is 0.02–0.19 $M_\odot$, with 2σ error bars $\sim$ factor 10 (and detection efficiencies not taken into account), in agreement with the estimation made through the method of mass moments[39] $<M> \sim 0.08\ M_\odot$. Recently, Bahcall et al.[40] used the Wide Field Camera 2 on board the repaired Hubble Space Telescope to establish a color-magnitude diagram in a small region in the general direction of the LMC. They concluded that low-mass stars do not contribute significantly (<6%) to the observed candidate microlensing events toward the LMC (see also Hu et al.[41]). There is still the remote possibility that the LMC events are not due to halo objects, but rather to microlensing by faint stars in the LMC itself[42]. Concerning deflectors of smaller masses, the EROS team did not detect any event in their CCD data. They are thus able to exclude at 90% confidence level the mass range 5 10^{-8}–7 10^{-4} $M_{\odot}$; the galactic halo is not dominated by such objects[43].

A third group started in 1992 to search for microlensing events. A polish-US collaboration, OGLE (Optical Gravitational Lensing Experiment) operates a 1 m telescope from Las Campanas in Chile. They began to use a only 1 CCD camera giving a $0.25^{o} \times 0.25^{o}$ field and they observe only toward the galactic bulge through Baade's window. Also in 1993, the OGLE team published its first event[44]. At the end of 1994, they had detected ten events[45]. Also monitoring the bulge, the MACHO group has now identified more than 45 microlensing candidates (with at least three in common with OGLE), well distributed over the HR diagram[46]. With similar error bars as for the LMC events, the estimated mass range for deflectors toward the bulge seems to be higher than in the halo, around 0.28 $M_{\odot}$[39].

To be complete, one has to mention the french DUO (Disk Unseen Objects) project who started in 1993 a Schmidt survey of the galactic bulge from ESO-La Silla in Chile. Monitoring 15 million stars, they have already detected several microlensing candidates[47]. Last, at least two other projects (one french called AGAPE[48], one US[49]) are underway, both aiming at microlensing toward the Andromeda galaxy (see Kaplan et al., in these proceedings). But instead of the usual monitoring of resolved stars, these expriments work through pixel monitoring to look for stronger amplifications of unseen objects raising them above the detection threshold of an instrumental system.

Advances in data processing now allow the OGLE and MACHO projects (and soon the EROS one) to detect in real time microlensing events ("early warning system" first implemented by OGLE[50]) and immediately alert the community, providing thus the opportunity to look in detail at the fine structure of the light curves, especially at caustic crossings. This will give key insight into the nature of the source star and the lensing system; in particular, a better sampling offers the presently most powerful tool to search for planets orbiting stellar deflectors toward the galactic bulge[51,52,53]. Already, a light curve from a binary lens event may have been recorded (OGLE 7th event[54], followed also by MACHO[46]). Also, it will be possible to distinguish between microlensing by a point mass and by an hydrogen cold dark cloud[55]. Moreover, thanks to this alert system, an ESO team have confirmed that the spectral features of a microlensed star toward the bulge did not change during the event (IAUC 6069 and 6071).

Last but not least, more precise light curves will allow to see parallax events, such as the first one reported with MACHO data[46]. This latter event is the longest one yet detected (about 100 days), during which the light curve has begun to be distorted by the effect of the motion of the Earth. It is then possible to determine the transverse velocity of the deflector projected at the position of the Earth – 54 km/s in the observed case – and thus to infere its mass (still depending on its location) – a low mass star in the Galactic disk or a brown dwarf in the bulge. Finally, it is clear that all the projects will end up with huge unprecedented catalogues of all kinds of variable stars, of extreme astrophysical potentiality too. The first published exemples are for eclipsing binaries[56] or Cepheids[57] in the LMC, a galaxy less evolved than ours.

CONCLUSIONS

The preliminary conclusions that can be drawn with respect to unseen baryonic matter from the actual results (as for February 1995) of the different on-going projects searching for microlensing events are the following:

i) The observed microlensing optical depth toward the Large Magellanic Cloud seems to appear well below the prediction[38] ($\sim$ 5 10^{-7}) expected for a standard halo entirely made of MACHOS, but significantly above the maximum possible value[40,58] ($< 10^{-8}$) as due to known distributions of stars. Furthermore, low mass halo objects within the range $\sim$ 5 10^{-8}–7 10^{-4} $M_\odot$ seem excluded[43].

ii) The observed microlensing optical depth toward the galactic bulge seems to be far above (by a factor 4?[59]) the prediction[60] ($\sim$ 8.5 10^{-7}) computed with known stellar populations of the bulge. This might strengthen plenty of earlier evidences[61,62] that the bulge is in fact barlike, elongated toward us[63,64]. Nevertheless, a more heavy ("maximum") disk[65] could still be an alternative explanation.

A first very preliminary implication for the composition of the halo of our Galaxy might be that the halo mass fraction in MACHOS in the range $\sim$ 10^{-1}–10^{-4} $M_\odot$ is less than $\sim$20% (see Bennett et al. and Magneville et al. in these proceedings), apparently consistent with expectation for a Universe whose primary component is cold dark matter[66]. However, one has to be cautious; uncertainties are such that a larger MACHOS fraction cannot yet be excluded. If indeed MACHOS do exist, one will have then to understand their origin and distribution[67,68].

The phenomenon of gravitational microlensing has now been convincingly proved to be a manageable important new tool for addressing a wide variety of scientific questions, from star formation to galaxy evolution and cosmology. The preliminary results concerning the Galactic baryonic dark matter need to be enforced with much more data. Also, other lines of sight, like for instance toward the SMC, should be monitored to explore the shape of the dark halo. All on-going experiments will continue for several years; in particular, the EROS one is planning an instrumental upgrade to a 1 m telescope and a 2 $\times$ 8 CCDs 2K $\times$ 2K in the course of 1995. They also have a common satellite project in order to exploit the parallax events and search for extrasolar planets. Whatever will be the future results, one should finally get the answer to the dark matter problem!

REFERENCES

1. Bouchet, F. & M. Lachièze-Rey Eds. 1994. Proceedings of the 9th IAP Astrophysics Meeting, Paris, France, July 12-16, 1993. Editions Frontières.
2. Dekel, A., Bertschinger, E., Yahil, A., Strauss, M., Davis, M. & J. Huchra. 1993. Astrophys. J. **412**: 1-21.
3. Zwicky, F. 1933. Helv. Phys. Acta. **6**: 110.
4. Elbaz, D., Arnaud, M. & H. Bohringer. 1995. Astron. Astrophys. **293**: 337-346.
5. Refsdal, S. & J. Surdej. 1994. Rep. Prog. Phys. **56**: 117-185.
6. Mamon, G. 1995. *In* XIVth Moriond Astrophysics Meeting on Clusters of Galaxies, p. 291. Editions Frontières.
7. Faber, S. & J. Gallagher. 1979. Annu. Rev. Astron. Astrophys. **17**: 135-187.
8. Battaner, E., Garrido, J., Membrado, M. & E. Florido. 1992. Nature. **360**: 652-653.
9. Persic, M. & P. Salucci. 1992. Month. Not. R. Astr. Soc. **258**: 14p-18p.
10. Peebles, P. 1993. *In* Principles of Physical Cosmology, p. 476. Princeton Series in Physics.
11. Vangioni-Flam, E. & M. Cassé. 1995. Astrophys. J. **441**: 471-476.
12. L3 Collab., Adeva, B. et al. 1989. Phys. Letters B. **231**: 509. ALEPH Collab., Decamp, D. et al. 1989. Phys. Letters B. **231**: 519. OPAL Collab., Akrawy, M.Z. et al. 1989. Phys. Letters B. **231**: 530. DELPHI Collab., Aarnio, P. et al. 1989. Phys. Letters B. **231**: 539.
13. Walker, T., Steigman, G., Schramm, D., Olive, K. & H. Kang. 1991. Astrophys. J. **376**: 51-69.
14. Smith, M., Kawano, L. & R. Malaney. 1993. Astrophys. J. Suppl. **85**: 219-247.
15. Ferlet, R. 1992. *In* Astrochemistry of Cosmic Phenomena. IAU Symp. **150**: 85-90.
16. Songaila, A., Cowie, L. L., Hogan, C. & M. Rugers. 1994. Nature. **368**: 599-604.
17. Carswell, R., Rauch, M., Weymann, R., Cooke, A. & J.K. Webb. 1994. Month. Not. R. Astr. Soc. **268**: L1-L4.
18. Pierce, M., Welch, D., McClure, R., van den Bergh, S., Racine, R. & P. Stetson. 1994. Nature **371**: 385-389.
19. Tytler, D. 1994. Paper presented at the ESO Workshop on QSOs Absorption Systems, Garching, Germany, November 24-28.

20. Oort, J. 1932. Bull. Astron. Inst. Neth. **6**: 249.
21. Bahcall, J. 1987. *In* Dark Matter in the Universe. IAU Symp. **117**: 17-27.
22. Fich, M. & S. Tremaine. 1991. Annu. Rev. Astron. Astrophys. **29**: 409-445.
23. Robin. A., Crézé, M. & V. Mohan. 1992. Astron. Astrophys. **265**: 32-39.
24. Ostriker, J., Peebles, P. & A. Yahil. 1975. Astrophys. J. **193**: L1-L4.
25. Trimble, V. 1987. Annu. Rev. Astron. Astrophys. **25**: 425-472.
26. Carr, B. 1990. Comments Astrophys. **14**: 257-280.
27. Liebert, J. & A. Burrows. 1993. Rev. Mod. Phys. **65**: 301.
28. de Rújula, A., Jetzer, P. & E. Massó. 1992. Astron. Astrophys. **254**: 99-104.
29. Pfenniger, D. & F. Combes. 1994. Astron. Astrophys. **285**: 94-118. 30. Pfenniger, D., Combes, F. & L. Martinet. 1994. Astron. Astrophys. **285**: 79-93.
31. Einstein, A. 1936. Science. **84**: 506.
32. Paczyński, B. 1986. Astrophys. J. **304**: 1-5.
33. Arnaud, M. et al. 1994. Exp. Astron. **4**: 265-276.
34. Arnaud, M. et al. 1994. Exp. Astron. **4**: 279-296.
35. Aubourg, E. et al. 1993. Nature **365**: 623-625.
36. Beaulieu, J.P. et al. 1995. Astron. Astrophys. In press.
37. Alcock, C. et al. 1993. Nature **365**: 621-623.
38. Griest, K. 1991. Astrophys. J. **366**: 412-421.
39. Jetzer, P. 1994. Astrophys. J. **432**: L43-L45.
40. Bahcall, J., Flynn, C., Gould, A. & S. Kirhakos. 1994. Astrophys. J. **435**: L51-L54.
41. Hu, E., Huang, J., Gilmore, G. & L. L. Cowie. 1995. Nature. In press.
42. Sahu, K. 1994. Nature **370**: 275-276.
43. Aubourg, E. et al. 1995. Astron. Astrophys. In press.
44. Udalski, A. et al. 1993. Acta Astron. **43**: 289-294.
45. Udalski, A. et al. 1994. Astrophys. J. Letters **426**: L69-L72.
46. Bennett, D. et al. 1994. Preprint.
47. Alard, C. 1994. Private communication.
48. Baillon, P., Bouquet, A., Giraud-Héraud, Y. & J. Kaplan. 1993. Astron. Astrophys. **277**: 1-9.
49. Crotts, A. 1992. Astrophys. J. Letters. **399**: L43-L46.
50. Udalski, A. et al. 1994. Acta Astron. **44**: 227.
51. Gould, A. & A. Loeb. 1992. Astrophys. J. **396**: 104-114.
52. Sazhin, M. & A. Cherepashchuk. 1994. Astron. Letters **20**: 523-528.
53. Bolato, A. & E. Falco. 1994. Astrophys. J. **436**: 112-116.
54. Udalski, A. et al. 1994. Astrophys. J. **436**: L103-L106.
55. Henriksen, R. & L. Widrow. 1995. Astrophys. J. **441**: 70-76.
56. Grison, P. et al. 1995. Astron. Astrophys. Suppl. **109**: 447-469.
57. Beaulieu, J.P. et al. 1995. Astron. Astrophys. In press.

58. Gould, A., Miralda-Escudé, J. & J. Bahcall. 1994. Astrophys. J. **423**: L105-L108.
59. Udalski, A. et al. 1994. Acta Astron. **44**: 165.
60. Kiraga, M. & B. Paczyński. 1994. Astrophys. J. **430**: L101-L104.
61. de Vaucouleurs, G. 1964. *In* The Galaxy and the Magellanic Clouds. IAU Symp. **20**: 195-199.
62. Holt, S. & F. Verter Eds. 1993. AIP Conf. Proc. 278. Back to th Galaxy.
63. Paczyński, B. et al. 1994. Astrophys. J. **435**: L113-L116.
64. Zhao, H.S., Spergel, D. & M. Rich. 1995. Astrophys. J. **440**: L13-L16.
65. Alcock, C. et al. 1994. Astrophys. J. In press.
66. Gates, E., Gyuk, G. & M. Turner. 1995. Phys. Rev. Letters. Submitted.
67. Wasserman, I. & E. Salpeter. 1994. Astrophys. J. **433**: 670-686.
68. De Paolis, F., Ingrosso, G., Jetzer, P. & M. Roncadelli. 1995. Astron. Astrophys. In press.

Clusters of Galaxies

HANS BÖHRINGER

Max-Planck-Institut für extraterrestrische Physik
85740 Garching, FRG

INTRODUCTION

Clusters of galaxies are as indicated by their name recognized as density enhancements in the galaxy distribution in the sky. In three dimensional space one finds that the galaxy density is higher by a factor of the order of 100 on average over the cluster and reaches on overdensity of up to a factor of 10^4 in the central regions of rich clusters. But there is more to clusters of galaxies than that. They are very interesting astrophysical laboratories as well as important tracers of the large scale structure of the Universe as will be illustrated in this contribution.

In addition to the galaxies, clusters also contain a large amount of dark matter as recognized already by Zwicky (1934) in his dynamical study of the Coma cluster, when it just became possible to obtain redshifts for the cluster galaxies. The dark matter is about an order of magnitude more massive than the estimated mass of the galaxies if the galaxies are to be gravitationally bound.

The origin of these galaxy clusters is best understood by considering the present knowledge about the formation of structure in the Universe. In this scenario the astronomical objects have formed from inhomogeinities in the density distribution in the early Universe. These density fluctuations have grown by the action of gravity. When the growth of some overdense regions reaches an amplitude comparable to the mean density of the Universe they decouple from the cosmic Hubble flow and recollapse. When part of the potential energy released in the collapse is transformed into kinetic energy (in proportions governed by virial equilibrium) dynamically stable astronomical objects form.

The power spectrum describing the density fluctuations in the early Universe can be inferred from our present knowledge of structure (e.g. see the lecture given by Margaret Geller in this Symposium), and one concludes that over the relevant length scale for the important astronomical objects the amplitude of the fluctuations decreases with wavelength. Thus small objects reach the critical amplitude for collapse first and these objects may be identified with the galactic cores. Subsequently galaxies form, they aggregate into groups and finally even larger regions become gravitationally unstable which leads to the formation of rich galaxy clusters. This sequence does not yet continue beyond the rich cluster scale (characterized by masses around 10^{15} $M_\odot$), since larger regions in the Universe have not reached the critical amplitude for collapse.

Therefore clusters of galaxies occupy a very special place in the cosmic hierarchy. They are the largest objects in the Universe that have collapsed to a nearly virial state and are thus characterized by a proper dynamical equilibrium configuration. The ideal equilibrium form is for example tentatively described by the King model[1]. One finds that this model describes many of the observed clusters reasonably well, while others are still in a critical intermediate state of formation and have a more peculiar structure.

This particular cosmological role of clusters can also be illustrated in another way by looking at the amplitudes of the gravitational potential in the Universe. If we could measure this potential throughout present day space, neglecting here very local "anomalies" in the potential caused by stars or galaxy nuclei, we would find that the depth of the potential wells associated with astronomical objects increases with their mass. Approximating the objects very roughly by homogeneous spheres one gets the simple parameterized relation, $\phi = 8.7 \cdot 10^{-9} \quad \Delta M R^{-1}$, where ϕ is the modified potential measured in $(\mathrm{km/s})^2$, ΔM the mass of the object (minus the mean density times the occupied volume) in solar mass units, and R the radius in Mpc. This gives us a means to roughly approximate the depth of the potential for various known objects. Now let us assume that this potential is filled with a collisionless tracer gas (as given by stars on small and galaxies on large scale) which is in virial equilibrium. Then the local velocity dispersion of the gas particles is the square root of half the value of the potential. Using these tracer particles to measure the potential depth one would find values of 100 - 300 km s^{-1} for field galaxies, values of 300 - 500 km s^{-1} for groups of galaxies, and values of 800 - 1500 km s^{-1} for rich galaxy clusters. Turning to superclusters this trend does not continue. In a supercluster we will essentially observe the potentials of the individual clusters.

Thus in our tracer particle picture, clusters of galaxies are marked by regions of the highest velocity dispersions. If we now fill this model Universe with gas that thermalizes in these potentials so that the main mass component, the protons, gain the same specific energy as the tracer particles, we will find that the hottest places are the centers of galaxy clusters again. The gas temperatures of the hot gas in rich clusters would be of the order of 3 - 13 keV. This gas would have very strong emission in soft X-rays and thus we would see the distribution of clusters and the large scale structure that they are outlining in an X-ray picture of this Universe.

Being the largest clearly defined objects makes clusters of galaxies superlatives in several ways and explains their importance in cosmology. They are the largest astrophysical laboratories for which the physical conditions can be fairly well defined and they are the largest single building blocks that can be used to probe the large scale structure of the Universe. In addition clusters of galaxies are relatively primitive objects to first order. They were formed by the gravitational collapse of matter without much dissipation and their formation process is mostly characterized by virialization. Thus their formation is a much simpler process than galaxy formation which involves dissipation and very complicated astrophysics. That galaxy clusters are so much more easy to model makes them even more valuable probes for cosmology.

The above sketched picture of clusters in the gravitational potential in the Universe gives us also a recipe for their study. One can either study the spatial and redshift distribution of the galaxies in the cluster in optical astronomy, or one can observe the hot intracluster gas by means of X-ray astronomy to study the dynamical structure of the clusters.

It was in particular X-ray astronomy with the development of powerful, space based X-ray telescopes, that has opened a new window for the study of the physics of galaxy clusters in the last years. Therefore this review is to a large part devoted to illustrate these new results. Section 2 provides examples of the observed X-ray morphology of clusters, while section 3 reviews the results on the mass determination and on the investigation of the matter composition in clusters. The sections 4 and 5 deal with further effects that have to do with the superlative role of clusters: they provide the largest masses for observed gravitational deflection of light rays in the Universe giving rise to gravitational lensing (section 4) and the giant hot gas halos of the clusters lead to measurable distortions in the microwave background, producing the so-called Sunyaev-Zel'dovich effect (section 5). In section 6 the use of clusters as cosmological probes for the large scale structure of the Universe is considered, and the last section provides a conclusion.

X-RAY MORPHOLOGY OF GALAXY CLUSTERS

The detailed study of the X-ray morphology of clusters of galaxies requires an X-ray telescope which, due to the absorption of the Earth's atmosphere, has to be placed in space. The first satellite that carried an X-ray telescope on board, was the EINSTEIN observatory[2]. It provided good images for of the order of 200 clusters and gave first deep insight into the physics of these objects. The German ROSAT observatory[3] followed with well improved sensitivity and spatial resolution. It is still operative since its launch in 1990. A third important instrument that has been launched recently is the Japanese satellite ASCA[4] with a low spatial resolution X-ray telescope, but with simultaneous good spectral resolution thanks to the newly employed X-ray CCD detectors.

Here we will mostly report on results obtained with the ROSAT observatory that provides by far the most detailed view on the X-ray morphology of clusters. Its observing window covers the energy range 0.1 to 2.4 keV. The telescope in connection with the position sensitive proportional counter (PSPC) reaches a spatial resolution of 20 arcsec at the optical axis and allows simultaneously for an energy resolution of the detected photons of about 45% at around 1 keV[5]. The second focal plane instrument that can be used with the telescope, the high resolution imager (HRI), has a higher spatial resolution of 5 arcsec on axis but no energy discrimination and lower sensitivity.

As an illustration of the morphology and physics of clusters that can be studied with the ROSAT observatory we will discuss results on the cluster of galaxies in the constellation Virgo. It is the most prominent, fairly rich cluster of galaxies in our immediate neighborhood, with a distance of about 20 Mpc. Due to its proximity it has been subject to many studies at optical and radio wavelength mostly concerning its galaxy population[6], and more than 400 cluster galaxy redshifts have been

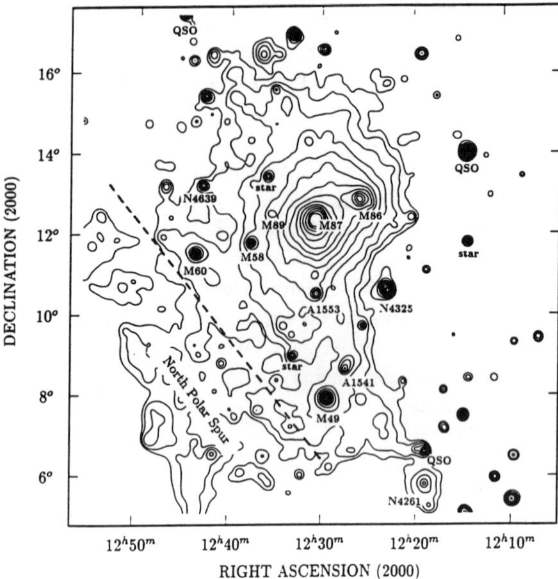

Figure 1: X-ray image of the Virgo cluster from the ROSAT All Sky Survey in the energy band 0.4 - 2 keV[10]. The image has been smoothed by a variable Gaussian filter with a maximum σ of 24 arcmin for the low surface brightness regions. Some cluster galaxies as well as foreground and background X-ray sources are labled.

measured[7]. With its large mass concentration so nearby, the cluster is responsible for part of the peculiar motion of our galaxy and the cluster is an important milestone in establishing the cosmological distance ladder. Thus an understanding of the morphology of this cluster and a mass determination is quite important. Unfortunately the Virgo cluster is a very complex, unrelaxed system and does not easily lend itself to modeling and to a virial mass estimate. The ROSAT observation of the cluster in the All Sky Survey gives the first detailed X-ray picture of the cluster as a whole and provides the means to model the large scale structure of this object.

The galaxy field of the cluster extends over more than 10 degrees in the sky and therefore only selected fields were imaged with the EINSTEIN observatory[8]. Scans with the non-imaging collimated detector on GINGA revealed extended emission on scales of several degrees[9]. In the ROSAT All Sky Survey the Virgo region was scanned almost homogeneously with an average exposure time of 450 sec. Fig. 1 shows a contour plot of the X-ray surface brightness in the energy band from 0.4 to 2.4 keV[10]. The most prominent feature is the giant almost spherically symmetric X-ray halo around the bright elliptical galaxy M87. The X-ray halos around the ellipticals M86 and M49 are also clearly visible and M58 and M60 show up as X-ray sources. Faint non-symmetrical X-ray emission from the cluster extends over the whole region including both M87 and M49 with dimensions of more than 8 and 5

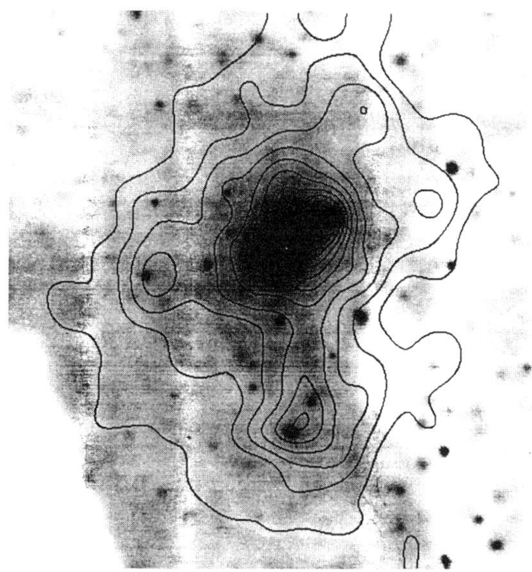

Figure 2: Number density of the galaxies in the Virgo cluster from the photometric survey of Binggeli et al.[11]. The density contours are superposed on the X-ray image of the Virgo cluster shown in Fig. 1 (from Schindler et al.[12]).

degrees in north-south and east-west direction, respectively.

Fig. 2 shows the density distribution of the galaxies in Virgo from the photometric survey of Binggli, Tammann, and Sandage[11] in form of a contour plot superposed on the X-ray image[12]. The images are surprisingly similar. In particular the asymmetric extension to the east and the sharp edge in the west are well visible in both contour maps. The similarity of the distribution of the galaxies and the intracluster plasma supports the picture that both cluster components follow the overall (mostly dark) matter distribution of the cluster very closely. This is also indicated in other cluster observations, but can only be seen in such detail in the nearby Virgo cluster.

The data from the ROSAT All Sky Survey observation of Virgo can also be analyzed spectroscopically in concentric rings around M87 out to 3 degrees radius[10]. The resulting temperature profile shows a significant decrease of the temperature towards the centre inside a radius of $\sim$ 100 kpc consistent with predictions from earlier studies on the cooling flow structure in M87[13]. Outside the cooling flow radius the temperature data are consistent with nearly isothermal plasma and with earlier measurements with the X-ray spectrometer on board of GINGA[14]. The combined data indicate a temperature of 2 - 2.4 keV for the outer halo of M87.

The total X-ray luminosity of the Virgo cluster with the extent shown in Fig. 1 amounts to an X-ray luminosity of about $8 \cdot 10^{43}$ erg s^{-1}in the ROSAT band (0.1 to 2.4 keV). By subtracting a spherically symmetric model X-ray image for the core region of the Virgo cluster from the observed image shown in Fig. 1 one can show that most of the X-ray emission comes from a symmetric cluster core around M87[10]. The two X-ray halos around M86 and M49 provide X-ray fluxes that are about one order of magnitude smaller. The residual image obtained after the subtraction is very irregular and contains about 10 - 15% of the X-ray flux. There are two prominent extended regions, one north of M87 and an irregular extended region around the more compact M49 halo as well as a connecting region between M87 and M49.

The fairly symmetric core region can actually be traced out to a radius of 1.5 to 1.8 Mpc except for the western region with the sharp edge and it can quite well be fit by a surface brightness distribution of a β-model[15] with a value of β of 0.46. With the given temperature and gas density distribution the integrated radial mass profile can be determined as explained in the next section. For the core region of Virgo to a radius of 1.8 Mpc one finds a gravitational mass of $1.5 - 5.5 \cdot 10^{14}\ M_{\odot}$. The large uncertainty of this value comes from the large errors in the measurement of the temperature profile. The gas density distribution can be determined with much higher precision - assuming again spherical symmetry of the cluster core - and one finds a gas mass of $4 - 5.5 \cdot 10^{13}\ M_{\odot}$. Thus about 7 to 36% of the inferred gravitational mass can be accounted for by the X-ray luminous gas (A more detailed temperature and mass profile for the inner region of the M87 halo has been derived from a ROSAT pointed observation[16]).

The gravitational effect of the Virgo cluster is responsible for a peculiar acceleration of our galaxy and the local group. The "Virgo infall velocity" so produced is estimated to be about 250 km s^{-1} for which a Virgo mass of about $10^{15}\ M_{\odot}$ is required[17]. The core region of Virgo which is now well seen in X-rays accounts already for a major fraction of 15 - 55% of this inferred mass.

The study described above shows, that in the Virgo cluster only the central part is dynamically old, while the large surrounding region is not virialized and still infalling. This demonstrates that clusters like Virgo are still growing by accretion of material at present and that the accretion process is generally fairly inhomogeneous and unisotropic.

The detailed ROSAT X-ray images have revealed other interesting examples of cluster evolution by accretion. In particular there are frequent cases where the accretion occurs by collisions of major cluster components. Two prime examples are the Coma cluster[18] and the cluster Abell 2256[19] shown in Fig. 3. While the main body of the Coma cluster is a dynamically old, nearly relaxed system, one notes a group or poor cluster of galaxies in the south west which is just infalling into the main cluster body. In A2256 a similar event happens with a smaller mass ratio of the components (probably around 1:4). Here the merger event has already progressed further and the heating of the intracluster gas by the collision event can be studied.

In a temperature distribution map recently derived for the deep observation of A2256[20] one notes two hot regions in the inner part of the cluster at position angles perpendicular to the collision axis. This effect is consistent with the findings of n-body+hydrodynamical simulations showing that a major shock wave is set off in the collision that starts with a lenticular shape strongly elongated perpendicular to

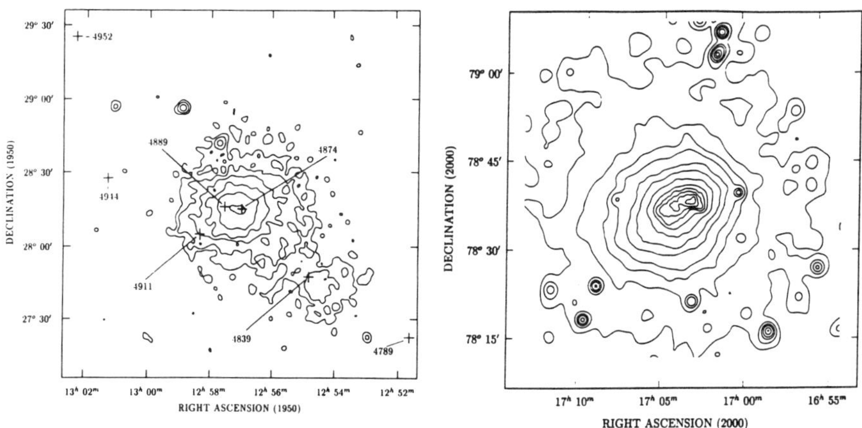

Figure 3: ROSAT X-ray images of the Coma cluster (left, from the ROSAT All Sky Survey)[18] and of Abell 2256 (right, ROSAT deep pointed observation)[19] in the energy band 0.5 - 2 keV. Both clusters show the infall of a smaller cluster component into the main cluster at different stages of the merging process.

the infall direction[21,22]. One is thus beginning to see details of the physics of cluster collisions in the X-ray observations.

The growth of clusters by accretion is also interesting from a more general point of view. If the Universe has critical density it is more likely that the surroundings of clusters are gravitationally bound and are accreted in the course of time than in a low density Universe. In the low density case cluster growth is stopped quite rapidly. Therefore the accretion rate of clusters observed at present can be used as a measure of the mean density of the Universe, Ω_0, as shown e.g. by Richstone et al.[23]. Different studies have started to relate the observable cluster morphology to an accretion rate and to Ω_0 [24–26], and first results indicate that the density is close to critical.

MASS AND COMPOSITION OF GALAXY CLUSTERS

It is very important to have a precise mass measurement of galaxy clusters. First it is necessary to get a correct account of the matter composition of the cluster and to establish the existence of "dark matter". Second, for the use of clusters to understand the formation of large scale structure and to obtain constraints on the primordial power spectrum, we need to know their mass. The mass of clusters can be determined from data of optical observations using the spatial and redshift distribution of the cluster galaxies assuming that the cluster is in virial equilibrium[27,28]. Apart from the problem that many clusters are not in virial equilibrium and that many hundreds of galaxies are needed for good statistics (which is rarely available), a major uncertainty in this mass determination is introduced by the fact that the results are very dependent on the assumed isotropy or anisotropy of the galaxy orbits (which cannot be determined from present observations)[29].

Using X-ray observations of the hot gas in clusters therefore provides an advantage. The gas is clearly isotropic and the statistics is determined by the number of photons received, which is only restricted by the length of the observations. Presently the most severe limitation is the imprecision with which the temperature distribution of the intracluster gas can be determined. But this is rapidly improving with the application of X-ray CCDs.

The mass determination through X-ray observations is based on the hydrostatic equation and requires the knowledge of the gas density and temperature distribution. The cluster mass profile for a spherically symmetric cluster is then given by:

$$M(r) = -\frac{kT_g(r)\, r^2}{m_h \mu G}\left(\frac{d\log T_g(r)}{dr} + \frac{d\log \rho(r)}{dr}\right) \tag{1}$$

where $\rho(r)$ and $T_g(r)$ are the density and temperature profile of the gas and the other parameters have their usual meaning.

The gas density distribution and gas mass profile can directly be determined from the X-ray surface brightness profile by geometrical deprojection assuming spherical symmetry. The temperature of the intracluster plasma is determined by fitting model spectra[30] to the observed photon spectra. Apart from the new results obtained with the ASCA satellite, observations with previous instruments either had good spectral resolution but no imaging capability or in the case of ROSAT good imaging but low simultaneous spectral resolution. Therefore at present the temperatures determined as a function of radius still contain large uncertainties which translate into corresponding uncertainties in the mass profile of the cluster.

Since the X-ray emissivity of the thermal gas in the ROSAT energy band is only weakly dependent on the temperature in the relevant temperature range, the observed X-ray surface brightness directly provides a value for the emission measure in the line of sight. The count rates vary less than 6% over the energy range from 2 to 10 keV. Therefore a precise knowledge of the temperature is not required for calculations of the gas density distribution.

One of the great advantages of ROSAT is the good sensitivity and the low internal background which allows to map the X-ray emission in some clusters out to the virial radius for the first time and thus allows a mass estimate of the clusters in total. An example of such a case is the Perseus cluster as observed in the ROSAT All Sky Survey. Here the X-ray emission could be traced out to a radius of 3 h_{50}^{-1} Mpc which is close to the estimated virial edge of the cluster[31,32].

Fig. 4 shows the gravitational mass, the gas mass, and the integrated galaxy mass profile for the Perseus cluster (for $h_{50} = 1$, where h_{50} is the Hubble constant measured in units of 50 km s^{-1} Mpc^{-1}). The gas mass is larger than the galaxy mass by more than a factor of 5. The gas mass cannot account for the binding mass,

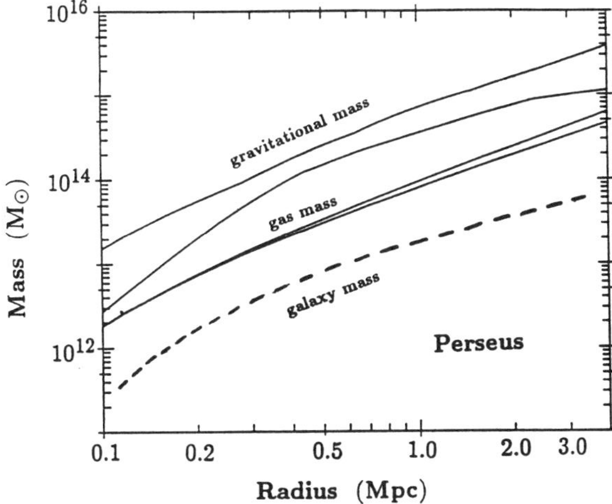

Figure 4: Integrated radial profiles for the gas and gravitational mass in the Perseus cluster of galaxies as determined from ROSAT X-ray data. The upper two curves indicate the constraint for the gravitational mass profile while the lower two curves bracket the gas mass profile.

however, and the major mass component has to be attributed to "dark matter". The gas density profile has a shallower form than the distribution of the galaxies.

Similar results for the gas mass and gravitational mass were derived from ROSAT data for some other nearby clusters[16,19,33–38]. The general results are summarized in Table 1. Particularly interesting are also the results for the low temperature clusters and for small groups of galaxies. For these objects the low temperature allows a better temperature determination with the data from ROSAT. A case worth noting is for example the group N5044 with a mean temperature around 1 keV for which the temperature has been determined with very low uncertainties ($\sim$ 5%)[36].

These results for the cluster masses derived from X-ray observations rest on the assumptions of hydrostatic equilibrium and spherical symmetry of the clusters. The reliability of this way of mass determination was recently tested for realistic, still evolving clusters by simulating the X-ray emission and data analysis process in model clusters obtained from n-body+hydrodynamic simulations[39,40]. The results show that the above simplifying assumptions lead on average to errors not larger than about 15 - 20%, but can be well as high as a factor of two at the location of thermalizing shock waves in merging clusters. Thus the mass estimates should in general be reliable - in particular on average for a cluster sample - but severe deviations can occur in peculiar cases (often recognizable by their X-ray morphology).

The large values for the ratio of the gas mass to the gravitational mass in rich clusters of galaxies is quite surprising. If we consider a hierarchical scenario for the formation of structure in the Universe - like the CDM model - clusters of galaxies are formed from overdensities in the mass distribution purely by the action of gravitation. Since in a strictly hierarchical model no dissipation occurs at scales larger

Table 1: Mass and composition of galaxy clusters

Rich clusters :	$M_{grav} \sim 5 \cdot 10^{14} - 5 \cdot 10^{15} M_\odot$
Galaxies :	2 - 7% $\cdot h_{50}^{-1}$ of M_{grav}
Intracluster gas :	10 - 30% $\cdot h_{50}^{-1.5}$ of M_{grav}
Dark matter :	60 - 85% of M_{grav}
Iron abundance :	$0.35 \pm (0.15)$ solar
Groups of galaxies :	$M_{grav} \sim 3 \cdot 10^{13} - 5 \cdot 10^{14}\ M_\odot$
Galaxies :	5 - 10% $\cdot h_{50}^{-1}$ of M_{grav}
Intracluster gas :	3 - 25% $\cdot h_{50}^{-1.5}$ of M_{grav}
Iron abundance :	$0.25 \pm (0.15)$ solar

than clusters before their formation baryonic matter and nonbaryonic matter should be collected indiscriminantly into the newly forming cluster potentials. Therefore the two mass fractions throughout the entire cluster should reflect the ratio of the mass components in the Universe in general. The gas mass fraction of 10 - 30% constitutes a lower limit to the baryon mass but may actually comprise by far the major component of all baryons in clusters. Extrapolating this fraction of 10 - 30% for the baryonic mass to the composition of the matter in the Universe in general has some very interesting cosmological consequences, in particular compared to the current understanding of the primordial nucleosynthesis [41]. A comparison of results from nucleosynthesis models with observed elemental abundances leads to constraints of the baryon density in the Universe of $\Omega_B = 0.02 - 0.1 h_{50}^{-1}$ [42]. These results are only marginally consistent with a mean density corresponding to $\Omega_o = 1$. This has been discussed in detail by White et al. [43].

The chemical composition of the baryonic matter can also be investigated by X-ray spectroscopy through the K-shell emission lines of the most abundant, heavy metals. The very impressive cluster X-ray spectra recently obtained with the ASCA satellite imply an intracluster abundance of iron of about 0.35 ± 0.15 the solar value deduced from the most prominent iron emission lines[44]. But also the emission lines of O, Mg, Si, and S are present in some spectra[45], giving a first indication that most of the intracluster heavy elements are produced by supernovae type II.

GRAVITATIONAL LENSING

The discovery of "gravitational arcs" in clusters of galaxies in 1986/87[46,47] opened another way to determine the mass of clusters. These arcs or smaller arclets are gravitationally distorted images of background galaxies. The large mass of the cluster acts as a gravitational lens bending the light rays from the source on their path to the observer. Fort and Mellier recently gave an excellent review on this subject[48].

As a guiding line to distinguish different regions where image distortion occurs, one generally refers to an ideal configuration that leads to the so-called Einstein ring. Here a background galaxy is directly in the line of sight with the observer and the center of the lens. If the lensing cluster is symmetric and exceeds the critical surface mass density (~ 1 g cm^{-2}) the light rays from the source can be bend at all azimuthal position angles around the cluster and the observer sees a circular image. The strongest distortions occur near this configuration. Thus the longest curved arcs are usually observed near the Einstein radius. The area inside and near the Einstein radius is called the region of strong lensing, while weak lensing effects are observed outside this radius leading to arclets not too far out and to very small elliptical distortions further away, which can only be detected statistically for a larger ensemble of background galaxies.

From the lens equation one can calculate the following radius for the Einstein ring

$$\Theta_E = 28.8'' \left(\frac{\sigma_r}{1000 km/s}\right)^2 \frac{D_{ls}}{D_s} = 28.8'' \left(\frac{T_g}{6.25 keV}\right) \frac{D_{ls}}{D_s} \tag{2}$$

for typical velocity dispersions or X-ray gas temperatures of very rich clusters. The parameter D_{ls}/D_s is the ratio of the lense-source and the observer-source distance which is generally of the order of unity but smaller than 1. Since clusters of galaxies provide the deepest gravitational potentials on large scale - as outlined above - the largest bending angles in gravitational lensing are observed for these objects.

The geometry of the arcs and arclets in the strong lensing region have been used to infer the masses of cluster cores[49] and also to reconstruct the shape of the gravitational potential[50–52], e.g. for the clusters A370 and A2218. For A370 a double potential is found from lens reconstruction and confirmed by a high resolution ROSAT image[48]. For A2218, which also needs a double potential, the X-ray image seems to imply a hydrodynamically disturbed cluster core[51], which may also explain why the X-ray mass of the cluster core is smaller by a factor of 2 - 2.5 compared to the lensing mass[49]. In another case, A963, which appears undisturbed, the lensing and X-ray masses are in good agreement[53].

The mass over larger parts of a cluster can only be investigated by studying the weak lensing effects further outside the Einstein radius[54]. Kaiser and Squires[55] have pioneered an non-parametric mass distribution reconstruction technique based on a statistical study of the weak gravitational shear in the background galaxy images over a large region of a CCD image of a cluster. Applying this technique a few surprising results were obtained, in particular the result by Fahlman et al.[56] on the cluster MS1224 where the lensing mass is higher by a factor of 2 - 3 than the virial mass estimate based on the galaxy velocity dispersion. Also the lensing mass to light ratio for this cluster of $\sim 800 h_{100} \frac{M_\odot}{L_\odot}$ (with h_{100} being the Hubble constant measured in units of 100 km s^{-1} Mpc^{-1}), which is astonishingly high compared to the results from virial mass or X-ray mass estimates. For two other clusters investigated by Smail et al.[57], MS1455+22 and MS0016+16, two of the brightest clusters from the Einstein Medium Sensitivity Survey, the lensing mass and X-ray masses are in good agreement. Another interesting case of a weak lensing study concerns Cl0024+17 by Bonnet et al.[58] which shows a gravitational shear effect out to an Abell radius (3 h_{50}^{-1} Mpc), providing the means to study the mass distribution out to the edge of the virialized cluster system. The restrictions in these observations at present come

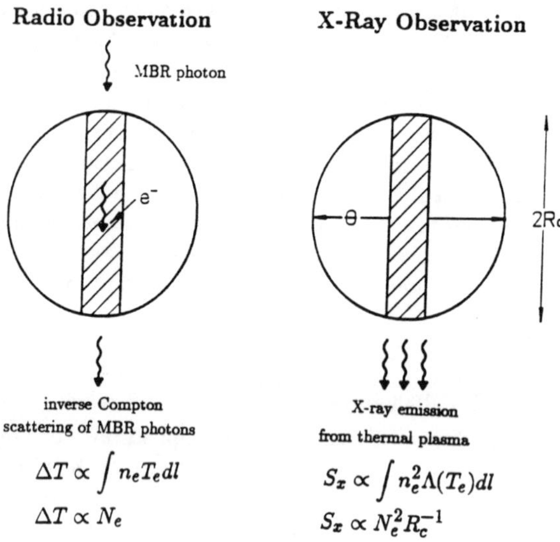

Figure 5: Sketch of the Sunyaev-Zel'dovich effect

from the size of the CCD field and the seeing limit in the observations. Very large field-of-view CCD cameras for this purpose are under construction promising even more interesting results for the future. At present it is not perfectly clear if the high lensing masses observed for some cases are really in conflict with the results from the conventional methods or if this problem has to be attributed to the sum of disturbing effects: substructure in clusters, boundary conditions in the lens reconstruction, etc. But this question should soon be clarified by more and better data.

THE SUNYAEV-ZEL'DOVICH EFFECT

The hot electron plasma of the intracluster medium of galaxy clusters produces a distortion in the microwave background, known as the "Sunyaev-Zel'dovich-Effect". It can be used for the determination of absolute distances to clusters of galaxies independent of the conventional cosmological distance ladder. The effect is caused by the fact that microwave background photons passing through the cluster are inverse-Compton scattered to higher energies. This leads to a decrement in the microwave background in the Rayleigh-Jeans part of the microwave spectrum and an increment in the Wien part[59]. Since clusters of galaxies contain the by far the largest hot plasma halos known in the Universe this is the only measurable effect of a distortion of the microwave background between the last scattering surface at the epoch of recombination and the observer.

A sketch of the effect is given in Fig. 5. The clusters are optically very thin for the scattering and the effect is small and linear. As shown in the figure the decrement (increment) is proportional to the electron temperature (measurable by X-ray spectroscopy) and to the column density of the electrons in the plasma in the line of sight. The same thermal plasma is also responsible for the X-ray emission. But the X-ray surface brightness is now proportional to the emission measure, which can be expressed as the square of the column density divided by a characteristic length scale. That the two observable effects depend on the electron density with a different power is the key to the application of this effect to distance measurements. Knowing the column density of the electrons from the microwave decrement (increment) measurement and the shape of the plasma density distribution, the characteristic radius can be calculated as absolute quantity. The corresponding apparant characteristic radius is obtained from X-ray imaging (for clusters of galaxies that are believed to be fairly spherically symmetric). From the comparison of the absolute and apparant radius the distance of the cluster can be calculated.

In practice the radioastronomical observation of the microwave decrement (increment) is rather difficult since the expected effect is of the order of $\frac{\Delta I}{I} \sim 10^{-4}$ for even the most massive and X-ray luminous clusters. This small effect has to be observed in the presence of much larger contamination effects from the Earth's atmosphere, the ground, and from unrelated radio sources in the clusters. Birkinshaw et al.[60,61] have pioneered these studies and observed this effect in Abell 2218, Abell 665, and Cl0016+16. Recently a larger number of research teams have entered the field and the effect is now detected in of the order of 20 galaxy clusters.

The cases analyzed so far which are documented in the literature are A665[61] where a Hubble constant of 45 ± 17 km s^{-1} Mpc^{-1} was deduced, A2218 with $H_0 = 65 \pm 25$ km s^{-1} Mpc^{-1} [62] and $H_0 = 38(+18, -16)$ km s^{-1} Mpc^{-1} [63], and Coma with $H_0 = 74 \pm 29$ [64]. The results still span a wide range. But the progress in the experimental techniques is very encouraging.

The recently achieved experimental improvements include the start of the operation of the Ryle telescope array at Cambridge[65] which can map the Suyaev-Zel'dovich effect two-dimensionally and the use of microwave Bolometers at 1.2 and 2.2 mm wavelength with which the effect can be measured both as decrement and increment. This latter technique has been applied successfully by two groups recently, by a team at CalTech[66] and by Andreani et al. at ESO, La Silla[67]. Thus a larger number of more accurate and reliable measurements should soon become available.

PROBING THE LARGE SCALE STRUCTURE WITH CLUSTERS

Clusters of galaxies are good tracers of the large scale structure in the Universe. As mentioned above clusters form from the regions with the highest amplitudes in the primordial density fluctuation field in the early Universe (at comoving scales of $\sim 6 - 16\ h_{100}^{-1}$ Mpc radius). If the primordial density fluctuations are described by a Gaussian process - e.g. if they originate in quantum fluctuations at the beginning of the inflationary epoch - the statistics of the occurance of protoclusters can be derived theoretically from the power spectrum that describes the primordial density fluctuation field statistically. For cosmological models that do not lead to a Gaussian fluctuation field, e.g. models with cosmic strings or textures - more complex statistics has to be used.

In this context there are in particular two statistical measures of the observed

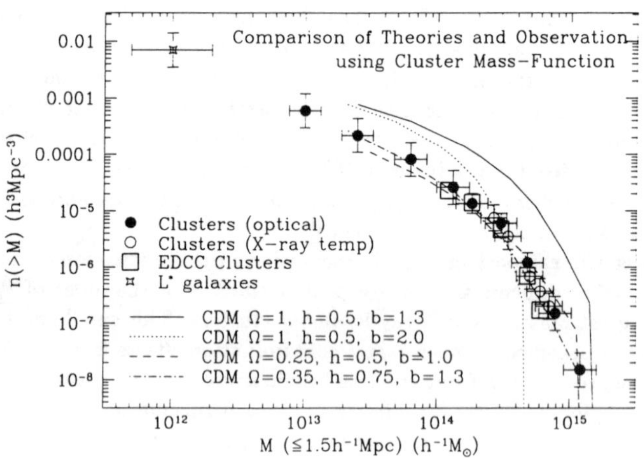

Figure 6: Cluster mass function derived from optical and X-ray data by Bahcall and Cen[74]. Predictions of different cosmological models with CDM type primordial power spectra are compared to the observational data.

cluster population that are most useful to constrain the primordial power spectrum. These are the cluster mass spectrum describing the spatial density of clusters as a function of their size and the two-point-correlation function describing the statistics of the spatial distribution of clusters. The first method provides information on the primordial power spectrum on comoving scales corresponding to the size of proto-clusters (6 - 16 Mpc). And the two-point-correlation function gives constraints on the power spectrum from scales of the intercluster distances to the size of the cluster surveys (10 - several hundred Mpc). To relate the primordial power spectrum to the observed cluster distribution two approaches are used: with some simplifications analytical methods as the Press-Schechter statistics[68] or Peak statistics[69] can be used to calculate the mass or correlation function[70] of clusters and alternatively n-body models are used to simulate the cluster formation for different cosmological scenarios[71–73].

Cluster mass function

So far there are not yet enough clusters for which the total mass has been determined reliably from X-ray data as described above to allow the construction of a mass function from these data. Therefore one has to rely on more simple mass estimates. Bahcall and Cen[74] have, for example, constructed a cluster mass function from known global parameters of clusters like velocity dispersions, X-ray luminosities, and temperatures. The resulting function shown in Fig. 6 can best be fit by a cosmological model with the parameters $\Omega_0 = 0.25 - 0.35$, $h_{100} = 0.5 - 0.75$, $\sigma_8 = 0.77 - 1.0$, and a cold dark matter (CDM) type primordial power spectrum. (The parameter σ_8 gives the normalization for the power spectrum, where σ_8 is the variance (rms) of the density fluctuation field smoothed with a filter radius of 8 h_{100}^{-1} Mpc). The mass function is extended down to elliptical galaxies. If this function is integrated (down to galaxy masses of $\sim 10^{12}$ $M_\odot$) one finds that these objects comprise already 20% of a critical density Universe (for $h_{100} = 0.5$). White et al.[75] used in a similar way the spatial density of richness class 1 Abell clusters and a typical mass estimate for these objects of $4 - 5 \cdot 10^{14}$ $M_\odot$ to get constraints on the primordial power spectrum. In this analysis more care was taken in the assignment of the masses to the clusters in order to include the complete virialized cluster system. The general results are similar to those of Bahcall and Cen, but the mass scale is shifted to higher masses by 40%. This results in a normalization parameter of $\sigma_8 = 0.5 - 0.6$.

The direct observable property of clusters that comes closest to the mass is the X-ray temperature, which effectively probes the depth of the cluster potential. The temperature function[76,77] has been used in the past to get information on the primordial power spectrum e.g. by Henry and Arnaud[77]. We can use the present results on clusters masses obtained by detailed modeling of the ROSAT data to construct a mass-temperature relation[78]. Using the known X-ray temperature functions[76,77] one gets with large statistical errors a mass function that is consistent with the results of Bahcall and Cen[74].

Cluster correlation function

While the mass function of clusters gives the first moment of the cluster distribution in space, more detailed information on the structural coherence on large scales is contained in the second moment of the spatial distribution of clusters, the two-point-correlation function. Several attempts have been made to determine the correlation function of galaxy clusters and the results were nicely compiled some years ago in the review by Bahcall [79]. The interesting point about the correlation function of clusters compared to that of galaxies is the fact that it has a higher amplitude and that the relation of the amplitude of the correlation function of clusters to the correlation function of the underlying matter distribution can be calculated by theoretical considerations [70] (contrary to the galaxy correlation function which depends on the so-called biasing parameter, that cannot be calculated ab initio from astrophysical principles).

Studies on the spatial correlation of Abell clusters[80,81] gave a correlation length of $r_{cc} = 20(\pm 5)h_{100}^{-1}$ Mpc and thus a correlation amplitude for cluster that is about a factor of ~ 7 higher than that of galaxies. More recent results based on machine produced cluster catalogues obtained from the digitized data of the UK Schmidt survey (COSMOS and APM) gave smaller values for the correlation length of $r_{cc} \sim 15$ h_{100}^{-1} Mpc[82,83] (see Fig. 7). The difference has been attributed to the difference in richness of the selected clusters[84], and more importantly to selection biases in the construction of the cluster catalogues[85]. The selection effects in the Abell cluster sample manifest themselves in giving different values for the correlation length in

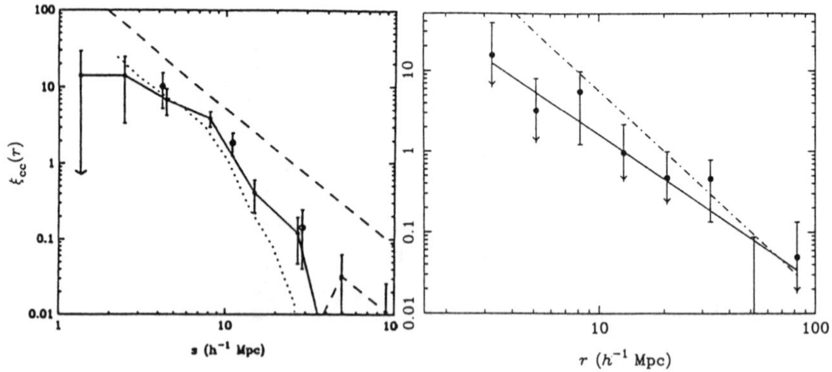

Figure 7: Two-point-correlation function of clusters of galaxies for the machine produced, optical APM cluster sample[83] (left) and for the X-ray selected ROSAT cluster sample[86] (right).

the line-of-sight (redshift) direction and perpendicular to it. Selecting the clusters primarily by their X-ray emission promises the avoidance of these selection biases.

Fig. 7 also shows the cluster correlation function determined for a flux limited sample of ROSAT All Sky Survey clusters around the South Galactic Pole [86]. The clusters have been selected primarily by their X-ray emission (plus some additional optical criteria that were used to identify the X-ray cluster candidates on plates and to later verify the cluster nature by optical follow-up observations). In a second sample the correlation function was determined for a sample of Abell clusters with X-ray emission in the region of the North Galactic Pole [87]. The two results are in good agreement with the results from the COSMOS and APM cluster samples. Bahcall and Cen [72] have attributed the difference of the optical Abell and X-ray samples to the fact that sampling by an X-ray flux limit is different from the optical sampling while Dalton et al.[88] have made biased selection effects in the optical Abell cluster sample responsible for the disagreement. The overall result on the cluster correlation function is that it has too much power on large scales to be consistent with a primordial CDM power spectrum and $\Omega_0 = 1$, but it agrees with a low density model or other models that provide extra power on scales between $\sim 30 - 100$ Mpc.

A large scale redshift survey for the ROSAT All Sky Survey clusters is under way in the form of an ESO key programme which will deliver a much larger and better defined data set of about 700 southern clusters[89]. This data set will be large and significant enough to provide a reliable X-ray cluster correlation function and to allow the study of the dependence of the correlation amplitude on different selection criteria.

CONCLUSIONS

The examples of research on clusters of galaxies shown above illustrate that clusters have become increasingly important cosmological probes. In particular X-ray astronomy has advanced the detailed study of the physics of clusters in recent years. By probing the X-ray emitting intracluster gas it is now possible to obtain reliable gravitational masses of clusters. With the new results from the ASCA satellite one can now derive cluster mass profiles with errors as small as $\sim$ 20%. These observations clearly confirm the existence of unseen matter in clusters of galaxies and imply also that the missing mass cannot only be concentrated in the galaxies or in a central black hole but the major part has to be distributed smoothly throughout the cluster.

The more detailed X-ray images of clusters now available clearly show a high frequency of substructure in clusters giving evidence that clusters are still evolving by accretion. This sets a lower limit to the mean density of the Universe; and it is one of the interesting tasks for the near future to better quantify this lower limit to Ω_0.

The discovery of gravitational lensing has provided an alternative way of cluster mass determination. Some of the present results showing a disagreement with the virial and X-ray masses are quite puzzling; but the fast development in this area will soon lead to the resolution of this question.

Clusters are also very important tracers of the large-scale structure of the matter distribution in the Universe. It was shown how the results from ROSAT - notably from the ROSAT All Sky Survey - provide a unique possibility to construct cluster mass functions and cluster two-point correlation functions based on X-ray data, which are in many ways more reliable than the purely optical data used in the past. First preliminary results were shown. The large redshift surveys for ROSAT detected clusters that are currently carried out will provide good results from large samples for the statistical measures of the cluster population and spatial distribution.

ACKNOWLEDGMENTS

I thank my colleagues S. Schindler, D. M. Neumann, W. Voges, and U.G. Briel for the collaboration in the study of galaxy clusters with ROSAT. I especially also thank the ROSAT team for providing the ROSAT data and for the analysis software.

References

[1.] King, I., AJ, **67:** 471.

[2.] Giacconi, R., et al., 1979, ApJ, **230:** 540.

[3.] Trümper, J., 1993, Science, **260:** 1769.

[4.] Tanaka, Y., Inoue, H., and Holt, S.S., 1994, Publ. Astron. Soc. Japan, **46:** L37.

[5.] Briel, U.G., Pfeffermann, E., Hartner, G., and Hasinger, G., 1988, Proc. SPIE, **982:** 401.

[6.] Richter,O.G. and Binggeli, B. (eds.), 1985, *The Virgo Cluster*, ESO Workshop Proc. No. 20.

[7.] Binggeli, B., Popescu, C.C., and Tammann, G.A., 1993, A&AS, **98:** 275.

[8.] Forman, W., Jones, C., and DeFaccio, M., 1985, ESO Workshop Proc. No. 20: *The Virgo Cluster*, O.-G.Richter, B. Binggeli (eds.), p. 323.

[9.] Takano, S., Awaki, H., Koyama, K., Kunieda, H., Tawara, Y., Yamauchi, S., Makishima, K., & Ohashi, T., 1989 Nat, **340:** 289.

[10.] Böhringer, H., Briel, U.G., Schwarz, R.A., Voges, W., Hartner, G., and Trümper, J., 1994, Nat, **368:** 828.

[11.] Binggeli, B., Tammann, G.A., and Sandage, A., 1987, AJ, **94:** 251.

[12.] Schindler, S.C. et al., 1995, Proccedings of the 17th Texas Symposium on Relativistic Astrophysics and Cosmology, Voges, W. et al. (eds), Extended Poster Abtracts, MPE Report, MPE, Garching, (in press)

[13.] Stewart, G.C., Canizares, C.R., Fabian, A.C., and Nulsen, P.E.J., 1984, ApJ, **278:** 536.

[14.] Koyama, K., Takano, S., and Tawara, Y., 1991, Nat, **350:** 135.

[15.] Jones, C. and Forman, W., 1984, ApJ, **276:** 38.

[16.] Nulsen, P.E.J. and Böhringer, H., 1995, MNRAS, (in press).

[17.] Davis, M., Tonry, J., Huchra, J.P., and Latham, D.W., 1980, ApJ, **238:** L113.

[18.] Briel, U.G., Henry, J.P., Schwarz, R.A., Böhringer, H., Ebeling, H., Edge, A.C., Hartner, G., Schindler, S., Trümper, J., and Voges, W., 1991, A&A**246:** L10.

[19.] Briel, U.G., Henry, J.P., and Böhringer, H., 1991, A&A, **259:**, L31.

[20.] Briel, U.G. and Henry, J.P., 1994 Nat, **372:** 439.

[21.] Schindler, S. and Müller, E., 1993, A&A, **272:** 137.

[22.] Evrard, G. , 1990, in Clusters of Galaxies, Oegerle, W.R., Fitchett, M.J., Danley, L. (eds.), Cambridge Univ. Press, p. 287.

[23.] Richstone,D., Loeb,A., and Turner, E.L., 1992, ApJ, **393:**, 477.

[24.] Mohr, J.J., Fabricant, D.G., and Geller, M.J., 1993, ApJ, **413:** 492.

[25.] Mohr, J.J., Evrard, A.E., Fabricant, D., and Geller, M., 1995, ApJ, (in press).

[26.] Neumann, D.M. and Böhringer, H., in "Clustering in the Universe", Proc. of the Rencontre de Moriond, Les Arcs, March 1995, Balkowski et al. (eds.), Edition Frontiere, (in press).

[27.] Kent, S.M. and Gunn, J.E., 1982, AJ, **87:** 945.

[28.] Kent, S.M. and Sargent, W.L.W., 1983, AJ, **88:** 697.

[29.] The, L.S. and White, S.D.M., 1986, AJ, **92:** 1248.

[30.] Raymond, J.C. and Smith, B.W., 1977, ApJS, **35:** 419.

[31.] Schwarz, R.A., Edge, A.C., Voges, W., Böhringer, H., Ebeling, H., and Briel, U.G., 1992, A&A, **256:** L11.

[32.] Böhringer, H., 1994, in *Cosmological Aspects of X-ray Clusters of Galaxies*, W.C. Seitter (ed.), Kluwer Publ., p. 123.

[33.] Henry, J.P., Briel, U.G., and Nulsen, P.E.J., 1993, A&A, **271:** 413.

[34.] Neumann, D.M. and Böhringer, H., 1995, A&A, (in press).

[35.] David, L.P., Jones, C., Forman, W., and Daines, S., 1994, ApJ, **428:** 544.

[36.] Allen, S.W., Fabian, A.C., Johnstone, R.M., White, D., Daines, S.W., Edge, A.C., and Stewart, G.C., 1993, MNRAS, **262:** 901.

[37.] Ponman, T.J. and Bertram, D. , 1993, Nat, **363:** 51.

[38.] Mulchaey, J.S., Davis, D.S., Mushotzky, R.F., and Burstein, D., 1993, ApJ, **404:** L9.

[39.] Schindler, S., 1995, A&A, (submitted)

[40.] Evrard, A.E., 1994, in *Clusters of Galaxies*, Proc. of the XIVth Moriond Astrophysics Meeting, Durret, F., Mazure, A., Tran Thanh Van, J. (eds.), Edition Frontiere, p. 241.

[41.] White, S.D.M. and Frenk, C.S., 1991, ApJ, **371:** 52.

[42.] Walker, T.P., Steigman, G., Schramm, D.N., Olive, K.A., and Kang, H-S., 1991, ApJ, **376:** 51.

[42.] White, S.D.M., Navarro, J.F., Evrard, A.E., Frenk, C.S., 1993, Nat, **366:** 429.

[44.] Ohashi,T., 1995, this volume

[45] Fukazawa, Y., Ohashi, T., Fabian, A.C., Canizares, R.C., Ikebe, Y., Makishima, K., Mushotzky, R.F., and Yamashita, K., 1994, Publ. Astron. Soc. Japan, **46:** L55.

[46.] Soucail, G., Fort, B., Mellier, Y., and Picat, J.P., 1987, A&A, **172:** L14.

[47.] Lynds, R. and Petrosian, V., 1986, B.A.A.S., **18:** 1014.

[48.] Fort, B. and Mellier, Y., 1994, A&AR, **5:** 239.

[49.] Miralda-Escude, J. and Babul, A., 1995, ApJ, (submitted).

[50.] Kneib, J.-P., Mellier, Y., Fort, B., and Mathez, G., 1993, A&A**273:** 367.

[51.] Kneib, J.P., Mellier, Y., Pello, R., Miralda-Escude, J., Le Borgne, J.-F., Böhringer, H., and Picat, J.-P., 1995, A&A, (in press).

[52.] Hammer, F., Gioia, I., Le Fevre, O., and Luppino, G., this volume.

[53.] Bautz, M., Poster presented at the 17th Texas Symposium.

[54.] Tyson, J.A., Valdes, F., and Wenk, R.A., 1990, ApJ, **349:** L1.

[55.] Kaiser, N. and Squires, G., 1993, ApJ, **404:** 441.

[56.] Fahlman, G.G., Kaiser, N., Squires, G., and Woods, D., 1995, ApJ, **437:** 56.

[57.] Smail, I., Ellis, R.S., Fitchett, M.J., and Edge, A.C., 1995, MNRAS, **273:** 277.

[58.] Bonnet, H., Mellier, Y., and Fort, B., 1994, ApJ, **427:** L83.

[59.] Sunyaev, R.A. & Zel'dovich, Y.B., 1972, Comm. Astrophys, Space Sci., **4:** 173.

[60.] Birkinshaw, M., Gull, S.F., Hardebeck, H.E., 1984, Nat, **309:** 34.

[61.] Birkinshaw, M., Hughes, J.P, & Arnaud, K.A., 1991, ApJ, **379:** 466.

[62.] Birkinshaw, M., and Hughes, 1994, ApJ, **420:** 33.

[63.] Jones, M., et al., 1993, Nat, **365:** 320.

[64.] Herbig, T., Lawrence, C.R., Readhead, A.C.S., and Gulkis, S., 1995, (CalTech preprint)

[65.] Lasenby, A.N., 1992, in *Clusters and Superclusters of Galaxies*, A.C. Fabian (ed.), Kluwer, p219.

[66.] Wilbanks, T.M., Ade, P.A.R., Fischer, M.L., Holzapfel, W.L., and Lange, A.E., 1994, ApJ, **427:** L75.

[67.] Andreani, P., Böhringer, H., Booth, R., Dall'Oglio, G., Nyman, L.-A., Pizzo, L., Shaver, P., and Whyborn, N., 1995, The Messenger, **78:** 41.

[68.] Press, W.H. and Schechter, P., 1974, ApJ**187:** 425.

[69.] Bardeen, J.M., Bond, J.R., Kaiser, N., and Szalay, A.S., 1986, ApJ, **304:** 15.

[70.] Kaiser, N., 1984, ApJ, **284:** L9.

[71.] Frenk, C.S., White, S.D.M., Efstathiou, G., and Davis, M., 1990, ApJ, **351:** 10.

[72.] Bahcall, N.A. and Cen, R., 1994, ApJ, **426:** L15

[73.] Bahcall, N.A., Cen, R., and Gramann, M., 1994, ApJ, **430:** L13

[74.] Bahcall, N.A., and Cen,R., 1992, ApJ, **407:** L49.

[75.] White, S.D.M., Efstathiou, G., and Frenk, C.S., 1993, MNRAS, **262:** 1023.

[76.] Edge, A.C., Stewart, G.C., Fabian, A.C., and Arnaud, K.A., 1990, MNRAS, **245:** 599.

[77.] Henry, J.P., and Arnaud, K.A., 1991, ApJ, **372:** 410.

[78.] Böhringer, H., 1995, in *Large Scale Structure of the Universe*, 10th Potsdam Cosmology Workshop, Mücket, J., et al. (eds.), World Scientific, Singapore, (in press).

[79.] Bahcall, N.A., 1988, ARA&A, **26:** 631.

[80.] Bahcall, N.A. and Soneira, R.M., 1983, ApJ, **270:** 20.

[81.] Postman, M., Geller, M.J., and Huchra, J.P., 1992, **384:** 404.

[82.] Nichol, R.C., Collins, C.A., Guzzo, L., and Lumsden, S.L., 1992, MNRAS, **255:** 21p.

[83.] Dalton, G.B., Efstathiou, G., Maddox, S.J., and Sutherland, W., 1992, ApJ, **390:** L1.

[84.] Bahcall, N.A. and West, M.J., 1992, ApJ, **392:** 419.

[85.] Sutherland, W.J., and Efstathiou, G., 1991, MNRAS, **248:** 159.

[86.] Romer, A.K., Collins, C.A., Böhringer, H., Cruddace, R.G., Ebeling, H., MacGillivray, H.T., Voges, W., 1994, Nat, **372:** 75.

[87.] Nichol, R.C., Briel, U.G., and Henry, J.P., 1994, MNRAS, **267**, 771.

[88.] Dalton, G.B., Croft, R.A.C., Efstathiou, G., Sutherland, W.J., Maddox, S.J., Davis, M., 1995, MNRAS, (in press).

[89.] Guzzo, L., et al., 1995, in *Wide-Field Spectroscopy and the Distant Universe*, Maddox, S.J. and Aragon-Salamanca, A. (eds.), World Scientific, Singapore, (in press).

High Redshift Quasars

P.A. Shaver

European Southern Observatory
Karl-Schwarzschild-Str. 2
85748 Garching bei München
Germany

1. INTRODUCTION

The discovery of quasars was the origin of the Texas Symposium. The invitation to the first Texas Symposium, held in December 1963, included the following text: *In February, Fred Hoyle and William Fowler suggested that the energies which lead to the formation of radio sources could be supplied through the gravitational collapse of a superstar. Such an object, with a mass between one hundred thousand and one hundred million solar masses, would be located in the center of the galaxy. The gravitational collapse of this supersun could supply the necessary energy if it were to shrink down close to the Schwarzschild radius. Last March, astronomers and radio-astronomers in Australia and California identified two extragalactic radio sources with galaxies of a type never seen before. The source 3C273B seems to be a superstar......The intriguing new discoveries and the theory put forward by Hoyle and Fowler open up the discussion of a wealth of exciting questions....*" (Robinson *et al.*, 1965).

Thus, the primary issue at first was the origin of the extraordinary luminosities, and over the years it has become widely accepted that supermassive black holes do indeed power quasars, as they allow maximum efficiency and compactness within the framework of conventional physics, and readily explain various phenomena such as the remarkable stability in the orientations of radio jets over very long timescales. The recent HST and VLBA observations demonstrating the likely presence of black holes in the nuclei of M87 and NGC 4258 respectively (Ford *et al.*, 1994; Harms *et al.*, 1994; Miyoshi *et al.*,

1995) have added further strength to that interpretation. However, as Rees (1993b) remarked, *"quasars themselves, despite strong circumstantial evidence that they involve supermassive black holes, have not yet offered any distinctive insights into strong-field gravity. Relativistic astrophysics has bloomed..... primarily through pulsars and X-ray sources."*. Quasar studies over the years have focussed more on their use as beacons and probes of the distant Universe.

Given their great luminosities, quasars are detectable out to very high redshifts, and for decades they provided the only clues to the high redshift Universe: (1) The epoch of galaxy formation, and the evolution of the space density of quasars. The very existence of quasars at high redshifts imposes constraints on the primordial power spectrum and theories of galaxy formation. (2) The evolution of large scale structure (clustering of and with quasars). (3) Probes of intervening matter using quasar absorption lines: the distribution and nature of absorbing matter, the evolution of chemical abundances, the evolution of structure, etc. (4) Probes of intervening matter using gravitational lensing, providing a measure of the masses of intervening galaxies and clusters, a probe of dark matter, and determination of the Hubble constant.

In this review I will briefly describe strategies and recent searches for high-redshift quasars, evidence for the decline in space densities beyond the "quasar epoch" at $z \sim 2$-3, properties of high-redshift quasars and implications for theories of galaxy formation, and the use of high-redshift quasars as beacons and probes. $H_\circ = 50$ km s^{-1} Mpc^{-1} and $q_\circ = 0.5$ are used throughout.

2. SEARCHES FOR HIGH-REDSHIFT QUASARS

Quasars are now seen out to $z = 4.9$. Fig. 1 shows the history of record redshifts for quasars since they were discovered in 1963. $z = 2$ was reached very quickly, and then progress was slower. As shown below, this was undoubtedly due to the fact that the actual space density of quasars increases rapidly up to $z \sim 2$-3, and then falls off at higher redshifts. It can also be seen in fig. 1 that all the record redshifts up to the mid-80s came from *radio-selected* quasars, and thereafter from optically-selected quasars. This can be readily understood, as follows. By the mid-60s, all known quasars were still radio-selected. In 1965 the much larger population of radio-quiet quasars was discovered (Sandage, 1965). These were found to be blue objects, and searches were made using an "ultraviolet excess" criterion. But this technique is only sensitive up to $z \sim 2$, and the record redshift from radio quasars was already beyond that redshift. It was only in the late-80s that new and very effective optical selection techniques based on *red* colours were introduced; these made it possible for the first time to find quasars at much higher redshifts. The fact that, in spite of determined large-scale searches using appropriate selection criteria, it has been so difficult to find quasars at redshifts $z > 5$, suggests that there may really be very few of them.

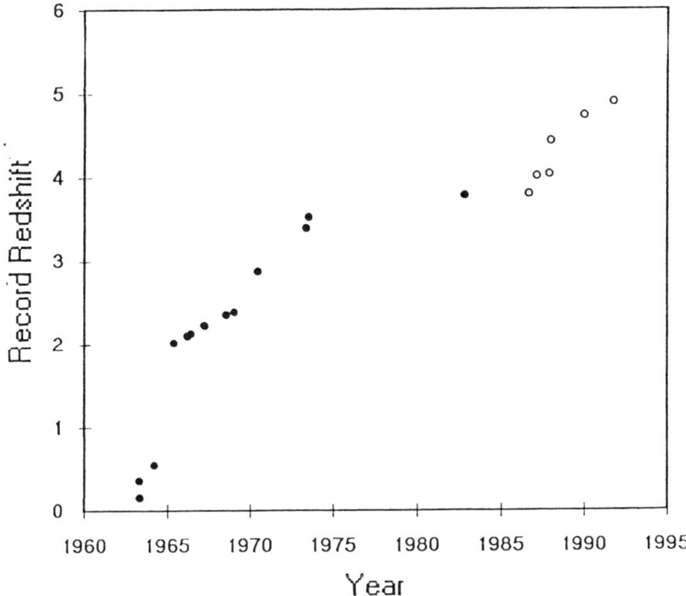

FIGURE 1. Record redshifts of quasars (radio (•) and optical (○)) as a function of time (refs. from Schneider *et al.*, 1992).

To what redshifts *should* quasars be observable? There are two obvious possible limitations. (1) Quasar luminosities: but quasars are so luminous that they should be detectable out to very high redshifts. A flat-spectrum radio-loud quasar with absolute magnitude $M = -27$ and radio power $P = 10^{27}$ W Hz^{-1} sr^{-1} would still have an apparent magnitude of ~ 21 (in the absence of intervening absorption) and radio flux density of $\sim$ 100 mJy at $z = 10$. (2) Opacity of the Universe: Thomson scattering and dust obscuration along the line of sight. Thomson scattering affects radiation at all frequencies $\ll mc^2/h$ equally (hence also radio emission), but even if all baryons in the Universe were in the form of an ionized intergalactic medium, and assuming a high baryonic density, $\Omega_B \sim 0.3$ (*cf.* Steigman & Felten (1995) for a discussion of uncertainties regarding the baryonic density), the opacity due to Thomson scattering would still only be ~ 0.2 at $z = 10$. Absorption by intervening dust and gas, discussed below, could obscure distant quasars at optical wavelengths, but not at infrared or radio wavelengths. Thus, quasars should in principle be detectable to redshifts well in excess of 5, if they are there.

2.1 Techniques

Several effective techniques have been developed over the years to search for quasars (*cf.* the reviews by Warren & Hewett (1991) and Hewett & Foltz (1994)). The original approach, as mentioned above, was based on radio emission. However, only ~5-10% of quasars are radio-loud, and at low flux densities confusion with radio galaxies is large, especially for steep-spectrum sources.

A number of optical selection techniques based on the characteristic properties of quasars, illustrated in fig. 2, have been developed over the years. The first of these properties to be exploited was the fact that quasar continua are blue, and it was found that quasars could be identified efficiently as ultraviolet excess (UVX) objects from their $U - B$ colours. This technique works up to $z \sim 2.2$, where the Lyα emission line moves into the B-band. Another approach is based on slitless spectroscopy, in which objective prism or grating/prism (grism) surveys are used to select objects with unusual spectra: emission lines, absorption troughs, blue continuum, Lyman decrement. These are usually biased towards strong-line objects. An advance over the single-colour UVX method is the multicolour method, in which several optical colours are used to segregate quasars from the stellar locus in a multidimensional colour space. This has been used especially in searches for higher-redshift quasars (*e.g.* Warren, Hewett, & Osmer, 1994). At the highest redshifts ($z > 4$) the Lyα forest and the Lyman continuum begin to severely depress the flux in the B-band, and a single colour (*e.g.* $B - R$) again suffices to identify quasars.

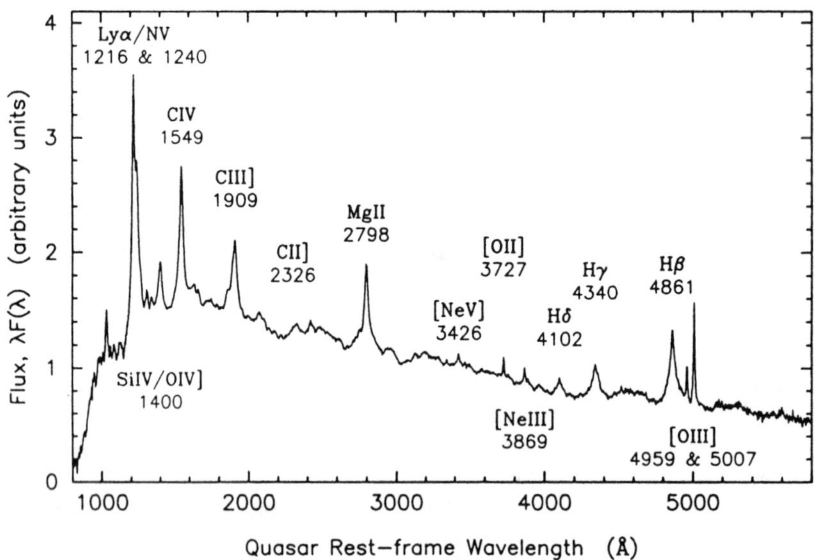

FIGURE 2. Composite quasar spectrum (Francis *et al.*, 1991).

Still other search techniques have been employed, but so far they have been used mostly to check the completeness of optical samples. Quasars are X-ray sources, and high-latitude X-ray surveys contain a high fraction of quasars; this approach will be increasingly productive as the positional accuracy and sensitivity of X-ray surveys increase. Most quasars ($\sim$ 80-90%) are variable, so this is also an efficient selection criterion (Usher *et al.*, 1983; Hawkins, 1986), although there may be a luminosity-dependent selection bias (Véron & Hawkins, 1995). Finally, proper motion studies can be used to separate extragalactic objects from galactic stars (Kron & Chiu, 1981); there is still considerable contamination by (slow) halo stars, but again this is a useful method to check the completeness of existing samples.

2.2 Recent High-Redshift Searches

Recent optical searches have been very successful in finding quasars at $4 < z < 5$, and in assembling samples sufficiently complete that space densities can be estimated. Warren, Hewett, & Osmer (1994) conducted a multicolour survey in 6 bands (u, b_j, v, or, r, i) to $m_{or} = 20$ over 43 deg^2. 86 quasars were found in the range $2.2 < z < 4.5$. About 50 would be expected in the range $3.5 < z < 4.5$ if there were no evolution at $z > 3.5$, but only 8 were found, indicating that, while the comoving space density continues to increase to $z = 3.3$, it then declines by a factor > 3 to $z = 4$. Schmidt, Schneider, & Gunn (1991, 1995) made a grism search using the Palomar 5m telescope to an equivalent width limit of $EW_{obs} = 50Å$ over 62 deg^2. 90 quasars were found at $2.75 < z < 4.75$, but only 9 of them were at $z > 4$. They conclude that at $z > 2.75$ the comoving space density decreases by a factor of 3 per unit redshift. Irwin, McMahon, & Hazard (1991, 1995) made a multicolour survey in B, R, and I (selected by $B - R$) to $R = 19$ over 1600 deg^2. This survey was highly successful in finding high-redshift quasars: 27 were found at $z > 4$. These authors concluded that the comoving space density of bright quasars is the same at $z = 2$ and $z = 4$, in apparent contradiction with the results of Warren *et al.* and Schmidt *et al.*, but Warren *et al.* pointed out that there is no inconsistency if the space density of bright quasars increases from $z = 2$ to $z = 3.3$ and then declines to higher redshifts.

Radio samples have also been used to estimate space densities of quasars at high redshifts. Dunlop & Peacock (1990) studied a sample of 171 flat-spectrum and 353 steep-spectrum radio sources, and although some of the redshifts were estimates and 15-20% of the sources were uncertain insofar as they could be at high redshifts, they were able to conclude that a flattening or turnover in comoving space density was likely. Hook (1994) conducted a search on the basis of identifications of large samples of flat-spectrum radio sources with accurate positions, incorporating optical colour information (POSS O-E) to select high-redshift objects, and found 26 new quasars at $z > 3$. It was again concluded that the space density of such objects falls off steeply above $z \sim 3$, although in this case there was some uncertainty because of 20% incompleteness due to the plate limit, below which the high-redshift quasars might be expected.

Thus, with appropriate selection techniques these new surveys have been quite successful, both in finding high-redshift quasars and in estimating space densities. This success is illustrated in Table 1, which lists all of the known quasars with $z > 4$; of these 50 objects, 33 were found using optical colour techniques, 10 using slitless spectroscopy, two were radio-loud quasars, two variable quasars, one X-ray quasar, and two were found serendipitously. All of the above studies, with the possible exception of that by Irwin *et al.*, indicate that the space density of quasars decreases above $z \sim 3$, in agreement with earlier suggestions (*e.g.* Sandage, 1972; Osmer, 1982).

However, it has been suggested that the turnover in space density may be merely illusory. Ostriker & Heisler (1984) proposed that dust in intervening galaxies could completely obscure distant quasars and explain the apparent turnover in space density (see also Heisler & Ostriker, 1988; Wright, 1990). It was suggested that the effect may be all or nothing, some quasars being completely obscured and others totally unaffected, so that there would be little or no evidence for reddening of *visible* quasars as a function of redshift, as observed (*e.g.*, Wright & Malkan, 1987; Schneider *et al.*, 1989a; Storrie-Lombardi, 1994). A related possibility is obscuration by dust in damped Lyα systems. Fall & Pei (1993) summarized evidence for the presence of dust in such systems: (1) the continuum spectra of quasars with damped systems are redder than those of quasars without damped systems (Pei *et al.*, 1991), and (2) chromium is more depleted than zinc in damped systems, as in our Galaxy, consistent with selective depletion onto dust grains (Pettini *et al.*, 1990). These results suggest a dust to gas ratio ~ 0.1 at $z \sim 2$. The fraction of missing quasars due to such systems is "modest" ($\sim 10 - 70\%$ at $z = 3$ and up to 90% at $z = 4$), but may be enough to flatten the quasar turnover at high redshifts and explain all of the high-redshift UV background by quasars alone. (Pei (1995a,b) has also considered the opposite effect, a magnification of quasars by gravitational lensing, and concludes that a true turnover in space density is likely). Finally, Rieke *et al.* (1979; 1982) and Webster *et al.* (1995) have found that some radio-selected quasars are so extremely red ($R - K \sim 5$) that they may have been missed in optical searches (although this effect appears to be independent of redshift, so it is most likely instrinsic and should not affect the turnover). The question therefore remains as to whether there is a real turnover in the space density of quasars at high redshifts.

TABLE 1. Quasars at $z > 4$

Name	z	m(R)	Survey Technique	Reference
PC 1247+34	4.90	20.4	Colour 5m+CCD	Schneider *et al.* (1991b)
PC 1158+46	4.73	20.2	Grism 5m+CCD	Schneider *et al.* (1989b)
APM 1202−07	4.70	18.7	Colour UKST+APM	Irwin *et al.* (1991)
APM 2237−06	4.55	18.3	Colour UKST+APM	Irwin *et al.* (1991)
APM 0019−15	4.52	19.0	Colour UKST+APM	Irwin *et al.* (1991)
APM 1033−03	4.50	18.5	Colour UKST+APM	Irwin *et al.* (1991)
APM 1114−08	4.50	19.4	Colour UKST+APM	Irwin *et al.* (1991)
PKS 1251−40	4.46	19.9I	Radio 3.6m+CCD	Shaver *et al.* (1995)
PC 0953+47	4.46	19.5	Grism 5m+CCD	Schneider *et al.* (1991a)
PC 1233+47	4.45	20.6	Grism 5m+CCD	Schneider *et al.* (1991a)
APM 1335−04	4.45	19.4	Colour UKST+APM	Irwin *et al.* (1991)
Q 0051−27	4.43	20.0	Colour UKST+APM	Warren *et al.* (1987b)
APM 0952−01	4.43	18.7	Colour UKST+APM	Irwin *et al.* (1991)
APM 0103+00	4.43	18.6	Colour UKST+APM	Irwin *et al.* (1991)
SGP 0051−27	4.40	20.0	Colour UKST+APM	Warren *et al.* (1987b)
Q 2203+29	4.40	20.8	Serendipity	McCarthy *et al.* (1988)
APM 1013+00	4.38	18.8	Colour UKST+APM	Irwin *et al.* (1991)
PC 0307+02	4.38	20.4	Grism 5m+CCD	Schneider *et al.* (1989a)
Q 2134−45	4.36	20.5	Variability UKST+COSMOS	Hawkins *et al.* (1995)
APM 0351−10	4.36	18.7	Colour UKST+APM	Irwin *et al.* (1991)
APM 0951−04	4.35	19.0	Colour UKST+APM	Irwin *et al.* (1991)
RXJ 1759+66	4.32	20.1	X-ray ROSAT	Henry *et al.* (1994)
APM 0111−28	4.30	18.7	Colour UKST+APM	Irwin *et al.* (1991)
GB 1508+57	4.30	19.0	Radio POSS+APM	Hook *et al.* (1995)
APM 1050−00	4.29	18.6	Colour UKST+APM	Irwin *et al.* (1991)
PC 0751+56	4.28	19.9	Grism 5m+CCD	Schneider *et al.* (1989a)
Q 2133−43	4.26	20.5	Variability UKST+COSMOS	Hawkins *et al.* (1995)
APM 2235−03	4.25	18.2	Colour UKST+APM	Irwin *et al.* (1991)
APM 0401−17	4.23	18.7	Colour UKST+APM	Irwin *et al.* (1991)
PC 0027+05	4.21	21.9	Serendipity	Schneider *et al.* (1994)
APM 0151−00	4.20	18.9	Colour UKST+APM	Irwin *et al.* (1991)
APM 1328−04	4.20	19.0	Colour UKST+APM	Irwin *et al.* (1991)
APM 0245−06	4.20	18.6	Colour UKST+APM	Irwin *et al.* (1991)
APM 0945−04	4.18	18.8	Colour UKST+APM	Irwin *et al.* (1991)
PC 0104+02	4.17	19.7	Grism 5m+CCD	Schneider *et al.* (1989a)
APM 2248−12	4.16	18.5	Colour UKST+APM	Irwin *et al.* (1991)
APM 1144−07	4.16	18.6	Colour UKST+APM	Irwin *et al.* (1991)
APM 0046−24	4.15	18.9	Colour UKST+APM	Irwin *et al.* (1991)
HM 0000−26	4.10	17.5	Objective Prism UKST	Hazard *et al.* (1991)
PC 2331+02	4.09	20.0	Grism 5m+CCD	Schneider *et al.* (1989a)
SGP 0101−30	4.07	19.3	Colour UKST+APM	Warren *et al.* (1987b)
Q 0111−28	4.04	19.8	Colour Schmidt+CCD	Smith *et al.* (1994)
APM 0241−01	4.04	18.2	Colour UKST+APM	Irwin *et al.* (1991)
APM 1302−14	4.04	18.6	Colour UKST+APM	Irwin *et al.* (1991)
PC 0910+56	4.04	20.8	Grism 5m+CCD	Schmidt *et al.* (1987)
SGP 0046−29	4.01	19.0	Colour UKST+APM	Warren *et al.* (1987a)
APM 1346−03	4.01	18.8	Colour UKST+APM	Irwin *et al.* (1991)
APM 2212−16	4.00	18.1	Colour UKST+APM	Irwin *et al.* (1991)
PC 1301+47	4.00	21.3	Grism 5m+CCD	Schneider *et al.* (1991a)
APM 1117−13	4.00	18.0	Colour UKST+APM	Irwin *et al.* (1991)

2.3 A Search for $z > 5$ Radio-Loud Quasars

In searching for high-redshift quasars, one would ideally like to be unrestricted by the selection technique and able to find objects at the highest possible redshifts. (a) Searches based on radio emission have no obvious redshift limit, and radio emission is unaffected by dust obscuration; only a small fraction of all quasars are radio-loud, but large homogeneous catalogues are available with accurate positions, so this is an attractive possibility. (b) Searches based on millimetre/submillimetre emission may be a strong possibility for the future. The prominent far-infrared bump would be redshifted into these bands at high redshifts, so that even radio-quiet quasars would become "radio-loud"; fig. 3 illustrates this feature, and shows that the ratio of 300 GHz to 5 GHz flux density for a steep-spectrum radio-loud quasar would be a strong indicator

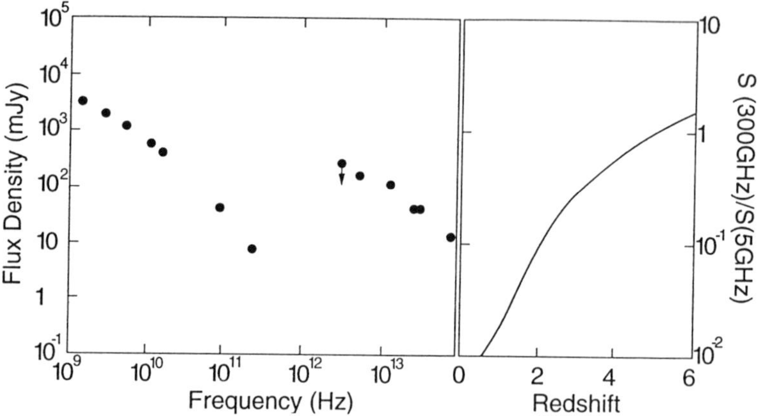

FIGURE 3. (*left*) Spectrum of a steep-spectrum radio-loud quasar (Q1704+608, from Chini *et al.*, 1989). (*right*) Ratio of flux densities at 300 and 5 GHz for such an object as it is moved to higher redshifts.

of redshift. Large-scale surveys with sensitive millimetre telescopes will be required to realize this possibility. (c) Near-infrared searches would be sensitive out to $z > 10$ in the K-band, but the present arrays are small and the sky background is high. (d) Optical searches in the far red are sensitive out to $z \sim 7$, but they rely on small CCD arrays, and again the sky background is high. (e) X-Ray searches would have no obvious redshift limit, but improved sensitivity and positional accuracy will be required before this becomes a very productive approach (the few $z \sim 3 - 4$ X-ray quasars from ROSAT all come from very deep exposures in small regions). At present, therefore, a radio-based strategy seems the best choice for an *unrestricted* search for quasars at $z > 5$.

Such a search has been made recently, based on a sample of flat-spectrum radio sources from the Parkes catalogue coupled with an optical criterion (absence from the B-band) to select $z > 5$ objects (Shaver, Wall, Kellermann, & Hawkins, 1995). There were two basic assumptions. (a) flat radio spectrum: a very high fraction of flat-spectrum sources are quasars (83% of the identified sources in the Parkes catalogue with spectral index between 2.7 and 5 GHz $\alpha > -0.4$ are quasars). But spectra steepen at high frequencies, so there is a possibility that high-redshift sources could be selectively omitted from the sample. Analysis of the quasi-simultaneous radio spectra of radio-loud quasars and BL Lacs from Gear *et al.* (1994) shows that this should not become important until $z > 10$ (fig. 4). (b) $z > 5$ objects should be totally obscured in the optical B band by Lyα forest and Lyman limit absorption. This assumption is supported by observed quasar spectra at $4 < z < 5$; simulations at higher

FIGURE 4. (*below, left*) Expected fraction of intrinsically flat-spectrum radio-loud quasars (solid histogram) and BL Lac objects (dashed histogram) for which the *observed* spectral index at 6 cm is flatter than -0.4, as a function of redshift, based on the quasi-simultaneous measurements of Gear *et al.* (1994).

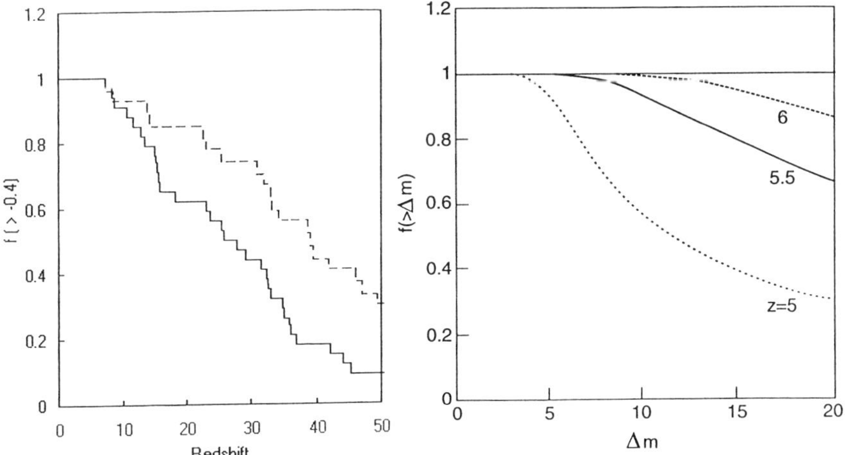

FIGURE 5. (*above, right*) Expected fraction of quasars at redshifts 5, 5.5, and 6 for which the absorption in the B band due to the Lyα forest and Lyman limit is greater than Δm, as a function of Δm. These extrapolations, by Møller (private communication; *cf.* Møller & Warren, 1991), assume $dn/dz \propto (1+z)^{2.75}$ and $f(N)dN \propto N^{-1.60}$.

redshifts based on (extrapolated) absorber properties indicate that there should be at least 5 and 10 magnitudes of absorption in the B band at $z = 5$ and 5.5 respectively (fig. 5).

The sample was comprised of 896 flat-spectrum sources ($\alpha > -0.4$ between 2.7 and 5 GHz) from the Parkes catalogue with $S_{2.7} \gtrsim 0.25$ Jy in the range $+2.5° > \delta > -80°$, of which 581 were already identified, 83% with quasars. VLA and Australia Telescope positions were obtained for the unidentified sources, accurate to < 1 arcsec, and UKST/COSMOS B_J identifications were made for 185 sources. CCD imaging observations with EFOSC on the ESO 3.6m telescope in the B, Gunn-i, and Gunn-z bands were made for the remaining 130 sources. Tentatively, it appears that all sources can be identified either with galaxies or with stellar objects which are present in the B-band (and therefore not at $z > 5$). Only one very red stellar identification has been found, which was marginally present in the B-band (hence at high redshift, but not at $z > 5$): PKS 1251-407 (fig. 6). Its redshift was found to be $z = 4.46$, making it the highest-redshift radio source presently known. Objects at higher redshifts could have been found with similar ease. If, as now seems likely, there are no unidentified sources left in the sample which could be associated with quasars at $z > 5$, this gives a firm upper limit on the space density at $z > 5$ *which is independent of any optical magnitude limit.*

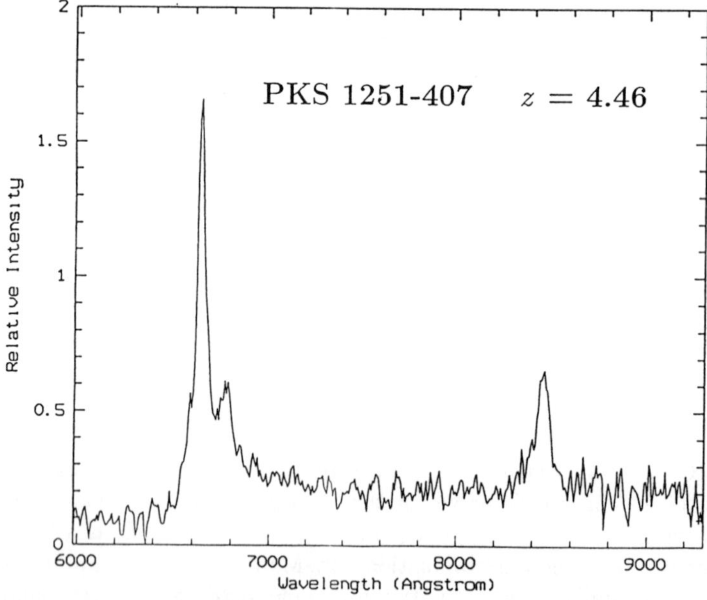

FIGURE 6. Spectrum of the highest-redshift known radio-loud quasar, PKS 1251-407 (Shaver, Wall, & Kellermann, 1995).

3. THE QUASAR EPOCH

Comparison of the upper limit on the space density of radio quasars at high redshift (considering just the range $5 < z < 7$) with the measured space density for similar objects at lower redshift gives a straightforward measure of the turnover, which is free of any possible complications due to optical obscuration. A flux density limit of 0.25 Jy was adopted, and a selection was made of similar objects ($\alpha > -0.4$, and radio power corresponding to 0.25 Jy at $z = 7$) at lower redshifts from the Parkes catalogue in the declination range $+10° > \delta > -45°$ where redshift information is 70% complete. This provided *lower* limits on the space density at $z < 5$, to be compared with the *upper* limit at $5 < z < 7$. A clear turnover is present, as shown in fig. 7.

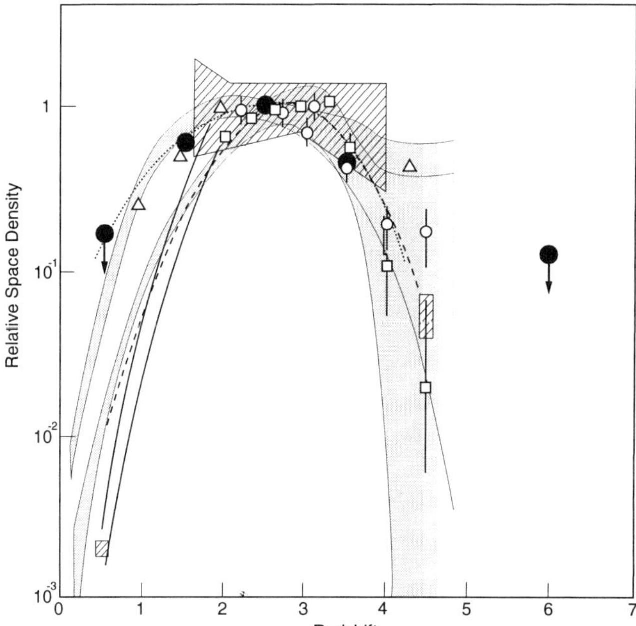

FIGURE 7. Space densities, normalized to $z \sim 2.5$ and plotted as a function of redshift, for the Parkes flat-spectrum radio-loud quasars (●), the optically-selected quasar samples of Warren *et al.*, 1994 (□), Schmidt *et al.*, 1991, 1995 (○), and Irwin *et al.*, 1991, 1995 (combined and normalized with Miller *et al.*, 1993) (△), the flat and steep spectrum sources studied by Dunlop & Peacock, 1990 (shaded areas), and the flat-spectrum radio-loud quasar sample of Hook, 1994 (dotted line). Also shown are normalized ultraviolet background radiation intensities (hatched areas) at $z \sim 0.5$ (Kulkarni & Fall, 1993), $z \sim 2-4$ (Bechtold, 1995), and $z \sim 4.5$ (Williger *et al.*, 1994; Storrie-Lombardi, 1994). The dashed line is a representative curve used in fig. 8.

This turnover cannot be due to obscuration. The radio emission is unaffected by dust, and if all sources are already accounted for, there are none left to be optically obscured quasars at $z > 5$. Could it be that some of the sources are actually at $z > 5$ but have been *misidentified* with the very galaxies that happen to lie along the line of sight and obscure them? This could not explain the turnover, as the positional agreement is 1 arcsec *r.m.s.* (the *same* for the identified galaxies and quasars). To remove the turnover would require several such coincidences; given the observed surface density of such galaxies the probability of *all* of the $z > 5$ quasars being obscured by these (visible) galaxies is very small. Could some of the sources be obscured by damped Lyα absorbers of exceptional column density (which are rich in dust, molecules, etc.)? Not if all sources are identified. The only possibility would be if sources are *completely* missed (also in the radio) due to both optical obscuration and free-free absorption in the radio. An optical depth of 0.1 at 2.7 GHz requires an emission measure of 10^6 pc cm^{-6} implying a very small size for typical excitation parameters, and a correspondingly small probability of lying along the line of sight. Large amounts of dust and ionized gas *intrinsic* to the quasar could conceivably do it, but this would presumably affect objects at all redshifts, so the turnover in space density remains.

Can it be argued now that this turnover applies to *all* quasars, and not just to the radio-loud quasars? The radio-loud fraction decreases up to $z \sim 1$ (La Franca et al., 1994; Schmidt et al., 1995b) (largely due to steep-spectrum quasars), and is then approximately constant (largely flat-spectrum quasars). There is no evidence that the radio-loud quasar population turns over first, and no theoretical reason to expect it. If the optical turnover were entirely due to obscuration, then at least some radio quasars should also be affected. If the apparent space density of optically-selected quasars were down by a factor of (say) ten at $z \sim 4.5$ due to obscuration, there would be only a 10% chance of PKS 1251-407 having been found rather than a blank field. Then, it seems unlikely to be just a *coincidence* that the radio and optical quasar populations appear to turn over together. Finally, and perhaps most significantly, the UV background also seems to exhibit a similar peak in redshift (fig. 7). It may be due to a large population of active galaxies, of which quasars are only a part (Madau, 1992; Loeb & Eisenstein, 1995), and the redshift distribution may reflect the epoch of galaxy formation and activity generally. It seems reasonable to conclude that the turnover *does* apply to all quasars, and is not strongly affected by dust obscuration. This "quasar epoch" is illustrated as a function of cosmic time in fig. 8. The narrowness of this peak in time implies an extraordinary synchronization of the evolution of AGN over the whole sky, which may be useful in constraining theories for the formation of structure (Turner, 1991a,b), perhaps eventually even cosmologies.

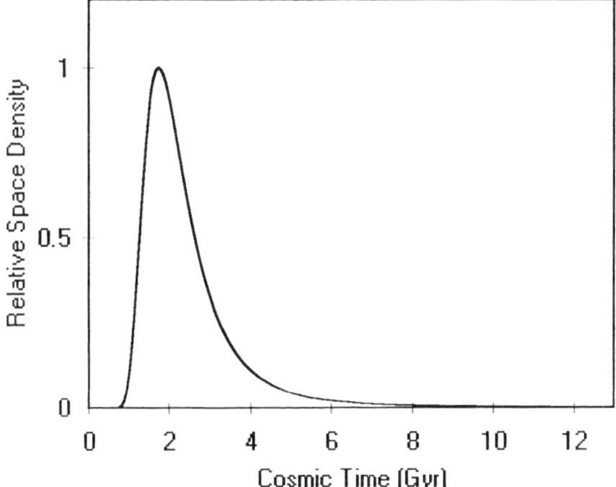

FIGURE 8. Relative space density as a function of cosmic time, computed from the dashed curve in fig. 7.

4. HIGH-REDSHIFT QUASARS AND COSMOLOGY

4.1 Properties of High-Redshift Quasars

It is significant that the properties of high-redshift quasars are not markedly different from those of quasars at lower redshifts. Apparently even in these exceptionally dense nuclear regions there is no huge enrichment over cosmological time scales. If anything, the N V/C IV ratios and the metallicities inferred from the broad emission lines seem to be *higher* at high redshifts than at low redshifts (Hamann & Ferland, 1992; Elston *et al.*, 1994), implying more rapid chemical evolution, more vigorous star formation, and deeper potential wells at the highest redshifts. Model calculations indicate that abundances in excess of solar can be reached in $\sim$ 1 Gyr, and are thereafter only weakly time-dependent (Matteucci & Padovani,1993) (see fig. 9). The fact that no quasars are found with obviously low abundances implies that considerable evolution to high metallicities occurs *before* the quasar becomes observable, even at high redshifts. One feature which is notable in the spectra of the highest-redshift radio-loud quasars is the narrowness of the emission lines. It is known that radio-loud quasars tend to have narrower emission lines than radio-quiet quasars (Baldwin *et al.*, 1988; Francis *et al.*, 1993); the $z > 4$ quasars are exceptional even for radio-loud quasars (see fig. 6). It has been argued (Francis *et al.*, 1993) that narrow-line quasars are unlikely to be "pole-on" quasars, as they have broad emission line components underlying the narrow cores which define the class. The reason for this difference thus remains unknown.

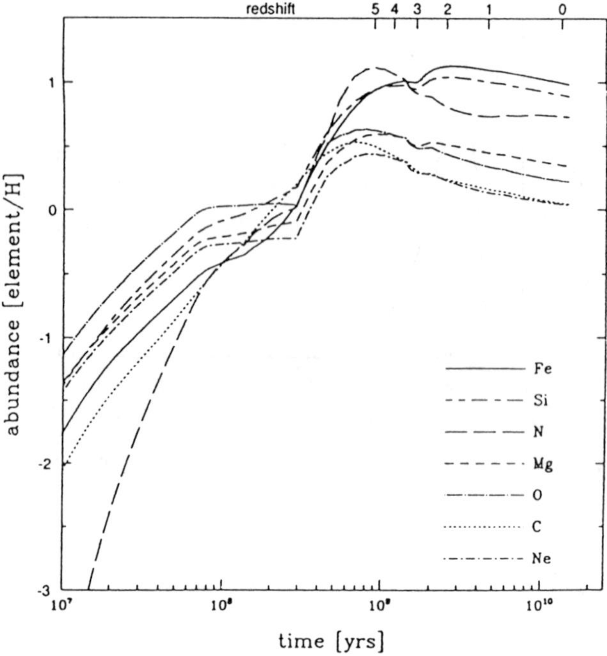

FIGURE 9. Time dependence of the abundances of various elements relative to solar for a galaxy with luminous mass 10^{12} $M_{\odot}$ and a Salpeter initial mass function (from Matteucci & Padovani, 1993). The steep increase at $t \sim 0.3-1$ Gyr is due to the maximum in the Type Ia supernova rate.

The X-ray properties of high-redshift quasars also do not appear to be exceptional. The $z = 4.3$ X-ray quasar found by Henry *et al.* (1994) has X-ray and optical properties comparable to those of 3C273 (2.7-10.6 keV X-ray luminosity of 2 10^{45} erg s^{-1}), and optical properties similar to those of other $z > 4$ quasars. The ratio of X-ray to optical luminosity does not appear to be correlated with redshift (Avni & Tananbaum, 1986; Pickering *et al.*, 1994).

If young quasars are the sites of active star formation, there may be copious amounts of gas and dust, as in starburst galaxies. Recent millimetre-wavelength detections of dust and CO in a lensed quasar at $z = 2.5$ (Barvainis *et al.*, 1992, 1994) and dust in quasars at redshifts up to 4.7 (Andreani *et al.*, 1993; McMahon *et al.*, 1994) support this interpretation and indicate dust and gas masses of 10^{8-9} and 10^{11} $M_{\odot}$ respectively, implying galactic scale total masses. Again, however, there is no indication that these properties are a strong function of redshift.

4.2 Quasar and Galaxy Formation

The very existence of quasars at $z \sim 5$ already provides important constraints on the epoch of galaxy formation and on the power spectrum of primordial fluctuations. Several arguments based on the existence and properties of quasars at high redshifts (their high luminosities, the extraordinary synchronized evolution implied by the narrowness of the "quasar epoch", evidence for large structures and peculiar velocities, heavy elements in high-redshift absorption systems) suggest that active galaxy formation must have taken place at very early epochs ($z > 5 - 10$) (*e.g.* Turner, 1991a,b). As mentioned above (*cf.* fig. 9), the metallicities inferred from quasar emission lines (especially Fe enrichment due to Type Ia supernovae) take ~ 1 Gyr to develop. At $z = 5$ the age of the Universe is just 0.9 Gyr (for $H_\circ = 50$ km s^{-1} Mpc^{-1} and $q_\circ = 0.5$), so the timescale is already tightly constrained. This problem would be eased with smaller values of $H_\circ$, $q_\circ$, or both, and indeed it has been proposed (Hamann & Ferland, 1993; Matteucci & Padovani, 1993; Elston *et al.*, 1994) that these abundances can be used as a "cosmic clock".

This timescale problem had been thought to present difficulties for hierarchical models such as CDM (Efstathiou & Rees, 1988), in which galaxy formation comes late, but recent simulations by Katz *et al.* (1994) have shown that concentrations suitable for quasars (virialized masses $> 8\ 10^{11} M_\odot$) could in fact have formed by $z \sim 8$. CDM models in which quasar lifetimes are assumed short ($\ll 10^9$ yr) can reproduce the "rise and fall" of the quasar population as a superposition of ~ 100 generations of short-lived quasars (Efstathiou & Rees, 1988; Haehnelt & Rees, 1993). Such models imply that the redshift-dependence of AGN is related to the formation of the galaxies themselves, and that most galaxies today host remnant black holes.

COBE data provide a normalization for the amplitude of the initial power spectrum on large scales, and the existence of collapsed objects at $z \sim 5$ gives a lower limit to $\delta\rho/\rho$ on small scales (Blanchard *et al.*, 1993; Haehnelt, 1993; Rees, 1993b). Thus, the quasar luminosity function at high redshift provides a strong constraint on the spectral index of the power spectrum of initial density fluctuations and/or hot dark matter fraction; the simple adiabatic (pancake) models dominated by neutrinos and some other HDM models have serious difficulties with this constraint (Blanchard *et al.*, 1993; Haehnelt, 1993). As the high-redshift quasars are thought to be associated with high peaks in the density field, they should be clustered *ab initio.* They might have affected the large-scale distribution of galaxies if they dominated the UV radiation field in a patchy way and so modulated the environment in which galaxies formed (Rees, 1993a; Miralda-Escudé & Rees, 1994).

4.3 High-Redshift Quasars as Beacons and Probes

The clustering of quasars and of galaxies around them provides a powerful handle on the evolution of large-scale structure. Quasars serve as beacons, marking the peaks in the large-scale density distribution, the deepest potential wells and the earliest-forming bound systems of galactic scale. At intermediate redshifts, it has been found that quasars are indeed located in regions of high galaxy density, radio-loud quasars more so than radio-quiet quasars (*e.g.* Yee & Green, 1987). At higher redshifts the only direct indications of clustering of luminous objects come from studies of quasar-quasar clustering (Boyle, 1990, and references therein). It is found that quasars cluster somewhat like galaxies today; new studies soon to be undertaken (at the AAT in the south, and with the Sloane survey in the north) based on very large samples should make it possible to determine the evolution of this clustering. Whether quasar clustering is really typical of galaxy clustering generally, and whether it is related in a simple way to the evolution of large-scale structure in the linear regime, remains to be determined (quasars are exotic and rare objects, presumably located preferentially in high density regions which are the first to go nonlinear). It is also important to understand whether the "quasar epoch" was the formation epoch of galaxies themselves, or whether it was caused by increasing interaction and merging of pre-existing galaxies followed by depletion of accessible fuel in the merged systems.

Distant quasars also serve as probes of the intervening medium. Absorption lines in their spectra convey a wealth of information about non-luminous matter distributed along the lines of sight. Correlations of absorbers with each other and with quasars give sensitive information about the nature of the absorbers and the evolution of large scale structure. Up to now, theories of large scale structure have only been constrained at $z \sim 1000$ (fluctuations in the the microwave background) and $z \sim 0$ (the distribution of galaxies today), but theories which satisfy these constraints can differ greatly from each other at intermediate redshifts (Cen *et al.*, 1993). For the first time, the Lyα forest properties are being used as powerful new constraints on such theories (Cen *et al.*, 1994; Petitjean *et al.*, 1995). Furthermore, the advent of 8-10m class telescopes now makes possible correlations in the absorption spectra of *groups* of quasars, so that three-dimensional (spatial) clustering can be investigated. Studies of absorption lines in pairs of quasars indicate very large sizes for the Lyα clouds (Bechtold *et al.*, 1994; Dinshaw *et al.*, 1994, 1995; Smette *et al.*, 1995), and possible evidence for large flattened structures, which are a generic feature of collapse in some models of structure formation. Studies of the properties and redshift dependence of damped Lyα systems may eventually provide a comprehensive view of galaxy formation and the history of the baryonic component of the Universe. Sensitive studies can also be made of the chemical composition of the absorbers, eventually revealing the history of heavy-element formation and enrichment. The properties of the diffuse intergalactic medium can be studied, including the Gunn-Peterson test (Giallongo *et al.*, 1994) and measures of the intergalactic UV radiation field as a function of redshift as mentioned above. It has been suggested that a cosmic rest frame may be provided by the high-redshift Lyα forest (Rauch, 1994). Finally, quasar absorption lines make possible certain critical cosmological tests, such as *in situ* measurements of the microwave background temperature at high redshifts, which should increase with redshift as $1+z$ (Songaila *et al.*, 1994b), and the primordial deuterium abundance (Carswell *et al.*, 1994; Songaila *et al.*, 1994a; Tytler, 1995).

High-redshift quasars serve as probes of intervening matter also through the effects of gravitational lensing: their images can be amplified, distorted and split by the gravitational fields of intervening massive objects (*cf.* Schneider, Ehlers, & Falco, 1992). These effects provide a means to determine the total masses of intervening galaxies and clusters, and so provide another handle on the fraction of mass which is dark. In principle it is also possible to determine the Hubble constant from the time delays between variations in the multiple images of a lensed quasar (*e.g.*, Roberts *et al.*, 1991), and perhaps even to constrain the value of q_o (Narayan, 1991).

A variety of other possible applications of high-redshift quasars to cosmology have been suggested over the years. The "Baldwin Effect" (a correlation of the C IV emission line equivalent width and continuum absolute luminosity) may provide one of the few possibilities of using quasars for "classical" cosmological tests; this phenomenon is observed in samples of radio-loud quasars (Baldwin *et al.*, 1989), but it is weak in optically-selected samples (Zitelli *et al.*, 1991). Hawkins (1993, 1995) has proposed that the variability of quasars may be used to study the possible presence of collapsed dark matter along the line of sight as a result of gravitational lensing effects, and suggests that the existence of a large population of Jupiter-mass lensing objects sufficient to make up the cosmological critical density is indicated. It has been found that the observed proper motions of superluminal radio sources are anticorrelated with redshift, as expected in a standard Friedmann cosmology (Cohen *et al.*, 1988) (fig. 10). It has been proposed that such observations can be used, in conjunction with models having a "standard velocity", to estimate H_o and q_o; while at the moment they are probably more useful in testing the models themselves, they may indeed eventually provide constraints on the cosmological parameters (Yahil, 1979; Cohen *et al.*, 1988; Rust *et al.*, 1989; Cohen & Vermeulen, 1992). One of the most promising recent possibilities is that the angular sizes of radio sources may be used to determine q_o. Hoyle made this suggestion many years ago, but attempts to employ the angular separation between the radio lobes for this purpose were inconclusive because it became apparent that any evolution in the density of the intergalactic medium could seriously affect the linear separation for a given luminosity. The compact radio cores in the nuclei of quasars, however, should be unaffected by the outer environment, and may come closer to serving as a "standard rod". Kellermann (1993) has made a preliminary study using this approach, and finds that the results are best matched by a $q_o = 0.5$ cosmology (fig. 11).

FIGURE 10. (*below, left*) Proper motion of VLBI components in jets of radio galaxies and quasars, plotted against redshift (Cohen *et al.*, 1988).

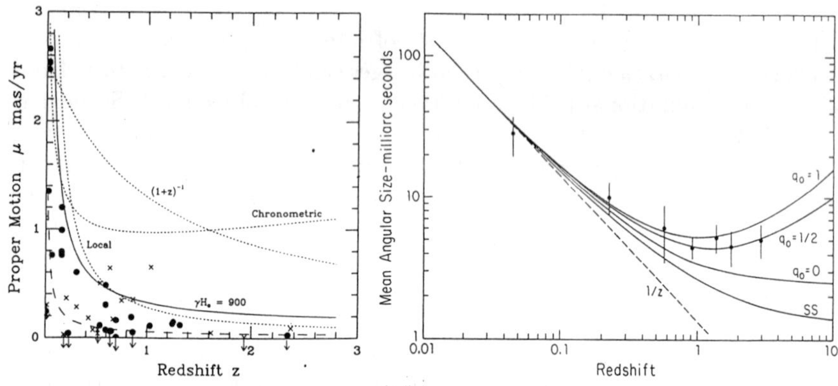

FIGURE 11. (*above, right*) Angular size of compact radio sources as a function of redshift (Kellermann, 1993).

5. CONCLUSIONS

Much of the excitement about quasars at the first Texas Symposium concerned their extraordinary energies and their possible relevance to strong-field gravity. While they may not have contributed as much in that area over the years as some other classes of objects, they have certainly led to impressive developments in our understanding of the distant Universe and cosmology by virtue of their use as beacons and probes. Searches for and studies of quasars up to $z = 5$ have provided important constraints on theories of structure and galaxy formation, and probes of the intervening medium which are revealing an increasingly rich wealth of detail on the nature and evolution of the Universe.

In principle we might expect to detect quasars out to redshifts of 10 or more. Their absence in open-ended radio searches indicates that the downturn in comoving space density is real, and not merely due to obscuration. $z > 5$ quasars must be rare, and we will need other tools to study the "dark ages" beyond. At $z \sim$ 4-5 the density of intergalactic neutral hydrogen is increasing rapidly with redshift and the ultraviolet background seems to be decreasing, suggesting that this was the epoch of rapid galaxy formation. Still, the existence of quasars at $z \sim 5$ with typical metallicities and the presence of heavy elements in absorption systems at similar redshifts indicates that considerable and distributed star formation must have taken place still earlier. Searches for thermal dust emission at millimetre wavelengths from highly redshifted young galaxies or HI line emission at metre wavelengths from neutral gas in protoclusters (Sunyaev & Zel'dovich, 1972; Scott & Rees, 1990) may open the way to studies of the earliest phases of structure formation in the Universe - the next frontier.

ACKNOWLEDGEMENTS

I am grateful to Palle Møller for running the lengthy simulations of Lyman absorption, to Richard McMahon for providing much of the information in Table 1, to several others for stimulating discussions and helpful information – Jill Bechtold, Bob Carswell, Mike Fall, Martin Haehnelt, Fabio La Franca, Martin Rees, Maarten Schmidt, Ed Turner, and Steve Warren – and especially my colleagues in the high-redshift quasar search, Jasper Wall, Ken Kellermann, and Mike Hawkins.

REFERENCES

Andreani, P., La Franca, F., Cristiani, S. 1993, *Mon. Not. R. astr. Soc* **261**, L35

Avni, Y., Tannenbaum, H. 1986, *Astrophys. J.* **305**, 83

Baldwin, J.A., McMahon, R., Hazard, C., Williams, R.E. 1988, *Astrophys. J.* **327**, 103

Baldwin, J.A., Wampler, E.J., Gaskell, C.M. 1989, *Astrophys. J.* **338**, 630

Barvainis, R., Antonucci, R., Coleman, P. 1992, *Astrophys. J.* **399**, L19

Barvainis, R., Tacconi, L., Antonucci, R., Alloin, D., Coleman, P. 1994, *Nature* **371**, 586

Bechtold, J. 1995, in *ESO Workshop on QSO Absorption Lines* (ed. G. Meylan), (in press)

Bechtold, J., Crotts, A.P.S., Duncan, R.C., Fang, Y. 1994, *Astrophys. J.* **437**,

L83

Blanchard, A., Buchert, T., Klaffl, R. 1993, *Astron. Astrophys.* **267**, 1

Boyle, B.J. 1990, *Aust. J. Phys.* **43**, 251

Carswell, R.F., Rauch, M., Weymann, R.J., Cooke, A.J. and Webb, J.K. 1994, *Mon. Not. R. astr. Soc.* **268**, L1

Cen, R., Ostriker, J.P., Peebles, P.J.E. 1993, *Astrophys. J.* **415**, 423

Cen, R., Miralda-Escudé, J., Ostriker, J.P., Rauch, M. 1994, *Astrophys. J.* **437**, L9

Chini, R., Biermann, P.L., Kreysa, E., Gemünd, H.-P. 1989, *Astron. Astrophys.* **221**, L3

Cohen, M.H., Barthel, P.D., Pearson, T.J., Zensus, J.A. 1988, *Astrophys. J.* **329**, 1

Cohen, M.H., Vermeulen, R.C. 1992, in *Extragalactic Radio Sources - From Beams to Jets* (Proc. 7th IAP Astrophys. Meeting), p. 98

Dinshaw, N., Impey, C.D., Foltz, C.B., Weymann, R.J., Chaffee, F.H. 1994, *Astrophys. J.* **437**, L87

Dinshaw, N., Foltz, C.B., Impey, C.D., Weymann, R.J., Morris, S.L. 1995, *Nature* **373**, 223

Dunlop, J.S., Peacock, J.A. 1990, *Mon. Not. R. astr. Soc.* **247**, 19

Efstathiou, G., Rees, M.J. 1988, *Mon. Not. R. astr. Soc.* **230**, 5p

Elston, R., Thompson, K.L., Hill, G.J. 1994, *Nature* **367**, 250

Fall, S.M., Pei, Y.C. 1993, *Astrophys. J.* **402**, 479

Ford, H.C., Harms, R.J., Tsvetanov, Z.I., Hartig, G.F., Dressel, L.L., Kriss, G.A., Bohlin, R.C., Davidsen, A.F., Margon, B., Kochhar, A.K. 1994, *Astrophys. J.* **435**, L27

Francis, P.J., Hewett, P.C., Foltz, C.B., Chaffee, F.H., Weymann, R.J., Morris, S.L. 1991, *Astrophys. J.* **373**, 465

Francis, P.J., Hooper, E.J., Impey, C.D. 1993, *Astron. J.* **106**, 417

Gear, W.K., Stevens, J.A., Hughes, D.H., Litchfield, S.J., Robson, E.I., Teräsranta, H., Valtaoja, E., Steppe, H., Aller, M.F., Aller, H.D. 1994, *Mon. Not. R. astr. Soc.* **267**, 167

Giallongo, E., D'Odorico, S., Fontana, A., McMahon, R.G., Savaglio, S., Cristiani, S., Molaro, P., Trevese, D. 1994, *Astrophys. J* **425**, L1

Haehnelt, M.G. 1993, *Mon. Not. R. astr. Soc.* **265**, 727

Haehnelt, M.G., Rees, M.J. 1993, *Mon. Not. R. astr. Soc.* **263**, 168

Hamann, F., Ferland, G. 1992, *Astrophys. J.* **391**, L53

Hamann, F., Ferland, G. 1993, *Astrophys. J.* **418**, 11

Harms, R.J., Ford, H.C., Tsvetanov, Z.I., Hartig, G.F., Dressel, L.L., Kriss, G.A., Bohlin, R., Davidsen, A.F., Margon, B., Kochhar, A.K. 1994, *Astrophys. J.* **435**, L35

Hawkins, M.R.S. 1986, *Mon. Not. R. astr. Soc.* **219**, 417

Hawkins, M.R.S. 1993, *Nature* **366**, 242

Hawkins, M.R.S. 1995, *Mon. Not. R. astr. Soc.* (in press)

Hawkins, M.R.S., Shaver, P.A., Clements, D. 1995, *Mon. Not. R. astr. Soc.* (in preparation)
Hazard, C. *et al.* 1991 (*cf.* Schneider, Schmidt, Gunn 1991a)
Heisler, J., Ostriker, J.P. 1988, *Astrophys. J.* **332**, 543
Henry, J.P., Gioia, I.M., Böhringer, H., Bower, R.G., Briel, U.G., Hasinger, G.H., Aragon-Salamanca, A., Castander, F.J., Ellis, R.S., Huchra, J.P., Burg, R., McLean, B. 1994, *Astron. J.* **107**, 1270
Hewett, P.C., Foltz, C.B. 1994, *Publ. Astr. Soc. Pacific* **108**, 113
Hook, I.M. 1994, *Quasars at High Redshift* (Ph.D. Thesis, Cambridge Univ.)
Hook, I.M., McMahon, R.G., Patnaik, A.R. Browne, I.W.A., Irwin, M.J. 1995, *Mon. Not. R. astr. Soc.* (in press)
Irwin, M.J., McMahon, R.G., Hazard, C. 1991, in *The Space Distribution of Quasars* (ASP Conference Series 21, ed. D. Crampton),p. 117
Irwin, M.J., McMahon, R.G., Hazard, C. 1995 (in preparation)
Katz, N., Quinn, T., Bertschinger, E., Gelb, J.M. 1994, *Mon. Not. R. astr. Soc.* **270**, L71
Kellermann, K.I. 1993, *Nature* **361**, 134
Kron, R.G., Chiu, L.-T.G. 1981, *Publ. Astron. Soc. Pacific* **93**, 397
Kulkarni, V.P., Fall, M. 1993, *Astrophys. J.* **413**, L63
La Franca, F., Gregorini, L., Cristiani, S., de Ruiter, H., Owen, F. 1994, *Astron. J.* **108**, 1548
Loeb, A., Eisenstein, D.J. 1995, *Astrophys. J.* (in press)
Madau, P. 1992, *Astrophys. J.* **389**, L1
Matteucci, F., Padovani, P. 1993, *Astrophys. J.* **419**, 485
McCarthy, P.J., Dickinson, M., Filippenko, A.V., Spinrad, H., van Breugel, W.J.M. 1988, *Astrophys. J.* **328**, L29
McMahon, R.G., Omont, A., Bergeron, J., Kreysa, E., Haslam, C.G.T. 1994, *Mon. Not. R. astr. Soc.* **267**, L9
Miller, L., Goldschmidt, P., La Franca, F., Cristiani, S. 1993, in *Observational Cosmology* (ASP Conf. Ser. **21**, eds. G. Chincarini, A. Iovino, T. Maccacaro, D. Maccagni), p. 614
Miralda-Escudé, J., Rees, M.J. 1994, *Mon. Not. R. astr. Soc.* **266**, 343
Miyoshi, M., Moran, J., Herrnstein, J., Greenhill, L., Nakal, N., Diamond, P., Inoue, M. 1995, *Nature* **373**, 127
Møller, P., Warren, S. 1991, in *The Space Distribution of Quasars* (ASP Conf. Ser. 21, ed. D. Crampton), p. 96
Narayan, R. 1991, *Astrophys. J.* **378**, L5
Osmer, P.S. 1982, *Astrophys. J.* **253**, 28
Ostriker, J.P., Heisler, J. 1984, *Astrophys. J.* **278**, 1
Pei, Y.C. 1995a, *Astrophys. J.* **438**, 623
Pei, Y.C. 1995b, *Astrophys. J.* **440**, 485
Pei, Y.C., Fall, S.M., Bechtold, J. 1991, *Astrophys. J.* **378**, 6
Petitjean, P., Mücket, J.P., Kates, R.E. 1995, *Astron. Astrophys.* **295**. L9
Pettini, M., Boksenberg, A., Hunstead, R.W. 1990, *Astrophys. J.* **348**, 48

Pickering, T.E., Impey, C.D., Foltz, C.B. 1994, *Astron. J.* **108**, 1542

Rauch, M. 1994, *Mon. Not. R. astr. Soc.* **271**, 13

Rees, M.J. 1993a, *Proc. Natl. Acad. Sci. USA* **90**, 4840

Rees, M.J. 1993b, *Quart. J. Roy. astr. Soc.* **34**, 279

Rieke, G.H., Lebofsky, M.J., Kinman, T.D. 1979, *Astrophys. J.* **232**, L151

Rieke, G.H., Lebofsky, M.J., Wisniewski, W.Z. 1982, *Astrophys. J.* **263**, 73

Roberts, D.H., Lehar, J., Hewitt, J.N., Burke, B.F. 1991, *Nature* **352**, 43

Robinson, I., Schild, A., Schucking, E.L. (eds.) 1965, *Quasi-Stellar Sources and Gravitational Collapse* (Univ. of Chicago Press)

Rust, B.W., Nash, S.G., Geldzahler, B.J. 1989, *Astrophys. Space Sc.* **152**, 141

Sandage, A. 1965, *Astrophys. J.* **141**, 1560

Sandage, A. 1972, *Astrophys. J.* **178**, 25

Schmidt, M., Schneider, D.P., Gunn, J.E. 1987, *Astrophys. J.* **321**, L7

Schmidt, M., Schneider, D.P., Gunn, J.E. 1991, in *The Space Distribution of Quasars* (ASP Conference Series 21, ed. D. Crampton),p. 109

Schmidt, M., Schneider, D.P., Gunn, J.E. 1995 (in preparation)

Schmidt, M., van Gorkom, J.H., Schneider, D.P., Gunn, J.E. 1995b, *Astron. J.* **109**, 473

Schneider, D.P., Schmidt, M., Gunn, J.E. 1989a, *Astron. J.* **98**, 1507

Schneider, D.P., Schmidt, M., Gunn, J.E. 1989b, *Astron. J.* **98**, 1951

Schneider, D.P., Schmidt, M., Gunn, J.E. 1991a, *Astron. J.* **101**, 2004

Schneider, D.P., Schmidt, M., Gunn, J.E. 1991b, *Astron. J.* **102**, 837

Schneider, D.P., Schmidt, M., Gunn, J.E. 1994, *Astron. J.* **107**, 880

Schneider, D.P., van Gorkom, J.H., Schmidt, M., Gunn, J.E. 1992, *Astron. J.* **103**, 1451

Schneider, P., Ehlers, J., Falco, E.E. 1992, *Gravitational Lenses* (Springer-Verlag, Berlin)

Scott, D., Rees, M.J. 1990, *Mon. Not. R. astr. Soc.* **247**, 510

Shaver, P.A., Wall, J.V., Kellermann, K.I. 1995, *Mon. Not. R. astr. Soc.* (in preparation)

Shaver, P.A., Wall, J.V., Kellermann, K.I., Hawkins, M.R.S. 1995 (in preparation)

Smette, A., Robertson, J.G., Shaver, P.A., Reimers, D., Wisotzki, L., Köhler, Th. 1995, *Astron. Astrophys.* (in press)

Smith, J.D., Djorgovski, S., Thompson, D., Brisken, W.F., Neugebauer, G., Matthews, K., Meylan, G., Piotto, G., Suntzeff, N.B. 1994, *Astron. J.* **108**, 1147

Songaila, A., Cowie, L.L., Hogan, C.J., Rugers M. 1994a *Nature* **368**, 599.

Songaila, A., Cowie, L.L., Vogt, S., Keane, M., Wolfe, A.M., Hu, E.M., Oren, A.L., Tytler, D.R., Lanzetta, K.M. 1994b, *Nature* **371**, 43

Steigman, G., Felten, J.E. 1995, in *Proc. of the St. Petersburg Gamow Seminar* (eds. A.M. Bykov & R.A. Chevalier); *Space Science Rev.* (Kluwer, Dordrecht), in press

Storrie-Lombardi, L.J. 1994, *Absorption in the Highest Redshift Quasars* (Ph.D Thesis, Cambridge Univ.)

Sunyaev, R.A., Zel'dovich, Ya.B. 1972, *Astron. Astrophys.* **20**, 189

Turner, E.L. 1991a, *Astron. J.* **101**, 5

Turner, E.L. 1991b, in *The Space Distribution of Quasars* (ASP Conf. Ser. Vol. 21, ed. D. Crampton), p. 361

Tytler, D. 1995, in *ESO Workshop on QSO Absorption Lines* (ed. G. Meylan), (in press)

Usher, P.D., Warnock, A., Green, R.F. 1983, *Astrophys. J.* **269**, 73

Véron, P., Hawkins, M.R.S. 1995, *Astron. Astrophys.*, in press

Warren, S.J., Hewett, P.C. 1991, *Rep. Prog. Phys.* **54**, 243

Warren, S.J., Hewett, P.C., Irwin, M.J., McMahon, R.G., Bridgeland, M.T., Bunclark, P.S., Kibblewhite, E.J. 1987a, *Nature* **325**, 131

Warren, S.J., Hewett, P.C., Osmer, P.S., Irwin, M.J. 1987b, *Nature* **330**, 453

Warren, S.J., Hewett, P.C., Osmer, P.S. 1994, *Astrophys. J.* **421**, 412

Webster, R., *et al.* 1995, in preparation (*cf.* Nov. 1994 *Sky & Telescope* p. 10)

Williger, G.M., Baldwin, J.A., Carswell, R.F., Cooke, A.J., Hazard, C., Irwin, M.J., McMahon, R.G., Storrie-Lombardi, L.J. 1994, *Astrophys. J.* **428**, 574

Wright, E.L. 1990, *Astrophys. J.* **353**, 411

Wright, E.L., Malkan, M.A. 1987, *Bull. Amer. Astron. Soc.* **19**, 699

Yahil, A. 1979, *Astrophys. J.* **233**, 775

Yee, H.K.C., Green, R.F. 1987, *Astrophys. J.* **319**, 28

Zitelli, V., Zamorani, G., Marano, B., Mignoli, M., Boyle, B. 1991, in *The Space Distribution of Quasars* (ed. D. Crampton), ASP Conf. Ser. **21**, 105

THE EXTRAGALACTIC X– AND γ–RAY BACKGROUNDS

GIANCARLO SETTI

Dipartimento di Astronomia, Universita' di Bologna
Istituto di Radioastronomia CNR, via Gobetti 101, I-40129 Bologna, Italy

INTRODUCTION

The observations of the X– and γ–ray backgrounds cover an extremely wide energy range from 0.1 keV to 30 GeV. The extragalactic components are not identified with the same degree of confidence across the energy spectrum. Only in the hard X–ray regime above $\sim$ 3 keV, at which photon energies the entire Galaxy becomes fully transparent, early measurements made with UHURU, Ariel V and HEAO–1 A2 have shown that the large scale intensity distribution is highly isotropic (to within 1%, once a small galactic contribution is subtracted), thus providing very strong observational support in favour of its extragalactic origin[1,2]. Below 3 keV galactic absorption and emission components become progressively more important, while in the γ–ray domain the Galaxy contributes a significant fraction of the total background.

Following its discovery more than three decades ago[3], it was soon realized that the relatively large energy flux carried by the X–ray background cannot be accounted for in terms of sources and processes confined to the present epoch, but must involve the X–ray properties of the Universe at earlier times, say at redshifts $z > 1$. Countless papers and review articles have been written on the origin of the extragalactic X–ray background (XRB). But even now, after many years of a large research effort, most authors refer to the origin of the XRB as one of the main unsolved problems. Nevertheless, I believe that the results obtained with the ROSAT and GINGA missions, combined with other achievements in the field of extragalactic astronomy, support the earlier hypothesis[4] that the bulk of the XRB is due to the summed contributions of known classes of active galactic nuclei (AGN). Similarly, the exciting findings which are being obtained with the Compton Gamma Ray Observatory (CGRO) indicate that also the extragalactic γ–ray background (GRB) is due to a sub–class of AGN which emit a large fraction of their energy in the γ–ray domain.

SPECTRAL CHARACTERISTICS

The spectral energy distribution over the whole energy range is shown in Fig. 1. The associated energy density is $w_{X,\gamma} \simeq 10^{-4}$ eV cm^{-3}. As one can see the distribution is double-peaked, characterizing two spectral regions

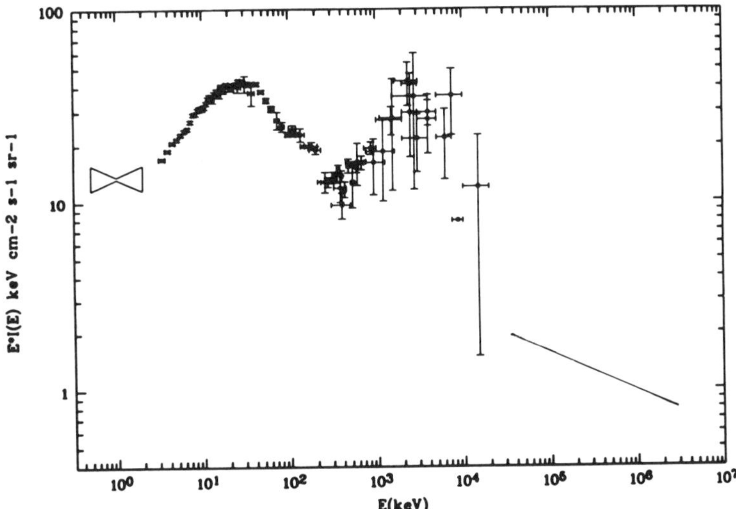

FIGURE 1. A representation of the spectrum of the X– and γ–ray extragalactic backgrounds. The bow–tie shape represents the 0.5 – 2 keV measurements from ROSAT[6]. In the interval 3 keV – 10 MeV one has plotted a selection of data from various experiments[14]. The solid line above 30 MeV has been plotted with the slope derived from the EGRET measurements[31] and the normalization from SAS–2[30].

which carry a large fraction of the total energy flux. For the purpose of the present discussion it is then convenient to take 400 keV as the dividing line between the X– and γ–ray domains. With the fits to the data described in the following sub–sctions one finds $\sim$ 60% of the total intensity in the XRB ($\sim$ 13% in the soft X–ray domain) and $\sim$ 40% in the GRB (4% in high energy γ–rays, > 35 MeV).

The soft XRB

The spectral properties in the soft X–ray domain (E < 3 keV) are still somewhat uncertain. At the very soft end, where the shadowing effects due to the interstellar absorption are extremely severe, the measured background is dominated by the thermal emission due to a local hot (10^6 K) gas. Below $\sim$ 1 keV one has isolated a second thermal component due to the emission of a somewhat hotter (a few 10^6 K) gas whose nature (galactic or intergalactic associated with the Local Group) is still unclear[5]. The results obtained with the ROSAT–PSPC[6] indicate that the XRB spectrum in the 0.5 – 2.0 keV interval can be represented by a power law of energy spectral index

$\alpha = 1.12 \pm 0.12$ (flux $\propto E^{-\alpha}$) and a normalization 13.0 ± 0.2 keV cm^{-2} s^{-1} sr^{-1} keV^{-1} at 1 keV. This result is reproduced in Fig. 1 and, as one can see, it implies that the XRB spectrum steepens up between 2 and 3 keV with respect to a downward extrapolation from higher energies. On the other hand it has been recently reported[7] that preliminary results from ASCA–SIS, which is sensitive in the 0.4–12 keV band, are described by a single power law with energy index $\alpha = 0.4$ in the 1 to 10 keV energy interval and a normalized intensity of 9.6 keV cm^{-2} s^{-1} sr^{-1} keV^{-1}, in complete agreement with previous measurements above 3 keV, while the spectrum steepens below 1 keV. Taken at their face value these findings appear to disagree with those from ROSAT. The rather complex situation has been discussed in great detail by Hasinger[8] who concludes that agreement among the various observational results can be obtained by assuming a normalized intensity of 11 keV cm^{-2} s^{-1} sr^{-1} keV^{-1} and a turn up of the spectrum only below 1 keV.

Most importantly, the ROSAT–PSPC pointed observations of the Lockman Hole have enabled an extension of the source counts down to 2×10^{-15} erg cm^{-2} s^{-1} in the 0.5 – 2 keV band[6], a flux more than one order of magnitude fainter than the Einstein Observatory Deep Survey (EDS) limit. The resulting Log N – Log S, obtained by adding the data from other shallower survey fields, is fully consistent with the EDS [9] and with the Einstein Observatory EMSS total counts in the region of overlap. It is found that about 60% of the ROSAT X–ray background is already resolved into discrete sources, whilst this fraction increases to at least 75% on the basis of an extension of the Log N – Log S relationship down to a flux $\simeq 10^{-16}$ erg cm^{-2} s^{-1} from a fluctuation analysis performed in the same Lockman field[6].

What's the nature of the sources? A large fraction of those detected in five deep ROSAT fields with fluxes $> 10^{-14}$ erg cm^{-2} s^{-1} have been spectroscopically identified. The results indicate that type 1 AGNs, mostly quasars, are by far the dominant population ($\sim$ 60%), followed by galaxies and galaxy clusters; foreground stars account for about 15% of the sources and their fraction should decrease at fainter X–ray fluxes[8,10,11].

The hard XRB (3 – 400 keV)

The shape of the XRB spectrum in the 3 – 100 keV energy interval closely follows that of an optically thin thermal *bremsstrahlung* at a temperature kT $\simeq$ 40 keV [12,13], while above $\sim$ 100 keV the spectral data exceed the thermal fit. An analysis of the combined observed spectra has been made by Gruber[14] who concludes that, while a thermal *bremsstrahlung* spectrum with an e–folding energy = 41.13 keV accurately fits the data up to 60 keV, above this energy the sum of two power laws is required, one with an exponent – 2.0 (an extension of the *bremsstrahlung* form) and the other with exponent – 0.7, and with normalizations such that at 60 keV the differential energy spectral slope is about – 1.6, gradually flattening to about – 0.7 at

MeV energies. It should also be noted that below 10 keV the XRB energy spectrum is well represented by a power law of index $\alpha = 0.4$.

The fact that the XRB spectrum is so well reproduced by a 40 keV thermal *bremsstrahlung* was considered by some to support an earlier proposal[15] that the bulk of the XRB is produced by a hot ($> 10^8$ K) diffuse intergalactic gas (IGG). Subsequent investigations [16–18] have shown that a viable model, in the framework of the big–bang cosmology, can only be obtained by assuming that the IGG has been suddenly reheated at a redshift $z_r = 3 - 5$ to a temperature T $= 40 \times (1 + z_r)$ keV and that the corresponding density parameter is $\Omega > 0.2$, in excess of the baryonic matter density allowed by the primordial nucleosynthesis argument. These parameters would cause a large Compton distortion of the CMB: the very precise black-body shape of the CMB, as measured by the FIRAS instrument on board COBE, imposes that the contribution of a hot diffuse IGG to the XRB cannot be more [19,20] than 10^{-5}.

In principle, strong clumping can help in circumventing these difficulties. Nevertheless, to avoid violation of the upper limits on the temperature fluctuations of the CMB, one would have to hypothesize the existence of a large number of unrealistically small size clumps (< few tens kpc)[21]. Likewise, the interesting idea[22] of linking the origin of the XRB to the formation of the large scale structure of the Universe, via the thermal *bremsstrahlung* emission from the gas heated to the required temperature by infall into deep potential wells of massive (10^{15-16} $M_\odot$) condensates at large redshifts, is severely constrained by the smoothness of the XRB and of the CMB. Indeed, several authors[23] have pointed out that the close resemblance of the XRB spectrum to that of a 40 keV thermal *bremsstrahlung* provides in itself a strong argument against any interpretation involving a hot IGG as the main contributor to the XRB intensity. This is because any reasonable subtraction of the integrated contributions from known classes of extragalactic X–ray sources would destroy the almost perfect thermal shape which, therefore, must be considered as accidental.

Presently, there are no other models which may satisfactorely explain the hard XRB in terms of diffuse processes taking place in the intergalactic medium. As a consequence, the only alternative possibility is that of the summed contribution from extragalactic sources. Among these the AGNs, in particular Seyfert galaxies and quasars, are known for some time[4] to be the most likely candidates. However, unlike the situation registered in the soft X–rays, only a small fraction of the hard XRB has been actually resolved into sources: the HEAO–1 A2 all–sky survey has resolved $\sim 1\%$ of the 2 – 10 keV XRB down to a flux limit $\simeq 3 \times 10^{-11}$ erg cm^{-2} s^{-1} and determined a Euclidean Log N – Log S [24]. At these bright fluxes AGNs and clusters of galaxies are the two dominant classes of sources . The counts to fainter fluxes in the 2 – 10 keV band have been constrained from the analyses of the fluctuations in the XRB: the results of the P(D) fitting techniques applied to the GINGA measurements[25,26] and to those of the HEAO–1 A2 experiment[27] are consistent with a Euclidean slope of the source number counts extrapolated from the HEAO–1 A2 Log N – Log S down to a flux $\sim 10^{-13}$ erg cm^{-2} s^{-1}, at which flux about 15% of the 2 – 10 keV XRB is resolved.

The main difficulty with the idea that the AGNs could supply the bulk of the XRB has been that their hard X–ray spectra (2 – 20 keV) are well represented by a power law of mean energy spectral index $< \alpha > = 0.7$, much steeper than that of the XRB[28]. Recent developments have permitted to overcome this obstacle (see the next Section).

The GRB (E > 400 keV)

The results of a number of observations made with various instruments from high–altitude balloons and space platforms in the 0.4 – 10 MeV energy range are displayed in Fig. 1. The large dispersion in the data and the large error bars indicate the intrinsic difficulty of the measurments at these photon energies, mainly due to the instrumental background induced by ambient variable sources. This has lead some to even question the reality of the "MeV bump". We assume the Gruber's fit, already discussed in the preceding sub–section, as the most reliable approximation to the present observational knowledge. At 6 MeV the Gruber's fit gives an intensity (νI_ν) of 25 keV cm^{-2} s^{-1} sr^{-1}, and then the spectrum must slope down (very sharply, indeed!) to meet the derived intensity of the GRB at 30 MeV.

Observations made with the SAS–2 satellite in the 35 – 200 MeV band have shown a signal strongly correlated with the intertellar gas column densities, but persistent even at high galactic latitudes. This has been interpreted as an isotropic component probably extragalactic[29]. Therefore, the studies of the GRB above several tens MeV are difficult not only because of the instrumental background produced by local cosmic rays, but also because of the need to subtract the diffuse gamma radiation from the Galaxy, mainly due to the interaction of cosmic rays with the interstellar medium. From the analysis of the SAS–2 data, together with an improved estimate of the galactic contribution based on the interstellar absorption from Galaxy counts, one has obtained[30] a GRB intensity above 100 MeV of $(1.3 \pm 0.5) \times 10^{-5}$ ph cm^{-2} s^{-1} sr^{-1}, and $(5.5 \pm 1.3) \times 10^{-5}$ ph cm^{-2} s^{-1} sr^{-1} above 35 MeV, and a power law fit to the spectrum with energy spectral index $\alpha = 1.3 \pm 0.4$ in the 30 – 150 MeV energy interval. At the poles the Galaxy contributes a photon flux about 50% that of the extragalactic component.

Preliminary results obtained with the CGRO–EGRET experiment in the 30 MeV – 30 GeV band, which have been analysed taking into account very detailed models of the galactic emission, confirm the power law spectrum extended over a much wider energy interval with an energy index $\alpha = 1.2 \pm 0.2$ and an integrated intensity above 100 MeV in agreement within the errors with that found from SAS–2 [31,32]. Most importantly, EGRET has revealed a large number of powerful extragalactic sources all identified with compact flat spectrum radio quasars and BL Lac objects. Presently, 33 such sources have been detected at $> 5\sigma$ and another 11 at $4 - 5\sigma$[33]. By considering only the sources detected at $> 5\sigma$, and the maximum observed flux from each, one finds that they contribute a total flux of 1.93×10^{-5} ph cm^{-2} s^{-1}, or $\sim$ 15% of the GRB in 10 sr outside the Galactic plane. Since many sources show a high degree of variability on timescales from days to months or longer [33], it can be safely estimated that at least 10% of the high energy GRB has been resolved into descrete sources pertaining to a particular class of AGNs.

AGNs AND THE HARD XRB SPECTRAL PROBLEM

One of the achievements of the GINGA mission has been the discovery that the X–ray spectra of a large fraction of a sample of Seyfert 1 galaxies flatten beyond $\sim$ 10 keV, the mean spectral index in the $10 - 18$ keV interval being $< \alpha > = 0.28$ with $\sigma = 0.15$ [34]. This fact has been interpreted either as photoelectric absorption by thick cold matter partial coverage of an underlying power law continuum source or as a hump produced by reprocessed X–rays (reflection) from thick cold ($< 10^6$ K) matter surrounding the central source, possibly in an accretion disk around a massive black hole[35–37]. The Compton reflection model is generally thought to be more representative of the physical conditions prevailing in the AGNs.

It has been conjectured[38] that a flattening of the average AGN spectrum above about 10 keV might provide a solution to the AGN vs. XRB spectral problem, simply because the redshifted contributions from the flat portion of the source spectra would dominate the $3 - 30$ keV XRB spectral shape, if the AGN volume emissivity adequately increases as a function of look-back time out to a redshift cut-off $z = 3$. The rather sharp decline of the XRB spectrum above 30 keV could be modeled by introducing a high energy cut-off at $\sim$ 100 keV. Following this recipe, and by adopting a partial coverage absorption model on a power law continuum ($\alpha = 0.7$), one has obtained[39] a good fit of the $3 - 100$ keV XRB spectrum with spectral source parameters typical of the Sy 1 spectra measured by GINGA, assumed to be exponentially cut-off at a computed energy of $\sim$ 115 keV, and with a reasonable cosmological evolution of the local AGN luminosity function derived from the HEAO–1 bright sample[24].

Fabian *et al.*[40] emphasized the physical nature of the relatively sharp break of the XRB spectrum at $\sim$ 30 keV and noted that this feature can be adequately reproduced by the redshifted Compton reflection spectrum from sources at a redshift $z \sim 2$, thus avoiding the need of introducing artificial cut-offs. It was found that the fit of the XRB spectrum from several keV to $\sim$ 1 MeV requires a new class of strong ($\lesssim 10^{45}$ erg s^{-1}) sources with more than 90% of the emitted flux in the reflected component, and a strong cosmological evolution such that the comoving volume emissivity of the sources increases as $(1 + z)^4$ out to a redshift $z \sim 5$. 'Reflection' models of this type have been further investigated with similar conclusions on the strength of the reflected component[41,42]. However, while a new class of AGNs, such as that postulated by Fabian *et al.*, could not have been hitherto detected, other models [42] are based on the evolution of the local XLF of AGNs and, therefore, the conclusion that $\sim$ 80% of the source flux is channeled in the reflected component is at variance with the average value actually observed in the Sy 1 spectra. A general difficulty with all these models is that the predicted source counts and/or spectra in the soft X–ray band are inconsistent with those of the EMSS and ROSAT surveys[43]. It has also been pointed out[44] that these models do not adequately fit the position and the width of the peak of the XRB intensity. Recently, one has proposed another 'reflection' model [45] where the typical Sy 1 spectrum is made by Compton reflection of $\sim$ 50% of a primary X–ray flux produced by thermal comptonization of seed photons in a hot plasma cloud with a Thompson depth of a few and a temperature kT = 40 keV, which means that the spectrum is exponentially cut-off at $\sim$ 100 keV. By adopting the cosmological evolution derived[11] from the analysis of the combined AGN sample from the EMSS and ROSAT surveys, it is found that a very good fit of the 3 – 100 keV XRB spectrum can be obtained with the evolution extended up to $z = 4$, but with a local volume emissivity about 2 times that derived from the above mentioned source samples. No details are given about the expected source counts. However, the assumed source spectral shape, in combination with the high redshifts involved in the model, is likely to produce too many sources with too flat spectra in the soft X–rays.

In conclusion, it appears that pure 'reflection' models are rather inefficient in producing the XRB, unless a completely new class of AGNs is postulated.

Alternatively, it has been proposed that the synthesis of the XRB may be obtained by assuming the existence of a large population of heavily absorbed AGNs[46,47]. In particular, Setti & Woltjer [46] have discussed a model based on the X–ray properties of AGN unified schemes, first introduced by Antonucci & Miller[48] for the Seyfert galaxies, in which the central source can be hidden (depending on the viewing angle) by a thick torus of surrounding absorbing matter. Radio–loud quasars and strong radio galaxies can be similarly unified, the quasar phenomenon showing up whenever the source axis is aimed toward us within a specified angle[49], whilst the existence of a hidden population of radio-quiet quasars is still uncertain, although the luminous IR galaxies have been proposed as likely candidates. Accordingly, it has been shown that the X–rays can be absorbed up to > 20 keV whenever the lines of sight are intercepted by the tori, while above the absorption cut-offs the X–ray properties should be the same as those of the 'unabsorbed' AGNs, unless the X–ray emission itself is largely beamed. X–ray observations of Seyfert 2 galaxies and strong radiogalaxies support this hypothesis[50–52]. However, a combination of data from GINGA and CGRO-OSSE (2 – 500 keV) on a small sample of Sy galaxies indicate that, while the Sy 2 X-ray spectra are substantially harder than those of Sy 1s, there are differences which may not be explained by the simplest version of the AGN unified model[53]. By adopting source spectra with the canonical 2 – 10 keV slope ($\alpha = 0.7$), very simple fractional distributions of the absorption cut-offs and an internally consistent cosmic evolution of the AGN XLF out to z = 3, it was demonstrated that the 3 – 30 keV XRB spectral intensity can be accurately reproduced with a number of 'absorbed' AGNs about equal to that of the 'unabsorbed' ones. Very good fits of the XRB have been obtained by more detailed investigations, thus confirming the basic validity of this scenario[54,55].

We have seen that different models are able to reproduce the hard XRB as a superposition of the emission from AGNs. We have also underlined that it is critically important to check these models against other observational constraints, in particular the predicted source counts and spectra in the soft X-rays. One important feature of Setti & Woltjer's proposal is that the 'absorbed' AGNs, essential to explain the XRB above several keV, may not show up significantly in the soft X-rays, so that these two spectral regions are to some extent indepedent. Therefore, the next step is the construction of a model consistent with the main statistical properties of AGN X-ray samples[56,57]. A specific attempt in this direction will be briefly described in the following Section.

A BASELINE MODEL OF THE XRB FROM AGNs

The basic assumptions of the model[57], aimed at reproducing the XRB spectrum in the 3 – 100 keV interval, can be summarized as follows:

a) A typical primary AGN continuum spectrum with spectral indices $\alpha = 1.3$ below 1.5 keV and $\alpha = 0.9$ above this energy, exponentially cut-off at an e-folding energy of ~ 320 keV. A Compton reflection component (50% of the primary flux) has been added to the continuum of the low luminosity AGNs ($< 7 \times 10^{43}$ erg s^{-1} in the 0.3 – 3.5 keV interval). This is consistent with the broad spectral characteristics observed in Sy 1s and quasars[57].

b) The AGNs are surrounded by tori of absorbing material (solar composition) with 45° half opening angle, such that hydrogen column densities N_H of up to 10^{25} cm^{-2} can be intercepted depending on the source characteristcs and orientation with respect to the line of sight to the central source.

c)The adopted (z = 0) XLF is that determined by Boyle et al. [11] in the 0.3–3.5 band, evolved in luminosity by $(1+z)^{2.6}$ up to z = 2.25, thereafter constant and cut-off at z = 4. It covers a wide luminosity range from 10^{42} to 10^{47} erg s^{-1}.

The number of 'absorbed' sources as a function of $N_H(> 10^{21}$ cm$^{-2})$ is essentially the only adjustable parameter of the model. This distribution has

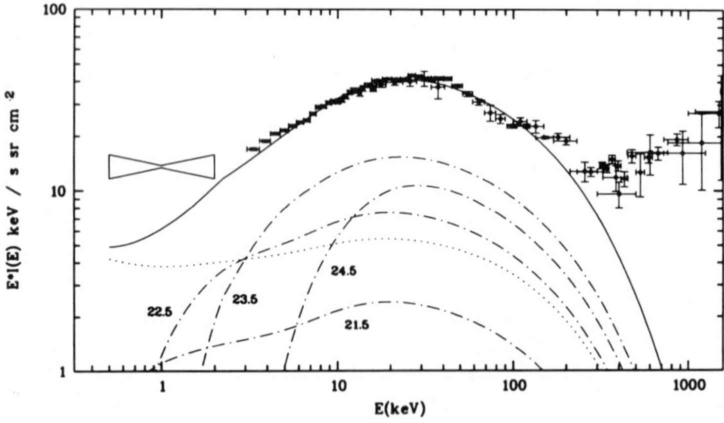

FIGURE 2. The fit (solid line) to the X-ray background of the AGN baseline model discussed in the text. Also plotted are the contributions from unabsorbed AGNs (dotted line) and unabsorbed ones (dot-dashed lines; labels = Log N_H).

been parameterized by dividing the N_H range into four equal intervals on a log scale and by assigning to the absorbed objects the mean N_H value of the corresponding interval. It is found that the observational constraints are met by a number distribution of 'absorbed' AGNs, normalized to the 'unabsorbed' ones, given by 0.35, 1.10, 2.30, 1.65, in order of increasing N_H.

Specifically, the following constraints have been identified and checked for consistency:

1) The XRB from several keV to about 100 keV is well approximated (Fig. 2) to better than 5% accuracy with respect to the best fit curve of Gruber[14]. Below $\sim$ 5 keV the fit underestimates the background intensity by 5–10% to allow for possible contributions from other classes of objects[58].

2) The source counts in the 2 – 10 keV band (Fig. 3) are consistent with the AGN surface density of the bright sample from the HEAO–1 A2 all-sky survey and with the constraints from the fluctuation analysis of GINGA fields[26] down to a flux of $\sim 10^{-13}$ erg cm^{-2} s^{-1}. The local X-ray volume emissivity (3.8×10^{38} erg s^{-1} Mpc^{-3}) is fully consistent with that obtained from a cross-correlation of the HEAO–1 A2 maps with the IRAS galaxies[59].

3) The redshift and absorption (N_H) distributions of bright AGNs are in good agreement with those of the HEAO–1 A2 AGN sample[24].

4) The predicted source counts in the soft X-rays (Fig. 4) are in good agreement with the EMSS Log N – Log S [60] and with the ROSAT deep survey counts[6]. An extrapolation of the predicted counts down to a zero flux accounts for $\sim$ 74% of the 1 – 2 keV ROSAT XRB. Moreover, and most

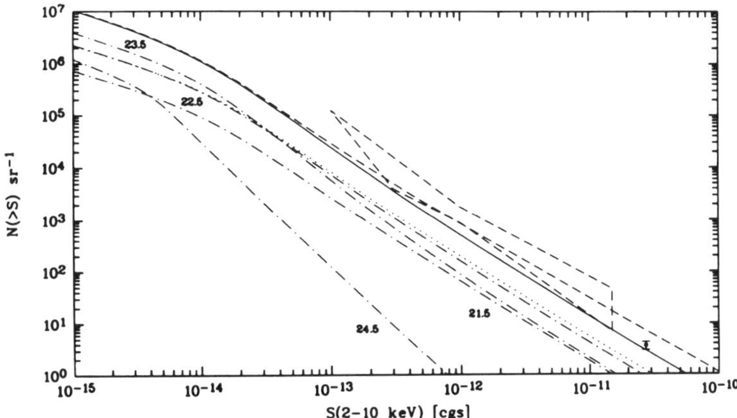

FIGURE 3. The predicted source counts in the 2 – 10 keV energy interval of the baseline model, AGN (solid line) and AGN plus galaxy clusters (upper dashed line), compared with the AGN surface density from the HEAO–1 A2 AGN complete sample (dot w/bar)[24] and the P(D) projected counts from GINGA (dashed bow-tie)[26]. Also shown the contributions from unabsorbed AGNs (dotted) and absorbed ones (dot-dashed; labels = Log N_H).

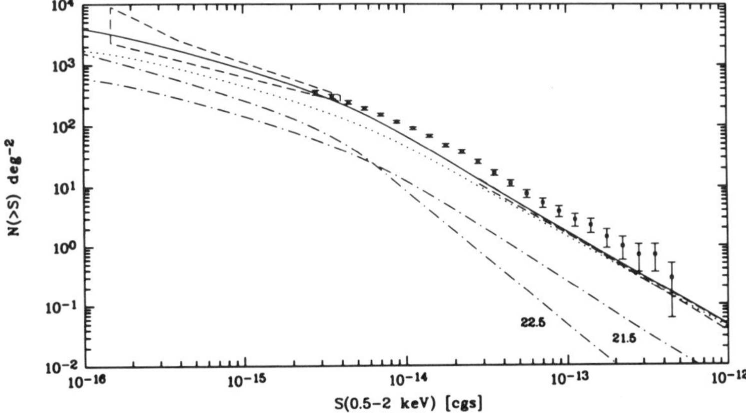

FIGURE 4. The AGN baseline model predicted counts in the 0.5 – 2 keV energy interval (solid line) compared with the AGN counts from the EMSS[60] (dashed line) and with the ROSAT total counts and their extension from a fluctuation analysis (dashed contour)[6]. Also shown are the contributions to the counts from unabsorbed AGN (dotted line) and absorbed ones (dot-dashed lines; labels = Log N_H). The contributions from absorbed AGNs with Log $N_H > 10^{23}$ are negligeable.

importantly, the source spectra at different flux levels have mean slopes in agreement with those found in the EMSS and ROSAT surveys.

5) The redshift distributions of the EMSS AGN sample and of the AGNs so far identified in five deep ROSAT fields are well reproduced by the model.

The ratios between absorbed and unabsorbed objects of the baseline model cannot be immediately associated with the Sy 2 / Sy 1 ratio. This is because objects with intrinsic absorptions corresponding to N_H values of up to several times 10^{22} cm^{-2} have been frequently classified as Sy 1 galaxies and, viceversa, Sy 2 galaxies sometimes have $N_H < 10^{22}$ cm^{-2}. If one takes column densities in the range $10^{22} - 10^{22.5}$ cm^{-2} as the dividing line between the two types, then the model predicts a ratio Sy 2 / Sy 1 = 2.4 – 3.7, in good agreement with that found from a complete sample of optically selected Seyfert galaxies[61].

THE ORIGIN OF THE γ–RAY BACKGROUND

The high energy GRB

The attempts made to explain the high energy GRB have not met the same type of difficulties, at least from the standpoint of the energy supply, as for the XRB. Despite the very limited amount of information, it was clear that the problem was more that sources might overproduce the background intensity rather than the opposite. For instance, following the COS–B detection of 3C 273, it was noted that no more than 10% of the quasars, a fraction close to that of the radio-loud quasar population, could emit the same γ–ray flux (> 100 MeV) compared to the optical emission otherwise the GRB would be exceeded[62]. Similar considerations have been used to constrain the γ–ray emission from other type of AGNs[63]. Alternatively, it has been proposed that the SAS-2 GRB might be due to diffuse processes, mainly matter–antimatter annihilation in connection with some cosmological models[64]. The findings with the EGRET instrument have now placed the question of the origin of the GRB on a much firmer basis.

As already mentioned, all high galactic latitude EGRET sources have been identified with compact radio–loud quasars and BL Lac objects having 5 GHz radio fluxes $S_5 > 0.5$ Jy and flat radio spectra ($\alpha < 0.5$ between 2.7 and 5 GHz, for $S \propto \nu^{-\alpha}$). Out of 44 sources detected above 4σ there are 35 quasars and 9 BL Lacs. Padovani *et al.*[65], on the basis of a much smaller sample then available (12 quasars and 4 BL Lacs), argued that there is a strong correlation between the 5 GHz and 100 MeV luminosities of these objects, which enables the estimate of their contribution to the GRB via the knowledge of the radio luminosity functions and of their time evolution. It is concluded that the total contribution to the 100 MeV GRB is 66^{+31}_{-24} per cent, about equally distributed between the two classes of objects. A basically similar approach has been followed by Stecker *et al.*[66] using the same sample (with the addition of 5 sources detected in the $4 - 5\ \sigma$ range), but performing a "survival analysis" to take into account the non-detection of flat spectrum radio sources with $S_5 > 1$ Jy in the sky area covered by CGRO. The computed intensity at 100 MeV is $\sim 30\%$ that of Padovani *et al.*, or about 20% that of the SAS–2 background. Because the average source spectrum appeared to be harder ($\alpha \sim 1$) than that of the SAS–2 background, it was claimed that these sources may dominate over other components only above ~ 3 GeV. The lower estimate has been attributed to the fact that Padovani *et al.* assume that all sources produce γ–rays in proportion to their radio emission without taking into account the non-detections.

A different approach has been followed by Setti & Woltjer[67]. Since the identified sources represent a γ–ray selected sample the mean ratio between the γ–ray and radio fluxes (F_γ/S_5) should be in excess of the true average. On the other hand, if one considers all flat–spectrum radio sources in a complete sample down to some specified radio flux limit and sets $F_\gamma = 0$ for all the sources not yet detected, then the F_γ/S_5 must be less than the true average. By making use of a sample of 24 identified γ–ray sources and of several complete radio samples down to $S_5 = 2$ Jy it is found that the average F_γ/S_5 should be in the interval $0.2 - 0.6$ (F_γ, the flux above 100 Mev, in units of 10^{-7} ph cm^{-2} s^{-1} and S_5 in Jy). From the radio source counts one finds that the flat–spectrum sources with $S_5 > 1$ mJy contribute 360 Jy sr^{-1} with an uncertainty of $\pm 10\%$. The 1 mJy flux is representative of the separation between quasar–like nuclei and normal (including star burst) galaxies. Thus the predicted GRB should be in the range $(0.7 - 2.2) \times 10^{-5}$ ph cm^{-2} s^{-1} sr^{-1} . Note that the value of the SAS-2 GRB is nicely placed in the middle of this interval. Since not all flat–spectrum radio sources are quasars or BL Lacs, the predicted background must be corrected downward by a somewhat uncertain amount, unlikely to exceed 20%. Therefore, one can place a rather firm lower limit ($\sim 40\%$) to the contribution of flat-spectrum radio sources to the high energy GRB. A large proportion of the intensity is contributed by the quasar population: among the 33 ($> 5\sigma$) EGRET sources[33] there are 27 quasars and 6 BL Lacs, the last ones contributing only about 11% of the total observed source flux. Contrary to quasars, BL Lacs do not evolve with cosmic time and, therefore, their percentage contribution to the GRB should be less than 10% that of the quasars. It is interesting to note that, whilst the spectral indices of the EGRET sources are spread over a wide interval ($\alpha \sim 0.4 - 2.0$)[33], the average spectral index of the quasars is steeper ($< \alpha > \simeq 1.2$) than that of the small sample of detected BL Lacs ($< \alpha > = 0.8$) and is in perfect agreement with the EGRET spectrum of the GRB.

As a corollary to these findings it immediately follows that flat–spectrum radio quasars cannot be at distances much smaller than indicated by their redshifts, otherwise their summed contributions, integrated out to the Hubble radius, would easily exceed the observed GRB (unless the Galaxy is accidentally placed near the centre of a large excess in the distribution of these objects)[67]. It should be noted that these sources, if placed at the cosmological distances indicated by their redshifts, are precisely those which show the occurrence of large superluminal motions (at least 7 of the detected EGRET sources are superluminal) and in which inverse Compton catastrophes would be most serious. As a consequence the above argument provides direct evidence that superluminal motions do occur and that the associated relativistic effects do solve the Compton problems.

In addition to the AGNs, normal and starburst galaxies may contribute a sizable fraction of the high energy GRB. The estimate can be made in a way similar to that followed above for the AGNs. It can be argued that in these objects there should be a rough proportionality between the radio and γ–ray emission: a large fraction of the galactic radio power is due to the synchrotron process from the cosmic ray electron component while, at the same time, γ–ray photons are largely produced by the interactions of cosmic rays ($p - p$ collisions) with the interstellar gas. By converting the integrated contribution of normal and starburst galaxies to the radio background at 1400 MHz into a γ–ray intensity, with the γ–ray to radio flux ratio calibrated on our Galaxy and the LMC, it is found that these objects may contribute about 10% of the GRB intensity[67]. Obviously this estimate is uncertain in view of our ignorance about the relevant galaxy parameters at earlier epochs: the γ–ray luminosity of young galaxies may be enhanced both because they are richer in gas content and possibly in cosmic ray sources, e.g. supernovae, which entails an increased rate of $p - p$ collisions; on the other hand the (non-thermal) radio luminosity depends on the (unknown) evolution of the strength of the average galactic magnetic fields and, of course, on the relativistic electrons whose ratio with respect to the nuclear cosmic rays may change with cosmic time.

The "MeV" GRB

The question of the origin of the "MeV" background, which carries most of the flux in the γ–ray regime (accordindg to the Gruber's fit), is still very much uncertain. Following earlier detections of two Seyfert galaxies from balloon experiments, it was commonly thought that Seyferts would be very bright sources at MeV energies. Likewise it has been frequently argued that the integrated contribution from Seyfert galaxies could easily account for the intensity of the MeV bump in the GRB spectrum. Up to now the COMPTEL instrument on board CGRO, which is sensitive in the $0.75 - 30$ MeV band, has failed to detect any Seyfert galaxy, thus setting a rather stringent upper limit to the average flux from this class of objects, at least an order of magnitude fainter than the fluxes quoted in the claimed detetections[68].

Instead, COMPTEL has detected a number of EGRET sources at flux levels generally in agreement with those expected on the basis of an extrapolation of the EGRET spectra down to a few MeV[69]. These sources are characterized by high time variability at all frequencies, but in most cases time correlated observations in the different spectral domains are not available. However, (quasi)simultaneous observations of the quasar 3C 273[70] and a compilation of spectral data taken at different epochs for other objects, such as the quasars CTA 102 and 3C 454.3[71], clearly indicate that the multiwavelength energy spectra (νf_ν) of these sources are peaked at MeV energies. Recently, COMPTEL has detected a very bright "MeV" source, probably coincident with the EGRET quasar PKS 0506−612, whose spectrum (νf_ν) is strongly peaked at a few MeV and must break sharply to meet the EGRET flux above 30 MeV[72]. It is pointed out that the "MeV" background intensity could be readily explained if a sizable fraction of the EGRET type of sources (mostly quasars), accounting for the bulk of the high energy GRB, has a γ-ray spectrum like that of PKS 0506−612.

CONCLUSIONS

On the basis of present observational evidence it seems reasonable to conclude that both the XRB and the GRB are mostly due to the summed contributions of different types of AGNs.

With reference to the XRB it has been demonstrated that it is possible to construct a baseline model which accurately explains the spectral intensity observed in the $5 - 100$ keV energy interval and is consistent with all available data in the soft and hard X-rays. A key feature of the model is represented by the contribution of "absorbed" AGNs, which are predicted by the simplest version of the AGN unified schemes. Both low-luminosity AGNs (Seyfert nuclei) and quasars are assumed to equally evolve in cosmic time out to a redshift $z = 2 - 2.5$. The model predicts that the AGNs account for about 75% of the soft XRB measured by ROSAT, or as much as 90% if the lower normalization of the XRB intensity indicated by the preliminary ASCA results applies.

It appears that at least 40% of the high energy GRB (> 30 MeV) can be readily explained as the summed contribution from flat-spectrum radio quasars and BL Lac objects, the largest fractional contribution being that of the quasars. The background intensity due to BL Lacs is probably less than $\sim 10\%$ that of the quasars. While these objects may easily account for almost all the intensity of the > 30 MeV GRB, about 10% may still be contributed by normal galaxies.

A complete understanding of the "MeV" GRB is not available yet. Recent results from the COMPTEL instrument suggest that the compact flat spectrum quasars observed by EGRET might have the correct spectral shape to significantly contribute to the "MeV" bump. However, it is premature to draw any definitive conclusion: more data on the sources and, most importantly, a much better knowledge of the background spectral intensity are required before one can attempt a detailed modeling of the "MeV" GRB.

ACKNOWLEDGMENTS

It is a pleasure to thank Dr A. Comastri for informative discussions.

REFERENCES

1. Warwick, R.S., J.P. Pye & A.C. Fabian. 1980. Mon. Not. R. astr. Soc. 190: 243-260.
2. Iwan, D. *et al.* 1982. Astrophys. J. 260: 111–123.
3. Giacconi, R. *et al.* 1962. Phys. Rev. Letters. 9: 439–443.
4. Setti, G. & L. Woltjer. 1973. *In* Proceedings of the IAU Symposium 55 on X- and Gamma-Ray Astronomy. H. Bradt & R. Giacconi, Eds.: 208–211. Reidel Publ. Co.. Dordrecht.
5. Wang, Q.D. & R. McCray. 1993. Astrophys. J. 409: L37–L40.
6. Hasinger, G. *et al.* 1993. Astron. Astrophys. 275: 1–15.
7. Gendreau, K.C. *et al.* 1995, PASJ in press
8. Hasinger, G. 1994. *In* Proceedings of the IAU Symposium 168 on Examining the Big Bang and Diffuse Background Radiations. M. Kafatos, Ed. Kluwer Academic Press. Dordrecht. In press.
9. Primini *et al.* 1991. Astrophys. J. 374: 440–455.
10. Shanks, T., I. Georgantopoulos & G.C. Stewart. 1991. Nature. 353: 315–320.
11. Boyle, B.J. *et al.* 1993, Mon. Not. R. astr. Soc. 260: 49–58.
12. Marshall, F.E. *et al.* 1980. Astrophys. J. 235: 4–10.
13. Boldt, E. 1987. Phys. Reports. 146: 215–257.
14. Gruber, D.E. 1992. *In* The X-ray background. X. Barcons & A.C. Fabian, Eds.: 44–53. Cambridge Univ. Press. Cambridge.
15. Cowsik, R. & E.J. Kobetich. 1972. Astrophys. J. 177: 585–593.
16. Field, G.B. & S.C. Perrenod. 1977. Astrophys. J. 215: 717–722.
17. Guilbert, P.W. & A.C. Fabian. 1986. Mon. Not. R. astr. Soc. 220: 439–452.
18. Taylor, G.B. & E.L. Wright. 1989. Astrophys. J. 339: 619–628.
19. Wright, E.L. *et al.* 1994, Astrophys. J. 420: 450–456.
20. Mather, J. 1994. These Proceedings.
21. Barcons, X. & A.C. Fabian. 1988. Mon. Not. R. astr. Soc. 230: 189–206.
22. Daly, R.A. 1987. Astrophys. J. 322: 20–33.
23. Giacconi, R. & Zamorani G. 1987. Astrophys. J. 313: 20–27.
24. Piccinotti, G. *et al.* 1982. Astrophys. J., 253: 485–503.
25. Warwick, R.S. & G.C. Stewart. 1989. *In* Proceedings of the 23rd ESLAB Symposium on Two-Topics in X-Ray Astronomy. Vol.**2**: 727–731. ESA SP-296.
26. Hayashida, K. 1990. Ph.D. Thesis. Tokyo University. ISAS RN 466.

27. Shafer, R.A. 1983. Ph.D. Thesis. University of Maryland. NASA Tech.Mem. 85029.
28. Turner, T.J. & K.A. Pounds. 1989. Mon. Not. R. astr. Soc. 240: 833–880.
29. Fichtel, C.E., G.A. Simpson & D.J. Thompson. 1978. Astrophys. J. 222: 833–849.
30. Thompson, D.J. & C.E. Fichtel. 1982. Astron. Astrophys. 109: 352–354.
31. Sreekumar, P. & D.A. Kniffen. 1994. _In_ Proceedings of the IAU Symp. No.168 on Examining the Big Bang and Diffuse Background Radiations. M. Kafatos, Ed. Kluwer Academic Press. Dordrecht. In press.
32. Osborne, J.L., A.W. Wolfendale & L. Zhang. 1994. Phys. G: Nucl. Part. Phys. 20: 1089–1101.
33. von Montigny, C. _et al._ 1994. Astrophys. J. In press.
34. Nandra, K., & K.A. Pounds. 1994. Mon. Not. R. astr. Soc. 268: 405–429.
35. Piro, L., M. Yamauchi & M. Matsuoka. 1989. _In_ Proceedings of the 23rd ESLAB Symposium on Two-Topics in X-Ray Astronomy. Vol.**2**: 819–823. ESA SP-296.
36. Matsuoka, M. _et al._. 1990. Astrophys. J. 361: 440–458.
37. Pounds, K.A. _et al._ 1990. Nature. 344: 132–133.
38. Schwartz, D.A. & W.H. Tucker. 1988. Astrophys. J. 332: 157–162.
39. Morisawa, K. _et al._. 1990. Astron. Astrophys. 236: 299–304.
40. Fabian, A.C. _et al._. 1990, Mon. Not. R. astr. Soc. 242: 14p–16p.
41. Rogers, R.D. & G.B. Field. 1991. Astrophys. J. 378: L17–L20.
42. Terasawa, N. 1991. Astrophys. J. 378: L11–L16.
43. Setti, G. 1992. _In_ The X-ray background. X. Barcons & A.C. Fabian, Eds.: 187–200. Cambridge Univ. Press. Cambridge.
44. Zdziarski, A.A. _et al._ 1993. Astrophys. J. 405: 125–129.
45. Zdziarski, A.A., P.T. Zycki & J.H. Krolik. 1993. Astrophys. J. 414: L81–L84.
46. Setti, G. & L. Woltjer. 1989. Astron.Astrophys. 224: L21–L23.
47. Grindlay, J.E. & M. Luke. 1990. _In_ Proceedings of the IAU Coll.No.115 on High Resolution X-ray Spectroscopy of Cosmic Plasmas. P. Gorenstein & M. Zombeck, Eds.: 276–280. Kluwer Academic Press. Dordrecht.
48. Antonucci, R.R.J. & J.S. Miller. 1985. Astrophys. J. 297: 621–632.
49. Barthel, P.D. 1989. Astrophys. J. 336: 606–611.
50. Awaki, H. _et al._ 1991. Pub. Astr. Soc. Japan. 43: 195–212.
51. Koyama, K. 1992, _In_ X-Ray Emission from Active Galactic Nuclei and the Cosmic X-Ray Background. W. Brinkmann & J. Trümper, Eds.: 74–80. MPE Report 235.
52. Allen, S.W. & A.C. Fabian. 1992. Mon. Not. R. astr. Soc. 258: 29p–32p.
53. Zdziarski, A.A. _et al._ 1995. Astrophys. J. 438: L63–L66.

54. Madau, P., G. Ghisellini & A.C. Fabian. 1993. Astrophys. J. 410: L7–L10.
55. Matt, G. & A.C. Fabian. 1994. Mon. Not. R. astr. Soc. 267: 187–192.
56. Madau, P., G. Ghisellini & A.C.Fabian. 1994. Mon. Not. R. Astr. Soc. 270: L17–L21.
57. Comastri, A., G. Setti, G. Zamorani & G. Hasinger. 1995. Astron. Astrophys. 296: 1–12.
58. Setti, G. 1990. *In* Proceedings of the IAU Symposium 139 on The Galactic and Extragalactic Background Radiation. S. Bowyer & Ch. Leinert, Eds.: 345–356. Kluwer Academic Press. Dordrecht.
59. Miyaji, T., O. Lahav, K. Jahoda & E. Boldt. 1994. Astrophys. J. In press.
60. Della Ceca, R. *et al.*. 1992. Astrophys. J. 389: 491–498.
61. Huchra, J. & R. Burg. 1992. Astrophys. J. 393: 90–97.
62. Setti, G. & L. Woltjer. 1979. Astron. Astrophys. 76: L1–L2.
63. Bignami, G.F., *et al.* 1979. Astrophys. J. 232: 649–658.
64. Stecker, F.W. 1985. Nucl. Phys. **B**. 252: 25–36.
65. Padovani, P. *et al.* 1993. Mon. Not. R. astr. Soc. 260: L21–L24.
66. Stecker, F.W., M.H. Salamon & M.A. Malkan. 1993. Astrophys. J. 410: L71–L74.
67. Setti, G. & L. Woltjer. 1994. Astrophys. J. Suppl. 92: 629–631.
68. Maisack, M. *et al.* 1994. Astron. Astrophys. In press.
69. Williams, O.R. *et al.* 1995. Astron.Astrophys. In press.
70. Lichti, G.G. *et al.* 1994. Astron. Astrophys. In press.
71. Blom, J.J. *et al.* 1995. Astron. Astrophys. 295: 330-334.
72. Bloemen, H. *et al.* 1995. Astron. Astrophys. 293: L1-L4.

Gravitational Radiation

KIP S. THORNE

California Institute of Technology
Pasadena, CA 91125 USA

INTRODUCTION

1 Introduction

According to general relativity theory, compact concentrations of energy (e.g., neutron stars and black holes) should warp spacetime strongly, and whenever such an energy concentration changes shape, it should create a dynamically changing spacetime warpage that propagates out through the Universe at the speed of light. This propagating warpage is called *gravitational radiation*—a name that arises from general relativity's description of gravity as a consequence of spacetime warpage.

There is an enormous difference between gravitational waves, and the electromagnetic waves on which our present knowledge of the Universe is based:

- Electromagnetic waves are oscillations of the electromagnetic field that propagate through spacetime; gravitational waves are oscillations of the "fabric" of spacetime itself.
- Astronomical electromagnetic waves are almost always incoherent superpositions of emission from individual electrons, atoms, or molecules. Cosmic gravitational waves are produced by coherent, bulk motions of huge amounts of mass-energy—either material mass, or the energy of vibrating, nonlinear spacetime curvature.
- Since the wavelengths of electromagnetic waves are small compared to their sources (gas clouds, stellar atmospheres, accretion disks, ...), from the waves we can make pictures of the sources. The wavelengths of cosmic gravitational waves are comparable to or larger than their coherent, bulk-moving sources, so we cannot make pictures from them. Instead, the gravitational waves are like sound; they carry, in two independent waveforms, a stereophonic, symphony-like description of their sources.
- Electromagnetic waves are easily absorbed, scattered, and dispersed by matter. Gravitational waves travel nearly unscathed through all forms and amounts of intervening matter [1, 2].

- Astronomical electromagnetic waves have frequencies that begin at $f \sim 10^7$ Hz and extend on *upward* by roughly 20 orders of magnitude. Astronomical gravitational waves should begin at $\sim 10^4$ Hz (100-fold lower than the lowest-frequency astronomical electromagnetic waves), and should extend on *downward* from there by roughly 20 orders of magnitude.

These enormous differences make it likely that:

- The information brought to us by gravitational waves will be very different from (almost "orthogonal to") that carried by electromagnetic waves; gravitational waves will show us details of the bulk motion of dense concentrations of energy, whereas electromagnetic waves show us the thermodynamic state of optically thin concentrations of matter.
- Most (but not all) gravitational-wave sources that our instruments detect will not be seen electromagnetically, and conversely, most objects observed electromagnetically will never be seen gravitationally. Typical electromagnetic sources are stellar atmospheres, accretion disks, and clouds of interstellar gas—none of which emit significant gravitational waves; while typical gravitational-wave sources are the cores of supernovae (which are hidden from electromagnetic view by dense layers of surrounding stellar gas), and colliding black holes (which emit no electromagnetic waves at all).
- Gravitational waves may bring us great surprises. In the past, when a radically new window has been opened onto the Universe, the resulting surprises have had a profound, indeed revolutionary, impact. For example, the radio universe, as discovered in the 1940s, 50s and 60s, turned out to be far more violent than the optical universe; radio waves brought us quasars, pulsars, and the cosmic microwave radiation, and with them our first direct observational evidence for black holes, neutron stars, and the heat of the big bang [3]. It is reasonable to hope that gravitational waves will bring a similar "revolution".

In this lecture I shall review the present status of attempts to detect gravitational radiation and plans for the future, and I shall describe some examples of information that we expect to garner from the observed waves. I shall begin, in Section 2, with an overview of all the frequency bands in which astrophysical gravitational waves are expected to be strong, the expected sources in each band, and the detection techniques being used in each. Then in subsequent sections I shall focus on the two bands whose sources are largely in the present-day Universe: the "high" and "low" frequency bands.

2 Frequency Bands, Sources, and Detection Methods

Four gravitational-wave frequency bands are being explored experimentally: the high-frequency band (HF; $f \sim 10^4$ to 1 Hz), the low-frequency band (LF; $f \sim 1$ to 10^{-4} Hz), the very-low frequency band (VLF; $f \sim 10^{-7}$ to 10^{-9} Hz), and the extremely-low-frequency band (ELF; $f \sim 10^{-15}$ to 10^{-18} Hz).

2.1 High-Freqency Band, 1 to 10^4 Hz

A gravitational-wave source of mass M cannot be much smaller than its gravitational radius, $2GM/c^2$, and cannot emit strongly at periods much smaller than the light-travel time around this gravitational radius, $4\pi GM/c^2$; correspondingly, the frequencies at which it emits are

$$f \lesssim \frac{1}{4\pi GMc^2} \sim 10^4 \mathrm{Hz} \frac{M_\odot}{M} , \tag{1}$$

where $M_\odot$ is the mass of the Sun. To achieve a size of order its gravitational radius and thereby emit near this maximum frequency, an object presumably must be heavier than the Chandrasekhar limit, about the mass of the sun, $M_\odot$. Thus, the highest frequency expected for strong gravitational waves is $f_{max} \sim 10^4$ Hz. This defines the upper edge of the high-frequency gravitational-wave band.

The high-frequency band is the domain of Earth-based gravitational-wave detectors: laser interferometers (on which I shall focus in this article because of their great promise), and resonant-mass detectors (which may play special roles at the highest frequencies; cf. the end of Sec. 4.1.4).

At frequencies below about 1 Hz, Earth-based detectors face nearly insurmountable noise (i) from fluctuating Newtonian gravity gradients (due, e.g., to the gravitational pulls of inhomogeneities in the Earth's atmosphere which move overhead with the wind), and (ii) from Earth vibrations (which are extremely difficult to filter out mechanically below ~ 1 Hz). This defines the 1 Hz lower edge of the High-frequency band; to detect waves below this frequency, one must fly one's detectors in space.

A number of interesting gravitational-wave sources fall in the high-frequency band: the stellar collapse to a neutron star or black hole in our Galaxy and distant galaxies, which sometimes triggers supernovae; the rotation and vibration of neutron stars (pulsars) in our Galaxy; the coalescence of neutron-star and stellar-mass black-hole binaries ($M \lesssim 1000 M_\odot$) in distant galaxies; and possibly such sources of stochastic background as vibrating loops of cosmic string, phase transitions in the early Universe, and the big bang in which the Universe was born.

I shall discuss the high-frequency band in Sections 3 and 4, focusing on sources in the present-day universe (i.e., ignoring, for lack of space, stochastic background from the early universe).

2.2 Low-Frequency Band, 10^{-4} to 1 Hz

The low-frequency band, 10^{-4} to 1 Hz, is the domain of detectors flown in space: microwave-frequency Doppler tracking of spacecraft at present, and optical tracking of spacecraft by each other in the future—a technique on which I shall focus in Sec. 5 because of its great promise.

The 1 Hz upper edge of the low-frequency band is defined by the gravity-gradient and seismic cutoffs on Earth-based instruments; the $\sim 10^{-4}$ Hz lower edge is defined by expected severe difficulties at lower frequencies in isolating spacecraft from the buffeting forces of fluctuating solar radiation pressure, solar wind, and cosmic rays.

The low-frequency band should be populated by waves from short-period binary stars in our own Galaxy (main-sequence binaries, cataclysmic variables, white-dwarf binaries, neutron-star binaries, ...); from white dwarfs, neutron stars, and small black holes spiraling into massive black holes ($M \sim 3 \times 10^5$ to $3 \times 10^7 M_\odot$) in distant galaxies; and from the inspiral and coalescence of supermassive black-hole binaries ($M \sim 100$ to $10^8 M_\odot$). The upper limit, $\sim 10^8 M_\odot$, on the masses of black holes that can emit in the low-frequency band is set by Eq. (1) with $f \gtrsim 10^{-4}$ Hz. There should also be a low-frequency stochastic background from such early-universe processes as vibrating cosmic strings, phase transitions, and the big-bang itself.

I shall discuss the low-frequency band in Sections 5 and 6.

2.3 Very-Low-Frequency Band, 10^{-7} to 10^{-9} Hz

Joseph Taylor and others have achieved a remarkable gravity-wave sensitivity in the very-low-frequency band (VLF) by the timing of millisecond pulsars: When a gravitational wave passes over the Earth, it perturbs our rate of flow of time and thence the ticking rates of our clocks relative to clocks outside the wave. Such perturbations will show up as apparent fluctuations in the times of arrival of the pulsar's pulses. If no fluctuations are seen at some level, we can be rather sure that neither Earth nor the pulsar is being bathed by gravitational waves of the corresponding strength. If fluctuations with the same time evolution are seen simultaneously in the timing of several different pulsars, then the cause could well be gravitational waves bathing the Earth.

By averaging the pulses' times of arrival over long periods of time (months to tens of years), a very high timing precision can be achieved, and correspondingly tight limits can be placed on the waves bathing the Earth or the pulsar. The upper edge of the VLF band, $\sim 10^{-7}$ Hz, is set by the averaging time, a few months, needed to build up high accuracy; the lower edge, $\sim 10^{-9}$ Hz, is set by the time, ~ 20 years, since very steady millisecond pulsars were first discovered.

Strong gravitational-wave sources are generally compact, not much larger than their own gravitational radii. The only compact bodies that can radiate in the VLF band or below, i.e., at $f \lesssim 10^{-7}$ Hz, are those with $M \gtrsim 10^{11} M_\odot$ [cf. Eq. (1)]. Conventional astronomical wisdom suggests that compact bodies this massive do not exist, and that therefore the only strong waves in the VLF band and below are a stochastic background produced by the same early-universe processes as might radiate at low and high frequencies: cosmic strings, phase transitions, and the big bang.

Of course, conventional wisdom could be wrong. Nevertheless, it is conventional to quote measurement accuracies in the VLF band and below in the language of a stochastic background: the fraction $\Omega_g(f)$ of the energy required to close the Universe that lies in a bandwidth $\Delta f = f$ centered on frequency f. The current 95%-confidence limit on Ω_g from pulsar timing in the VLF band is $\Omega_g < 6 \times 10^{-8} H^{-2}$, where H is the Hubble constant in units of 100 km sec^{-1} Mpc^{-1} [4]. This is a sufficiently tight limit that it is beginning to cast doubt on the (not terribly popular) suggestion, that the Universe contains enough vibrating loops of cosmic string for their gravitational pulls to have seeded galaxy formation.

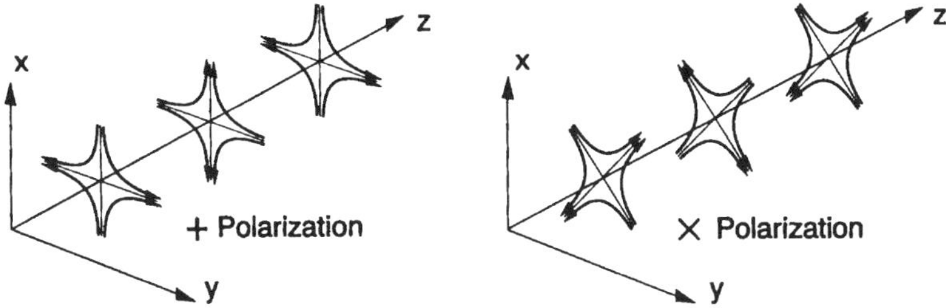

Figure 1: The lines of force associated with the two polarizations of a gravitational wave. (From Ref. [7].)

2.4 Extremely-Low-Frequency Band, 10^{-15} to 10^{-18} Hz

Gravitational waves in the extremely-low-frequency band (ELF), 10^{-15} to 10^{-18} Hz, should produce anisotropies in the cosmic microwave background radiation. The tightest limit from microwave observations comes from the lower edge of the ELF band $f \sim 10^{-18}$ Hz, where the gravitational wavelength is about π times the Hubble distance, and the waves, by squeezing all of the space inside our cosmological horizon in one direction, and stretching it all in another, should produce a quadrupolar anisotropy in the microwave background. The quadrupolar anisotropy measured by the COBE satellite, if due primarily to gravitational waves (which some or even most of it *could* be [5, 6]), corresponds to an energy density $\Omega_g(10^{-18}\mathrm{Hz}) \sim 10^{-9}$.

3 Ground-Based Laser Interferometers

3.1 Wave Polarizations, Waveforms, and How an Interferometer Works

According to general relativity theory, a gravitational wave has two linear polarizations, conventionally called + (plus) and × (cross). Associated with each polarization there is a gravitational-wave field, h_+ or $h_\times$, which oscillates in time and propagates with the speed of light. Each wave field produces tidal forces (gravitational stretching and squeezing forces) on any object or detector through which it passes. If the object is small compared to the waves' wavelength (as is the case for ground-based interferometers and resonant mass antennas), then relative to the object's center, the forces have the quadrupolar patterns shown in Fig. 1. The names "plus" and "cross" are derived from the orientations of the axes that characterize the force patterns [1].

A laser interferometer gravitational wave detector ("interferometer" for short) consists of four masses that hang from vibration-isolated supports as shown in Figure 2, and the indicated optical system for monitoring the separations between the masses [1, 7]. Two masses are near each other, at the corner of an "L", and one mass is at the end of each of the L's long arms. The arm lengths are nearly equal, $L_1 \simeq L_2 = L$. When a gravitational wave, with frequencies high compared to the masses' $\sim$ 1 Hz pendulum frequency, passes through the detector, it pushes the

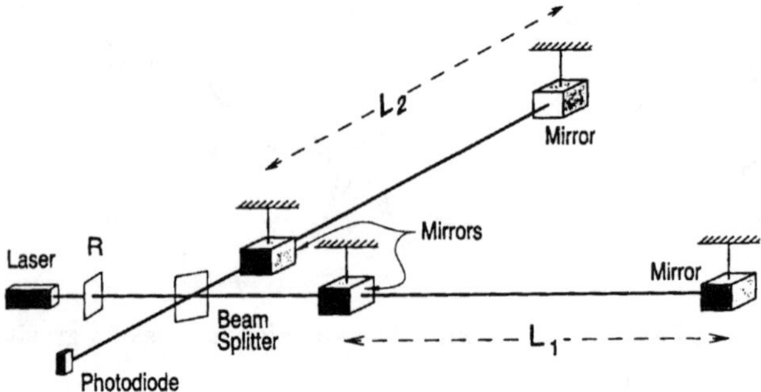

Figure 2: Schematic diagram of a laser interferometer gravitational wave detector. (From Ref. [7].)

masses back and forth relative to each other as though they were free from their suspension wires, thereby changing the arm-length difference, $\Delta L \equiv L_1 - L_2$. That change is monitored by laser interferometry in such a way that the output of the photodiode (the interferometer's output) is directly proportional to $\Delta L(t)$.

If the gravitational waves are coming from overhead or underfoot and the axes of the + polarization coincide with the arms' directions, then it is the waves' + polarization that drives the masses, and $\Delta L(t)/L = h_+(t)$. More generally, the interferometer's output is a linear combination of the two wave fields:

$$\frac{\Delta L(t)}{L} = F_+ h_+(t) + F_\times h_\times(t) \equiv h(t) \ . \tag{2}$$

The coefficients F_+ and $F_\times$ are of order unity and depend in a quadrupolar manner on the direction to the source and the orientation of the detector [1]. The combination $h(t)$ of the two h's is called the *gravitational-wave strain* that acts on the detector; and the time evolutions of $h(t)$, $h_+(t)$, and $h_\times(t)$ are sometimes called *waveforms.*

3.2 LIGO, VIRGO, and the International Interferometric Network

Interferometers are plagued by non-Gaussian noise, e.g. due to sudden strain releases in the wires that suspend the masses. This noise prevents a single interferometer, by itself, from detecting with confidence short-duration gravitational-wave bursts (though it might be possible for a single interferometer to search for the periodic waves from known pulsars). The non-Gaussian noise can be removed by cross correlating two, or preferably three or more, interferometers that are networked together at widely separated sites.

The technology and techniques for such interferometers have been under development for nearly 25 years, and plans for km-scale interferometers have been developed over the past 13 years. An international network consisting of three km-scale interferometers, at three widely separated sites, is now in the early stages of construction. It includes two sites of the American LIGO Project ("Laser Interferometer Gravitational Wave Observatory") [7], and one site of the French/Italian VIRGO Project (named after the Virgo cluster of galaxies) [8].

LIGO will consist of two vacuum facilities with 4-kilometer-long arms, one in Hanford, Washington (in the northwestern United States) and the other in Livingston, Louisiana (in the southeastern United States). These facilities are designed to house many successive generations of interferometers without the necessity of any major facilities upgrade; and after a planned future expansion, they will be able to house several interferometers at once, each with a different optical configuration optimized for a different type of wave (e.g., broad-band burst, or narrow-band periodic wave, or stochastic wave). The LIGO facilities and their first interferometers are being constructed by a team of about 80 physicists and engineers at Caltech and MIT, led by Barry Barish (the PI) and Gary Sanders (the Project Manager). Substantial contributions are also being made by scientists at other institutions.

The VIRGO Project is building one vacuum facility in Pisa, Italy, with 3-kilometer-long arms. This facility and its first interferometers are a collaboration of more than a hundred physicists and engineers at the INFN (Frascati, Napoli, Perugia, Pisa), LAL (Orsay), LAPP (Annecy), LOA (Palaiseau), IPN (Lyon), ESPCI (Paris), and the University of Illinois (Urbana), under the leadership of Alain Brillet and Adalberto Giazotto.

Both LIGO and VIRGO are scheduled for completion in the late 1990s, and their first gravitational-wave searches are likely to be performed in 2000 or 2001.

LIGO alone, with its two sites which have parallel arms, will be able to detect an incoming gravitational wave, measure one of its two waveforms, and (from the time delay between the two sites) locate its source to within a $\sim 1^{\circ}$ wide annulus on the sky. LIGO and VIRGO together, operating as a *coordinated international network*, will be able to locate the source (via time delays plus the interferometers' beam patterns) to within a 2-dimensional error box with size between several tens of arcminutes and several degrees, depending on the source direction and on the amount of high-frequency structure in the waveforms. They will also be able to monitor both waveforms $h_{+}(t)$ and $h_{\times}(t)$ (except for frequency components above about 1kHz and below about 10 Hz, where the interferometers' noise becomes severe).

The accuracies of the direction measurements and the ability to monitor more than one waveform will be severely compromised when the source lies anywhere near the plane formed by the three LIGO/VIRGO interferometer locations. To get good all-sky coverage will require a fourth interferometer at a site far out of that plane; Japan and Australia would be excellent locations, and research groups there are carrying out research and development on interferometric detectors, aimed at such a possibility. A 300-meter prototype interferometer called TAMA is under construction in Tokyo, and a 400-meter prototype called AIGO400 has been proposed for construction north of Perth.

Two other groups are major players in this field, one in Britain led by James Hough, the other in Germany, led by Karsten Danzmann. These groups each have two decades of experience with prototype interferometers (comparable experience to the LIGO team and far more than anyone else) and great expertise. Frustrated by inadequate financing for a kilometer-scale interferometer, they are constructing, instead, a 600 meter system called GEO600 near Hanford, Germany. Their goal is to develop, from the outset, an interferometer with the sort of advanced design that LIGO and VIRGO will attempt only as a "second-generation" instrument, and thereby achieve sufficient sensitivity to be full partners in the international network's first gravitational-wave searches; they then would offer a variant of their interferometer as a candidate for second-generation operation in the much longer arms of LIGO and/or VIRGO. It is a seemingly audacious plan, but with their extensive experience and expertise, the British/German collaboration might pull it off successfully.

4 High-Frequency Gravitational-Wave Sources

4.1 Coalescing Compact Binaries

The best understood of all gravitational-wave sources are coalescing, compact binaries composed of neutron stars (NS) and black holes (BH). These NS/NS, NS/BH, and BH/BH binaries may well become the "bread and butter" of the LIGO/VIRGO diet.

The Hulse-Taylor [9, 10] binary pulsar, PSR 1913+16, is an example of a NS/NS binary whose waves could be measured by LIGO/VIRGO, if we were to wait long enough. At present PSR1913+16 has an orbital frequency of about 1/(8 hours) and emits its waves predominantly at twice this frequency, roughly 10^{-4} Hz, which is in the low-frequency band—far too low to be detected by LIGO/VIRGO. However, as a result of their loss of orbital energy to gravitational waves, the PSR1913+16 NS's are gradually spiraling inward. If we wait roughly 10^8 years, this inspiral will bring the waves into the LIGO/VIRGO high-frequency band. As the NS's continue their inspiral, the waves will then sweep upward in frequency, over a time of about 15 minutes, from 10 Hz to $\sim 10^3$ Hz, at which point the NS's will collide and coalesce. It is this last 15 minutes of inspiral, with $\sim 16,000$ cycles of waveform oscillation, and the final coalescence, that LIGO/VIRGO seeks to monitor.

4.1.1 Wave Strengths Compared to LIGO Sensitivities

Figure 3 compares the projected sensitivities of interferometers in LIGO [7] with the wave strengths from the last few minutes of inspiral of BH/BH, NS/BH, and NS/NS binaries at various distances from Earth. The two solid curves at the bottoms of the stippled regions (labeled $h_{\rm rms}$) are the rms noise levels for broad-band waves that have optimal direction and polarization. The tops of the stippled regions (labeled $h_{\rm SB}$ for "sensitivity to bursts") are the sensitivities for highly confident detection of randomly polarized, broad-band waves from random directions (i.e., the sensitivities for high confidence that any such observed signal is not a false alarm due to Gaussian noise). The upper stippled region and its bounding curves are the expected performances of the first interferometers in LIGO; the lower stippled region and curves are performances of more advanced LIGO interferometers.

As the NS's and/or BH's spiral inward, their waves sweep upward in frequency (left to right in the diagram). The dashed lines show their "characteristic" signal strength h_c (approximately the amplitude h of the waves' oscillations multiplied by

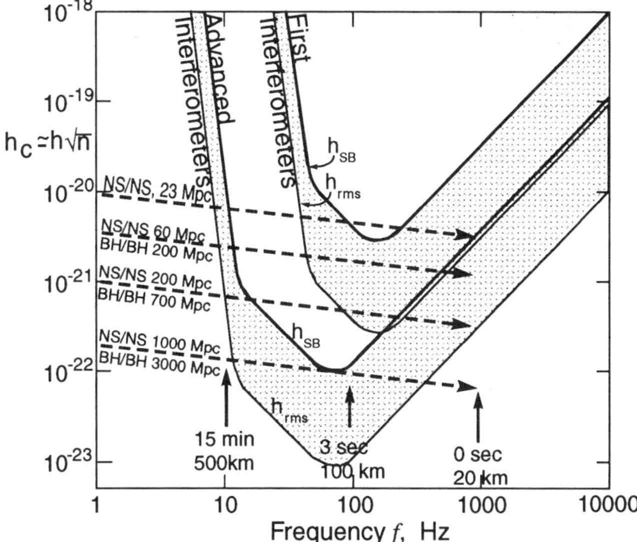

Figure 3: LIGO's projected broad-band noise $h_{\rm rms}$ and sensitivity to bursts $h_{\rm SB}$ [7] compared with the strengths of the waves from the last few minutes of inspiral of compact binaries. The signal to noise ratios are $\sqrt{2}$ higher than in Ref. [7] because of a factor 2 error in Eq. (29) of Ref. [1].

the square root of the number of cycles spent near a given frequency, $\sqrt{n}$); the signal-to-noise ratio is this h_c divided by the detector's $\sqrt{5}h_{\rm rms}$, $S/N = h_c/(\sqrt{5}h_{\rm rms})$, where the $\sqrt{5}$ converts $h_{\rm rms}$ from "optimal direction and polarization" to "random direction and polarization") [7, 1]. The arrows along the NS/NS inspiral track indicate the time until final coalescence and the separation between the NS centers of mass. Each NS is assumed to have a mass of 1.4 suns and a radius $\sim$ 10 km; each BH, 10 suns and $\sim$ 20 km.

4.1.2 Coalescence Rates

Phinney [11], taking account of all present knowledge about neutron-star binaries and all plausible ways in which our knowledge might be flawed, has placed "ultraconservative" bounds on how far LIGO/VIRGO must look to see several coalescences of NS/NS binaries per year: a lower limit of 23 Mpc; an upper limit of 1000 Mpc.

Several "best" estimates within these bounds have been made. Experts in the evolution of binary systems have combined the theory of binary evolution with the population of main sequence binaries observed in our galaxy, to obtain an estimate of $\sim$ 60Mpc for the distance to several coalescences per year [12, 13, 14]. Phinney [11], and independently Narayan, Piran, and Shemi [15] have used the statistics of binary pulsar searches in our galaxy to estimate a distance of $\sim$ 200 Mpc for several coalescences per year for those classes of NS/NS binaries that have been found in our galaxy as binary pulsars. Bailes [16], on the basis of more recent binary pulsar searches, has argued that this distance should be increased by a factor 2.

By comparing these estimates with the signal strengths in Fig. 3, we see that (i) the first interferometers in LIGO/VIRGO have a possibility but not a high probability of seeing NS/NS coalescences; (ii) advanced interferometers are almost certain of seeing them (the requirement that this be so was one factor that forced the LIGO/VIRGO arm lengths to be so long, several kilometers); and (iii) they are most likely to be discovered roughly half-way between the first and advanced interferometers—which means by an improved variant of the first interferometers several years after LIGO operations begin.

We have no good observational handle on the coalescence rate of NS/BH or BH/BH binaries. However, estimates based on our galaxy's main-sequence binary population and evolutionary considerations suggest a distance $\sim$ 200 Mpc for several NS/BH or BH/BH coalescences per year [11, 15, 12, 14]. This estimate should be regarded as a plausible upper limit on the event rate and lower limit on the distance to look [11, 15]. If this estimate is correct, then NS/BH and BH/BH binaries will be seen before NS/NS, and might be seen by the first LIGO/VIRGO interferometers or soon thereafter; cf. Fig. 3. However, this estimate is far less certain than the (rather uncertain) NS/NS estimates!

Once coalescence waves have been discovered, each further improvement of sensitivity by a factor 2 will increase the event rate by $2^3 \simeq 10$. Assuming a rate of several NS/NS per year at 200 Mpc, the advanced interferometers of Fig. 3 should see $\sim$ 100 per year.

4.1.3 Inspiral Waveforms and the Information They Can Bring

Neutron stars and black holes have such intense self gravity that it is exceedingly difficult to deform them. Correspondingly, as they spiral inward in a compact binary, they do not gravitationally deform each other significantly until several orbits before their final coalescence [17, 18]. This means that the inspiral waveforms are determined to high accuracy by only a few, clean parameters: the masses and spin angular momenta of the bodies, and the initial orbital elements (i.e. the elements when the waves enter the LIGO/VIRGO band).

Though tidal deformations are negligible during inspiral, relativistic effects can be very important. If, for the moment, we ignore the relativistic effects—i.e., if we approximate gravity as Newtonian and the wave generation as due to the binary's oscillating quadrupole moment [1], then the shapes of the inspiral waveforms $h_+(t)$ and $h_\times(t)$ are as shown in Fig. 4.

The left-hand graph in Fig. 4 shows the waveform increasing in amplitude and sweeping upward in frequency (i.e., undergoing a "chirp") as the binary's bodies spiral closer and closer together. The ratio of the amplitudes of the two polarizations is determined by the inclination ι of the orbit to our line of sight (lower right in Fig. 4). The shapes of the individual waves, i.e. the waves' harmonic content, are determined by the orbital eccentricity (upper right). (Binaries produced by normal stellar evolution should be highly circular due to past radiation reaction forces, but compact binaries that form by capture events, in dense star clusters that might reside in galactic nuclei [19], could be quite eccentric.) If, for simplicity, the or-

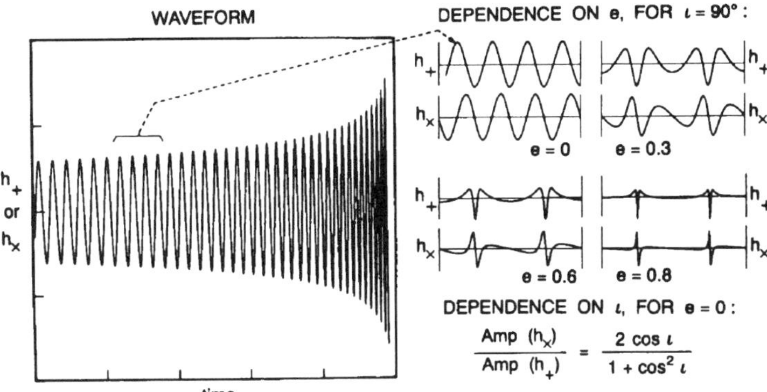

Figure 4: Waveforms from the inspiral of a compact binary, computed using Newtonian gravity for the orbital evolution and the quadrupole moment approximation for the wave generation. (From Ref. [7].)

bit is circular, then the rate at which the frequency sweeps or "chirps", df/dt [or equivalently the number of cycles spent near a given frequency, $n = f^2(df/dt)^{-1}$] is determined solely, in the Newtonian/quadrupole approximation, by the binary's so-called *chirp mass*, $M_c \equiv (M_1 M_2)^{3/5}/(M_1 + M_2)^{1/5}$ (where M_1 and M_2 are the two bodies' masses). The amplitudes of the two waveforms are determined by the chirp mass, the distance to the source, and the orbital inclination. Thus (in the Newtonian/quadrupole approximation), by measuring the two amplitudes, the frequency sweep, and the harmonic content of the inspiral waves, one can determine as direct, resulting observables, the source's distance, chirp mass, inclination, and eccentricity [20, 21].

As in binary pulsar observations [10], so also here, relativistic effects add further information: they influence the rate of frequency sweep and produce waveform modulations in ways that depend on the binary's dimensionless ratio $\eta = \mu/M$ of reduced mass $\mu = M_1 M_2/(M_1 + M_2)$ to total mass $M = M_1 + M_2$ [22] and on the spins of the binary's two bodies [23]; These relativistic effects are reviewed and discussed at length in Refs. [24, 25]. Two deserve special mention: (i) As the waves emerge from the binary, some of them get backscattered one or more times off the binary's spacetime curvature, producing wave *tails*. These tails act back on the binary, modifying its inspiral rate in a measurable way. (ii) If the orbital plane is inclined to one or both of the binary's spins, then the spins drag inertial frames in the binary's vicinity (the "Lense-Thirring effect"), this frame dragging causes the orbit to precess, and the precession modulates the waveforms [24, 26, 27].

Remarkably, the relativistic corrections to the frequency sweep will be measurable with very high accuracy, even though they are typically $\lesssim 10$ per cent of the Newtonian contribution, and even though the typical signal to noise ratio will be only ~ 9 even after optimal signal processing. The reason is as follows [28, 29, 24]:

The frequency sweep will be monitored by the method of "matched filters"; in other words, the incoming, noisy signal will be cross correlated with theoretical templates. If the signal and the templates gradually get out of phase with each other by more than $\sim 1/10$ cycle as the waves sweep through the LIGO/VIRGO band, their cross correlation will be significantly reduced. Since the total number of cycles spent in the LIGO/VIRGO band will be $\sim 16,000$ for a NS/NS binary, ~ 3500 for NS/BH, and ~ 600 for BH/BH, this means that LIGO/VIRGO should be able to measure the frequency sweep to a fractional precision $\lesssim 10^{-4}$, compared to which the relativistic effects are very large. (This is essentially the same method as Joseph Taylor and colleagues use for high-accuracy radio-wave measurements of relativistic effects in binary pulsars [10].)

Preliminary analyses, using the theory of optimal signal processing, predict the following typical accuracies for LIGO/VIRGO measurements based solely on the frequency sweep (i.e., ignoring modulational information) [30, 28, 29, 31, 24]: (i) The chirp mass M_c will typically be measured, from the Newtonian part of the frequency sweep, to $\sim 0.04\%$ for a NS/NS binary and $\sim 0.3\%$ for a system containing at least one BH. (ii) *If* we are confident (e.g., on a statistical basis from measurements of many previous binaries) that the spins are a few percent or less of the maximum physically allowed, then the reduced mass μ will be measured to $\sim 1\%$ for NS/NS and NS/BH binaries, and $\sim 3\%$ for BH/BH binaries. (Here and below NS means a $\sim 1.4 M_\odot$ neutron star and BH means a $\sim 10 M_\odot$ black hole.) (iii) Because the frequency dependences of the (relativistic) μ effects and spin effects are not sufficiently different to give a clean separation between μ and the spins, if we have no prior knowledge of the spins, then the spin/μ correlation will worsen the typical accuracy of μ by a large factor, to $\sim 30\%$ for NS/NS, $\sim 50\%$ for NS/BH, and a factor ~ 2 for BH/BH [30, 28]. These worsened accuracies might be improved somewhat by waveform modulations caused by the spin-induced precession of the orbit [26, 27], and even without modulational information, a certain combination of μ and the spins will be determined to a few per cent. Much additional theoretical work is needed to firm up the measurement accuracies.

LIGO/VIRGO observations of compact binary inspiral have the potential to bring us far more information than just binary masses and spins:

- They can be used for high-precision tests of general relativity. In scalar-tensor theories (some of which are highly attractive alternatives to general relativity [32]), radiation reaction due to emission of scalar waves places a unique signature on those waves that LIGO/VIRGO would detect—a signature that can be searched for with high precision [33].

- They can be used to measure the Hubble constant, deceleration parameter, and cosmological constant [20, 21, 34, 35]. The keys to such measurements are that (i) advanced interferometers in LIGO/VIRGO will be able to see NS/NS out to cosmological redshifts $z \sim 0.3$, and NS/BH out to $z \sim 2$. (ii) The direct observables that can be extracted from the observed waves include the source's luminosity distance r_L (measured to accuracy ~ 10 per cent in a large fraction of cases), and its direction on the sky (to accuracy ~ 1 square degree)—accuracies good enough that only one or a few electromagnetically-observed clusters of galaxies should fall within the 3-dimensional gravitational error boxes, thereby giving promise to joint gravitational/electromagnetic statistical studies. (iii) Another direct gravitational observable is $(1 + z)M$ where z is redshift and M is any mass in the system (measured to the accuracies quoted above). Since the masses of NS's in binaries seem to cluster around $1.4M_{\odot}$, measurements of $(1 + z)M$ can provide a handle on the redshift, even in the absence of electromagnetic aid.

- For a NS or small BH spiraling into a massive ~ 50 to $500M_{\odot}$ BH, the inspiral waves will carry a "map" of the spacetime geometry around the big hole—a map that can be used, e.g., to test the theorem that "a black hole has no hair" [36]; cf. Sec. 6.3 below.

4.1.4 Coalescence Waveforms and their Information

The waves from the binary's final coalescence can bring us new types of information.

BH/BH Coalescence: In the case of a BH/BH binary, the coalescence will excite large-amplitude, highly nonlinear vibrations of spacetime curvature near the coalescing black-hole horizons—a phenomenon of which we have very little theoretical understanding today. Especially fascinating will be the case of two spinning black holes whose spins are not aligned with each other or with the orbital angular momentum. Each of the three angular momentum vectors (two spins, one orbital) will drag space in its vicinity into a tornado-like swirling motion—the general relativistic "dragging of inertial frames," so the binary is rather like two tornados with orientations skewed to each other, embedded inside a third, larger tornado with a third orientation. The dynamical evolution of such a complex configuration of spacetime warpage (as revealed by its emitted waves) may well bring us surprising new insights into relativistic gravity. Moreover, if the sum of the BH masses is fairly large, ~ 40 to $200M_{\odot}$, then the waves should come off in a frequency range $f \sim 40$ to 200 Hz where the LIGO/VIRGO broad-band interferometers have their best sensitivity and can best extract the information the waves carry.

To get full value out of such wave observations will require having theoretical computations with which to compare them [37]. There is no hope to perform such computations analytically; they can only be done as supercomputer simulations. The development of such simulations is a major effort within the world's relativity community.

NS/NS Coalescence: The final coalescence of NS/NS binaries should produce waves that are sensitive to the equation of state of nuclear matter, so such coalescences have the potential to teach us about the nuclear equation of state [24]. In essence, we will be studying nuclear physics via the collisions of atomic nuclei that have nucleon numbers $A \sim 10^{57}$—somewhat larger than physicists are normally accustomed to. The accelerator used to drive these nuclei up to the speed of light is the binary's self gravity, and the radiation by which the details of the collisions are probed is gravitational.

A number of research groups [38, 39, 40, 41, 17, 18, 42, 43] are engaged in numerical astrophysics simulations of NS/NS coalescence, with the goal not only to predict the emitted gravitational waveforms and their dependence on equation of state, but also (more immediately) to learn whether such coalescences might power the γ-ray bursts that have been a major astronomical puzzle since their discovery in the early 1970s. If advanced LIGO interferometers were now in operation, they could report definitively whether or not the γ-bursts are produced by NS/NS binaries; and if the answer were yes, then the combination of γ-burst data and gravitational-wave data could bring valuable information that neither could bring by itself. For example, we could determine when, to within a few msec, the γ-burst is emitted relative to the moment the NS's first begin to touch; and by comparing the γ and gravitational times of arrival, we might test whether gravitational waves propagate with the speed of light to a fractional precision of $\sim 0.01\text{sec}/3 \times 10^9\,\text{lyr} \sim 10^{-19}$.

Unfortunately, the final NS/NS coalescence will emit its gravitational waves in the kHz frequency band ($800\text{Hz} \lesssim f \lesssim 2500\text{Hz}$) where photon shot noise will prevent them from being studied by the standard, "workhorse," broad-band interferometers of Fig. 3. However, a specially configured ("dual-recycled") interferometer invented by Brian Meers [44], which could have enhanced sensitivity in the kHz region at the price of reduced sensitivity elsewhere, may be able to measure the waves and extract their equation of state information, as might massive, spherical bar detectors [24, 45]. Such measurements will be very difficult and are likely only when the LIGO/VIRGO network has reached a mature stage.

4.2 Stellar Core Collapse and Supernovae

Several features of the stellar core collapse, which triggers supernovae, can produce significant gravitational radiation in the high-frequency band. We shall consider these features in turn, the most weakly radiating first.

4.2.1 Boiling of the Newborn Neutron Star

Even if the collapse is spherical, so it cannot radiate any gravitational waves at all, it should produce a convectively unstable neutron star that "boils" vigorously (and nonspherically) for the first ~ 0.1 second of its life [46]. The boiling dredges up high-temperature nuclear matter ($T \sim 10^{12}\text{K}$) from the neutron star's central regions, bringing it to the surface (to the "neutrino-sphere"), where it cools by neutrino emission before being swept back downward and reheated. Burrows estimates [47, 48] that the boiling should generate $n \sim 10$ cycles of gravitational waves with frequency $f \sim 100\text{Hz}$ and amplitude $h \sim 3 \times 10^{-22}(30\text{kpc}/r)$ (where r is the distance to the source), corresponding to a characteristic amplitude $h_c \simeq h\sqrt{n} \sim 10^{-21}(30\text{kpc}/r)$; cf. Fig. 5. LIGO/VIRGO will be able to detect such waves only in the local group of galaxies, where the supernova rate is probably no larger than ~ 1 each 10 years. However, neutrino detectors have a similar range, and there could be a high scientific payoff from correlated observations of the gravitational waves emitted by the boiling's mass motions and neutrinos emitted from the boiling neutrino-sphere.

4.2.2 Axisymmetric Collapse, Bounce, and Oscillations

Rotation will centrifugally flatten the collapsing core, enabling it to radiate as it implodes. If the core's angular momentum is small enough that centrifugal forces do not halt or strongly slow the collapse before it reaches nuclear densities, then the

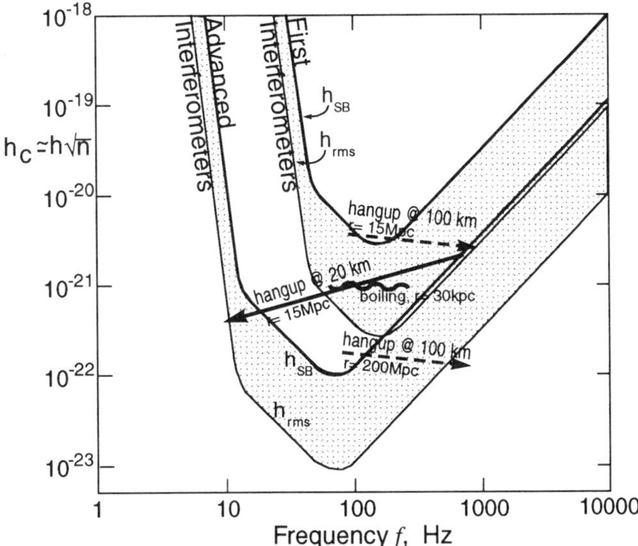

Figure 5: Characteristic amplitudes of the gravitational waves from various processes accompanying stellar core collapse and supernovae, compared with projected sensitivities of LIGO's interferometers.

core's collapse, bounce, and subsequent oscillations are likely to be axially symmetric. Numerical simulations [49, 50] show that in this case the waves from collapse, bounce, and oscillation will be quite weak: the total energy radiated as gravitational waves is not likely to exceed $\sim 10^{-7}$ solar masses (about 1 part in a million of the collapse energy) and might often be much less than this; and correspondingly, the waves' characteristic amplitude will be $h_c \lesssim 3 \times 10^{-21}(30\text{Mpc}/r)$. These collapse-and-bounce waves will come off at frequencies ~ 200 Hz to ~ 1000 Hz, and will precede the boiling waves by a fraction of a second. Like the boiling waves, they probably cannot be seen by LIGO/VIRGO beyond the local group of galaxies and thus will be a very rare occurrence.

4.2.3 Rotation-Induced Bars and Break-Up

If the core's rotation is large enough to strongly flatten the core before or as it reaches nuclear density, then a dynamical and/or secular instability is likely to break the core's axisymmetry. The core will be transformed into a bar-like configuration that spins end-over-end like an American football, and that might even break up into two or more massive pieces. In this case, the radiation from the spinning bar or orbiting pieces *could* be almost as strong as that from a coalescing neutron-star binary, and thus could be seen by the LIGO/VIRGO first interferometers out to the distance of the Virgo cluster (where the supernova rate is several per year) and by advanced interferometers out to several hundred Mpc (supernova rate $\sim 10^4$ per year); cf. Fig. 5. It is far from clear what fraction of collapsing cores will have enough angular momentum to break their axisymmetry, and what fraction of those will actually radiate at this high rate; but even if only $\sim 1/1000$ or $1/10^4$ do so, this could ultimately be a very interesting source for LIGO/VIRGO.

Several specific scenarios for such non-axisymmetry have been identified:

Centrifugal hangup at $\sim$ 100km radius: If the pre-collapse core is rapidly spinning (e.g., if it is a white dwarf that has been spun up by accretion from a companion), then the collapse may produce a highly flattened, centrifugally supported disk with most of its mass at radii $R \sim 100$km, which then (via instability) may transform itself into a bar or may bifurcate. The bar or bifurcated lumps will radiate gravitational waves at twice their rotation frequency, $f \sim 100$Hz — the optimal frequency for LIGO/VIRGO interferometers. To shrink on down to $\sim$ 10km size, this configuration must shed most of its angular momentum. *If* a substantial fraction of the angular momentum goes into gravitational waves, then independently of the strength of the bar, the waves will be nearly as strong as those from a coalescing binary. The reason is this: The waves' amplitude h is proportional to the bar's ellipticity e, the number of cycles n of wave emission is proportional to $1/e^2$, and the characteristic amplitude $h_c = h\sqrt{n}$ is thus independent of the ellipticity and is about the same whether the configuration is a bar or is two lumps [21]. The resulting waves will thus have h_c roughly half as large, at $f \sim 100$Hz, as those from a NS/NS binary (half as large because each lump might be half as massive as a NS), and they will chirp upward in frequency in a manner similar to those from a binary.

It is rather likely, however, that most of the excess angular momentum does *not* go into gravitational waves, but instead goes largely into hydrodynamic waves as the bar or lumps, acting like a propeller, stir up the surrounding stellar mantle. In this case, the radiation will be correspondingly weaker.

Centrifugal hangup at R $\sim$ 20km: Lai and Shapiro [51] have explored the case of centrifugal hangup at radii not much lager than the final neutron star, say $R \sim 20$km. Using compressible ellipsoidal models, they have deduced that, after a brief period of dynamical bar-mode instability with wave emission at $f \sim 1000$Hz (explored by Houser, Centrella, and Smith [52]), the star switches to a secular instability in which the bar's angular velocity gradually slows while the material of which it is made retains its high rotation speed and circulates through the slowing bar. The slowing bar emits waves that sweep *downward* in frequency through the LIGO/VIRGO optimal band $f \sim 100$Hz, toward $\sim$ 10Hz. The characteristic amplitude (Fig. 5) is only modestly smaller than for the upward-sweeping waves from hangup at $R \sim 100$km, and thus such waves should be detectable near the Virgo Cluster by the first LIGO/VIRGO interferometers, and at distances of a few 100Mpc by advanced interferometers.

Successive fragmentations of an accreting, newborn neutron star: Bonnell and Pringle [53] have focused on the evolution of the rapidly spinning, newborn neutron star as it quickly accretes more and more mass from the pre-supernova star's inner mantle. If the accreting material carries high angular momentum, it may trigger a renewed bar formation, lump formation, wave emission, and coalescence, followed by more accretion, bar and lump formation, wave emission, and coalescence. Bonnell and Pringle speculate that hydrodynamics, not wave emission, will drive this evolution, but that the total energy going into gravitational waves might be as large as $\sim 10^{-3}M_{\odot}$. This corresponds to $h_c \sim 10^{-21}(10\text{Mpc}/r)$.

4.3 Spinning Neutron Stars; Pulsars

As the neutron star settles down into its final state, its crust begins to solidify (crystalize). The solid crust will assume nearly the oblate axisymmetric shape that centrifugal forces are trying to maintain, with poloidal ellipticity $\epsilon_p \propto$(angular velocity of rotation)^2. However, the principal axis of the star's moment of inertia tensor may deviate from its spin axis by some small "wobble angle" θ_w, and the star may deviate slightly from axisymmetry about its principal axis; i.e., it may have a slight ellipticity ϵ_e in its equatorial plane.

As this slightly imperfect crust spins, it will radiate gravitational waves [54]: ϵ_e radiates at twice the rotation frequency, $f = 2f_{\text{rot}}$ with $h \propto \epsilon_e$, and the wobble angle couples to ϵ_p to produce waves at $f = f_{\text{rot}} + f_{\text{prec}}$ (the precessional sideband of the rotation frequency) with amplitude $h \propto \theta_w \epsilon_p$. For typical neutron-star masses and moments of inertia, the wave amplitudes are

$$h \sim 6 \times 10^{-25} \left(\frac{f_{\text{rot}}}{500\text{Hz}}\right)^2 \left(\frac{1\text{kpc}}{r}\right) \left(\frac{\epsilon_e \text{ or } \theta_w \epsilon_p}{10^{-6}}\right) \tag{3}$$

The neutron star gradually spins down, due in part to gravitational-wave emission but perhaps more strongly due to electromagnetic torques associated with its spinning magnetic field and pulsar emission. This spin-down reduces the strength of centrifugal forces, and thereby causes the star's poloidal ellipticity ϵ_p to decrease, with an accompanying breakage and resolidification of its crust's crystal structure (a "starquake") [55]. In each starquake, θ_w, ϵ_e, and ϵ_p will all change suddenly, thereby changing the amplitudes of the star's two gravitational "spectral lines" $f = 2f_{\text{rot}}$ and $f = f_{\text{rot}} + f_{\text{prec}}$. After each quake, there should be a healing period in which the star's fluid core and solid crust, now rotating at different speeds, gradually regain synchronism. By monitoring the amplitudes, frequencies, and phases of the two gravitational-wave spectral lines, and by comparing with timing of the electromagnetic pulsar emission, one might learn much about the physics of the neutron-star interior.

How large will the quantities ϵ_e and $\theta_w \epsilon_p$ be? Rough estimates of the crustal shear moduli and breaking strengths suggest an upper limit in the range $\epsilon_{\max} \sim 10^{-4}$ to 10^{-6}, and it might be that typical values are far below this. We are extremely ignorant, and correspondingly there is much to be learned from searches for gravitational waves from spinning neutron stars.

One can estimate the sensitivity of LIGO/VIRGO (or any other broad-band detector) to the periodic waves from such a source by multiplying the waves' amplitude h by the square root of the number of cycles over which one might integrate to find the signal, $n = f\hat{\tau}$ where $\hat{\tau}$ is the integration time. The resulting effective signal strength, $h\sqrt{n}$, is larger than h by

$$\sqrt{n} = \sqrt{f\hat{\tau}} = 10^5 \left(\frac{f}{1000\text{Hz}}\right)^{1/2} \left(\frac{\hat{\tau}}{4\text{months}}\right)^{1/2} \tag{4}$$

This $h\sqrt{n}$ should be compared (i) to the detector's rms broad-band noise level for sources in a random direction, $\sqrt{5}h_{\text{rms}}$, to deduce a signal-to-noise ratio, or (ii) to

$h_{\rm SB}$ to deduce a sensitivity for high-confidence detection when one does not know the waves' frequency in advance [1]. Such a comparison suggests that the first interferometers in LIGO/VIRGO might possibly see waves from nearby spinning neutron stars, but the odds of success are very unclear.

The deepest searches for these nearly periodic waves will be performed by narrow-band detectors, whose sensitivities are enhanced near some chosen frequency at the price of sensitivity loss elsewhere—e.g., "dual recycled" interferometers or resonant bars. With "advanced-detector technology," dual-recycled interferometers might be able to detect with confidence all spinning neutron stars that have [1]

$$(\epsilon_e \text{ or } \theta_w \epsilon_p) \gtrsim 3 \times 10^{-10} \left(\frac{500\text{Hz}}{f_{\rm rot}}\right)^2 \left(\frac{r}{1000\text{pc}}\right)^2 \tag{5}$$

There may well be a large number of such neutron stars in our galaxy; but it is also conceivable that there are none. We are extremely ignorant.

5 LISA and the Low-Frequency Band

In the 2014 time frame, the European Space Agency (ESA) and/or NASA is likely to fly a *Laser Interferometer Space Antenna* (LISA) which will achieve remarkably good sensitivities in the low-frequency band, 10^{-4}Hz to 1 Hz.

5.1 Mission Status

LISA is largely an outgrowth of 15 years of studies by Peter Bender and colleagues at the University of Colorado. Unfortunately, the prospects for NASA to fly such a mission have not looked good in the early 1990s. By contrast, prospects in Europe have looked much better, so a largely European consortium was put together in 1993 with Bender's participation but under the leadership of Karsten Danzmann (Hannover) and James Hough (Glasgow), to propose LISA to the European Space Agency. This proposal has met with considerable success; LISA might well achieve approval to fly as an ESA Cornerstone Mission around 2014 [56]. Members of the American gravitation community hope that NASA will join together with ESA in this endeavor, and that working jointly, ESA and NASA will be able to fly LISA considerably sooner than 2014.

5.2 Mission Configuration

As presently conceived, LISA will consist of six compact, drag-free spacecraft (i.e. spacecraft that are shielded from buffeting by solar wind and radiation pressure, and that thus move very nearly on geodesics of spacetime). All six spacecraft would be launched simultaneously by a single Ariane rocket. They would be placed into the same heliocentric orbit as the Earth occupies, but would follow $20^{\rm o}$ behind the Earth; cf. Fig. 6. The spacecraft would fly in pairs, with each pair at the vertex of an equilateral triangle that is inclined at an angle of $60^{\rm o}$ to the Earth's orbital plane. The triangle's arm length would be 5 million km (10^6 times larger than LIGO's arms!). The six spacecraft would track each other optically, using one-Watt

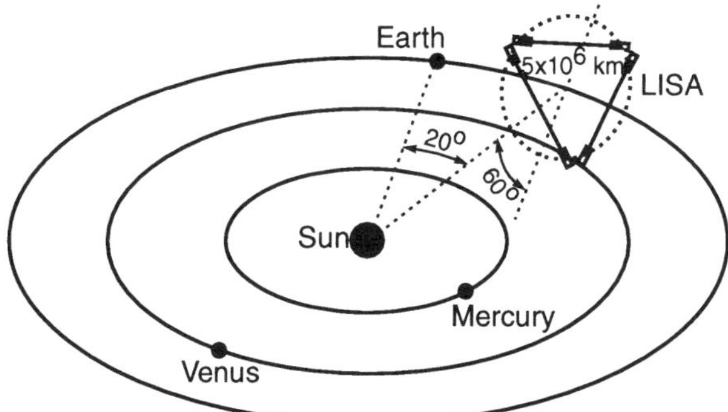

Figure 6: LISA's orbital configuration.

YAG laser beams. Because of diffraction losses over the 5×10^6km arm length, it is not feasible to reflect the beams back and forth between mirrors as is done with LIGO. Instead, each spacecraft will have its own laser; and the lasers will be phase locked to each other, thereby achieving the same kind of phase-coherent out-and-back light travel as LIGO achieves with mirrors. The six-laser, six-spacecraft configuration thereby functions as three, partially independent but partially redundant, gravitational-wave interferometers.

5.3 Noise and Sensitivity

Figure 7 depicts the expected noise and sensitivity of LISA in the same language as we have used for LIGO (Fig. 3). The curve at the bottom of the stippled region is $h_{\rm rms}$, the rms noise, in a bandwidth equal to frequency, for waves with optimum direction and polarization. The top of the stippled region is $h_{\rm SB} = 5\sqrt{5}h_{\rm rms}$, the sensitivity for high-confidence detection ($S/N = 5$) of a broad-band burst coming from a random direction, assuming Gaussian noise.

At frequencies $f \gtrsim 10^{-3}$Hz, LISA's noise is due to photon counting statistics (shot noise). The noise curve steepens at $f \sim 3 \times 10^{-2}$Hz because at larger f than that, the waves' period is shorter than the round-trip light travel time in one of LISA's arms. Below 10^{-3}Hz, the noise is due to buffeting-induced random motions of the spacecraft that are not being properly removed by the drag-compensation system. Notice that, in terms of dimensionless amplitude, LISA's sensitivity is roughly the same as that of LIGO's first interferometers (Fig. 3), but at 100,000 times lower frequency. Since the waves' energy flux scales as f^2h^2, this corresponds to 10^{10} better energy sensitivity than LIGO.

5.4 Observational Strategy

LISA can detect and study, simultaneously, a wide variety of different sources scattered over all directions on the sky. The key to distinguishing the different sources is the different time evolution of their waveforms. The key to determining each source's direction, and confirming that it is real and not just noise, is the manner

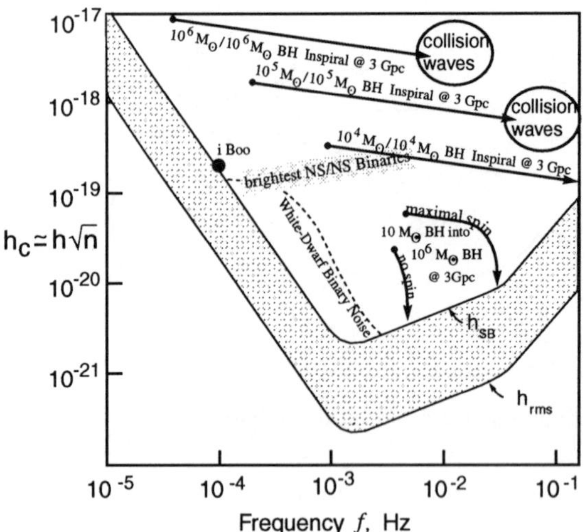

Figure 7: LISA's projected broad-band noise h_{rms} and sensitivity to bursts h_{SB}, compared with the strengths of the waves from several low-frequency sources. [*Note:* When members of the LISA team plot curves analogous to this, they show the sensitivity curve (top of stippled region) in units of the amplitude of a periodic signal that can be detected with $S/N = 5$ in one year of integration; that sensitivity to periodic sources is related to the h_{SB} used here by $h_{SP} = h_{SB}/\sqrt{f \cdot 3 \times 10^7 \text{sec.}}$]

in which its waves' amplitude and frequency are modulated by LISA's complicated orbital motion—a motion in which the interferometer triangle rotates around its center once per year, and the interferometer plane rotates around the normal to the Earth's orbit once per year. Most sources will be observed for a year or longer, thereby making full use of these modulations.

6 Low-Frequency Gravitational-Wave Sources

6.1 Waves from Binary Stars

LISA has a large class of guaranteed sources: short-period binary stars in our own galaxy. A specific example is the classic binary 44 i Boo (HD133640), a $1.35M_\odot/0.68M_\odot$ system just 12 parsecs from Earth, whose wave frequency f and characteristic amplitude $h_c = h\sqrt{n}$ are depicted in Fig. 7. (Here h is the waves' actual amplitude and $n = f\hat{\tau}$ is the number of wave cycles during $\hat{\tau}$ =1 year of signal integration.) Since 44 i Boo lies right on the h_{SB} curve, its signal to noise ratio in one year of integration should be $S/N = 5$.

To have an especially short period, a binary must be made of especially compact bodies—white dwarfs (WD), neutron stars (NS), and/or black holes (BH). WD/WD binaries are thought to be so numerous that they might produce a stochastic background of gravitational waves, at the level shown in Fig. 7, that will hide some other interesting waves from view [57]. Since WD/WD binaries are very dim optically, their actual numbers are not known for sure; Fig. 7 might be an overestimate.

Assuming a NS/NS coalescence rate of 1 each 10^5 years in our galaxy [11, 15], the shortest period NS/NS binary should have a remaining life of about 5×10^4 years, corresponding to a gravitational-wave frequency today of $f \simeq 5 \times 10^{-3}$Hz, an amplitude (at about 10kpc distance) $h \simeq 4 \times 10^{-22}$, and a characteristic amplitude (with one year of integration time) $h_c \simeq 2 \times 10^{-19}$. This is depicted in Fig. 7 at the right edge of the region marked "brightest NS/NS binaries". These brightest NS/NS binaries can be studied by LISA with the impressive signal to noise ratios $S/N \sim 50$ to 500.

6.2 Waves from the Coalescence of Massive Black Holes in Distant Galaxies

LISA would be a powerful instrument for studying massive black holes in distant galaxies. Figure 7 shows, as examples, the waves from several massive black hole binaries at 3Gpc distance. The waves sweep upward in frequency (rightward in the diagram) as the holes spiral together. The black dots show the waves' frequency one year before the holes' final collision and coalescence, and the arrowed lines show the sweep of frequency and characteristic amplitude $h_c = h\sqrt{n}$ during that last year. For simplicity, the figure is restricted to binaries with equal-mass black holes: $10^4 M_\odot/10^4 M_\odot$, $10^5 M_\odot/10^5 M_\odot$, and $10^6 M_\odot/10^6 M_\odot$.

By extrapolation from these three examples, we see that LISA can study much of the last year of inspiral, and the waves from the final collision and coalescence, whenever the holes' masses are in the range $3 \times 10^4 M_\odot \lesssim M \lesssim 10^8 M_\odot$. Moreover, LISA can study the final coalescences with remarkable signal to noise ratios: $S/N \gtrsim 1000$. Since these are much larger S/N's than LIGO/VIRGO is likely to achieve, we can expect LISA to refine the experimental understanding of black-hole physics, and of highly nonlinear vibrations of warped spacetime, which LIGO/VIRGO initiates—*provided* the rate of massive black-hole coalescences is of order one per year in the Universe or higher. The rate might well be that high, but it also might be much lower.

By extrapolating Fig. 7 to lower BH/BH masses, we see that LISA can observe the last few years of inspiral, but not the final collisions, of binary black holes in the range $100 M_\odot \lesssim M \lesssim 10^4 M_\odot$, out to cosmological distances.

Extrapolating the BH/BH curves to lower frequencies using the formula (time to final coalescence) $\propto f^{-8/3}$, we see that equal-mass BH/BH binaries enter LISA's frequency band roughly 1000 years before their final coalescences, more or less independently of their masses, for the range $100 M_\odot \lesssim M \lesssim 10^6 M_\odot$. Thus, if the coalescence rate were to turn out to be one per year, LISA would see roughly 1000 additional massive binaries that are slowly spiraling inward, with inspiral rates df/dt readily measurable. From the inspiral rates, the amplitudes of the two polarizations, and the waves' harmonic content, LISA can determine each such binary's luminosity distance, redshifted chirp mass $(1+z)M_c$, orbital inclination, and eccentricity; and from the waves' modulation by LISA's orbital motion, LISA can learn the direction to the binary with an accuracy of some tens of arcminutes.

6.3 Waves from Compact Bodies Spiraling into Massive Black Holes in Distant Galaxies

When a compact body with mass μ spirals into a much more massive black hole with mass M, the body's orbital energy E at fixed frequency f (and correspondingly at fixed orbital radius a) scales as $E \propto \mu$, the gravitational-wave luminosity $\dot{E}$ scales as $\dot{E} \propto \mu^2$, and the time to final coalescence thus scales as $t \sim E/\dot{E} \propto 1/\mu$. This means that the smaller is μ/M, the more orbits are spent in the hole's strong-gravity region, $a \lesssim 10GM/c^2$, and thus the more detailed and accurate will be the map of the hole's spacetime geometry, which is encoded in the emitted waves.

For holes observed by LIGO/VIRGO, the most extreme mass ratio that we can hope for is $\mu/M \sim 1M_\odot/300M_\odot$, since for $M > 300M_\odot$ the inspiral waves are pushed to frequencies below the LIGO/VIRGO band. This limit on μ/M seriously constrains the accuracy with which LIGO/VIRGO can hope to map out the spacetime geometries of black holes and test the black-hole no-hair theorem (Sec. 4.1). By contrast, LISA can observe the final inspiral waves from objects of any mass $M \gtrsim 0.5M_\odot$ spiraling into holes of mass $3 \times 10^5 M_\odot \lesssim M \lesssim 3 \times 10^7 M_\odot$.

Figure 7 shows the example of a $10M_\odot$ black hole spiraling into a $10^6 M_\odot$ hole at 3Gpc distance. The inspiral orbit and waves are strongly influenced by the hole's spin. Two cases are shown [58]: an inspiraling circular orbit around a non-spinning hole, and a prograde, circular, equatorial orbit around a maximally spinning hole. In each case the dot at the upper left end of the arrowed curve is the frequency and characteristic amplitude one year before the final coalescence. In the nonspinning case, the small hole spends its last year spiraling inward from $r \simeq 7.4GM/c^2$ (3.2 Schwarzschild radii) to its last stable circular orbit at $r = 6GM/c^2$ (3 Schwarzschild radii). In the maximal spin case, the last year is spent traveling from $r = 6GM/c^2$ (3 Schwarzschild radii) to the last stable orbit at $r = GM/c^2$ (half a Schwarzschild radius). The $\sim 10^5$ cycles of waves during this last year should carry, encoded in themselves, rather accurate values for the massive hole's lowest few multipole moments [59]. If the measured moments satisfy the "no-hair" theorem (i.e., if they are all determined uniquely by the measured mass and spin in the manner of the Kerr metric), then we can be sure the central body is a black hole. If they violate the no-hair theorem, then (assuming general relativity is correct), either the central body was not a black hole, or an accretion disk or other material was perturbing its orbit [60]. ¿From the evolution of the waves one can hope to determine which is the case, and to explore the properties of the central body and its environment.

Models of galactic nuclei, where massive holes reside, suggest that inspiraling stars and small holes typically will be in rather eccentric orbits [61]. This is because they get injected into such orbits via gravitational deflections off other stars, and by the time gravitational radiation reaction becomes the dominant orbital driving force, there is not enough inspiral left to fully circularize their orbits. Such orbital eccentricity will complicate the waveforms and complicate the extraction of information from them. Efforts to understand the emitted waveforms are just now getting underway.

The event rates for inspiral into massive black holes are not at all well understood. However, since a significant fraction of all galactic nuclei are thought to contain massive holes, and since white dwarfs and neutron stars, as well as small black holes, can withstand tidal disruption as they plunge toward the massive hole's horizon, and since LISA can see inspiraling bodies as small as $\sim 0.5M_{\odot}$ out to 3Gpc distance, the event rate is likely to be interestingly large.

7 Conclusion

It is now 35 years since Joseph Weber initiated his pioneering development of gravitational-wave detectors [62] and 25 years since Forward and Weiss initiated work on interferometric detectors. Since then, hundreds of talented experimental physicists have struggled to improve the sensitivities of these instruments. At last, success is in sight. If the source estimates described in this lecture are approximately correct, then the planned interferometers should detect the first waves in 2001 or several years thereafter, thereby opening up this rich new window onto the Universe. One payoff should be deep new insights into compact astrophysical bodies and their roles in binary systems.

8 Acknowledgments

My group's research on gravitational waves from compact bodies and the waves' relevance to LIGO/VIRGO and LISA is supported in part by NSF grants AST-9417371 and PHY-9424337 and by NASA grant NAGW-4268. This written version of my lecture has been adapted from a longer review article that I am writing for the Proceedings of the Snowmass '94 Summer Study on Particle and Nuclear Astrophysics and Cosmology in the Next Millenium.

References

[1] K. S. Thorne. In S. W. Hawking and W. Israel, editors, *Three Hundred Years of Gravitation*, pages 330–458. Cambridge University Press, 1987.

[2] K. S. Thorne. In N. Deruelle and T. Piran, editors, *Gravitational Radiation*, page 1. North Holland, 1983.

[3] W. J. Sullivan. *The Early Years of Radio Astronomy.* Cambridge University Press, 1984.

[4] V. M. Kaspi, J. H. Taylor, and M. F. Ryba. *Astrophys. J.*, 428:713, 1994.

[5] L. M. Krauss and M. White. *Phys. Rev. Lett.*, 69:969, 1992.

[6] R. L. Davis, H. M. Hodges, G. F. Smoot, P. J. Steinhardt, and M. S. Turner. *Phys. Rev. Lett.*, 69:1856, 1992.

[7] A. Abramovici et. al. *Science*, 256:325, 1992.

[8] C. Bradaschia et. al. *Nucl. Instrum. & Methods*, A289:518, 1990.

[9] R. A. Hulse and J. H. Taylor. *Astrophys. J.*, 324:355, 1975.

[10] J. H. Taylor. *Rev. Mod. Phys.*, 66:711, 1994.

[11] E. S. Phinney. *Astrophys. J.*, 380:L17, 1991.

[12] A. V. Tutukov and L. R. Yungelson. *Mon. Not. Roy. Astron. Soc.*, 260:675, 1993.

[13] H. Yamaoka, T. Shigeyama, and K. Nomoto. *Astron. Astrophys.*, 267:433, 1993.

[14] V. M. Lipunov, K. A. Postnov, and M. E. Prokhorov. *Astrophys. J.*, 423:L121, 1994. and related, unpublished work.

[15] R. Narayan, T. Piran, and A. Shemi. *Astrophys. J.*, 379:L17, 1991.

[16] M. Bailes. In J. van Paradijs, E. van den Heuvel, and E. Kuulkers, editors, *Proceedings of I. A. U. Symposium 165, Compact Stars in Binaries*. Kluwer Academic Publishers, 1995. in press.

[17] C. Kochanek. *Astrophys. J.*, 398:234, 1992.

[18] L. Bildsten and C. Cutler. *Astrophys. J.*, 400:175, 1992.

[19] G. Quinlan and S. L. Shapiro. *Astrophys. J.*, 321:199, 1987.

[20] B. F. Schutz. *Nature*, 323:310, 1986.

[21] B. F. Schutz. *Class. Quant. Grav.*, 6:1761, 1989.

[22] C. W. Lincoln and C. M. Will. *Phys. Rev. D*, 42:1123, 1990.

[23] L. E. Kidder, C. M. Will, and A. G. Wiseman. *Phys. Rev. D*, 47:3281, 1993.

[24] C. Cutler, T. A. Apostolatos, L. Bildsten, L. S. Finn, E. E. Flanagan, D. Kennefick, D. M. Markovic, A. Ori, E. Poisson, G. J. Sussman, and K. S. Thorne. *Phys. Rev. Lett.*, 70:1984, 1993.

[25] C. M. Will. In M. Sasaki, editor, *Relativistic Cosmology*, pages 83–98. Universal Academy Press, 1994.

[26] T. A. Apostolatos, C. Cutler, G. J. Sussman, and K. S. Thorne. *Phys. Rev. D*, 49:6274, 1994.

[27] L. E. Kidder. *Phys. Rev. D*, 1995. in press.

[28] C. Cutler and E. E. Flanagan. *Phys. Rev. D*, 49:2658, 1994.

[29] L. S. Finn and D. F. Chernoff. *Phys. Rev. D*, 47:2198, 1993.

[30] E. Poisson and C. M. Will. *Phys. Rev. D*, 1995. submitted.

[31] P. Jaranowski and A. Krolak. *Phys. Rev. D*, 49:1723, 1994.

[32] T. Damour and K. Nordtvedt. *Phys. Rev. D*, 48:3436, 1993.

[33] C. M. Will. *Phys. Rev. D*, 50:6058, 1994.

[34] D. Markovic. *Phys. Rev. D*, 48:4738, 1993.

[35] D. F. Chernoff and L. S. Finn. *Astrophys. J. Lett.*, 411:L5, 1993.

[36] F. Ryan, L. S. Finn, and K. S. Thorne. *Phys. Rev. Lett.*, 1995. in preparation.

[37] E. E. Flanagan and S. A. Hughes. *Phys. Rev. D*, 1995. in preparation.

[38] M. Shibata, T. Nakamura, and K. Oohara. *Prog. Theor. Phys.*, 88:1079, 1992.

[39] M. Shibata, T. Nakamura, and K. Oohara. *Prog. Theor. Phys.*, 89:809, 1993.

[40] F. A. Rasio and S. L. Shapiro. *Astrophys. J.*, 401:226, 1992.

[41] T. Nakamura. In M. Sasaki, editor, *Relativistic Cosmology*, page 155. Universal Academy Press, 1994.

[42] X. Zhuge, J. M. Centrella, and S. L. W. McMillan. *Phys. Rev. D*, 50:6247, 1994.

[43] M. B. Davies, W. Benz, T. Piran, and F. K. Thielemann. *Astrophys. J.*, 431:742, 1994.

[44] B. J. Meers. *Phys. Rev. D*, 38:2317, 1988.

[45] D. Kennefick, D. Laurence, and K. S. Thorne. Phys. Rev. D. in preparation.

[46] H. A. Bethe. *Rev. Mod. Phys*, 62:801, 1990.

[47] A. Burrows, J. Hayes, and B. A. Fryxell. *Astrophys. J.*, 1995. in press.

[48] A. Burrows, 1995. private communication.

[49] L. S. Finn. *Ann. N. Y. Acad. Sci.*, 631:156, 1991.

[50] R. Mönchmeyer, G. Schäfer, E. Müller, and R. E. Kates. *Astron. Astrophys.*, 256:417, 1991.

[51] D. Lai and S. L. Shapiro. *Astrophys. J.*, 442:259, 1995.

[52] J. L. Houser, J. M. Centrella, and S. C. Smith. *Phys. Rev. Lett.*, 72:1314, 1994.

[53] I. A. Bonnell and J. E. Pringle. *Mon. Not. Roy. Astron. Soc.*, 273:L12, 1995.

[54] M. Zimmermann and E. Szedenits. *Phys. Rev. D*, 20:351, 1979.

[55] S. L. Shapiro and S. A. Teukolsky. Wiley: Interscience, 1983. Section 10.10 and references cited therein.

[56] P. Bender, A. Brillet, I. Ciufolini, K. Danzmann, R. Hellings, J. Hough, A. Lobo, M. Sandford, B. Schutz, and P. Touboul. *LISA, Laser interferometer space antenna for gravitational wave measurements: ESA Assessment Study Report.* R. Reinhard, ESTEC, 1994.

[57] D. Hils, P. Bender, and R. F. Webbink. *Astrophys. J.*, 360:75, 1990.

[58] L. S. Finn and K. S. Thorne. *Phys. Rev. D*, 1995. in preparation.

[59] F. Ryan. *Phys. Rev. D*, 1995. in preparation.

[60] D. Molteni, G. Gerardi, and S. K. Chakrabarti. *Astrophys. J.*, 436:249, 1994.

[61] D. Hils and P. Bender, 1995. preprint.

[62] J. Weber. *Phys. Rev.*, 117:306, 1960.

Inflation: From Theory To Observation and Back

MICHAEL S. TURNER

Departments of Physics and of Astronomy & Astrophysics
Enrico Fermi Institute, The University of Chicago, Chicago, IL 60637-1433
and
NASA/Fermilab Astrophysics Center
Fermi National Accelerator Laboratory, Batavia, IL 60510-0500

OVERVIEW

Alan Guth introduced cosmologists to inflation at the 1980 Texas Symposium. Since, inflation has had almost as much impact on cosmology as the big-bang model itself. However, unlike the big-bang model, it has little observational support. Hopefully, that situation is about to change as a variety and abundance of data begin to test inflation in a significant way. The observations that are putting inflation to test involve the formation of structure in the Universe, especially measurements of the anisotropy of the cosmic background radiation. The cold dark matter models of structure formation motivated by inflation are holding up well as the observational tests become sharper. In the next decade inflation will be tested even more significantly, with more precise measurements of CBR anisotropy, the mean density of the Universe, the Hubble constant, and the distribution of matter, as well as sensitive searches for the nonbaryonic dark matter predicted to exist by inflation. As an optimist I believe that we may be well on our way to a standard cosmology that includes inflation and extends back to around 10^{-32} sec, providing an important window on the earliest moments and fundamental physics.

1 BEYOND THE BIG BANG MODEL

The hot big-bang cosmology is a remarkable achievement. It provides a reliable account of the Universe from about 10^{-2} sec to the present. Further, it together with modern ideas in particle physics—the Standard Model, supersymmetry, grand unification, and superstring theory—provides a sound framework for sensible speculation all the way back to the Planck epoch and perhaps even earlier.[1]

These speculations have allowed cosmologists to address a deeper set of questions: What is the nature of the ubiquitous dark matter that is the dominant component of the mass density? Why does the Universe contain only matter? What is the origin of the tiny inhomogeneities that seeded the formation of structure, and

[1]Before the advent of the Standard Model (point-like quarks and leptons with "weak interactions" at short distances) cosmology hit a wall at about 10^{-5} sec. At this time the Universe was a strongly interacting gas of overlapping hadrons with the number of "fundamental particles" increasing exponentially with mass.

how did that structure evolve? Why is the portion of the Universe that we can see so flat and smooth? What is the value of the vexing cosmological constant? How did the expansion begin—or was there a beginning?

In the past fifteen years much progress has been made, and many believe that the answers to all these questions involve events that took place during the earliest moments and involved physics beyond the Standard Model [1]. For example, the matter-antimatter asymmetry, quantified as a net baryon number of about 10^{-10} per photon, is believed to have developed through interactions that do not conserve baryon number and C, CP (matter-antimatter symmetry) and occurred out of thermal equilibrium. Until recently it was believed that "baryogenesis" involved unification-scale physics and occurred around 10^{-34} sec; recent work suggests that baryogenesis might have occurred at the weak scale ($T \sim 300$ GeV and $t \sim 10^{-11}$ sec) and involved baryon-number and C, CP violation within the Standard Model [2].

The most optimistic early-Universe cosmologists (of which I am one) believe that we are on the verge of solving all of the above problems and extending our knowledge of the Universe back to around 10^{-32} sec after "the bang." The key to this is inflation. Among other things, inflation has led to the cold dark matter models of structure formation, which are characterized by scale-invariant density perturbations and dark matter whose composition is primarily slowly moving elementary particles (e.g., axions or neutralinos). The cold dark matter theory is crucial to testing inflation, and if it proves correct, would complete the standard cosmology by connecting the theorist's early Universe which is smooth and formless to the astronomer's Universe which is inhomogeneous and abounds with structure.

1.1 Evidence

Four pillars provide the observational support on which the hot big-bang model rests: (1) The uniform distribution of matter on large scales and the isotropic expansion that maintains this uniformity; (2) The existence of a nearly uniform and accurately thermal cosmic background radiation (CBR); (3) The abundances (relative to hydrogen) of the light elements D, ^{3}He, ^{4}He, and ^{7}Li; and (4) The existence of small fluctuations in the temperature of the CBR across the sky at the level about 10^{-5}. The Hubble expansion supports the general notion of an expanding Universe; the CBR provides almost indisputable evidence of a hot, dense beginning. The agreement between the light-element abundances predicted by primordial nucleosynthesis and those observed tests the model back to about 10^{-2} sec and leads to the most accurate determination of the baryon density, $\Omega_B \simeq 0.009h^{-2} - 0.022h^{-2}$ [3]. The small fluctuations in the temperature of CBR across the sky indicate the existence of primeval density perturbations of a similar size which amplified by gravity over the age of the Universe has led to the abundance of structure seen today (galaxies, clusters of galaxies, superclusters, voids, and great walls).

At present, observational support for inflation is fragmentary at best. However, a wealth of diverse observations are beginning to seriously test inflation, and even someone who is not an early-Universe optimist would have to concede that inflation is likely to be tested in a significant way within the next five years or so. The crucial tests include measurements of CBR anisotropy, the present degree of inhomogeneity as probed by redshift surveys and peculiar-velocity measurements, x-ray studies of clusters of galaxies, increasingly accurate measurements of the Hubble constant, the study of the Universe at high redshift by large ground-based telescopes and the Hubble Space Telescope, the mapping of dark matter through gravitational lensing, the search for baryonic dark matter through microlensing and particle dark matter through both direct and indirect techniques, and on and on.

While the observational support for inflation is not overwhelming—yet!—there are a number of observations which are very encouraging. The evidence that the mass density of the Universe is significantly larger than that which baryons can account for continues to grow: measurements based upon peculiar velocities of the Milky Way and other galaxies indicate that $\Omega_{\rm matter} \gtrsim 0.3$ [4, 5] and x-ray and weak-gravitational lensing measurements of the masses of rich clusters continue to indicate mass to light ratios consistent with $\Omega_{\rm matter} \gtrsim 0.2$ or greater. (For a Hubble constant of greater than $60\,{\rm km\,s^{-1}\,Mpc^{-1}}$, baryons can contribute at most 5% of the critical density.) Further, the mass fraction of rich clusters that can be readily identified as baryonic (mostly hot gas) is only about $0.04h^{-3/2} - 0.1h^{-3/2}$ [6]. Likewise, the fraction of the dark halo of our galaxy that can be accounted for by baryons (faint stars and MACHOs) is only about $0.05 - 0.3$ [7]. CBR anisotropy has now been measured on angular scales from about $0.5°$ to $90°$ [8, 9], probing the spectrum of metric perturbations on scales from about $30h^{-1}\,$Mpc to almost $10^4h^{-1}\,$Mpc. The measurements are consistent with the scale-invariant prediction of inflation; see Fig. 1. Likewise, measurements of the spectrum of inhomogeneity on smaller scales from redshift surveys, say less than about $100h^{-1}\,$Mpc, are generally consistent with the predictions of cold dark matter model (see Fig. 2). The success of cold dark matter is all the more striking given that the only other models (Peebles' baryons only PBI model and topological defects + nonbaryonic dark matter) are on the verge of being ruled out.

2 INFLATIONARY THEORY

2.1 Generalities

As successful as the big-bang cosmology it suffers from a dilemma involving initial data. Extrapolating back, one finds that the Universe apparently began from a very special state: A slightly inhomogeneous and very flat Robertson-Walker spacetime. Collins and Hawking showed that the set of initial data that evolve to a spacetime that is as smooth and flat as ours is today of measure zero [10]. (In the context of simple grand unified theories, the hot big bang suffers from another serious problem: the extreme overproduction of superheavy magnetic monopoles; in fact, it was an attempt to solve the monopole problem which led Guth to inflation.)

The cosmological appeal of inflation is its ability to lessen the dependence of the present state of the Universe upon the initial state. Two elements are essential to doing this: (1) accelerated ("superluminal") expansion and the concomitant tremendous growth of the scale factor; and (2) massive entropy production [11]. Together, these two features allow a small, smooth subhorizon-sized patch of the early Universe to grow to a large enough size and contain enough heat (entropy in excess of 10^{88}) to easily encompass our present Hubble volume. Provided that

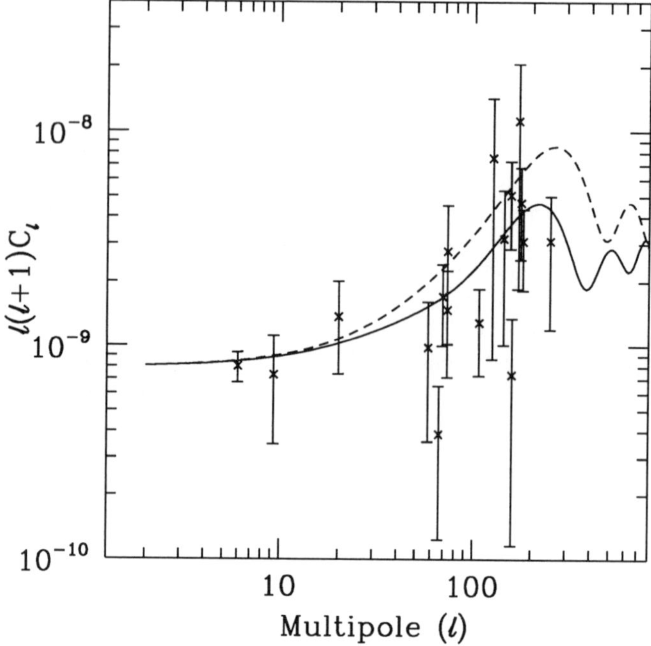

Figure 1: Summary of current measurements of CBR anisotropy in terms of a spherical-harmonic decomposition, $C_l \equiv \langle |a_{lm}|^2 \rangle$. The rms temperature fluctuation measured between two points separated by an angle θ is roughly given by: $(\delta T/T)_\theta \simeq \sqrt{l(l+1)C_l}$ with $l \simeq 200°/\theta$. The curves are the cold dark matter predictions, normalized to the COBE detection, for Hubble constants of $50\,\mathrm{km\,s^{-1}\,Mpc}$ (solid) and $35\,\mathrm{km\,s^{-1}\,Mpc^{-1}}$ (broken). (Figure courtesy of M. White.)

the region was originally small compared to the curvature radius of the Universe it would appear flat then and today (just as any small portion of the surface of a sphere appears flat).

While there is presently no standard model of inflation—just as there is no standard model for physics at these energies (typically 10^{15} GeV or so)—viable models have much in common. They are based upon well posed, albeit highly speculative, microphysics involving the classical evolution of a scalar field. The superluminal expansion is driven by the potential energy ("vacuum energy") that arises when the scalar field is displaced from its potential-energy minimum, which results in nearly exponential expansion. Provided the potential is flat, during the time it takes for the field to roll to the minimum of its potential the Universe undergoes many e-foldings of expansion (more than around 60 or so are required to realize the beneficial features of inflation). As the scalar field nears the minimum, the vacuum energy has been converted to coherent oscillations of the scalar field, which correspond to nonrelativistic scalar-field particles. The eventual decay of these particles into lighter particles and their thermalization results in the "reheating" of the Universe and

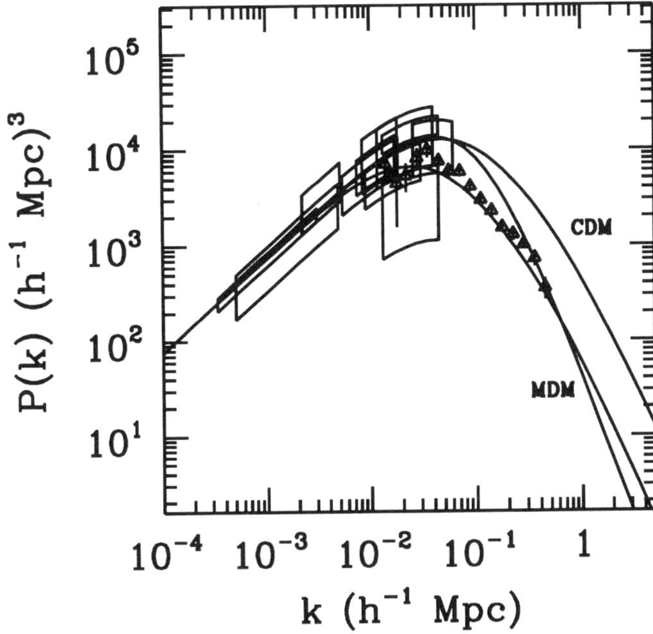

Figure 2: Comparison of the cold dark matter perturbation spectrum with CBR anisotropy measurements (boxes) and the distribution of galaxies today (triangles). Wavenumber k is related to length scale, $k = 2\pi/\lambda$; error flags are not shown for the galaxy distribution. The curve labeled MDM is hot + cold dark matter ("5 eV" worth of neutrinos); the other two curves are cold dark matter models with Hubble constants of 50 km s^{-1} Mpc (labeled CDM) and 35 km s^{-1} Mpc. (Figure courtesy of M. White.)

accounts for all the heat in the Universe today (the entropy production event).

Superluminal expansion and the tremendous growth of the scale factor (by a factor greater than that since the end of inflation) allow quantum fluctuations on very small scales ($\lesssim 10^{-23}$ cm) to be stretched to astrophysical scales ($\gtrsim 10^{25}$ cm). Quantum fluctuations in the scalar field responsible for inflation ultimately lead to an almost scale-invariant spectrum of density perturbations [12], and quantum fluctuations in the metric itself lead to an almost scale-invariant spectrum of gravity-waves [13]. Scale invariance for density perturbations means scale-independent fluctuations in the gravitational potential (equivalently, density perturbations of different wavelength cross the horizon with the same amplitude); scale invariance for gravity waves means that gravity waves of all wavelengths cross the horizon with the same amplitude. Because of subsequent evolution, neither the scalar nor the tensor perturbations are scale invariant today.

2.2 Metaphysical implications

Inflation alleviates the "specialness" problem greatly, but does not eliminate all dependence upon the initial state [14]. All open FRW models will inflate and become flat; however, many closed FRW models will recollapse before they can inflate. If one imagines the most general initial spacetime as being comprised of negatively and positively curved FRW (or Bianchi) models that are stitched together, the failure of the positively curved regions to inflate is of little consequence: because of exponential expansion during inflation the negatively curved regions will occupy most of the space today. Nor does inflation solve the smoothness problem forever; it just postpones the problem into the exponentially distant future: We will be able to see outside our smooth inflationary patch and Ω will start to deviate significantly from unity at a time $t \sim t_0 \exp[3(N - N_{\rm min}]$, where N is the actual number of e-foldings of inflation and $N_{\rm min} \sim 60$ is the minimum required to solve the horizon/flatness problems.

Linde has emphasized that inflation has changed our view of the Universe in a very fundamental way [15]. While cosmologists have long used the Copernician principle to argue that the Universe must be smooth because of the smoothness of our Hubble volume, in the post-inflation view, our Hubble volume is smooth because it is a small part of a region that underwent inflation. On the largest scales the structure of the Universe is likely to be very rich: Different regions may have undergone different amounts of inflation, may have different laws of physics because they evolved into different vacuum states (of equivalent energy), and may even have different numbers of spatial dimensions. Since it is likely that most of the volume of the Universe is still undergoing inflation and that inflationary patches are being constantly produced (eternal inflation), the age of the Universe is a meaningless concept and our expansion age merely measures the time back to the end of our inflationary event!

2.3 Specifics

In Guth's seminal paper [16] he introduced the idea of inflation, sung its praises, and showed that the model that he based the idea upon did not work! Thanks to very important contributions by Linde [17] and Albrecht and Steinhardt [18] that was quickly remedied, and today there are many viable models of inflation. That of course is both good news and bad news; it means that there is no standard model of inflation. Again, the absence of a standard model of inflation should be viewed in the light of our general ignorance about fundamental physics at these energies.

Many different approaches have taken in constructing particle-physics models for inflation. Some have focussed on very simple scalar potentials, e.g., $V(\phi) = \lambda\phi^4$ or $= m^2\phi^2/2$, without regard to connecting the model to any underlying theory [19, 20]. Others have proposed more complicated models that attempt to make contact with speculations about physics at very high energies, e.g., grand unification [21], supersymmetry [22, 23, 24], preonic physics [25], or supergravity [26]. Several authors have attempted to link inflation with superstring theory [27] or "generic predictions" of superstring theory such as pseudo-Nambu-Goldstone boson fields [28]. While the scale of the vacuum energy that drives inflation is typically of order $(10^{15}\,\mathrm{GeV})^4$, a model of inflation at the electroweak scale, vacuum energy $\approx (1\,\mathrm{TeV})^4$, has been proposed [29]. There are also models in which there are multiple epochs of inflation [30].

In all of the models above gravity is described by general relativity. A qualitatively different approach is to consider inflation in the context of alternative theories of gravity. (After all, inflation probably involves physics at energy scales not too different from the Planck scale and the effective theory of gravity at these energies could well be very different from general relativity; in fact, there are some indications from superstring theory that gravity in these circumstances might be described by a Brans-Dicke like theory.) Perhaps the most successful of these models is first-order inflation [31, 32]. First-order inflation returns to Guth's original idea of a strongly first-order phase transition; in the context of general relativity Guth's model failed because the phase transition, if inflationary, never completed. In theories where the effective strength of gravity evolves, like Brans-Dicke theory, the weakening of gravity during inflation allows the transition to complete. In other models based upon nonstandard gravitation theory, the scalar field responsible for inflation is itself related to the size of additional spatial dimensions, and inflation then also explains why our three spatial dimensions are so big, while the other spatial dimensions are so small.

All models of inflation have one feature in common: the scalar field responsible for inflation has a very flat potential-energy curve and is very weakly coupled. This typically leads to a very small dimensionless number, usually a dimensionless coupling of the order of 10^{-14}. Such a small number, like other small numbers in physics (e.g., the ratio of the weak to Planck scales $\approx 10^{-17}$ or the ratio of the mass of the electron to the W/Z boson masses $\approx 10^{-5}$), runs counter to one's belief that a truly fundamental theory should have no tiny parameters, and cries out for an explanation. At the very least, this small number must be stabilized against quantum corrections—which it is in all of the previously mentioned models.[2] In some models, the small number in the inflationary potential is related to other small numbers in particle physics: for example, the ratio of the electron mass to the weak scale or the ratio of the unification scale to the Planck scale. Explaining the origin of the small number that seems to be associated with inflation is both a challenge and an opportunity.

Because of the growing base of observations that bear on inflation, another approach to model building is emerging: the use of observations to constrain the underlying inflationary potential—and hence the title of this paper. In the next Section I focus on just this. Before I do, I want to emphasize that while there are many varieties of inflation, there are robust predictions which are crucial to sharply testing inflation.

[2]It is sometimes stated that inflation is unnatural because of the small coupling of the scalar field responsible for inflation; while the small coupling certainly begs explanation, these inflationary models are not unnatural in the rigorous technical sense as the small number is stable against quantum fluctuations.

2.4 Three robust predictions

Inflation makes three robust[3] predictions:

1. **Flat universe.** Because solving the "horizon" problem (large-scale smoothness in spite of small particle horizons at early times) and solving the "flatness" problem (maintaining Ω very close to unity until the present epoch) are linked geometrically [1, 11], this is the most robust prediction of inflation. Said another way, it is the prediction that most inflationists would be least willing to give up. (Even so, models of inflation have been constructed where the amount of inflation is tuned just to give Ω_0 less than one today [33].) Through the Friedmann equation for the scale factor, flat implies that the total energy density (matter, radiation, vacuum energy, ...) is equal to the critical density.

2. **Nearly scale-invariant spectrum of gaussian density perturbations.** Essentially all inflation models predict a nearly scale-invariant spectrum of gaussian density perturbations. Described in terms of a power spectrum, $P(k) \equiv \langle |\delta_k|^2 \rangle = Ak^n$, where δ_k is the Fourier transform of the primeval density perturbations, and the spectral index $n \approx 1$ is equal to unity in the scale-invariant limit. The overall amplitude A is model dependent. Density perturbations give rise to CBR anisotropy as well as seeding structure formation. Requiring that the density perturbations are consistent with the observed level of anisotropy of the CBR (and large enough to produce the observed structure formation) is the most severe constraint on inflationary models and leads to the small dimensionless number that all inflationary models have.

3. **Nearly scale-invariant spectrum of gravitational waves.** These gravitational waves have wavelengths from $\mathcal{O}(1\,\mathrm{km})$ to the size of the present Hubble radius and beyond. Described in terms of a power spectrum for the dimensionless gravity-wave amplitude at early times, $P_T(k) \equiv \langle |h_k|^2 \rangle = A_T k^{n_T - 3}$, where the spectral index $n_T \approx 0$ in the scale-invariant limit. Once again, the overall amplitude A_T is model dependent (varying as the value of the inflationary vacuum energy). Unlike density perturbations, which are required to initiate structure formation, there is no cosmological lower bound to the amplitude of the gravity-wave perturbations. Tensor perturbations also give rise to CBR anisotropy; requiring that they do not lead to excessive anisotropy implies that the energy density that drove inflation must be less than about $(10^{16}\,\mathrm{GeV})^4$. This indicates that if inflation took place, it did so at an energy well below the Planck scale.[4]

There are other interesting consequences of inflation that are less generic. For example, in models of first-order inflation, in which reheating occurs through the nucleation and collision of vacuum bubbles, there is an additional, larger amplitude,

[3] Because theorists are so clever, it is not possible nor prudent to use the word immutable. Models that violate any or all of these "robust predications" can and have been constructed.

[4] To be more precise, the part of inflation that led to perturbations on scales within the present horizon involved subPlanckian energy densities. In some models of inflation, the earliest stages, which do not influence scales that we are privy to, involve energies as large as the Planck scale.

but narrow-band, spectrum of gravitational waves ($\Omega_{\rm GW}h^2 \sim 10^{-6}$) [34]. In other models large-scale primeval magnetic fields of interesting size are seeded during inflation [35].

3 STRUCTURE FORMATION: A WINDOW TO THE EARLY UNIVERSE

The key to testing inflation is to focus on its robust predictions and their implications. Earlier I discussed the prediction of a flat Universe and its bold implication that most of the matter in Universe exists in the form of particle dark matter. Much effort is being directed at determining the mean density of the Universe and detecting particle dark matter.

The scale-invariant scalar metric perturbations lead to CBR anisotropy on angular scales from less than 1° to 90° and seed the formation of structure in the Universe. Together with the nucleosynthesis determination of Ω_B and the inflationary prediction of a flat Universe, scale-invariant density perturbations lead to a very specific scenario for structure formation; it is known as cold dark matter because the bulk of the particle dark matter is comprised of slowly moving particles (e.g., axions or neutralinos) [36].[5] A large and rapidly growing number of observations are being brought to bear in the testing of cold dark matter, making it the centerpiece of efforts to test inflation.

Finally, there are the scale-invariant tensor perturbations. They lead to CBR anisotropy on angular scales from a few degrees to 90° and a spectrum of gravitational waves. The CBR anisotropy arising from the tensor perturbations can in principle be separated from that arising from scalar perturbations. However, because the sky is finite, sampling variance sets a fundamental limit: the tensor contribution to CBR anisotropy can only be separated from that of the scalar if T/S is greater than about 0.14 [37] (T is the contribution of tensor perturbations to the variance of the CBR quadrupole and S is the same for scalar perturbations). It is also possible that the stochastic background of gravitational waves itself can be directly detected, though it appears that the LIGO facilities being built will lack the sensitivity and even space-based interferometery (e.g., LISA) is not a sure bet [38].

Before going on to discuss how cold dark matter models are testing inflation I want to emphasize the importance of the tensor perturbations. The attractiveness of a flat Universe with scale-invariant density perturbations was appreciated long before inflation. Verifying these two predictions of inflation, while important, will not provide a "smoking gun." The tensor perturbations are a unique feature of inflation. Further, they are crucial to obtaining information about the scalar potential responsible for inflation.

[5]The simpler possibility, that the particle dark matter exists in the form of 30 eV or so neutrinos which is known as hot dark matter, was falsified almost a decade ago. Because neutrinos move rapidly, they can diffuse from high density to low density regions damping perturbations on small scales. In hot dark matter large, supercluster-size objects must form before galaxies, and thus hot dark matter cannot account for the abundance of galaxies, damped Lyman-α clouds, etc. that is observed at high redshift.

3.1 Vanilla Cold Dark Matter: almost, but not quite?

The simplest version of cold dark matter, vanilla cold dark matter if you will, is characterized by: (1) $\Omega_B \sim 0.5$ and $\Omega_{\rm CDM} \sim 0.95$; (2) Hubble constant of $50\,{\rm km\,s^{-1}\,Mpc^{-1}}$; (3) Precisely scale-invariant density perturbations ($n = 1$); and (4) No contribution of tensor perturbations to CBR anisotropy. In cold dark matter models structure forms hierarchically, with small objects forming first and merging to form larger objects. Galaxies form at redshifts of order a few, and rarer objects like QSOs form from higher than average density peaks earlier. In general, cold dark matter predicts a Universe that is still evolving at recent epochs. N-body simulations are crucial to bridging the gap between theory and observation, and several groups have carried out large numerical studies of vanilla cold dark matter [39].

There are a diversity of observations that test cold dark matter; they include CBR anisotropy and spectral distortions, redshift surveys, pairwise velocities of galaxies, peculiar velocities, redshift space distortions, x-ray background, QSO absorption line systems, cluster studies of all kinds, studies of evolution (clusters, galaxies, and so on), measurements of the Hubble constant, and on and on. I will focus on how these measurements probe the power spectrum of density perturbations, emphasizing the role of CBR-anisotropy measurements and redshift surveys.

Density perturbations on a (comoving) length scale λ give rise to CBR anisotropy on an angular scale $\theta \sim \lambda/H_0^{-1} \sim 1^\circ(\lambda/100h^{-1}\,{\rm Mpc})$.[6] CBR anisotropy has now been detected by more than ten experiments on angular scales from about 0.5° to 90°, thereby probing length scales from $30h^{-1}\,{\rm Mpc}$ to $10^4h^{-1}\,{\rm Mpc}$. The very accurate measurements made by the COBE DMR can be used to normalize the cold dark matter spectrum (the normalization scale corresponds to about 20°). When this is done, the other ten or so measurements are in agreement with the predictions of cold dark matter (see Fig. 1).

The COBE-normalized cold dark matter spectrum can be extrapolated to the much smaller scales probed by redshift surveys, from about $1h^{-1}\,{\rm Mpc}$ to $100h^{-1}\,{\rm Mpc}$. When this is done, there is general agreement. However, on closer inspection the COBE-normalized spectrum seems to predict excess power on these scales (about a factor of four in the power spectrum; see Fig. 2). This conclusion is supported by other observations, e.g., the abundance of rich clusters and the pairwise velocities of galaxies. It suggests that cold dark matter has much of the truth, but perhaps not all of it [40], and has led to the suggestion that something needs to be added to the simplest cold dark matter theory.

There is another important challenge facing cold dark matter. X-ray observations of rich clusters are able to determine the ratio of hot gas (baryons) to total cluster mass (baryons + CDM) (by a wide margin, most of the baryons "seen" in clusters are in the hot gas). To be sure there are assumptions and uncertainties; the data at the moment indicate that this ratio is $0.04h^{-3/2} - 0.1h^{-3/2}$ [6]. If clusters provide a fair sample of the universal mix of matter, then this ratio should equal $\Omega_B/(\Omega_B + \Omega_{\rm CDM}) \simeq (0.009 - 0.022)h^{-2}/(\Omega_B + \Omega_{\rm CDM})$. Since clusters are large objects they should provide a pretty fair sample. Taking the numbers at face

[6] For reference, perturbations on a length scale of about 1 Mpc give rise to galaxies, on about 10 Mpc to clusters, on about 30 Mpc to large voids, and on about 100 Mpc to the great walls.

value, cold dark matter is consistent with the cluster gas fraction provided either: $\Omega_B + \Omega_{\rm CDM} = 1$ and $h \sim 0.3$ or $\Omega_B + \Omega_{\rm CDM} \sim 0.3$ and $h \sim 0.7$. The cluster baryon problem has yet to be settled, and is clearly an important test of cold dark matter.

Finally, before going on to discuss the variants of cold dark matter now under consideration, let me add a note of caution. The comparison of predictions for structure formation with present-day observations of the distribution of galaxies is fraught with difficulties. Theory most accurately predicts "where the mass is" (in a statistical sense) and the observations determine where the light is. Redshift surveys probe present-day inhomogeneity on scales from around one Mpc to a few hundred Mpc, scales where the Universe is nonlinear ($\delta n_{\rm GAL}/n_{\rm GAL} \gtrsim 1$ on scales $\lesssim 8h^{-1}$ Mpc) and where astrophysical processes undoubtedly play an important role (e.g., star formation determines where and when "mass lights up," the explosive release of energy in supernovae can move matter around and influence subsequent star formation, and so on). The distance to a galaxy is determined through Hubble's law ($d = H_0^{-1}z$) by measuring a redshift; peculiar velocities induced by the lumpy distribution of matter are significant and prevent a direct determination of the actual distance. There are the intrinsic limitations of the surveys themselves: they are flux not volume limited (brighter objects are seen to greater distances and vice versa) and relatively small (e.g., the CfA slices of the Universe survey contains only about 10^4 galaxies and extends to a redshift of about $z \sim 0.03$). Last but not least are the numerical simulations which link theory and observation; they are limited in dynamical range (about a factor of 100 in length scale) and in microphysics (in the largest simulations only gravity, and in others only a gross approximation to the effects of hydrodynamics/thermodynamics). Perhaps it would be prudent to withhold judgment on vanilla cold dark matter for the moment and resist the urge to modify it—but that wouldn't be as much fun!

3.2 The many flavors of cold dark matter

The spectrum of density perturbations today depends not only upon the primeval spectrum (and the normalization on large scales provided by COBE), but also upon the energy content of the Universe. While the fluctuations in the gravitational potential were initially (approximately) scale invariant, the Universe evolved from an early radiation-dominated phase to a matter-dominated phase which imposes a characteristic scale on the spectrum of density perturbations seen today; that scale is determined by the energy content of the Universe, $k_{\rm EQ} \sim 10^{-1}h\,{\rm Mpc}^{-1}\,(\Omega_{\rm matter}h/\sqrt{g_*})$ (g_* counts the relativistic degrees of freedom, $\Omega_{\rm matter} = \Omega_B + \Omega_{\rm CDM}$). In addition, if some of the nonbaryonic dark matter is neutrinos, they reduce power on small scales somewhat through freestreaming (see Fig. 2). With this in mind, let me discuss the variants of cold dark matter that have been proposed to improve its agreement with observations.

1. **Low Hubble Constant + cold dark matter (LHC CDM)** [41]. Remarkably, simply lowering the Hubble constant to around $30\,{\rm km\,s^{-1}\,Mpc^{-1}}$ solves all the problems of cold dark matter. Recall, the critical density $\rho_{\rm crit} \propto H_0^2$; lowering H_0 lowers the matter density and has precisely the desired effect. It has two other added benefits: the expansion age of the Universe is comfortably

consistent with the ages of the oldest stars and the baryon fraction is raised to a value that is consistent with that measured in x-ray clusters. Needless to say, such a small value for the Hubble constant flies in the face of current observations [42]; further, it illustrates that the problems of cold dark matter get even worse for the larger values of H_0 that are favored by recent observations.

2. **Hot + cold dark matter (νCDM) [43].** Adding a small amount of hot dark matter can suppress density perturbations on small scales; adding too much leads back to the longstanding problems of hot dark matter. Retaining enough power on very small scales to produce damped Lyman-α systems at high redshift limits Ω_ν to less than about 20%, corresponding to about "5 eV worth of neutrinos" (i.e., one species of mass 5 eV, or two species of mass 2.5 eV, and so on). This admixture of hot dark matter rejuvenates cold dark matter provided the Hubble constant is not too large, $H_0 \lesssim 55\,\mathrm{km\,s^{-1}\,Mpc^{-1}}$; in fact, a Hubble constant of closer to $45\,\mathrm{km\,s^{-1}\,Mpc^{-1}}$ is preferred.

3. **Cosmological constant + cold dark matter (ΛCDM) [44].** (A cosmological constant corresponds to a uniform energy density, or vacuum energy.) Shifting 50% to 80% of the critical density to a cosmological constant lowers the matter density and has the same beneficial effect as a low Hubble constant. In fact, a Hubble constant as large as $80\,\mathrm{km\,s^{-1}\,Mpc^{-1}}$ can be accommodated. In addition, the cosmological constant allows the age problem to solved even if the Hubble constant is large, addresses the fact that few measurements of the mean mass density give a value as large as the critical density (most measurements of the mass density are insensitive to a uniform component), and allows the baryon fraction of matter to be larger, which alleviates the cluster baryon problem. Not everything is rosy; cosmologists have invoked a cosmological constant twice before to solve their problems (Einstein to obtain a static universe and Bondi, Gold, and Hoyle to solve the earlier age crisis when H_0 was thought to be $250\,\mathrm{km\,s^{-1}\,Mpc^{-1}}$). Further, particle physicists can still not explain why the energy of the vacuum is not at least 50 (if not 120) orders of magnitude larger than the present critical density, and expect that when the problem is solved the answer will be zero.

4. **Extra relativistic particles + cold dark matter (τCDM) [45].** Raising the level of radiation has the same beneficial effect as lowering the matter density. In the standard cosmology the radiation content consists of photons + three (undetected) cosmic seas of neutrinos (corresponding to $g_* \simeq 3.36$). While we have no direct determination of the radiation beyond that in the CBR, there are at least two problems: What are the additional relativistic particles? and Can additional radiation be added without upsetting the successful predictions of primordial nucleosynthesis which depend critically upon the energy density of relativistic particles? The simplest way around these problems is an unstable tau neutrino (mass anywhere between a few keV and a few MeV) whose decays produce the radiation. This fix can tolerate a larger Hubble constant, though at the expense of more radiation.

5. **Tilted cold dark matter (TCDM)** [46]. While the spectrum of density perturbations in most models of inflation is very nearly scale invariant, there are models where the deviations are significant ($n \approx 0.8$) which leads to smaller fluctuations on small scales. Further, if gravity waves account for a significant part of the CBR anisotropy, the level of density perturbations can be lowered even more. A combination of tilt and gravity waves can solve the problem of too much power on small scales, but seems to lead to too little power on intermediate and very small scales.

In evaluating these better fit models, one should keep the words of Francis Crick in mind (loosely paraphrased): A model that fits all the data at a given time is necessarily wrong, because at any given time not all the data are correct(!). ΛCDM provides an interesting/confusing example. When I discussed it in 1990, I called it the best-fit Universe, and quoting Crick, I said that ΛCDM was certain to fall by the wayside [47]. In 1995, it is still the best-fit model [48].

Let me end by defending the other point of view, namely, that to add something to cold dark matter is not unreasonable, or even as some have said, a last gasp effort to saving a dying theory. Standard cold dark matter was a starting point, similar to early calculations of big-bang nucleosynthesis. It was always appreciated that the inflationary spectrum of density perturbations was not exactly scale invariant [20] and that the Hubble constant was unlikely to be exactly $50\,\mathrm{km\,s^{-1}\,Mpc}$. As the quality and quantity of data improve, it is only sensible to refine the model, just as has been done with big-bang nucleosynthesis. Cold dark matter seems to embody much of the "truth." The modifications suggested all seem quite reasonable (as opposed to contrived). Neutrinos exist; they are expected to have mass; there is even some experimental data that indicates they do have mass. It is still within the realm of possibility that the Hubble constant is less than $50\,\mathrm{km\,s^{-1}\,Mpc^{-1}}$, and if it is as large as $70\,\mathrm{km\,s^{-1}\,Mpc^{-1}}$ to $80\,\mathrm{km\,s^{-1}\,Mpc^{-1}}$ a cosmological constant seems inescapable based upon the age problem alone. There is no data that can preclude more radiation than in the standard cosmology and deviations from scale invariance were always expected.

3.3 Reconstruction

If inflation and the cold dark matter theory is shown to be correct, then a window to the very early Universe ($t \sim 10^{-34}$ sec) will have been opened. While it is certainly premature to jump to this conclusion, I would like to illustrate one example of what one could hope to learn. As mentioned earlier, the spectra and amplitudes of the the tensor and scalar metric perturbations predicted by inflation depend upon the underlying model, to be specific, the shape of the inflationary scalar-field potential. If one can measure the power-law index of the scalar spectrum and the amplitudes of the scalar and tensor spectra, one can recover the value of the potential and its first two derivatives around the point on the potential where inflation took place [49]:

$$V = 1.65 T m_{\rm Pl}{}^4, \tag{1}$$

$$V' = \pm\sqrt{\frac{8\pi r}{7}}\, V/m_{\rm Pl}, \tag{2}$$

$$V'' = 4\pi\left[(n-1)+\frac{3}{7}r\right]V/{m_{\rm Pl}}^2, \tag{3}$$

where $r \equiv T/S$, a prime indicates derivative with respect to ϕ, $m_{\rm Pl} = 1.22\times10^{19}\,{\rm GeV}$ is the Planck energy, and the sign of V' is indeterminate. In addition, if the tensor spectral index can be measured a consistency relation, $n_T = -r/7$, can be used to further test inflation. Reconstruction of the inflationary scalar potential would shed light both on inflation as well as physics at energies of the order of $10^{15}\,{\rm GeV}$.

4 The Future

The stakes for cosmology are high: if correct, inflation/cold dark matter represents a major extension of the big bang and our understanding of the Universe, which can't help but shed light on the fundamental physics at energies of order 10^{15} GeV.

What are the crucial tests and when will they be carried out? Because of the many measurements/observations that can have significant impact, I believe the answer to when is sooner rather than later. The list of pivotal observations is long: CBR anisotropy, large redshift surveys (e.g., the Sloan Digital Sky Survey will have 10^6 redshifts), direct searches for nonbaryonic in our neighborhood (both for axions and neutralinos) and baryonic dark matter (microlensing), x-ray studies of galaxy clusters, the use of back-lit gas clouds (quasar absorption line systems) to study the Universe at high redshift, evolution (as revealed by deep images of the sky taken by the Hubble Space Telescope and the Keck 10 meter telescope), measurements of both H_0 and q_0, mapping of the peculiar velocity field at large redshifts through the Sunyaev-Zel'dovich effect, dynamical estimates of the mass density (using weak gravitational lensing, large-scale velocity fields, and so on), age determinations, gravitational lensing, searches for supersymmetric particles (at accelerators) and neutrino oscillations (at accelerators, solar-neutrino detectors, and other large underground detectors), searches for high-energy neutrinos from neutralino annihilations in the sun using large underground detectors, and on and on. Let me end by illustrating the interesting consequences of several possible measurements.

A definitive determination that H_0 is greater than $55\,{\rm km\,s^{-1}\,Mpc^{-1}}$ would falsify LHC CDM and νCDM. A definitive determination that H_0 is $75\,{\rm km\,s^{-1}\,Mpc^{-1}}$ or larger would necessitate a cosmological constant. A flat Universe with a cosmological constant has a very different deceleration parameter than one dominated by matter, $q_0 = -1.5\Omega_\Lambda + 0.5 \sim -(0.4-0.7)$ compared to $q_0 = 0.5$, and this could be settled by galaxy number counts or numbers of lensed quasars. The level of CBR anisotropy in τCDM and LHC CDM on the 0.5° scale is about 50% larger than the other models, which should be easily discernible. If neutrino-oscillation experiments were to provide evidence for a neutrino of mass 5 eV (or two of mass 2.5 eV) νCDM would seem almost inescapable.

Many more CBR measurements are in progress and there should many interesting results in the next few years. In the wake of the success of COBE there are proposals, both in the US and Europe, for a satellite-borne instrument to map the CBR sky with a factor of ten better resolution. A map of the CBR with $0.5^\circ - 1^\circ$ resolution could separate the gravity-wave contribution to CBR anisotropy and provide evidence for the third robust prediction of inflation, as well as determining other important parameters [50], e.g., the scalar and tensor indices, Ω_Λ, and even Ω_0 (the position of the "Doppler" peak scales as $0.5^\circ/\sqrt{\Omega_0}$) [51]).

Acknowledgments

This work was supported in part by the DOE (at Chicago and Fermilab) and the NASA (at Fermilab through grant NAG 5-2788).

References

[1] See e.g., E.W. Kolb and M.S. Turner, *The Early Universe* (Addison-Wesley, Redwood City, CA, 1990).

[2] A. Cohen, D. Kaplan, and A. Nelson, *Annu. Rev. Nucl. Part. Sci.* **43**, 27 (1992).

[3] C. Copi, D.N. Schramm, and M.S. Turner, *Science* **267**, 192 (1995).

[4] A. Dekel, *Annu. Rev. Astron. Astrophys.* **32**, 371 (1994).

[5] M.A. Strauss et al., *Astrophys. J.* **397**, 395 (1992); N. Kaiser et al., *Mon. Not. R. astron. Soc.* **252**, 1 (1990).

[6] S.D.M. White et al., *Nature* **366**, 429 (1993); U.G. Briel et al., *Astron. Astrophys.* **259**, L31 (1992); D.A. White and A.C. Fabian, *Mon. Not. R. astron. Soc.*, in press (1995).

[7] E. Gates, G. Gyuk, and M.S. Turner, *Phys. Rev. Lett.* **74**, 3724 (1995).

[8] M. White, D. Scott, and J. Silk, *Ann. Rev. Astron. Astrophys.* **32**, 319 (1994).

[9] D. Wilkinson, in these proceedings (1995).

[10] C.B. Collins and S.W. Hawking, *Astrophys. J.* **180**, 317 (1973).

[11] Y. Hu, M.S. Turner, and E.J. Weinberg, *Phys. Rev. D* **49**, 3830 (1994).

[12] A. H. Guth and S.-Y. Pi, *Phys. Rev. Lett.* **49**, 1110 (1982); S. W. Hawking, *Phys. Lett. B* **115**, 295 (1982); A. A. Starobinskii, *ibid* **117**, 175 (1982); J. M. Bardeen, P. J. Steinhardt, and M. S. Turner, *Phys. Rev. D* **28**, 697 (1983).

[13] V.A. Rubakov, M. Sazhin, and A. Veryaskin, *Phys. Lett. B* **115**, 189 (1982); R. Fabbri and M. Pollock, *ibid* **125**, 445 (1983); A.A. Starobinskii *Sov. Astron. Lett.* **9**, 302 (1983); L. Abbott and M. Wise, *Nucl. Phys. B* **244**, 541 (1984).

[14] M.S. Turner and L.M. Widrow, *Phys. Rev. Lett.* **57**, 2237 (1986); L. Jensen and J. Stein-Schabes, *Phys. Rev. D* **35**, 1146 (1987); A.A. Starobinskii, *JETP Lett.* **37**, 66 (1983).

[15] A.D. Linde, *Inflation and Quantum Cosmology* (Academic Press, San Diego, CA, 1990).

[16] A.H. Guth, *Phys. Rev. D* **23**, 347 (1981).

[17] A.D. Linde, *Phys. Lett. B* **108**, 389 (1982).

[18] A. Albrecht and P.J. Steinhardt, *Phys. Rev. Lett.* **48**, 1220 (1982).

[19] A.D. Linde, *Phys. Lett. B* **129**, 177 (1983).

[20] P.J. Steinhardt and M.S. Turner, *Phys. Rev. D* **29**, 2162 (1984).

[21] Q. Shafi and A. Vilenkin, *Phys. Rev. Lett.* **52**, 691 (1984); S.-Y. Pi, *ibid* **52**, 1725 (1984).

[22] R. Holman, P. Ramond, and G.G. Ross, *Phys. Lett. B* **137**, 343 (1984).

[23] K. Olive, *Phys. Repts.* **190**, 309 (1990).

[24] H. Murayama et al., *Phys. Rev. D(RC)* **50**, R2356 (1994).

[25] M. Cvetic, T. Hubsch, J. Pati, and H. Stremnitzer, *Phys. Rev. D* **40**, 1311 (1990).

[26] E.J. Copeland et al., *Phys. Rev. D* **49**, 6410 (1994).

[27] See e.g., M. Gasperini and G. Veneziano, *Phys. Rev. D* **50**, 2519 (1994); R. Brustein and G. Veneziano, *Phys. Lett. B* **329**, 429 (1994); T. Banks et al., hep-th/9503114.

[28] K. Freese, J.A. Frieman, and A. Olinto, *Phys. Rev. Lett.* **65**, 3233 (1990).

[29] L. Knox and M.S. Turner, *Phys. Rev. Lett.* **70**, 371 (1993).

[30] J. Silk and M.S. Turner, *Phys. Rev. D* **35**, 419 (1986); L.A. Kofman, A.D. Linde, and J. Einsato, *Nature* **326**, 48 (1987).

[31] D. La and P.J. Steinhardt, *Phys. Rev. Lett.* **62**, 376 (1991).

[32] E.W. Kolb, *Physica Scripta* **T36**, 199 (1991).

[33] M. Bucher A.S. Goldhaber, and N. Turok, hep-ph/9411206 (1994).

[34] M.S. Turner and F. Wilczek, *Phys. Rev. Lett.* **65**, 3080 (1990); A. Kosowsky, M.S. Turner, and R. Watkins, *ibid* **69**, 2026 (1992).

[35] M.S. Turner and L.M. Widrow, *Phys. Rev. D* **37**, 2743 (1988); B. Ratra, *Astrophys. J.* **391**, L1 (1992).

[36] For an overview of the cold dark matter scenario of structure formation see e.g., G. Blumenthal et al., *Nature* **311**, 517 (1984).

[37] L. Knox and M.S. Turner, *Phys. Rev. Lett.* **73**, 3347 (1994).

[38] M.S. Turner, J. Lidsey, and M. White, *Phys. Rev. D* **48**, 4613 (1993).

[39] See e.g., C. Frenk, *Physica Scripta* **T36**, 70 (1991).

[40] See e.g., J.P. Ostriker, *Ann. Rev. Astron. Astrophys.* **31**, 689 (1993); A. Liddle and D. Lyth, *Phys. Rep.* **231**, 1 (1993).

[41] J. Bartlett et al., *Science* **267**, 980 (1995).

[42] W. Freedman et al., *Nature* **371**, 757 (1994); M. Fukugita, C.J. Hogan, and P.J.E. Peebles, *ibid* **366**, 309 (1993).

[43] Q. Shafi and F. Stecker, *Phys. Rev. Lett.* **53**, 1292 (1984); S. Achilli, F. Occhionero, and R. Scaramella, *Astrophys. J.* **299**, 577 (1985); S. Ikeuchi, C. Norman, and Y. Zahn, *Astrophys. J.* **324**, 33 (1988); A. van Dalen and R.K. Schaefer, *Astrophys. J.* **398**, 33 (1992); M. Davis, F. Summers, and D. Schlegel, *Nature* **359**, 393 (1992); J. Primack et al., *ibid* **74**, 2160 (1995); D. Pogosyan and A.A. Starobinskii, astro-ph/9502019.

[44] M.S. Turner, G. Steigman, and L. Krauss, *Phys. Rev. Lett.* **52**, 2090 (1984); M.S. Turner, *Physica Scripta* **T36**, 167 (1991); P.J.E. Peebles, *Astrophys. J.* **284**, 439 (1984); G. Efstathiou et al., *Nature* **348**, 705 (1990); L. Kofman and A.A. Starobinskii, *Sov. Astron. Lett.* **11**, 271 (1985).

[45] S. Dodelson, G. Gyuk, and M.S. Turner, *Phys. Rev. Lett* **72**, 3578 (1994); J.R. Bond and G. Efstathiou, *Phys. Lett. B* **265**, 245 (1991).

[46] R. Cen, N. Gnedin, L. Kofman, and J.P. Ostriker, *ibid* **399**, L11 (1992); R. Davis et al., *Phys. Rev. Lett.* **69**, 1856 (1992); F. Lucchin, S. Mattarese, and S. Mollerach, *Astrophys. J.* **401**, L49 (1992); D. Salopek, *Phys. Rev. Lett.* **69**, 3602 (1992); A. Liddle and D. Lyth, *Phys. Lett. B* **291**, 391 (1992); J.E. Lidsey and P. Coles, *Mon. Not. R. astron. Soc.* (1993); T. Souradeep and V. Sahni, *Mod. Phys. Lett. A* **7**, 3541 (1992).

[47] M.S. Turner, *Physica Scripta* **T36**, 167 (1991).

[48] L. Krauss and M.S. Turner, astro-ph/9504003.

[49] E.J. Copeland, E.W. Kolb, A.R. Liddle, and J.E. Lidsey, *Phys. Rev. Lett.* **71**, 219 (1993); *Phys. Rev. D* **48**, 2529 (1993); M.S. Turner, *ibid*, 3502 (1993); *ibid* **48**, 5539 (1993).

[50] L. Knox, astro-ph/950454 (submitted to *Phys. Rev. D*).

[51] M. Kamionkowski et al., *Astrophys. J.* **426**, L57 (1994).

Particle Physics and Cosmology

John ELLIS

Theoretical Physics Division, CERN
CH - 1211 Geneva 23

ABSTRACT

The agreement between the observed light element abundances and calculations of homogeneous cosmological nucleosynthesis constrains inhomogeneous models, and suggests that most of the matter in the Universe is invisible Dark Matter. This could be in the form of neutrinos, lightest supersymmetric particles (LSPs) or axions. Interactions between LSPs or axions and nuclear matter are controlled by the spin decomposition of the nucleon.

1 Cosmological Nucleosynthesis and Non-Baryonic Dark Matter

One of the most important pieces of evidence for the Big Bang model is provided by the abundances of the light nuclei. For example, all stars, galaxies, etc., which we can measure contain at least 24% by mass of ^{4}He. It is believed that this and the smaller abundances of other light elements (D, ^{3}He, ^{7}Li) were cooked by the Universe when it was about 10^9 times hotter and smaller than it is today. The temperature of the Universe would have been about 10^9 K when its age was about 100 secs. During the early history of the Universe, when its expansion was dominated by relativistic matter, the age t would have been related to the temperature T by the following approximate relation

$$t(\text{sec}) \simeq \left[T(\text{MeV})\right]^{-2} \tag{1}$$

with the numerical coefficient depending on the number of particle species. As seen in Fig. 1, the consistency with observation of calculations of nucleosynthesis in a homogeneous Big Bang requires there to be only a small number of neutrino species, in agreement with the determination from LEP

$$N_\nu = 2.987 \pm 0.016 \tag{2}$$

and the baryon density ρ_B today to be less than the critical density [1] ρ_C:

$$\Omega_B \equiv \rho_B/\rho_C \sim 0.01 \text{ to } 0.1 \tag{3}$$

which is rather more than that seen shining in stars, etc, ...

Going back in time, the next interesting event in the history of the Universe which we encounter is the transition from quarks and gluons to protons and neutrons. Just as atoms ionize at high temperature (cf. the microwave background radiation), we expect quarks to "ionize" or deconfine at high temperatures, above about 200 MeV or 2×10^{12} K. Above this temperature, hadronic matter is believed to exist in the form of a plasma of quarks and gluons. As the Universe expanded and cooled, the quarks and gluons would have combined to form protons, neutrons and other strongly-interacting particles, perhaps leading also to inhomogeneities in the early Universe, if the quark-hadron transition was first order. It has been suggested [2] such these inhomogeneities might relax the above upper bound on the density of strongly-interacting matter in the Universe today, and the possibility has even been raised that the baryons could provide the critical density and all the dark matter beloved of astrophysicists. This possibility seems to be disfavoured by recent theoretical estimates of the parameters of the quark-hadron phase transition [3], which suggest that the quark-hadron transition is at most weakly first-order. The existence of the quark-gluon plasma and the nature of the phase transition are being explored by experimental programmes colliding heavy nuclei at Brookhaven and at CERN.

Inhomogeneities could have arisen during the quark-gluon phase transition, and possibly affected Big Bang nucleosynthesis if the transition was first order. In a first-order transition, bubbles of the new vacuum form separated by regions of the old phase. If the distance between nucleation sites of the new phase is larger than the characteristic diffusion length of protons in the early Universe

$$d \gtrsim 0.5 \text{ m} \tag{4}$$

then neutrons diffuse more than protons, and the n/p ratio as well as the $\eta = n_B/n_\gamma$ could vary through space. This could lead to different light element abundances, and hence possibly a relaxation of the cosmological upper bound on the present baryon density [2]. However, estimates of the QCD phase transition parameters suggest for the bubble surface tension σ and the latent heat L:

$$\sigma/T_c^3 \sim 0.1 \,, \qquad L/T_c^4 \sim 10$$

which corresponds to $d \sim$ (1 to 10) cm.

Moreover, calculations of this scenario suggest that if $\Omega_B = 1$ then there is an overabundance of ^{7}Li. This problem could be avoided only if there was some special process able to deplete the cosmological ^{7}Li abundance, but none is known. Recent

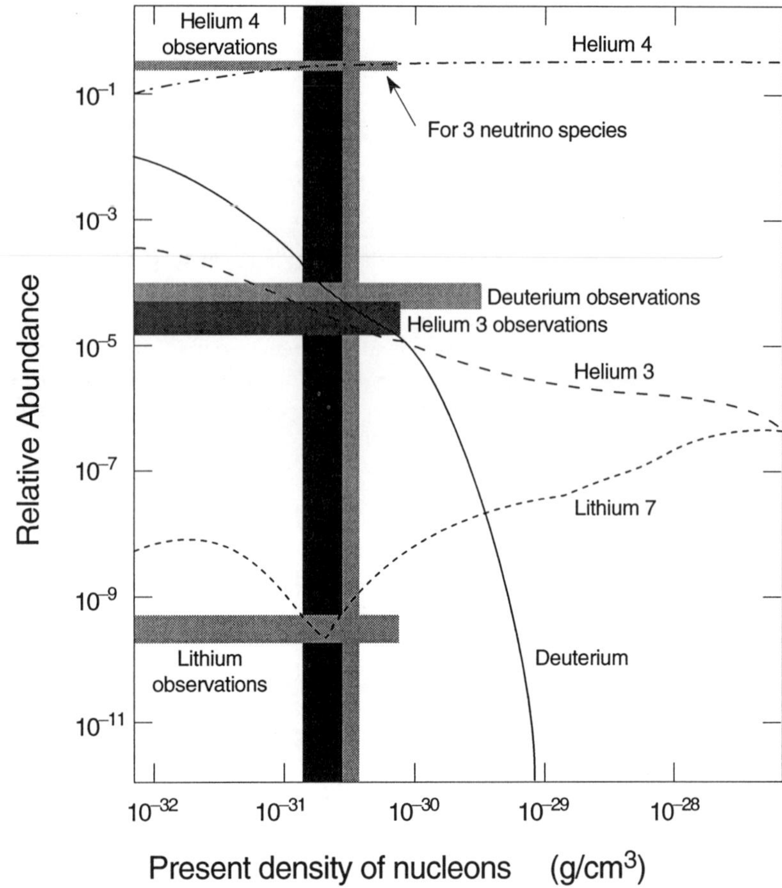

Figure 1: Abundances of the light elements: confrontation between data and predictions based on homogeneous Big Bang nucleosynthesis [1].

studies [4] indicate that there is also a problem with the abundance of ^{4}He, which is also overproduced, and with deuterium D, which tends to be underproduced in such an inhomogeneous nucleosynthesis scenario. As seen in Fig. 2, the general conclusion is that $\Omega_B = 1$ is excluded, that there is very little room for any substantial deviation from the standard homogeneous Big Bang nucleosynthesis scenario, and hence that

$$n_B/n_\gamma \simeq 3 \times 10^{-10} \tag{5}$$

It seems that a strong first-order quark-hadron transition is disfavoured by astrophysical as well as theoretical considerations.

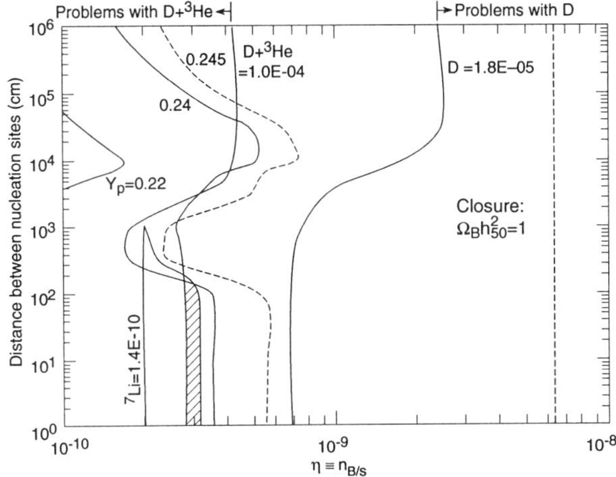

Figure 2: Constraints on inhomogeneous Big Bang nucleosynthesis models imposed by the observed light element abundances [4].

2 How Much Dark Matter ?

Inflation tells us to expect that the density of the Universe as a whole is indistinguishable from the critical density: $\Omega = 1$. However, observations only provide $\Omega_{\text{visible}} \lesssim 0.01$, whereas Big Bang nucleosynthesis apparently restricts $\Omega_B \lesssim 0.1$. Measurements of rotation curves suggest that galactic haloes contain $\Omega_{\text{halo}} \sim 0.1$. Thus the halo could in principle be made out of baryons, although arguments have been given that these could not be in the form of gas, dust or snowballs [5]. The possibility that they could be brown dwarfs has attracted much interest recently, and some microlensing events caused by brown dwarfs have apparently been seen [6]. However, the best estimate is that they only contribute a fraction [7]

$$f = 0.20^{+0.33}_{-0.14} \tag{6}$$

of a simple spherical halo. Therefore, it seems that the dominant contribution to the local halo density, estimated to be

$$\rho_{halo} \sim 0.3\,\text{GeV}\,\text{cm}^{-3} \sim 0.01\,m_{\odot}\,pc^{-3} \tag{7}$$

is from non-baryonic dark matter.

3 Hot or Cold Dark Matter ?

Theories of galaxy formation seem to require some non-baryonic dark matter to enhance initial density perturbations via gravitational instability. Theorists of structure formation in the early Universe distinguish two categories of dark matter: hot, which was relativistic when galaxy-sized perturbations came within the event horizon and began to grow, and cold, which was non-relativistic at that epoch. The one you favour depends on your pet theory of galaxy formation. If you believe these formed from a Gaussian random field of perturbations laid down during inflation, then you should prefer cold dark matter. This is because it enables perturbations to grow on all scales from galaxies upwards, whereas hot dark matter escapes from small galaxy-sized perturbations, retarding their growth. Thus, in a pure hot dark matter scenario, galaxies only form late as a result of larger structures breaking up. This scenario seemed to be disfavoured by the observation of galaxies and quasars at high redshifts, as well as by other considerations. However, if galaxies formed from seeds, such as cosmic strings, hot dark matter might be preferred.

Assuming some type of inflationary scenario, the comparison of COBE [8] and other large-scale data with other data on structures at smaller scales seems to require a modification of the pure cold dark matter scenario [9]. One possibility is that there is a non-zero cosmological constant, but this seems very unlikely from the point of view of particle physics, since there is no known natural reason why the cosmological constant should be in the range of cosmological utility. The two prime candidates for modifying the standard cold dark matter scenario seem to be a mixed dark matter with an admixture of hot dark matter

$$\Omega_{\text{Cold}} \sim 0.7, \ \Omega_{\text{Hot}} \sim 0.2, \ \Omega_{\text{Baryons}} \lesssim 0.1 \tag{8}$$

or a tilt in the spectrum of primordial density perturbations. The parameters of these two models can be fixed by galaxies and by the COBE data, which fix the overall normalization of $\delta\rho/\rho$: the question is then what the models predict at intermediate scales, in particular for the peculiar motions of clusters. At least some astrophysicists [10] will tell you that the cold dark matter with tilt model predicts peculiar motions on the scale of clusters that are too small by comparison with the data. Moreover, at least in simple models one expects very small amount of tilt. Therefore, my own preference at the moment is for a mixed dark matter model as in Eq. (8).

4 Massive Neutrinos ?

The relic energy density in neutrinos can be in the range of interest to astrophysicists and cosmologists for three regions of the neutrino mass, as seen in Fig. 3. When the neutrino mass is much less than 1 MeV, its number density is independent of its mass, so its energy density increases linearly with the mass and, because of their very large abundance similar to that of photons (2), becomes comparable to the critical density when the neutrino weighs about 30 eV. A neutrino in this mass range would constitute hot dark matter. For larger neutrino masses, the neutrino density Ω_ν is too large, until the neutrino mass increases to a few GeV. In this region, the neutrino number density becomes suppressed by a Boltzmann factor exp $-m_\nu/T$, and the relic massive neutrino density can be again of astrophysical and cosmological interest [11]. Neutrinos and many other neutral particles can annihilate through the Z^0, so their annihilation cross-sections are maximal when $m_\nu = m_Z/2$, resulting in a local minimum of the neutrino energy density. As the neutrino mass is further increased, the annihilation cross-section decreases, resulting in an increase in the number density and hence the energy density. There is thus a second region of large neutrino masses around 1 TeV where the neutrino could constitute cold dark matter [12]. Similar features are exhibited by the relic density of other neutral non-baryonic candidates for dark matter.

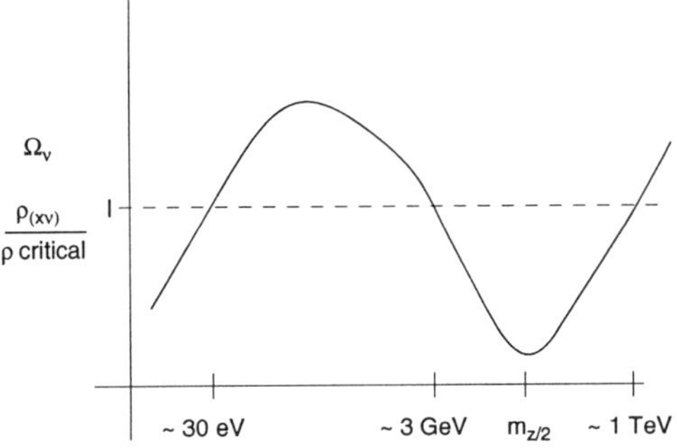

Figure 3: Qualitative picture of the neutrino relic density ρ_ν as a function of the neutrino mass m_ν.

Data on solar neutrinos [13] may provide extra motivation for the possibility that one of the neutrinos, most probably the ν_τ, contributes significantly to dark matter. As is well known, all four experiments searching for solar neutrinos report a deficit compared with the standard solar model. This could be due to astrophysical reasons, but there is a very elegant explanation in terms of matter-enhanced neutrino oscillations. The preferred fit to the combined data on solar neutrinos requires

$$\delta m_\nu^2 \sim 10^{-5}\, eV^2,\ sin^2 2\theta \sim 10^{-2} \tag{9}$$

Very light neutrino masses such as these find a natural explanation within GUTs, via the seesaw mass matrix mechanism [14]. According to this, there is mixing between the light left-handed neutrinos of the Standard Model and massive right-handed singlet neutrinos, via a mass matrix of the form

$$(\nu_L, \bar{\nu}_R) \begin{pmatrix} \sim 0 & m_q \\ m_q & M_{\rm GUT} \end{pmatrix} \begin{pmatrix} \nu_L \\ \bar{\nu_R} \end{pmatrix} \tag{10}$$

where $M_{\rm GUT}$ should be within a few orders of magnitude of the grand unification mass scale $m_X \sim 10^{16}$ GeV. This mass matrix suggests the hierarchy

$$m_{\nu_e} : m_{\nu_\mu} : m_{\nu_e} \simeq m_u^2 : m_c^2 : m_t^2 \tag{11}$$

which, when combined with Eq. (9), suggests that

$$m_{\nu_e} << m_{\nu_\mu} \sim 3 \times 10^{-3}\, \mathrm{eV} \tag{12}$$

Scaling up the ν_μ mass by m_t^2/m_c^2, we arrive at the estimate

$$m_{\nu_e} \sim 10\, \mathrm{eV} \tag{13}$$

offering the possibility [15] that the ν_τ could constitute the hot dark matter discussed in the previous section.

More detailed modelling of neutrino mixing angles finds that the value of $\sin^2 2\theta$ in Eq. (9) is very natural, and suggests that $\sin^2 \theta_{\mu\tau}$ could be in the range detectable in a new round of accelerator experiments. Two such experiments, NOMAD and CHORUS, have started taking data at CERN [16], and may be able to tell us within a couple of years whether the ν_τ has a mass in the range of Eq. (13) and a mixing angle $\sin^2 2\theta_{\mu\tau} \gtrsim 10^{-4}$, as seen in Fig. 4. In the longer term, ideas are being discussed [17] at CERN for sending a neutrino beam to the Gran Sasso laboratory, a distance of 732 km. The neutrino beam would be generated along a beam transfer line from the SPS to the LHC, and would provide a known flux of neutrinos that could be used to test suggestions of oscillations in atmospheric neutrinos.

Could the non-baryonic dark matter, argued above to dominate the galactic halo, consist of neutrinos? It has long been argued that the neutrino density in the halo is significantly restricted by phase-space arguments. Updating these arguments with the ranges of neutrino masses and other prameters currently favoured, we find [18] that at most a very small fraction of our galactic halo density could be made up of neutrinos. This and the estimate [7] of the brown dwarf density mentioned earlier together suggest that most of the halo material is cold dark matter, to which the rest of this talk is devoted.

5 Lightest Supersymmetric Particle ?

Most supersymmetric theories contain one stable particle, which should be around today as a cosmological relic from the Big Bang [19]. This is because most supersymmetric theories have a multiplicatively-conserved quantum number called R parity, that takes the value $+1$ for all particles and -1 for all sparticles. Its conservation is linked to B and L conservation, since R parity may be represented as

$$R = (-1)^{3B+L+2S} \tag{14}$$

This expression also shows how R parity may be violated, for example, by a violation of L either in the vacuum through a vacuum expectation value for some sneutrino field $\tilde{\nu}$, or explicitly, for example by a coupling between a Higgs field and a lepton field. However, these possibilities are severely constrained by laboratory limits on lepton-number-violating interactions, as well as by cosmological considerations. There are three important implications of R parity conservation. (1) Sparticles must be produced in pairs, for example $e^+e^- \to \tilde{\mu}^+\tilde{\mu}^-$ or $\bar{p}p \to \tilde{q}\tilde{g}X$. (2) Heavier sparticles must decay into lighter sparticles, for example $\tilde{e} \to e\tilde{\gamma}$, $\tilde{q} \to q\tilde{g}$. (3) The lightest supersymmetric particle must be stable, because it has no available decay mode.

If the stable relic particle had electromagnetic charge or strong interactions, then it would have presumably condensed along with ordinary matter into galaxies, stars and planets, and hence be detectable as an anomalous heavy isotope, with a calculated abundance

$$\frac{n(\text{relic})}{n(p)} \sim \frac{10^{-10}}{\alpha^2} \begin{array}{ll} \nearrow \; 10^{-10} & \text{(strongly interacting)} \\ \searrow \; 10^{-6} & \text{(weakly interacting)} \end{array} \tag{15}$$

These estimates conflict with experimental upper limits [20] on the abundances of anomalous heavy isotopes relative to protons

$$\frac{n(\text{relic})}{n(p)} \lesssim 10^{-15} \quad \text{to} \quad 10^{-30} \quad \text{for} \quad 1\,GeV \lesssim m_{\text{relic}} \lesssim 10\,TeV \tag{16}$$

Hence we conclude [24] that the stable relic particle must be electromagnetically neutral and only have weak interactions.

Supersymmetric candidates include the sneutrinos $\tilde{\nu}$ of spin zero, some mixture of $\tilde{\gamma}$, $\tilde{H}$ and $\tilde{Z}$, called a neutralino, of spin 1/2, and the gravitino of spin 3/2. The most plausible of these is the lightest neutralino. There are in the minimal supersymmetric extension of the Standard Model four neutralinos, whose masses and mixing are parametrized by three parameters: $m_{1/2}$, the umixed gaugino mass, μ, a Higgs mixing parameter, and $\tan\beta$, the ratio of supersymmetric Higgs vacuum expectation values. The composition of the lightest supersymmetric particle simplifies in the limits $m_{1/2} \to 0$, in which it is approximately a photino $\tilde{\gamma}$, and $\mu \to 0$, in which

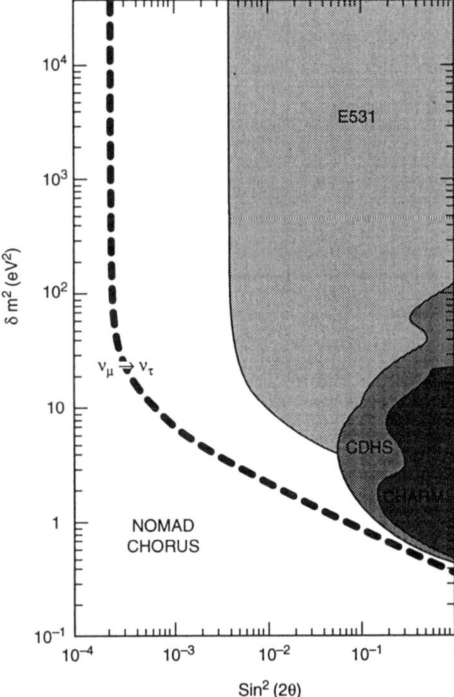

Figure 4: Present experimental limits on (ν_μ, ν_τ) mixing in the $(\sin^2\theta, \delta m^2)$ plane, and the planned sensitivity of the CHORUS and NOMAD experiments [15].

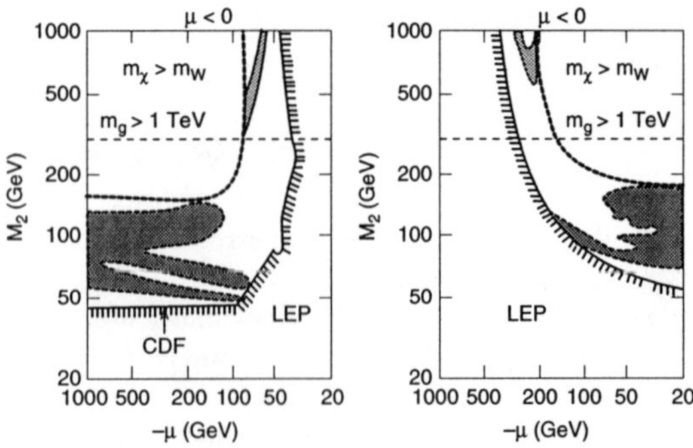

Figure 5: Regions of the $(\mu, m_{1/2})$ plane excluded by LEP and CDF searches, together with regions where the lightest neutralino mass $m_\chi < m_W$, and the relic neutralino density is in an "interesting" range $0.1 < \Omega_{LSP}h^2 < 1$.

limit it is a higgsino $\tilde{H}$. However, experiments rule out either of these limits [21], and the generic lightest neutralino is a complicated mixture of $\tilde{\gamma}$, $\tilde{Z}$ and $\tilde{H}$. Constraints on neutralinos and their charged and strongly-interacting partners constrain the lightest supersymmetric particle

$$m_\chi \gtrsim (10 \text{ to } 20)\,\text{GeV} \tag{17}$$

As seen in Fig. 5, there are large regions of the neutralino parameter space where $\Omega_{\text{LSP}} = \rho_{\text{LSP}}/\rho_{\text{critical}} \simeq 1$ is possible. We therefore conclude that the LSP is a good candidate for the non-baryonic cold dark matter.

6 Searches for Dark Matter Particles

Some of the search strategies discussed in this section may also apply to other candidates for cold dark matter, but we will focus on neutralinos.

6.1 Annihilation in the Galactic Halo

If the galactic halo is made of non-baryonic dark matter, then most of its density would consist of Cold Dark Matter particles. Every once in a while, these will find each other and annihilate in the halo, producing mainly $\bar{l}l$ and $\bar{q}q$ final states, which eventually yield stable particles such as $\bar{p}, p, e^+, e^-, \gamma$ and ν in the cosmic rays [22]. Searches for antiprotons, positrons and γ's have been used [23] to set upper limits on the density of cold dark matter particles in the halo which are several times larger than the expected halo density. Since the fluxes are quadratically dependent on the relic density, and since the fluxes of antiprotons and positrons are uncertain because of uncertainties in the lifetime of cosmic ray particles within the galactic magnetic field, it is unlikely that this search strategy will be able to prove or disprove soon the hypothesis that the halo is made out of neutralinos.

6.2 Annihilation in the Sun or Earth

Here the idea [24] is that every once in a while a Cold Dark Matter particle will pass through the Sun and collide elastically with a nucleus inside it. This will cause it to lose recoil energy, which may transfer it from a hyperbolic orbit to an elliptic orbit with a perihelion below the solar surface. Repeated collisions with solar nuclei will then cause it to fall into an approximately isothermal distribution whose temperature is fixed by energy balance with the core of the Sun.[1] The population of cold dark matter particles within the Sun is controlled by evaporation from the solar surface and/or annihilation in its core. The latter could give rise to an observable flux of high-energy solar neutrinos, which could be detected either directly in proton decay experiments or indirectly via the upward-going muons produced by their collisions in rock beneath the detectors.

The rate Γ_T for neutralino trapping by the Sun is proportional to the elastic neutralino-nucleus scattering cross-section $\sigma(\chi A \to \chi A)$, and the annihilation rate Γ_A is proportional to the trapping rate: $\Gamma_A = \Gamma_{T/2}$ if the neutralino mass exceeds about 3 GeV, so that evaporation of neutralinos from the surface of the Sun can be neglected. The dominant nuclear species inside the Sun is just the proton p, so the annihilation rate is essentially proportional to $\sigma(\chi p \to \chi p)$. The spin-dependent scattering matrix element M is given by Z and squark $\tilde{q}$ exchange. In the simplified case where χ is a pure $\tilde{\gamma}$,

$$M \propto \sum_q Q_q^2 \Delta q \tag{18}$$

where the Δq are the fractions of the proton spin carried by different species of quark q and we have assumed, for the sake of simplicity, degenerate $\tilde{q}$ masses. Neutron β-decay tells us the value of $g_A/g_V = \Delta u - \Delta d$, and hyperon β-decay together with flavour $SU(3)$ fixes $\Delta u + \Delta d - 2\Delta s$. Early estimates of the individual Δq assumed [25] that $\Delta s = 0$, as in the naïve quark model, but this is disfavoured by data on polarized lepton-nucleon scattering. *Deep-inelastic polarized ep or μp scattering* measures exactly the photino-proton scattering combination (18), and the early EMC experiment at CERN indicated [26] that $\Delta s \neq 0$. More recent data from the SMC experiment at CERN [27] and the E142, E143 experiments at SLAC [28] are highly consistent if higher-order perturbative QCD corrections are taken into account, and highly compatible with the Bjorken sum rule, yielding $\alpha_s(M_Z^2) = 0.122^{+0.005}_{-0.009}$ [29]. A global fit to all available data yields [29]

$$\begin{aligned} &\Delta u = 0.83 \pm 0.03 \ , \quad \Delta d = -0.42 \pm 0.03 \ , \quad \Delta s = -0.10 \pm 0.03 \\ &\Delta\Sigma \equiv \Delta u + \Delta d + \Delta s = 0.31 \pm 0.07 \end{aligned} \tag{19}$$

A small value of $\Delta\Sigma$ is expected within Skyrmion-like soliton models of the nucleon [30], but some authors have suggested that the axial $U(1)$ anomaly may play a rôle in suppressing $\Delta\Sigma$ [31].

The EMC and subsequent results indicate that $\sigma(\tilde{\gamma} p \to \tilde{\gamma} p)$, and hence the solar trapping rate, is smaller than the naïve quark model would have indicated. As a

[1] The LSP is unable to cool the core of the Sun sufficiently to have an effect on the low-energy solar neutrino flux.

result, assuming supersymmetric model parameters that give a relic density close to the critical density, and making other conservative assumptions, the upper limit on the local neutrino density is several times the expected halo density [32]. In this case, there is reasonable hope that a new round of experiments will be able to explore the range of densities expected for Cold Dark Matter particles that constitute most of the galactic halo.

6.3 Scattering in the Laboratory

The most direct way to search for cold dark matter particles is via their elastic scattering on heavy nuclei in the laboratory [33]. The Cold Dark Matter particles are expected to have a typical velocity $v \simeq 300$ km s^{-1}, and to deposit a recoil energy

$$\Delta E < m_\chi v^2 = 10 \left(\frac{m_\chi}{10\,\mathrm{GeV}} \right) \mathrm{keV} \tag{20}$$

The effective interaction between neutralinos and quarks is given by

$$\mathcal{L}_{eff} = \sum_q \left[(\bar{\chi}\gamma_\mu\gamma_5\chi) \left(\bar{q}\gamma^n (A_q(\frac{1-\gamma_5}{2}) + B_q(\frac{1+\gamma_5}{2}))q \right) + \bar{\chi}\chi\, C_q m_q \bar{q}q \right] \tag{21}$$

The first term is due to $\tilde{f}$ and Z exchange, and is spin-dependent. The second term is due to H and $\tilde{f}$ mixing, and is spin-independent. The proton matrix element of the spin-dependent interaction may be written in the form

$$M_{SD}(\chi p \to \chi p) = 4\,\mathbf{s}_\chi \cdot \mathbf{s}_p \sum_{q\epsilon p} (B_q - A_q)\Delta q \tag{22}$$

and similarly for the neutron, where the s are spin vectors, and the Δq are the different quark contributions to the proton spin. In what follows, we will assume [34]

$$\Delta n = 0.77 \pm 0.08,\ \Delta d = -0.49 \pm 0.08,\ \Delta s = -0.15 \pm 0.08 \tag{23}$$

which comes from a previous fit to data on polarized μp scattering, and is consistent with more recent data on polarized μD, μp, e^3He and ep scattering[29]. The proton matrix elements of the spin-independent interaction may be written as

$$M_{SI}(\chi p \to \chi p) = \left[\frac{\hat{f} m_u C_u + m_d C_d}{m_u + m_d} + f C_s + \frac{2}{27}(1 - f - \hat{f})(C_c + C_b + C_t) \right] \tag{24}$$

where the fractions of the proton mass carried by quarks are estimated to be [34]

$$\hat{f} \simeq 0.05,\ f \simeq 0.2 \tag{25}$$

on the basis of extractions of the $\pi - N$ σ-term from, e.g., *low-energy* πN *scattering*. The spin-dependent interaction is dominant for light nuclei, whilst the spin-independent interaction is coherent and dominant for heavy nuclei.

Figure 6 displays figures of merit for spin-dependent LSP-nucleus scattering, as calculated [34] in a naive single-particle shell model, and as corrected by the more successful odd-group model [35]. We see that Fluorine is likely to be the best nucleus for monitoring the spin-dependent interaction. LSP-nucleus elastic scattering rates are given by

$$R = (R_{SD}+R_{SI})Y\left(\frac{4m_\chi m_N}{(m_\chi+m_N)^2}\right)\left(\frac{\rho_\chi}{0.3\,\mathrm{GeV\,cm^{-3}}}\right)\left(\frac{v}{320\,\mathrm{kms^{-1}}}\right)\text{ events } kg^{-1}d^{-1} \tag{26}$$

where Y is the isotopic abundance, ρ_{LSP} is the local density of cold dark matter particles, and v is the velocity of the LSP relative to the Earth. The quantity R_{SD} is the spin-dependent interaction factor

$$R_{SD} = 5.5\left(\frac{10^4\,GeV^2}{m_0^2-m_\chi^2}\right)\lambda^2 J(J+1)\zeta_s\left(\frac{m_0^2-m_\chi^2}{e^2}\sum_q(B_q-A_q)\Delta q\right)^2 \tag{27}$$

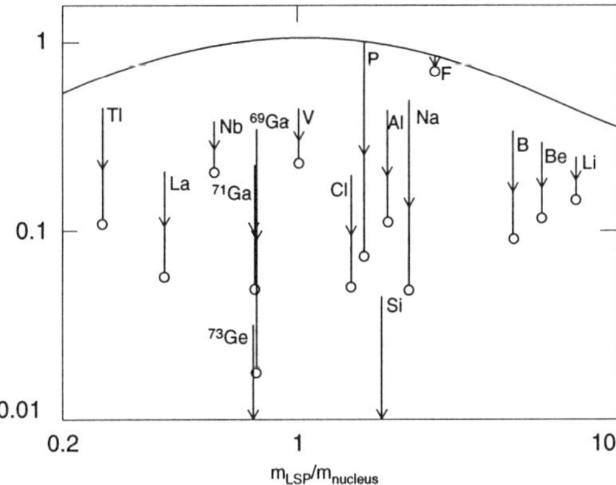

Figure 6: Figures of merit for spin-dependent neutralino-nucleus scattering. The tops of the arrows are the naive shell-model predictions, the bottoms are those found [32] in the odd-group model [33].

where λ is a spin factor obtained using the odd-group model, J is the nucleon spin, ζ_s is a spin form factor, and the Δq were given in Eq. (23). The quantity R_{SI} is the spin-independent interaction rate given by

$$R_{SI} = 210\, m_N^2\, m_\chi^4\, \zeta_c\left(\hat{f}\frac{m_u C_u + m_d C_d}{m_u+m_d} + f\,C_s + \frac{2}{27}(1-f-\hat{f})\,(C_c+C_b+C_t)\right)^2 \tag{28}$$

where ζ_c is the charge form factor of the nucleus, and the $\hat{f}, f$ were given in Eq. (25). We adjust the supersymmetry model parameters so as to give critical density for the LSP for a present-day Hubble expansion rate $h_0 = 50$ km /second/MPc. The resulting elastic scattering rates depend on the masses of the supersymmetric Higgs particles, as seen in Fig. 7. We see there that Fluorine and Germanium are useful in complementary regions of the LSP parameter space, whilst Thallium may be useful in regions similar to those of Germanium. Figure 8 shows in the case of Germanium the "probability" of detection, defined to be the fraction of the parameter space where $\Omega_{\rm LSP} = 1$ is possible, in which the elastic scattering rate is "observable", in the sense of being above 0.1 events per kilogram per day [34]. We see that there is quite a good probability of detection, although this cannot be guaranteed. We also see that the regions of supersymmetric parameter space that are detectable in cold dark matter searches complement and extend those accessible to LEP II and the LHC.

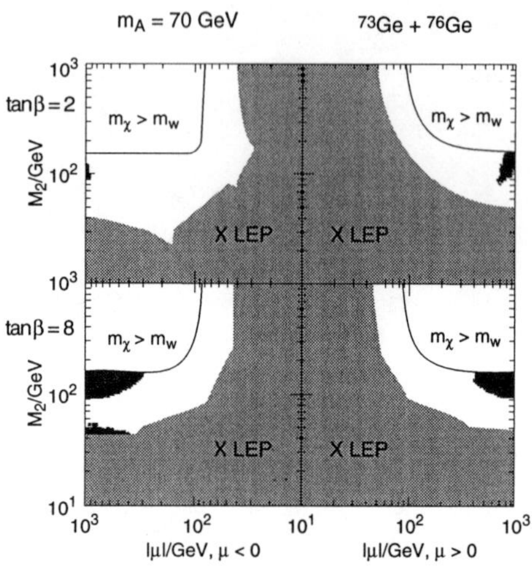

Figure 7: Elastic scattering rates for $^{73}Ge + ^{76}Ge$, for the pseudoscalar Higgs mass $m_A = 70\,GeV$ [32]. The rates are above 0.1 events $kg^{-1}d^{-1}$ in the shaded regions.

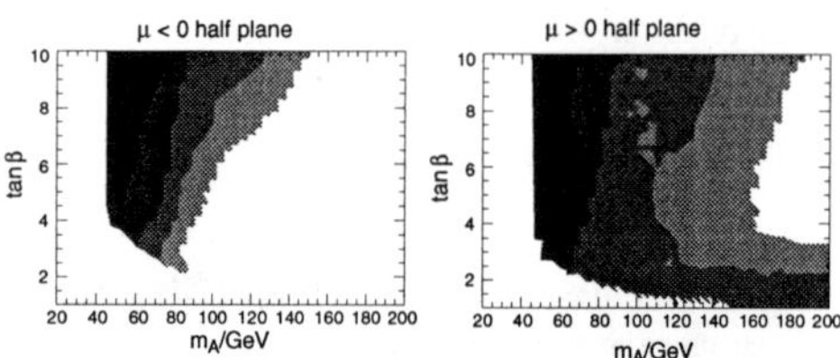

Figure 8: The "probability" of detecting neutralino dark matter, as defined by the fraction of the parameter space where the critical density $\Omega_{LSP} = 1$ can be attained in which the $\chi - Ge$ elastic scattering rate exceeds 0.1 events $kg^{-1}d^{-1}$ [32].

7 Axions ?

As you know, the axion was introduced to ensure CP conservation in the strong interactions [36]. It is a light spin-0 boson whose mass and couplings are determined by a model-dependent decay constant f_a, analogous to the pion decay constant f_π:

$$m_a \sim \frac{m^2\pi}{f_a}\ , \quad g_{a\bar{f}f} \sim \frac{m_f}{f_a}\ , \quad g_{a\gamma\gamma} \sim \frac{1}{f_a} \tag{29}$$

The absence of the axion in many laboratory experiments tells us that

$$f_a \gtrsim 1 \text{ TeV} \tag{30}$$

Astrophysics places more stringent lower bounds on f_a [37]. For example, the fact that the Sun shines photons, not axions, and unsuccessful searches for the solar axio-electric effects: $a+$ (atomic e) $\rightarrow$ (unbound e) tell us that [37]

$$f_a \gtrsim 10^7 \text{ GeV} \tag{31}$$

On the other hand, cosmologists tell us that axions should be present as relics from the Big Bang in the form of coherent waves that would act as Cold Dark Matter, despite the axion's very low mass. These coherent waves would have an acceptable energy density [38]

$$\Omega_a \equiv \frac{\rho_a}{\rho_{crit}} \lesssim 1 \quad \text{if} \quad f_a \lesssim 10^{12} \text{ GeV} \tag{32}$$

and the density would be astrophysically intersting ($\Omega_a \gtrsim 0.1$) if $f_a \gtrsim 10^{11}$ GeV. Much of the range of values of f_a between (31) and the cosmological upper bound (32) is excluded by considerations of the energetics of Red Giants and White Dwarfs:

$$f_a \gtrsim \text{few} \times 10^9 \text{ GeV} \tag{33}$$

One possibility for extending this bound is provided by the supernova 1987a.

According to the standard theory of Type-II supernovae, 99% of the binding energy $\sim 3 \times 10^{53}$ ergs of the remnant neutron star is released as neutrinos, with at most 1% released as electromagnetic radiation. Since the radius of the neutron star is much larger than the mean free path of the neutrinos, they are thermalized and emitted from a neutrinosphere close to the surface with a temperature $T \sim$ 5 MeV, over a period of about 10 s. This theory is in good agreement with the observations by the Kamiokande [39] and IMB [40] experiments. If axions were sufficiently strongly coupled to nucleons, they would escape freely from the core of the neutron star, reduce the energy carried away by the neutrinos, and shorten the neutrino pulse. The absence of such an effect gives in principle a lower bound on f_a.

To evaluate this bound, we need a model for axion-nucleon couplings, which are related by chiral symmetry to the Δq:

$$g_{aN} = C_{aN}\frac{m_N}{f_a} \tag{34}$$

where [41]

$$C_{ap} = 2[-2.76\Delta n - 1.13\Delta d - 0.98\Delta s - \cos(2\beta)\ (\delta n - \Delta d - \Delta s)] \tag{35a}$$

$$C_{an} = 2[-2.76\Delta d - 1.13\Delta u - 0.98\Delta s - \cos(2\beta)\ (\delta d - \Delta u - \Delta s)] \tag{35b}$$

where $\cos 2\beta$ is related to the ratio of two unknown Higgs vacuum expectation values. The latest determination of the *nucleon-spin decomposition* enables $C_{ap,an}$ to be estimated [29] with relatively small errors, as seen in Fig. 9. The total rate of axion emission by the supernova is approximately proportional to

$$C_{an}^2 + 0.83(C_{an} + C_{ap})^2 + 0.47C_{ap}^2 \tag{36}$$

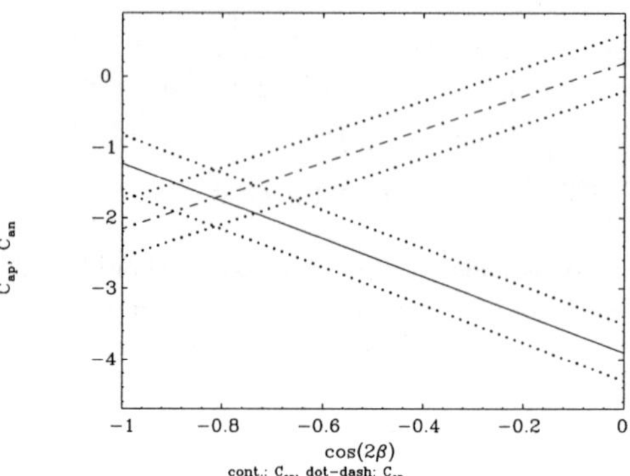

Figure 9: The axion-nucleon coupling coefficients $C_{ap,an}$ (35), shown for different values of $\cos(2\beta)$, including the uncertainties in the nucleon-spin decomposition [29].

The spin uncertainty in this combination is shown in Fig. 10. We see that this uncertainty is much less than that associated with finite-density nuclear effects, that may be as large as an order of magnitude. Thus, polarized lepton-nucleon scattering experiments have made their contribution to bounding f_a. The agreement of supernova 1987a observations with theory indicates that

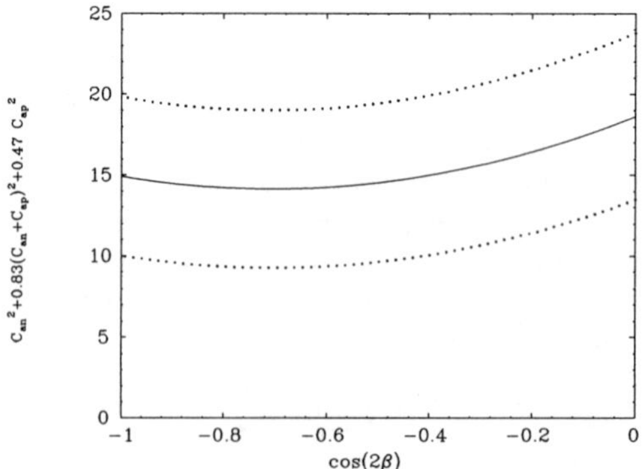

Figure 10: The quantity in (36) which controls the axion emission rate from nucleons in a supernova core.

$$f_a \gtrsim \text{few} \times 10^{10} \text{ GeV} \tag{37}$$

indicating that the interesting cosmological window is still open.

8 A Pre-Revolutionary Situation ?

The current situation in particle astrophysics and cosmology reminds me of that in particle physics just prior to the establishment of the Standard Model of particle physics. The observation by COBE of fluctuations $\delta T/T$ in the microwave background radiation reminds me of the initial discovery of neutral currents in Gargamelle. Both observations were rapidly confirmed by many other experiments, and the prospective mapping of the inflaton potential reminds me of the long effort to determine $\sin^2\theta_W$ precisely. For many particle physicists, the crucial paradigm shift came with the discovery of the J/ψ and charm: the analogue now might be the discovery of a non-zero neutrino mass large enough to provide interesting Hot Dark Matter, for example via neutrino oscillation measurements. The subsequent discovery of the W and Z might be parallelled by the discovery of a massive Cold Dark Matter particle, such as the neutralino or the axion. The discovery of any of these particles would take us beyond the Standard Model, perhaps into the realm of GUTs or supersymmetry. Thus, the coming revolution in particle astrophysics and cosmology might also trigger a new revolution in particle physics itself.

References

[1] W. Fowler and F. Hoyle, *Ap.J.Supp.* **9** (1964) 201;
R. Wagoner, *Ap.J.* **179** (1973) 343.
J. Yang, M.S. Turner, G. Steigman, D.N. Schramm and K.A. Olive, *Ap.J.* **281** (1984) 493.
B. Pagel, *Physica Scripta* **T36** (1991) 7.

[2] J.H. Applegate and C.J. Hogan, *Phys.Rev.* **D31** (1985) 3037;
J.H. Applegate, C.J. Hogan and R. Scherrer, *Phys.Rev.* **D35** (1987) 1151;
C. Alcock, G.M. Fuller and G.J. Mathews, *Ap.J.* **320** (1987) 439.

[3] J. Ignatios, K. Kajantie, H. Kurki-Sainio and M. Laine, Helsinki Preprint HU-TFT-93-43 (1993).

[4] D. Thomas, D.N. Schramm, K.A. Olive and B.D. Fields, *Ap.J.* **406** (1993) 569.

[5] D. Hegyi and K.A. Olive, *Astrophys. J.* **303** (1986) 56.

[6] C. Alcock et al., *Nature* **365** (1993) 621;
E. Aubourg et al., *Nature* **365** (1993) 623.

[7] MACHO Collaboration, C. Alcock et al., astro-ph 9501091 (1995);

[8] G.F. Smoot et al., *Ap. J. Lett.* **396** (1992) L1;
E.L.Wright et al., *Ap. J. Lett.* **396** (1992) L13.

[9] See, for example:
G. Efstathiou, J.R. Bond and S.D.M. White, *Mon. Not. R. Ast. Soc.* **258** (1992) 1P;
M.S. Turner, FNAL preprint Conf-92/313-A (1992).

[10] G. Efstathiou, private communication (1993).

[11] B.W. Lee and S. Weinberg, *Phys. Rev. Lett.* **39** (1977) 165;
P. Hut, *Phys. Lett.* **69B** (1977) 85.

[12] S. Dimopoulos, *Phys. Lett.* **B246** (1990) 347.

[13] P. Anselmann et al., *Phys. Lett.* **B285** (1992) 390, and references therein.

[14] T. Yanagida, Proc. Workshop on the Unified Theory and the Baryon Number in the Universe (KEK, Japan, 1979);
R. Slansky, talk at the Sanibel Symposium, Caltech preprint CALT-68-709 (1979).

[15] J. Ellis, J. Lopez and D.V. Nanopoulos, *Phys. Lett.* **B292** (1992) 189.

[16] CHORUS Collaboration, N. Armenise et al., CERN-SPSC/90-42 (1990);
NOMAD Collaboration, P. Astier et al., CERN-SPSC/91-21 (1991).

[17] C. Rubbia and J.P. Revol, private communications (1993).

[18] J. Ellis and P. Sikivie, *Phys.Lett.* **B321** (1994) 390.

[19] H. Goldberg, *Phys. Rev. Lett.* **50** (1983) 1419;
J. Ellis, J.S. Hagelin, D.V. Nanopoulos, K.A. Olive and M. Srednicki, *Nucl. Phys.* **B238** (1984) 453.

[20] J. Rich, M. Spiro and J. Lloyd-Owen, *Physics Reports* **151** (1987) 239.

[21] J. Ellis, G. Ridolfi and F. Zwirner,*Phys. Lett.* **B237** (1990) 423.

[22] J. Silk and M. Srednicki, *Phys. Rev. Lett.* **53** (1984) 624.

[23] J. Ellis et al., *Phys. Lett.* **B214** (1988) 403.

[24] J. Silk, K.A. Olive and M. Srednicki, *Nucl. Phys.* **B279** (1987) 804.

[25] J. Ellis and R. Jaffe, *Phys.Rev.* **D9** (1974) 1444; **D10** (1974) 1669.

[26] J. Ashman et al., *Phys.Lett.* **B206** (1988) 364 and *Nucl.Phys.* **B328** (1989) 1.

[27] B. Adeva et al., *Phys.Lett.* **B302** (1993) 533;
D. Adams et al., *Phys.Lett.* **B329** (1994) 399.

[28] P.L. Anthony et al., *Phys.Rev.Lett.* **71** (1993) 959;
K. Abe et al., SLAC preprint PUB-6508 (1994).

[29] J. Ellis and M. Karliner, CERN Preprint TH. 7324/94 (1994).

[30] S.J. Brodsky, J. Ellis and M. Karliner, *Phys.Lett.* **B206** (1988) 309.

[31] A.V. Efremov and O.V. Teryaev, Dubna preprint JIN-E2-88-287 (1988);
G. Altarelli and G. Ross, *Phys.Lett.* **B212** (1988) 391;
R.D. Carlitz, J.D. Collins and A.H. Mueller, *Phys.Lett.* **B214** (1988) 229.

[32] J. Ellis, R.A. Flores and S. Ritz, *Phys.Lett.* **B198** (1987) 493.

[33] M. Goodman and E. Witten, *Phys. Rev.* **D30** (1985) 3059.

[34] J. Ellis and R.A. Flores, *Phys. Lett.* **B263** (1991) 259; *Nucl. Phys.* **B400** (1993) 25;
J. Ellis and R.A. Flores, *Nucl. Phys.* **B307** (1988) 883.

[35] J. Engel and P. Vogel, *Phys. Rev.* **D40** (1989) 3132.

[36] R.D. Peccei and H.R. Quinn, *Phys.Rev.Lett.* **38** (1977) 1440, and *Phys.Rev.* **D16** (1977) 1791.

[37] F.T. Avignone et al., *Phys.Rev.* **D35** (1987) 1490.

[38] L. Abbott and P. Sikivie, *Phys.Lett.* **120B** (1983) 133;
J. Preskill, M. Wise and F. Wilczek, *Phys.Lett.* **120B** (1983) 127;
M. Dine and W. Fischler, *Phys.Lett.* **120B** (1983) 137.

[39] K. Hirata et al., *Phys.Rev.Lett.* **58** (1987) 1490.

[40] R.M. Bionta et al., *Phys.Rev.Lett.* **58** (1987) 1494.

[41] R. Mayle et al., *Phys.Lett.* **B203** (1988) 188 and **B219** (1989) 515;
G. Raffelt, *Physics Reports* **198** (199) 1.

HST Highlights: New Views of the Universe[a]

H. S. STOCKMAN

Space Telescope Science Institute
3700 San Martin Drive
Baltimore, Maryland 21218

INTRODUCTION

The First Servicing Mission to the *Hubble Space Telescope (HST)* has restored completely the observatory's capabilities for unexcelled optical resolution and ultraviolet sensitivity[1,2]. Since December 1993, the international community has utilized *HST* to observe a wide variety of astronomical targets, from the impact of Comet Shoemaker-Levy 9 on Jupiter[3] to some of the most distant known galaxies. While the breadth of this research is too great to review other than statistically, Dr. Freedman and I highlight four major topics of extragalactic research which have been enabled by the servicing mission and are especially germane to this 17th Texas Symposium on Relativistic Astrophysics: the Distance Scale, supernovae, the nuclei of radio galaxies, and high redshift galaxies. The first topic, the determination of the Hubble constant utilizing distant Cepheid variables is designated as one of the *HST* Key Projects and is addressed by Dr. Freedman in these proceedings[4].

EXTRAGALACTIC SUPERNOVAE

The study of distant supernovae with *HST* will dramatically improve our understanding of these remarkable phenomena and ultimately may allow an independent means of determining the Hubble Constant. Indeed, measurements of supernovae at cosmological distances hold the promise of determining the cosmological deceleration parameter, q_o[5]. Ironically, the closest supernova in modern history, SN1987A in the Large Magellanic Cloud, has provided one of *HST*'s most mysterious images[6]. Images of SN1987A obtained in a narrow Hα(6563Å) and [NII](6548Å) filter with the corrected *Wide Field Planetary Camera 2* (*WFPC2*) reveal two ring structures in addition to the well known ring which circles the supernova remnant. All three rings appear to have approximately the same eccentricity and axial symmetry. The widths of the two "outer arcs" are unresolved in Hubble images, whereas the inner ring can be resolved into individual knots. Comparison of the images to those predicted for thin shells and cylinders indicate that these structures are true rings of material which has been photoionized by the recent explosion. No single theory for such structures appears to be satisfactory. The proximity of SN1987A offers the unique opportunity to study the interaction of a supernovae with the wind structures created during the last stages of late-star evolution.

[a] Based on observations with the NASA/ESA *Hubble Space Telescope*, obtained at the Space Telescope Science Institute, which is operated by AURA, Inc. under NASA Contract No. NAS 5-26555.

THE NUCLEI OF RADIO GALAXIES

Before the servicing mission, ground-based and *HST* observations found evidence that several nearby radio galaxies possessed both massive cores and internal disk-like structures: the early work of Young[7] and Sargent [8]on M87, the disk of gas and dust in the nucleus of NGC 4261[9] and more than a dozen galaxies with ionization cones in presumably caused by obscuration of ultraviolet light by a central torus[10]. Further study of these inner regions promises to address three major areas in our understanding of Active Galactic Nuclei (AGN). Can we obtain direct kinematic evidence of the putative "black holes" which power AGN through accretion? And can we discern the outer parts of the accretion disk which is the most likely mechanism to remove angular moment from the accreting material. Are the axes of these gas and dust disks aligned with the relativistic jets which are a common feature of bright radio sources? Since the optical emission from the nuclei of radio galaxies is relatively faint compared to typical quasars, they make ideal targets for such studies.

Images of the nucleus of M87 taken with the *WFPC2* in Hα and [NII] have resolved a disk-like structure with a diameter of approximately 74 pc[11] . The derived inclination of the disk is consistent with an axis aligned with the well-known optical and radio jet. The estimated mass of the disk, M_d = 3.9 x 10^3 solar masses, is comparable to that observed in the nucleus of NGC 4261. Using the *Faint Object Camera (FOS)*, Ford and collaborators[12] have mapped the emission line velocities in the disk to within 18 pc of the core. They find 1000 km s^{-1} peak-to-peak radial velocity variations, corresponding to a central mass of 2.4 x 10^9 solar masses. This value is consistent with the estimates made by earlier workers but these *HST* measurements probe an order of magnitude closer to the center and thereby exclude the possibility that a stellar cusp may be responsible for the deduced mass. With these new measurements, the ratio of M87's core mass to light is $M/L > 170$ in solar units (normal stellar mixtures have $M/L < 10$). This is the strongest direct evidence that a massive black hole resides at the center of M87.

The *HST* observations of the M87, NGC 4261, and NGC 4774, show that the inner disk structures have symmetry axes similar to those of the large scale radio jets. It is tempting to infer from these data that the disks, which have accretion lifetimes of about 10^6 yr, must be feed by structures with lifetimes considerably longer than those of the larger radio lobes , or 10^{7-8} yr. Otherwise, we might expect greater angular discrepancies between the older radio structure and the axes of the nuclear dust/accretion disks. Indeed, a broader *HST* survey of 3CR radio galaxies is finding examples of significant misalignments[13]. Internal dust structures due to recent merger events have been found in M84 and 3C84 and are perpendicular with the radio structures, as one might expect. However, similar images of the more distant sources 3C449 and 3C305 show internal dust lanes which are nearly parallel to the radio structures in these galaxies. Further *HST* imaging and high angular resolution spectroscopy of the inner regions of radio galaxies and normal galaxies will be crucial to our better understanding of these sources and their evolutionary status.

HIGH REDSHIFT GALAXIES

If it is not an irony of nature that the structure of high redshift galaxies has a scale comparable to the best seeing obtained for ground-based observatories, it is evidence of how limiting capabilities directly limit our understanding of the Universe. As an example, it has been known for over two decades that clusters of galaxies with redshifts $z > 0.1$ often contain larger populations of bluer galaxies than would be deduced by reversing the aging process in the clusters which we observe in our neighborhood. This effect has generally been interpreted to be due to a higher proportion of young galaxies at those early times, probably unresolved spirals. Indeed, this was dramatically confirmed by deconvolved *HST* images of the $z = 0.4$ cluster CL0939+4713 obtained by Dressler and collaborators prior to the servicing mission[14]. In these *WFPC* images, it is possible to classify the morphologies of ellipticals and spirals to high accuracy. Recently, Dressler has reobserved CL0939+4713 with *WFPC2* and has confirmed the earlier classifications[15]. These new data show considerably more detail than those of the uncorrected *WFPC* and also provide further evidence for mergers and interactions between galaxies in this cluster. Such mergers are a likely mechanism for the relative reduction in spiral galaxies in high density settings between the epoch of CL0939+4713 and today.

Dickinson and others have obtained[16] even deeper images of the cluster which surrounds the bright radio galaxy 3C 324 at $z = 1.2$. At this redshift, the R band used by Dickinson corresponds to U in the rest frame of the cluster. Partly for this reason, many of the resolved structures are not the typical elliptical, smooth spiral, and spiral shapes which characterize galaxies today. FIGURE 1 provides examples of ordinary appearing E/S0 galaxies (in the lower right panel) and a large fraction of blue irregular shapes (in the left panel). The host galaxy of 3C 324 is shown in the upper right panel and its elongated shape is aligned with the radio structure in the *HST* images. When observed from the ground in the 2.2μm (K) band, it appears smoother, more similar to a classical elliptical galaxy. Some of the galaxies which are classified as elliptical in the *WFPC2* images are also bright in the K band (R-$K \approx 6$), thus confirming the existence of an older, established stellar population in these objects.

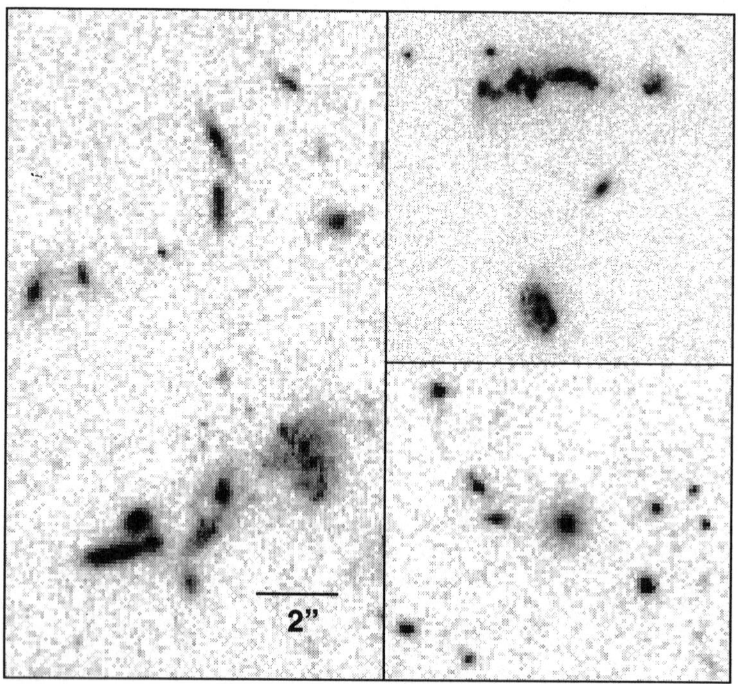

FIGURE 1. *WFPC2 R* band images of galaxies in the field of the radio galaxy 3C 324[16].

Clearly, the combination of the *WFPC2* and the high resolution & low background IR imaging provided by the 1997 installation of the *Near Infrared Camera and Multi-Object Spectrometer (NICMOS)* will yield unique information about the structure of these galaxies and the currently undetectable dwarfs.

Perhaps the most tantalizing *HST* images are those of normal galaxies at redshifts in excess of $z > 2$. The images of CL0939+4713, for instance, have revealed a probable cluster of faint, resolved sources surrounding a previously known QSO at $z = 2.055$. Over two dozen sources lie within a projected distance of 100 kpc of the QSO. Although no spectroscopic data are available to confirm the high redshift of these objects, their general appearance and faintness are unlike the brighter, E/S0 galaxies in the 3C 324 field. Dressler[15] also notes that the more accurate data obtained with *WFPC2* does not indicate that they are part of larger, more diffuse galaxies. Rather they are probably galactic fragments, in the process of forming larger structures through mergers and interactions.

The most distant cluster of galaxies which has been imaged[17] by *HST* is a group of faint sources near the line-of-sight to the bright QSO Q0000-263 ($z = 4.11$). The field was chosen based upon the evidence in the QSO spectrum of an intervening optically-thick absorption system with $z = 3.390$. The redshifts for one of the nearby galaxies (G2, $z = 3.428$) has been confirmed by the detection of Ly-α emission. The probable absorbing galaxy (G1) does not show emission lines but has the correct colors and apparent luminosity to place it at such a high redshift. The inner portions of G2 appear consistent with a relaxed system while the outer regions show some evidence of structure, suggesting ongoing star formation. The importance of these data, apart from their usefulness in understanding early galactic morphologies, is that Cold Dark Matter (CDM) theories predict galactic formation to have occurred relatively recently, $z \leq 2.0$. The very existence of relaxed systems at redshifts of $z = 3$-4 would have serious implications about the CDM picture and the geometry and fate of the expanding universe.

ACKNOWLEDGMENTS

We deeply appreciate the assistance and access to unpublished material provided by Stefi Baum, Chris Burrows, Mark Dickinson, Alan Dressler, & Holland Ford.

REFERENCES

1. TRAUGER, J.T. *et al.* 1994, Astrophys. J. **435**: L3.
2. JEDRZEJEWSKI, R.I. *et al.* 1994, Astrophys. J. **435**: L7.
3. HAMMEL, H.B. *et al.* 1995, Science, in press.
4. FREEDMAN, W. L. 1995, *In* Annals for 17th Texas Symp. on Relativistic Astrophys.
5. PERLMUTTER, S. *et al.* 1995, Astrophys. J. **440**: L41.
6. BURROWS, C.J. *et al.* 1995, Astrophys. J. in press.
7. YOUNG, P.J. *et al.* 1978, Astrophys. J. **221**: 721.
8. SARGENT, W.L.W. *et al.* 1978, Astrophys. J. **221**: 731.
9. JAFFE, W. *et al.* 1993, Nature **364**: 213.
10. WILSON, A.S. & Z. TSETVANOV1994, Astron. J. **107**: 1227.
11. FORD, H.C. *et al.* 1994, Astrophys. J. **435**: L27.
12. HARMS, R. J. *et al.* 1994, Astrophys. J. **435**: L35.
13. BAUM, S. 1994, private communication.
14. DRESSLER, A. , A. OEMLER, J.E. GUNN & H. BUTCHER 1992, Astrophys. J. **404**: L45.
15. DRESSLER, A. , A. OEMLER, W.B. SPARKS & R.A. LUCAS, 1994, Astrophys. J. **435**: L23.
16. DICKINSON, M. *et al.* 1995, in preparation.
17. GIAVALISCO, M. , F.D. MACCHETTO, P. MADAU & W.B. SPARKS 1995, Astrophys. J. **441**: L13.

HST Highlights: The Extragalactic Distance Scale[a,b]

WENDY L. FREEDMAN[c]
(for the H_0 Key Project Team)

[c]*Carnegie Observatories*
813 Santa Barbara Street
Pasadena, California 91101

INTRODUCTION

Early in this century, astronomer Edwin Hubble[1] discovered a correlation between the distance to a galaxy and its recession velocity. Interpreted subsequently as evidence for an overall expansion of the Universe, these and subsequent data form part of the observational basis for the Big Bang cosmological model. The expansion rate or Hubble constant, H_0, determines both the expansion time scale and the size scale of the Universe, in addition to constraining the amount of dark matter, the density of baryons produced in the Big Bang, and early structure formation in the Universe. Unfortunately, however, a measurement of a precise value for the Hubble constant has turned out to be more difficult than anticipated and remains an outstanding problem in observational cosmology to the present day.

In the mid-1970s, providing a solution to this problem was the motivation for setting the final aperture size for the Hubble Space Telescope (HST): the aperture size was chosen to allow the discovery of Cepheid variable stars in galaxies as distant as the Virgo cluster. A decade or so later, the Space Telescope Science Institute Working Group on galaxies specified the measurement of the Hubble constant as one of the top priority or 'Key Projects' to be carried out by the HST. The goal of the Key Project on the extragalactic distance scale is to provide a measurement of the Hubble constant to an accuracy of 10%.

The underlying basis of the Key Project is the measurement of accurate distances to galaxies using the period-luminosity (PL) relation for Cepheid variables. (For recent reviews of the Cepheid distance scale see Feast and Walker[2] and Madore and Freedman[3].) The strategy adopted by our Key Project team has three primary objectives:

(1) discovery of Cepheids and measurement of distances to about 20 nearby spiral galaxies

[a] Based on observations with the NASA/ESA Hubble Space Telescope, obtained at the Space Telescope Science Institute, which is operated by AURA, Inc. under NASA Contract No. NAS 5-26555.

[b] Support for this work was provided by NASA through grant number 2227-87A from the Space Telescope Science Institute which is operated by the Association of Universities for Research in Astronomy Inc. under NASA Contract NAS5-26555.

with distances in the approximate range of 4 < d < 20 Mpc. The distances of these primary galaxies will be used to calibrate several secondary distance techniques which extend out to distances in the range of 20 to 100 Mpc or more (for example, the Tully-Fisher relation, surface brightness fluctuations, the planetary nebula luminosity function, and supernovae of both types Ia and II).

(2) discovery of Cepheids and measurement of the distances to galaxies in both the Virgo and Fornax clusters.

And finally,

(3) to test for systematic effects in the measurement of extragalactic distances; i example, to provide external checks on the zero point of the Cepheid period-luminos relation, to undertake tests to determine if there is a dependence of the Cepheid zeı point on heavy element abundance, and to provide an external comparison of severa independent secondary distance indicators (as a result of (1) above).

Highlights of this Key Project to date are briefly described below. Results from other recent HST programs are described by Stockman in another entry in this volume.

RECENT RESULTS

Progress is now actively being made along all three of the lines described above. Prior to the refurbishment of the telescope, observations were made in two fields in the nearby galaxy M81. This galaxy is a useful calibrator for several secondary methods including the Tully-Fisher relation, the planetary nebula luminosity function, surface brightness fluctuations and potentially for type II supernovae. Thirty Cepheids were discovered and a distance to this galaxy has recently been published.[4,5] In addition, a field in the outer regions of the face-on spiral galaxy M101 was observed and 29 Cepheids discovered.[6] The field observed is one of two chosen to allow a test of the metallicity sensitivity of the Cepheid PL relation. Since the refurbishment mission in December of 1993, the pace of the program has begun to increase rapidly. Data have been acquired for several galaxies: M101 (an inner field), the Virgo cluster galaxy M100, and three inclined spiral galaxies: NGC 925 (a member of the NGC 1023 group), NGC 7331, and NGC 3351 (a member of the Leo Group.) Most recently we have published a distance to the Virgo cluster galaxy M100,[7] and these results are summarized briefly below. Further analysis of these fields is proceeding.[8,9]

Twelve V-band (5500 Angstroms) and 4 I-band (8000 Angstroms) observations of M100 were made using the HST Wide Field and Planetary Camera (WFPC2). The observations were made over a two-month period in the spring of 1994. To facilitate the discovery of Cepheids with a range of periods, the observing epochs were spaced following a power-law distribution.[4,10] A search was conducted using wide-field camera chips 1, 2, and 4, and twenty Cepheids having periods in the range of 20 to 65 days were identified.[7] The V and I PL relations for these 20 stars are plotted in Figure 1. With two independent wavelength bands, a correction can be made for the effects of interstellar dust extinction (both foreground obscuration due to dust in our own Galaxy, in addition to that internal to M100).

Apparent distance moduli at V and I for M100 were obtained by minimizing the residuals in the combined PL relations for the two galaxies, and determining the offset with respect to the calibrating LMC sample. The latter sample is used to define the slope of the relation. The

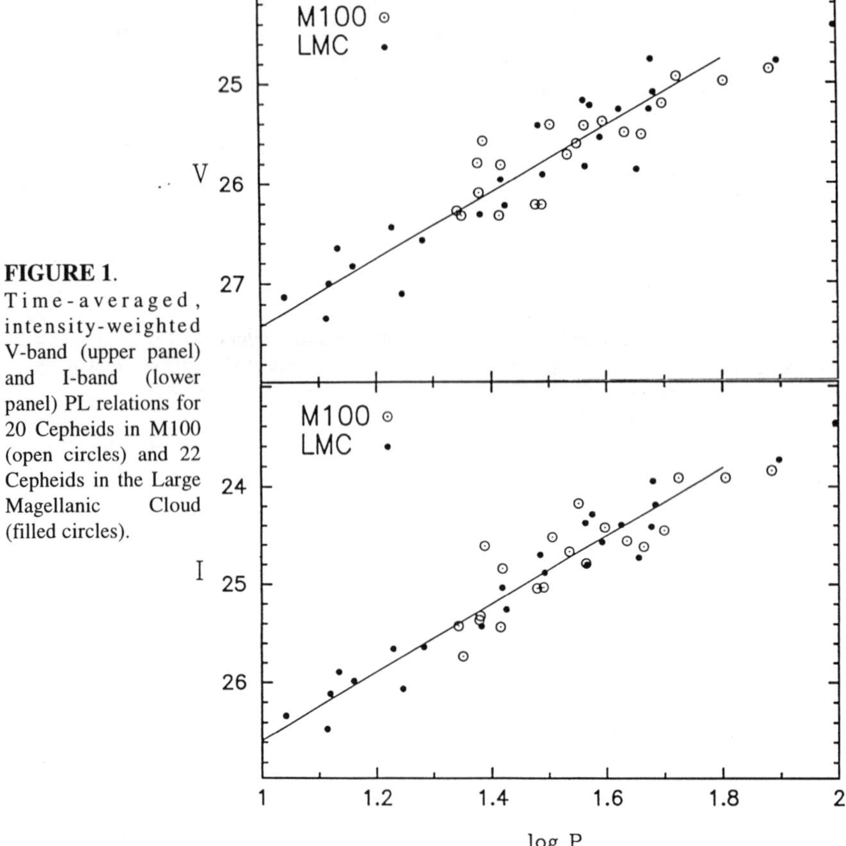

FIGURE 1. Time-averaged, intensity-weighted V-band (upper panel) and I-band (lower panel) PL relations for 20 Cepheids in M100 (open circles) and 22 Cepheids in the Large Magellanic Cloud (filled circles).

true (reddening-corrected) distance modulus adopted for the LMC is 18.5 mag with a mean reddening for the LMC Cepheids of E(B-V) =0.10 mag. Based on this sample of 20 Cepheids, the V-band apparent modulus for M100 is measured to be 31.31 mag. At I the apparent modulus is 31.25 mag, yielding a mean visual extinction of 0.15 ± 0.17 mag for the M100 Cepheid sample. The true distance modulus to M100 is determined to be 31.16 ± 0.20 mag, corresponding to a linear distance of 17.1 ± 1.8 Mpc.

Given a measurement of the distance and the Hubble velocity of the Virgo cluster, a value of H_0 can be determined. However, at the present time, the largest uncertainty in this determination is due to the fact that the distribution of spiral galaxies in the Virgo cluster is both extended and complex. Hence, the distance to M100 alone cannot define the mean distance to the Virgo cluster to an accuracy of better than 15-20%.[7,11] Adopting a recession velocity for the Virgo cluster of 1404 ± 80 km/sec[12] and a Virgo distance of 17.1 Mpc yields a value of $H_0 = 82 \pm 6$ (random) $\pm$ 16 (systematic) km/sec/Mpc. The dominant sources of uncertainty in this estimate are systematic and are due to the reddening correction, the zero point of the Cepheid PL relation, the position of M100 with respect to the center of the cluster, and the adopted recession velocity of the cluster.

An estimate of H_0 can also be made using the measured relative distance between the Virgo cluster and the more distant Coma cluster.[7] This determination avoids the uncertainty in the velocity of the Virgo cluster. Adopting a distance of 17.1 Mpc for the Virgo cluster, a relative Virgo-Coma distance of 77 Mpc, and a recession velocity for Coma of 7200 km/sec, yields a value of $H_0 = 77 \pm 6$ (random) $\pm$ 15 (systematic) km/sec/Mpc. These results indicate that the value of the Hubble constant is ~80 km/sec/Mpc out to a distance of ~100 Mpc, with an accuracy of $\pm 20\%$.

A value of $H_0 = 80 \pm 17$ km/sec/Mpc is consistent with a low-density ($0.1 < \Omega < 0.3$) Universe and an age of 12 Gyr. An expansion age of 12 Gyr is consistent with other measured age estimates based on stellar evolution theory applied to globular clusters, white-dwarf cooling estimates for the Galactic disk, and radioactive dating of elements, which give 14 $\pm$ 2 Gyr. However, for the standard (Einstein-de Sitter, $\Omega = 1$, $\Lambda = 0$) cosmological model, the expansion age is 8(+2.3,-1.3) Gyr for $H_0 = 80 \pm 17$ km/sec/Mpc. This expansion age is below the age estimates from other methods listed above. This well-known age conflict highlights the importance of decreasing the uncertainties in all of these age estimates.

Our analysis has shown that the remaining uncertainty in the value of H_0 is still dominated by systematic errors. An accuracy of 10% or better will only be reached when we have measured distances to a larger sample of galaxies so that the magnitude of these systematic errors can be assessed directly. Hence, the remaining two years of the Key Project will be critical. Our results to date, however, suggest that the measurement of H_0 to an accuracy of 10% is now in fact a feasible goal.

ACKNOWLEDGMENTS

All of this work was carried out in collaboration with the other members of the HST Key Project team and I sincerely acknowledge their contributions: R. Kennicutt, J. Mould, F. Bresolin, L. Ferrarese, H. Ford, J. Graham, M. Han, P. Harding, J. Hoessel, R. Hill, J. Huchra, S. Hughes, G. Illingworth, D. Kelson, B. Madore, R. Phelps, A. Saha, N. Silbermann, P. Stetson, and A. Turner. We also acknowledge the substantial contributions and participation of the late Marc Aaronson who led the team when it was formed.

REFERENCES

1. HUBBLE, E. P. 1929. Proc. Nat. Acad. Sci. USA. **15**: 168.
2. FEAST, M. & A. WALKER. 1987. Ann. Rev. Astron. Astrophys. **25**: 345.
3. MADORE, B. F. & W.L. FREEDMAN. 1990. Astrophys. J. **365**: 186.
4. FREEDMAN, W. L., ET AL. 1994. Astrophys. J. **427**: 628.
5. HUGHES, S.M.G., ET AL. 1994. Astrophys. J. **428**: 143.
6. KELSON, D., ET AL. 1995. Astrophys. J. Submitted.
7. FREEDMAN, W. L., ET AL. 1994. Nature. **371**: 757.
8. FERRARESE, L., ET AL. 1995. In preparation.
9. HILL, R., ET AL. 1995. In preparation.
10. MADORE, B. F. 1995. In preparation.
11. MOULD, J. R., ET AL. 1995, Astrophys. J. Submitted.
12. HUCHRA, J. 1994. *In* Extragalactic Distance Scale. S. van den Bergh & C.J. Pritchet, Eds. Publs. Astron. Soc. Pacif. Conf. Series. **4**: 257.

Discovery of Explosion Fragments of the Vela Supernova

BERND ASCHENBACH
Max-Planck-Institut für Extraterrestrische Physik
D-85740 Garching, Germany

In our Galaxy, there are about 200 supernova remnants (SNR) known, primarily from their radio emission[1]. In the ROSAT all-sky X-ray survey about 80 remnants have been detected and a large fraction has been mapped[2]. Additionally, more than 100 remnant candidates have been discovered, out of which currently 10 objects have been confirmed as SNRs by follow-up radio and/or optical observations. Only 13 SNRs seem to contain a pulsar, but the physical association is quite uncertain in a number of cases, e.g. that of the Vela SNR and the Vela pulsar PSR 0833–45.

The Vela SNR has been X-ray imaged by ROSAT as well[3], and the ROSAT image shows for the first time the full extent of the remnant with a diameter of 8.3° or 73 pc adopting a distance of 500 pc (Fig. 1). Over more than 3/4 of the circumference the remnant's periphery is well matched by a circle, the center of which is offset from the pulsar's current position by (25±5) arcmin. Fig. 1 also shows at least 6 extended features, labelled A – F, well outside of the remnant's boundary. Five (B, C, D/D′, F) of the protruding objects show a distinct 'boomerang' type structure which opens towards the SNR center. The other two objects A & E look like truncated cones, opening towards the remnant's centre as well. They extend from the general SNR boundary by 1.2° and 2.4°, or for a distance of 500 pc, by ~10 pc and ~22 pc, respectively. A magnified view of object A is shown in Fig. 2, illustrating the pronounced symmetric delineation of the trailing X-ray emission. The emission region of four of the protruding objects (A, B, D, F) is sufficiently well defined to construct the symmetry axis of each. The four symmetry axes intersect each other close to the remnant's geometric center. The intersection of their symmetry axes with the known proper motion vector of the pulsar[4], gives the objects' origin at (14.9±7.2) arcmin away from the present pulsar position along the proper motion axis towards the south-east. The proximity of the remnant's geometric center and the origin of the six protruding objects to the pulsar position strongly favors their common origin in the Vela supernova event.

The residual angular offset divided by the measured proper motion velocity of 0.049 arcsec yr^{-1} gives a new and independently determined age of the Vela SNR of (18000±9000) years, assuming the objects' origin mark the explosion site of the progenitor star. This datum is consistent with the pulsar's spin-down age[5] $\frac{P}{2 \cdot \dot{P}}$ = 11400 years.

We suggest that the X-ray emission associated with the protruding objects is produced by shock-heating of the ambient medium by supersonic motion of the objects. This is strongly supported by the recent observation of radio emission from

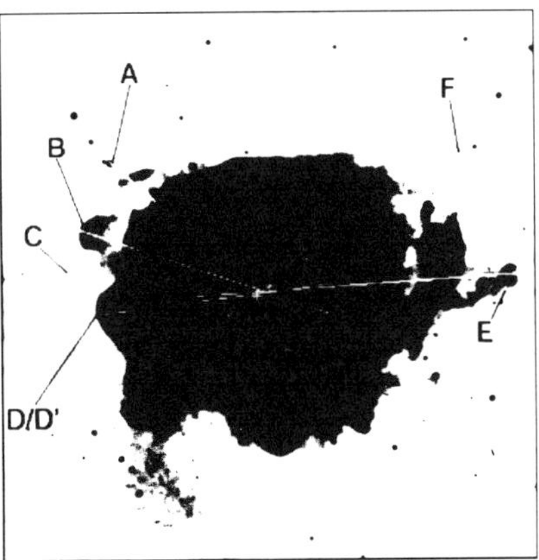

Figure 1. ROSAT image (14°×14°) of the Vela SNR. The small white cross marks the present position of the Vela pulsar PSR 0833–45; the arrow indicates the direction of its proper motion. The bigger white cross identifies the geometric center of the SNR. The objects A – F outside the remnant's boundary are suggested to be explosion fragments; D/D′ is a superposition of two objects. The symmetry axes of the emission regions trailing objects A, B, D, F are drawn towards the SNR center. They intersect in a very narrow region – suggesting the Vela supernova site.

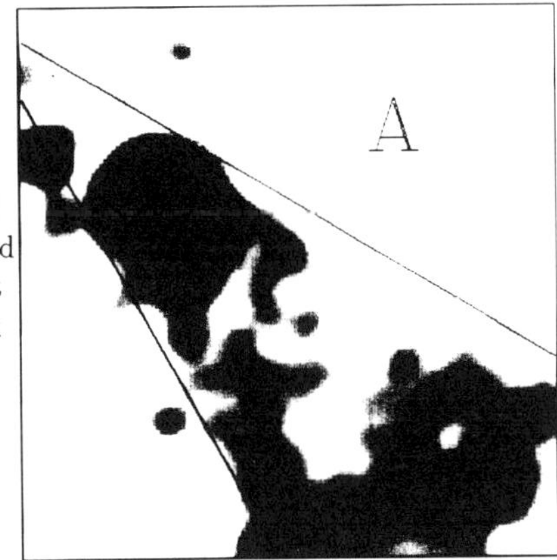

Figure 2. Detailed (0.8°×0.8°) ROSAT image of fragment A. The emission is bound to the region delineated by two straight lines. The bisecting line extrapolates back to the explosion site (c.f. Fig. 1).

the leading edge of object A[6].

Four of the protruding objects are trailed by wakes of X-ray emission, which we propose to be Mach cones, extending back to the blast wave front of the Vela SNR. The Mach number Ma is related to the opening angle of the cone and Mach numbers between 2.4 and 4.0 are found. These low Mach numbers imply that the temperature of the ambient medium through which the objects are moving must be high, close to X-ray temperatures. Outside the general Vela SNR boundary we have found in the ROSAT all-sky survey data such an excess emission of almost constant surface brightness in a circular area with a diameter of 20° and the center within the Vela SNR boundary. The analysis of the ROSAT PSPC spectrum shows the temperature of this region to be $kT_{amb} = 0.10 \pm 0.02$ keV.

Given Ma and kT_{amb} we can determine the current velocity v, which ranges between 390 ± 60 km s^{-1} and 650 ± 110 km s^{-1}, and the expected temperature kT_{exp}

Figure 3. 14°×14° ROSAT X-ray colour image of the Vela SNR. Hardness of the X-ray spectrum increases from dark to light. A cellular structure, defined by dark, i.e. spectrally soft loops appears. The loops are suggested to be the soft seams of fragments seen almost head-on, rather than sideways like the fragments A – F. The soft seams of fragments D/D′ are clearly seen.

of each of the objects from standard shock equations[7]

$$\frac{T_{\rm exp}}{T_{\rm amb}} = \frac{[2\gamma Ma^2 - (\gamma - 1)][(\gamma - 1)Ma^2 + 2]}{(\gamma + 1)^2 Ma^2}$$

with the result of kT_{exp} between 0.27 keV and 0.60 keV. A comparision of kT_{exp}, inferred primarily from geometry, with kT, determined from measured spectra, shows very good agreement for each of the objects, which strongly supports the view that we see shock-heated gas confined in Mach cones.

We suggest that the protruding objects represent blobs of matter formed in the collapse of the progenitor star and expelled in the subsequent supernova explosion. They will travel largely unaffected through the hot, tenuous, post-shock plasma behind the blast wave and will pass its front at a late stage, when the blast wave has significantly decelerated. The suggestion for such a scenario, in which the explosion of some supernovae might resemble more that of a splinter bomb than that of a pressure bomb - giving rise to long-living confined explosion fragments -, has been made earlier[8]. Only recently, two-dimensional hydrodynamical calculations of supernova explosions of massive stars reveal that massive stars do not explode in a spherical symmetric way, but highly asymmetric due to the formation of Rayleigh-Taylor instabilities in the outer layers of the progenitor star near the H/He interface and the He/CO interface[9,10] as well as by convective instabilities developing close to the mantle of the nascent neutron star[11,12]. Highly heterogeneous clumps of matter so-called 'mushrooms' are formed by these instabilities.

Whereas the angular size of the 'mushrooms' in the chemical transition zones seem to be relatively small, the angular size of the matter clumps close to the neutron star[13] is about 30°. If such a pattern will be maintained throughout the explosion - such late-time hydrodynamical calculations are not yet available - about 40 – 50 explosion fragments are likely to dominate the X-ray appearance of the remnant. It is interesting to note that the spectrally resolved ROSAT image of the

Vela SNR shows a network of about such a number of cells (c.f. Fig. 3), which are defined by a closed seam-like loop of intensive soft X-rays. This is the structure expected for the Mach cone shrapnels not looked at sideways but nearly head-on or slightly inclined.

The fact that we have observed shrapnel like objects in an SNR which is associated with a neutron star lends support to the presence of matter instabilities in type Ib or type II supernova explosions of massive stars. Whether we observe the effects which went on in the outer or deep inner layers remains to be seen. If on the other hand the creation of such fragments is bound to the birth of a neutron star, their occurrence might be as frequent as or as rare as that of neutron stars in SNRs, and the Vela SNR is one of these rare cases. In turn, the discovery of shrapnels around other SNRs might point to the existence of a hidden, hitherto unknown neutron star.

References

[1.] Green, D.A., PASP **103:** 209.

[2.] Aschenbach, B., 1995, in *New Horizon of X-RAY Astronomy - First Results from ASCA*, Makino, F. and Ohashi, T. (eds.), Universal Academy Press, Inc., Tokyo, p103.

[3.] Aschenbach, B., Egger, R. & Trümper, J., 1995, Nature, (in press).

[4.] Bailes, M., Reynolds, J.E., Manchester, R.N., Kesteven, M.J. & Norris, R.P., 1989, ApJ **343:** L53.

[5.] Taylor, J.H., Manchester, R.N. & Lyne, A.G., 1993, ApJS **88:** 529.

[6.] Strom, R., Johnston, H.M., Verbunt, F. & Aschenbach, B., 1995, Nature, (in press).

[7.] Landau, L.D. & Lifschitz, E.M., 1959, *Fluid Mechanics*, Pergamon Press, Oxford, p329.

[8.] Kundt, W.,1988, in *Lecture Notes in Physics* **316:** *245, 'Supernova Shells and Their Birth Events'*, Kundt, W. (ed.), Springer Verlag, Berlin.

[9.] Herant, M. & Benz, W., 1991, ApJ **370:** L81.

[10.] Müller, E., Fryxell, B. & Arnett, D., 1991, A&A **251:** 505.

[11.] Burrows, A. & Fryxell, B. A., 1992, Science **258:** 430.

[12.] Janka, H.-T. & Müller, E., 1994, in *IAU Colloquium 145: Supernovae and Supernova Remnants* , McCray, R. & Wang Zhenru (eds.), Cambridge University Press, Cambridge, (in press).

[13.] Janka, H.-T. & Müller, E., 1994, A&A **290:** 496.

THE X–RAY BACKGROUND

GÜNTHER HASINGER
Astrophysikalisches Institut Potsdam, An der Sternwarte 16
14482 Potsdam, Germany
and
Universität Potsdam, Am Neuen Palais 10, 14469 Potsdam, Germany

ABSTRACT

The progress in the measurement and understanding of the X-ray background is reviewed here with particular emphasis on a discussion of their implication on large-scale structure. New and important constraints on large-scale structure are obtained from measurements of the smoothness of the XRB. Recently the first discovery of a signal in the angular correlation function of the XRB could be announced. Finally, new information on the evolution of clusters of galaxies could be obtained from medium-deep ROSAT surveys.

1. Introduction

The X-ray background (XRB), discovered as the first cosmic background radiation[1] well in advance of the Cosmic Microwave Background (CMB), presented one of the long-standing puzzles of modern astrophysics. At higher energies and on scales larger than about 10 degrees its celestial distribution is very isotropic, apart from a weak dipole anisotropy[2], indicating its cosmological origin. Major steps have been taken in the past few years towards an understanding of its nature. Originally the XRB spectrum in the range 3-40 keV, which resembles very closely a thermal bremsstrahlung model with a temperature of $\sim 40\ keV$[3] led the way to an interpretation in terms of a hot, diffuse intergalactic medium. Such a truly diffuse hot plasma would, however, produce a severe Compton distortion on the CMB spectrum, which has not been observed by the COBE satellite[4]. This puts stringent constraints on the fraction of the XRB originating from hot gas and leaves the alternative interpretation of the XRB in terms of discrete sources the only feasible one. Nevertheless, the existence of cooler and/or significantly clumped hot plasma has not been ruled out by the COBE measurements.

At higher X-ray and soft gamma-ray energies (above 4 keV) measurements until recently have been performed only using collimated X-ray detectors with relatively coarse angular resolution (degrees). Consequently, while these measurements yielded very reliable estimates of the intensity and shape of the **total** X-ray background, they were only able to resolve a small fraction ($\sim 3\%$) of the XRB into discrete sources. The situation in the soft X-ray band (0.1-3 keV), where grazing incidence focussing optics can be used, is diametrically opposite. Due to the high sensitivity and angular resolution (below 1 arcmin) a substantial fraction of the X-ray background could already be resolved into discrete sources here. Deep surveys with the ROSAT satellite were able to resolve about 60% of the 1-2 keV background into discrete sources amounting to a surface density of $> 400\ deg^{-2}$ at a flux of $2.5 \cdot 10^{-15}\ erg\ cm^{-2}\ s^{-1}$ (ref. 5,6). Direct optical follow-up studies could identify a substantial fraction of these objects as active galactic nuclei (QSOs and Seyfert galaxies, see e.g. ref.7 and references therein). However, for several reasons the total extragalactic X-ray background in this energy range is very hard to measure, so that its detailed shape and intensity are still a matter of debate.

While the explanation of the X-ray background in terms of the summed X-ray emission of discrete objects (e.g. AGN), integrated in Olber's sense over cosmic distance and time, is a very attractive one (see eg. ref. 8), there were two puzzles in recent years which made this interpretation questionable, in particular in the hard X-ray band: the first one is the "spectral paradox", i.e. the fact that no single class of objects known so far has a spectrum resembling that of the XRB (e.g. ref. 9). The second one is the "logN-logS" paradox, i.e. the fact that fluctuation analyses in the "hard" (2-10 keV) band indicate a surface density of objects a factor of 2-3 higher than that in the "soft" (0.5-2 keV) band (e.g. ref. 10). Another complication is, that the number counts of the faintest objects in the ROSAT deep surveys exceed the predictions from the most recent determination of the AGN X-ray luminosity function[11] by about a factor of two[5,6], so that the possibility of a "new class of sources" had to be invoked.

However, a better understanding of the various classes of active galactic nuclei in terms of the "Unified Model", where differences between the classes are mainly due to orientation effects (e.g. ref. 12) has shed new light on these questions. Recent detailed spectral observations of bright, nearby AGN in the X-ray and soft gamma-ray band, as well as ROSAT deep surveys have now obtained new ingredients that led the way to a solution of the two puzzles above in a complete and self-consistent way, assuming only known objects with measured properties[13,14].

In this *review* I discuss the angular correlation function of the XRB, which on one hand provides stringent constraints on the evolution of clustering in the universe and on the other hand for the first time reveals significant large-scale structure in the XRB. I summarize the major soft X-ray surveys and follow-up optical identifications leading to new constraints about the luminosity evolution of clusters of galaxies.

2. The angular correlation function

2.1. ROSAT pointed observations

Since the X-ray background is made up largely from discrete sources one would expect some variance in the background due to those sources. A signal in the angular correlation function (ACF) can give strong constraints on the clustering properties of the sources contributing to the X-ray background. However, the XRB is remarkably smooth. Until recently no signal could be detected in the XRB ACF neither at soft X-ray nor at hard X-ray energies (see ref. 15 for a review and figure 2). A first signal could be found in a 1-2 keV analysis of 50 deep ROSAT pointed observations in about 10% of the fields[16]. This signal could be clearly associated with a few extended, very-low X-ray surface brightness clusters or groups of galaxies at moderate redshift. These objects are now termed "blotches"[17]. In trying to obtain an upper limit on structure in the background due to the clustering of sources producing the bulk of the emission, Soltan & Hasinger excluded the fields with significant cluster emission. Indeed, once those fields were excluded, only

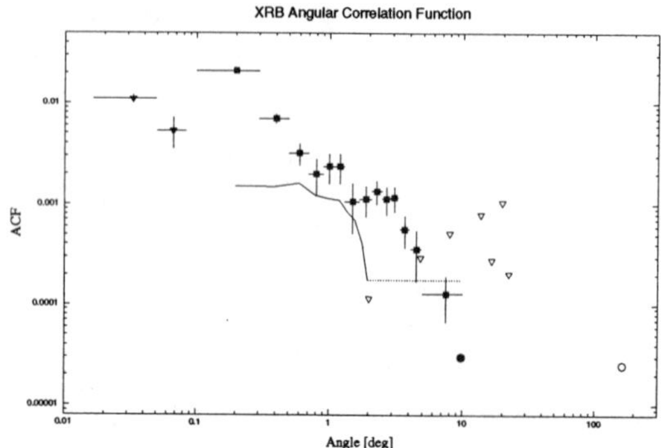

Fig. 1. XRB angular correlation function (after ref. 15) Open circle: XRB dipole moment[2]. Filled circle: upper limit from HEAO-1 A2[18]. Open triangles: upper limits from Ginga pointings[19]. Continuous line: 2-sigma upper limits from Ginga scan data[20]. Filled squares: ROSAT all-sky survey[21]. Filled triangles: upper limits to the ACF from 50 ROSAT pointed observations[16].

upper limits for the ACF could be obtained, however, those limits strongly constrain the nature and clustering properties of the sources contributing to the residual X-ray background. According to this analysis, less than 35% of the residual background can be due to objects with clustering properties similar to QSOs. The objects which make up the remainder of the background must have a clustering length smaller than normal galaxies and/or show very strong cosmologic evolution of their clustering[16]. In figure 2 the angular correlation function at 2 and 4 arcmin is shown, determined from the average of all 50 ROSAT pointings, i.e. including the "blotchy" fields. A significant signal is detected there.

2.2. ROSAT survey data

In a recent analysis of the 0.9-1.3 keV background from a "clean" region of 1 sr in the ROSAT all-sky survey, Soltan et al.[21] detected a very significant signal in the ACF of the X-ray background, extending out to about 10° (see figure 2). The angular resolution of this measurement is 12'. At angles of $0.5 - 5^o$ this signal corresponds to roughly 3% fluctuations of the X-ray background. The authors could show convincingly that a large fraction of this signal must be extragalactic, actually galactic contributions to the fluctuations could be largely removed using the angular correlation function in a softer energy band (0.7-0.9 keV). Therefore this measurement represents the first discovery of the long-sought signal in the extragalactic X-ray background flux. Soltan et al. correlated the X-ray background fluctuations with the Abell catalogue of clusters of galaxies and find a significant crosscorrelation signal, not only with the direct cluster emission (within 20') but also an extended component reaching out to $\sim 5^o$. The magnitude of this effect, which corresponds roughly to one third of the total background fluctuations, excludes the possibility that this signal is due to the cluster-cluster or the galaxy-cluster correlation and the tentative assumption put forward by the authors is, that typical Abell clusters are surrounded by large ($\sim 10\ Mpc$) haloes of diffuse hot gas with an average luminosity of $\sim 10^{43.4}\ erg\ s^{-1}$. This could be the first detection of hot diffuse supercluster gas emission. Another possibility yet to be tested is the explanation by a number of discrete sources associated with groups of galaxies concentrated around rich clusters.

Similarly, the ROSAT data are being correlated to catalogues of other object classes (e.g. nearby groups of galaxies), work which will presented elsewhere. A crosscorrelation between the ROSAT all-sky survey map and the 2nd-year COBE DMR map[24] is in progress, but the results are still inconclusive.

3. X-ray surveys

X-ray surveys are important for our understanding of the object classes contributing to the X-ray background as well as their cosmological evolution. Because of the requirement to obtain optical spectroscopic identifications for large, statistically complete samples of X-ray selected objects, X-ray survey work is quite tedious and time consuming. For many years the two X-ray surveys from HEAO-1[22] and the Einstein observatory Extended Medium Sensitivity Survey (EMSS[23]) were the only available workhorses. With the advent of the ROSAT X-ray observatory substantial progress has been made both in depth and in coverage of X-ray surveys. The optical follow-up work is, however, still a problem. There are two almost complete ROSAT surveys substantially fainter than the EMSS, which contribute most to our understanding of the sources of the XRB: the RIXOS project and the ROSAT Deep Survey project. RIXOS, the ROSAT International X-ray Optical Survey is large program to completely optically identify serendipitous sources from PSPC pointings longer than 8 ksec down to a flux limit of $3 \cdot 10^{-14}\ erg\ cm^{-2}\ s^{-1}$. For the purpose of optical identifiations 1.5 years of the CCI international time on the Canary Island telescopes was awarded to a large consortium (PI Keith Mason) in the years 1993-1994. In a solid angle of $\sim 15\ deg^2$ (60 fields) the identifications are 100% complete, the remaining ~ 20 fields have some small incompleteness.

A small number of PSPC pointings reach sensitivities substantially below $10^{-14}\ erg\ cm^{-2}\ s^{-1}$, the so called ROSAT deep surveys. Optical counterparts in these pointings typically have magnitudes in the range $m_R = 19-24$, so that optical identifications become very time consuming. In this review data from a total of 5 ROSAT PSPC fields with optical identifications largely complete down to an X-ray flux of $10^{-14}\ erg\ cm^{-2}\ s^{-1}$ have been combined (see also ref 25): the Lockman Hole[5,6,26], the Marano Field[27], the North-ecliptic pole field[28,29] and the QSF3 and GSP4 fields[11,30].

At fainter X-ray fluxes the typical magnitude of optical counterparts increases correspondingly and the surface density of faint galaxies rises dramatically. While ROSAT PSPC position errors can be as small as 2-3" for brighter sources, the X-ray error boxes at detection threshold are still quite large (10-15" radius), so that the likelihood of spurious associations of faint galaxies with faint X-ray sources increases substantially. In order to overcome these difficulties and to be able to push the limit for secure optical identifications an order of magnitude deeper we have started observations for an Ultradeep ROSAT HRI survey inside the PSPC survey of the Lockman Hole. A total of 1Msec of HRI observations and a new "X-ray speckle" technique to correct for aspect errors are forseen to obtain arcsecond positions for about 70 objects down to a flux limit of $7 \cdot 10^{-16}$ $erg\ cm^{-2}\ s^{-1}$. At the extremely faint optical magnitudes expected, probably only new telescopes of the 8-10m class with multislit spectroscopy capabilities over a large field can achieve completeness in a reasonable exposure time. Nevertheless this is the only possibility to obtain secure information about the role of faint galaxies and clusters for the X-ray background.

Table 1: Summary of X-ray Surveys

Survey	S_{lim} $[erg/cm^2s]$	Area $[deg^2]$	Total [%]	Unid. [%]	AGN [%]	Gal. [%]	Clus. [%]	Stars
EMSS	10^{-13}	430	835	4	55	2	13	26
RIXOS	$3 \cdot 10^{-13}$	14.9	285	14	51	4	11	20
DEEP	10^{-14}	0.76	61	8	62	7	5	15
ULTRA	$7 \cdot 10^{-16}$	0.06	70?					

4. Optical identifications and cluster evolution

Table 1 gives a summary of the survey papameters and optical identification content for the EMSS, RIXOS and ROSAT Deep Surveys, which span roughly a decade in limiting flux. Active galactic nuclei, mainly QSOs, represent the majority in all three surveys and are expected to contribute a large fraction of the X-ray background. Foreground stars, galaxy clusters and apparently normal or weak emission line galaxies contribute only $5 - 25\%$ each. Of particular interest is the relative behaviour of galaxy clusters on one hand and apparently normal or weak emission line galaxies on the other hand. The sub-Euclidean slope for clusters of galaxies found in the Einstein Medium Survey[31] seems to continue to much fainter X-ray fluxes: for the three samples discussed here a power law slope of -1.1 is found for the integral source counts. This is consistent with the strong negative evolution found for the X-ray luminosity of galaxy clusters[32,33]. An independent estimate of the cluster evolution, taking into account the redshift distribution of a completely identified set of medium survey clusters, could be obtained from the RIXOS sample[34]. Also this result confirms the negative luminosity evolution of clusters.

5. Acknowledgements

At this location I want to thank my collaborators in quite a number of projects for the fruitful cooperation over many years and the permission to show some new material in advance of publication (representative for many more): R. Bower, R. Burg, R. Ellis, R. Giacconi, K. Mason, R. McMahon, M. Schmidt, A. Soltan, J. Trümper, G. Zamorani.

6. References

1. Giacconi R., Gursky H., Paolini F.R. & Rossi B.B., *Phys.Rev.Letters*, **9**, 439 (1962).

2. Shafer R.A. & Fabian A.C. in: *Early Evolution of the Universe and its Present Structure*, G.O. Abell, G. Chincarini, eds., Reidel:Dordrecht, p.333 (1983).
3. Marshall F.E. et al., *Astrophys. J.*, **235**, 4 (1980).
4. Mather J. et al. *Astrophys. J. (Letters)*, **354**, L37 (1990).
5. Hasinger G., Burg R., Giacconi R., Hartner G., Schmidt M., Trümper J. & Zamorani G., *Astr. Astroph.*, **275**, 1 (1993).
6. — *Astr. Astroph.*, **291**, 348 (1994).
7. Georgantopoulos I. et al. *Mon. Not. R. astr. Soc.*, (submitted) (1994).
8. Setti G. & Woltjer L., *Astr. Astroph.*, **224**, L21 (1989).
9. Boldt E. *Phys. Rep.*, **146**, 215 (1987).
10. Mushotzky R. in: *Frontiers of X-ray Astronomy*, Y. Tanaka, K. Koyama eds., Universal Academy Press: Tokyo, p.657 (1992).
11. Boyle B.J., Griffiths R.E., Shanks T., Stewart G.C. & Georgantopoulos I., *Mon. Not. R. astr. Soc.*, **260**, 49 (1993).
12. Antonucci R.R. *Ann. Rev. Astr. Astrophys. Astrophys.*, **31**, 473 (1993).
13. Matt G. (preprint), (1994).
14. Comastri A., Setti G., Zamorani G. & Hasinger G. *Astr. Astroph.*, (in press) (1995).
15. Fabian A.C. & Barcons X. *Ann. Rev. Astr. Astrophys. Astrophys.*, **30**, 429 (1992).
16. Soltan A. & Hasinger G. *Astr. Astroph.*, **288**, 77 (1994).
17. Hasinger G., Schmidt M. & Trümper *Astr. Astroph.*, **246**, L2 (1991).
18. Jahoda K. *Adv. Space. Res.* **Vol. 13, No. 12**, p.(12)231 (1993).
19. Carrera F.J., Barcons X., Butcher J., Fabian A.C., Stewart G.C., Warwick R.S., Hayashida K. & Kii T. *Mon. Not. R. astr. Soc.*, **249**, 698 (1991).
20. Carrera F.J., Barcons X., Butcher J.A., Fabian A.C., Stewart G.C., Toffolatti L., Warwick R.S., Hayashida K., Inoue H. & Kondo H. *Mon. Not. R. astr. Soc.*, **260**, 376 (1993).
21. Soltan A., Hasinger G., Egger R., Snowden S. & Trümper J. *Astr. Astroph.*, (submitted) (1995).
22. Piccinnotti G., Mushotzky R.F., Boldt E.A., et al. *Astrophys. J.*, **253**, 485 (1982).
23. Gioia I.M., Maccacaro T., Schild R.E., et al., *Astrophys. J. Suppl.*, **72**, 567 (1990).
24. Smoot G.F. et al. *Astrophys. J.*, **396**, L1 (1992).
25. Branduardi-Raymont G., Mason K.O., Warwick R.S. et al. *Mon. Not. R. astr. Soc.*, **270**, 947 (1994).
26. deRuiter H. (in preparation) (1995).
27. Zamorani G. et al., (in preparation) (1994).
28. Henry J.P. et al. *Astron. J.*, **107**, 1270 (1994).
29. Bower R. et al. (in preparation) (1994).
30. Shanks T., Georgantopoulos I., Stewart G.C., Pounds K.A., Boyle B.J. & Griffiths R.E. *Nature*, **353**, 315 (1991).
31. Gioia I.M., Maccacaro T., Schild R.E., Stocke J.T., Liebert J.W., Danziger I.J., Kunth D. & Lub J. *Astrophys. J.*, **283**, 495 (1984).
32. Edge A.C., Stewart G.C., Fabian A.C. & Arnaud K.A. *Mon. Not. R. astr. Soc.*, **245**, 559 (1990).
33. Henry J.P., Gioia I.M., Maccacaro T., Morris S.L., Stocke J.T. & Wolter A. *Astrophys. J.*, **386**, 408 (1992).
34. Castander F.J., Bower R.G., Ellis R.S., Aragon-Salamanca A., Mason K.O., Hasinger G., McMahon R.G., Carerra F.J., Mittaz J.P.D., Perez-Fournon I. & Barcons X. *Nature*, (subm.) (1994).

An Overview of the ASCA Mission

Yasuo Tanaka

Institute of Space and Astronautical Science
Sagamihara, 229 Kanagawa-ken, Japan
and
Max-Planck Institut für Extraterrestrische Physik
85748 Garching, Germany

ASCA is the fourth Japanese X-ray astronomy satellite launched on February 20, 1993. While it weighs only 417 kg, *ASCA* is a high-throughput X-ray astronomy observatory which is capable of simultaneous imaging and spectroscopic measurements over a wide energy range 0.5 - 10 keV.

ASCA carries four identical grazing-incidence X-ray telescopes utilizing multi-nested thin foil conical optics, each equipped with an imaging spectrometer at its focal plane. The angular resolution is modest, with a half-power diameter of $\sim 3'$. However, the point spread function has a cusp-shaped peak which produces a sharp image core and allows us to separate two sources separated by $1'$.

The focal plane detectors are two CCD cameras (named SIS) and two gas scintillation imaging spectrometers (named GIS). All four detectors are operated simultaneously, and data obtained from each of them are separately available.

The SIS has a superior energy resolution, with resolving power of $\Delta E/E$ of $\sim$50 at 6 keV and is sensitive down to 0.5 keV. The resolving power of GIS is $\sim$ 13 at 6 keV and sensitive down to 1 keV. The GIS has a higher detection efficiency than the SIS above $\sim$4 keV. The SIS has a square field of view of $20' \times 20'$, whereas the GIS has a much larger field of view of a $50'$-diameter circle. Therefore, the SIS and GIS have complementary characteristics.

As compared with the two high-sensitivity imaging missions prior to *ASCA*, the spatial resolution of *ASCA* is comparable to the IPC of *the Einstein Observatory*, whereas the *ROSAT* PSPC has substantially higher spatial resolution and source detection capability than *ASCA*. On the other hand, *ASCA* has a coverage of much wider energy range, 0.5 - 10 keV, and high spectroscopic capability. These are the unique capabilities of *ASCA*.

The *ASCA* instruments cover the most important energy band for the spectroscopy, because the K-emission lines and the K-absorption edges from oxygen through nickel (and also L-line complex of iron) at various ionization stages are all in this band. The SIS can individually resolve all major lines. Motion of plasma of the order or greater than 1000 km/s can also be measured. Together with capability of imaging, spatially-resolved spectroscopy has become possible for the first time with *ASCA*.

The non X-ray background rates were confirmed to be very low. The source detection limit can be estimated from the observations of the "blank sky" (dominated by the cosmic X-ray background). For a typical exposure of 40 ks, the 5σ detection limit for a single detector is $\sim 4 \times 10^{-14}$ ergs/cm^{2}s in the range 2 - 10 keV, for a Crab-like spectrum. The detection limits of the SIS and GIS are about the same. For fainter sources, the source confusion will become a serious limiting factor. From the results so far obtained, detailed spectroscopic studies are possible for the sources of flux $> 10^{-12}$ ergs/cm^{2}s, and coarse spectra can be obtained down to a flux level of several times 10^{-14} ergs/cm^{2}s.

After completing the in-orbit performance verification phase with observations of about 150 sources, we have been operating *ASCA* for observations of selected proposals from general users since October, 1993. All instruments and satellite systems are functioning normally.

Many new findings are emerging. The following papers will show some of the significant results obtained from the early *ASCA* observations on supernova remnants, active galactic nuclei, and clusters of galaxies.

ASCA Observations of Supernova Remnants

ROBERT PETRE
NASA/Goddard Space Flight Center
Greenbelt, MD 20771 USA

AND THE *ASCA* TEAM

Supernova remnants (SNR's) are an attractive target for *ASCA* because their properties facilitate usage of all of *ASCA's* unique attributes.[1] Their spectra are usually dominated by lines from highly ionized metals (usually He- and H-like ions). In young remnants the metal abundances are enriched by ejecta, resulting in very strong lines. *ASCA's* broad band allows simultaneous measurements of plasma diagnostics from oxygen to iron. Galactic remnants can be spatially resolved using *ASCA's* imaging capabilities, facilitating measurements of plasma properties as a function of distance from the shock and spectral mapping of regions of particular interest. Finally, it is possible to combine *ASCA's* spectral and spatial resolution to produce the first X-ray images of SNR's in the light of a particular element, or in the line-free continuum, thereby allowing searches for stratification and clumping which might reveal important clues about the nature of the progenitor star and the explosion mechanism. Below we present via examples three fundamental new insights on SNR's provided by *ASCA*.

SPECTRAL COMPLEXITY

The SNR E0102-72 is the brightest remnant in the Small Magellanic Cloud. Its nearly symmetrical, limb-brightened X-ray morphology suggests a symmetric explosion into a relatively uniform interstellar medium.[2] The prominent line emission from O, Ne, and Mg detected in the *ASCA* SIS spectrum (Fig. 1) suggests it arose from a Type II supernova. Given the apparent lack of influence of major ISM inhomogeneities, and the fact that the *ASCA* beam samples its entire 40" angular diamter (40"), E0102-72 offers an opportunity for comparison of its X-ray

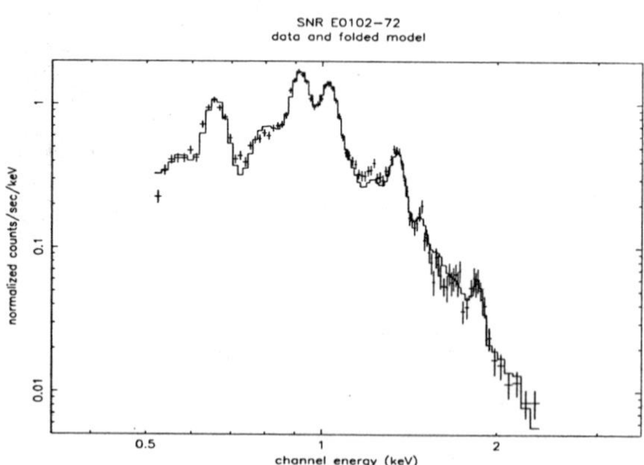

Figure 1: *ASCA* spectrum of SNR E0102-72, showing prominent O, Ne and Mg lines.

spectrum the results of hydrodynamic modeling of SNR's that take into account time-dependent ionization.[3,4] These models are characterized by two parameters: T_e, the electron temperature,and *nt*, the product of the number density *n* and the time *t* since the material was shocked. In E0102-72, no single component NEI model provides a satisfactory representation of the data. In fact, in order to achieve an acceptable fit, at least three different (T, *nt*) combinations are required, one for each prominent atomic species. This unanticipated spectral complexity appears consistently in the SNR's observed by *ASCA*. In E0102-72 it indicates that each element is encountering different shock conditions, and that therefore the supernova ejecta are stratified.[5]

STRATIFICATION OF EJECTA AND COMPLEX STRUCTURE

More direct evidence for stratification is found in the remnant W49B.[6] Here, the very strong Si, S, and especially Fe lines suggest a Type Ia origin. The high signal in the various lines allow construction of narrow-band images. Also, as this remnant is about 4' across, slightly larger than the *ASCA* beam, it is amenable to image deconvolution. As shown in Fig. 2, deconvolved images of the Si and S lines reveal a clear shell structure, whereas the Fe emission appears centrally concentrated. As in E0102-72, spectral fitting indicates that each atomic species requires its own set of NEI conditions.

The Type II remnant Cas A offers evidence for a different sort of structure.[7] Its high X-ray surface brightness makes it possible to examine spectral variations on a spatial scale of an arc minute. Doing so has revealed a systematic pattern of energy shifts in the peaks of all the prominent lines. If this pattern is interpreted as arising from bulk motion, then the southern half of the remnant is moving toward us and the northern half away, with a maximum velocity of ~1,000 km s^{-1}. This pattern can be best explained if the expanding material is confined to an inclined ring, as originally suggested by Markert et al.[8]. Cas A is also amenable to the construction of deconvolved narrow band images. The images in the strong Si and S lines look very similar to the *Einstein* HRI image of Cas A,[9] blurred to 12" resolution. On the other hand, an image constructed using high energy continuum photons, from the 4.0-6.5 keV band, has a qualitatively different appearance. The overall shell structure is maintained, but the knots of highest surface brightness are different from the line maps. In particular, the brightest knot is on the the western edge of the remnant, where the line maps show a deficit of surface brightness. The continuum map shows an uncanny resemblance to the radio image, suggesting that non-thermal processes provide a significant contribution to the high energy continuum.

POSSIBLE COSMIC RAY ACCELERATION SITES

Perhaps the most exciting *ASCA* SNR result to date comes from SN1006, the remnant of the brightest optical supernova ever recorded,[10] thought to be a type Ia.[11] Previous X-ray observations yielded the quizzical result that the 2-20 keV spectum appeared featureless.[12] Two explanations were advanced, one involving a combination of extreme NEI conditions and X-ray continuum emission from fully ionized carbon (which challenges the prevailing models of Type Ia composition),[11] and another invoking synchrotron emission from electrons accelerated via Fermi processes by magnetic fields compressed by the shock front.[13] *ASCA* spectra, shown in Fig. 3, clearly reveal that the X-ray emission from most of the remnant is

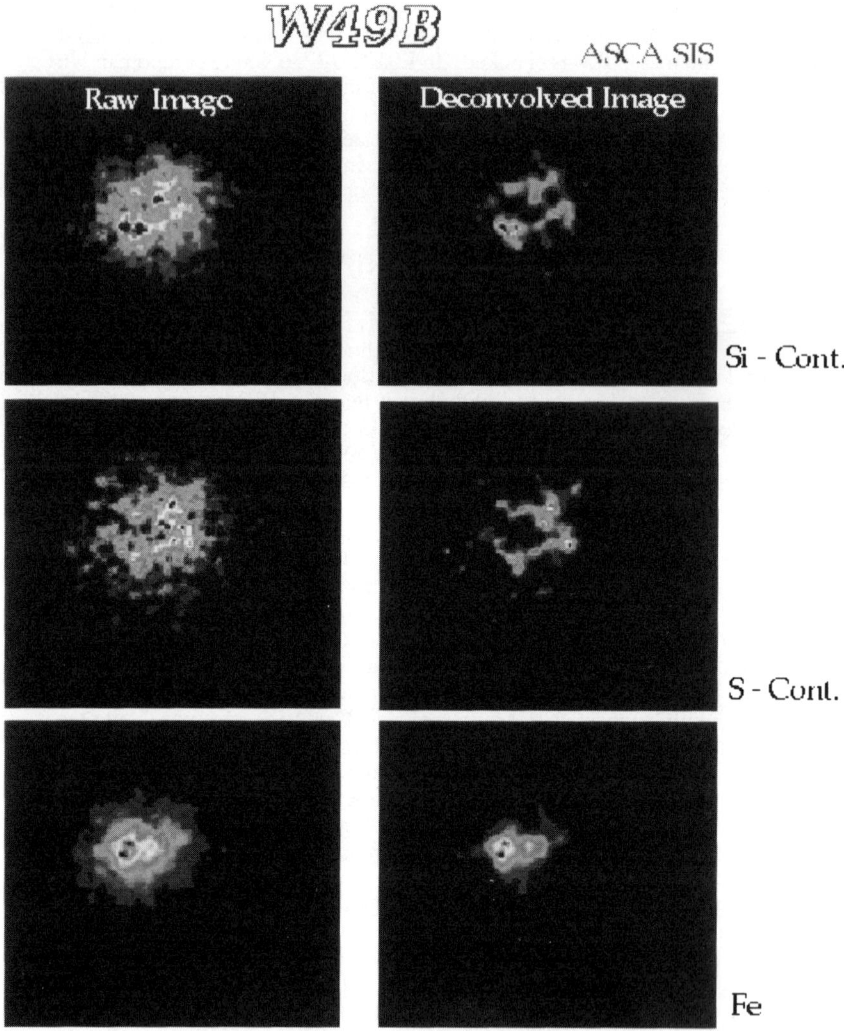

Figure 2: *ASCA* SIS images of SNR W49B in Si, S, and Fe K lines. Left panels show raw images; right panels show images deconvolved using Lucy-Richardson algorithm. The Si and S emission is confined to a ring, while the Fe emission is centrally concentrated, suggesting stratification of ejecta. Figure courtesy of R. Fujimoto, ISAS.

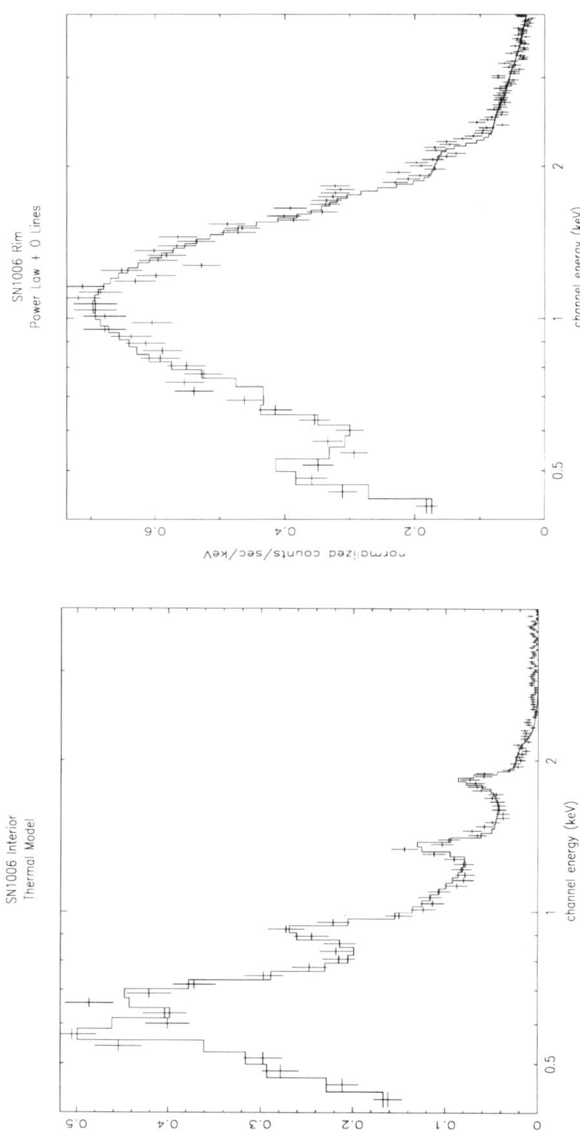

Figure 3: *ASCA* SIS spectra of regions of SN1006. Left panel shows spectrum of interior, which is clearly dominated by line emission. Right panel shows spectrum of NW rim, which is featureless above 800 eV.

thermal, and that the non-thermal emission is confined to the two highest surface brightness quadrants of the rim, diametrically opposite each other.[14] The presence of a strong oxygen line in these rims suggests that thermal emission is present, but that it is rendered undetectable by the order-of-magnitude brighter non-thermal component. Such a high surface brightness, featureless spectrum cannot be explained by any plausible NEI model, and we therefore conclude that the spectrum is truly non-thermal. If the X-ray spectrum is produced by synchrotron emission from shock accelerated electrons, then the maximum electron energy required exceeds 200 TeV. Thus in the bright rims of SN1006 we have found the first strong evidence for a site of high energy cosmic ray acceleration. In light of this result, it is possible to interpret the strong radio/X-ray continuum correlation in Cas A as evidence for for cosmic ray acceleration there as well.

(1) Tanaka, Y., Inoue, H., & Holt, S.S. 1994, PASJ, 46, L37
(2) Hughes, J.P. 1994, in The Soft X-Ray Cosmos, eds. E.M. Schlegel & R. Petre (New York: AIP), p. 144
(3) Hamilton, A.J.S., Sarazin, C.L., & Chevalier, R.A. 1983, ApJS,
(4) Hughes, J.P., & Helfand, D.J. 1985, ApJ, 291, 544
(5) Hayashi, I., Koyama, K., Ozaki, M., Miyata, E., Tsunemi, H., Hughes, J.P., & Petre,.R. 1994, PASJ, 46, L121
(6) Fujimoto, R., et al. 1995, PASJ, submitted
(7) Holt, S.S., Gotthelf, E.V., Tsunemi, H., & Negoro, H. 1994, PASJ, 46, L151
(8) Markert, T.H., Canizares, C.R., Clark, G.W., & Winkler, P.F. 1983, ApJ, 268, 134
(9) Murray, S.S., Fabbiano, G., Fabian, A.C., Epstein, A., & Giacconi, R. 1979 ApJ, 234, L69
(10) Stephenson, F.R., Clark, D.H., & Crawford, D.F. 1977, MNRAS, 180, 567
(11) Hamilton, A.J.S. Sarazin, C.L. & Szymkowiak, A.E. 1986, Astrophys. J. 300, 698
(12) Becker, R.H., Szymkowiak, A.E., Boldt, E.A., Holt, S.S., and Serlemitsos, P.J. 1980, Astrophys. J., 240, L33
(13) Reynolds, S.P., & Chevalier, R.A. 1981, ApJ, 245, 912
(14) Koyama, K., Petre, R., Gotthelf, E.V., Matsuura, M., Ozaki, M., & Holt, S.S. 1995, Nature, submitted

ASCA results on Active Galactic Nuclei

K. MITSUDA
Institute of Space and Astronautical Science
3-1-1 Yoshinodai, Sagamihara, Kanagawa 229, Japan

1 Introduction

The high energy resolution and the wide energy coverage are the major advantages in ASCA observations of Active Galactic Nuclei (AGN). From detailed study of emission lines and/or absorption structures, physical state of material near central engines of AGN have been extensively investigated. On the other hand, the high sensitivity in the energy range above 2 keV is crucial for detecting AGN which are hidden in the lower energy range because of absorptions. In this paper these results are reviewed with emphasis on the latter kind of studies.

2 Diagnostics of Matter around AGN

Existence of a warm absorber near the AGN was indicated by previous satellites [1]. The ASCA data showed clear evidence for the existence of warm material near the central engine. K-absorption edges of highly ionized oxygen (O VII and O VIII) have been reported for MCG-6-30-15 [2], NGC4151 [3], NGC3227, NGC5548, Mkn 841, MR2251-178 [4]. Such warm absorber demonstrates existence of intermediately photo-ionized plasma in the AGN systems. From six observations of MR2251-178 performed in a time span of about two months, Otani et al. [4] investigated the relation between the ionization state and X-ray flux. The source was in a very bright state and the X-ray flux varied by about 50 %. The depth of the O VII and O VIII absorption edges were studied as functions of X-ray intensity and only the O VIII absorption showed clear correlations. These results strongly constrain the size the distance from the central engine of the warm material.

Existence of warm material is indicated also from emission lines. Inoue et al. [5] decomposed the energy spectrum from the central region of Cen A into spatially-extended and point-like components. The latter component has a hard absorbed spectrum but at the same time shows emission lines whose energies are consistent with the K emission lines from Mg XI and Si XIII.

Broad and/or double peaked iron K emission lines were detected from ASCA observations of the AGN, NGC 5548, MCG 6-30-15, NGC 4151, and Fairall 9. The line profiles of such AGN are consistent with the accretion disk line emission models. If we interpret them in the frame work of the models, we can pause strong constraints on the location of the line emitting regions ([6], [4]).

3 Searches for hidden AGN

In this section, the results of searches for AGN in two different classes of objects, infrared luminous galaxies, and nearby normal galaxies are summarized.

3.1 AGN in infrared luminous galaxies

Ultra-luminous infrared galaxies emit $10^{12}L_{\odot}$ in the far infrared wavelengths. This luminosity is comparable to the total luminosity of Quasars. Most of the ultra-luminous infrared galaxies show morphological evidences of strong interactions, namely merging, such as double nuclei and/or tidal tails ([7]). Two scenarios have been considered as the energy source of luminous infrared emissions; the starburst activity and AGN, both fueled by large amount of gas in molecular clouds. The characteristics of the optical lines from the center of these object are similar to those of typical AGN. Sanders et al. [7] proposed merging of two gas rich spirals as the origin of Quasars. According to the scenarios of merging, the ultra-luminous infrared galaxies hosts an AGN highly obscured by the dense molecular clouds. Thus the X-ray measurement, in high energy range in particular, is one of the best opportunities to directly detect emissions from such AGN.

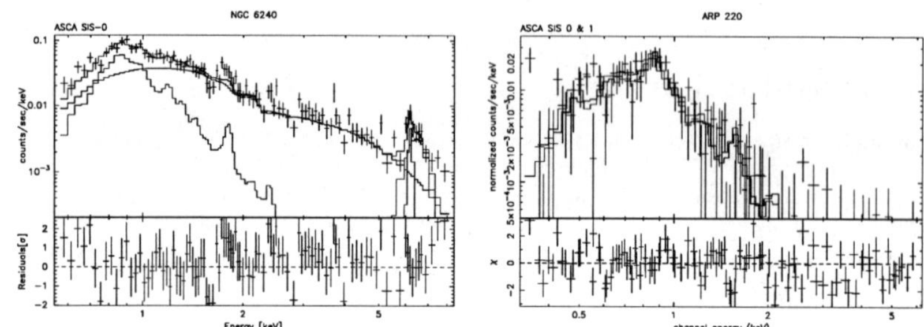

Figure 1: Energy spectra of two infrared luminous galaxies, NGC 6240 (left) and Arp 220, obtained with ASCA SIS (CCD camera).

NGC 6240 is an on-going merger. Two nuclei separated by about 0.8 kpc are detected in infrared wavelengths. The luminosity of the galxy is $5 \times 10^{11} L_{\odot}$. NGC6240 was clearly detected with ASCA ([8]). The energy spectrum of this galaxy is shown in Fig. 1. The spectrum above about 2 keV is characterized by a power-law spectrum with a photon index of 1.7. We notice that there is a strong, complex iron K emission line feature, which is consistent with being a mixture of separate narrow lines from nearly neutral to highly-ionized irons. In the low energy range, we find an excess flux above the model and indication of several line features. These suggest existence of a thin thermal emission (kT=0.6 keV) in low energy range. These spectral features are strikingly similar to those of the famous Seyfert 2 galaxy, NGC 1068 [9]. The existence of the power-law component strongly suggests that NGC 6240 hosts Seyfert nucleus. On contrary to these results, from Arp 220, one of the most famous advanced mergers, we failed to detect emissions in high energy range [10]. From this galaxy we only detected soft emissions which can be represented by a thin thermal model.

From NGC 6240 we have detected complex iron emission lines, which are most likely to come from obscured AGN as in the cases of some Seyfert II galaxies. Since the equivalent width of the line is about 2 keV, the continuum emission from the central engine is thought to be blocked and only emissions scattered by a thin hot plasma are visible. From the intensity of the iron lines, we can place an lower limit on the luminosity of the power-law component, which is 1.2×10^{43} erg/sec. This is the lower limit of the X-ray luminosity of the AGN. Thus the emission from the AGN is a candidate of the heat source of dust grains. On the other hand, only upper limit (3×10^{-13} erg sec^{-1} cm^{-2} in 2-10 keV) was obtained for Arp 220. This could be interpreted to be because the AGN is obscured by thick matter completely surrounding the AGN. The upper limit of the luminosity corrected for the absorption depends on the column density to the center of the galaxy, which we consider is in the range 10^{23} to 10^{24} cm^{-2} ([11] [12]). Then the upper limit of X-ray emission in 2-10 keV range is in the range $(0.3 - 2) \times 10^{42}$ erg/sec assuming a photon index of 1.7. This indicates that Arp 220 does not host a bright AGN; even if it does, it is only a low-luminosity AGN. An AGN does not play an important role in heating dust grains.

3.2 Low-luminosity AGN in normal galaxies

HEAO-A2 survey suggests that the luminosity function of Seyfert galaxies in 2-10 keV band show flattening at the luminosity of around 10^{42} erg/sec [13]. Then it is very important to understand how the luminosity function is related to the activity of nucleus of normal galaxies. This could be related to the evolutions of galaxies and active galactic nucleus, and the origin of cosmic X-ray background.

Name	Optical	L_X	Photon Index	N_H	Fe K Lines	Ref
M106	LINER–Sy2	4×10^{40}	1.85 ± 0.05	1.5×10^{23}	6.4 keV	[14]
M51	LINER–Sy2	1×10^{40}	1.4	--	6.4 keV	[15]
M81	LINER	1×10^{40}	1.85 ± 0.05	1.4×10^{21}	Complex	[16]
NGC3079	LINER	5×10^{40}	1.0 ± 0.5	$< 1 \times 10^{23}$	-	[18]
M33	HII	2×10^{39}	($E_{cut} = 2$ keV)	1.6×10^{21}	-	[19]
NGC1313	HII	4×10^{39}	$1.8 \pm 0.0?$	1.6×10^{21}	-	[20]
GC		$\sim 10^{36}$	~ 1.2	4×10^{22}	Complex	[21] [22]

In the Table, hard X-ray spectral parameters of the nuclear region of the galaxies observed in ASCA PV observations are summarized. Some spectra requires additional soft thermal emissions. With the spatial resolution of ASCA telescope, extended emissions around the some of the nuclei may not be resolved. We, thus, focus on the hard power-law component which is likely to be from the nucleus.

The results of M106 were already published by Makishima et al. [14]. This galaxy is optically classified as intermediate between LINER and Seyfert. The mass of the central engine has recently been estimated to be $3.6 \times 10^7 M_\odot$ from water maser observations [23]. The ASCA spectrum of this galaxy strongly supports the existence of an AGN obscured by dense matter of $N_H = 1.5 \times 10^{23}$. Makishima et al. also noticed iron K emission line with center energy of 6.4 keV.

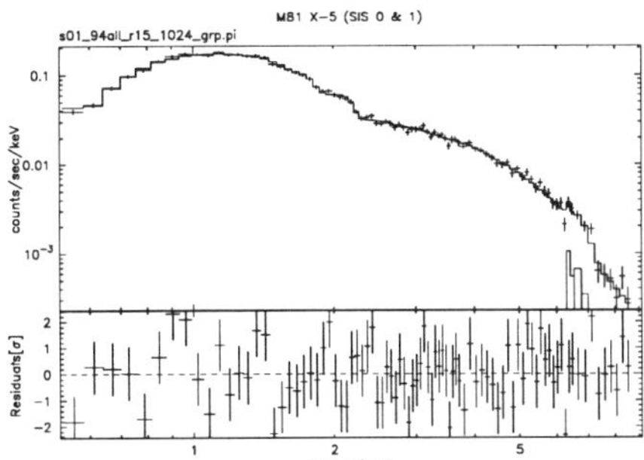

Figure 2: Energy spectrum of the nuclear source of M81. The spectrum can be represented by a power law function with a photon index of 1.9. A significant line feature is found between 6 to 7 keV. In this figure, it is fitted with a sum of three narrow lines.

The nucleus of the galaxy M81 is classified as a LINER. This galaxy was observed many times with ASCA for the observations of SN1993J, to obtain a long exposure on the nucleus of this galaxy. The energy spectrum of the nuclear source, X-5 (Fig. 2, [16]) can be represented by a power-law function with a photon index of 1.9. Ishisaki et al. [16] reported temporal variations of about 10 % on time scales of a few tens of hours. This is a clear signature of an AGN. In the energy spectrum, we notice a spectral structure around 7 keV, which can be represented either by a broad emission line centered at about 6.7 keV or a blend of narrow lines of different ionization states; nearly neutral to Hydrogen like.

The results of M106 and M81 indicate that the low luminosity AGN in a luminosity range of 10^{40-41} erg/sec have X-ray characteristics which are very similar to those of brighter classes of AGN. They have a power law spectrum with an photon index of about 1.8 and exhibits iron K emission lines with an equivalent width of a few hundred eV. Two results also indicate that there are wide variety in the degree of absorption; some are highly obscured like Seyfert 2, but some have nearly no absorption.

On contrary to these results, the bright source at the center of M33 has an energy spectrum which requires an exponential cut off with an e-folding energy of 2 keV ([19]). Takano et al. [19] discussed the possibility of a binary X-ray source containing a stellar mass-black hole. The energy spectrum of the bright source near (45") the center of NGC1313 can be fitted with a power-law function (photon index = 1.8). However, from these bright sources, ASCA failed to detect line emissions within the limit of the statistics.

Emission line from neutral iron is one of the most important signatures of AGN. Then the ASCA results of the galactic center could be interpreted in the frame work of AGN ([21] [22]). If the 6.4 keV emission line detected by ASCA is induced by emissions from an AGN which had been bright in the past or which is hidden behind a dense cloud, the luminosity of the AGN is estimated to be $\sim 1 \times 10^{39}$ erg/sec.

The author would like express his thanks to T.Kii for valuable discussions and allowing the author to use his results before publications. He is also grateful to T.Nakagawa, C.Otani, Y.Terashima, Y.Ishisaki, Y.Fukazawa, and S.Uno, who provided him with their unpublished results. He also would like to thank Prof. Y.Tanaka, the organizer of the ASCA session, and all the members of the ASCA team.

References

[1] Nandra,K. and Pounds,K.A. 1992, *Nature*, **359**, 215.

[2] Fabian,A.C. et al. 1994, *PASJ*, **46**, L59.

[3] Mihara,T. et al. 1994, *PASJ*, **46**, L137.

[4] Otani,C. 1995, Ph.D.thesis, University of Tokyo.

[5] Inoue,H. et al. 1995, in preparation.

[6] Fabian,A.C. 1995, this volume.

[7] Sanders,D.B. et al. 1988, *ApJ*, **325**, 74.

[8] Nakagawa,T. et al. 1995, in preparation.

[9] Ueno,S. et al. 1994, *PASJ*, **46**, L71.

[10] Kii,T. et al. 1995, in preparation.

[11] Scoville,N.Z. et al. 1986, *ApJ.Let*, **311**, L47.

[12] Neugebauer,G. et al. 1987, *AJ*, **93**, 1057.

[13] Piccinotti,G. et al. 1982, *ApJ*, **253**, 485.

[14] Makishima,K. et al. 1994, *PASJ*, **46**, L77.

[15] Makishima,K. 1994, in *New Horizon of X-Ray Astronomy*, Univ.AcademyPress, 171.

[16] Ishisaki,Y. et al. 1995, in preparation.

[17] Terashima,Y. 1995, Master Thesis, Nagoya University.

[18] Fukazawa,Y. et al. 1995, in preparation.

[19] Takano,M. et al. 1994, *ApJ.Let*, **436**, L47.

[20] Petre,R. et al. 1994, *PASJ*, **46**, L115.

[21] Koyama,K. 1994, in *New Horizon of X-Ray Astronomy*, Univ.AcademyPress, 181.

[22] Sonobe,T. 1995, Master Thesis, Gakushuin University.

[23] Miyoshi,M. et al., *Nature*, **373**, 127.

ASCA Results on Clusters of Galaxies

TAKAYA OHASHI
Department of Physics, Tokyo Metropolitan University
1-1 Minami-Ohsawa, Hachioji, Tokyo 192-03, Japan
ohashi@phys.metro-u.ac.jp

INTRODUCTION

Recent results on clusters of galaxies obtained from the ASCA observations are reported. The performances of ASCA[1] such as the imaging capability up to 10 keV with extremely good energy resolution (about 2% FWHM at 6 keV with the SIS and and 8% with the GIS) are particularly suited for the study of galaxy clusters. However, characteristics of the X-ray telescopes have much to be further invesitigated. It has been realized that the point spread function is energy dependent and may cause a spurious temperature profile if one performs, for example, a simple ring-cut analysis. The modelling of the telescope characteristics based on the in-orbit data is under way, and I shall report here the results from these recent efforts.

ABUNDANCE GRADIENTS

First, we show the results from the ring-cut analysis (slicing the image as a function of projected radius into several rings) for the GIS[2,3] data. This method has a problem of producing wrong temperature profiles in particular for hot clusters. The effect is not more than about 20% for clusters with $kT \lesssim 3.5$ keV[4], therefore the results presented here can be taken as a measure of the abundance gradients. The measured radial distributions of iron abundance are plotted in Fig. 1. The profiles are not deconvolved, so the true feature would be narrower than the "raw" profiles shown in Fig. 1. A large enhancement in the abundance by more than a factor of 3 is seen in the central region of the Centaurus cluster[5]. This level of iron concentration pushes up the apparent mean abundance (such as that measured with non-imaging detectors) by 30% compared with the level in the outside region. A smaller but significant enhancement of the abundance is also seen from AWM7[6] (and possibly from PKS2354-35). Similar feature has been reported for the Virgo cluster around M87[7,8]. The Perseus cluster shows no significant iron concentration in the center[6,9].

The abundance gradients are seen from poor clusters ($kT \lesssim 4$ keV) hosting cD galaxies. The sedimentation of iron in the intracluster medium (ICM) is too slow to account for the metal concentration[10]. The ram-pressure stripping of the metal-rich intestellar matter is thought to be efficient in the cluster center, however, this mechanism does not naturally explain the association with cD galaxies. A reasonable way is to assume that the excess iron is produced in the cD galaxies. Since optical luminosities of cD galaxies do not differ much from cluster to cluster within a factor of several, cDs may inject similar amount of excess iron into the intergalactic space. Then, such a contribution would be relatively larger in gas-poor clusters. However, it is not clear how cDs can produce excess iron than ordinary elliptical galaxies. Also, strong cooling flows, which transport the surrounding metal-poor gas into the center, would suppress the iron concentration.

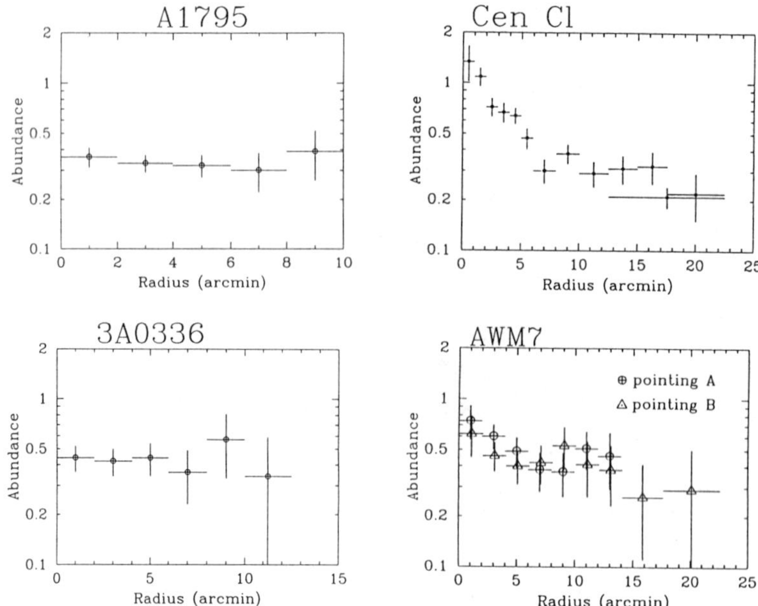

Figure 1: Iron abundance versus radius for 4 clusters

AVERAGE ABUNDANCES

The relation between the average metal abundance and the ICM temperature has been studied by excluding the central region ($r \lesssim 300$ kpc) where abundance enhancement is possible. The results by Y. Fukazawa are shown in Fig. 2. Although the number of points is not large, most of the iron abundances lie between 0.3 and 0.5 solar for the ICM temperatures 2.5 − 7 keV with no systematic correlation. However, very hot clusters ($kT > 8$ keV) tend to show a larger scatter between 0.1 and 0.5 times solar[11].

If one may assume that iron mass (M_{Fe}) in ICM is very roughly proportional to the stellar mass (M_{stellar}) because stars produce iron[12,13], then the constant iron abundance would imply that the gas injected from galaxies in to ICM is equally diluted by the primordial gas regardless of the cluster richness. In other words, the ratio of the gas mass (M_{gas}) to the stellar mass would be roughly constant among various clusters. Then, the efficiency of the galaxy formation may not be greatly dependent on the cluster richness.

The abundance of silicon may have larger systematic errors than iron, such as due to an effect of non-uniform temperature. However, it is interesting that for all clusters the silicon abundance turns out to be equal to or larger than the iron abundance. If Type-1 supernovae are the dominant source of metals in ICM, the iron abundance should be several times higher than the silicon one[14]. Therefore, we may safely infer that Type-1 SNe are a minor contributor to the metals in the ICM.

MASS ESTIMATION

We try to constrain the total mass M_{total} and gas mass M_{gas} in the ICM. In the analysis, we carried out a fitting for the radial profile data with model profiles based on Monte-Carlo simulations[15]. This method is developped by the SimASCA team.

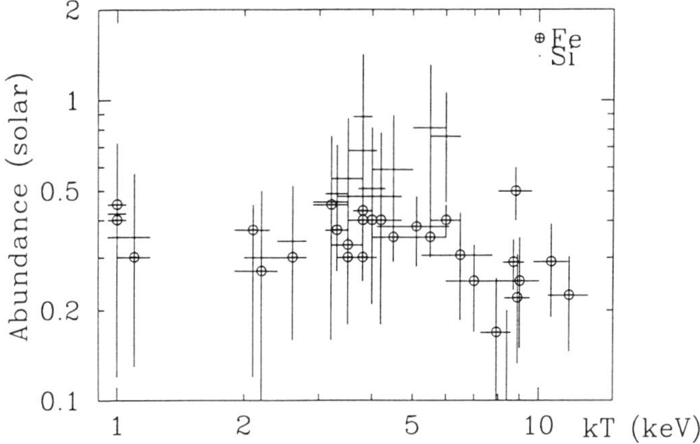

Figure 2: Average abundance of Fe and Si for various clusters

The results reported here are preliminary since there is a siginificant uncertainty in the off-axis point-spread function of the X-ray telescope and the effect of scatterings. Important information in estimating the ICM mass, apart from the beta-model parameters, is the maximum radius $r_{\max}$. We report here results for 2 clusters (A1795 with $z = 0.062$ and 3A0336+098 with $Z = 0.0349$ in collaboration with C. Sarazin) which are small enough (or distant enough) to be covered by the GIS f.o.v. and bright enough (or close enough) regarding the photon statistics. These clusters are reported to have cooling flows, but simple β models can fit the GIS data well if one excludes the cluster center ($r < 6'$). Spatially uniform temperature with no abundance gradient is assumed here, and these values are fixed to those observed in the outer regions in clusters. Core radius is also fixed. For a given β value, we generate models with various $r_{\max}$ values and compare the 3.5–10 keV radial profile at $r > 6'$ with the model. It is found that the χ^2 values bcome unacceptably large ($\chi^2_\nu > 2$) for $r_{\max}$ smaller than 15 arcmin. The χ^2_ν is close to unity for larger values of $r_{\max}$ even when $r_{\max}$ exceeds the radius of the GIS f.o.v. (about 25 arcmin). The properties of the ICM for the two clusters are summarized in Table 1.

Table 1: Parameters of the ICM for the 2 clusters

Cluster	kT (keV)	$Z_{\rm Fe}$ (solar)	$r_{\rm core}$ (′)/(kpc)	β	$r_{\max}$ (′)/)Mpc)
A1795	5.8	0.35	1.7/180	0.73	$\gtrsim$ 20/2.1
3A0336+098	3.5	0.45	1.3/80	0.65	$\gtrsim$ 20/1.2

Based on these reulsts, the total mass is calculated assuming the hydrostatic equilibrium[16]. The radial profies of the gas mass and the total mass thus obtained are plotted in Fig. 3. As seen in other clusters[17], the ratio of $M_{\rm gas}$ to $M_{\rm total}$ becomes larger as a function of radius. Requiring $r_{\max}$ to be greater than ~ 20 arcmin, $M_{\rm gas}$ turns out be at least 15% of $M_{\rm total}$ for $H_0 = 50$ km s^{-1} Mpc^{-1}. Further refinement in the model by including the second cool emission component is under way.

These results suggest that the amount of baryon may be as large as about 20% of the gravitational mass in these 2 clusters. It has been discussed recently that the baryon fraction of about 30% observed in the Coma cluster is too high if the

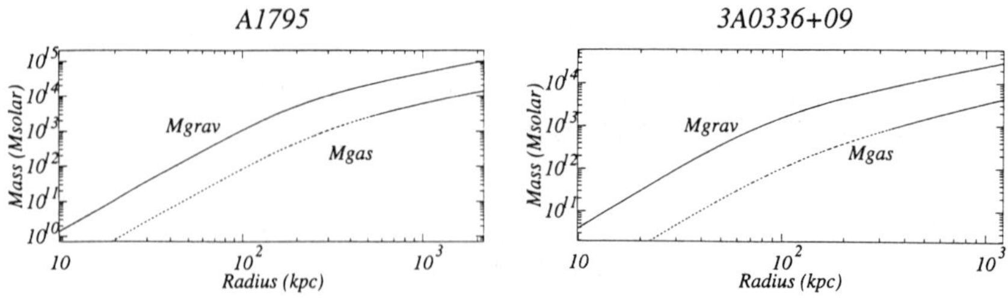

Figure 3: $M_{\rm gas}$ and $M_{\rm total}$ as a function of radius for A1795 and 3A0336+098

university has the critical density[17]. The big-bang nucleosynthesis predicts that baryon density should be about 6% of the critical density for $H_0 = 50$ km s^{-1} Mpc^{-1}, and numerical simulations show that the baryon density, averaged over a few Mpc, cannot exceed the cosmic average. The ASCA results suggest that the baryon content of clusters may be similarly high even for clusters with lower richness.

Acknowledgements The results reported here are analyzied by ASCA ANL, and SimASCA software development teams. Particular thanks are due to, Yasushi Ikebe, Tadayuki Takahashi, Yasushi Fukazawa, Yoshitaka Ishisaki, Ken'ichi Kikuchi, Noriko Yamasaki, and Yuzuru Tawara. Intriguing discussion with Kazuo Makishima, and Fumio Takahara are also gratefully acknowledged.

References

1. Tanaka, Y., H. Inoue & S.S. Holt 1994, PASJ, **46:** L37.
2. Ohashi, T., K. Makishima, M. Ishida, T. Tsuru, M. Tashiro, T. Mihara & Y. Kohmura 1991, SPIE, **1549:** 9.
3. Kohmura, Y. et al 1993, SPIE, **2006:** 78
4. Takahashi, T. et al 1994, ASCA Team internal report.
5. Fukazawa, Y., T. Ohashi, A.C. Fabian, C.R. Canizares, Y. Ikebe, K. Makishima, R.F. Mushotzky & K. Yamashita 1994, PASJ, **46:** L55.
6. Ohashi, T., Y. Fukazawa, Y. Ikebe, H. Ezawa, T. Tamura & K. Makishima 1994, *In* "New Horizon of X-Ray Astronomy", F. Makino & T. Ohashi eds.: 273, Universal Academy Press, Tokyo.
7. Koyama, K., S. Takano & Y. Tawara 1991, Nature, **350:** 135.
8. Matsumoto, H., T. Tsuru, H. Awaki, K. Koyama & ASCA Team 1994, *In* "New Horizon of X-Ray Astronomy", F. Makino & T. Ohashi eds.: 511, Universal Acad. Press, Tokyo.
9. Arnaud, K.A. et al 1994, ApJL, **436:** L67.
10. Rephaeli, Y. 1978, ApJ, **225:** 335.
11. Yamashita, K. 1995, *In* Proc. Les Arcs Meeting on Clusters of Galaxies, Méribel, France.
12. Tsuru, T. 1992, Ph. D. Thesis, University of Tokyo (ISAS RN-528).
13. Arnaud, M., R. Rothenflug, O. Boulade, L. Vigroux & E. Vangioni-Flam 1982, A&A, **254:** 49.
14. Nomoto, K., F-K. Thieleman & K. Yokoi 1984, ApJ, **286:** 644.
15. Ikebe, Y. 1995, Ph. D. Thesis, University of Tokyo.
16. Sarazin, C. 1986, "X-Ray Emission from Clsuters of Galaxies", Cambirdge Univ. Press., Cambirdge.
17. Böhringer, H. 1994, *In* " Cosmological Aspects of Clusters of Galaxies", W.C. Seitter ed.: 123, Kluwer Academic Publ.
18. White, S.D.M., J.F. Navarro, A.E. Evrard & C.S. Frenk 1993, Nature, **366:** 429.

Recent Highlights of the Compton Gamma Ray Observatory

CARL E. FICHTEL

NASA/Goddard Space Flight Center
Greenbelt, MD 20771, U. S. A.

The Compton Gamma Ray Observatory was launched on April 5, 1991 and continues to produce data of fundamental significance to several astrophysical problems. The satellite carries four experiments whose locations on the spacecraft are shown in the figure below. The Compton Telescope (COMPTEL) and the Energetic Gamma Ray Experiment Telescope (EGRET) are both wide field instruments and point along the z axis. The former covers the energy range from 1 to 30 MeV and the latter from 20 MeV to 30 GeV. The Oscillating Scintillation Spectrometer Experiment (OSSE) has a smaller field of view but may point anywhere in a plane from approximately the z axis to the x axis. It covers the energy range from 0.1 to 10 MeV. The Burst and Transient Source Experiment (BATSE) consists of eight instruments located on the corners of the spacecraft, and, as its name implies, is primarily designed for gamma ray bursts and transients in the 0.02 to 30 MeV range. The following four talks will summarize the recent scientific highlights from these four experiments.

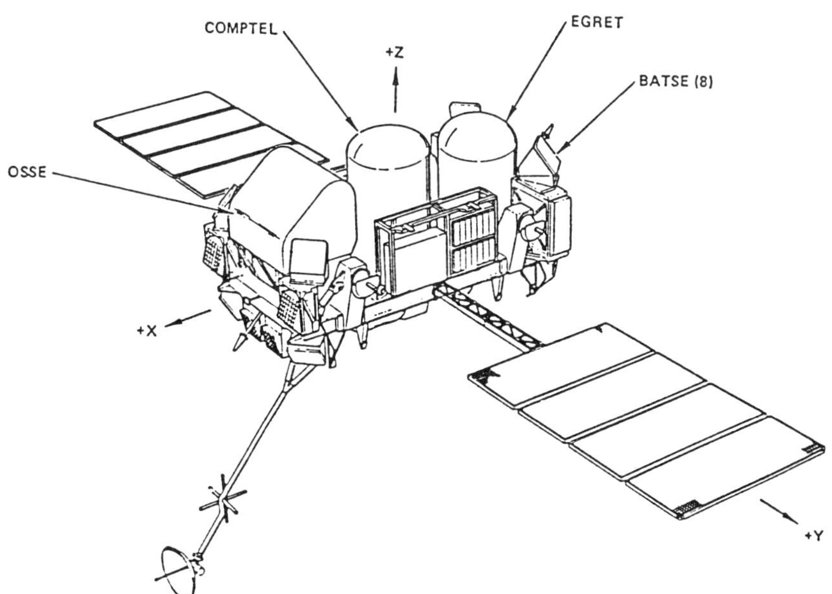

Fig. 1. The Compton Gamma Ray Observatory

Highlights of the Energetic Gamma Ray Experiment Telescope

CARL E. FICHTEL

NASA/Goddard Space Flight Center
Greenbelt, MD 20771, U. S. A.

1. INTRODUCTION

The significant potential contribution of high energy gamma ray astronomy to astrophysics had been recognized for a long time due to its direct connection to the dynamic high energy processes in the universe, but building a successful high energy gamma ray telescope qualified for space flight is difficult. The results of early small telescopes, especially those of SAS-2 and COS-B, gave support to the hopes for a rich return when a more sensitive instrument was flown. The Energetic Gamma Ray Experiment Telescope (EGRET) on the Compton Gamma Ray Observatory (GRO) has over ten times the sensitivity of these two earlier satellite telescopes just mentioned. It covers the energy range from about 2×10^7 eV to approximately 3×10^{10} eV. A description of the instrument is given by Kanbach et al.[1], and details of the calibration by Thompson et al.[2]. As anticipated, EGRET has indeed produced exciting and important new results on a broad range of astrophysical topics. In this short summary, some of the highlights will be given together with references to more extended treatments of these and other EGRET findings.

2. ACTIVE GALAXIES

One of the most striking observations of EGRET is the detection of high energy gamma-rays from active galaxies whose emission at most wavelengths is dominated by non thermal processes. These objects, called blazars, are highly variable at most frequencies and are bright, hard spectrum radio sources. At least 33 of these objects have strong gamma-ray emission as seen by EGRET, and another 11 are very likely sources of high energy gamma rays. Moreover, it is not just the close ones that are being seen, but there are a significant number of the sources at large z values. The distribution of the blazars that are observed in high energy gamma rays, called grazars, is in fact similar within the rather limited statistics to that observed in the radio wavelength range. A remarkable aspect of the high energy gamma ray emission is seen when the the intensity of the gamma ray emission expressed as νf_ν is compared to other wavelengths as a function of ν. For those that have been seen, the gamma ray luminosity generally exceeds that at other wavelengths, in one case by about two orders of magnitude (von Montigny et al.[3]).

For those grazars for which adequate observations exist, time variability is observed with timescales from months down to a day. On the basis of light travel time arguments, the size of the gamma-ray emitting region is of the order of the radius of a 10^{10} $M_\odot$ black hole multiplied by the relativistic factor--a small region on a galactic scale. This time variation observation is consistent with the concept that the radiation originates in the knots observed in blazars that are seen in the radio range to be moving outward at superluminal apparent speeds. The spectrum of the observed blazars may be represented in general by a power in energy with a differential spectral index of between 1.6 and 2.6. The average index is 2.1, and this average is independent of z within uncertainties. For a further discussion of the blazar results of EGRET, see von Montigny et al.[3].

Finally, it should be noted that, whereas there are these very impressive results for blazars, not a single Seyfert galaxy has been seen by EGRET in the high energy gamma-ray range.

3. GAMMA RAY BURSTS

Rivaling the AGN results in their surprise and scientific importance are the gamma ray bursts detected by EGRET. Prior to the EGRET results, these bursts had been thought to be a low energy gamma-ray phenomenon of short duration. EGRET has imaged 5 bursts above 30 MeV out of the few 100 gamma ray bursts detected by the BATSE instrument within EGRET's field of view. Gamma rays have been observed with energies well above a GeV. In the burst of 17 February 1994 (Hurley et al.[4]), the high-energy gamma-ray emission extended for an hour or more beyond the sub MeV emission detected by BATSE. The highest energy gamma ray was 18 GeV. It occurred almost an hour after the sub MeV emission became undetectable.

The existence and long time delay of high-energy gamma rays add many constraints to the nature of the source of gamma-ray bursts. The spectra extending up to GeV energies rule out thermal spectrum of most types. Therefore, particles must be accelerated to energies even higher than the energy of these gamma rays. High-energy gamma rays early in the burst require particles to be accelerated almost simultaneously with the keV gamma-ray emission. Shock acceleration is one such possibility.

The lack of attenuation of GeV gamma rays by pair production with the much more plentiful lower energy photons in the source requires beaming and/or bulk Lorentz factors in order to reduce the photon-photon opacity. The Lorentz factors required are much higher than those observed from other astrophysical phenomena, such as the superluminal motion in active galactic nuclei with bulk Lorentz factors less than 10. The photon density in burst sources is especially large because of the small size of the source determined from the rapid time variability and the large distances to bursts.

4. PULSARS

The realization that pulsars were neutron stars occurred almost immediately after their discovery in the radio portion of the electromagnetic spectrum. With the results of EGRET that now exist, the number of known high energy gamma-ray pulsars has risen from two to five. Except for the Crab pulsar, which has much the same appearance across the electromagnetic spectrum, the pulsars show a wide variety of pulse shapes and relative pulse arrival times. As an example, the Vela pulsar has a single radio pulse, a double optical pulse, a double X-ray pulse, and a double gamma ray pulse, none of which are in phase. What they do have in common is that all radiate far more energy in the form of gamma rays than in radio waves. In addition, there is a consistent trend of the gamma ray efficiency, in terms of the energy radiated in gamma rays as a fraction of the rotational energy, as a function of age of these five pulsars, specifically that the efficiency increases with age in the range of a thousand to a million years (Fichtel and Thompson[5] and references therein).

5. COSMIC RAYS AND THE MAGELLANIC CLOUDS

The study of the Magellanic Clouds in Gamma Rays has led to the clear resolution of the long standing question of whether cosmic rays are galactic or metagalactic in favor of the expected answer that they are galactic. The deciding measurement was the upper limit on the Small Magellanic Cloud (SMC) which was well below the flux that would have been present if the cosmic rays were metagalactic (Sreekumar and Fichtel[6]; Sreekumar et al.[7]). The upper limit for the SMC is also below the flux expected for quasi-stable equilibrium and is consistent with the SMC being in a disrupted state. The flux measured from the Large Magellanic Cloud is consistent with that expected from quasi-stable equilibrium, although possibly a bit on the low side (Sreekumar et al.[8]).

6. THE GALACTIC DIFFUSE RADIATION

A detailed map of the high energy gamma ray galactic diffuse radiation measured by EGRET now exists (Hunter et al.[9]). It is consistent in some detail with that expected from the interaction of cosmic rays with matter and photons and quasi-stable equilibrium of cosmic rays on a broad scale of about two kiloparsecs (Bertsch et al.[10]).

7. MICROSECOND BURSTS

Hawking[11] and Page and Hawking[12] investigated theoretically the possibility of detecting high-energy gamma rays produced by the quantum-mechanical decay of a small black hole created in the early Universe. They concluded that, at the very end of the life of the small black hole, it would radiate a burst of gamma rays peaked near 250 MeV with a total energy of about 10^{34} ergs in the order of a microsecond or less. The characteristics of a black hole are determined by laws of physics beyond the range of current particle accelerators; hence, the search for these short bursts of high-energy gamma rays provides at least the possibility of being the first test of this region of physics. A search of the EGRET data by Fichtel et al.[13] has led to an upper limit of 5×10^{-2} black hole decays per pc^3 yr^{-1}, placing constraints on this and other theories predicting microsecond high-energy gamma-ray bursts.

8. ENERGETIC SOLAR PARTICLES

Because the solar cycle is currently in a low state, the most significant results on solar physics remain those from June 1991. In the 11 June 1991 large solar event, high energy gamma rays were seen for 10 hours after the solar flare began (Kanbach et al.[14]). This unexpected result could either be due to acceleration over this period of time or trapping of the particles at the sun and slow release and interaction with the solar surface. The latter seems to be the easier explanation due to the other forms of radiation being observed only during the first hour.

9. EXTRAGALACTIC DIFFUSE RADIATION

There does appear to be an approximately uniform high energy gamma radiation on the basis of there being a uniform excess when the high energy gamma-ray galactic component is removed. This component is assumed to be extragalactic on the basis of its approximate uniformity. The spectrum is found to be well approximated by a differential power law in gamma-ray energy with a slope of 2.2 ± 0.2 (Sreekumar et al.[15]) and an intensity above 100 MeV that is consistent with that reported by SAS-2. It extends at least to 4 GeV (Sreekumar et al.[16]). Using the AGN high energy gamma-ray data only, Chiang et al.[17] derived the evolution funtion and the luminosity function and deduced that, within a relatively large uncertainty, the observed presumed extragalactic radiation is consistent with its being due to the sum of blazars.

10. SUMMARY

Highlights of the results from the Energetic Gamma Ray Experiment Telescope flown on the Compton Gamma Ray Observatory include: (1) The finding of a new class of objects--high energy gamma-ray emitting blazars or grazars, (2) the emission of high energy gamma-rays from a gamma ray burst for over an hour, with some gamma rays having energies over a GeV and two having energies over 10 GeV, (3) the observation of an increased fraction of pulsar electromagnetic radiation going into gamma rays as the age of the pulsar increases to a million years, (4) the determination with high certainty that cosmic rays are galactic, (5) the detailed mapping of the galactic diffuse radiation and the measurement of the pion bump in the high energy gamma-ray energy spectrum, (6) the absence of microsecond bursts and its implication for certain unification theories, (7) the long trapping time of over 10 hours for energetic solar particles following a flare, and (8) a measurement of the extragalactic diffuse radiation and, within relatively large uncertainty, its consistency with that expected from AGN emission.

REFERENCES

1. Kanbach, G., *et al.*, 1988, Space Sci. Rev. **49**, 69
2. Thompson, D., *et al.*, 1993, ApJ. Suppl., **86**, 629
3. von Montigny, C., *et al.*, 1995, to be published in the February issue of ApJ.
4. Hurley, K., *et al.*, 1994, Nature, **372**, 652
5. Fichtel, C. E., and Thompson, D. J., 1994, "High Energy Gamma Ray Astronomy", *In* High Energy Astrophysics, Models and Observations from MeV to EeV, ed. James M. Matthews, World Scientific Publishing Co.
6. Sreekumar, P., and Fichtel, C. E., 1991, A&A **251**, 446
7. Sreekumar, P., *et al.*, 1993, Phys. Rev. Letters **70**, 127
8. Sreekumar, P., *et al.*, 1992, ApJ. Letters **400**, L67
9. Hunter, S. D., *et al.*, 1995, submitted to ApJ
10. Bertsch, D. L., *et al.*, 1993, ApJ **416**, 587
11. Hawking, S. W., 1974, Nature, **248**, 30
12. Page, D. N. and Hawking, S. P., 1976, ApJ, **206**, 1
13. Fichtel, C. E., *et al.*, 1994, ApJ **434**, 557
14. Kanbach, G., *et al.*, 1993, A&A Suppl. **97**, 349
15. Sreekumar, P., *et al.*, 1994, January 1994 AAS Meeting
16. Sreekumar, P., *et al.*, 1995, submitted to the ApJ
17. Chiang, J., Fichtel, C. E., von Montigny, C., and Nolan, P. L., 1995, submitted to ApJ

Highlights from the COMPTEL 1 to 30 MeV Sky Survey

V. SCHÖNFELDER[1], K. BENNETT[4], H. BLOEMEN[2], W. COLLMAR[1], R. DIEHL[1], W. HERMSEN[2], L. KUIPER[2], G. LICHTI[1], M. MCCONNELL[3], J. RYAN[2], A. STRONG[1], C. WINKLER[4]

[1] *Max-Planck-Institut für extraterrestrische Physik, D-85740 Garching, Germany*
[2] *SRON - Utrecht, NL-3584 CA Utrecht, The Netherlands*
[3] *University of New Hampshire, Space Science Center, Durham, NH 03824, USA*
[4] *Astrophysics Division of ESA/ESTEC, NL-2201 AZ Noordwijk, The Netherlands*

INTRODUCTION

COMPTEL is the first gamma-ray telescope that has performed a complete survey of the sky at MeV-energies (1 to 30 MeV). Before the launch of the Compton Observatory, only a very few celestial objects had been seen in this energy range. COMPTEL has now opened this new window to astronomy. This article summarizes highlights from the first three phases of the Compton mission from April 1991 to August 1994.

THE COMPTEL ALL-SKY MAPS

COMPTEL all-sky maps in three energy ranges (1-3, 3-10, and 10-30 MeV) are shown in Fig. 1. These maps are based on the full COMPTEL database up to mid 1994. The maps are made using the maximium entropy method to deconvolve the data using the COMPTEL response function. The concentration of emission along the Galactic plane is the most striking characteristic, as expected. Much of this emission is attributable to the diffuse Galactic component, although the imaging method used for these images, is chosen to emphasize point sources and partly suppresses extended emisison.

Many known COMPTEL point sources are visible in Fig. 1 as indicated in Fig. 2. Galactic sources include the Crab nebula and pulsar (the strongest COMPTEL source in all 3 energy bands), the Vela pulsar, Cyg X-1, the Gregory-Taylor radio source GT 0236+610 (counterpart of the COS-B source 2C 135+01), and a strong unidentified source at $1 \approx 18°$. Extragalactic sources include 3C 273 and Cen A; also the region of the Large Magellanic Cloud shows enhanced emission. Other extragalactic COMPTEL detections (marked in Fig. 2) are not visible in Fig. 1, since they flare up only occasionally: on average they are too weak to become visible in the all-sky maps.

Although these maps are quite new and care must be taken in their interpretation, they clearly represent a major step forward being the first full-sky maps in this energy range.

GALACTIC GAMMA-RAY SOURCES

The few identified Galactic COMPTEL sources of continuum emission are pulsars and X-ray binaries. Up to now, in total, 6 spin-down pulsars are known to be gamma-ray pulsars; 4 of them have been seen by COMPTEL, namely Crab, Vela, PSR 1509-58, and Geminga. The properties of the first 3 pulsars are summarized in ref. 1; the fourth pulsar, Geminga is very weak at MeV-energies (in spite of being one of the brightest sources in the sky above 100 MeV). Its COMPTEL detection was only possible by combining data collected during three years of the Compton mission up to mid 1994.

A comparison of the energy spectra of Vela and Geminga is made in Fig. 3. Both pulsars have their peak luminosity near 1 GeV. The other two gamma-ray pulsars (PSR 1706-44 and PSR 1055-52) have very hard spectra and are only seen by EGRET above a few hundred MeV.

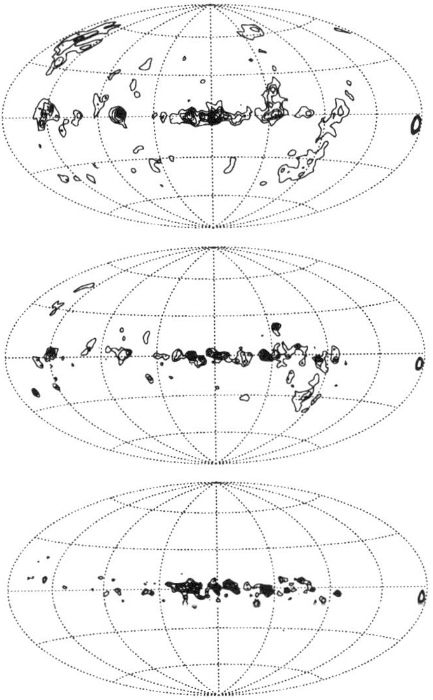

Fig. 1: COMPTEL full-sky images in the energy ranges 1-3, 3-10, and 10-30 MeV, using data from 1991 to 1994. Coordinate system is Galactic, centered on l = 0°, b = 0°.

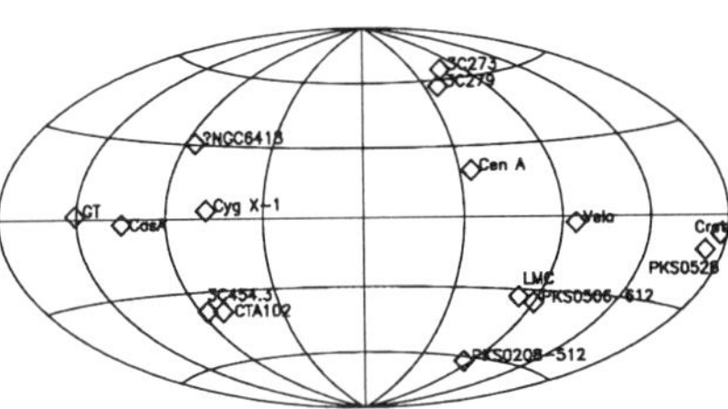

Fig. 2: Gamma-ray sources detected by COMPTEL; some of these (see text) are visible in the full-sky maps.

Among the Galactic X-ray binary sources Cyg X-1 is the most promising stellar black-hole candidate. The COMPTEL energy spectrum of Cyg X-1 (Fig. 4) extends to at least 5 MeV. This is a surprise, because temperatures with kT-values as high as about 200 keV would be needed to explain the high-energy end of the spectrum in the context of classical comptonisation models. The existence of an additional non-thermal component at MeV energies cannot be excluded at this time.

INTERSTELLAR CONTINUUM GAMMA-RAY EMISSION

In the COMPTEL energy range, the most important processes for the production of interstellar continuum gamma-ray emission are bremsstrahlung and inverse Compton scattering from relativistic electrons interacting with matter and the interstellar radiation field, respectively. The COMPTEL bremsstrahlung emissivity spectrum (Fig. 5) nicely connects to the COS-B emissivity spectrum at higher energies, dominantly by π°- decay emission. Calculations of the sum of the electron-bremsstrahlung and the π°-decay components match the overall shape of the measured spectrum from about 1 MeV to 1 GeV (ref. 8).

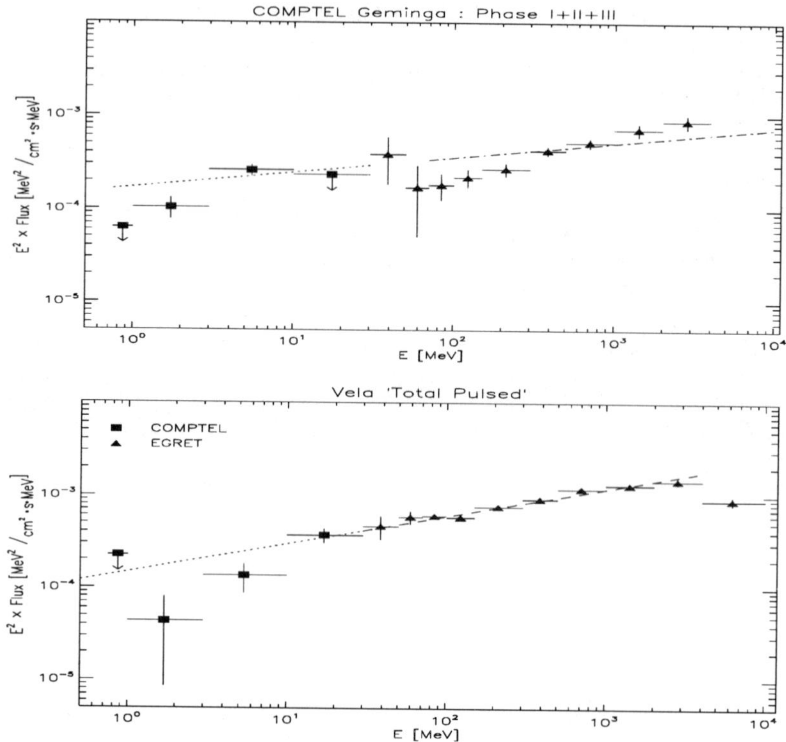

Fig. 3: Energy spectra of the Vela and Geminga pulsars. The EGRET data are taken from ref. 2 (Vela), and ref. 3 (Geminga), the COS-B spectra (dashed-dotted line) are taken from ref. 4.

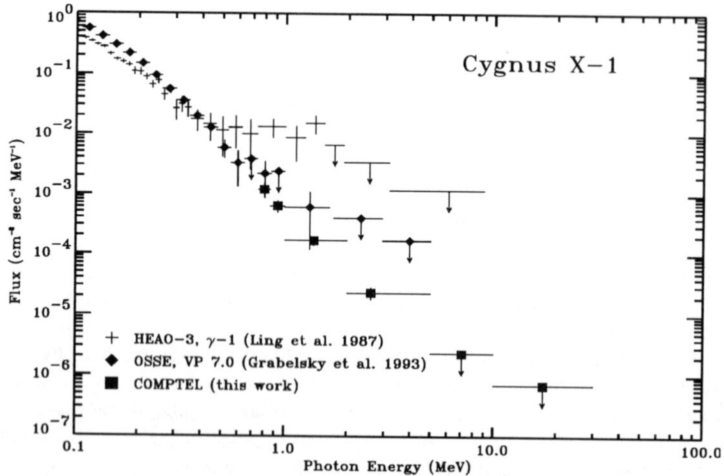

Fig. 4: Cyg X-1 energy spectrum. The COMPTEL summed spectrum from phase-1,-2, and -3 observations (ref. 5) is compared with the OSSE-spectrum from viewing period 7 (ref. 6), and with the HEAO-C spectrum (ref. 7).

EXTRAGALACTIC GAMMA-RAY SOURCES

In the field of extragalactic astronomy COMPTEL has made two major findings. First, no single Seyfert galaxy (including NGC 4151) has so far been found to show emission above 750 keV (ref. 10). This is a disappointment. Seyfert galaxies are hard X-ray emitters, but apparently not intense gamma-ray emitters. Second, a few gamma-ray blazars - discovered by EGRET above 100 MeV (ref. 11) - have been seen by COMPTEL at MeV-energies. In Fig. 2 in total 8 blazars are marked, from which at least occasional gamma-ray emission was detected by COMPTEL. However, for two of the COMPTEL sources that can be identified with blazars, two potential counterparts are present (3C 454.3 / CT 102 and PKS 0506-612 / PKS 0522-611), so six COMPTEL γ-ray blazar sources can be seen in Fig. 2. Spectral information exists for most of these objects. Fig. 6 shows energy spectra of 6 candidate blazars. It is clear from this figure that the COMPTEL fluxes or flux limits do not allow extrapolations of the EGRET power-law spectra towards lower energies. Spectral breaks do exist in all six cases somewhere in the range 1 to 30 MeV. The change of the spectral power-law index in some cases is definitely larger than 0.5 (e.g. in case of 3C 273, see ref. 12). The spectra of PKS 0506-612 / 0522-611 (ref. 13) and PKS 0208-512 (ref. 14) suggest that for these objects even spectral bumps exist rather than spectral breaks. AGNs with spectral bumps at MeV-energies may be of importance to explain the well-known MeV-bump of the diffuse cosmic background radiation.

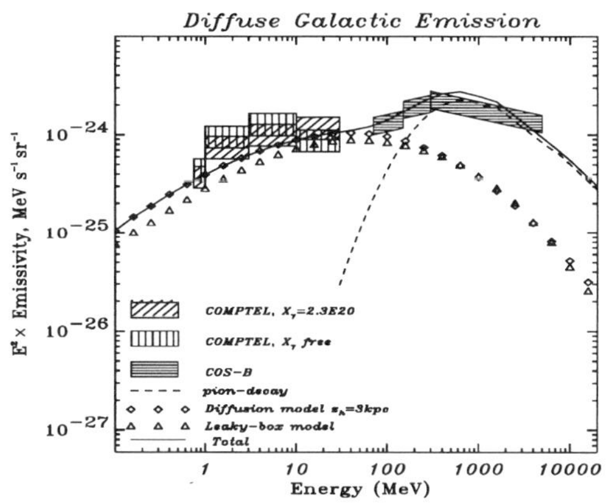

Fig. 5: Emissivity spectrum of the diffuse Galactic continuum emission based on COMPTEL (ref. 8) and COS-B (ref. 9) data. Calculated spectra from bremsstrahlung and π^0-decay are shown for comparison.

GAMMA-RAY LINE SPECTROSCOPY

Perhaps the most exciting COMPTEL results are in the field of gamma-ray line spectroscopy. COMPTEL has generated the first map of the entire galactic plane in the light of the 1.809 MeV line from radioactive ^{26}Al (ref. 19). It has for the first time detected the 1.156 MeV line from radioactive ^{44}Ti from a supernova remnant (namely Cas A, ref. 20). In addition, it has - again for the first time - detected MeV-emission from the Orion complex region, which may be due to nuclear interaction lines from exited ^{12}C and ^{16}O nuclei (ref. 21). Finally, after steady improvements in the COMPTEL data analysis technique, hints were found for the existance of the ^{56}Ni $\Rightarrow$ ^{56}Co $\Rightarrow$ ^{56}Fe-decay chain in the type Ia SN 1991T: the data show indications of the two ^{56}Co-lines at 847 keV and

1.238 MeV (ref. 22). These gamma-ray line detections were subject of the Mini-Symposium on Gamma Ray Line Spectroscopy at this Conference.

The appearance of the 1.809 MeV map from radioactive ^{26}Al was a surprise: the emission along the plane is not smooth, but clumpy. The reason may be that nearby sources or source regions contribute significantly to the observed distribution. A possible example is the Vela supernova remnant (ref. 23). Other localized source regions may be the Cygnus-region, the Carina region, and local spiral arm regions.

The detection of the 1.156 MeV line from radoactive ^{44}Ti from Cas A illustrates the potential of gamma-ray line spectroscopy for revealing previously undetected Galactic supernova remnants from the last several hundred years. This method may be the only way to perform a direct measurement of the Galactic supernova rate.

The possible detection of nuclear interaction lines from the Orion complex was a surprise, as well. Much weaker line fluxes had been expected. It appears that efficient nucleonic particle accelerators exist in the Orion complex.

The two ^{56}Co-lines at 847 keV and 1.238 MeV from SN 1991T have only marginally been detected. Still, the derived flux of the 847 keV line of (5.3 ± 2.0 stat. ± 1.6 syst.) $\cdot 10^{-5}$ cm^{-2} sec^{-1} is consistent with most of the theoretical estimates (see ref. 24).

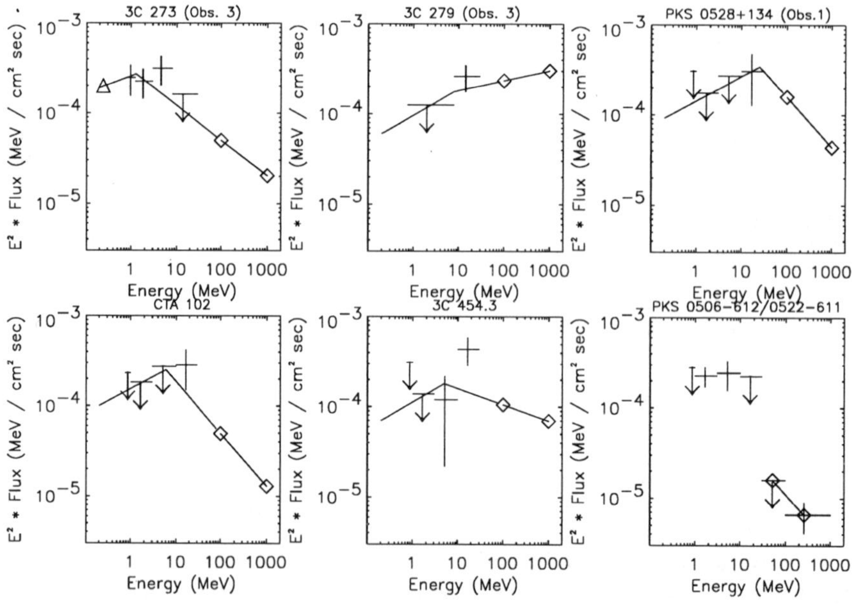

Fig. 6: Illustration of spectral breaks at MeV-energies found by COMPTEL for 6 AGNs. The solid lines (framed with diamonds) represent the spectral shapes of the sources as measured by EGRET (ref. 15). The spectral points represent the COMPTEL measurements or upper limits. Eye fits to the breaks in the combined COMPTEL/EGRET spectra are indicated. The break of the 3C 273 spectrum is required from a combination of OSSE and COMPTEL data (ref. 12). References for COMPTEL results are: (ref. 16) for 3C 279, (ref. 17) for PKS 0528+134 (ref. 18) for CTA 102 and 3C 454.3 and (ref. 13) for PKS 0506-612 / PKS 0522-611.

CONCLUSIONS

In this brief summary article, only part of the COMPTEL results could be included. Observations of cosmic gamma-ray bursts (see ref. 25) and solar flares were not addressed at all. Nevertheless, the results presented have illustrated that the sky is rich in phenomena that can be studied at MeV-energies.

ACKNOWLEDGEMENT

The COMPTEL project is supported by the German government through DARA grant 50 Q 9096.8, by NASA under contract NAS5-26645, and by the Netherlands Organization for Scientific Research NWO.

REFERENCES

1. Bennett, K. et al., 1994 ApJ Suppl. 90, 823
2. Kanbach, G. et al., 1994, A&A 289, 855
3. Mayer-Hasselwander, H.A., 1994, ApJ 421, 276
4. Grenier, I.A. et al., 1994, ApJ Suppl. 90, 813
5. McConnell, M. et al., 1995, COSPAR Hamburg, Adv. Sp. Res., in press
6. Grabelsky, D.A. et al., 1994, AIP-Conf. Proc. 280, 345, eds: M. Friedlander, N. Gehrels, and D.J. Macomb
7. Ling, J.C. et al., 1987, ApJ 321, L117
8. Strong, A.W. et al., 1994, A&A 292, 82
9. Strong, AW. et al., 1988, A&A 207, 1
10. Maisack, M. et al., 1995, A&A, in press
11. Fichtel, C.E., 1993, AfIP Conf. Proc. 280, p. 461, eds: M. Friedlander, N. Gehrels, and D.J. Macomb
12. Lichti, G.G. et al., 1995, A&A, in press
13. Bloemen, H. et al., 1995, A&A, 293, L1
14. Blom, H. et al., 1995, A&A, in press
15. Hunter, S. et al., 1993, ApJ 409, 134
16. Williams, O.W. et al., 1995, A&A, in press
17. Collmar, W. et al., in preparation for submission to A&A
18. Blom, H. et al., 1995a, A&A, in press
19. Diehl, R. et al., 1995, A&A, in press
20. Iyudin, A. et al., 1994, A&A 284, L1
21. Bloemen, H. et al., 1994, A&A 281, L5
22. Morris, D. et al., 1995, this volume
23. Oberlack, U. et al., 1994, ApJ Suppl. 92, 433
24. Ruiz-Lapuente, P., Lichti, G.G., Lehoucq, R., Canal, R., and Casse, M., 1993, ApJ 417, 547
25. Kippen, M. et al., 1995, this volume

BATSE Highlights

GERALD J. FISHMAN

Space Sciences Laboratory
NASA-Marshall Space Flight Center
Huntsville, Al USA

INTRODUCTION

Since the launch of the Compton Gamma Ray Observatory in April 1991, the Burst And Transient Source Experiment (BATSE) has been serving the high-energy astronomy community as an all-sky monitor at energies >20 keV. The hard x-ray sky is extremely variable, replete with gamma ray bursts, pulsars and a wide variety of other highly variable and transient sources. Below I highlight only a small sample of the widely diverse phenomena accessible to BATSE.

GAMMA-RAY BURSTS

Gamma-ray bursts represent the ultimate in high-luminosity, low duty cycle sources. During their brief appearance, these objects often outshine the combined flux of all other objects in the sky in the energy of the BATSE Large Area Detectors (LADs), from 20 keV to 2 MeV. This phenomenon has no analog in other fields of astronomy. Recent BATSE observations have produced a wealth of new temporal, spectral and global properties of gamma-ray bursts (cf. Fishman, Brainerd and Hurley 1994). In spite of this, the present state of uncertainty in the distance scale, the energy source and the emission mechanism of gamma-ray bursts is as great as ever.

The mini-symposium on gamma-ray bursts in these proceedings provides a good summary of the current state of observations, interpretations and theory; I will not try to summarize those papers. There is a continued feeling that the identification of a counterpart in another wavelength region may be the key to our understanding these enigmatic objects. A new capability exists for near real-time searches of gamma-ray burst sky regions using data from BATSE. The BAtse COordinate DIstribution NEtwork (BACODINE), developed at NASA-GSFC in conjunction with the BATSE team (Barthelmy et al. 1994), is now in full operation. It notifies over twenty sites worldwide of the occurrence and location of a gamma-ray burst, often while the burst is still in progress.

The distribution of over one thousand gamma-ray bursts on the sky shows no anisotropy, confirming previous BATSE observations will smaller samples of bursts. Recent papers using BATSE data (Meegan et al 1995; Briggs et al. 1995, and Hartmann et al. 1995) describe the constraints on any observed anisotropy and also set limits on the fraction of bursts which may repeat. These limits, combined with the observed inhomogeneity of gamma-ray bursts (as required by the deficiency of weak bursts), are continuing to cause more theorists to consider cosmological models of gamma-ray bursts.

TRANSIENT SOURCES

The high sensitivity and all-sky capabilities of BATSE make it ideal for the discovery and nearly continuous monitoring of hard x-ray transients. Most of these observations are accomplished using Earth occultation: observing a change in the background as the sources being studied rise above and set behind the earth's limb twice each spacecraft orbit. The all-sky monitoring of sources by BATSE has been greatly augmented by a newly developed transform imaging technique (Zhang et al. 1993). Through this technique, sources can be much better separated and located. Typical source location accuracies attained are a few tenths of a degree. Figure 1 shows representative samples of a field with three sources taken at different times.

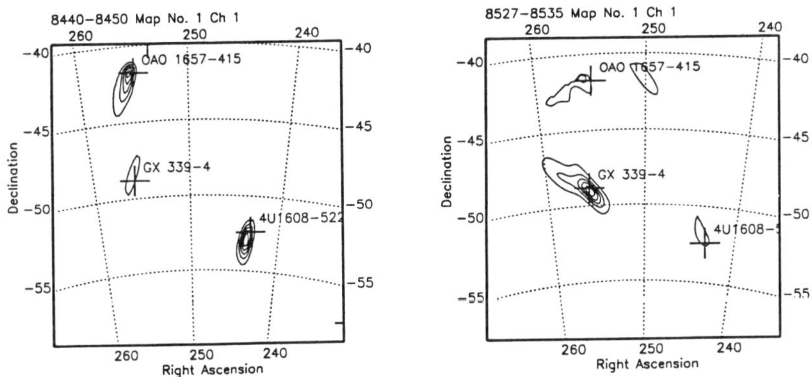

FIGURE 1. A region of the sky imaged by the BATSE occultation transform imaging technique (Zhang et al. 1993) in the energy range 20-50 keV. It shows three sources at two different times, separated by about 10 days. The high variability of these sources is evident.

Within the past three years, radio and x-ray observations have been made of two remarkable galactic objects, GRS 1915+105 and GRO J1655-40. Both objects show high variability with repeated outbursts over several months, hard x-ray spectra, and radio jet structures expanding from a central source with superluminal velocities. These galactic hard x-ray/radio jet sources may be more common than previously thought. They have not previously been observed due to their low duty-cycle, the lack of rapid, high resolution radio imaging, and the lack of all-sky hard x-ray coverage. In addition to their intrinsic interest, these unusual objects may also serve to shed light on high-energy jet phenomena in AGNs.

BATSE discovered X-ray Nova Scorpii (GRO J1655-40) in July 1994. Subsequent observations were made in the optical and radio regions, as described in a flurry of IAU Circulars in the last half of 1994. Some of the early BATSE observations of GRO J1655-40 are presented in more detail in the black-hole mini-symposium in these proceedings by W. Paciesas and they will soon be in print (Harmon et al. 1995). In that same session, F.

Mirabel gives an invited talk on radio observations of both this object and the first "micro-quasar", GRS 1915+105.

PULSARS AND QPO SOURCES

Sources with unique temporal structures can be separated from the background in the all-sky detector data by epoch-folding, FFT and power spectral studies. Thus far, about twenty pulsed sources have been observed. Similarly, the folding of occultation data has provided the detection of several sources, including LMC X-4, by their orbital or precessional periods. Many of the x-ray binaries are only observable by BATSE for less than a few percent of the time. Two new binary pulsars, GRO J1008-57 (Wilson et al. 1994) and GRO J1948+32 (Finger et al 1994a) were discovered by BATSE during routine searches of all-sky FFT data .

In 1994, a very intense outburst of the Be binary x-ray source A0535+26 occurred. The pulsed flux from this source rose to about 5xCrab in the energy range 20-100 keV. This is perhaps the highest flux ever observed from any pulsar in this energy range. The high

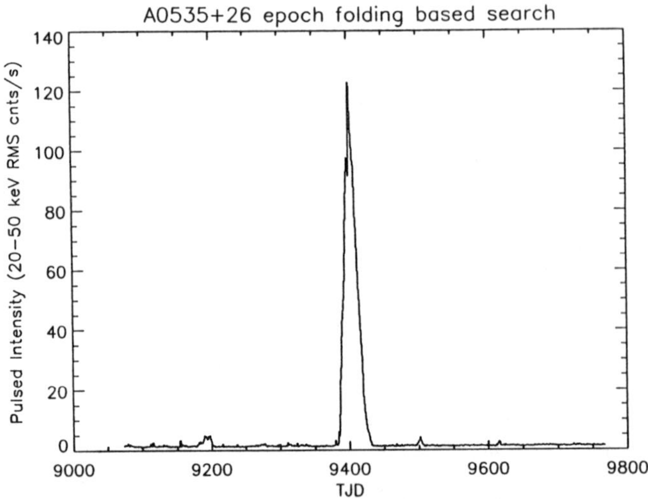

FIGURE 2. Over two years of nearly continuous observations of the Be binary x-ray system A0535+26 by BATSE, showing several outbursts, as measured by its pulsed hard x-ray flux. These flares occur at intervals near the binary orbital period, 111 days. The large outburst, lasting about 50 days is seen near TJD 9400 (17 February 1994).

intensity, combined with the continuous viewing throughout the outburst, has allowed some unique diagnostics of this system to be made from the BATSE data. In particular, quasi-periodic oscillations (QPO) were observed in this source for the first time (Finger et al. 1994b). Simultaneous observations of the QPO frequency changes, the intensity changes and the period changes (torque) on the neutron star of the A0535+26 system permits modeling of an accreting binary system as never before possible (Finger et al. 1995). The observation of a cyclotron line feature (discovered by OSSE, confirmed by

BATSE and HEXE-MIR) will allow further diagnostics of the system by directly measuring the magnetic field of the neutron star. Another transient, discovered simultaneously by BATSE and WATCH-GRANAT in 1992 (GRO 1719-24= GRS 1716-249), also shows QPO behavior with curious changes of the QPO frequency (van der Hooft et al. 1995).

SUMMARY

These recent highlights from BATSE illustrate the diverse objects and phenomena under continuous observation by the BATSE all-sky detectors. The high-energy sky has provided BATSE with other phenomena to study, from distances as close as the Earth's the upper atmosphere (Fishman et al. 1994) to the monitoring of the quasar 3C273 (Paciesas et al. 1994). Of special note, at intermediate distances, has been BATSE contribution to the study of solar flares (e.g. Ramaty et al. 1994; Aschwanden et al. 1995) and soft gamma-ray repeaters (SGR's) (Kouveliotou et al. 1994).

In addition, BATSE has been providing a valuable monitoring function for many other experiments and observatories. The BATSE team is involved in about one hundred collaborations and guest investigations. The experiment continues to operate extremely well and we anticipate many additional years of exciting results.

REFERENCES

Aschwanden, M.J. et al. 1995. Ap. J. 447, in press (July 10)
Barthelmy S. 1994. In "Gamma-Ray Bursts-The Second Huntsville Workshop" AIP Conf. Proc # 307
Briggs, M. et al. 1995 Ap. J. in press
Finger, M. et al. 1994a IAU Circ. #5977
Finger, M. et al. 1994b IAU Circ. #5934
Finger, M. et al. 1995. *In* Proc. IAU Symp. #165, The Hague, in press
Fishman, G.J., Brainerd, J.J. and Hurley, K.C., eds. 1994. "Gamma-Ray Bursts-The Second Huntsville Workshop" AIP Conf. Proc # 307
Fishman, G.J. et al. 1994. Science 264, 1313
Harmon, B.A. et al. 1995. Nature - in press
Hartmann, D. et al. 1995. Ap. J., in press
Kouveliotou, C. 1994. et al. Nature 368, 125
Meegan, C.A. et al. 1995. Ap J. Lett., in press
Paciesas et al. 1994. *In* The Second Compton Symposium, C.E. Fichtel et al., eds. AIP Conf. Proc # 304, p. 674
Ramaty, R. et al. 1994. Ap. J. 436, 941
Van der Hooft et al. 1995. *In* Proc. IAU Symp. #165, The Hague, in press
Wilson et al. 1994. *In* The Second Compton Symposium, C.E. Fichtel et al., eds. AIP Conf. Proc # 304, p. 390
Zhang et al. 1993. Nature 366, 248

Highlights from OSSE on the Compton Observatory[a]

J. D. KURFESS
Naval Research Laboratory
Washington, DC 20375

INTRODUCTION

Recent highlights from the Oriented Scintillation Spectrometer Experiment (OSSE) on the *COMPTON* Observatory are presented. These include results on the galactic center region, on galactic X-ray sources, the average spectrum of Seyfert galaxies, spectra of quasars, and observations of several supernovae. OSSE covers the energy range from 50 keV to 10 MeV (Johnson et al. 1993). The 3.8° x 11.4° FWHM non-imaging field-of-view of the four identical OSSE detectors is a compromise between discrete and diffuse source observational requirements. Offset pointing is used to estimate the background for the source observations. The OSSE limiting sensitivities are obtained through observations which normally consist of 2-3 week viewing periods.

GALACTIC CENTER REGION

The galactic center region has been observed for a total of 30 weeks during the initial 2-1/2 years of the mission. The objectives of these observations including mapping the distribution of the 0.511 MeV emission, mapping the diffuse galactic continuum emission, searching for evidence of a variable point source(s) of 0.511 MeV emission, and the detection and study of other discrete sources. Purcell et al. (1993a, 1993b, 1994) discuss the galactic center observations. Briefly, the 0.511 MeV emission is adequately described by two components. The more intense component, with an intensity of $1.5 \times 10^{-3}\ \gamma\ cm^{-2}s^{-1}$ in the inner radian is a spheroid centered on the galactic center and with scale size of about 1.2 kpc. There is also a narrow disk component that is observed out to ±60° galactic longitude, but whose intensity at the galactic center is only ~20% that of the spheroidal component. No evidence for discrete sources of 0.511 MeV emission has been observed. The upper limit for steady 0.511 MeV emission from 1E 1740.7-2942 is $\sim 10^{-4}\ \gamma\ cm^{-2}s^{-1}$. Future work is aimed at producing maps of the galactic center region to facilitate improved modeling of the origin of the positrons though correlations with the several candidate sources (novae, supernovae, pulsars, black holes).

Several correlative observations have been undertaken with the imaging SIGMA experiment (Cordier et al. 1994) to extract the diffuse and discrete components in the galactic center region. We have, for the first time, obtained the continuum spectrum of the central region of our Galaxy from which the contributions of the stronger discrete sources have been removed. Figure 1 shows a preliminary spectrum of the diffuse flux from the galactic center with the long axis of the OSSE collimator parallel to the galactic plane. For comparison, the spectrum at galactic longitude +25° is also shown. The continuum emission at the two positions is very similar. The main difference in the two spectra is the increased flux of 0.511 MeV and associated positronium continuum emission from the

[a] This work was supported under NASA DPR S-10987C.

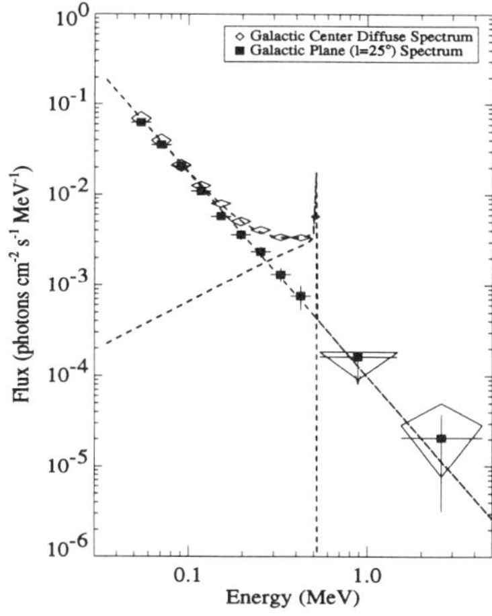

galactic center region. This indicates that the continuum emission is extended in galactic longitude, similar to that seen in high-energy gamma rays. Combining these results with X-ray, COMPTEL, and EGRET data will provide an improved understanding of both the electronic and nucleonic components of cosmic rays, their contributions to production of the galactic diffuse gamma rays, and their energy input to the interstellar medium.

Figure 1. The diffuse emission from the galactic center region obtained by subtracting from OSSE observations the discrete source contributions from simultaneous SIGMA observations. The spectrum at +25° longitude is also shown.

GALACTIC SOURCES

A0535+26 is a high mass X-ray binary which has a recurrent, transient pulsar with a pulse period of 103 s. Kendziorra et al. (1994), reporting on results obtained with the HEXE instrument on MIR, suggested evidence for cyclotron absorption features at 50 and 100 keV. OSSE observed A0535+26 during an unusually intense outburst from 8-17 Feb 1994. Fig. 2a shows the phase-averaged spectrum summed over the entire OSSE observation. The spectrum shows a clear deficit near 100 keV which demands spectral features or multi-component spectra to adequately characterize the emission. Grove et al. (1995) find that the best fit is obtained with an absorption feature at 110 keV. The optical depth of the feature is large (τ=1.8) as seen in Fig. 2c. The OSSE data for the phase-averaged spectrum is consistent with no absorption feature at 55 keV; however, the lower limit of the OSSE energy range (45 keV) makes a definitive statement about a 55 keV line difficult. The most likely interpretation (fundamental at 110 keV), implies a magnetic field of 1.1 x 10^{13} gauss, the most intense for an accreting source.

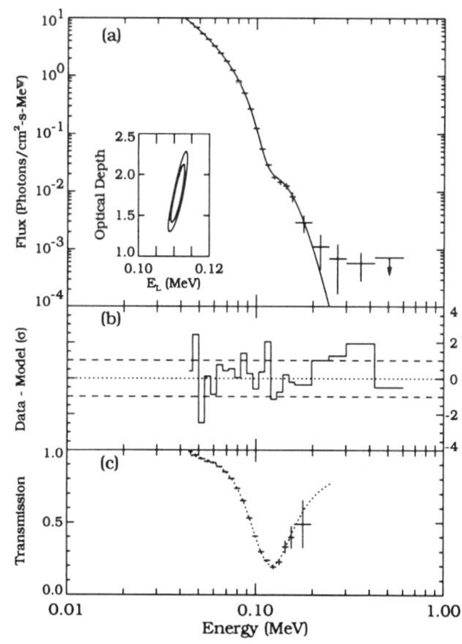

Figure 2. (a) Energy spectrum for A0535+26. The best fit model is a power law times exponential with an absorption feature at 110 keV. (b) residuals from the best fit model. (c) Transmission function for the absorption line feature.

1E 1740.7-2942 is a variable hard X-ray source about 1° from the galactic center. Opposing radio jets emanating from the location of 1E 1740.7-2942 have been discovered by Mirabel et al. (1991). SIGMA has reported several episodes of enhanced emission above 200 keV which have been interpreted as red-shifted positron annihilation radiation. The most compelling evidence occurred on 13-14 Oct 1990 (Bouchet et al. 1991). A similar 1-day transient, but of lower significance, occurred on 20-21 Sept 1992 (Cordier et al. 1993). This unusual character led to speculation that it may be the source of variable 0.511 MeV positron annihilation radiation which has been reported from the galactic center region (Riegler et al. 1985; Leventhal et al. 1989). Ramaty et al. (1992) suggest a model wherein outbursts of 1E 1740.7-2942 produce electron-positron pairs near a black hole. Some of the positrons annihilate on local material and produce the broad and gravitationally red-shifted emission seen by SIGMA. The remainder form part of the jet and eventually annihilate in a nearby molecular cloud (Bally and Leventhal 1991) producing the narrow 0.511 MeV line.

OSSE also has observations for the 1-day event of 20-21 Sept 1992. Jung et al. (1995) do not confirm an excess emission above 200 keV in the OSSE spectrum shown in Figure 3. The 3σ upper limit obtained by OSSE for a broad red-shifted positron feature is 2.4×10^{-3} γ $cm^{-2}s^{-1}$. For the reported SIGMA flux, OSSE should have seen a 5-13σ increase. This, coupled with the lack of variability in the 0.511 MeV emission from the galactic center region in ~200 days of OSSE observations, and the lack of evidence for SIGMA-like events in the SMM data (Harris et al. 1995), suggests that the case for pair production associated with galactic black hole candidates is not yet compelling.

Cygnus X-1 In June, 1993 Harmon et al. (1993) reported a decrease in the intensity of Cyg X-1 which corresponded to a transition from the γ_2 state to the γ_1 state. These states were defined by Ling et al. (1987) who had reported strong, broad MeV emission associated with the low intensity γ_1 state observed by the HEAO-C1 spectrometer in 1979-1980. The OSSE data in the 45-140 keV energy range show gradual intensity changes but no clear bipolar intensity levels, similar to those suggested by HEAO. However, Cyg X-1 does appear to exhibit temperature-intensity correlations, wherein episodes of lower effective temperature of the emission are observed as the overall X-ray luminosity is reduced. This correlation is shown in Figure 4. Many of the OSSE observations have been at intensity levels which are similar to the HEAO γ_1 level. We have seen no evidence in the OSSE observations for excess MeV emission similar to that reported by Ling et al. (1987).

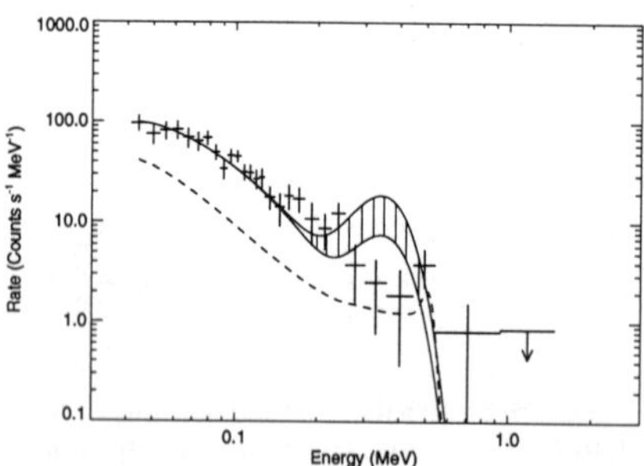

Figure 3. OSSE Spectrum of the galactic center region on 20-21 Sept 1992. The SIGMA spectrum (Cordier et al.) is shown by the hatched region.

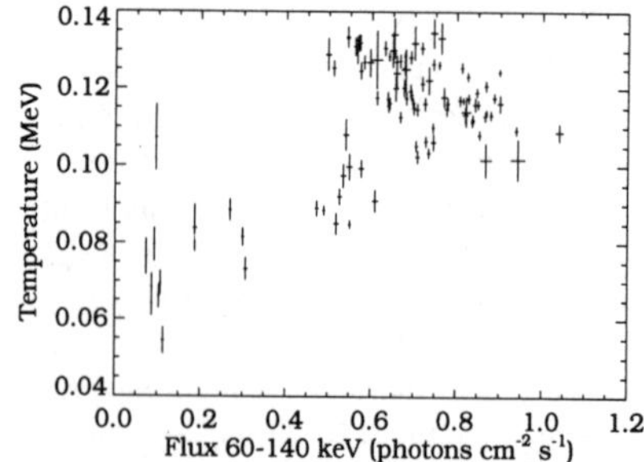

Figure 4. Temperature-Intensity correlation for Cyg X-1 observations when the OSSE spectra are fit with a thermal bremsstrahlung model.

ACTIVE GALAXIES

Johnson et al. (1994) have measured the average spectrum for Seyfert galaxies by summing the spectra of weak Seyferts observed by OSSE during the first 2+ years of the mission. This spectrum is shown in Figure 5 and is best modeled as a thermal rather than a power-law spectrum. A spectral break is required between the X-ray data and the low-energy γ-ray data in the average Seyfert spectrum. This average Seyfert spectrum follows an early OSSE measurement of a thermal-like spectrum from the Seyfert galaxy NGC 4151 (Maisack et al. 1992). An early observation of NGC 4151 by Perotti et al. (1981) had indicated a power-law spectrum which extended up to the MeV region. Assuming that this was typical of most Seyferts limited their contribution to the diffuse γ-ray background because extrapolating the hard X-ray spectra overproduced the γ-ray background in the several hundred keV region. The OSSE spectra of Seyfert galaxies removes this problem.

This spectrum is in contrast to the spectra of blazars observed by GRO. The discovery of blazars as a class of high-energy gamma ray sources (Fichtel et al. 1994) has been one of the highlights of the *COMPTON* Observatory. Blazar emission, which is believed to originate in relativistic jets, exhibit spectra which extend to 1 GeV and which often exhibit spectral breaks in the MeV region (McNaron-Brown et al. 1995).

SUPERNOVAE

OSSE has observed four supernovae with the primary objective of testing models for heavy element nucleosynthesis. OSSE observations of SN1987a provided the first detection of ^{57}Co γ-rays (Kurfess et al. 1992). These data constrained the late time energy input into the supernova remnant and, coupled with the previous observations of ^{56}Co, the nucleosynthetic processes in Type II supernovae (Clayton et al 1992).

Gamma radiation from a Type II supernova at the distance of M81 would not be expected to be detectable by GRO. However, OSSE did observe hard X-radiation from SN1993j following the explosion in March 1993 (Leising et al. 1994). Enhanced emission was observed in the periods 10-15 and 24-37 days after the explosion with an $\sim E^{-2.2}$ spectrum. This emission is not due to scattered nuclear line gamma radiation which would have a much harder spectrum below 100 keV. The radiation observed by OSSE is believed to arise from matter in the pre-supernova stellar wind which is heated by the supernova shock to temperatures of $\sim 10^9$ °K a rather surprising result.

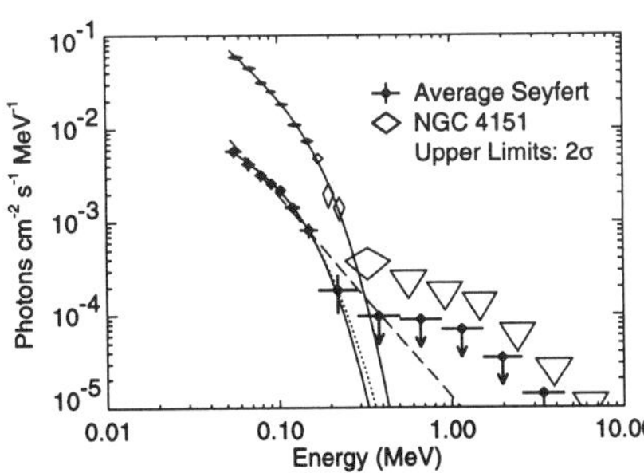

Figure 5. Average spectrum of weak Seyfert galaxies. The spectrum of NGC 4151 is also shown.

Iyudin et al. (1994) have recently reported the COMPTEL detection of ^{44}Ti ($\tau_{1/2}$= 46.4 - 66.6 yrs) in the supernova remnant Cassiopeia A by the observation of 1.156 MeV gamma rays at a flux of $7.0 \pm 1.7 \times 10^{-5}$ γ cm^{-2}s^{-1}. This is the first detection of ^{44}Ti gamma rays and demonstrates the capability for gamma ray astronomy to discover recent galactic supernovae. At a distance of 2.8 ± 0.2 kpc, the

COMPTEL observation implies the production of 1.5-3 x 10^{-4} M_o of ^{44}Ti, at the upper end of the expected range for Type II (core collapse) supernova. OSSE observed Cas A for a period of three weeks during 1992. Analysis of the OSSE data has been undertaken (The et al. 1994) by jointly fitting the OSSE spectrum for the three ^{44}Ti decay lines at 68, 78 and 1156 keV. They find a 3σ upper limit for each of these lines of ~6 x 10^{-5} γ $cm^{-2}s^{-1}$. This result is in marginal disagreement with the COMPTEL result. A flux of 3-5 x 10^{-5} γ $cm^{-2}s^{-1}$ would be compatible with the two separate results with ~2% probability. Additional OSSE-COMPTEL observations to be undertaken in 1995 should achieve a sensitivity adequate to detect ~1 x 10^{-4} M_o of ^{44}Ti in Cas A with high confidence.

SN1991t, a Type Ia supernovae in NGC 4527, was observed early in the mission. At a distance of ~13 Mpc, ^{56}Co might be marginally detectable by OSSE and COMPTEL. OSSE does not detect the ^{56}Co emission (Leising et al. 1995). The OSSE data, combined with limits obtained by COMPTEL (Lichti et al. 1994), appear to rule out those models for Type 1a supernovae which produce close to 1 M_o of ^{56}Ni.

REFERENCES

Bally, J., and Leventhal, M., 1991, Nature, **353**, 234.
Bouchet, L., et al., 1991, ApJ. (Letters), **383**, L45.
Clayton, D.D. et al. 1992, ApJ (Letters), **399**, L141.
Cordier, B. et al., 1993, Astron. & Astrophys., **275**, L1.
Cordier, B., et al., 1994, AIP Conference Proceedings, **304**, 446.
Fichtel, C.E., et al., 1994, ApJ. Supp., **94**, 551.
Grove, J.E. et al., 1995, ApJ (Letters) **438**,L25.
Harmon, B.A., et al., 1993, IAUC 5813.
Harris,, M.J., Share, G.H., and Leising, M.D. 1994, ApJ **433**,87.
Iyudin, A.F., et al., 1994, AIP Conference Proceedings, **304**, 156.
Johnson, W.N., et al., 1993. ApJ. Supp., **86**,693.
Johnson, W.N., et a;. 1994, AIP Conference Proceedings, **304**, 515.
Jung, G.V., et al. 1995,to be published in Astron. Astrophys.
Kendziorra, E., et al., 1994, Astron. Astrophys. **291**,L31.
Kurfess, J. D., et al., 1992, ApJ (Letters) **399**, L137.
Leising, M.D., et al. 1994, ApJ. (Letters) **431**, L95.
Leising, M.D.,et al. 1995, submitted to ApJ.
Leventhal, M, et al. 1989, Nature, **339**, 36.
Lichti, G.G., et al., 1994, Astron. Astrophys. **292**, 569.
Ling, J.C., et al., 1987, Ap.J., **321**, L117.
Maisack, M., et al., 1992, ApJ (Letters), **407**, 167.
McNaron-Brown, K., et al., 1995, submitted to ApJ.
Mirabel, I.F., et al., 1991, Nature, **358**, 215.
Perotti, F., 1981, ApJ (Letters), **247**, L63.
Phlips, B. et al., 1995, submitted to ApJ.
Purcell, W.R., et al., 1993a, AIP Conference Proceedings, **280**, 107.
Purcell, W.R., et al., 1993b, ApJ (Letters), **413**, L85.
Purcell, W.R., et al., 1994, AIP Conference Proceedings, **304**, 403.
Ramaty R., et al., 1992, ApJ (Letters), **392**, L63.
Riegler, G.R., et al., 1985, ApJ, **294**, L13.
The, L.-S., et al., 1995, to be published in ApJ.

Detections of Neutron Stars in the Extreme Ultraviolet

Stuart Bowyer
Center for EUV Astrophysics, 2150 Kittredge Street,
University of California, Berkeley, CA 94720–5030

Neutron stars were not expected to be observable in the extreme ultraviolet (EUV) and hence their detection in this band is a surprise. More important is that EUV observations provide unique new information on these objects. In the accompanying table, I list the neutron stars detected in the EUV with the Extreme Ultraviolet Explorer (EUVE). The binary object Her X-1 is the only object detected in the all-sky survey. This is not surprising since it is the only neutron star system with sufficient intensity to be detectable in the 500–2000 second sky integrations, which are typical for the EUVE all-sky survey. All the neutron stars were detected in the 100 Å (Lexan) bandpass. A more complete description of EUVE and the all-sky survey are provided in Bowyer & Malina [1] and Bowyer et al. [2].

A phased analysis of the flux from Her X-1 [6] shows that the EUV flux emanates from the accretion disc and not from the neutron star itself. While these results help elucidate the character of the disc, they are not the topic of this symposium and will not be discussed further.

Timing information from EUVE is routinely available to 8 ms accuracy, and with an easily implemented alternate mode is available to better than 1 ms. This allows the discrimination of pulse and whole-body emission from virtually all pulsars.

The pulsar B0656+14 was observed for over 100,000 seconds and these data have recently been released for general dissemination via the EUVE Public Archive. A preliminary examination of these data shows the data are not pulsed, and hence the EUV flux is most likely thermal emission from the neutron star surface.

Table 1: Neutron Stars Detected with EUVE

Object	Deep Survey (counts s^{-1})	Lexan Survey (counts s^{-1})	Age (yr)	DM (cm^{-3})	d (pc)	Ref.
J0437-4715	0.038		5×10^9	2.6	140	[3]
B0656+14	0.024		1×10^5	14.0	760	[4]
Geminga	0.02		5×10^5		< 500	[5]
Her X-1		0.18	$< 10^7$		< 5000	[6]

The X-ray and gamma-ray (but not radio!) pulsar, Geminga, has been observed for 300,000 seconds. These data have just been released to the scientific community via the EUVE Public Archive, but as yet no analysis of the data has been published.

The millisecond pulsar J0437-4715 was observed with EUVE in January 1994 for about 70,000 seconds. An unrelated EUV source about 3.′8 away was well resolved with the $\sim 1'$ point-spread function of the telescope. Because the data were not taken in the high time resolution mode, a pulse vs. whole-body deconvolution is not possible for these data. Nonetheless, Edelstein, Foster, & Bowyer [3] have shown that a substantial amount of information can be derived from this observation.

The observed dispersion measure of 2.65 pc cm^{-3} implies a distance of 140 pc for this pulsar. This value is consistent with optical data on the white dwarf companion star. An estimate of the hydrogen column density to this system can be obtained from the compilation of Fruscione et al. [7]; a reasonable value is $N_{\mathrm{HI}} = 10^{19}$ cm^2. Combining the response of the 100 Å bandpass of the Deep Survey telescope with absorption by the interstellar medium (ISM) associated with this hydrogen column [8] yields an effective area weighted wavelength of 116 Å. This is a substantially lower energy than the X-ray data obtained for this object and leads to new insights regarding this source.

Becker & Truemper [9] modeled the X-ray emission of J0437-4715 with a power-law model having $\alpha = -2.60$ and with a composite model. The composite model consisted of a blackbody "hot spot" ($T \approx 1.7 \times 10^6$ K, area = 0.05 km^2), responsible for a possible feature at 0.8 keV, combined with a power-law source with $\alpha = -2.85$ that dominates at lower energies. The predicted EUV flux for both power laws was consistent with the observed EUV luminosity only if the absorbing column was in the narrow range of $N_{\mathrm{HI}} = 2.5 \pm 0.2 \times 10^{20}$ cm^{-2}. This column exceeds the hydrogen column out of the Galaxy [10], and hence both of these models can be ruled out as the source of the EUV flux.

The EUV data were modeled as a blackbody to determine if this could produce the ROSAT observed 0.8 keV feature. This could be accomplished only if the hot spot emitting area was from 3 times (for $N_{\mathrm{HI}} = 10^{19}$ cm^{-2}) to 25 times (for $N_{\mathrm{HI}} = 10^{20}$ cm^{-2}) larger than the area required by the X-ray data. Edelstein et al. concluded that the observed EUV flux was not consistent with a blackbody hot spot capable of producing the soft X-ray feature [3].

Edelstein et al. examined whether blackbody emission could provide an alternative explanation to the 0.1–0.4 keV ROSAT flux that had been attributed to power-law emission. They concluded that both the EUV flux and the X-ray flux could be explained by a blackbody with a temperature $\sim 5.7 \times 10^5$ K, an emitting area of ~ 3 km^2, and ISM absorption associated with a hydrogen column of $N_{\rm HI} = 5 \times 10^{19}$ cm^{-2}.

Alternatively, the EUV flux could be independent of the X-ray flux and originate from the entire surface of the pulsar. In this case, a minimum surface temperature of $T_{\rm min} = 1.6 \times 10^5$ K was found for a distance of 140 pc and a stellar radius of 10 km. An upper limit to the pulsar's temperature, $T_{\rm max} = 4.0 \times 10^5$ K, was established by using the known hydrogen column out of the Galaxy [10].

Thermal EUV emission from the surface of a $\sim 5 \times 10^9$ year old neutron star is inconsistent with standard cooling models for neutron stars [11] and would require some form of reheating. Four sources for reheating were considered: (1) frictional heating at the crust-core interface through the unpinning of internal vortex lines; (2) accretion from the ISM; (3) accretion from the white dwarf companion onto the neutron star; and (4) nucleon decay catalyzed by heavy magnetic monopoles.

Theoretical work [11, 12] predict that for neutron stars older than 10^9 yr, the thermal surface temperatures from frictional heating will be less than 10^5 K. Hence, standard frictional heating models can be ruled out.

The gravitational accretion of ISM could heat the neutron star if the velocity of the star is less than ~ 10 km s^{-1} [13, 14, 15]. A velocity of 91 ± 3 km s^{-1} has been measured for PSR 0437-4715 using timing measurements [16] and of 63 ± 30 km s^{-1} using scintillation observations [17]. These velocity estimates are sufficiently large that significant heating by accretion from the ISM can be ruled out.

Stellar material abated from the neutron stars' white dwarf companion by a high energy particle wind from the pulsar could accrete on the neutron star and produce heating. An upper limit to mass accretion by a high-energy particle wind from the pulsar is:

$$\dot{M} \leq f L_{\rm SD}/\gamma c^2 ,$$

where f is the fraction of the total flux from the neutron star intercepted by the companion, $L_{\rm SD}$ is the pulsar spin-down luminosity, γ is the Lorentz factor of the particles, and c is the speed of light. For parameters appropriate to this system, the maximum luminosity from this accretion is an insignificant fraction of the observed EUV luminosity.

It has been suggested that neutron stars could be heated by magnetic "monopole-catalyzed" nucleon decay [18, 19]. Magnetic monopoles that hit the surface of a neutron star will be captured by the star at a rate proportional to the monopole flux. The combination of the upper limit to the surface temperature of PSR 0437-4715 and the object's extreme age results in a value for the Galactic monopole flux 3 orders of magnitude lower than existing limits.

Four neutron stars (or binaries with a neutron star as one component) have been detected in the EUV to date. In three of these systems the observed emission is believed to be from the neutron star itself, but a detailed analysis has been carried out for only one of these objects. In this case substantial new information on neutron star physics has been obtained. We can expect more progress in this field when analysis of the existing data is completed and as new neutron stars are found to be emitting in the EUV.

This work is supported by NASA contract NAS5-30180.

References

1. Bowyer S., & Malina R. F. 1991, in Extreme Ultraviolet Astronomy, ed. R. F. Malina & S. Bowyer, (New York: Pergamon Press), 397.
2. Bowyer S., Lieu R., Lampton M., Lewis J., Wu X., Drake J. J., & Malina R. F. 1994, The First EUVE Source Catalog, ApJS, 93 (2), 569.
3. Edelstein J., Foster R. S., & Bowyer S. 1995, ApJ (submitted).
4. Finley J. P., Ogelman H., & Edelstein J. 1994, BAAS, 26(2), 870.
5. Halpern J. 1994, private communication.
6. Vrtilek S. D., Mihara T., Primini F. A., Kahabka P., Marshal H., Agerer F., Charles P. A., Cheng F. H., Dennerl K., la Dous C., Hu E. M., Rutten R., Serlemitsos P., Soong Y., Stull J., Trumper J., Voges W., Wagner R. M., & Wilson R. 1994, Observations of Hercules X-1/HZ Hercules, ApJ, 436, L9.
7. Fruscione A., Hawkins I., Jelinsky P., & Wiercigroch A. 1994, The Distribution of Neutral Hydrogen in the Interstellar Medium. I. The Data, ApJS, 94, 127.
8. Rumph T., Bowyer S., & Vennes S. 1994, Interstellar Medium Continuum, Autoionization, and Line Absorption in the Extreme Ultraviolet, AJ, 107 (6), 2108.
9. Becker W., & Trumper, J. 1993, Nature, 365, 528.
10. Dickey J. M., & Lockman F. J. 1990, ARA&A, 28, 215.
11. Shibazaki N., & Lamb, F. K. 1989, ApJ, 346, 808.
12. Umeda H., Shibazaki N., Nomoto K., & Tsuruta S. 1993, ApJ, 408, 186.

13. Paczyński, B. 1990, ApJ, 348, 485.

14. Hartmann D., Epstein R. I., & Woosley S. E. 1990, ApJ, 348, 625.

15. Treves A., & Colpi M. 1991, A&A, 241, 107.

16. Bell J. F., Bailes M., Manchester R. N., Weisberg J. M., & Lyne A. G. 1995, ApJ (in press).

17. Johnston S., & Nicastro L. 1995, preprint.

18. Freese K., Turner M., & Schramm D. N. 1983, Phys. Rev. Lett., 51, 1625.

19. Kolb E., & Turner M. 1984, ApJ, 286, 702.

Isolated Neutron Stars and Their Emission Throughout the Electromagnetic Spectrum

Patrizia A. Caraveo
Istituto di Fisica Cosmica del CNR,
Via Bassini,15 - 20133 MILANO

INTRODUCTION

As of today, less than 3% of the radio pulsar population has been seen at other wavelengths. 16 objects have been detected in soft X-rays, of these 6 have been detected in the optical, 3 in hard X-rays and 5 in high-energy gamma-rays (see (1) and (2) for a summary of the observational data). Luckily enough, this small sample encompasses young as well as middle-age and old pulsars. Young objects are favorite by their high rotational energy losses but, unfortunately, they are very few. Older objects are more numerous but, relying on smaller energy reservoirs, are intrinsecally dimmer and only the nearby ones can be detected. The comparison of the multifrequency behaviour of pulsars of different age is, indeed, essential to understand the emission mechanisms at work and, in particular, to disentangle the thermal from the non-thermal ones. This is especially true in the optical, UV and soft X-ray domains, where the still hot surface of an isolated neutron star can shine through thermal processes. Moreover, the comparison of the efficiencies at different wavelengths, i.e. the fraction of the rotational energy loss going into the different photon energy domains, is a powerful tool to understand how the multiwavelength behaviour of Isolated Neutron Stars is driven by the pulsar parameters (see (2) and (3) for a discussion).

THE DATA

Table 1 gives the up-to-date summary on the database available for doing multiwavelength astronomy of Isolated Neutron Stars (INS). The energy ranges have been chosen to match the sensitivity of the current or past generation of instruments. The column "UV" refers to EUVE data, "X_{soft}" refers to ROSAT and Einstein data, "X_{hard}" refers to SIGMA, as well as BATSE/GRO and OSSE/GRO data, "γ_{soft}" refers to COMPTEL/GRO data and "γ_{hard}" to EGRET/GRO data.
In order to be included in Table 1 an isolated neutron star must have been seen in at least one energy band beside radio. The objects have been ordered as a function of their age, τ = P/2Pdot, and the symbols refer to the method used to attribute the radiation detected to the INS. **P** means that the identification is based on pulsation,

while **D** stands for detection and reflects a good positional coincidence between the radio pulsar and a point source detected at other wavelengths. In the soft X-ray band some of the pulsed emission can be recognized to be of thermal origin, these cases are idendified with a **T**. Finally, non-detections are identified with a -.

TABLE 1

PULSAR	*Radio*	*Optical*	*UV*	X_{soft}	X_{hard}	γ_{soft}	γ_{hard}
0531+21	P	P		P	P	P	P
0540-69	P	P		P			-
1509-58	P	D		P	P	D	-
0833-45	P	P		P,T	-	P	P
1706-44	P	-		D	-	-	P
1800-21	P			D,T			-
1823-13	P			D,T			-
2334+61	P			D,T			-
1951+32	P			P			-
0656+14	P	D	D	P,T			-
GEMINGA	-	D	D	P,T		-	P
1055-52	P	-		P,T	-	-	P
0355+54	P			D,T			-
1929+10	P		D	P,T			-
0823+26	P			D,T			-
0950+08	P	-	D	D,T			-

LOW ENERGIES: THERMAL vs. NON-THERMAL

While high-energy (hard X-ray and γ-ray) radiation can be produced only through magnetospheric non-thermal processes, radiation softer than few keV can indeed be produced by the hot surface of the neutron star. According to standard cooling theories, temperatures of $<10^{6}$°K are to be expected after the initial rapid cooling.Thus, Isolated Neutron Stars should be seen as thermal emitters in the optical, UV and soft X-ray domains, provided that their non-thermal emission is not overwhelming. This happens in young objects like the Crab, PSR 0540-69, PSR 1509-58, PSR 1951+32 where the soft X-ray and/or optical emissions are clearly of non-thermal nature.

PSR 0833-45, the Vela pulsar, marks a turning point since its soft X-ray emission is, at least partially, of thermal nature while the optical emission most probably is not.

For pulsars older than 10^5y the thermal emission is clearly dominating the soft X-ray domain (4) and also the fluxes recorder from the proposed optical counterparts of Geminga (5) and PSR 0656+14 (6) can be explained as the Rayleigh-Jeans tail of the black-body detected in X-rays.

The thermal origin of the X-ray radiation does not prevent it from being pulsed (7). Owing to the influence of the magnetic field on the thermal conductivity, the neutron star surface is not at the same temperature, so that shallow modulation, not exceeding 15-20%, is usually seen from these middle-age pulsars (see 4 for a complete discussion of the ROSAT data). Moreover, the non-thermal processes producing high-energy γ radiation in the case of Geminga and 1055-52 can give a significant contribution to the thermal emission, providing extra heating to small regions of the surface of the neutron star, most probably the polar caps. These show up as a second, higher temperature component in the soft X-ray emission (8,9,4).

For neutron stars older than 10^6y only this hotter component is detectable so that PSR 1929+10 (8) and PSR 0950+08 (9) are seen in the soft X-ray domain as thermal emitters characterized by relatively high temperatures but exceedingly small emitting area. The small area involved imply very small fluxes and make it possible to detect only the very near objects.

HIGH ENERGIES: THE ANTICORRELATION

For the high-energy side, although always working with a grand total of 6 objects, we face a very interesting situation.

The high energy γ-ray emission appears to be anticorrelated with the hard X-ray one: the higher the output in high energy γ-rays, the lower the hard X-ray production. This can be better seen if one computes the emissivity,i.e. the ratio between the pulsar luminosity, as measured in a given spectral domain, and its total energy loss Edot, assuming, for sake of semplicity, isotropic emission.

Fig 1 shows the hard X-ray and high energy γ-ray emissivities as a function of the pulsar period derivative. The choice of Pdot is based on the results of Goldoni et al.(2) and Caraveo (3) who have identified Pdot as the parameter effectively driving the high energy behaviour of the Isolated Neutron Stars detected so far.

The trend in fig 1 is as clear as it is undisputable: the hard X-ray efficiency goes up, at the expenses of the high energy γ-ray output, in close correlation with the value of the objects Pdot, hence their magnetic field. The higher the magnetic field, the higher the output in hard X-ray and the lower the high energy γ-ray throughput. The non detections of PSR 1509-58 by EGRET on one side and of Geminga by SIGMA and COMPTEL on the other, provide further confirmation of this intriguing anticorrelation.

A NS magnetic field, an essential ingredient to build up the electric field needed to accelerate particles, is also responsible for the absorption of the high energy photons generated by the same particles through curvature radiation.

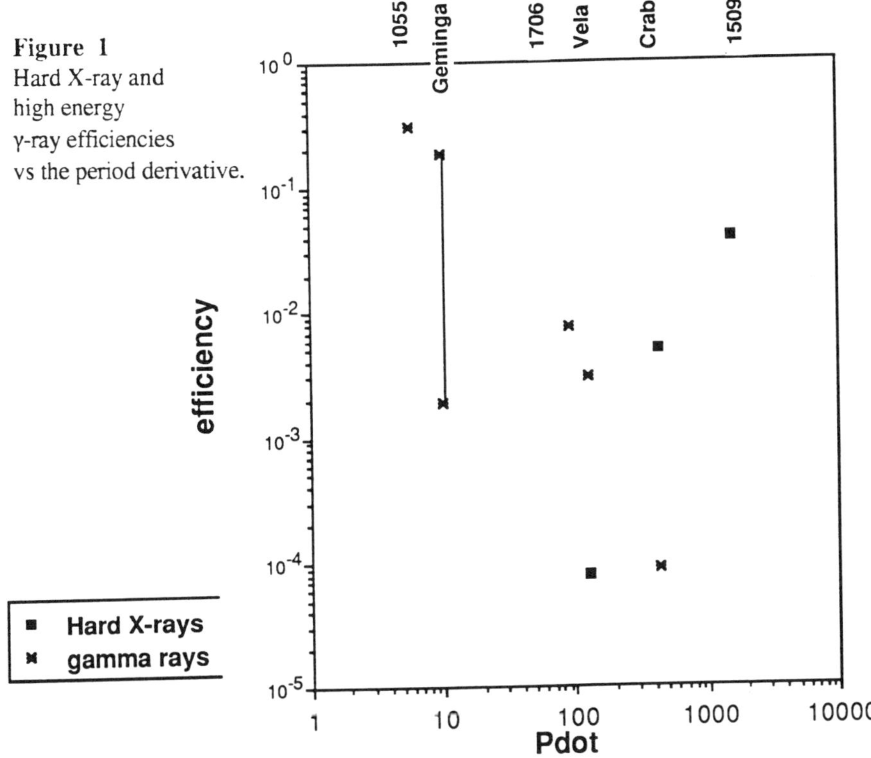

Figure 1 Hard X-ray and high energy γ-ray efficiencies vs the period derivative.

Thus the driving mechanism is not only the production of high energy photons, but also their capability to escape. Lu et al. (12) have generalized the magnetic absorption mechanism proposed by Hardee (13) and have introduced the generation number parameter, i.e. the number of times the original high energy photon must be absorbed and re-emitted in order to be able to escape the pulsar magnetosphere. Accordingly, one can compute the maximum energy allowable for a photon in order not to be magnetically absorbed. While the generation number goes up with the magnetic field, the maximum photon energy decreases, explaining the anticorrelation between the emissivities measured in hard X-ray and high energy γ-rays.

REFERENCES

1- Mereghetti. S., Caraveo, P.A., Bignami G.F. Ap.J.Suppl **92**,521 (1994)
2- Goldoni, P., Musso, C., Caraveo, P. A., Bignami G.F. A&A in press (1995)
3- Caraveo P. A. "Advances in Space Research" ed. N. Gehrels in press (1995)
4- Ogelman H. in "The Lives of Neutron Stars" Ed.A.Alpar et al., p. 101 (1995)
5- Bignami G.F. et al ApJ. **319**,358 (1987)
6- Caraveo, P.A., Bignami,G.F., Mereghetti, S., Ap.J.Lett **422**,L87(1994)
7- Greenstein G. and Hartke G.J. Ap.J. **271**, 283 (1983)
8- Halpern J.P. and Ruderman M. Ap.J. **415**, 286 (1993)
9- Ögelman H. and Finley, J.P. Ap.J. Letters **413**, L31 (1993)
10-Yancopoulous S, Hamiltin, T.T., Helfand D. Ap.J. **429**,832(1993)
11- Manning, R.A. and Willmore A.P. MNRAS (1994)
12-Lu, T., Wei, D.M., Song L.M. A&A (1994)
13-Hardee, P.E., ApJ. **216**,873 (1977)

ROSAT Observations of Neutron Stars

WERNER BECKER

Max-Planck-Institut für extraterrestrische Physik
Giesenbachstraße 1, 85740 Garching, Germany
Email: web@mpe-garching.mpg.de

INTRODUCTION

Neutron stars are among the most fascinating astronomical objects in the universe. Born in the imploding core of a supernova, they provide a unique class of stellar objects with properties that make them nearly ideal probes for investigating a wide variety of physical problems. One primary goal addressed with ROSAT was to search for thermal X-ray emission from cooling neutron stars[1,2]. The close link between the thermal evolution of neutron stars and the physical characteristics of neutron star material at supernuclear densities provides an important starting point for the empirical study of matter at extreme energies and baryon densities. The possibility of altered hadronic interactions, the existence of stable pions, kaons, hyperons or free quarks in the core of a neutron star and the influence of superfluid neutrons in the inner crust all yield different cooling rates and theoretical predictions, so that the neutron star surface temperature as a function of the neutron star age reflects the stellar composition[3–5]. Comparing the neutron star temperatures and temperature upper limits measured by ROSAT with the theoretical predictions based on different equations of state thus provides the empirical basis essential for the verification of neutron star models and cooling theories.

Another point of interest for observing rotation-powered pulsars with ROSAT was that of magnetospheric X-ray emission. The exact physical mechanisms that operate in the pulsar magnetosphere to produce the intense, broad-band radiation are complex and we know very little about them. Experimental X-ray results to date have been sparse; before the launch of ROSAT, X-rays had been observed from only nine rotation-powered pulsars[6], and only the youngest, most powerful pulsars, such as the Crab and PSR 1509-58, yielded enough photons for a detailed spectral and timing analysis needed to probe the origin of the radiation. ROSAT, with a significantly larger collecting area and more sensitive instruments than previous X-ray satellites, has made it possible to further the study of pulsars detected by EINSTEIN and EXOSAT and to detect X-ray emission from nine[1] more pulsars.

In the following we will give a brief overview of the X-ray detected rotation-powered pulsars and draw a picture of their emission properties as it is suggested by ROSAT. A summary of the characteristic pulsar parameters is given in Table 1.

[1]as of June 1995

Characteristics of the radio, optical, X-ray and γ-ray detected rotation-powered pulsars (as of June 1995)

Pulsar	Comment	detected						$\dot{E}/(4\pi d^2)$	lg $\dot{E}$	lg L_{xp}^{tot}	lg L_{xp}^{puls}	lg L_x^{pn}	lg$(P/2\dot{P})$	P	lg $\dot{P} \times 10^{-15}$	D	lg B	Ref.
		R	O	X_s	X_h	γ_s	γ_h	erg/s/cm^2	erg/s	erg/s	erg/s	erg/s	Jahre	ms	s s^{-1}	kpc	Gauss	
B0531 + 21	Crab	p	p	p	p	p	p	$9.3 \cdot 10^{-7}$	38.65	36.2	36.1	37.5	3.10	33.40	420.96	2.00	12.58	1
B0833 − 45	Vela	p	p	p	p	p	p	$2.3 \cdot 10^{-7}$	36.84	32.7	31.7	33.4	4.05	89.29	124.68	0.50	12.53	11
J0633 + 17	Geminga	-	d	p	-	-	p	$1.2 \cdot 10^{-8}$	34.51	31.7	29.6		5.53	237.09	10.97	0.15	12.21	14,28
J0437 − 47	ms Pulsar	p	-	p	-	-	-	$1.1 \cdot 10^{-8}$	34.40	30.7	30.1		8.88	5.75	$1.2 \cdot 10^{-4}$	0.14	8.93	23
B1706 − 44	G343.1-02.3	p	-	d	-	-	p	$8.6 \cdot 10^{-9}$	36.53	33.1			4.24	102.45	93.04	1.82	12.49	9
B1509 − 58	MSH 15-52	p	d	p	p	p	-	$7.7 \cdot 10^{-9}$	37.25	34.3	34.3	35.3	3.19	150.23	1540.19	4.40	13.19	1,6
B1951 + 32	CTB 80	p	-	p	-	-	p	$5.0 \cdot 10^{-9}$	36.57	33.4	33.0	34.0	5.03	39.53	5.85	2.50	11.69	11
J0751 + 18	ms Pulsar	p	-	d	-	-	-	$2.4 \cdot 10^{-9}$	35.87	31.8			7.83	3.47	$7.9 \cdot 10^{-4}$	1.60	9.23	21
B1823 − 13		p	-	d	-	-	-	$1.3 \cdot 10^{-9}$	36.45	33.1			4.33	101.45	74.95	4.12	12.45	29
B1800 − 21	G8.7-0.1	p	-	d	-	-	-	$1.2 \cdot 10^{-9}$	36.35				4.30	133.61	134.32	3.94	12.63	30
B1929 + 10		p	-	p	-	-	-	$1.1 \cdot 10^{-9}$	33.59	30.1	29.5		6.49	226.51	1.16	0.17	11.71	18
B1957 + 20	ms Pulsar	p	-	d	-	-	-	$5.7 \cdot 10^{-10}$	35.20	31.3			9.18	1.60	$1.7 \cdot 10^{-5}$	1.53	8.22	23,24
B0656 + 14		p	d	p	-	-	-	$5.5 \cdot 10^{-10}$	34.58	32.5	31.7		5.05	384.87	55.03	0.76	12.67	12
B0540 − 69	SNR in LMC	p	p	p	-	-	-	$5.1 \cdot 10^{-10}$	38.17	36.3	36.2	37.3	3.22	50.37	479.06	49.4	12.70	31
B0950 + 08		p	-	d	-	-	-	$3.2 \cdot 10^{-10}$	32.75	28.9			7.24	253.06	0.23	0.12	11.39	20
B1055 − 52		p	-	p	-	-	p	$1.1 \cdot 10^{-10}$	34.48	32.5	31.5		5.73	197.10	5.83	1.53	12.03	15
B0355 + 54		p	-	d	-	-	-	$8.8 \cdot 10^{-11}$	34.66	31.9			5.75	156.38	4.39	2.07	11.92	32
B2334 + 61	G114.3+0.3	p	-	d	-	-	-	$8.6 \cdot 10^{-11}$	34.79	31.8			4.61	495.24	191.91	2.46	12.99	10
B0823 + 26		p	-	d	-	-	-	$2.6 \cdot 10^{-11}$	32.66	29.6			6.69	530.66	1.72	0.38	11.99	19

TABLE 1. A list of pulsars that have been detected in the radio, optical, X- and γ-ray wavebands, ordered according to their spin-down flux density at Earth $\dot{E}/4\pi d^2$. The individual columns are as follows: 1) Pulsar name; 2) Comment, e.g. association with a SNR, alternative pulsar name, or millisecond pulsar; 3-8) Energy ranges in which pulsed (p) or unpulsed (d) radiation has been detected: R – radio, O – optical, X_s – soft X-rays ($E_\gamma \sim 1$keV), X_h – hard X-rays ($E_\gamma \sim 10$keV), γ_s – soft γ-rays ($E_\gamma \sim 1$MeV) and γ_h – hard γ-rays ($E_\gamma > 100$MeV). lg $\dot{E}$ is the decimal logarithm of the pulsar spin-down power $I\Omega\dot{\Omega}$; L_{xp}^{tot} is the sum of the pulsed and unpulsed X-ray luminosities assuming isotropic emission; L_{xp}^{puls} is the pulsed luminosity alone; L_x^{pn} is the total X-ray luminosity including the contribution from the pulsar's synchrotron nebula. Each of the luminosities is calculated for the ROSAT energy range, 0.1 – 2.4 keV. Columns headed $P/2\dot{P}$, P, $\dot{P}$, D and B contain the dynamic age, period, period derivative, distance and magnetic field $B = 3.3 \times 10^{19}(P\dot{P})^{1/2}$ for a neutron star of moment of inertia $I = 10^{45}$g cm^2 and radius 10 km. Radio pulsar parameters from Taylor et al. (1995).

X-ray Emission Properties of Rotation-Powered Pulsars

Nearly 700 neutron stars are currently known as rotation-powered pulsars[7]. In the course of the ROSAT mission, about 10% of them have been observed in detailed pointed observations, leading to the detection of pulsed X-ray emission from 10 of them while 9 more pulsars are identified only by positional coincidence with a radio pulsar[1] (c.f. Table 1). The collective X-ray emission properties of these pulsars suggest that their X-ray emission characteristics depend mainly on the object's age. Young pulsars with a characteristic age of about $1-2\times 10^3$ years emit sharp X-ray pulses with a pulsed fraction of up to 100%. Their X-ray emission is thought to be connected with the acceleration of electrons/positrons in the pulsar magnetosphere and shows spectra well described by a power-law $dN/dE \propto E^{-\alpha}$ with photon index $\alpha \approx 2$. No thermal X-ray emission from the surface of these young pulsars could be detected. The observation of thermal X-rays is complicated by the fact that the youngest and hottest neutron stars like the Crab, PSR 0540-69 and PSR 1509-58 are also those with the strongest magnetospheric emission which buries the thermal component[8]. The steady X-ray emission from the pulsar powered synchrotron nebula is found to be harder than the spin modulated pulsar contribution.

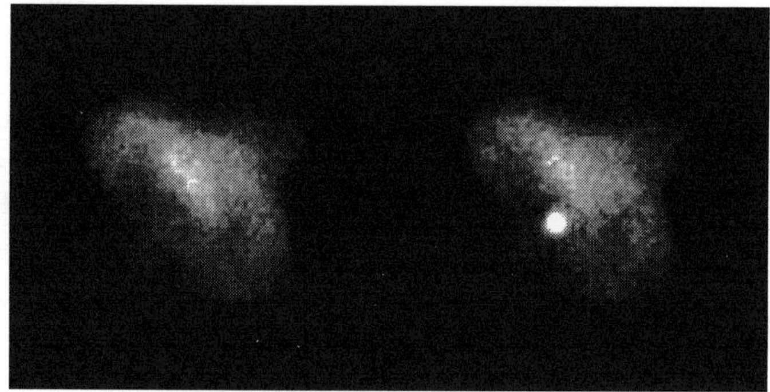

FIGURE 1. Crab pulsar and nebula as seen with the ROSAT HRI in the energy range 0.1-2.4 keV. Shown is the 2 arcmin synchrotron nebula during the pulse-off (left) and pulse-on phase.

Different from the Crab-like pulsars are pulsars having an age between 10^4-10^5 years. They show smoother lightcurves and were found to resemble more the X-ray emission properties of the Vela-pulsar; i.e. strong steady emission from a pulsar powered synchrotron nebula combined with a small pulsed contribution from magnetospheric or thermal emission[9–11]. For the Vela pulsar, the spectrum of the pulsed X-ray emission is of thermal origin[12] and contributes only below ~ 1.2 keV. The emission from the synchrotron nebula dominates the Vela pulsar spectrum above ~ 0.5 keV and follows a power-law with photon index $\alpha \approx 2$.

Three of the nineteen detected pulsars have ages in the range 10^5-10^6 years. They form the group of so-called 'cooling' neutron stars[13–16] and include PSR 0656+14, Geminga and the 197 ms pulsar PSR 1055-52. For all of them, the

ROSAT data suggest that the main component of the detected X-radiation is of thermal origin and a relic of the initial heat content in the neutron stars' birth event. The X-ray emission characteristic of these cooling neutron stars is found to follow a dichotomy, i.e. the spectra are best described by a two-component model, in which the low-energy radiation is represented by a black-body spectrum and the high-energy component either by a thermal spectrum (polar cap emission) or a power law (magnetospheric emission). The existence of two spectral components is also backed up by phase-resolved analysis. All three pulsars show a phase shift and a change in the pulsed fraction from $\sim 15-30\%$ below a transition energy of $\sim 0.5-0.6$ keV, rising to up to $\sim 60\%$ above. The X-ray lightcurves show a sinusoidal pulse profile. The modulation of the neutron star's cooling emission can be explained by non-uniformities in the surface temperature due to the presence of the strong magnetic field which gives rise to an anisotropic heat flow in the neutron star's outer layers[17]. A comparison between the empirically derived neutron star surface temperatures and temperature upper limits and the cooling curves as predicted by the FP model[18] is shown in figure 2. The predicted curves are for standard and accelerated cooling. The latter is caused by an enhanced neutrino emissivity due to the presence of a pion condensation in the neutron star core.

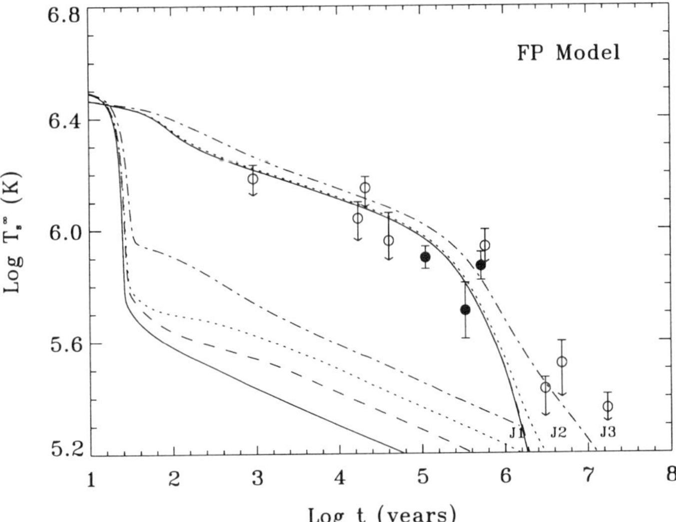

FIGURE 2. Predicted cooling curves and empirically derived neutron star surface temperatures and temperature upper limits. The dotted and dashed lines represent the thermal evolution of the neutron star with internal frictional heating for strong (J3), weak (J2) and superweak (J1) pinning of crustal superfluid vortex lines. The filled circles are temperatures measured by ROSAT[12–15]. The open circles are upper limits, in which a thermal spectrum for the detected X-rays has been assumed[1]. The arrow lengths indicates the effects of uncertainties in the pulsar distances and column densities. The individual points belong to the Crab, PSR 1706-44, PSR 1823-13, PSR 2334+61, PSR 0656+14, Geminga, PSR 1055-52, PSR 0355+54, PSR 1929+10, PSR 0823+16 and PSR 0950+08 (from left to right).

Figure 2 clearly shows that the neutron star surface temperatures derived from the soft energy part of the spectrum (filled circles) and the temperature upper limits (open circles) are in all cases compatible with the predictions of standard cooling models, i.e. there is no indication of a progressive neutron star cooling as would be expected by the presence of exotic matter in the neutron star core region or by the direct URCA processes, respectively. This is the first time that empirical results constrain neutron star cooling models.

Besides the young pulsars and those whose surface cooling is visible in the X-ray waveband, ROSAT has detected X-ray emission from a further six pulsars: PSR 1929+10[19], PSR 0823+26[20], PSR 0950+08[21], PSR J0751+18[22], PSR J0437-4715[23] and PSR 1957+20[24,25]. All are characterized by a high spin-down age of $\sim 5 \times (10^6 - 10^9)$ years and a relatively close distance. Pulsed X-ray emission, however, could only be detected from the two closest of these pulsars, the 5.75 ms binary pulsar PSR J0437-4715[23] and from PSR 1929+10[19]. The discovery of pulsed X-rays from PSR J0437-4715 by Becker & Trümper (1993) represents the first conclusive detection of a millisecond pulsar outside the radio band. The X-ray pulse profile is very broad with a single pulse stretching across almost the entire phase cycle. The relative strength of the X-ray pulse changes with energy, from 30 ± 3 % at 0.1–0.6 keV, rising to 53 ± 6 % at 0.6–1.1 keV, and sinking again to 27 ± 9 % at 1.1–2.4 keV. The energy averaged pulsed fraction over the whole ROSAT energy band is found to be 33 ± 3 %. No phase shift exists between the pulses at different energies in contrast to the behaviour seen in PSR 0656+14, PSR 1055-52 and Geminga. The results of the spectral analysis for PSR J0437-4715 and PSR 1929+10 strongly indicate that the pulsed X-ray emission is of thermal origin ($T \sim 2$–4×10^6 K, $R_{bb} \sim 50 - 150$ m) and comes from a hot spot on the neutron star surface which is probably located at the magnetic poles[19,23]. Since detectable thermal X-ray emission was expected before ROSAT only from neutron stars younger than $\sim 10^6$ years, the ROSAT results on the millisecond pulsar PSR J0437-4715 and PSR 1929+10 show for the first time that in rotation-powered pulsars continuous heating takes place and yields an important contribution to the soft X-ray emission from these objects even for very old pulsars. An appropriate heating mechanism proposed by most magnetospheric pulsar models is the bombardment of the polar caps by energetic particles, streaming back to the surface from the pulsar magnetosphere and heating the polar cap region up to a few million degrees. The X-ray luminosities observed by ROSAT are in good agreement with the predictions made by these models[26,27]

PROSPECTS

ROSAT has brought important progress for the neutron star and pulsar astronomy. Its significantly larger collecting area and higher sensitivity compared to previous X-ray satellites allowed for the first time to distinguish between various X-ray emission processes. Follow-up experiments such as the ESA cornerstone XMM and NASA's AXAF, which are designed to have a higher sensitivity and a better spectral, timing and spatial performance than ROSAT, will be well suited to further the study of neutron stars and to enhance our understanding of these objects in many aspects.

REFERENCES

1. Becker, W., 1995, *Investigation of rotation powered pulsars with ROSAT. A search for cooling neutron stars*, Ph.D. thesis, Ludwig-Maximilians-Universität München, available as MPE-Report 260.

2. Ögelman, H., 1995, *X-ray observations of cooling neutron stars*, in *The Lives of Neutron Stars*, eds A. Alpar, U. Kilizóglu & J. van Paradijs, Kluwer Academic Publishers.

3. Umeda, H., Shibazaki, N., Nomoto, K. & Tsuruta, S., 1993, *Thermal evolution of neutron stars with internal frictional heating*, Astrophys. Journ., **408**, 286-293.

4. Haensel, P. & Gnedin, O.Yu., 1994, *Direct URCA processes involving hyperons and cooling of neutron stars*, Astrophys. & Astron., **290**, 458.

5. Pethick, C.J., 1992, *Cooling of neutron stars*, Reviews of Modern Physics, **64**, 1133.

6. Seward, F.D., Wang, Z.R., 1988, *Pulsars, X-ray synchrotron nebulae and guest stars*, Astrophys. Journ., **332**, 199.

7. Taylor, J., Lyne, A.G., Manchester, R.M., 1995, *Princeton-Pulsar-Catalogue.*

8. Becker W. & Aschenbach B., 1995, *ROSAT HRI observations of the Crab pulsar: An improved temperature upper limit for PSR 0531+21*, in *The Lives of Neutron Stars*, eds A. Alpar, U. Kilizóglu & J. van Paradijs, Kluwer Academic Publishers.

9. Becker W., Brazier K.T.S. & Trümper, 1995, *ROSAT observations of the radio and gamma-ray pulsar PSR 1706-44*, Astrophys. & Astron., **298**, 528.

10. Becker W., Brazier K.T.S. & Trümper, 1995, *ROSAT observations of PSR 2334+61 in the supernova remnant G114.3+0.3*, submitted to Astrophys. & Astron.

11. Safi-Harb, S. & Ögelman, H., 1995, *ROSAT observations of the unusual supernova remnant CTB 80*, in *The Lives of Neutron Stars*, eds A. Alpar, U. Kilizóglu & J. van Paradijs, Kluwer Academic Publishers.

12. Ögelman, H., Finley, J.P. & Zimmerman, H.U., 1993, *Pulsed X-rays from the Vela Pulsar*, Nature, **361**, 136-138.

13. Finley, J.P.,Ögelman, H. & Kiziloglu, Ü., 1992, *ROSAT observations of PSR 0656+14: A pulsating and cooling neutron star*, Astrophys. Journ., **394**, L21.

14. Halpern, J.P. & Holt, S., 1992, *Discovery of soft X-Ray pulsations from the γ-ray source Geminga*, Nature, **357**, 222-224.

15. Halpern, J.P. & Ruderman, M., 1993, *Soft X-ray properties of the Geminga pulsar*, Astrophys. Journ., **415**, 286-297.

16. Ögelman, H. & Finley, J.P., 1993, *ROSAT observations of pulsed soft X-ray emission from PSR 1055-52*, Astrophys. Journ., **413**, L31-L34.

17. Schaaf, M.E., 1990, *Surface-to-core temperature variation of homogeneously magnetized neutron stars*, Astrophys. & Astron., 61-70.

18. Umeda, H., Shibazaki, N., Nomoto, K. & Tsuruta, S., 1993, *Thermal evolution of neutron stars with internal frictional heating*, Astrophys. Journ., **408**, 286-293.

19. Yancopoulos, S., Hamilton, T.T. & Helfand, D.J., 1994, *The detection of pulsed X-ray emission from a nearby radio pulsar*, Astrophys. Journ., **429**, 832-843.

20. Sun, X., Trümper, J., Dennerl, K. & Becker, W., 1993, *Detection of soft X-rays from PSR 0823+26*, IAU Circular No. 5895

21. Manning, R. & Willmore, P., 1994, *ROSAT observations of PSR 0950+08*, Mon. Not. R. Astron. Soc., **266**, 635-639.

22. Becker, W., Brazier, K.T.S., Trümper, J., Lundgreen, S.C., Zepka, A.F., & Cordes, J., 1995, *ROSAT observations of the binary millisecond pulsar PSR J0751+18*, in preparation

23. Becker W. & Trümper J., 1993, *Detection of pulsed X-rays from the binary millisecond pulsar J0437-4715*, Nature, 365, 528-530.

24. Kulkarni, S.R., Phinny, E.S., Evans, C.R. & Hasinger, G., 1992, *X-ray detection of the eclipsing millisecond pulsar PSR 1957+20*, Nature, **359**, 300-302.

25. Fruchter, A.S., Bookbinder, J., Garcia, M.R. & Bailyn, C.D., 1992, *X-rays from the eclipsing pulsar 1957+20*, Nature, **359**, 303-304.

26. Cheng, K.S., Ho, C. & Ruderman, M.A., 1986, *Energetic radiation from rapidly spinning pulsars. I. Outer magnetosphere gaps*, Astrophys. Journ., **300**, 500-539

27. Chen, K. & Ruderman, M., 1993, *Origin and radio pulse properties of millisecond pulsars*, Astrophys. Journ., **408**, 179-185.

28. Becker, W., Brazier, K.T.S. & Trümper, J., 1993, *Geminga: relative phases of the X-ray and γ-ray pulses*, Astrophys. & Astron., **273**, 421-424.

29. Finley, J.P. & Ögelman, H., 1993, *Detection of soft X-ray emission from PSR 1823-13*, IAU Circular No. 5787

30. Finley, J.P., Ögelman, H., 1994, *The PSR 1800-21 / G8.7-0.1 association: a view from ROSAT*, Astrophys. Journ.**434**, L25-L28

31. Finley, J.P., Ögelman, H., Hasinger, G. & Trümper, J., 1993, *ROSAT observations of the LMC pulsar PSR 0540-69*, Astrophys. Journ., **410**, 323-327.

32. Slane, P., 1994, *X-ray emission from PSR 0355+54*, Astrophys. Journ., **437**, 458-464.

Plasma Configurations Around Rotating, Magnetised Objects - Particle Trapping by the Force-Free Surface[1]

K.O. THIELHEIM
Institut für Reine und Angewandte Kernphysik
Abteilung Mathematische Physik
University of Kiel
Otto Hahn Platz 3
24118 Kiel, Germany

MOTIVATION OF PRESENT WORK

In his lecture delivered on occasion of the presentation of the 1989 Nobel Prize in Physics, W. Paul outlines the principles of strong focussing, trapping and cooling of electrically charged particles in man-made machines [1]. Some of these mechanims can also been found in astrophysical configurations, specifically, in rotating magnetised neutron stars, for example. The present work is directed to the understanding of fundamental mechanisms at work in pulsar magnetospheres leading to the formation of plasma configurations[2].

'STAGE ONE' - PARTICLE DYNAMICS IN THE VACUUM FIELDS

The Force-Free Surface

The vacuum fields of a homogeneously magnetised sphere [2] rotating with its vector of angular velocity ω inclined by an angle $\chi \neq 0, \pi$ against its vector of magnetic dipole moment μ is known to have a non-trivial force-free surface (FFS), on which, by definition, the electric and magnetic vectors are orthogonal [3]. Figure 1 illustrates the FFS of an inclined rotator[3].

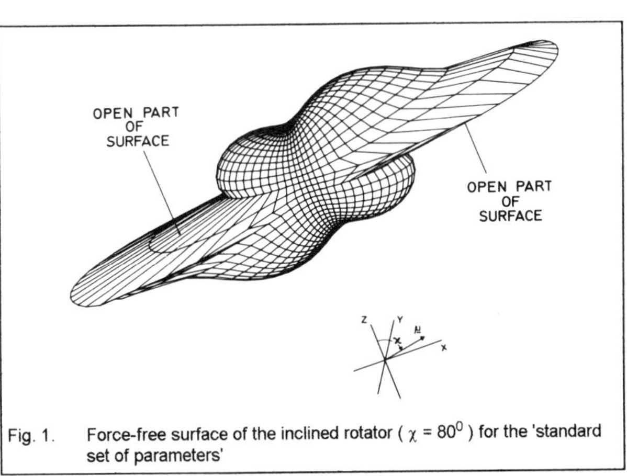

Fig. 1. Force-free surface of the inclined rotator ($\chi = 80^0$) for the 'standard set of parameters'

[1] Paper presented at the 17th Texas Symposium on Relativistic Astrophysics, München, Germany, December 11 - December 17, 1994.

[2] Different types of configurations are predicted for different regions of pulsar magnetospheres. In this contribution, due to limited space, we shall concentrate on configurations forming under the influence of the FFS.

[3] Numerical results presented here are for a 'standard set of parameters': $r_N = 10^{-6}$ cm, $\mu = 10^{30}$ G cm^3 and $\omega = 20\pi$ sec^{-1}, believed to be typical for pulsars.

The Equations of Motion

In preliminary studies, ('stage one',) we have considered the development of orbits and of energy for electrically charged particles, specifically for electrons and protons, in the vacuum fields of a rotating magnet[4]. These calculations were done by numerical integration of the equations of motion incorporating radiation reaction [5],

$$du_j / d\tau = \eta_0 F_{jk} u^k + \tau_0 G_{jk} u^k, \quad (1)$$

where u^k is the (four-) vector of velocity and τ is the eigen time. F_{jk} is the tensor of the external field[5] and η_0 = e/mc. The radiation force tensor is introduced in the form

$$G_{jk} = \eta_0 u^l \partial_l F^{EXT}{}_{jk} + G^T{}_{jk}, \quad (2)$$

where

$$G^T{}_{jk} = (u^{LL}{}_j u_k - u_j u^{LL}{}_k) /c^2 \quad (3)$$

is the tensor of ' Thomson force'. $\tau_0 = 2e^2/3mc^3$ is the radiation constant[6] and $u^{LL}{}_j = \eta_0{}^2 F^{EXT}{}_{jk} F^{EXTkl} u_l$ is an abbreviation used for the 'second Lorentz acceleration'.

The Trapping Mechanism

In the present context, we are interested in particle motion developing within a regime governed by the FFS [6].

Within the regime of the FFS, very near to the surface of the rotating magnet, well inside a sphere the radius of which equals the light radius $r_L = c/\omega$, particle trajectories essentially follow magnetic field lines. Somewhat further out, though still within that sphere, particle orbits develop a tendency to gyrate around magnetic field lines. Beyond that region, particle orbits more and more alienate from magnetic field lines. The systematics of particle acceleration near the FFS is demonstrated in Fig. 2.

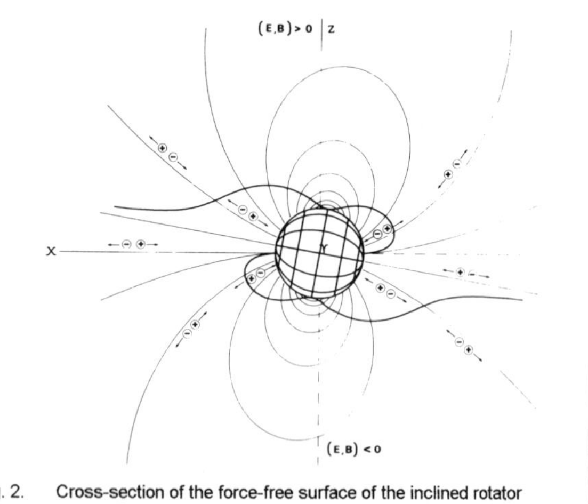

Fig. 2. Cross-section of the force-free surface of the inclined rotator ($\chi = 80^0$) with the ω-μ plane for the 'standard set of parameters' and systemactis of the direction of particle acceleration

[4] A review of results obtained on 'stage one' is given in [4].

[5] The 'external' field is understood as the field which is due to all other electromagnetically interacting particles around.

[6] Gaussian units will be used throughout this paper. Accordingly, the electric as well as the magnetic field strengths are measured in units of 1 G = 300 V/cm. $\eta_o = 1.76 \cdot 10^7$ $g^{-1/2}cm^{1/2}$ for the electron and $\eta_o = 9.58 \cdot 10^3$ $g^{-1/2}cm^{1/2}$ for the proton. $\tau_o = 6.27 \cdot 10^{-24}$ sec for the electron and $\tau_o = 5.12 \cdot 10^{-27}$ sec for the proton.

Three types of orbits, originating from positions very near to or on the surface of the magnet, have been found:

Particles moving monotonically in outward direction. Such particles can achieve very high energies and contribute to cosmic particle radiation.

Particles returning to the immediate neibourhood of the magnet on orbits resembling those known from classical Störmer theory [7].

Trapped particles oscillating on a magnetic field line about the FFS in a motion damped by radiation losses.

The latter type of orbits for which a number of examples are shown in figure 3 is of main interest here. Oscillation and damping ('cooling') is further illustrated by figure 4.

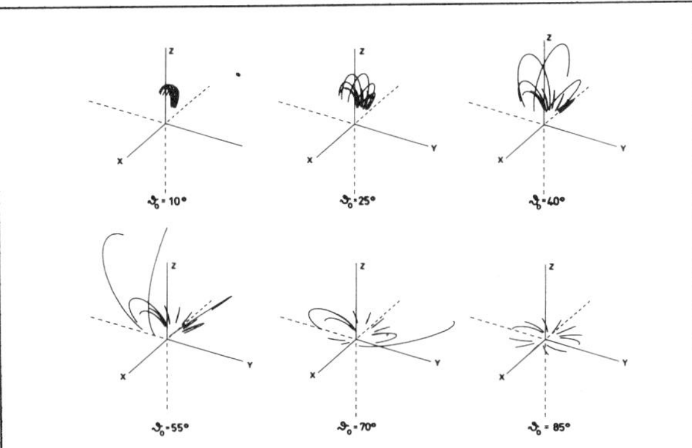

Fig. 3. Orbits of protons trapped by the force-free surface of the inclined rotator ($\chi = 80^0$, 'standard set of parameters'). ϑ_0 is the initial latitudinal angle and φ_0 is the initial longitudinal angle.

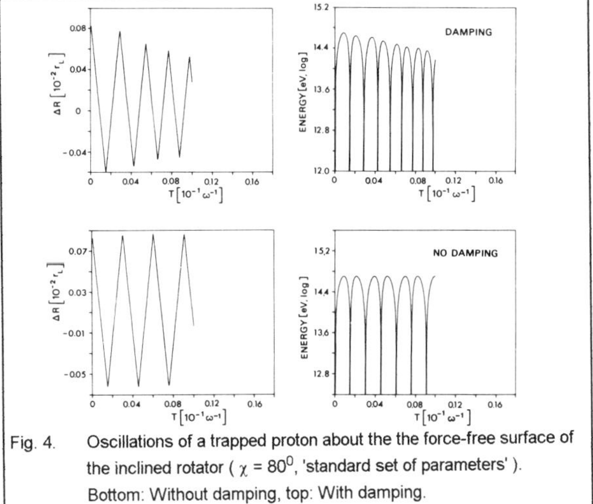

Fig. 4. Oscillations of a trapped proton about the the force-free surface of the inclined rotator ($\chi = 80^0$, 'standard set of parameters').
Bottom: Without damping, top: With damping.
Left: Distance from the FFS, right: Energy.

'STAGE TWO' - PARTICLES IN THE PLASMA FIELDS

The Iterative Procedure

H. Wolfsteller and K.O. Thielheim have developed a numerical iterative procedure for the generation of quasi-stable plasma configurations [8]. This preliminary work is intended to clarify conditions for the existence and structural features of plasma configurations. In this context, it was not necessary to perform numerical integrations of the equations of motion.

In each iteration step, a certain amount of charge was allowed to separate from each segment of the surface of the neutron star and to move along the adjacent magnetic field line (neglecting drift effects) and to settle down where the electro-static forces tangent to the magnetic field line vanish. The total electric charge dismissed locally from the surface in each iteration step was chosen to be proportional to the magnitude of the component of the electric vector tangent to the magnetic field line (given the appropriate sign of that component).

Modifications of the electric field (with respect to vacuum fields) due to space charges were taken into account in each iteration step. Modifications of the magnetic field were found to be insignificant. A minimum value for the tan-gent electric component was adopted as a current limiting condition to define the end of the iteration process, as is illustrated in figure 5.

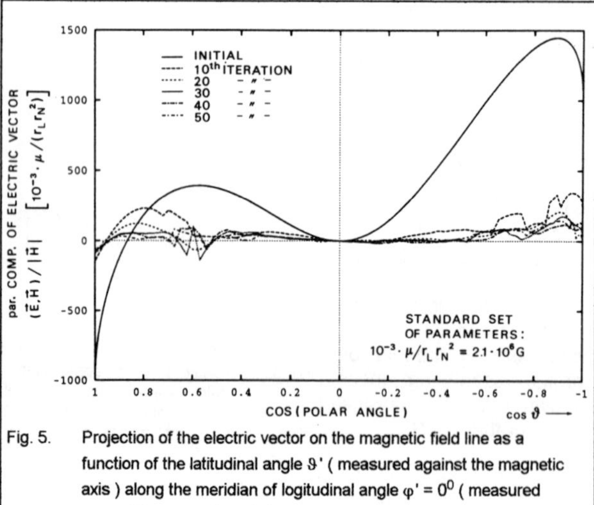

Fig. 5. Projection of the electric vector on the magnetic field line as a function of the latitudinal angle ϑ' (measured against the magnetic axis) along the meridian of logitudinal angle $\varphi' = 0^0$ (measured around the magnetic axis).

Corotating Charge Clouds

Figure 6 illustrates preliminary results for corotating charge clouds.

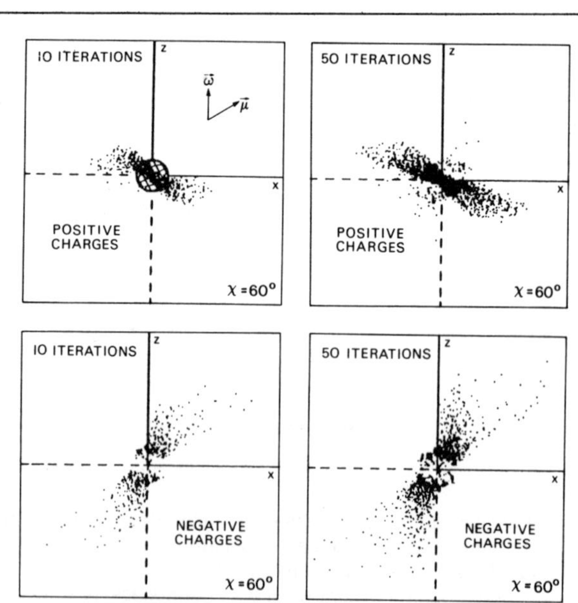

Fig. 6. Quasi-stable corotating charge clouds formed under the influence of the FFS of the inclined rotator (χ = 80^0, 'standard set of parameters'). Bottom: Negative charges, top: Positive charges. Left: After ten iterative steps, right: After fifty iterative steps.

References

[1] W. Paul, Rev. of Mod. Phys. **62**, 531 (1990).

[2] A.J. Deutsch, Ann. d'Astrophys. **18**, 1 (1955).

[3] E.A. Jackson, Ap.J. **222**, 675 (1978).

[4] K.O. Thielheim, Fund. of Cosm. Phys. **13**, 357 (1989).

[5] K.O. Thielheim, Phys. Scripta, **T52**, 123 (1994).

[6] K.O. Thielheim, and H. Wolfsteller, Ap. J. **71**, 583 (1989).

[7] C. Störmer, Z. Astrophys., **1**, 237 (1930).

[8] K.O. Thielheim and H. Wolfsteller, Ap. J. **431**, 718 (1994).

Boundary Effects in the Pulsar Magnetosphere

ESTELLE ASSEO

Centre de Physique Théorique
Ecole Polytechnique
91128 Palaiseau Cedex FRANCE

INTRODUCTION

The source of coherent pulsar radio emission has often been related to the development of plasma instabilities in pulsar polar cap models[1]. These instabilities have been studied either for a relativistic beam of e^- (or e^+), or for a relativistic plasma of e^- and e^+, or for the interaction between both, mainly by using straight geometry and an approximation of infinite and homogeneous beam and plasma.

For a tentative interpretation of radio observations, according to Rankin's classification of pulsar profiles which show a hollow cone or a full cone of radio emission[2], we include in our analysis curved geometry and finite extent of the beam[3]. Indeed, relativistic particles are constrained to move along the extremely strong pulsar magnetic field that is dipolar and thus curved. They form a flowing beam that has a finite extent, being limited to the bundle of open magnetic field lines present in the emission region. This beam is bounded by external plasmas that have different characteristics, namely the plasma that lie on closed field lines and eventually the plasma inside the hollow cone.

LOCAL CYLINDRICAL GEOMETRY

Locally, we use cylindrical geometry assuming that, at a given point, the curved magnetic field lines can be approximated by circles, each with a radius R_c, equal to the local radius of curvature. This cylindrical geometry is particularly appropriate to describe the hollow cone region of emission, that involves a group of open magnetic field lines with slightly different radii of curvature. As the bundle of open magnetic field lines forming the hollow cone diverges from the surface of the star and is issued from a relatively small area of this surface, these same field lines can be considered as forming the full cone region of emission closer to the surface of the star. With such an approximation, only a very small part of the emission region is not described. We are thus able to propose qualitative results for the full cone region of emission[3].

DISPERSION RELATIONS

The dispersion relation for the whole system of relativistic finite beam and external plasmas is obtained by matching the corresponding solutions for the electromagnetic field components at different beam-plasma interfaces[3]. Instead of the classical dispersion relation, $W = 0$, which describes a two-stream instability for infinite homogeneous relativistic beam and plasmas, assuming straight geometry, we obtain a dispersion relation modified by terms that account for geometric properties of the flow, namely curvature of the magnetic field through the radius of curvature R_c, finite width of the beam a and azimuthal wavenumber of the perturbation m, and for the presence of external media with dielectric constants W_i and W_e. The analysis of this dispersion relation in different asymptotic limits reveals a "new" instability, that we have called the "finite beam" instability, in addition to the radiative and two-stream instabilities[3].
A quite general dispersion relation is derived, describing a "finite beam" instability for a beam bounded by non-identical external media. It is obtained close to the resonant frequency $m\,\Omega_0$, at $\omega = m\,\Omega_0 + \delta\omega$ assuming that $Y = \delta\omega / m\,\Omega_0$, is very small. Considering a specific asymptotic limit, it can be written

$$\tanh\left[(2\,m\,W^{1/2}/3)\,\{(-Y)^{3/2} - (a/R_c - Y)^{3/2}\}\right] = W^{1/2}\,[W_i^{1/2} + W_e^{1/2}]\,/\,[W + W_i^{1/2}\,W_e^{1/2}]$$

It is quite a complicated dispersion relation due to the complex argument in the hyperbolic tangent and, except for the particular solution $W = 0$ describing the two-stream instability, it is very difficult to solve. We have solved it mainly numerically or by using specific approximations. For instance, for a beam bounded by identical external media, the available simplified dispersion relations[3] show dependence on geometric properties of the flow through the dimensionless parameters $\varepsilon = R_c / m\,a$ and $\tau = (1/m)\,(R_c/a)^{3/2}$, whereas the presence of external plasmas appears through the dielectric constant W_i.

PARAMETERS ε and τ. DIELECTRIC CONSTANTS

Estimates for ε and τ (with usual pulsar parameters) are very small for most open magnetic field lines, so that $\tanh[\;] \approx 1$. Approximate simplified dispersion relations take one of the two very simple forms, $W = W_i$ or $W = -W_i$, depending on the specified asymptotic limit. In the extreme case of almost straight field lines, large values for ε and τ have to be considered and the corresponding approximation $\tanh z \approx z$ leads to more complex dispersion relations.

Dielectric constants W and W_i, respectively for a beam of relativistic monoenergetic particles (γ_p) and for the interior plasma presumably at rest, write

$$W \approx 1 - P^2/Y^2 \qquad \text{and} \qquad W_i \approx 1 - P_i^2(1 - 2\,Y)$$

They depend on the relative variation of the frequency, Y, and on the ratio

between plasma and resonant frequencies, namely $P = \omega_p / \gamma_p^{3/2} m \Omega_0$ for the beam plasma and $P_i = \omega_{pi} / m \Omega_0$ for the interior plasma .
P and P_i depend on the density of the concerned plasma and vary with the distance to the center of the star. Thus, we may define a characteristic distance r_{lim} where $P_i = 1$. For pulsars which are not too fast ($P_{sec} \approx 10^{-1}$ s) and at radio frequencies (10^8-10^9 Hz), we obtain r_{lim} in the domain $10 R_* \leq r_{lim} \leq 60 R_*$.
W_i varying, according to the value of P_i , with the location in the emission region, is not the same close to the surface of the star ($r < r_{lim}$, $P_i >> 1$), at the transition zone ($r = r_{lim}$, $P_i \approx 1$) and beyond the transition zone ($r > r_{lim}$, $Pi << 1$). Thus, the behavior of radiative or "finite beam" instabilities differs according to the distance of the emission region relatively to the surface of the neutron star.

THE DIVERSE INSTABILITIES[3]

(1) The two-stream instability reaches maximum growth rate well beyond r_{lim}.

(2) An analysis of the general dispersion relation describing a radiative instability, for an extremely thin relativistic beam bounded by identical or non-identical external media[3], shows that this instability has significant growth rates only at and beyond the characteristic distance r_{lim} . Comparison with the growth rate of the two-stream instability indicates that the radiative instability is dominant beyond the distance r_{lim} , as long as P/P_i , equivalent to the ratio of densities, is not too small.

(3) With either a large or small argument for the tanh in the dispersion relation that describes the "finite beam" instability, one concludes that this instability is very efficient close to the surface of the star, below a critical distance comparable to r_{lim}. This is obvious in a few cases that manifest stability beyond the distance r_{lim} . A comparison between growth rates shows that the "finite beam" instability is easily dominant close to the surface of the star, for not too small values of the ratio P/P_i .

WAVE POLARIZATION. ELECTROMAGNETIC ENERGY.

An evaluation of radial and transverse polarization ratios for the unstable waves arising from the "finite beam", radiative or two-stream instabilities[3,4], shows that such waves are at the same time electrostatic and electromagnetic and thus are true candidates to account for pulsar radio emissions. Indeed, the polarization is essentially in the radial direction, perpendicular to the mean magnetic field. In some very special conditions, the "finite beam" instability may also have a small percentage of transverse polarization and account for some circular polarization.
Estimates for the maximum electromagnetic energy available through one or the other instability[3,4,5] indicate that energy is sufficient to account for the high level of coherent pulsar radio emissions. For instance, the maximum luminosity expected from the "finite beam" instability $L \leq L_{max} \approx 2\ 10^{29}$ erg/s , is comparable to observed fluxes in the range 10^{25} - 10^{30} erg/s.

INTERPRETATION OF THE RESULTS

We provide numerical estimates for the opening angle Θ of any unstable domain located between the surface of the star and the distance R and for its associated width W_c. Both estimates depend on P_i and are evaluated in the presumed emission region at $P_i =$ 1, 10 and 100, that is for emission locations below r_{lim}[3].

We compare our numerical estimates for the core widths to the core widths measured by Rankin[2], and later analysed by Blakiewicz et al[6], for two groups of pulsars that show core emission. For the first set of pulsars (with $\alpha \approx 90°$) the observed values correspond to emission domains located very close to the surface of the neutron star, P_i being very close to 100, whereas for the second group of pulsars (with $\alpha \neq 90°$) the measured core widths correspond to emission domains located slightly farther from the surface of the star, where $1 < P_i < 10$, i.e. at a distance just below or up to r_{lim}.

These results lead us to believe that the "finite beam" instability, which is most efficient in the region between the surface of the star and the transition zone, could account for core emission. This is consistent with recent suggestions by Rankin[2] that there are separate mechanisms for core and conal emissions.

SUMMARY

In this paper based on our most recent work[3],

(1) we show the possible existence of a "finite beam" instability, which differs from radiative or two-stream instabilities and develops from a one-component beam of relativistic particles bounded by external plasmas,

(2) we suggest that this "finite beam" instability could be related to core emission whereas two-stream or radiative instabilities would be related to conal emissions,

(3) we emphasize the importance of boundaries that delimit the finite relativistic beam and also of the nature of external plasmas which are essential to the determination of the instability process together with the curvature of the magnetic field, the finite extent of the beam and the azimuthal wavenumber of the studied perturbation.

REFERENCES

1 Ruderman, M.A., Sutherland, P.G., 1975, Astrophys.J, 196, 51

2 Rankin, J.M., 1990, Astrophys.J, 352, 258

3 Asseo, E., 1995, Mont. Not. R.Astron. Soc. (accepted)

4 Asseo, E., Pellat, R., Sol, H, 1983, Astrophys.J, 266, 201

5 Asseo, E., Pelletier, G., Sol, H, 1990, Mont. Not. R.Astron. Soc.,247, 529

6 Blaskiewicz, M., Cordes, J. M., Wasserman, I., 1991, Astrophys.J, 370, 643

SUPERNOVA REMNANTS vs PULSAR NEBULAE

WOLFGANG KUNDT
Institut für Astrophysik der Universität
D-53121 Bonn

SNRs COME IN TWO VARIETIES

Supernova Remnants (=SNRs) are observationally defined as extended sub-galactic emission regions with non-thermal spectra, often better recognizable at radio and/or X-ray frequencies than at optical ones. Their name implies that they are thought to be the (ejected) remnants of a (recent) SN-explosion. But this is not guaranteed in all cases: SNRs are often confused with Pulsar Nebulae, i.e. with density enhancements illuminated (non-thermally) by a not-too-weak pulsar (=PSR). The latter belong to the subclass of 'exotic' SNRs. They tend to be fainter than SNRs, and have bizarre morphologies (rabbit, bird, la paloma, mouse, tornado,...)[7].

There is a simple reason for the existence of two classes of SNRs: SNRs in the strict sense owe their relativistic electrons to the SN explosion that gave birth to the remnant (see below) - i.e. to an event of duration less than a day - whereas PSR nebulae are blown steadily by a PSR of typical age $\lesssim 10^5$ yr, during its causting away from its birth site; their bizarre morphologies are due to the different and varying densities of the medium[7] encountered by the drifting PSR.

A better understanding of (the nature of) SNRs is desirable for at least two reasons: (i) Two of the 'soft gamma-ray repeaters' (=SGRs) project onto SNRs (in the broad sense); their distances have been estimated as either $\gtrsim 10$ Kpc, or $\lesssim 50$ pc by different authors[16,8]. (ii) Peculiar velocities of several young PSRs have been estimated from their age and distance from the center of a nearby SNR; they may be large overestimates if the neutron star was born strongly off-center - as is suggested by their likely progenitor systems, the (runaway) Wolf-Rayet stars, cf. figure 1, and ref. 5.

SUPERNOVA EXPLOSIONS

Much of what has been claimed above depends on the correct SN model, hence I will repeat - and update - my earlier conclusions[6]: I restrict my considerations to core-collapse explosions of stars of initial mass $\gtrsim 5\ M_{\odot}$ (which I deem typical for SNe). Because of its extremely long runway, from the (length) scale of a neutron star (of $\gtrsim 10^6$ cm) to the scale of a supergiant ($10^{13 \pm 0.5}$ cm), the SN 'piston' cannot be non-relativistic; adiabatic cooling losses would reduce its temperature (in proportion to r^{-2}) way below the 10^{10} K required for pushing matter (radially) to velocities of $\gtrsim 10^4$ Kms^{-1}. The piston cannot consist of photons either because photons would mix with matter, and would thereby yield a non-relativistic plasma. Neutrinos do not qualify because of too low an opacity of the stellar envelope: their radial momentum would have to be transferred (to the envelope) at an efficiency of $\gtrsim 30\%$, an unrealistic demand. Remain magnetic fields and their decay product - a magnetized relativistic plasma consisting mainly of electron-positron pairs - as the expected[6] SN piston.

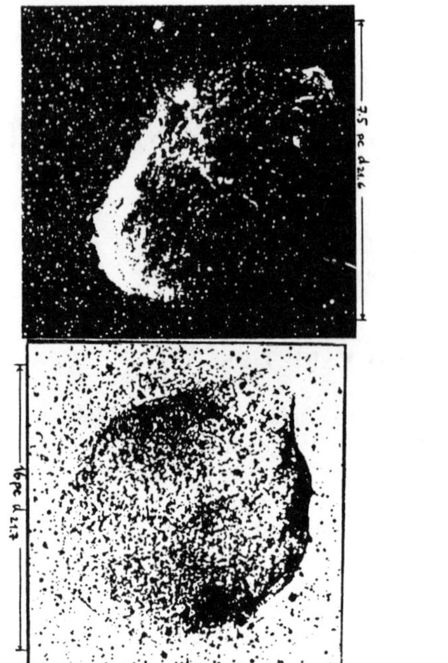

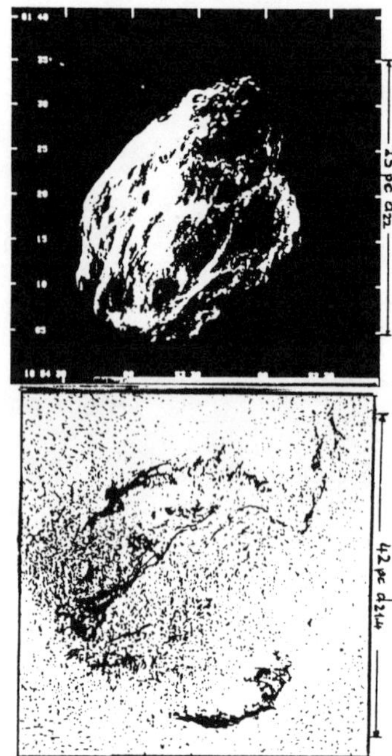

FIGURE 1. The Wolf-Rayet nebulae NGC 6888 (optical positive, rotated through 90^0; top left) and S 308 (at [OIII] 5007 Å, bottom left) confronted with the SNRs W 44 (at 20 cm; top right) and the Cygnus Loop (POSS red, rotated through 143^0; bottom right). Note the similar large-scale morphologies and comparable sizes, whereby the (older) SNRs are somewhat larger. In the non-spherical (runaway, NGC 6888) nebula, the exciting star (marked by a cross) is located distinctly off-center. When it goes SN, its stellar remnant will not be born at the shell's center.

In pressure contact with the relativistic piston, the envelope of a (supergiant) star is torn and squeezed into a huge number of filamentary fragments: a SN is a (thick-walled) splinter bomb, not a (thin-walled) pressure bomb (like conceived by Shklovskii[14]). This conclusion is confirmed by SN spectra which show a large range of velocities - from a few 10^2 Kms^{-1} to several 10^4 Kms^{-1}, see figure 2 - not a thin shell, of well-defined speed $10^{3.8\pm\ 0.3}$ Kms^{-1}. At the same time, pair-plasma SNe are expected to flare at radio frequencies as soon as they have reached the radial distance beyond which the surrounding windzone (of the progenitor star) gets transparent, in agreement with the observations[17]. (Most authors postulate in-situ acceleration to extremely relativistic energies in order to explain the radio detections, an assumption which other people consider a Münchhausen effect. In SNe, shock acceleration would have to function orders of magnitude faster than in other astrophysical situations for which it has been assumed).

The splinter (or shrapnel) structure of a SN explosion is not always obvious from its maps; SNRs often look like smooth, moderately thin shells; (see refs.12, 13, and 17 for surveys at IR, X-ray, and radio frequencies, respectively). Such maps (of shell remnants) do not, however, directly trace the ejecta. Rather, they show the illuminated edges of former windzones: we see the matter glow where its densest regions are traversed by the ejecta. Emission scales as the product of target density and density of the ejecta. Illustrative examples are (i) Cas A , where both the 'jet' (to the NE) and the 'fast-moving

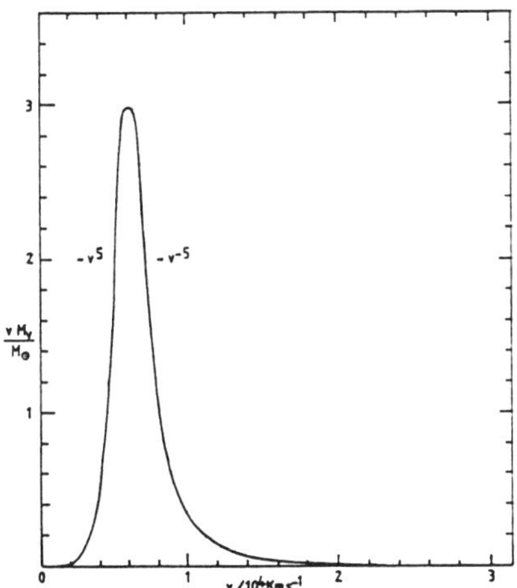

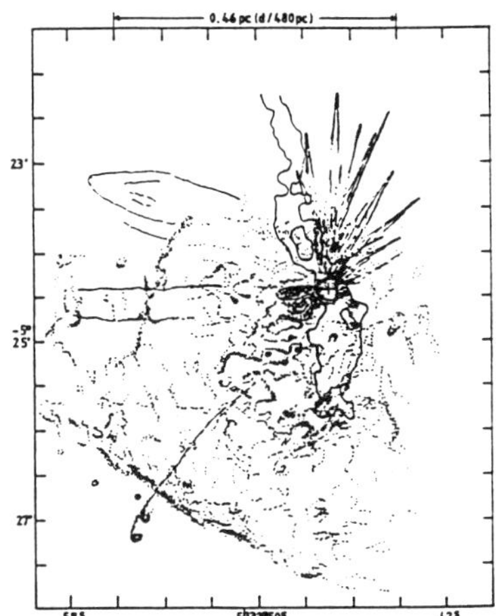

FIGURE 2. Typical radial mass distribution, $\Delta M =: \int M_v\, dv$, in the shell ejected by a supernova, as judged from SN spectra, and from young SN shells. v stands for the (radial) velocity which scales as radial distance during the epoch of free expansion - i.e. for a significant fraction of the SNR lifetime if the ejecta are of shrapnel type.

FIGURE 3. Sketch of the outflow in the Orion Nebula (M42) - with center one arcmin NW from the four trapezium stars - which has been called a 'bipolar flow' in recent literature but which may well be the youngest known SNR in the Galaxy, see ref. 9. The figure is a multiple-frequency map, combining radio continuum intensities with (optical) stars, optical emission lines, and CS contours. All the measured velocities in this region are consistent with a radial Hubble flow from (near) the IR source IRc2.

flocculi' are seen out to twice the distance of the sharp-edged shell[4], (ii) the Vela SNR mapped at X-rays by ROSAT, with its ≈ 50 protrusions[1] beyond the 'outer edge', and (iii) the $\gtrsim 10^2$ yr-old explosion[9] in Orion, see figure 3. In all (3) cases, what tends to be called "the shock wave" is by no means the front of the explosion. Of course, many SNRs show (so far) sharp outer edges; shrapnel-type morphology can be invisible when the ejected fragments are small (and numerous) enough.

Figure 1 compares the Wolf-Rayet windzone NGC 6888 (at optical frequencies) with the SNR W44 (at 20 cm), as a typical example of a runaway system: the former may well be a progenitor of the latter. (A similar interpretation has been proposed[2] for Kepler's SNR). But then we learn that the exploding star need not be located at the center of a SN shell. In this way, the peculiar velocity of a PSR can be systematically overestimated if based on a PSR-SNR association. On the other hand, a sufficiently slow progenitor - like S 308 (again figure 1) - will have the SN explosion occur at the center (of the windzone).

If in-situ acceleration (of the relativistic component) inside a SN, or SNR is not feasible, the observed relativistic electrons (and positrons) must all be provided by the SN explosion. In my understanding, this is implied by the explosion mechanism. Occasionally one can hear a worry, that such a pair-plasma cloud would lose all its energy in expanding to the size of the SNR. The worry does not apply, however, when the inertia of the CSM is small so that the pair plasma expands effectively into vacuum. Of course, some expansion losses cannot be avoided, and eventually there is enough CSM to halt the growth. From then on, the pair plasma will preferentially squeeze into the wakes of the ejecta - like insulating foam blown into the space between a door-framing and the wall. The knotty, time-variable radio structure of Cas A may thus find a plausible explanation, as well as the embarrassingly large radio sizes of SNe 1987A (ref. 15) and 1993J (ref. 10), a few months or years after the explosion.

PULSAR NEBULAE

Little is yet known about the luminosity function of PSR nebulae. Historically, the first identified non-trivial PSR nebula (after the nebulae surrounding the Crab, and Vela PSRs) was the 'bird', whose powering PSR is located at the tip of the beak; cf. ref. 7. It was followed by CTB 80 whose PSR 1951+32 had hidden for decades after the first study of its nebula. It is surrounded by a compact nebula (of size $\approx$ 0.3 pc) inside the much larger 'rabbit' ($\approx$ 20 pc). There is yet no consensus on whether the compact nebula owes its existence to a second (more recent) explosive event[11], or to the blowing[7] of the PSR.

Meanwhile, at least 12 PSR nebulae have been discovered[3], among them *one* inside of Puppis A. In all cases, the PSR is located on a (more or less) symmetry axis through the nebula. Its compact nebula (bow shock) and extended nebula ('remnant') have radii of $10^{17 \pm 1.2}$ cm and (15 ± 5) pc, respectively. They are probably at the high-luminosity end of their luminosity function, so that many more PSR nebulae will have to be extracted from the background in the future.

CONCLUSIONS

A distinction between stellar windzones, SN ejecta, and PSR nebulae can have strong implications on the determination of PSR peculiar velocities (from associations), and on the identification of the sources of the soft gamma-ray repeaters.

REFERENCES

1. ASCHENBACH, B. 1995. plenary contribution to this Symposium.
2. BANDIERA, R. 1987. Astrophys. J. **319:** 885-892.
3. CARAVEO, P.A. 1993. Astrophys. J. **415:** L 111-114.
4. FESEN, R.A., R.H. BECKER, & W.P. BLAIR. 1987. Astrophys. J. **313:** 378-388.
5. FRAIL, D.A., W.M. GOSS, & J.B.Z. WHITEOAK. 1994. Astrophys. J. **437:** 781-793.
6. KUNDT, W. 1990. In Neutron Stars and their Birth Events. W. Kundt, Ed.: 22-50. NATO ASI C **300**, Kluwer, Dordrecht.
7. KUNDT, W. & H.-K. CHANG. 1992. Astrophys. Sp. Sci. **193:** 145-154.
8. KUNDT, W. & H.-K. CHANG. 1994. In GAMMA-RAY BURSTS. G.J. Fishman, J.J. Brainerd & K. Hurley, Eds.: 596-599. AIP Conf. Proceedings **307**, New York.
9. KUNDT, W. & A. YAR. 1995. Astrophys. Sp. Sci.: in press.
10. MARCAIDE, J.M., et al (23 authors). 1995. Nature **373:** 44-45.
11. SAFI-HARB, S., H. ÖGELMAN, & J.P. FINLEY. 1995. Astrophys. J.: in press.
12. SAKEN, J.M., R.A FESEN, & J.M. SHULL. 1992. Astrophys.J. Suppl. **81:** 715-745.
13. SEWARD, F.D. 1990. Astrophys. J. Suppl. **73:** 781-819.
14. SHKLOVSKII, I.S. 1964. Astron. Zh. **41:** 176, 676.
15. STAVELEY-SMITH, L., D.S. BRIGGS, A.C.H. ROWE, R.N. MANCHESTER, J.E. REYNOLDS, A.K. TZLOUMIS, & M.J. KESTEVEN. 1993. Nature **366:** 136-139.
16. VASISHT, G., D.A. FRAIL, & S.R. KULKARNI. 1995. Astrophys. J. Lett: in press.
17. WEILER, K.W., & R.A. SRAMEK. 1988. Ann. Rev. Astron. Astrophys. **26:** 295-341.

Neutron Star Kicks and Multi-Dimensional Supernova Models[a]

H.-THOMAS JANKA[b,c] AND EWALD MÜLLER[c]

[c]*Max-Planck-Institut für Astrophysik*
Karl-Schwarzschild-Str. 1
D-85740 Garching, Germany

OBSERVATIONS

Radio interferometric measurements[1–4] and observations of the speed of the interstellar scintillation patterns[5,6] show that a large number of pulsars have proper motions between some ten and several hundred km/s. Reevaluation of older data based upon an improved distance scale for the pulsars together with new measurements that extended the sample of pulsars with known transverse velocities[4] reveal a significant number of objects with velocities close to 1000 km/s or even higher[7].

The complete sample of 99 pulsars considered by Lyne and Lorimer[7] includes 86 pulsars with interferometry data and another 13 with scintillation data. The sample has a mean transverse velocity of 300 ± 30 km/s. The increase of this number compared with previous analyses reflects both the change of the adopted distance scale and the greater number of young and higher velocity pulsars in the recent astrometric surveys. Considering only the 29 youngest pulsars of the sample (younger than 3 Myr), one finds a mean transverse velocity of 345 ± 70 km/s, implying an average space and birth velocity of pulsars of $\langle v_{\rm ns} \rangle = 450 \pm 90$ km/s. (The r.m.s. value is 535 km/s.) Associations between young supernova remnants and nearby pulsars[8–11] yield consistent results and might indicate even faster motions. Taking 13 reasonably convincing associations, Lyne and Lorimer[7] deduce a mean transverse velocity of about 530 ± 180 km/s, corresponding to an average space velocity of 690 ± 230 km/s.

The 10 oldest pulsars of the analysed sample have a mean transverse speed of only 105 ± 25 km/s, which confirms a suggested strong selection effect[6] that makes the observed velocities unrepresentative of those acquired at birth. Young pulsars appear to have higher velocities on average than

[a] This work was supported in part by the National Science Foundation under grant NSF AST 92-17969, by the National Aeronautics and Space Administration under grant NASA NAG 5-2081, and by an Otto Hahn Postdoctoral Scholarship of the Max-Planck-Society (H.-Th. J.).

[b] Present address: Department of Astronomy and Astrophysics, University of Chicago, 5640 S. Ellis Avenue, Chicago, Illinois 60637, U.S.A.

older ones not because pulsars slow down significantly as they age, but because young pulsars originate predominantly from Population I stars, close to the galactic plane, and move rapidly away from the plane. After about 10 Myr the mean distance of the fastest pulsars from an observer on the galactic plane has increased considerably and their detection probability is considerably reduced[7]. This explains the observed negative correlation between the transverse velocity and the characteristic pulsar age (spin down age $\tau = P/2\dot{P}$). Itoh and Hiraki[12] argue that also an apparent negative correlation between τ and the magnetic field of pulsars is mainly due to observational reasons. They show that even in the constant magnetic field model of pulsars without field decay a correlation results from the fact that the range of observed pulsar periods P is very much narrower than that of the observed pulsar period derivatives $\dot{P}$. Therefore a positive correlation between magnetic field strength and proper motions[13] seems to be merely a consequence of selection effects.

SUPERNOVA MODELS

Asymmetries during the early phase of the supernova explosion can provide a mechanism to accelerate the newly formed neutron star[14,15]. Comparatively small anisotropies of the matter ejection or of the emission of neutrinos are already sufficient to impart a kick of a few hundred km/s to the neutron star. With an anisotropy α (which quantifies the relative deviation from central symmetry), a neutron star mass $M_{\rm ns}$, an ejected mass $M_{\rm ej}$, and a velocity $v_{\rm ej}$ of the ejection, one can write for the recoil velocity of the neutron star

$$v_{\rm ns} = \alpha \cdot \frac{M_{\rm ej}}{M_{\rm ns}} \cdot v_{\rm ej} \,. \tag{1}$$

Considering the emission of neutrinos ($v_{\rm ej} = c$) and using a typical value of the gravitational binding energy of a $1.5\,M_\odot$ neutron star being released in neutrinos, $M_{\rm ej} \cong 3 \cdot 10^{53}\,{\rm erg}/c^2$, one finds that a 1% anisotropy, $\alpha \approx 0.01$, can already account for a velocity of $v_{\rm ns} \approx 330\,{\rm km/s}$.

Recent multi-dimensional hydrodynamical simulations of core collapse and the early phase of the supernova explosion show that convective processes occur in the nascent neutron star[16–22]. In the current models this leads to enhanced and anisotropic neutrino emission for a period $t_{\rm conv}$ of several 10 ms to several 100 ms after core bounce and shock formation. The associated kick imparted to the neutron star is given by

$$v_{\rm ns} = \alpha_1 \cdot \frac{c}{M_{\rm ns}} \cdot \int_0^{t_{\rm conv}} {\rm d}t\ L_\nu(t)/c^2 \,, \tag{2}$$

with the stochastic anisotropy $\alpha_1 \lesssim 1/\sqrt{N} \approx \sqrt{\pi\,(d/2)^2/(4\pi r^2)} = d/(4r) <$ 5%, when N is the number of convective eddies or cells and d the characteristic diameter of a rising blob as determined in the numerical simulations and the coverage factor of the sphere is assumed to be 1 (for details, see

Thompson and Duncan[23] and Janka and Müller[24]). Within a few hundred milliseconds after bounce only a minor fraction of the gravitational binding energy E_{gb} of the neutron star is radiated away in neutrinos, and $\int_0^{t_{conv}} dt\, L_\nu(t) \lesssim E_{gb}/10 \ldots E_{gb}/5 \lesssim (0.015 \ldots 0.03) \cdot M_{ns}c^2$, which yields an optimistic upper limit of $v_{ns} \lesssim 250 \ldots 450\,\mathrm{km/s}$. Significantly higher recoil velocities by this mechanism require a long duration of convection inside the protoneutron star, which, however, can not (yet?) be concluded from the models.

Of course, the nascent neutron star will obtain a net momentum when matter is accelerated and ejected anisotropically in the explosion wave, or, inverting this picture, when matter flows down and is accreted onto the neutron star in a non-spherical way. Indeed, hydrodynamical simulations[20–22,24–26] show that there is a period of several hundred milliseconds after core bounce, where violent, turbulent overturn takes place in the neutrino-heated region outside of the forming neutron star, below the supernova shock. Upflows and downflows of matter have velocities close to or even higher than the speed of sound. These overturn processes lead to a significant deformation of the shock front on large scales and an anisotropic distribution of outward moving matter behind the supernova shock. Again, this process is stochastic and rather than a single, jet-like structure one finds lumps and bubbles growing and rising into different directions. In two-dimensional simulations with a 180° grid these bubbles have a maximum angular diameter of about 45° to 60° (see Janka and Müller[21,22,24]). From this the anisotropy is determined to be less than 10%, $\alpha_2 < 0.1$. With a mass of $M_{ej} = 0.2 \ldots 0.4\, M_\odot$ being involved in the turbulent overturn and an explosion energy of $E_{ej} = (1 \ldots 2) \cdot 10^{51}\,\mathrm{erg}$ and assuming, optimistically, that all this energy is kinetic energy, one gets typical expansion velocities of $v_{ej} < \sqrt{2E_{ej}/M_{ej}} \approx (1.5 \ldots 3) \cdot 10^4\,\mathrm{km/s}$. Plugging the largest values of all quantities into Eq. (1) one finds a recoil velocity of about 800 km/s for this most extreme, and certainly too optimistic, case. Being a little more conservative, a more reliable upper limit of the possible kick velocity of the neutron star is $v_{ns} \lesssim 500\,\mathrm{km/s}$. The numerical evaluation of a set of two-dimensional computations by Janka and Müller[21,22] yields recoil velocities of 50 km/s, 75 km/s, and about 120–130 km/s, respectively. Due to projection effects the real velocities corresponding to the three-dimensional situation could possibly be up to a factor of 2 higher. Burrows, Hayes, and Fryxell[27] claim a value of about 300 km/s for their two-dimensional model.

Eq. (1) with $v_{ej} = \sqrt{2E_{ej}/M_{ej}}$ gives $v_{ej} \propto \alpha_2 \sqrt{M_{ej}}$ (since E_{ej} is a constant), which might suggest that a more favorable case develops at later times, when the mass behind the shock has increased. Note, however, that the convective shell gets decoupled from the protoneutron star after a few 100 ms and the anisotropy α_2 starts to *decrease* with time by the same relative amount as the square root of the mass enclosed by the supernova shock increases. The estimate given above applies to both the effects of accretion and anisotropic matter ejection during the phase of turbulence in the neutrino-heated zone between protoneutron star and supernova shock. It also limits the recoil velocities that can be obtained from a possible late, anisotropic fallback of clumpy, dense explosion fragments onto the neutron star[28]. Also in this case the momentum of the neutron star is equal to the negative of the net momentum of the ejected stellar material.

Anisotropic neutrino emission from accretion downflows could be another way to transfer momentum to the neutron star. Yet, with very optimistic values for the amount of accreted matter and the extreme assumption that the matter radiates away its total gravitational binding energy in neutrinos instantaneously during the very short, final stage of the infall (and thus with the largest possible anisotropy), this process can be identified[24] as being less important than the asymmetric accretion or ejection of matter, mainly because an energy equivalent of only about 5–10% of the rest mass is emitted in form of neutrinos.

DISCUSSION

The attainable neutron star kicks associated with turbulent processes during the early phase of the supernova explosion as observed in current hydrodynamical models are limited by the stochastic nature of the considered mechanisms and by their short duration or the small mass (energy) involved. Certainly, three-dimensional simulations (instead of two-dimensional with a 180° grid[21,22,24,26] or even with only a 90° grid[27]) have to be performed for a reliable determination of the size of the kick velocities. Yet, the estimates presented above demonstrate that a significantly more violent overturn associated with a larger anisotropy seems to be needed in the region between protoneutron star and supernova shock. Comparison of three-dimensional with two-dimensional simulations of the convective processes *inside* the protoneutron star indicate, however, that in three dimensions the turbulent overturn tends to be less violent rather than stronger, simply because the restriction to two dimensions is associated with an enforced conservation of (specific) angular momentum.

The occurrence of the $l = 1$ mode as claimed by Herant *et al.*[20,25] could not be confirmed by other simulations. In this respect rotation or strong magnetic fields might be of considerable interest and might lead to a more coherent and directed, if not even jet-like, flow pattern. Alternatively, Duncan and Thompson[29] and Thompson and Duncan[23] suggest that in highly magnetized, rapidly rotating neutron stars magnetic fields could affect the neutrino loss by locally suppressing convective energy transport, thus depressing the neutrino flux at the stellar surface and creating "dark" neutron star spots which induce a larger anisotropy of the neutrino emission. A similar effect could also be achieved, even without magnetic fields, by a long-lasting, quasi-stationary convection inside the protoneutron star, below the neutrino sphere[21,22] (compare Eq. (2)). In this context it might be interesting that galactic models for the gamma-ray bursters consider neutron stars with very strong magnetic fields, $B \sim \mathcal{O}(10^{15}\ \mathrm{G})$, as the possible source of the bursts. The neutron stars need to have velocities larger than 600–700 km/s to be able to leave the galactic disk and to populate the outer galactic halo nearly isotropically (Podsiadlowski, Rees, and Ruderman, presented at this conference). A positive correlation of magnetic field strength and pulsar proper motions might nevertheless not be an immediate consequence and might not provide a suitable observational constraint of these ideas, because the magnetic field present in the neutron star at its birth might be reprocessed and possibly amplified within a comparatively short period of the neutron star evolution (Wiebicke and Geppert, presented at this conference; Wiebicke, personal communication).

SUMMARY AND CONCLUSIONS

An increasing number of young pulsars are observed to have high space velocities. This suggests that the acceleration is likely to occur during the formation of the neutron star at the center of a supernova explosion. The current multi-dimensional supernova models were discussed with respect to their potential to explain the measured pulsar velocities. Anisotropic neutrino emission and non-spherical ejection of matter associated with turbulent processes in the nascent neutron star and during the onset of the explosion can yield recoil velocities up to several hundred km/s. Due to the stochastic nature and short duration of these acceleration mechanisms, however, they seem hardly able to account for proper motions close to or even in excess of 1000 km/s.

REFERENCES

1. Lyne, A.G., B. Anderson & M.J. Salter. 1982. Mon. Not. R. Astron. Soc. **201:** 503.
2. Bailes, M., R.N. Manchester, M.J. Kesteven, R.P. Norris & J.E. Reynolds. 1990. Mon. Not. R. Astron. Soc. **247:** 322.
3. Fomalont, E.B., W.M. Goss, A.G. Lyne, R.N. Manchester & K. Justtanont. 1992. Mon. Not. R. Astron. Soc. **258:** 497.
4. Harrison, P.A., A.G. Lyne & B. Anderson. 1993. Mon. Not. R. Astron. Soc. **261:** 113.
5. Lyne, A.G. & F.G. Smith. 1982. Nature **298:** 825.
6. Cordes, J.M. 1986. Astrophys. J. **311:** 183.
7. Lyne, A.G. & D.R. Lorimer. 1994. Nature **369:** 127.
8. Caraveo, P.A. 1993. Astrophys. J. **415:** L111.
9. Cordes, J.M., R.W. Romani & S.C. Lundgren. 1993. Nature **362:** 133.
10. Frail, D.A. & S.R. Kulkarni. 1991. Nature **352:** 785.
11. Stewart, R.T., J.L. Caswell, R.F. Haynes & G.J. Nelson. 1993. Mon. Not. R. Astron. Soc. **261:** 593.
12. Itoh, N. & K. Hiraki. 1994. Astrophys. J. **435:** 784.
13. Anderson, B. & A.G. Lyne. 1983. Nature **303:** 597.
14. Woosley, S.E. 1987. *In* The Origin and Evolution of Neutron Stars, D.J. Helfand & J.-H. Huang, Eds.: p. 255. Kluwer. Dodrecht.
15. Woosley, S.E. & T.A. Weaver. 1992. *In* The Structure and Evolution of Neutron Stars, D. Pines, R. Tamagaki & S. Tsuruta, Eds.: p. 235. Addison-Wesley. Redwood City, California.
16. Burrows, A. & B.A. Fryxell. 1992. Science **258:** 430.
17. Janka, H.-Th. & E. Müller. 1993. *In* Frontiers of Neutrino Astrophysics, Y. Suzuki & K. Nakamura, Eds.: p. 203. Universal Academy Press. Tokyo.

18. Müller, E. 1993. *In* Proc. of the 7th Workshop on Nuclear Astrophysics, Ringberg Castle, Germany, March 22–27, 1993. W. Hillebrandt & E. Müller, Eds. Report MPA/P7: p. 27. Max-Planck-Institut für Astrophysik. Garching.
19. Burrows, A. & B.A. Fryxell. 1993. Astrophys. J. **418:** L33.
20. Herant, M., W. Benz, W.R. Hix, C.L. Fryer & S.A. Colgate. 1994. Astrophys. J. **435:** 339.
21. Janka, H.-Th. & E. Müller. 1995. Phys. Rep. In press.
22. Janka, H.-Th. & E. Müller. 1995. Astron. Astrophys. Submitted.
23. Thompson, C. & R.C. Duncan. 1993. Astrophys. J. **408:** 194.
24. Janka, H.-Th. & E. Müller. 1994. Astron. Astrophys. **290:** 496.
25. Herant, M., W. Benz & S.A. Colgate. 1992. Astrophys. J. **395:** 642.
26. Janka, H.-Th. & E. Müller. 1993. *In* Proceedings of the IAU Colloquium 145, Xian, China, May 24–29, 1993. R. McCray & Wang Zhenru, Eds. Cambridge University Press. Cambridge, New York. MPA-Preprint 748.
27. Burrows, A., J. Hayes & B.A. Fryxell. 1994. Astrophys. J. Submitted.
28. Bethe, H.A. 1993. Astrophys. J. **419:** 197.
29. Duncan, R.C. & C. Thompson. 1992. Astrophys. J. **392:** L9.

A LONG TERM STUDY OF HIGH ENERGY GAMMA RAY EMISSION FROM THE VELA, GEMINGA AND CRAB PULSARS

P. V. RAMANAMURTHY[1,a], D. L. BERTSCH[1], C. E. FICHTEL[1], G. KANBACH[2], D. A. KNIFFEN[3], H. A. MAYER-HASSELWANDER[2], P. L. NOLAN[4], P. SREEKUMAR[1,b] AND D. J. THOMPSON[1]

[1] *NASA/Goddard Space Flight Center, Code 662, Greenbelt, MD 20771, U. S. A.*

[2] *Max-Planck Institut für Extraterrestrische Physik, D-85748 Garching, Germany*

[3] *Hampden-Sydney College, P.O. Box 862, Hampden-Sydney, VA 23943, U. S. A.*

[4] *Hansen Experimental Physics Laboratory, Stanford University, Stanford CA 94305, U. S. A.*

INTRODUCTION

High Energy Gamma Ray (HEGR, hereafter) emission in the energy range 30 MeV - 30 GeV from the various celestial sources has been studied by the Energetic Gamma Ray Experiment Telescope (*EGRET*) aboard the Compton Gamma Ray Observatory since its launch in April 1991. *EGRET* discovered a large number of celestial sources of various types; see Fichtel *et al.* [1] for a catalogue of sources seen during the first 18 months of observation. Among the detected sources are five pulsars: Vela (Kanbach *et al.* [2]), Geminga (Bertsch *et al.* [3]; Mayer-Hasselwander *et al.* [4]), Crab (Nolan *et al.* [5]), PSR B1706-44 (Thompson *et al.* [6]) and PSR B1055-52 (Fierro *et al.* [7]), the first three being much brighter than the latter two.

Theoretical models to explain HEGR production by isolated pulsars fall into two general categories: polar cap models (see Harding & Daugherty[8], and references quoted therein) and outer gap models (see Cheng *et al.* [9a, 9b]; Ho[10]; Chiang & Romani[11] and references quoted therein).

[a] NRC/NASA Senior Research Associate
[b] Universities Space Research Association

In this paper, we report very briefly our results on the long term variability of the three bright *EGRET* pulsars with respect to shape of the light curve, ratio of intensities of emission under the two peaks, flux and spectral index of total emission at $E > 100$ MeV; for details, see Ramanamurthy *et al.* [12].

OBSERVATIONS AND RESULTS

Each of the three pulsars was observed in several separate viewing periods (lasting for a few days to two weeks) spread over approximately three years. For each pulsar, gamma ray light curve consists of two peaks and a bridge region in between the two peaks where there is clear evidence of HEGR emission. The HEGR light curve for each pulsar is distinct and differs in shape from the others. Light curves of a given pulsar from the separate viewing periods were all combined to form a master template light curve. The template is then normalized and compared with light curves from the individual viewing periods. In all cases, there is no evidence for any time variability of the *overall* shapes of the light curves, with some exceptions. It is noticed, however, in the case of the Vela and Crab pulsars, occasionally the emission of high energy gamma rays under one of the two peaks is significantly less than the long term average. This effect seems to be more pronounced at higher energies.

We next consider the variability of the ratio of emission of HEGR under the two peaks (the P2/P1 ratio) in the case of the Crab pulsar over time scales of decades by considering the results from *SAS*-2 , *COS-B* and *EGRET* put together. Earlier, Kanbach[13], based on the *SAS*-2 and *COS-B* data, had given a sinusoidal fit to the ratio as a function of time. We have included the *EGRET* data points from the present analysis and the new fit is given by

$$P2/P1 = 0.92 - 0.5\sin(2\pi(T - 2653)/4907) \qquad (1)$$

where T = JD - 2440000 is the date of observation. The result is shown in Figure 1. Kanbach's and the present fits are shown by the dotted and solid sinusoidal lines respectively. The dashed horizontal line is the weighted mean (0.51) of all the data points. The reduced χ^2 values for the three fits (Kanbach's and present sinusoidal fits and no variability fit) are 1.43, 1.29 and 1.81, resulting in probabilities of 0.13, 0.20 and 0.03 respectively. While equation 1 gives the best fit, the null hypothesis (i.e. no variability) cannot be ruled out with confidence at the moment. More observations in the future may resolve the question.

Time-averaged fluxes of HEGR at $E > 100$ MeV from the Vela, Geminga and Crab pulsars are respectively 8.20, 3.47 and 2.45 in units of 10^{-6} $quanta\text{-}cm^{-2}\,s^{-1}$. In each case, there is some evidence for variability. Fits to the null (i.e. no variability) hypothesis resulted in *reduced* χ^2 values of 3.2 (for

5 D.O.F.), 3.4 (for 7 D.O.F.) and 1.9 (for 9 D.O.F.) for the Vela, Geminga and Crab pulsars respectively.

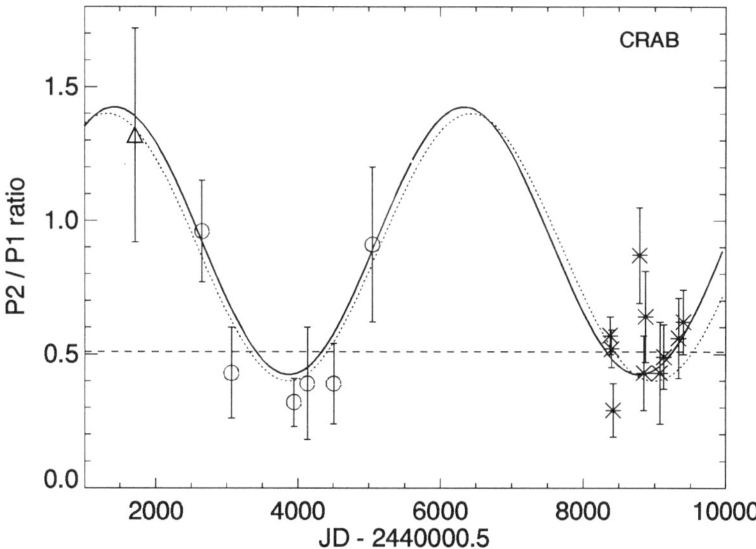

Fig. 1: see the text for details. Data are from *SAS*-2 (triangle), *COS-B* (circles) and *EGRET* (asterisks).

Spectral indices of the energy spectra showed a greater time variability than the integral ($E > 100$ MeV) fluxes from each of the pulsars; see Figure 2. Fits to the null (i.e. no variability) hypothesis resulted in *reduced* χ^2 values of 7.9 (for 5 D.O.F.), 4.9 (for 7 D.O.F.) and 2.1 (for 9 D.O.F.) for the Vela, Geminga and Crab pulsars respectively.

CONCLUSION

When all the results given above are viewed together, they indicate that the emission geometry in the pulsar magnetosphere appears to be constant in time but the $e^{\pm}$ beam intensities at the highest energies and/or the densities of low energy photon fields are variable over time scales of months to years; see Ramanamurthy *et al.* [12] for a more detailed discussion.

REFERENCES

1. Fichtel, C.E. *et al.* 1994, ***ApJS***, 94, 551
2. Kanbach, G. *et al.* 1994, ***A&A***, 289, 855
3. Bertsch, D.L. *et al.* 1992, ***Nature***, 357, 306

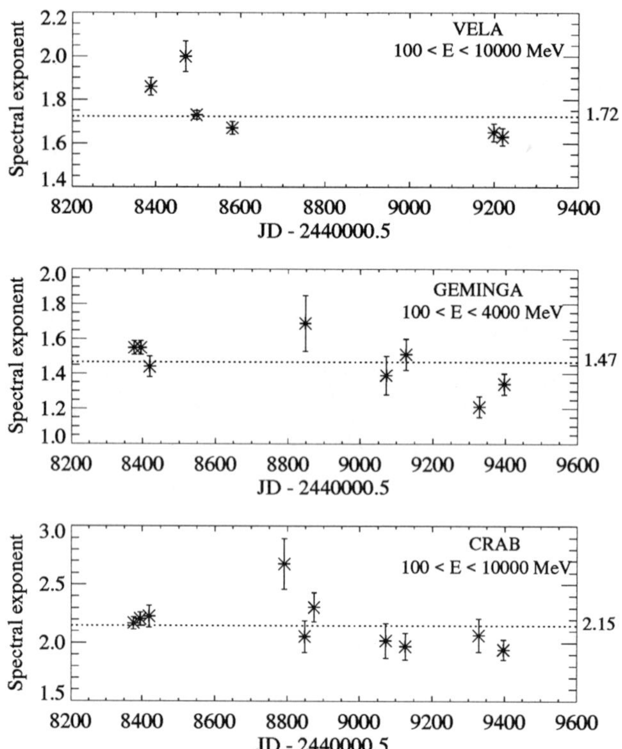

Fig. 2: Variation of gamma ray spectral index with time for the three pulsars. Dotted lines denote the weighted mean indices.

4. Mayer-Hasselwander, H.A. *et al.* 1994, ***ApJ***, 421, 276
5. Nolan, P.L. *et al.* 1993, ***ApJ***, 409, 697
6. Thompson, D.J. *et al.* 1992, ***Nature***, 359, 615
7. Fierro, J.M. *et al.* 1993, ***ApJ***, 413, L27
8. Harding, A.K. & Daugherty, J.K. 1993, in ***Isolated Pulsars***, Cambridge Univ. Press (ed. Van Riper, K.A., Epstein, R., & Ho, C.) , p. 279
9. (a) Cheng, K.S., Ho, C. & Ruderman, M. 1986, ***ApJ***, 300, p. 500 (b) Cheng, K.S., Ho, C. & Ruderman, M. 1986, ***ApJ***, 300, p. 522
10. Ho, C. 1993, in ***Isolated Pulsars***, Cambridge Univ. Press (ed. Van Riper, K.A., Epstein, R., & Ho, C.) , p. 271
11. Chiang, J. & Romani, R.W. 1993, in ***Isolated Pulsars***, Cambridge Univ. Press (ed. Van Riper, K.A., Epstein, R., & Ho, C.) , p. 287
12. Ramanamurthy, P.V. *et al.* , 1995 (submitted to ***ApJ***)
13. Kanbach, G. 1989, in ***Gamma Ray Observatory Science Workshop Proc.***, ed. W.N.Johnson, NASA Goddard Space Flight Center, p. 2-1

CGRO/OSSE Observations Of Pulsars: An Update

M. P. ULMER[a,b], P. C. SCHROEDER[a,b], J. M. CORDES[c], MATTHEW BAILES[d], AND JONATHON BELL[e]

[b]*Department of Physics and Astronomy*
Northwestern University
Evanston, IL 60208-3112

[c]*Astronomy Department*
Space Science Bldg.
Cornell University
Ithaca, NY 14853-6801

[d]*Australia Telescope National Facility*
P.O. Box 76
Epping NSW
2121, Australia

[e]*Mount Stromlo and Siding Spring Observatories*
Australian National University
Private Bag
Weston Creek
2611 ACT, Australia

INTRODUCTION

This is a progress report on *Compton Gamma-Ray Observatory* Oriented Scintillation Spectrometer Experiment (CGRO/OSSE) observations of radio pulsars. We report on observations of the Crab pulsar and PSR B0656+14 and PSR J0437−4715.

RESULTS AND DISCUSSION

Previous reports[1,2] have shown evidence for a ~ 13 year sinusoidal variation in the light curve of the Crab pulsar. The variation was quantified as the ratio of the intensity of the two peaks in the light curve, designated as P2/P1, where P2 is the intensity of the second peak and P1 the intensity of the first peak. The model that has been suggested to explain this variation is one in which the pulsar precesses and, as the view from Earth of the gamma-ray beam changes, so does the ratio of P2 to P1. The variations in different energy bands might track, although it is possible to evoke a more complicated model in which the beam profile width perpendicular to the plane scanning the Earth is different in different energy ranges. Our new data (taken over the energy range 50–200 keV, which is close enough to the previous 60–340 keV data in energy given that the low energy flux dominates) show, however, that the effect does not seem to be real, as can be seen in Figure 1. Most likely the high points in Figure 1 are simply statistical fluctuations as all the high points coincidently have large error bars.

[a]For the OSSE Team

We suggest that the apparent sinusoidal variation is not real, and note that another observation within a year should answer the question for our energy range. As noted above, however, this does not preclude an effect taking place at higher energies if the beam widths perpendicular to the plane vary with energy.

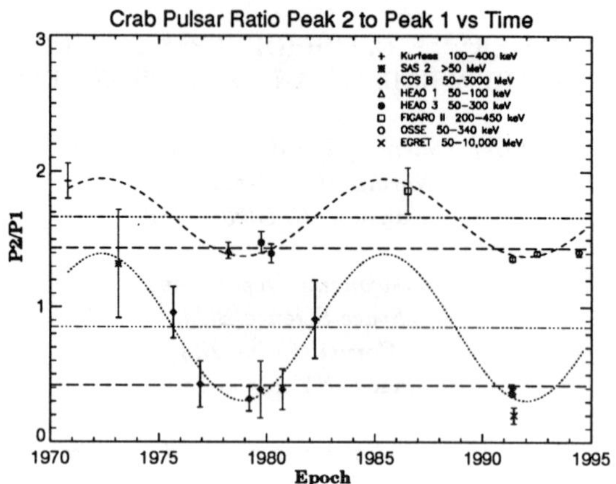

Figure 1: The ratio of the intensity of Peak 2/Peak 1 versus time at two different energies. (References for the data are: Kurfess[3]; Knight[4]; Mahoney, Ling, & Jacobson[5]; Agrinier et al.[6]; Thompson et al.[7]; Clear et al.[8]; and Nolan et al.[2]). The dot/dash horizontal lines represent the average of the sine wave fits, and the other horizontal lines are the best fit values based on the assumption of no variation in the ratio of P2/P1.

Keep in mind that we do not know the width of the beam of the pulsar perpendicular to the plane passing through the Earth as we turn to a discussion of the two other pulsars, PSR B0656+14 and PSR J0437−4715. We did not detect either of these pulsars in the 50–180 keV range, and the 2 sigma upper limits are 5.5×10^{-13} and 2.4×10^{-12} ergs cm^{-2} s^{-1}, where we have assumed a $1/E^2$ photon spectral shape and a 50% duty cycle. Before comparing with observations of other pulsars, we remark that for some comparisons it is necessary to either extend the hard X-ray spectrum and/or extrapolate the detected 100 MeV to 1 GeV spectrum down to the upper limit previously determined by OSSE. We have done this when we consider the total energy loss (Figure 2 below), but we have confined ourselves to the hard X-ray regions ($\sim$ 50–180 keV) when comparing with the soft X-ray (about 0.25–2.5 keV) flux in Figure 3 below.

We display our new upper limits in Figure 2 where we see that these two pulsars belie the apparent trend that the lower the energy loss for a pulsar, the more efficient the pulsar becomes in converting the energy loss to gamma rays. Note, however, the efficiency was calculated by assuming that all pulsars have the same width beam perpendicular to the plane passing through the earth. More likely, however, is the possibility that the widths are not all the same and it could very well be that pulsars which are losing energy more slowly have narrow beams such that their efficiency in converting energy loss into gamma-ray emission is not nearly as high as inferred in Figure 2. Since there are so many of these objects, the probability of the gamma-ray beam sweeping over the Earth from some pulsar is rather high and hence we detect some of these objects. The number of detections is too low, however, to make a quantitative calculation

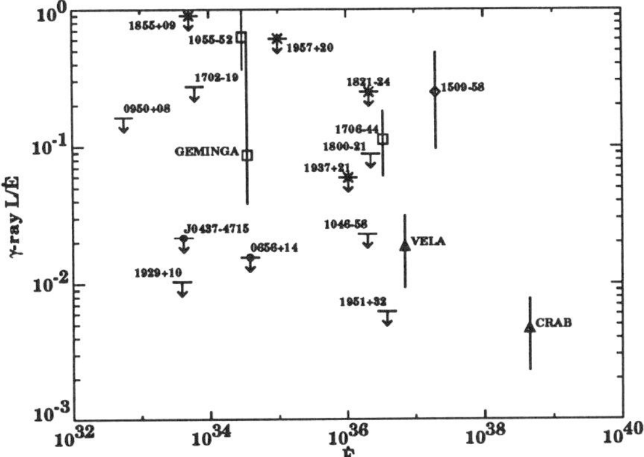

Figure 2: The gamma-ray production efficiency versus rate of energy loss. Upper limits from Schroeder et al.[9] are plotted only for those pulsars having luminosity limits less than their rotational energy loss rates. The diamonds represent pulsars which have been detected by OSSE, for which the luminosities were calculated for the energy range 0.1 MeV to 1 GeV. The squares represent pulsars which have been detected by EGRET, for which the luminosities were calculated for the energy range 1 MeV to 1 GeV. The triangles represent pulsars which have been detected by both OSSE and EGRET, for which the luminosities are determined for the energy range 0.1 MeV to 1 GeV. The new upper limits reported here are denoted with filled circles.

to derive a beam width versus energy loss rate. It may simply be, therefore, chance that determines whether or not a pulsar is detectable in the gamma-ray/hard X-ray region. In this regard, we point out that the beaming direction of the gamma-rays versus the radio emission of pulsars is another unknown that could account for some or all of the scatter in Figure 2.

Other potential predictors of a detectable gamma-ray flux are: (a) the ROSAT detections of a soft X-ray flux; and, (b) the energy loss divided by the square of the distance to the object ($\dot{E}/d^2$). In Figure 3 we see that, based on the ROSAT flux, it was very unlikely that we would have J0437−4715 and that B0656+14 should have been just detectable. Conversely, even if these pulsars were relatively inefficient (at the $\sim 3\%$ level) converters of energy loss into gamma-ray flux, we would have expected to detect these two pulsars. (For brevity we do not show a figure.) This suggests that for radio pulsars there is more to having a detectable gamma-ray flux than simply the energy loss and the distance from Earth.

Another observable that could be associated with a detectable gamma-ray flux is the shape of the radio pulse. We show in Figure 4 that these two pulsars are both single-pulsed, low-duty cycle pulsars. Note also the complex pulse shape of J0437−4715. The number of detections is so small, however, that it is premature to conclude anything definitive about pulsar shapes versus gamma-ray emission.

SUMMARY

In summary, then, the reason why only some pulsars are easily detectable in the

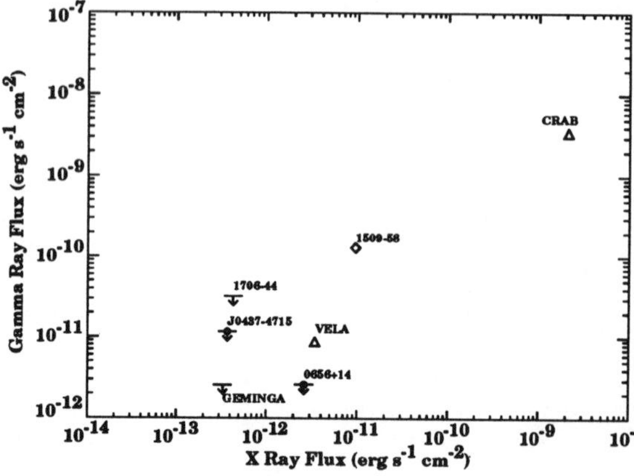

Figure 3: The ROSAT X-ray fluxes from Ögelman[10] versus the 50–180 keV detections and upper limits from OSSE. The triangles denote OSSE and EGRET detections, and the diamond denotes solely an OSSE detection. The OSSE (50–180 keV energy range) upper limits are from Schroeder et al.[9] and this work.

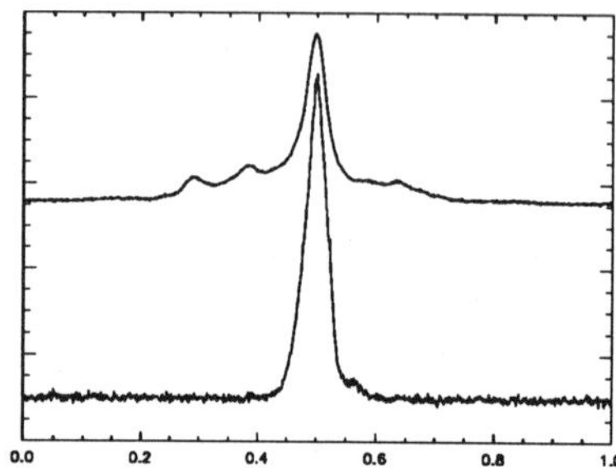

Figure 4: The radio light curves of PSR J0437−4715 (top) and PSR B0656+14 (bottom). The vertical scale is the pulsar radio intensity in arbitrary units. The horizontal scale is the phase of the intensity. The phase 0.5 has arbitrarily been chosen for the peak intensity.

hard X-ray or gamma-ray region remains a mystery. As the effect could be due to the pointing angle of the pulsar and not directly related to the soft X-ray flux or to the energy loss and distance to these objects, it behooves us to continue to expand our search for gamma-ray emission from radio pulsars. As OSSE has a narrow field of view, covering many pulsars in optimal sensitivity is a slow and time consuming process.

REFERENCES

1. Ulmer, M. P. 1994. Ap.J. Supp. **90**: 789–795.
2. Nolan, P. L., *et al.* 1993. Ap.J. **409**: 697.
3. Kurfess, J. D. 1971. Ap.J. **168**: L39.
4. Knight, F. K. 1982. Ap.J. **260**: 538.
5. Mahoney, W. A., J. C. Ling, & A. S. Jacobson. 1984. Ap.J. **278**: 784.
6. Agrinier, B., *et al.* 1990. Ap.J. **355**: 645.
7. Thompson, D. J., C. E. Fichtel, R. C. Hartman, D. A. Kniffen, & R. C. Lamb. 1977. Ap.J. **213**: 252.
8. Clear, J., K. Bennett, R. Buccheri, I. A. Grenier, W. Hermsen, H. A. Mayer-Hasselwander, & B. Sacco. 1987. A&A **174**: 85.
9. Schroeder, P. C., *et al.* 1995. Ap.J. Submitted.
10. Ögelman, H. 1995. *In* The Lives of the Neutron Stars. M. Alpar, Ü. Kiziloğlu, & J. van Paradijs, Eds.: 101–120. Kluwer Academic Publishers. Dordrecht.

Gamma-ray bursts and the structure of the Galactic halo

PHILIPP PODSIADLOWSKI[1], MARTIN J REES[1]
AND MALVIN RUDERMAN[2]

[1] *Institute of Astronomy*
Madingley Road,
Cambridge CB3 0HA

[2] *Department of Physics, Columbia University,*
538 West 120 Street, NY 10027

ABSTRACT

We have systematically investigated Galactic models for classical gamma-ray bursts (GRBs) in which gamma-ray bursters are ejected from the Galactic disk. It is relatively easy to construct models in which the predicted GRB sky distribution is isotropic and consistent with the present BATSE data if the outer halo is not spherically symmetric. The bursters must receive a velocity kick larger than $\sim 600-700\,\mathrm{km\,s^{-1}}$ at birth and remain active bursters for at least 10^9 yr. We suggest strongly magnetized neutron stars with magnetic fields of order 10^{15} G as possible sources of classical GRBs.

1 ISOTROPY OF A LONG-LIVED HIGH-VELOCITY HALO POPULATION

It may be premature to rule out Galactic models for the origin of classical GRBs. Indeed, a number of Galactic models have been constructed in which gamma-ray bursters are neutron stars escaping from the Galaxy (Li & Dermer 1992), two-population models (Lingenfelter & Higdon 1992; see, however, Paczynski 1992) or models in which bursters form an old halo population (Brainerd 1992; Salpeter & Wasserman 1993; Woosley 1993, Hartmann *et al.*1994b). Halo models have, however, fallen from favour because the observed isotropy seems to require significant fine-tuning of the models: see Hartmann (1994) for a thorough review.

We have investigated (Podsiadlowski *et al.*1995) halo models for GRBs in which bursters are born in the Galactic disk and are injected into the halo with a kick velocity received at birth (Shklovskii & Mitrofanov 1985). Our calculations are, in spirit, similar to those of Paczynski (1990) and Hartmann, Epstein & Woosley (1990); these earlier authors, however, used lower kick velocities, and focussed attention on cases where the bursts came from within a few kpc. It is relatively easy to construct halo models for GRBs if the halo is non-spherical (i.e. without requiring serious fine-tuning). However, a typical model also predicts some small anisotropies in the GRB sky distribution which reflect the asymmetries in the halo potential. This provides a potential test for verifying such models.

2 PHYSICS OF THE BURSTS

These results strengthen the motivation for investigating whether it would be physically plausible for neutron stars to generate bursts with the fluence, and at the rate, implied by the observations. There are three requirements.

(i) If a fraction f of neutron stars (those born with high velocities) are responsible for the bursts, then the ratio of the current burst rate to the estimated pulsar birthrate implies that each object must, over its lifetime, produce $10^6\,(f/0.01)^{-1}$ bursts (for a total Galactic population of $\sim 10^9$ neutron stars).

(ii) The energy must be released into a succession of bursts with characteristic energy $\sim 10^{41} - 10^{42}$ erg. This estimate assumes a characteristic fluence per burst of $10^{-6}\,\mathrm{erg\,cm^{-2}}$ (Fishman et al. 1994) and a characteristic burster distance of 50 – 100 kpc. Each burster must, therefore, be able to convert a total energy of $\sim 10^{47}\,(f/0.01)^{-1}$ erg into gamma-rays. If each burst were beamed, but in a random direction, then the energy per burst would go down, but the number of bursts would go up, leaving the overall energetic requirements unchanged.

(iii) The energy release must be spread over a timescale longer than $\sim 10^9$ yr. This means that each object must, on average burst about once every thousand years. However, the bursts from each star may not be independent 'poisson' events, so the detection of a series of repeated bursts from the same object, need not be incompatible with the model.

The stored magnetic energy inside the star can be substantially higher than inside, since the internal field is not uniform, but is concentrated in tubes of strength $B_c = 3 \times 10^{15}$ G occupying a fraction B/B_c of the volume.

The magnetic energy is

$$E_B \simeq \frac{BB_c}{8\pi}\frac{4\pi}{3}R^3 \simeq 3 \times 10^{47}\left(\frac{B}{6 \times 10^{14}\,\mathrm{G}}\right)\mathrm{erg}, \qquad (\text{for } B \leq B_c), \qquad (1)$$

for $R = 10$ km. For $B > B_c$, E_B is proportional to B^2, and for $B \gtrsim 10^{16}$ G superconductivity is quenched. Thus, for fields of order 10^{15} G, a neutron star can store enough energy to fulfil requirements (i) and (ii).

The answer the first part of question (ii) above — whether this energy can be released in appropriate 'quanta' to power a burst — depends on the properties of the neutron star's crust when it becomes stressed beyond its yield strength by changes in the core magnetic field (Ruderman 1991a). The core magnetic flux tubes move toward the core surface. There is a growing stress at the ends where they are pinned to the crust. During spindown, they drift outwards because they are linked to the outward movement of the superfluid neutron vortex array (Srinivasan et al. 1990; Ruderman 1991b). This rapid flux adjustment may be relevant to the frequent repetitions of soft repeaters, if these are indeed young highly-magnetised neutron stars (Duncan & Thompson 1992; Ruderman & Tavani 1994). But even if spin-down alone were to become less effective in moving flux tubes in old slowly rotating stars, flux tubes can still be pushed up to the base of the crust by their buoyancy. The key issues relevant to bursts are how much stress can build up in the crust, and what fraction of the stored stress is released in the 'quake' whereby the strain in the crust is reduced. The maximum stored energy is

$$E_s = \left(\frac{\mu\theta^2_{\rm max}}{2}\right) 4\pi\, R^2 d, \qquad (2)$$

where $d \sim 10^5$ cm is the crust thickness, μ the shear modulus ($\sim 3 \times 10^{29}$ dyne cm^{-2}), and $\theta_{\rm max}$ the maximum strain before breaking. (The likely value of $\theta_{\rm max}$ is 10^{-2}.) The maximum storable elastic stress energy in the crust is then about 2×10^{43} erg. However, a sudden stress-induced crust cracking would reduce the strain θ only by a much smaller $\Delta\theta$. If $\Delta\theta \simeq 10^{-4}$, crust-cracking events release $\sim 2(\Delta\theta/\theta_{\rm max})\, E_s \sim 4 \times 10^{41}$ erg. This estimate is consistent with the energy requirements for individual bursts provided that the conversion to gamma rays is very efficient (cf Blaes et al. 1989), and the internal field would indeed energise up to a million events of this type in each star.

The third question is the timescale over which the magnetic energy is released – can this be as long as 10^9 yr (cf (iii) above)? The effective electron viscosity which can limit flux tube response to moving neutron vortex arrays and to buoyancy forces is still poorly understood. For a single flux tube, Jones (1987) estimates $t_b = 8 \times 10^5$ yr; he notes that a dense array of flux tubes could take longer (maybe 10^7 yr), and that the time scales could be 10^9 yr if there were a non-superconducting baryonic component near the centre. Jones does not include the effect of proton superconductivity in screening the electron currents, which may reduce t_b. On the other hand, if the fields were as high as 10^{16} G, superconductivity would be quenched.

All we can claim is that a timescale in the range $10^9 - 10^{10}$ yr is not implausible. If the timescale were indeed in this range, we could then take seriously (cf (i) and (ii) above) the hypothesis that GRBs could come from a population of fast-moving neutron stars, with ultrastrong magnetic fields, originating in the Galactic disk. For these neutron stars to be the sources of the GRBs detected by BATSE, there would also have to be a strong correlation between being born with a huge magnetic field and acquiring a birth velocity well above average for radiopulsars. A variety of possible mechanisms for this have been suggested by Duncan and Thompson (1992).

3 CONCLUSIONS

Hakkila et al. (1994b) have shown that the isotropy of the bursts is hard to reconcile with any spherical halo distribution, where the halo is modelled as an isothermal sphere with a well-defined core radius. We found (Podsiadlowski *et al.*1995) perhaps surprisingly, that the constraints are eased when the trajectories are calculated in a more realistic non-spherical potential. On the simple (and indeed conservative) assumption that pulsars have formed at their present rate for the last few billion years, and come only from the disc, the bursts could be attributed to a fast-moving ($v \gtrsim 600-700\,\mathrm{km\,s^{-1}}$) subset of them. The considerations outlined here may significantly reduce the problems confronting a halo interpretation of classical gamma-ray bursts.

REFERENCES

Blaes O., Blandford R., Goldreich P., Madau P., 1989, ApJ, 343, 839
Brainerd J. J., 1992, Nat, 355, 522
Duncan R. C., Thompson C., 1992, ApJ, 329, L9
Fishman G. J. et al., 1994, ApJS, 92, 229
Hartmann D. H., 1994, in Fishman G. J., Brainerd J. J., Hurley K., eds, Gamma-Ray Bursts, Second Huntsville Workshop. AIP Press, New York, p. 562
Hartmann, D.H. *et al.*, 1994b, ApJS, 90, 893
Hakkila J. et al., 1994a, in Fishman G. J., Brainerd J. J., Hurley K., eds, Gamma-Ray Bursts, Second Huntsville Workshop. AIP Press, New York, p. 59
Hakkila J. et al., 1994b, ApJ, 422, 659
Jones P., 1987, MNRAS, 228, 513
Li H., Dermer C. D., 1992, Nat, 359, 514
Lingenfelter R. E., Higdon J. C., 1992, Nat, 356, 132
Paczynski, B., 1990, ApJ, 348, 485
Paczynski, B., 1992, Acta Astronomica. 42, 1
Podsiadlowski, P., Rees, M.J., Ruderman, M. MNRAS in press
Ruderman M., 1991a, ApJ, 382, 587
Ruderman M., 1991b, ApJ, 382, 576
Ryan J. M. et al., 1994, IAU Circ., 5950
Salpeter E. E., Wasserman I., 1993, in Phillips J. A., Thorsett S. E., Kulkarni S. R., eds, Planets around Pulsars. ASP Conference Series, San Francisco, p. 345
Shklovskii I. S., Mitrofanov I. G., 1985, MNRAS, 212, 545
Srinivasan G., Bhattacharya D., Muslimov A., Asygan A., 1990, Current Sci., 59, 31
Woosley S., 1993, in Phillips J. A., Thorsett S. E., Kulkarni S. R., eds, Planets around Pulsars. ASP Conference Series, San Francisco, p. 355

Accretion and Magnetic Field Decay in Neutron Stars Entering Binaries

U. GEPPERT[1)] and V. URPIN[1,2)]

1) MPI für Extraterrestrische Physik, Aussenstelle Berlin
Rudower Chaussee 5, D-12489 Berlin, Germany
2) A.F.Ioffe Institute, 194021 St.Petersburg, Russia

1. Introduction

The low magnetic fields of many binary pulsars strongly support the idea of magnetic field decay in neutron stars. Most of these pulsars are believed to be very old neutron stars passed through an accretion phase in binaries. For some binary systems, an accretion phase lasts as long as $10^7 - 10^8$ yr and the total amount of the accreted mass may reach $0.01 - 0.1 M_\odot$ (see, e.g., [1], [2]). A heavy mass transfer in these systems can play a crucial role in the evolution of the magnetic fields of neutron stars.

Recently Geppert & Urpin [3] have suggested a simple mechanism of an accelerated field decay in neutron stars undergoing accretion. Accretion can drastically change the thermal evolution of a neutron star if the accretion rate is sufficiently high and the mass transfer phase lasts a sufficiently long period (see, [4], [5]). Accretion heats the neutron star and decreases the crustal electrical conductivity since the latter depends on the temperature. A decrease of the conductivity accelerates the decay of the magnetic field if this field is supported by currents in the crust.

The present paper considers in more detail the influences of accretion rate and total amount of accreted mass on the final field strength at the end of the accretion phase. The suggested mechanism of the field decay may account for weak magnetic fields of many pulsars undergoing accretion in binary systems.

2. Description of the model

We consider the evolution of the magnetic field which occupies some fraction of the neutron star crust before a mass transfer phase. It has been argued (see [4], [5]) that accretion heating results in a drastical change of the thermal structure of neutron stars. The inward directed heat flux can rapidly heat the interior of a neutron star with standard *npe*-matter in the core since neutrino emission does not provide an effective mechanism of energy loss in this case. The characteristic time scale of this heating is probably of the order of $10^3 - 10^5$ yr. In the steady state, the crust is practically isothermal, with the exception of the very surface layers with densities $\rho \leq 10^6$ g/cm^3. We will address the field decay during this steady state phase.

Neglecting the anisotropy of conductivity, the induction equation governing the evolution of the magnetic field reads

$$\frac{\partial \vec{B}}{\partial t} = -\frac{c^2}{4\pi} \nabla \times \left(\frac{1}{\sigma} \nabla \times \vec{B} \right) + \nabla \times \left(\vec{v} \times \vec{B} \right), \tag{1}$$

where σ is the conductivity and $\vec{v}$ is the hydrodynamic velocity which is caused in the crust by the accreted mass flux.

In the neutron star crust, the most important scattering mechanisms of electrons are probably scattering on phonons and impurities. Scattering on phonons dominates the transport processes at not very high densities, whereas scattering on impurities gives the main contribution to the conductivity in deep crustal layers [6]. The total conductivity of the crystallized region is given by $\sigma^{-1} = \sigma_{ph}^{-1} + \sigma_{imp}^{-1}$, where σ_{ph} and σ_{imp} are the conductivities due to electron-phonon and electron-impurity scattering mechanisms, respectively. The quantity σ_{ph} depends on the temperature and density whereas σ_{imp} is independent of the temperature but depends on the impurity parameter Q characterizing the number density and charge of impurities.

The hydrodynamic velocity $\vec{v}$ entering equation (1) is caused by the flux of the accreted matter through the crust. This flow is certainly non–spherically symmetric in the surface layers where it can be strongly modulated by the magnetic field. In layers with a high density, however, deviations from the spherical symmetry are probably less pronounced and can be neglected. Then, one has from mass conservation $v = \dot{M}/4\pi r^2\rho$.

The value of the crustal temperature is strongly dependent on the accretion rate $\dot{M}$. In our computations, we used the dependence of the temperature on $\dot{M}$ obtained in the papers [4] and [5] for the accretion rate range $3 \times 10^{-10} M_{\odot}/\text{yr} \geq \dot{M} \geq 10^{-11} M_{\odot}/\text{yr}$.

3. Numerical results and discussion

We solved equation (1) with the corresponding boundary conditions numerically for various values of the accretion rates and durations of the accretion phase. It should be mentioned that the crustal field decay depends on the depth penetrated by the field before the accretion phase but is not sensible to the specific form of the initial field distribution (see [7]). In the present paper, the results are given for the initial depth corresponding to the density $\rho_0 = 10^{13}$ g/cm^3.

Fig.1 displays the time dependence of the surface magnetic field for different durations of the accretion phase. The field of a neutron star undergoing strong accretion decays much more rapidly in comparison with the field of an isolated star. So, the field of a cool non-accreting neutron star with impurity dominating the crustal conductivity with $Q = 0.01$ can be reduced only by a factor $3 - 4$ after 10^7 yr. The field decay in a neutron star entering a binary system is rapid during the accretion phase, but it slows down significantly when accretion stops. With the exception of a comparatively short initial stage, the field decay during the accretion phase can be described with high accuracy by a power law, $B \propto t^{-\gamma}$, with $\gamma \approx 0.8$.

If the accretion rate is sufficiently high and the accretion phase lasts sufficiently long, the surface field strength may be reduced by a few orders of magnitude. After accretion stops, the neutron star cools down due to both neutrino emission from the core and photon luminocity from the surface. When the neutron star becomes cool, the conductivity is mainly determined by the impurity scattering and may reach very high values depending on the parameter Q. At this stage, the field decay is as slow as in isolated neutron stars.

In Fig.2, the surface field at the end of the accretion phase is plotted against the total amount of accreted mass, ΔM, for different values of the accretion rate. It turns out that the final strength of the magnetic field, which is reached due to mass transfer, depends on both ΔM and accretion rate. At the same accretion rate, the magnetic field is obviously weaker for the neutron star that accreted more mass. This is because one needs a longer time at the same accretion rate in order

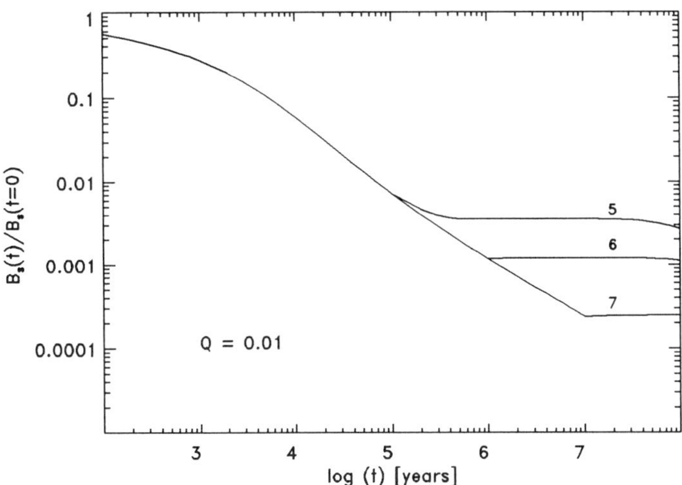

Figure 1: The time dependence of the surface magnetic field normalized to its initial value for different durations of the accretion phase and for $Q = 0.01$. The accretion rate is $\dot{M} = 3 \times 10^{-10} M_\odot/\text{yr}$. The field decay is shown for the initial densities penetrated by the field $\rho_0 = 10^{13}$ g/cm^3. Numbers near the curves indicate the logarithm of the duration of the accretion phase.

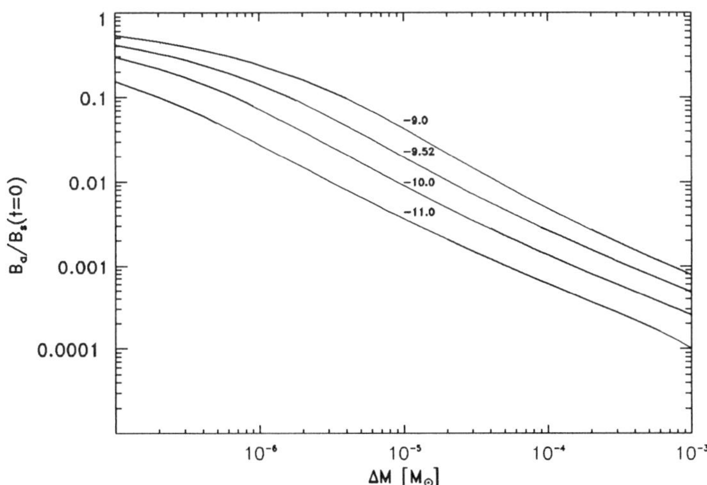

Figure 2: The dependence of the surface magnetic field at the end of the accretion phase, B_a, on the total amount of accreted mass, ΔM, for $\rho_0 = 10^{13}$ g/cm^3. The field is normalized to its strength before accretion starts. Numbers near the curves indicate the logarithm of the accretion rate in units $M_\odot/\text{yr}$.

to accrete more mass and, hence, the field may decay more significantly during this longer time. The computed dependence on ΔM may be fitted by a simple power law, $B \propto \Delta M^{-\gamma}$ with $\gamma \approx 0.8$, in a wide range of ΔM. The only exception is the region of a low accreted mass ($\Delta M < 10^{-6} M_{\odot}$) where the dependence is weaker. This simple and universal dependence follows directly from the time dependence of the magnetic field. As it was mentioned, the surface field is approximately proportional to $t^{-\gamma}$ during the accretion phase. Since at the end of this phase $t = \Delta M/\dot{M}$, we see that the final field strength should be proportional to $\Delta M^{-\gamma}$.

Regarding to the dependence on $\dot{M}$, it should be noted that an increase of $\dot{M}$ at a fixed total amount of the accreted mass results in a stronger field at the end of the accretion phase. The accretion rate influences the magnetic field in two ways. Firstly, at a fixed ΔM, an increase of $\dot{M}$ shortens a duration of the accretion phase and, hence, decreases a period of a rapid field decay. Secondly, an increase of $\dot{M}$ heats the crust to a higher temperature decreasing its conductivity, and, hence, it results in an acceleration of the accretion driven decay. The first effect is more pronounced and, therefore, the field strength after the accretion phase is greater for higher $\dot{M}$ at a fixed ΔM.

Thus, our computations clearly indicate that the neutron star magnetic fields after the accretion phase can be reduced by a factor $10^3 - 10^4$ in a comparison with the field of an isolated neutron star if the accretion phase is sufficiently extended and the accretion rate is sufficiently high. The suggested mechanism can be responsible for weak magnetic fields of millisecond pulsars and other pulsars which have undergone accretion with a high accretion rate.

References

1. Bhattacharya D. & van den Heuvel E.P.J. 1991, Phys.Rep., **203**, 1
2. Chanmugam G. 1992, ARA&A, **30**, 143
3. Geppert U. & Urpin V. 1994, MNRAS, **271**, 490
4. Fujimoto M.Y., Hanawa T., Icko Iben Jr. & Richardson M.B. 1984, ApJ, **278**, 813
5. Miralda-Escude J., Haensel P.& Paczynski B. 1990, ApJ, **362**, 572
6. Yakovlev D.G. & Urpin V.A. 1980, Sov. Astron., **24**,303
7. Urpin V., Chanmugam G. & Yeming Sang. 1994, ApJ, **433**, 780

Acknowledgements. One of the authors (V.U.) gratefully acknowledges the financial support under 94-02-06540 Grant of Russian Foundation of Fundamental Research, A-03-012 Grant within the ESO C&EE Programme and Research Grant of American Astronomical Society. This research was supported by the Deutsche Agentur für Raumfahrtangelegenheiten (DARA) GmbH under contract 50 QO 9201.

Anisotropic cooling and atmospheric radiation of neutron stars with strong magnetic field*

Yu. A. SHIBANOV[a], G. G. PAVLOV[b], V. E. ZAVLIN[c], L. QIN[d], and S. TSURUTA[d]

[a]*Ioffe Institute of Physics and Technology*
194021, St. Petersburg, Russia
[b]*Pennsylvania State University*
University Park, PA 16802, USA
[c]*MPI für Extraterrestrische Physik*
D-85740 Garching, Germany
[d]*Montana State University*
Bozeman, MT 59717, USA

INTRODUCTION

Soft-X-ray thermal-like radiation have been recently observed with *ROSAT* from at least four neutron stars (PSR 0833-45, PSR 1055-52, PSR 0656+14 and Geminga)[1]. A soft ($T \sim (3-10) \times 10^5$ K) component of the observed spectra is most probably emitted from all the visible neutron star (NS) surface. It pulses with the pulsar period, and the pulsed fraction, $\sim 10-30\%$, varies with photon energy. The spectra and light curves of these middle-age (10^4-10^6 yr) NSs contain important information on thermal evolution of NSs and on properties of superdense matter in their interiors.

An adequate analysis of thermal NS radiation should be based on theoretical models which are to include (i) nonuniformity of the surface magnetic field; (ii) anisotropy of the surface temperature distribution caused by anisotropic heat transport in magnetized NS envelopes[2,3]; (iii) strong anisotropy of radiative transfer in magnetic NS atmospheres and its effect on angular distribution and spectra of the emitted radiation[4]; (iv) gravitational redshift and deflection of trajectories (lensing) of emitted photons[3]. Some of these factors have been taken into account, in a simplified manner, in[3,5]. The results we are presenting here include all the ingredients (i)–(iv). We considered both dipolar and uniform distributions of the magnetic field, calculated corresponding distributions of the surface (effective) temperature for NSs of different ages, and used these temperatures as input parameters of local magnetic atmosphere models. The local specific spectral fluxes of the emerging radiation were further integrated over the visible NS surface with allowance for the gravitational effects to obtain the light curves and spectra observed by a distant observer. The presented results were obtained for a NS of mass $M = 1.4M_\odot$, radius $R = 10.7$ km, and the magnetic field at the pole $B_p = 1 \times 10^{12}$ G.

*This work was partialy supported by Russian Foundation of Fundamental Researches (grant 93-02-2916), by ESO C&EE (grant A-01-068), by ISF (grant R6A000), by NASA grants NAG5-2807 and NAGW-2208, and by Visiting Program of MPE.

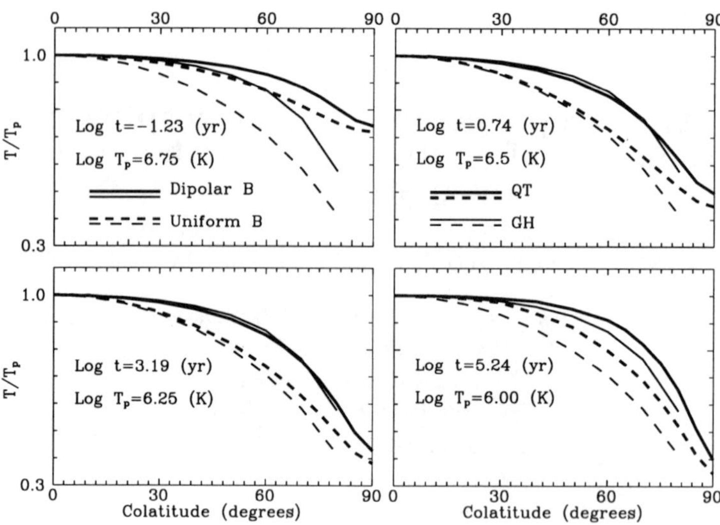

Figure 1: Surface temperature distributions for NSs of different ages with dipolar (solid) and uniform (dashed) magnetic fields. Thick lines (QT) show the results by Qin and Tsuruta[6]; thin lines (GH) correspond to the approximation $T/T_p = (\cos\theta_B)^{1/2}$.

RESULTS AND DISCUSION

Computations of NS cooling were carried out with the code described in[6]. Thermal conduction in the thin (a few meters) NS crust, which is responsible for anisotropy of the surface temperature distribution (Fig. 1), was considered in one-dimentional approximation (radial heat transport). The polar temperature T_p is practically the same as in the nonmagnetic case since the heat conductivity (determined mainly by electrons) along the magnetic field depends on B only slightly whereas the transverse conductivity is strongly suppressed due to rapid Larmor rotation of electrons[7]. Therefore the equatorial temperature is lower by a factor of 1.5 – 2.5 than the polar one. The difference is smaller for younger and hotter NSs since the suppression is less efficient at high temperatures. The temperature profiles for the uniform magnetic field are steeper than for the dipole one because the orientation of the latter is closer to the radial direction at intermediate polar angles θ. For comparison, simplified temperature profiles, $T/T_p = (\cos\theta_B)^{1/2}$, are shown ($\theta_B$ is the angle between the radial direction and the magnetic field). This approximate profile, which can be obtained assuming the transverse conductivity to be zero (cf.[2]), is fairly close to the computed profiles, for stars older than a few years, everywhere except for a narrow ($\sim 20^\circ$) belt near the magnetic equator. It should be noted that the one-dimentional approximation gives an upper limit to the temperature anisotropy which can be reduced by tangential heat transport or by meridional circulation.

The computed surface temperature distributions provide us with the effective temperature at a given surface element needed to compute the local specific intensity of radiation outgoing from the NS atmosphere[4]. Due to anisotropy of radiative transfer in strong magnetic fields, the angular distribution of the local emission has

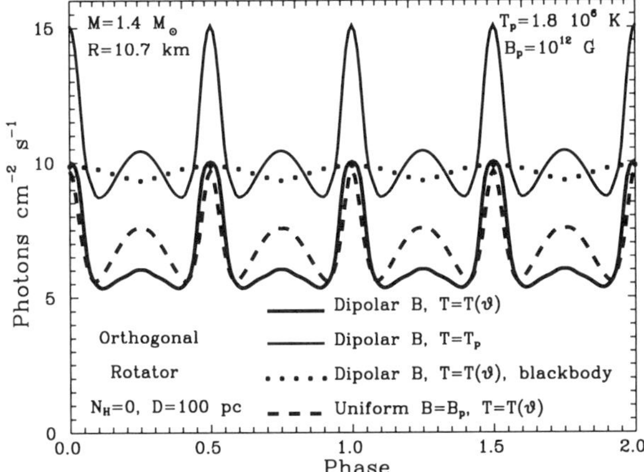

Figure 2: Energy-integrated light curves for rotating NS of the age 1.5×10^3 yr with different magnetic field and temperature distributions (from the left-bottom panel of Fig. 1). The NS rotation axis is perpendicular to both the magnetic axis and line of sight. For the dipolar field two more curves are added: for the atmospheric radiation with the isotropic temperature distribution, $T = T_p$, and for the blackbody-radiating surface with the anisotropic temperature distribution.

a high narrow peak along the field and a broad fan-like structure across the field. The local emission anisotropy causes pulsations of radiation from a rotating NS even for an isotropic temperature distribution[5].

The observable spectra and light curves are obtained by integration of the local fluxes over the visible NS surface with allowance for gravitational lensing. Similarity of the curves for the isotropic and anisotropic surface temperature distributions (see Fig. 2) means that their shapes are mainly determined by the anisotropic angular distribution of the local radiative fluxes emerging from the magnetic atmosphere. The reason is that the temperature anisotropy itself cannot produce substantial pulsed fraction because the main contribution comes from wide hotter polar regions which are seen at virtually any phase due to gravitational lensing. This conclusion is confirmed by the light curve computed under assumption that the NS surface radiates as a blackbody with temperature depending on θ (dotted curve in Fig. 2). To provide the observed modulation without allowance for the flux anisotropy, much smaller (and hotter) polar caps would be needed[8]. Replacement of the dipolar magnetic field with the uniform field does not affect essentially the pulsed fraction and the pulse shape, although the interpulses (secodary maxima at the phases $0.25, 0.75, \ldots$, corresponding to the transverse orientation of the magnetic axis to the line of sight) are stronger for the uniform field. The interpulses are due to the fan-like component of the emission from the polar regions enhanced by gravitaional lensing. Generally, the light curves for magnetic atmospheres with anisotropic temperature distributions are qualitatively similar to those obtained for the model of magnetic hot polar caps[9]. The corresponding light curves for different *ROSAT* PSPC energy bands are

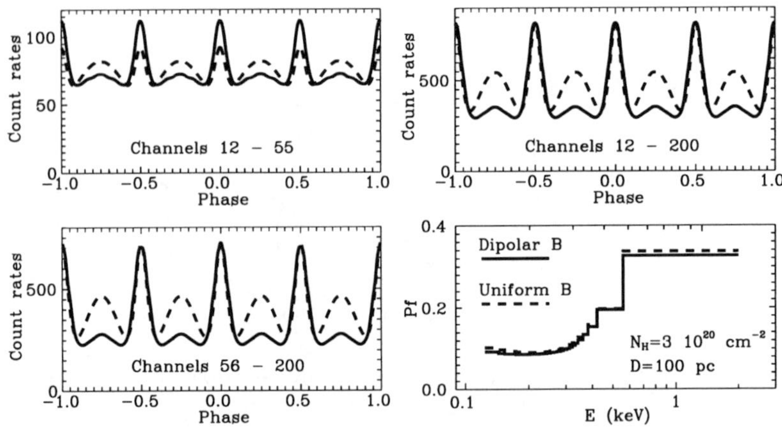

Figure 3: Light curves for different *ROSAT* PSPC output channels, and the output energy dependence of the pulsed fraction Pf for the case of magnetic atmosphere with the dipolar and uniform magnetic fields.

presented in Fig. 3 for a typical interstellar absorption. It is seen that the relative pulse heights and the pulsed fraction are high enough ($\sim 10 - 30\%$) and generally grow with energy.

The above results show that the model with all the ingredients (i)–(iv) included (see Introduction) can explain the obsereved values of the pulsed fraction of thermal radiation from NSs. The most important factors are the anisotropy of the local flux and gravitational lensing. If there were no strong anisotropy of the local flux, too large NS radii and nonrealistic equations of state would be needed to explain the relatively large pulsed fraction even with the anisotropic temperature distribution[3] unless much hotter polar caps are included[8]. At any field geometry main contribution to the NS radiation should come from (hotter) regions with radially directed magnetic field. Our model can be applied directly to relatively young NSs whose temperature remains $> 10^6$ K. Lower temperatures require additional analysis since tangential heat conduction is expected to be more important, and the atmospheric opacity is a more complicated function of direction and frequency.

REFERENCES

1. Ögelman, H. 1995. In Lives of Neutron Stars. A. Alpar, Ü. Kiziļoğlu & J. van Paradijs, Eds. Kluwer Academic Publishers. In press.
2. Greenstein, G., & Hartke, G. J. 1983. Astrophys J. **271:** 283.
3. Page, D. 1995. Astrophys. J. In press.
4. Pavlov, G. G. et al. 1994. Astron. Astrophys. **289:** 837.
5. Zavlin, V. E. et al., 1995. Astron. Astrophys. In press.
6. Qin, L., & Tsuruta, S. 1995. In preparation.
7. Urpin, V. A., & Yakovlev, D. G. 1980. Sov. Asron. **24:** 425.
8. Tsuruta, S., & Qin, L. 1995. This volume.
9. Zavlin, V. E. et al., 1995. This meeting, MPE reports. In press.

An Internal Friction Model for an Unusually Large Second Derivative of the Pulsar Rotation Rate

N. Shibazaki & S. Hirano

Department of Physics, Rikkyo University, Nishi-Ikebukuro, Tokyo 171, Japan

Abstract

The millisecond pulsar 1620 - 26 in the globular cluster M4 exhibits an unusually large second derivative of the pulse frequency[1, 2], which together with the other pulse timing parameters yields the braking index of 4×10^7. To understand this extreme behavior of the pulsar rotation, we consider the effect of a superfluid-crust coupling in the neutron star interior on the crust rotation. The frictional coupling between superfluid and crust becomes unstable when the temperature of the star falls down below a critical temperature. Then, the frictional instability develops a limit cycle[4]. The temperature increase in a part of the limit cycle enhances friction between the crust and the superfluid so largely and accelerates the angular momentum transfer from the superfluid to the crust so rapidly that the crust rotation is even spun up. This rapidly growing friction can produce an extremely large second derivative in the crust rotation rate as observed in PSR 1620 - 26.

1. Introduction

The braking index is a parameter which characterizes the spin-down behavior of a neutron star and defined by $n = \Omega_c \ddot{\Omega}_c / \dot{\Omega}_c^2$, where Ω_c is the crust angular velocity. Its canonical value, given by the magnetic dipole braking, is 3. The pulse timing parameters obtained for PSR 1620 - 26[1, 2] are $\Omega_c = 567$ rad s^{-1}, $\dot{\Omega}_c = -4.05 \times 10^{-14}$ rad s^{-2} and $\ddot{\Omega}_c = 1.18 \times 10^{-22}$ rad s^{-3}, which yield $n = 4 \times 10^7$. We suggest here that the cause for the extremely large $\ddot{\Omega}_c$ may be in the interior of the neutron star.

2. Frictional Instabilities

The superfluid in the inner crust rotates faster than the crust. The interaction between the inner crust superfluid and the crust creates friction, which yields internal torque and angular momentum transfer from the inner crust superfluid to the crust and dissipates the rotational energy of the superfluid into heat. The rotation of the crust is governed not only by the external braking torque but also by the internal accelerating torque. The internal torque in the vortex creep theory[3] is expressed as

$$N_{int} = -I_s \dot{\Omega}_s \approx \frac{2\Omega_s}{R} I_s v_0 \exp\left[-\frac{E_p}{kT}(1 - \frac{\omega}{\omega_{cr}})\right] , \quad (1)$$

where I_s and Ω_s are the moment of inertia and angular velocity of the inner crust superfluid, R is the neutron star radius, k is the Boltzmann constant, T is the internal temperature and ω is the angular velocity difference given by $\omega = \Omega_s - \Omega_c$, E_p is the pinning energy, v_0 ($\sim 10^7$ cm s^{-1}) is the typical velocity of microscopic motion of the vortex lines between pinning centers, and ω_{cr} is the critical angular velocity difference which, if exceeded, leads to sudden unpinning and outward motion of the vortex lines.

We have studied the thermal and rotational evolution of recycled pulsars after the neutron stars are spun up to the equilibrium rotation and the mass accretion is turned off. We find that the frictional coupling between the crust and the superfluid becomes unstable when the internal temperature falls down below a critical temperature. Then, the frictional instability develops a limit cycle[4], in which the temperature T and the angular velocity difference ω, which determine the strength of the friction between the crust and the superfluid, oscillate.

In order to understand the physical cause of the frictional instability consider the perturbation which gives the increase of the internal torque. The angular momentum is transferred from the superfluid to the crust, decreasing ω. The rotational energy of the superfluid is dissipated into heat, increasing T. The net change[4] in internal torque is given approximately by

$$\frac{\delta N_{int}}{N_{int}} \approx \left(\frac{1}{\tau_{HT}/\xi} - \frac{1}{\tau_{rr}}\right)\left(\frac{-\delta\omega}{|\dot{\Omega}_s|}\right) , \quad (2)$$

where τ_{HT} is the heating time given by $\tau_{HT} = C_V T/H$ with C_V being the heat capacity and H being the internal heating rate, ξ is the quantity which depends on $\Omega_s/|\dot{\Omega}_s|$ logarithmically and τ_{rr} is the rotational relaxation time given by $\tau_{rr} = (kT/E_p)(\omega_{cr}/|\dot{\Omega}_s|)$. When $\tau_{HT}/\xi < \tau_{rr}$, $\delta N_{int}/N_{int}$ is positive and hence the increase of N_{int} due to the increase of T overwhelms the decrease of N_{int} due to the decrease of ω. In consequence, the angular momentum transfer leads to the further increase of the internal torque, which in turn enhances the angular momentum transfer furthermore. The frictional coupling between the crust and the superfluid becomes unstable. The instability grows on a time of τ_{HT}/ξ. The critical temperature[4] is obtained from the condition $\tau_{HT}/\xi = \tau_{rr}$.

3. Results and Discussion

In a limit cycle the neutron star follows the two states, alternately, which are characterized in terms of the temperature and crust-superfluid coupling condition. Those are the high temperature-coupled state and the low temperature-decoupled state. The internal friction is large in the coupled state. The superfluid rotation is decelerated and its energy is dissipated into heat. The star evolves decreasing ω and increasing T in the coupled state. Note that just after the star enters into the coupled state, the internal torque on the crust grows rapidly and the angular momentum transfer from the superfluid to the crust is enhanced largely due to the frictional instability, while the external torque remains almost constant. In consequence, $\dot{\Omega}_c$ increases rapidly and largely. This variation of $\dot{\Omega}_c$ yields an enormously large $\ddot{\Omega}_c$ that can explain the observation of PSR 1620 - 26. When ω becomes sufficiently small and T becomes relatively high, the star enters into the decoupled state, where the internal friction is negligible. The star just cools and only the crust is decelerated by the external torque. The star evolves decreasing T and increasing ω in the decoupled state. When ω becomes sufficiently large and T becomes relatively low, the star enters into the coupled state, again. Note that the essence of the above results does not change for the different superfluid-crust coupling models as long as they depend sensitively on the temperature.

Table 1 shows the results of our calculations that simulate the thermal and rotational evolution of the old neutron star after recycled ($\Omega_{eq} \sim 650$ - 700 rad s^{-1}) in order to reproduce the observed timing parameters of PSR 1620 - 26. The critical angular velocity difference and the moment of inertia of the inner crust superfluid are constrained to be $\omega_{cr} \sim 10$ rad s^{-1} and $I_s \sim 3.6 \times 10^{43}$ g cm^2 from the comparison of our results with the timing parameters observed in PSR 1620 - 26. These values for ω_{cr} and I_s are in the range expected in current models. The strong pinning region is required in the inner crust from our model to explain $\ddot{\Omega}_c$ of PSR 1620 - 26. Weak and superweak pinning regions with $\omega_{cr} < 1$ rad s^{-1} are preferred in the vortex creep model for the post-glitch relaxation. These results indicate that there may be regions with different pinning strength in the inner crust. The contribution of the weak and superweak pinning regions to the frictional instability is negligible since their heating power ($I_s\omega_{cr}$) is small, while the strong pinning region does not participate in the glitch for some reason unknown right now.

TABLE 1 Timing Parameters Simulated for PSR 1620-26

ω_{cr} (rad s^{-1})	5	10	20	40
I_c (10^{45} g cm^2)	0.690	1.16	1.20	1.20
I_s (10^{43} g cm^2)	51.0	3.60	0.448	0.0575
μ (10^{27} G cm^3)	4.31	5.60	5.67	5.68
Ω_c (rad s^{-1})	567.4	566.8	567.8	568.0
$\dot{\Omega}_c$ (10^{-14} rad s^{-2})	-4.048	-4.048	-4.048	-4.048
$\ddot{\Omega}_c$ (10^{-22} rad s^{-3})	0.228	1.05	3.74	9.16
$\frac{\tau_{HT}}{\xi}$ (yrs)	104	18.1	4.44	1.32

We expect that PSR 1620 - 26 will increase its $\dot{\Omega}_c$ rapidly yielding increasingly large $\ddot{\Omega}_c$ and change its sign to positive in ~ 10 years. Maybe within the next several years the anomalous behavior of the pulse timing parameters may become evident. The continued observation of this very interesting pulsar can confirm the presence of a frictional instability in the interior of the neutron star. We also believe that there may be old pulsars which have a positive $\dot{\Omega}_c$ and a large $\ddot{\Omega}_c$ with both signs. If our interpretation is confirmed, we would have a new insight on the property and structure of neutron star interiors.

References

1. Backer, D. C., Foster, R. S. & Sallmen, S. 1993, Nature, 365, 817
2. Thorsett, S. E., Arzoumanian, Z. & Taylor, J. H. 1993, ApJ, 412, L33
3. Alpar, M. A., Anderson, P. W., Pines, D. & Shaham, J. 1984, ApJ, 276, 325
4. Shibazaki, N. & Mochizuki, Y. 1995, ApJ, 438, 288

Effects of magnetic fields on neutron star thermal evolution*

S. TSURUTA and L. QIN

Montana State University
Bozeman, MT 59717, USA

INTRODUCTION

In recent calculations of thermal evolution of neutron stars[1] the effects of magnetic fields have often been neglected, because the evolution of the total photon luminosity (or the equivalent effective temperaure) is thought to be insensitive to the presence of magnetic fields[2]. All thermal evolution calculations to date have been carried out as a one-dimensional (1D) problem (in the radial direction r). When the quantized magetic field is introduced, however, the electron and photon conductivities are no longer isotropic[2]. Due to this anisotropic nature of heat transport the 1D approach is no longer valid when we investigate the thermal evolution problems of magnetized neutron stars. Instead we have to solve the two-dimentional (2D) equations, to take into account the heat flow in the θ direction, where θ is the angle from the polar axis. Very recently, we have succeeded in constructing a realistic 2D code[3], with which we carried out the standard cooling of neutron stars under quantized magnetic fields. The anisotropic heat transport in magnetized plasmas also causes anisotropic radiation, which may be observable as pulsed thermal emission from these stars. Therefore, we also calculated the light curves for radiation emerging from the stellar atmosphere, which are fitted to the ROSAT data of some pulsars.

RESULTS

Standard Cooling of Magnetized Neutron Stars

Figure 1 shows the standard cooling of a magnetic neutron star obtained by using our 2D code[3], which solves the fully general relativistic thermal evolution equations in two-dimensions (r and θ) 'exactly' (meaning 'without making the isothermal approximation'). As a typical case we choose a neutron star with the FP type equation of state[4], stellar mass $M = 1.4M_\odot$, and radius R = 11 km. The total photon luminosity to be observed at infinity, L_γ^∞, is obtained by averaging the local fluxes over the whole stellar surface. The dipole magnetic field configuration is adopted. For younger stars of age $t <\sim 10^5$ years the total luminosity of a magnetic star is somewhat lower than the zero field case, but the effect is still relatively small. However, that is no longer the case for stars older than about 10^5 years. Instead of cooling quickly after about a millon years, the star keeps warm ($>\sim 10^5$ K) for 10^{8-9} years.

*This work was partialy supported by NSF grant AST-9013290 and NASA grant NAGW-2208.

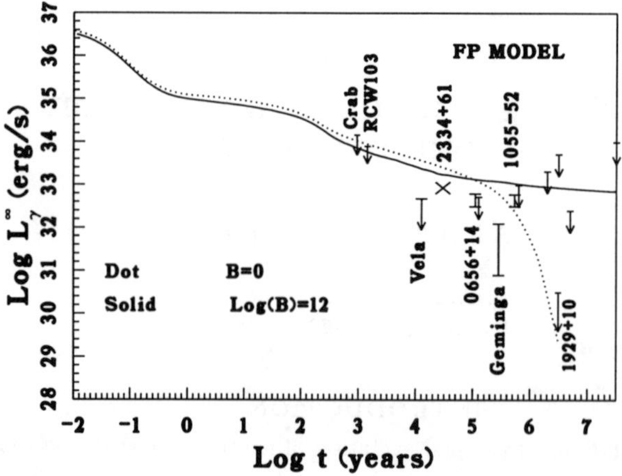

Figure 1: The standard cooling of a neutron star with the polar magnetic field strength of $B_p = 10^{12}$ Gauss (solid curve) and with the zero field (dotted curve). The total photon luminosity to be observed at infinity is shown as a function of the stellar age. The bars and a cross refer to the detection data of thermal emission from the respective pulsars as indicated, while the downward arrows refer to the upper limits, taken from Umeda et al.[5] and Becker et al.[6]. See the text for further details.

The main reason is that under quantized magnetic fields the electron conductivity is significantly reduced perpendicular to the magnetic field direction, while parallel to the field it becomes only slightly higher than the zero-field case[7]. For an isothermal stellar core, therefore, the polar surface temperature will be higher than the equatorial temperature. In previous works[8], magnetic cooling was obtained by carrying out 1D calculations along the polar direction. During the earlier 'neutrino cooling era' when the stellar energy is lost predominantly by neutrinos escaping from the interior this approximation may be still justified. However, as the star cools the neutrino cooling eventually becomes inefficient, and the star cools predominantly by photon radiation from the surface. During this later 'photon cooling era', since the star is a better insulator away from the polar direction, the interior temperature along the equatorial direction becomes much higher than along the polar direction, if heat flow in the θ direction is neglected. When the problem is treated realistically in 2D, the net effect of the heat flow from the equatorial to the polar direction is to keep the isothermal core warmer (than the zero field case), and hence it takes longer for a magnetic star to exhaust the residual energy.

Light Curves

Due to the anisotropic electron conductivity under strong magnetic fields the surface temperature is modulated along the θ direction. This temperature modulation is shown in our another paper in this volume[7]. Its effects may be observed as pulsation of the stellar radiation when viewed with an appropriate combination of the rotation and magnetic axes and the line of sight. The thermal radiation from some pulsars discovered by ROSAT generally exhibit pulsations of $\sim 10-30\%$[9]. We

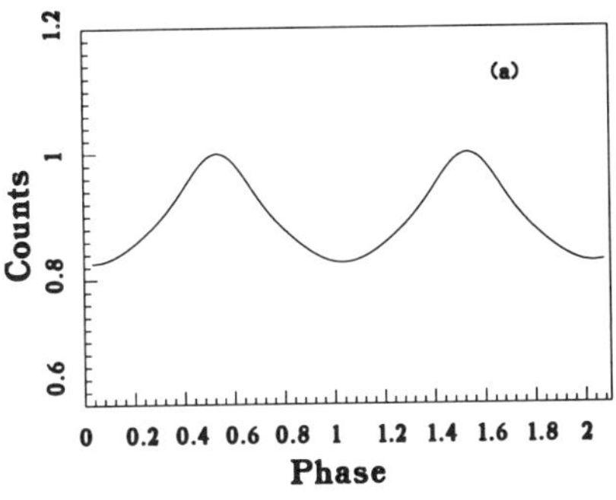

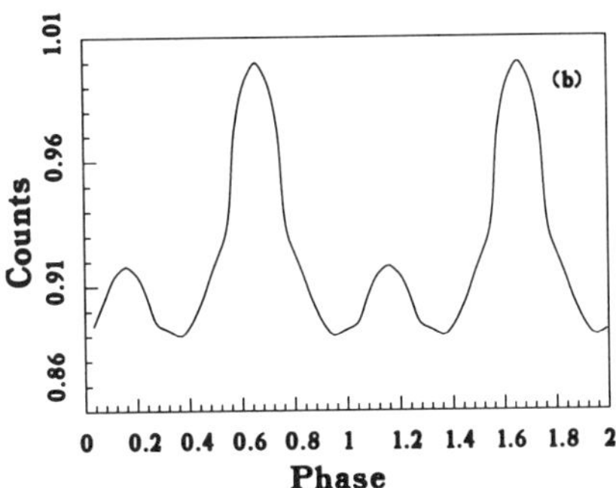

Figure 2: Light curves fitted to the ROSAT data on (a) PSR 0656+14, and (b) Vela pulsar.

calculated the light curves for radiation emerging from the stellar atmosphere, by integrating the local fluxes over the visible stellar surface and over the ROSAT energy band of 0.07 to 2.4 keV. The temperature modulation effect is included, but not the flux anisotropy. We found significant pulsations in our light curves. However, for a neutron star, due to its enormous gravity, the gravitational bending of the emitted light becomes important[10]. Its general effect is to wash out the modulation[7]. We find that when this gravitational effect is included the temperature modulation no longer can produce the observed amplitude of $\sim 10 - 30\%$.

On the other hand, the pulsars which emit thermal-like pulsed radiation generally appear to consist of two components, within ROSAT's soft X-ray band[9]. The higher energy component probably comes from hot spots, while the lower component from the whole stellar surface[9]. Therefore, we calculated light curves for radiation emerging from the stellar atmosphere, by including two hot spots, temperature modulation, and the gravitational bending (and redshift) effects. The pulse amplitude and shape depend on the relative angles among the rotation and magnetic axes and the line of sight. These angles are adjusted to fit our theoretical light curves to the ROSAT light curve data on PSR 0656+14 and Vela pulsar. The results are shown in Figures 2a and 2b, respectively. The reasonably good fits are obtained for the following parameters: (i) For PSR 0656+14, the hot spot temperature $T_h = 4.5\ T_s$, $\alpha = 25^o$, $\beta = 10^o$, and $\gamma = 20^o$, where α is the angle between the rotation axis and the line of sight, β is the angle between the rotation and magnetic axes, γ is the angular size of the hot spot, and Ts is the suface temperature. For the Vela pulsar, we obtained $T_h = 1.3\ T_s$, $\alpha = 5^o$, $\beta = 65^o$, and $\gamma = 60^o$. The hot spot temperatures and sizes thus deduced from the light curve fitting agree reasonably well with those obtained from spectral fitting.

DISCUSSION AND CONCLUDING REMARKS

We have shown that a magnetic neutron star may keep warm ($>\sim 10^5$ K) for much longer period ($\sim 10^{8-9}$ years) than predicted earlier. In all previous cooling calculations, on the other hand, a neutron star became cold very quickly after a few million years. Therefore, if warm older neutron stars are detected it was generally thought that some heating mechanism will have to be in operation. Our results may mean that such heating is no longer required, unless the magnetic field decay timescale is fairly short. Inverting the argument around, if no warm older neutron stars are found, then, our results may mean that the magnetic field decays fairly quickly, after a few million years.

In Figure 1 the magnetic cooling (solid curve) obtained with our 2D code is compared with the ROSAT data. When the effects of the equation of state and stellar mass are included the standard cooling appears to be consistent with some of the observational data (e.g., Crab, PSR 0656+14, and PSR 1055-52), but some other data (especially PSR 1929+10) are too low to be reconciled with the standard cooling. For these sources the non-standard fast cooling mechanisms may be required. Our results may imply that there are two classes of neutron stars, those cooling with the standard scenario and others cooling fast through a non-standard scenario. Such a possibility has been already suggested, and that is quite reasonable if the former class of neutron stars are somewhat less massive than the latter[5,11].

We have also shown that when the hot spots with the observed size and temperature are included the observed amount of pulsation amplitude is obtained for some appropriate combination of the relative angles α and β. In our work the effect of the anisotropic behavior of the photon flux under quantized magnetic field[7] was not included. This effect may change the final values of the relative angles, but probably not our conclusion, that the presence of hot spots is important to understand the light curves and spectra of the thermal-like pulsed radiation from some pulsars observed in the ROSAT window.

REFERENCES

1. see, e.g., Nomoto, K. & S. Tsuruta. 1987. Astrophys J. **312:** 711-726.
2. see, e.g., Hernquist, L. 1984. M.N.R.A.S. **213:** 313-336.
3. Qin, L. & S. Tsuruta. 1995. In preparation.
4. Friedman, B. & V. R. Pandharipande. 1981. Nucl. Phys. A. **361:** 502-520.
5. Umeda, H., K. Nomoto, S. Tsuruta, T. Muto & T. Tatsumi. 1994. Astrophys. J. **431:** 309-320.
6. Becker, W., J. Trümper & H. Ögelman. 1993. In Isolated Pulsars. K. A. Van Riper, R. I Epstein, & C. Ho, Eds.:104-109. Cambridge Univ. Press. Cambridge, UK.
7. Shibanov, Yu. A., G. G. Pavlov, V. E. Zavlin, L. Qin & S. Tsuruta. 1995. This volume.
8. see, e.g., Van Riper, K. A. 1991. Astrophys. J. Suppl. **75:** 444-462.
9. Ögelman, H. 1995. In Lives of Neutron Stars. A. Alpar, Ü., Kiziloğlu, & J. van Paradijs, Eds. Kluwer Academic Publishers, In press.
10. Pechenick, K. R., C. Ftaclas & J. M Cohen. 1983. Astrophys. J. **274:** 846-857.
11. Tsuruta, S. 1986. Comments on Astrophys. **11:** 151.

Fast Pulsars, Strange Stars, and Strange Dwarfs†

FRIDOLIN WEBER AND NORMAN K. GLENDENNING

Nuclear Science Division
Lawrence Berkeley Laboratory, MS: 70A-3307,
University of California
Berkeley, CA 94720, U.S.A.

INTRODUCTION

The hypothesis that strange quark matter may be the absolute ground state of the strong interaction (not ^{56}Fe) has been raised independently by Bodmer[1] and Witten[2]. Even to the present day there is no sound scientific basis on which one can either confirm or reject the hypothesis, so that it remains a serious possibility of fundamental significance for rare but exotic phenomena (see, for instance, Ref. [3]). One striking implication of the hypothesis would be that pulsars, which are conventionally interpreted as rotating neutron stars, almost certainly would be rotating strange stars (strange pulsars)[2,4,5,6,7]. This paper deals with an investigation of the properties of such objects. In addition to this, we develop the complete sequence of strange stars with nuclear crusts, which ranges from the compact members, with properties similar to those of neutron stars, to white dwarf-like objects (strange dwarfs) and discuss their stability against acoustical vibrations[8,9,10]. The properties with respect to which strange-matter stars differ from their non-strange counterparts, i.e., neutron stars and ordinary white dwarfs, are discussed, and observable signatures of strange stars are pointed out.

HADRONIC CRUST ON STRANGE STARS

The presence of electrons in strange quark matter stars is crucial for the possible existence of a nuclear crust on such objects. As shown in Ref. [6], the electrons, because they are bound to strange matter by the Coulomb force rather than the strong force, extend several hundred fermis beyond the surface of the strange star. Associated with this electron displacement is a electric dipole layer which can support, out of contact with the surface of the strange star, a crust of nuclear material, which it polarizes. The maximal possible density at the base of the crust (inner crust density, $\epsilon_{\rm crust}$) is determined by neutron drip, which occurs at about $\epsilon_{\rm drip} = 4.3\times 10^{11}$ g/cm^3. Any inner crust value smaller than that is possible. As an example, we show in Fig. 1 the equation of state of strange stars with inner crust densities $\epsilon_{\rm crust} = \epsilon_{\rm drip}$ and $\epsilon_{\rm crust} = 10^8$ g/cm^3. The nuclear crust is represented by

†This work was supported by the Director, Office of Energy Research, Office of High Energy and Nuclear Physics, Division of Nuclear Physics, of the U.S. Department of Energy under Contract DE-AC03-76SF00098.

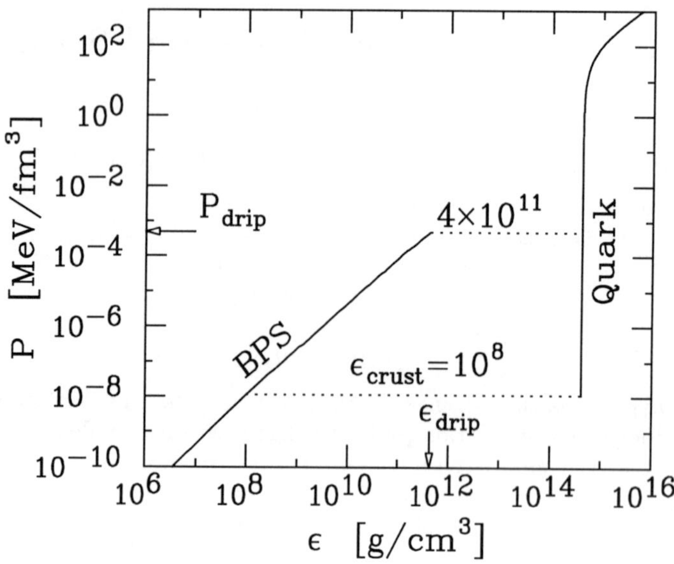

Figure 1: Equation of state of a strange star surrounded by a nuclear crust. $P_{\text{drip}}(\epsilon_{\text{drip}})$ is the pressure at the maximum possible inner crust density.

the equation of state of Baym, Pethick, and Sutherland (BPS)[11], the strange core is described by the bag model (details are given in Ref. [12]).

COMPLETE SEQUENCES OF STRANGE-MATTER STARS

The mass-radius relationship of strange-matter stars possessing maximally dense crusts (i.e., $\epsilon_{\text{crust}} = \epsilon_{\text{drip}}$), constructed for the equation of state exhibited in Fig. 1, is shown in Fig. 2 [8,12]. A value for the bag constant of $B^{1/4} = 145$ MeV has been chosen which represents strongly bound strange matter with an energy per baryon of ~ 830 MeV. The solid dots denote the maximum-mass stars, the arrows indicate the location of the minimum-mass star of each sequence. The sequence of strange stars has a minimum mass of $\sim 0.015\, M_\odot$ (radius of ~ 400 km) or about 15 Jupiter masses, which is about an order of magnitude smaller than that of the neutron star sequence. Smaller values of ϵ_{crust} lead to even lighter low-mass strange stars. For example, a value of $\epsilon_{\text{crust}} = 10^8$ g/cm^3 leads to objects as light as $M \sim 10^{-4}\, M_\odot$ [8]. Even such light objects, if existing abundantly enough in our Galaxy, could be seen in the presently performed gravitational microlensing experiments.

It is striking that the bulk properties of neutron and strange stars of masses that are typical for neutron stars, $1.1 \lesssim M/M_\odot \lesssim 1.8$, are relatively similar and therefore do not allow the distinction between the two different species of stars. The situation changes as regards the possibility of *fast rotation* of strange stars. This has its origin in the somewhat different

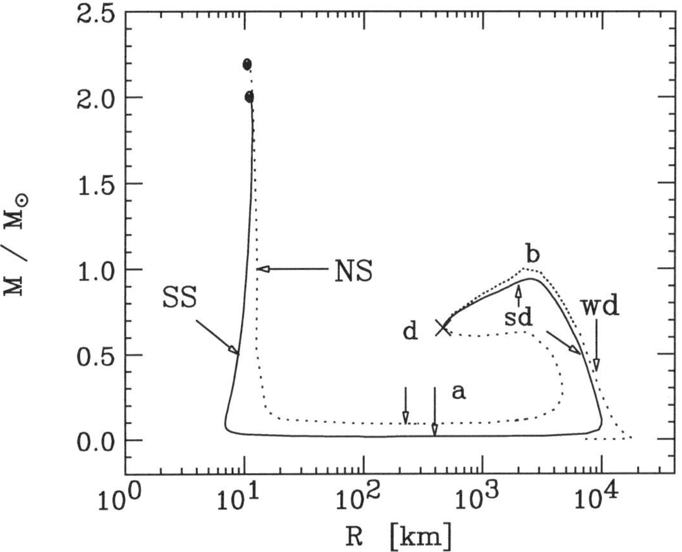

Figure 2: Mass versus radius of strange-star configurations with nuclear crusts (solid curve) and gravitationally bound stars (dotted curve). The following abbreviations are used: NS=neutron star, SS=strange star, wd=white dwarf, sd=strange dwarf.

mass-radius relationships of neutron stars and strange quark stars[5,13]. As a consequence of this, the *entire* family of strange stars can rotate rapidly, at Kepler periods in the range of ~ 0.4 to 0.8 msec (depending on star mass and bag constant), not just those near the limit of gravitational collapse to a black hole as is the case for neutron stars[8,12].

The strange-matter stars located in the right portion of Fig. 2 possess masses and radii similar to those of ordinary white dwarfs[8]. We refer to such strange stars as *strange dwarfs*. Because $\epsilon_{\rm crust} = 4 \times 10^{11}$ g/cm^3 the density at the base of their crusts is about 400 times higher than the density in the most massive white dwarf (at 'b'), which amounts $\sim 10^9$ g/cm^3, and about 4×10^4 times higher than in white dwarfs of typical mass, $M \sim 0.6\, M_\odot$. The stability of this new type of dense dwarfs will be shown next.

STABILITY OF STRANGE-MATTER STARS AGAINST RADIAL OSCILLATIONS

The normal modes of vibration of the strange-matter stars are computed from the Sturm-Liouville eigenequation (for details, see Ref. [8]). The four lowest-lying eigenfrequencies of massive strange stars and strange dwarfs, whose inner crust density is equal to neutron drip, are shown in Fig. 3. We find that the $n = 0$ mode becomes zero for the maximum- and minimum-

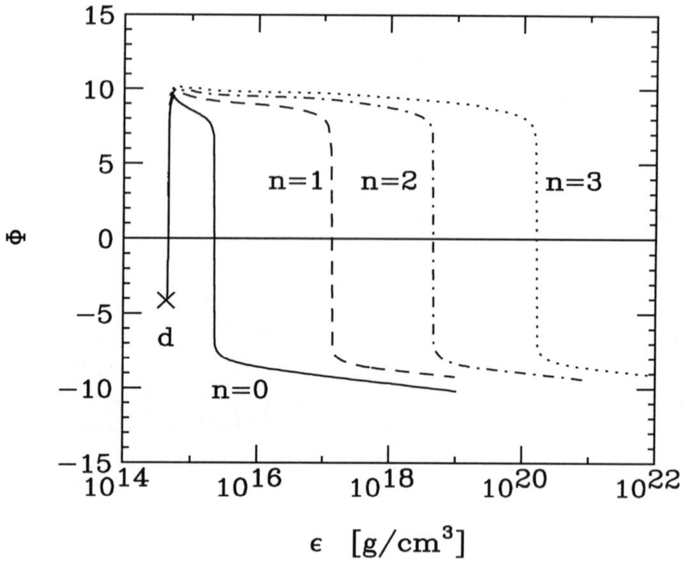

Figure 3: Pulsation frequencies, ω_n^2, of the lowest four ($n = 0, 1, 2, 3$) radial modes of strange stars with crust measured by $\Phi(x) \equiv \text{sign}(x) \log[1 + |x|]$, where $x \equiv (\omega_n/\text{sec}^{-1})^2$, as a function of central star density. The cross refers to the termination point of the strange dwarf sequence (cf. Fig. 2).

mass strange-star configurations in Fig. 2. Most importantly, the $n = 0$ mode passes through zero at 'b' and remains negative (i.e., $\omega_n^2 < 0$) for all densities down to the central density of the strange dwarf star at the termination point (cross). Since $\omega_0^2 < 0$ is associated with an exponentially growing mode of oscillation, all these strange dwarfs are *unstable* against radial oscillations. Thus, no stable strange dwarfs located between 'b' and the crossed termination point 'd' can exist stably. However all other members of the strange-star sequence, from the heaviest possible strange dwarf to the heaviest compact strange star (with the exception of the minimum-mass star at 'a' in Fig. 2), are stable and thus could exist in nature if the hypothesis is true.

SUMMARY

The main issues of this contribution can be stated as follows:

- The complete sequence of compact strange-matter stars can sustain extremely *rapid rotation* (considerably below 1 msec) and not just those close to the mass peak, as is the case for neutron stars!
- Strange stars can possess *nuclear crusts* of thickness ~ 1 km to ~

10^3 km, depending on central star density. This might be of great importance for their cooling behavior (presently under investigation)!

- Strange stars possess *masses* in the range $\sim 2\,M_\odot$ to $10^{-4}\,M_\odot$ and *radii* from several kilometers to $\sim 10^3$ km. Since masses and radii of $\sim 10^{-4} M_\odot$ and $\sim 10^3$ km are completely excluded for both neutron stars as well as white dwarfs, such star properties may serve as additional signatures for hypothetical strange stars!
- If the light, planetary-like strange stars exist and if they are abundant enough, then the gravitational *microlensing* experiments should see them!
- We find white-dwarf-like strange stars that owe their stability solely to the strange cores in their centers (*strange dwarfs*). They carry nuclear crusts whose density at the base is up to about 400 times higher than the central density in the most massive white dwarf. Hence such strange dwarfs constitute a possible new class of dense stars, if the strange matter hypothesis is correct!

REFERENCES

1. A. R. Bodmer, Phys. Rev. D **4** (1971) 1601.
2. E. Witten, Phys. Rev. D **30** (1984) 272.
3. C. Alcock and A. V. Olinto, Ann. Rev. Nucl. Part. Sci. **38** (1988) 161.
4. G. Chapline and M. Nauenberg, Ann. New York Academy of Sci. **302** (1977) 191.
5. N. K. Glendenning, Mod. Phys. Lett. **A5** (1990) 2197.
6. C. Alcock, E. Farhi, and A. V. Olinto, Astrophys. J. **310** (1986) 261.
7. P. Haensel, J. L. Zdunik, and R. Schaeffer, Astron. Astrophys. **160** (1986) 121.
8. N. K. Glendenning, Ch. Kettner, and F. Weber, *From Strange Stars to Strange Dwarfs*, (LBL-34869), November 1994.
9. Ch. Kettner, F. Weber, M. K. Weigel, and N. K. Glendenning, *Stability of Strange Quark Stars with Nuclear Crust against Radial Oscillations*, International Symposium on Strangeness and Quark Matter, September 1–5, 1994, Crete, Greece, to be published by World Scientific, (LBL-36209).
10. F. Weber, Ch. Kettner, M. K. Weigel, and N. K. Glendenning, *Strange-Matter Stars*, International Symposium on Strangeness and Quark Matter, September 1–5, 1994, Crete, Greece, to be published by World Scientific, (LBL-36210).
11. G. Baym, C. Pethick, and P. Sutherland, Astrophys. J. **170** (1971) 299.
12. N. K. Glendenning and F. Weber, Astrophys. J. **400** (1992) 647.
13. F. Weber and N. K. Glendenning, *Hadronic Matter and Rotating Relativistic Neutron Stars*, Proceedings of the Nankai Summer School, "Astrophysics and Neutrino Physics", p. 64–183, Tianjin, China, 17-27 June 1991, ed. by D. H. Feng, G. Z. He, and X. Q. Li, World Scientific, Singapore, 1993.

BATSE Observations of GRS 1915+105 and GRO J1655–40

W. S. PACIESAS[a,b], K. J. DEAL[a,b], B. A. HARMON[b], C. A. WILSON[b], S. N. ZHANG[b,c], AND G. J. FISHMAN[b]

[a] *University of Alabama in Huntsville, AL 35899 USA*

[b] *NASA/Marshall Space Flight Center Huntsville, AL 35812 USA*

[c] *Universities Space Research Association Huntsville, AL 35806 USA*

INTRODUCTION

The X-ray transients GRS 1915+105 and GRO J1655–40 are the only galactic sources so far found to have superluminally expanding radio jets. We compare and contrast CGRO/BATSE hard X-ray observations with the radio behavior of these two sources in an attempt to determine whether they constitute a distinct sub-class of X-ray transients.

GRS 1915+105 was discovered by WATCH/GRANAT[1] in August 1992 but subsequent analysis of BATSE data[2] showed it to be present as early as May 1992. Continued monitoring[2,3] showed that the outburst, which reached a typical (but highly variable) intensity of ~300 mCrab, lasted for more than 400 days until late July 1993, at which point the typical intensity dropped by at least a factor of 6. Sporadic later monitoring showed flaring activity below 20 keV in March and September 1994.[3] GRS 1915+105 was identified with a variable radio[4] and infrared[5,6] source, but no optical counterpart has been found, probably due to strong interstellar absorption.[7,8] VLA monitoring of the radio counterpart revealed a strong outburst in March/April 1994 which involved the ejection of a pair of bright condensations with large proper motion.[9] At the distance of 12 kpc derived from H I absorption,[8] the apparent velocity of the faster of these was superluminal. The inferred actual velocity of the jets was $0.92 \pm 0.08c$.

GRO J1655–40 (designated X-Ray Nova Scorpii 1994) was discovered by BATSE[10] in late July 1994. Continued hard X-ray monitoring showed a rather variable light curve with no clear trend during the first outburst, which lasted ~18 days. Two later outbursts were observed,[11,12] one in September lasting ~12 days, and another beginning in early November and still in progress as of this meeting. Identification of the optical counterpart[13,14] led to the discovery of strong, variable radio emission[15–17] from GRO J1655–40. The distance estimate from H I observations[18] is 3.5 kpc. VLBI and VLA observations of this source showed large proper motion ejection at apparently superluminal velocities.[19–21] Weaker radio outbursts were associated with the September and December X-ray outbursts.[22–24] The radio observations and the X-ray/radio correlations are discussed in more detail elsewhere.[25–27]

OBSERVATIONS

The Earth occultation technique enables BATSE to function effectively as an all-sky monitor between 20 keV to roughly 1 MeV.[28] Routine daily operations include occultation measurements of a catalog of known or suspected sources as well as a search, albeit less sensitive, of the data for new sources not included in the catalog. The standard analysis, or alternatives, may be performed on the archival data to measure previously unrecognized sources. This enabled us retroactively to detect emission from GRS 1915+105 for several months prior to its discovery.

We have attempted to quantify certain systematic errors in the BATSE occultation analysis by studying the measured deviations from zero flux in blank-sky regions. In general, we find a larger spread of the data than expected from statistics alone, the typical measured standard deviation being 20–30% greater. However, for sources in the Aquila region, which includes GRS 1915+105, the excess spread is even larger. When we performed such an analysis for the location of GRS 1915+105 prior to TJD 8730, which is consistent with zero average flux, the measured standard deviation was larger than the statistical prediction by a factor of 2.183.

The BATSE long-term intensity history of GRS 1915+105 is shown in Fig. 1, where we have increased the error bars by 2.183 in order to account for the systematic deviations. The hard X-ray emission from the source is confined primarily to two episodes: the original outburst lasting about 430 days from May 1992 to July 1993 and a secondary outburst lasting about 110 days from December 1993 to March 1994. Although quite variable on the ten-day timescale shown, the typical flux during both outbursts is similar: ~0.06 to 0.08 ph $cm^{-}2$ s^{-1} (20–100 keV), roughly 200–250 mCrab. There is also indication of persistent but variable lower-level emission of ~30 mCrab between and after the main outbursts. In particular, the flare beginning around TJD 9610 was also observed at lower energies.[3,29]

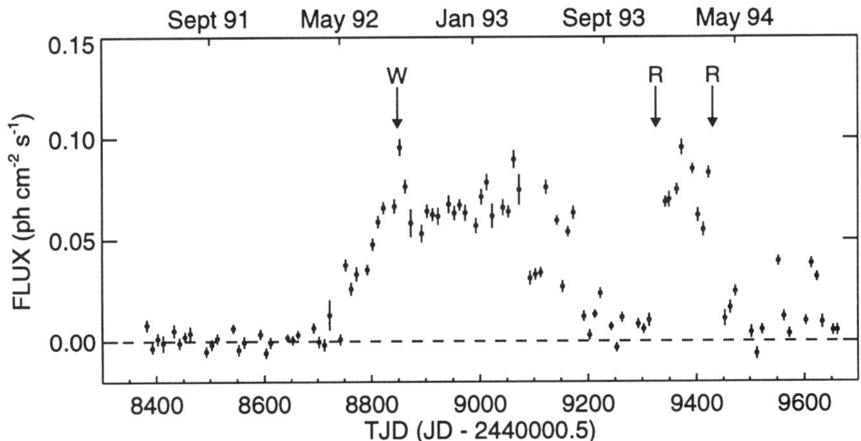

FIGURE 1. Intensity history of GRS 1915+105 in the energy range 20–100 keV. The points are averages of 10 days. The arrow marked W indicates the time of the WATCH discovery.[6] The arrows marked R indicate the times of radio flares.

Figure 2 shows the BATSE-determined intensity history of GRO J1655–40. The errors shown are statistical only, since we have not yet investigated the systematic fluctuations for this source. However, our previous experience indicates that systematic deviations in the Scorpius region are no greater, and probably much less, than those in Aquila. For this source, we identify three outburst episodes: the initial one of ~17 days which ended in mid-August, a slighly weaker one lasting ~10 days in September, and a longer, more intense one, beginning in early November and still in progress at the time of this symposium.[a] The typical flux during outburst is ~4 times that of GRS 1915+105, implying that the latter is ~3 times more luminous in hard X-rays than GRO J1655–40 if the distance estimates are correct.

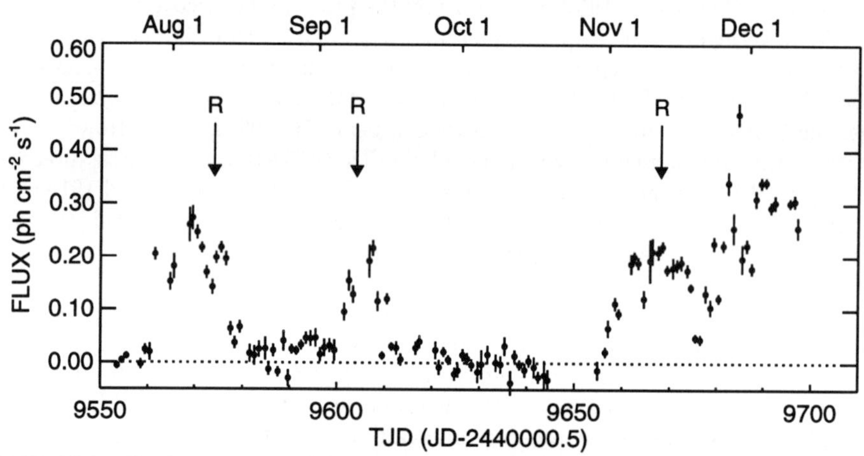

FIGURE 2. The intensity history of GRO J1655–40 in the energy range 20–100 keV. The points are averages over 1 day. Arrows marked R indicate the times of radio flares.

DISCUSSION

The radio counterpart of GRS 1915+105 was discovered in December 1992,[4] and thus radio coverage during the first hard X-ray outburst was incomplete. The two major radio flares of which we are aware occurred during the second hard X-ray outburst (indicated by arrows in Fig. 1). The first of these, in December 1993,[30] occurred near the hard X-ray rise and the second, including the VLA ejection event in March 1994,[9] coincides approximately with the hard X-ray decline. Unfortunately, the BATSE occultation analysis is affected by interference from other sources during the latter interval, and we cannot determine the time of the hard X-ray decrease more precisely without further study.

In GRO J1655–40, the radio flares are progressively less intense and their correspondence with hard X-ray outbursts is quasi-consistent, i.e., the first radio flare[17] coincides with the end of the first hard X-ray outburst, whereas the second radio flare[22,23] peaks around the middle of the second hard X-ray outburst and the third radio flare[24] peaks during the early part of the third outburst (an alternative interpretation is that the more intense hard X-ray emission after ~TJD 9680 is a fourth flare, in which case the third radio and X-ray flares coincide well while the inferred fourth radio flare may have been too weak to detect).

These preliminary results suggest some kind of correspondence between radio and hard X-ray activity in both sources; however, there is no simple one-to-one

[a] Subsequent BATSE data shows the outburst ending around TJD 9705 (19 Dec 1994).

relationship and the nature of the correlation is not consistent. We conclude that the radio and hard X-ray fluxes are affected in different ways by some more fundamental factor such as the local accretion rate in the inner disk.

It is interesting to consider whether superluminal radio jets are associated with a distinct subclass of X-ray novae. Although there are obvious differences in overall duration, the hard X-ray light curves of GRS 1915+105 and GRO J1655–40 have more in common with each other than with the "typical" X-ray nova. The latter may generally be described as having a smooth (on timescales of several days) decaying trend, usually exponential, with a weaker re-flare occurring a few months after the initial maximum. In contrast, Figs. 1 and 2 show strong variability within each outburst, with no clear maximum and no decaying trend, and the later outbursts are at least as strong, or stronger, than the first. Additionally, both sources have steeper hard X-ray spectra than other X-ray novae[2,25] and an unusually small difference in optical luminosity between outburst and quiescence was found for GRO J1655–40.[14] If this separate classification is supported by further studies, an obvious question to consider is whether the distinction is fundamental, such as a difference in the mass or nature of the compact object, or an accident of our observational perspective, as in the unified model for active galactic nuclei.

REFERENCES

1. CASTRO-TIRADO, A. *et al.* 1994. Ap. J. Suppl. Ser. **92**: 469–472.
2. HARMON, B. A. *et al.* 1994. *In* AIP Conf. Proc. No. 304. C. E. Fichtel, N. Gehrels & J. P. Norris, Eds.: 210–219. Am. Inst. Physics. New York, New York.
3. SAZONOV, S. YU. *et al.* 1994. Astronomy Letters **20**: 787–791.
4. MIRABEL, I. F. *et al.* 1993. IAU Circ. No. 5773.
5. MIRABEL, I. F. *et al.* 1993. IAU Circ. No. 5830.
6. CASTRO-TIRADO, A., S. A. BRANDT & N. LUND. 1993. IAU Circ. No. 5830.
7. GREINER, J. *et al.* 1994. *In* AIP Conf. Proc. No. 304. C. E. Fichtel, N. Gehrels & J P. Norris, Eds.: 260–264. Am. Inst. Physics. New York, New York.
8. MIRABEL, I. F. *et al.* 1994. Astron. Astrophys. **282**: L17–L20.
9. MIRABEL, I. F. & L. F. RODRIGUEZ. 1994. Nature **371**: 46–48.
10. ZHANG, S. N. *et al.* 1994. IAU Circ. No. 6046.
11. PACIESAS, W. S. *et al.* 1994. IAU Circ. No. 6075.
12. ZHANG, S. N. *et al.* 1994. IAU Circ. No. 6101.
13. BAILYN, C., S. JOGEE & J. OROSZ. 1994. IAU Circ. No. 6050.
14. BAILYN, C. D. *et al.* 1995. Nature, in press.
15. CAMPBELL-WILSON, D. & R. HUNSTEAD. 1994. IAU Circ. No. 6052.
16. HJELLMING, R. M. 1994. IAU Circ. No. 6055.
17. CAMPBELL-WILSON, D. & R. HUNSTEAD. 1994. IAU Circ. No. 6055.
18. MCKAY, D. & M. KESTEVEN. 1994. IAU Circ. No. 6062.
19. REYNOLDS, J. & D. JAUNCEY. 1994. IAU Circ. No. 6063.
20. HJELLMING, R. M. & M. RUPEN. 1994. IAU Circ. No. 6073.
21. HJELLMING, R. M. & M. RUPEN. 1994. IAU Circ. No. 6086.
22. HJELLMING, R. M. & M. RUPEN. 1994. IAU Circ. No. 6077.
23. CAMPBELL-WILSON, D., D. J. MCKAY & J. E. LOVELL. 1994. IAU Circ. No. 6078.
24. HJELLMING, R. M. & M. RUPEN. 1994. IAU Circ. No. 6107.
25. HARMON, B. A. *et al.* 1995. Nature, in press.
26. TINGAY, S. J. *et al.* 1995. Nature, **374**: 141–143.
27. HJELLMING, R. M. & M. RUPEN. 1995. Nature, submitted.
28. HARMON, B. A. *et al.* 1992. *In* The Compton Observatory Science Workshop. C. R. Shrader, N. Gehrels & B. Dennis, Eds.: 69–75. NASA CP-3137.
29. ALEXANDROVICH, N., K. BOROZDIN & R. SUNYAEV. 1994. IAU Circ. No. 6080.
30. RODRIGUEZ, L. F. & I. F. MIRABEL 1993. IAU Circ. No. 5900.

SIGMA Survey of the Galactic Bulge Region

J. Paul[a], A. Goldwurm[a], M. Vargas[a], J. Ballet[a],
J.-P. Roques[b], L. Bouchet[b], G. Vedrenne[b], P. Mandrou[b],
R. Sunyaev[c], E. Churazov[c], M. Gilfanov[c], A. Finogenov[c],
A. Vikhlinin[c], A. Dyachkov[c], N. Khavenson[c] and V. Kovtunenko[c]

[a] *CEA/DSM/DAPNIA/Service d'Astrophysique*
CE Saclay, 91191 Gif-sur-Yvette Cedex, France

[b] *Centre d'Etude Spatiale des Rayonnements*
BP 4346, 31029 Toulouse Cedex, France

[c] *Space Research Institute*
Profsoyuznaya, 84/32, Moscow, 117296, Russia

INTRODUCTION

The long/deep survey of the ~ 18° × 17° region around the GC performed by the hard X-ray and soft γ-ray (> 35 keV) SIGMA/*GRANAT* telescope[1] between 1990 March 24 and 1994 September 30 consists of 154 observations, grouped in 'campaigns' carried out on spring and fall of each year, for a total dead time corrected exposure of 2440 hours. This unprecedented survey, coupled to the high imaging performances of the telescope (angular resolution ~ 15′), has revealed that only a fraction of the many (> 60) X-ray sources of the region significantly radiate at higher energies. A total of 14 hard X-ray sources have been detected so far, and though they are usually variable, some of them are sufficiently strong and/or stable to appear in the averaged maps of the region, as obtained by the sum of all images recorded in the first 4 years[2]. A salient result of this first survey is that the > 35 keV emission from the one degree radius circle around the GC cannot be attributed only to 1E 1740.7−2942. The additional component can be resolved into the contribution of two other point sources : a known X-ray burster and a new source, GRS 1743−290, detected at ~ 9′ from Sgr A*, and which cannot be identified with the dynamic center of the Galaxy, observed to be remarkably silent above 35 keV in 1990-1993 (ref 2). We report here the salient results of the material collected in 1994, including the activity of the source identified with the Ter 1 Globular Cluster, the surprising behavior of the hard X-ray Nova 1993 in Ophiuchus, discovered by SIGMA in September 1993, which was active again in September 1994, and the discovery of GRS 1730−312, a new transient source detected at the end of September 1994.

OVERVIEW OF THE GALACTIC BULGE IN THE HARD X-RAYS

The contour of the source cluster detected by SIGMA is roughly compatible with that of the Galactic Bulge, the ~ 1 kpc radius spherical region which surrounds the GC and whose total mass, mostly in form of stars, is ~ 10^{10} $M_{\odot}$. It is not yet firmly established if the Galactic Bulge is a genuine component of the Galaxy, or a transition zone between the GC and the Halo. It features however a unique stellar population, with no star younger than 5 10^9 years, but still quite different from population II stars[3]. Contrary to the situation which prevails at higher energies, where it is well admitted that the galactic sites of localized γ-ray emission are associated with young population objects, the SIGMA survey reveals a population of hard X-ray and soft γ-ray sources clearly linked to old population stars. Six SIGMA sources have been identified with known X-ray sources belonging to different classes of neutron star binaries, i.e. one *accreting binary pulsar* : GX 1+4 (ref 4) and five *X-ray bursters* : X 1724−308 (in the Ter 2 globular cluster[5]), 4U/MXB 1728−34 (ref 6), KS 1731−26 (ref 7), X 1732−304 (in the Ter 1 globular cluster, see below) and A 1742−294 (ref 8). The two hardest sources, 1E 1740.7−2942 and GRS 1758−258, have been considered black hole candidates on the basis of their spectral properties[9], along with GRS 1716−249 (ref 10). Although precise estimates are not available, we can assume that all these sources, being clustered around the GC, are roughly at the same distance. Note that according to recent estimates of its distance[11], GRS 1716−249 should be considered as a foreground object, clearly distinct from the Galactic Bulge.

SALIENT RESULTS OF THE 1994 CAMPAIGNS

The source associated with the Ter 1 Globular Cluster

This source, occasionally detected by SIGMA before 1994 in short observation periods[12], was found in a higher state of activity during the 1994 campaigns. It then reaches a level > 5 σ in the summed 35-75 keV 1990-1994 image, at an average 35-75 keV flux of ~ 7 mCrab. This source is probably associated to the weak persistent X-ray source X 1732−304 which shows sporadic burst activity. But since the cluster angular size is much smaller than the SIGMA resolving power, this association, although likely, cannot be fully proven.

The activity of the hard X-ray Nova 1993 in Ophiuchus

GRS 1716−249 ≡ GRO J1719−24, the hard X-ray Nova 1993 in Ophiuchus, has been discovered by SIGMA/*GRANAT* and BATSE/*GRO*[10,11] on September 25, 1993. For several weeks after the outburst, it became one of the brighter hard X-ray and soft γ-ray source in the sky. Not detected in the spring of 1994 (2σ 35-150 keV upper limit ~ 10 mCrab), GRS 1716−249 featured a recrudescence of activity in September 1994 (ref 15). During the whole month, the source was present in the 35-150 keV, at a rather constant flux of ~ 100 mCrab. The $E^{-\alpha}$ power law approximation to the source spectrum gives $\alpha = 2.23 \pm 0.18$, a value to be compared to that obtained during the plateau of the 1993 outburst[10] ($\alpha = 2.32 \pm 0.02$). Such a temporal behavior, observed more than one year after the outburst, is markedly different from that of 'classical' hard X-ray Novæ (as e.g. Nova 1991 in Musca[16], and Nova 1992 in Perseus[17]). It suggests that GRS 1716−249 is a low mass X-ray binary system, caught during an early stage, a conclusion also inferred from the study of its optical and radio properties[11].

The discovery of the transient source GRS 1730−312

On September 22-23, 1994, SIGMA detected a stalwart hard X-ray emission from a previously unknown source, located ~ 3.5° from the GC (Fig. 1). Simultaneously, increasing 2-30 keV X-ray emission from this region was detected by the TTM coded mask instrument aboard the *MIR* Space Station[18]. The new source, GRS 1730−312 ≡ KS 1730−312, has been localized by SIGMA with 3.0′ accuracy (90% confidence error circle radius) at R.A. = $17^h30^m23^s.5$ and Dec. = −31°10′14″ (1950.0 Equinox), i.e. 1.4′ from the TTM position[18]. The

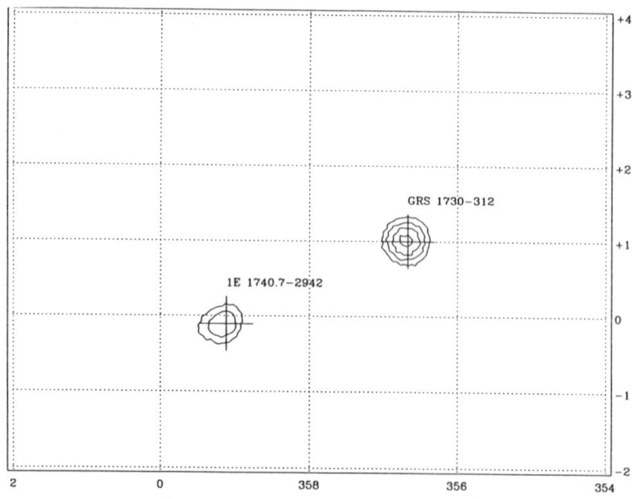

Fig. 1. Image of the sky region around GRS 1730−312 in the energy band 35-75 keV as seen by SIGMA in September 22-26, 1994. Contour levels are in unit of local standard deviations at 4.5 σ, 6.0 σ, 7.5 σ and 9.0 σ. The cross indicates the TTM position of GRS 1730−312.

35-75 keV light curve of the source (Fig. 2) is characterized by a sudden increase of the hard X-ray flux on September 22, followed by a ~ 3 day plateau, during which the 35-75 keV flux was nearly constant (mean value ~ 115 mCrab). The $E^{-\alpha}$ power law approximation to the source spectrum performed during the plateau gives $\alpha = 3.12 \pm 0.27$. The hardness ratio between the 75-150 keV and the

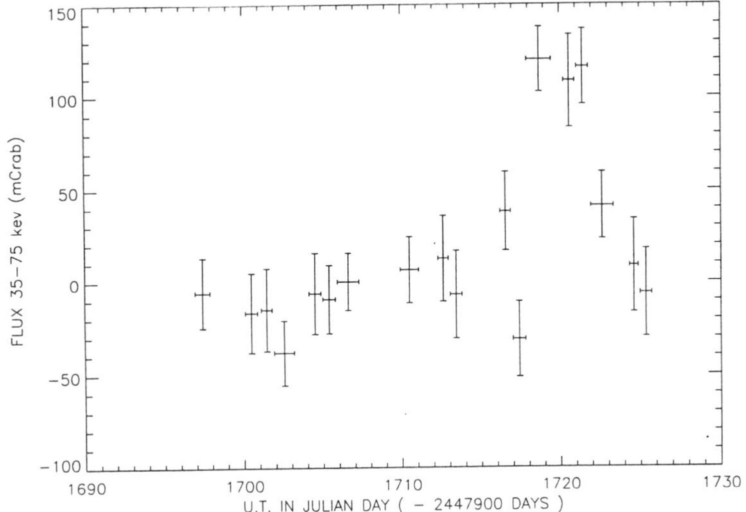

Fig. 2. Light curve of GRS 1730−312 in the 35-75 keV band.

35-75 keV source flux (in mCrab) was nearly constant during the whole event. Note that its mean value (0.63 ± 0.15) is reminiscent to that of all the GC X-ray bursters already detected by SIGMA[2], suggesting that GRS 1730−312 may be also such a binary system harboring a neutron star.

REFERENCES

1. Paul J., P. Mandrou, J. Ballet, *et al.* 1991. Adv. Space Res. **11**: 289-302.
2. Goldwurm, A., B. Cordier, J. Paul, *et al.* 1994. Nature **371**: 589-591.
3. Frogel, J.A. 1988. Ann. Rev. Astron. Astrophys. **26**: 51-91.
4. Laurent, P., L. Salotti, J. Paul, *et al.* 1993. Astron. Astrophys. **278**: 444-448.
5. Barret, D., S. Mereghetti, J.-P. Roques, *et al.* 1991. Astrophys. J. **379**: L21-L24.
6. Claret, A., A. Goldwurm, B. Cordier, *et al.* 1994. Astrophys. J. **423**: 436-440.
7. Barret, D., L. Bouchet, P. Mandrou, *et al.* 1992. Astrophys. J. **394**: 615-618.
8. Churazov, E., M. Gilfanov, R. Sunyaev, *et al.* 1994. Astrophys. J. Suppl. Ser. **92**: 381-385.
9. Sunyaev, R., E. Churazov, M. Gilfanov, *et al.* 1991. Astron. Astrophys. **247**: L49-L52.
10. Gilfanov, M., E. Churazov, R. Sunyaev, *et al.* 1994. *In* Proc. 4th Ann. Astr. Maryland Conf. In press.
11. Della Valle, M., I.F. Mirabel & L.F. Rodriguez. 1994. Astron. Astrophys. **290**: 803-806.
12. Goldwurm, A., B. Cordier, J. Paul, *et al.* 1994. *In* Proc. 2nd Compton Symp. C.E. Fichtel, N. Gehrels & J.P. Norris, Eds.: 421-425. AIP. New York, New York.
13. Ballet, J., M. Denis, M. Gilfanov & R. Sunyaev. 1994. IAU Cir. 5874.
14. Harmon, B.A., S.N. Zhang, W.S. Paciesas & G.J. Fishman. 1993. IAU Cir. 5874.
15. Churazov, E., M. Gilfanov, J. Ballet & E. Jourdain. 1994. IAU Cir. 6083.
16. Goldwurm, A., J. Ballet, P. Laurent, *et al.* 1993. Astron. Astrophys. Suppl. Ser. **97**: 293-297.
17. Roques, J.-P., L. Bouchet, E. Jourdain, *et al.* 1994. Astrophys. J. Suppl. Ser. **92**: 451-454.
18. Borozdin, K., N. Alexandrovich, V. Arefiev & R. Sunyaev. 1994. IAU Cir. 6083.

ASCA Observations of SS 433

NOBUYUKI KAWAI
The Institute of Physical and Chemical Research (RIKEN)
2-1 Hirosawa, Wako, Saitama 350-01, Japan

SUMMARY

The X-ray emission of SS 433 is considered to be generated in the vicinity of the base of the jet, at a distance probably two orders of magnitude or more smaller than the optical emission region[1]. It is therefore expected that the detailed study in X-ray would reveal the clearer information of the "central monster" who creates the antiparallel pair of the relativistic precessing jets. With the high resolution spectroscopic capability of ASCA[2], the thermal and dynamical properties of the X-ray emission region of the jets is starting to be studied in detail. Here we present the diagnosis of the jet using the X-ray emission lines observed in the early phase of the ASCA mission. The temperature at the base of the jet ($\sim 10\ keV$) and the kinetic luminosity ($\sim 10^{39}\ erg\ s^{-1}$) were derived.

We also report the result of the recent contemporaneous spectroscopic observations of SS 433 in X-ray and optical bands. The precession phase at the observation in October – November 1994 was significantly advanced compared with the prediction of the earlier model parameters[3]. It can be interpreted either as a result of random walk of the precession phase, or due to the secular decrease of the precession period.

X-RAY EMISSION LINE DIAGNOSIS

We detected numerous emission lines in the X-ray spectrum of SS 433 with ASCA. Most of these emission lines are thermal and emitted from the two antiparallel jets with a velocity of 0.26 c. In the observation on Apr 23, 1993, SS 433 was in a precession phase with relatively small difference between the Doppler shifts of the two jets. As a consequence the energy of the Kα line from He-like iron in the blue jet (the approaching jet) fell on very close to that of H-like iron in the red jet (the jet with positive radial velocity), and some lines could not be uniquely identified[4].

We attempted to decompose these complex lines into K-lines of H-like and He-like ions of Si, S, and Fe assuming the same temperature for the two jets. In this process, we further found that we need extra emission lines, most likely fluorescence K lines from the matter in the rest frame in low ionization states. It was later confirmed by observations at a different precession phase when the two jet components were widely separated[5]. In Table 1, the fluxes in the emission lines are shown for some of the prominent emission lines. The temperatures are derived assuming an origin in an

Table 1: Physical parameters derived from the Kα lines of Si, S and Fe ions.

Element	$\frac{H-like K\alpha}{He-like K\alpha}$ [1]	$kT_{iso}(keV)$ [2]	$kT_{base}(keV)$ [3]	$\dot{M}(g/s)$ [4]	$L_{kin}(erg/s)$ [5]
Si	1.40 ± 0.40	1.5	$5.5 \binom{+6}{-2}$	3.2×10^{19}	1×10^{39}
S	0.95 ± 0.35	2.0	$5.5 \binom{+5}{-2}$	2.5×10^{19}	0.8×10^{39}
Fe	0.33 ± 0.04	7.1	13 ± 2	2.5×10^{19}	0.8×10^{39}

[1] Ratio of photon numbers in the Kα lines
[2] Temperature derived from the line ratio for isothermal plasma
[3] Temperature at the base of the cooling jet
[4] Mass ejection rate for a single jet
[5] Kinetic energy flux in a single jet assuming $v = 0.26c$

isothermal plasm in ionization equilibrium[6]. The discrepancy of the temperatures derived for the three elements clearly demonstrates the multi-temperature nature of the emission, and is consistent with a picture of a cooling jet[1]. For example, the Fe K lines are more efficiently emitted at a higher temperature than the Si line emission region, and represent the hotter region, *i.e.* near the base of the jet.

The detailed analysis taking into account hydrodynamics and ionization processes will be presented elsewhere[5]. Here we estimate the physical parameters of the jet using a simplified picture. We assume that the jet starts at a certain point from the compact object, and travels at a constant velocity of 0.26 c. We assume that the jet cools only by the radiation. For the line emissivity we used Mewe et al[6]. We also assume solar abundance, the conservation of the number flux of electrons in the jet, and that the number of electrons is equal to that of protons. Then the temperature gradient along the jet is proportional to the product of the total volume emissivity and the electron density. The total line flux is found by integrating the emission line volume emissivity along the jet. The lower bound of the integral was set to 0.1 keV. The higher bound was determined from the ratio of the He-like Kα line and the H-like Kα line. It is the temperature at the innermost visible part of the X-ray jet. In table 1, we show the higher bound temperature and the mass ejection rate in units of $g\ s^{-1}$ and $erg\ s^{-1}$(so called kinetic luminosity) for a single jet. Although the upper bound temperature is somewhat different for Si, S and Fe, the mass ejection rate is consistent among them. The result will not change greatly when detailed hydrodynamics is taken into account, because adiabatic cooling has rather minor effect in the present situation.

The value is near the smallest extreme among the past estimates[1]. Still it exceeds the Eddington luminosity for neutron stars, and is probably good support for the idea that the relativistic jets are generated as a consequence of supercritical accretion.

CONTEMPORANEOUS OPTICAL & X-RAY SPECTROSCOPY

The optical emission lines are considered to be produced at a distance of $\sim$

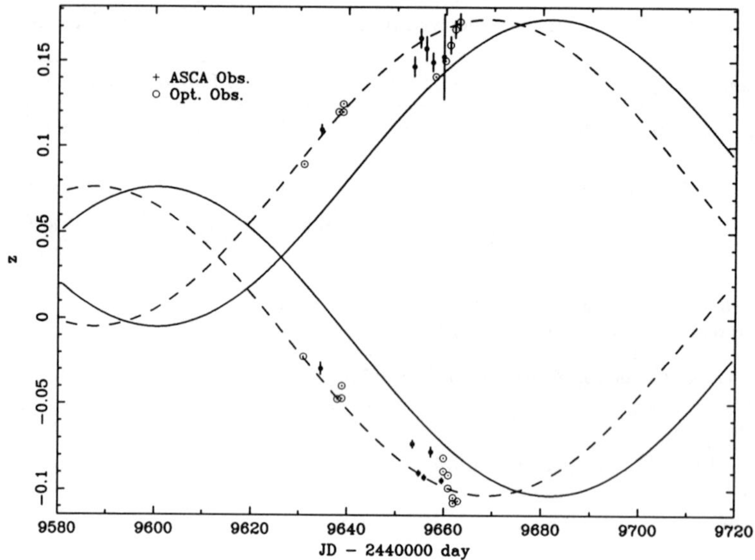

Figure 1: The Doppler shifts of the jets of SS 433 in October – November 1994 measured with X-ray lines (Fe Kα; filled dots) and with optical lines (Hα; open circles. The prediction of the kinematic model based on the observations in 1978 – 1988[3] is shown by the solid curves. The dashed curves shows the same model shifted by 13 days.

$10^{13} - 10^{15}$ cm from the base of the jets[7]. The X-rays too exhibit the emission lines with Doppler shifts generally consistent with those of the optical lines. The presence of the Fe K line suggests that the origin of the X-ray emission is in a hot plasma, probably in the vicinity ($< 10^{11} - 10^{12}$ cm) of the base of the jets[1]. The X-ray emitting plasma, as it travels in the jets, cools and then is observed as the source of the optical lines. Therefore the comparison of the kinematic properties of the jets in the optical and X-ray emission regions is quite important to understand the acceleration and propagation of the jets. Such comparison has become at last possible with ASCA.

We conducted a series of optical observations contemporaneously with ASCA in October – November 1994 at precession phases when the two jet Doppler shifts have rather large separation and the identification of the Fe K lines is relatively easy. Each X-ray observation is approximately 10 ks long. The optical spectra were taken with 1.88 m telescope at Okayama Astrophysical Observatory using the New Cassegrain Spectrograph. The optical Doppler shifts were determined from the moving Hα line. In some cases the red shifted component fell on the atmospheric feature and its wavelength could not be reliably determined. In Figure 1, the Doppler shifts of the two components were plotted against the dates observed. The dashed line is the prediction of the 5-parameter kinematic model based on 10 years' optical observations from 1978 to 1988[3]. The present observations are not strictly simultaneous, and the sampling is rather sparse. In addition it is known that the Doppler shifts of the lines exhibit significant scatter from the model prediction. And yet we obtained two major results.

First, the X-ray and the optical Doppler shifts are, in general, consistent with each other. This confirms, with the highest accuracy, the general consensus that the X-ray jets have same velocity as the optical jets[8,9]. There is no acceleration or deceleration while the jet travels the distance of $> 10^{13}$ cm. It justifies the assumption of the constant velocity in the calculation in section 1. The jet is accelerated to its full speed when it is first visible in X-rays.

Second, the precession phase is found to be ~13 days advanced from the earlier predictions[3]. The dashed line denotes the model curve shifted by 13 days, and it seems to fit the data much better. Baykal et al.[10] studied the fluctuation in the precession cycle ("164-day clock") , and characterized it as "random walk in 164d phase" with an expected rms variation of $\Delta P_{164} = \pm 2.2^d \Delta N^{1/2}$. The phase shift from the last data point[10] to the present work is $\sim 10 - 11d$ over 7 cycles, marginally consistent within 2σ deviation. The average precession period for the 7 cycles is $\sim$ 161 days. Alternatively, we could interpret the phase drift as mostly due to the secular change in the precession period $\dot{P} \sim -4 \times 10^{-4}$, *i.e.* $-P/\dot{P} \sim 1100 yrs$. Though it may look too short for an X-ray and optical source with persistent activity, it has been suggested that SS 433 is at a special phase in the evolution history of a close binary system which lasts for as short as $10^4 years$[11]. Further observations extending the base line can answer the true nature of the precession phase variation.

REFERENCES

1. Brinkmann W., Kawai N., Matsuoka M., Fink H.H. 1991, A&A 241, 112
2. Tanaka Y., Inoue H., Holt S.S. 1994, PASJ 46, L37
3. Margon B., Anderson S.F. 1989, ApJ 347, 448
4. Kotani T. et al. 1994, PASJ 46, L147
5. Kotani T., Kawai N., Brinkman W, et al. 1995, in preparation
6. Mewe R., Gronenschild E.H.B.M., van den Oord G.H.J. 1985, A&AS 62 197
7. Bodo G. et al. 1985, A&A 149, 246
8. Watson M.G., Stewart G.C., Brinkmann W., King A.R. 1986, MNRAS 222, 261
9. Yuan W. et al. 1995, A&A in the press
10. Baykal A., Anderson S.F., Margon B. 1993, AJ 106, 2359
11. van den Heuvel E.P.J., Ostriker J.P., Petterson J.A. 1980, A&A 81, L7

Globular Cluster X-ray Sources in M31

RODRIGO SUPPER

Max-Planck-Institut für extraterrestrische Physik
Postfach 1603
D-85740 Garching
Germany

INTRODUCTION

The giant spiral galaxy M31 is close enough for a detailed study of accreting X-ray sources with luminosities comparable to those in the Milky Way. With the *Einstein* observatory Trinchieri[1] and Fabbiano discovered 108 individual point sources within ∼ 100 ksec IPC and ∼ 200 ksec HRI observations, which had a limiting sensitivity of approximately 10^{37} erg/s. Long[2] and van Speybroeck detected 117 sources within these observations. A search for optical counterparts of these 117 sources by Crampton[3] et al. revealed 23 candidates for X-ray GC sources. The identifications were done by spatial correlation of the X-ray sources with GC lists of Hodge[4], Sargent[5] et al., Battistini[6] et al., and van den Bergh[7]. Among these 23 possible GC sources 19 were considered as being extremely strong candidates.

From these results Battistini[8] et al. calculated a luminosity distribution of the X-ray GC sources in M31 and compared it to that one of the Milky Way. They found a significant difference between both luminosity distributions and raised the question whether the GC sources in M31 are more luminous and/or more numerous compared with those in the Milky Way. Because these distributions are a function of X-ray GC sources as well as a function of optical GC sources, new X-ray and optical observations were expected to reveal more detailed results about the discrepancy of both luminosity distributions.

Within two ROSAT PSPC deep surveys of M31 563 individual X-ray sources have been discovered. The observations add up to ∼ 400 ksec of total exposure with a limited sensitivity of ∼ 10^{35} erg/s. Within these observations 38 X-ray GC sources have been identified by spatial correlations to optical GC lists of Battistini[6,9] et al. and Magnier[10]. The last one is the result of a new deep optical survey of M31. In the following the spectral properties of the brightest of these 38 X-ray GC sources will be discussed showing typical spectra for low mass X-ray binaries (LMXB). A new luminosity distribution will be presented and also a comparison to that one of the Milky Way, revealing that both distributions are comparable.

OBSERVATIONS

The first pointed M31 survey with the ROSAT PSPC was performed in July 1991. It consists of 6 contigous pointings covering the whole disk at equidistant positions and with ∼ 200 ksec of total observation time. Figure 1 shows an overlay of the central region of the 6 pointings over the D_{25} ellipse of M31 (Tully[11]). Each circle marks the boundary of the inner area of the PSPC with a radius of 20' in which the instrument has it's highest angular resolution, whereas the total field of view of the PSPC is 57' in radius.

The second pointed M31 survey with the ROSAT PSPC was performed in July/August 1992 and a few follow up observations in January 1993. It consists of 80 single pointings, each with 2.5 ksec observation time, covering the disk in 4 paths

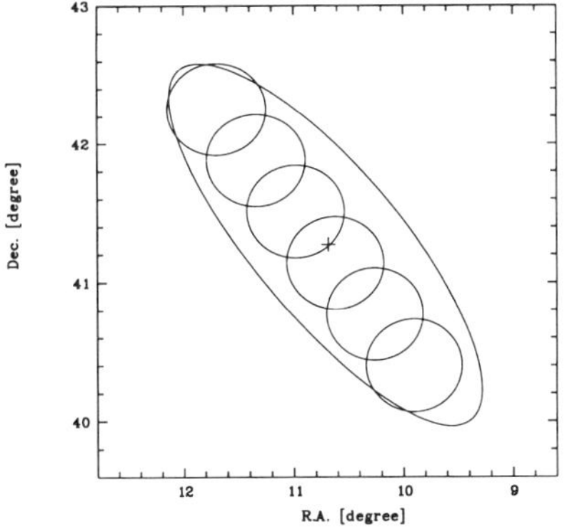

Figure 1: Locations of the central region of the 6 pointings of the first survey plotted over the D_{25} ellipse of M31. The cross marks the optical center of M31.

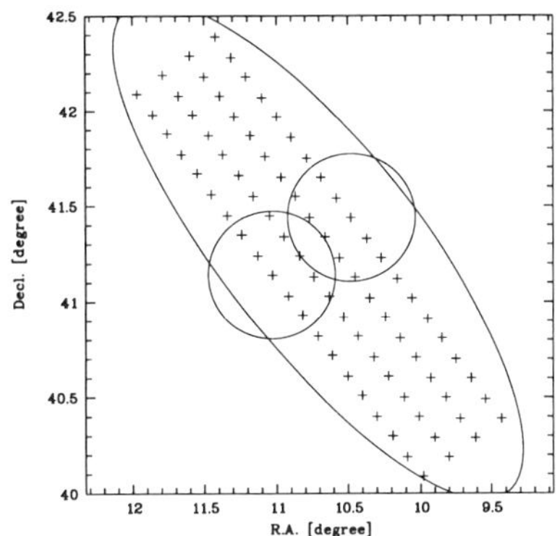

Figure 2: Locations of the on-axes pointing directions of the 80 pointings of the second survey, plotted over the D_{25} ellipse of M31. Two central regions are drawn representatively.

of 20 pointings each. Figure 2 shows the location of these 80 pointings marked with crosses representing the on-axes direction. For two pointings the circle of the inner area of the PSPC (20' radius) is drawn representatively. The D_{25} ellipse of M31 is also drawn for orientation. As can be seen from a comparison with Figure 1, the second survey is more homogenious than the fist one.

SPECTRAL PROPERTIES OF X-RAY GC SOURCES

The process of data preparation and the source detection strategy is explained in detail in Supper[12] et al. for the first survey. A description of the source identification technique can also be found there. For the second survey, the reader should refer to Supper[13]. Within the first survey 396 X-ray sources have been detected from which 29 correlate with GCs. The second survey adds 167 new sources and 9 new correlations with GCs. The main part of new sources in the second survey comes from the outer paths where this survey has a higher sensitivity than the first one.

For the brightest GC sources with more than 200 counts spectral fits have been carried out using a power law as spectral model. Since a detailed discussion of all spectra is beyond the scope of this paper, only a summary of the results can be given here:

For the mean power index $\Gamma = -1.6 \pm 0.3$ was found, which is at the hard end of the mean power index range of $\Gamma = -2.0 \pm 1.1$ for all bright sources independentely of any identification. The fitted N_H column densities correspond with the expected values from HI radio measurements. The spectrum of the brightest GC source is shown in Figure 3. Its spectrum is relatively hard ($\Gamma = 1.27 \pm 0.1$). In addition, this source shows variability on a time scale of $\sim$ 16 hours, which can be seen from its light curve in Figure 4. The hard spectra and the variability of the light curves is expected for LMXBs. Their calculated X-ray luminosities lie between $\sim 10^{36}$ and $\sim 10^{38}$ erg/s.

For the fainter sources hardness ratio colors have been calculated. The hardness

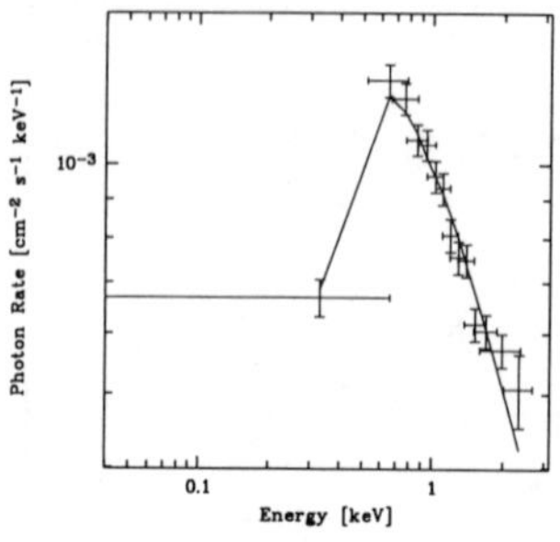

Figure 3: Spectrum and power law fit of the brightest GC source in M31.

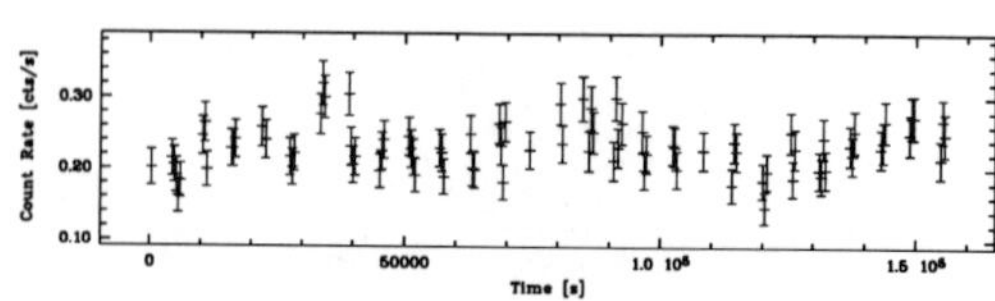

Figure 4: Light curve of the brightest GC source in M31. It shows variability on a time scale of ~ 16 hours.

ratio is defined as $HR_1 = (H - S)/(H + S)$, where H and S represent the count rates in the hard (0.5 - 2.0 keV) and soft (0.1 - 0.4 keV) energy band respectively. For all GC sources $0.0 \leq HR_1 \leq 1.0$ have been found which underlines their hard spectral properties (for comparison, foreground stars have hardness ratios between -0.85 and 0.7). A more detailed description of the spectral properties of all GC sources and fits with different spectral models can be found in Supper[12] et al. and Supper[13].

THE GC LUMINOSITY DISTRIBUTION

The spectral properties of the GC sources described above allow the discrimination against accidental correlations of X-ray sources with optical GCs and therefore to obtain a 'clean' sample of GC sources. As a result, 38 strong candidates for X-ray GC sources have been found with an error of ± 7. The error is the 1σ error level of accidentals comming out of the correlation process and having GC-like spectral properties.

27 of these GC sources are situated in the most sensitive region of the surveys. From the optical lists 473 uncorrelated GCs also lie within this region, for which an upper limit in count rate of 1.36×10^{-3} cts/s can be given from the ROSAT detection threshold within this area. With the number of detected and non-detected X-ray GC sources a luminosity distribution for GC sources in M31 can be determined by using the Kaplan Meier estimator to obtain a nonparametric estimate of the cumulative distribution function $\Phi(L)$ (Schmitt[14]). This distribution function $\Phi(L)$ gives the fraction of detected sources among all detectable sources at a given luminosity. To allow a straight foreward comparison with the results of Hertz[15] and Grindlay for the galactic GC sources, luminosities have been calculated by using an exponential spectral model with kT = 5 keV in the 0.5 - 4.5 keV energy band and assuming 690 kpc for M31 distance. The resulting distribution is shown in Figure 5 (thick line), where also the galactic distribution is shown (thin line), calculated in the same manner. Both distributions are consistent with being from the same parent distribution: a Kolmogorov-Smirnov test yields a probability of 86% for this hypothesis. Furthermore, the maximum luminosities are comparable.

This result suggests that the fraction of X-ray bright clusters among the total cluster population in both galaxies is similar and also their maximum luminosity is comparable, both against earlier reports. This is the result of the more complete detection of X-ray luminous GCs with ROSAT and the also more complete list of the optical detected GC population in M31. Applying a single power law $N(>L) = N_0 \cdot L^{\alpha}$ to the integral luminosity distribution of GCs in M31, yields $\alpha = -0.63 \pm 0.04$ including all data points or $\alpha = -1.02 \pm 0.05$ when only considering luminosities $\geq 1 \times 10^{37}$ erg/s. The difference indicates a flattening of the distribution, quite comparable to the result of the Milky Way.

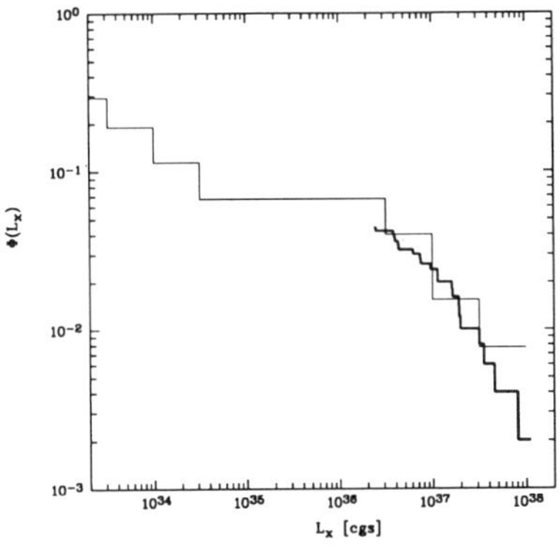

Figure 5: The integral luminosity distribution of the GC sources in M31 (thick line) together with the distribution in our own galaxy (thin line) reported by Hertz and Grindlay. Both presentations are normalized by the parameter-free estimator $\Phi(L)$ described by Schmitt.

ACKNOWLEDGEMENTS

I would like to thank my collaborators at MPE, MIT, and UvA. I would also like to thank Eugene Magnier for putting his list of GCs on my disposal. Parts of the analysis used the SASS and EXSAS data anlysis software. The ROSAT project has been supported by the Bundesministerium für Forschung und Technologie (BMFT) and the Max-Planck-Gesellschaft (MPG).

REFERENCES

1. Trinchieri G., Fabbiano G., 1991, ApJ 382, 82
2. Long K.S., Speybroeck L.P.van, 1983, X-ray emission from normal galaxies. In: Lewin W.H.G., Heuvel E.P.J.van den (eds.) Accretion driven stellar X-ray sources. Cambridge University Press, Cambridge, p. 117
3. Crampton D., Cowley A.P., Hutchings J.B., Schade D.J., Speybroeck L.P.van, 1984, ApJ 284, 663
4. Hodge P.W., 1981, Atlas of the Andromeda Galaxy (Seattle: University of Washington Press).
5. Sargent W., Kowal C., Hartwick F.D.A., Bergh S.van den, 1977, ApJ 82, 947
6. Battistini P., Bonoli F., Bracesi A., Fusi-Pecci F., Malagnini M., Marano B., 1980, A&AS 42, 357
7. Bergh S.van den, 1969, ApJS 19, 145
8. Battistini P., Bònoli F., Buonanno R., Corsi C.E., Fusi-Pecci F., 1982, A&A 113, 39
9. Battistini P.L., Bònoli F., Casavecchia M., et al., 1993, A&A 272, 77
10. Magnier E.A., 1993, PhD Thesis MIT
11. Tully R.B., 1988, Nearby Galaxies Catalog. Cambridge University Press
12. Supper R., Hasinger G., Pietsch W., et al. 1995, ROSAT PSPC survey of M31, A&A submitted
13. Supper R., 1995, PhD Thesis MPE
14. Schmitt J.H.M.M., 1985, ApJ 293, 178
15. Hertz P. Grindlay J.E., 1983, ApJ 275, 105

Observed Characteristics of Supersoft ROSAT Sources in the LMC and other Galaxies[a]

PETER KAHABKA[b]

[b] *Astronomical Institute*
University of Amsterdam, Kruislaan 403
NL-1098 SJ Amsterdam

INTRODUCTION

Supersoft sources (SSS) are soft ($T_{eff} \sim 1 - 10 \times 10^5 K$) and luminous ($L_{bol} \sim 10^{37} - 10^{38}\ erg\ s^{-1}$) X-ray objects. They are considered as a new and distinct class since *ROSAT* discovered seven candidates in the Magellanic Clouds (MCs) in addition to the four objects discovered by *Einstein.* Furthermore seven SSS were found in the Galaxy and at least 15 candidates in the Andromeda Galaxy. SSS can either be close or wide binaries but also hot central stars of planetary nebulae. Steady nuclear burning of hydrogen-rich matter on white dwarfs (WDs) is supposed to be the source of energy in these systems. This affords accretion rates of $\sim 1 - 4 \times 10^{-7} M_{\odot} yr^{-1}$ in close binary systems (orbital periods of $< 1^d - 2^d$) with high mass (0.7-1.2 $M_{\odot}$) WDs and supplied by more massive evolved main-sequence stars and accretion rates of $\sim 10^{-8} - 10^{-7}\ M_{\odot}\ yr^{-1}$ in wide binaries (orbital periods $> 100^d$) with low mass (0.5-0.7 $M_{\odot}$) WDs and supplied by giants with high mass loss. The latter systems are mainly found in the symbiotic binaries.

THE SAMPLE OF SUPERSOFT SOURCES

The number of supersoft systems presently known in the Magellanic Clouds, the Galaxy and the Andromeda Galaxy is ~30 (11 systems are found in the MCs, 7 in the Galaxy). The observations are summarized in review papers.[1,2,3] Optical identifications are known for 10 systems. Orbital periods range from 0.17 to 3.5 days in the close binary systems, for the wide (symbiotic) binaries orbital periods have mostly not been determined. The SSS show quite some variety of time variability, from persistent (since ~10 years), to transient, recurrent outbursts to eclipsing. About 3 systems show in X-rays modulation synchronous to the orbital modulation in the optical. Comparison of the orbital period distribution of SSS (c.f. Fig.1) with the predicted distribution of steady-state nuclear burning (of hydrogen-rich matter) high-mass (0.7-1.2 $M_{\odot}$) WDs shows, that the observed distribution is wider.[4]

But the number of objects is small and only two of six sources fall outside the predicted range. This does not consider the post-nova GQ Mus, which is not in a steady-state burning condition and even went off in X-rays recently. This may be an indication, that among the observed systems are recurrent SSS accreting at rates below the rate of steady-state burning or low-mass (~0.5-0.7 $M_{\odot}$) white dwarfs accreting at rates $\sim 10^{-8} M_{\odot}\ yr^{-1}$ as in the symbiotic binaries.[5]

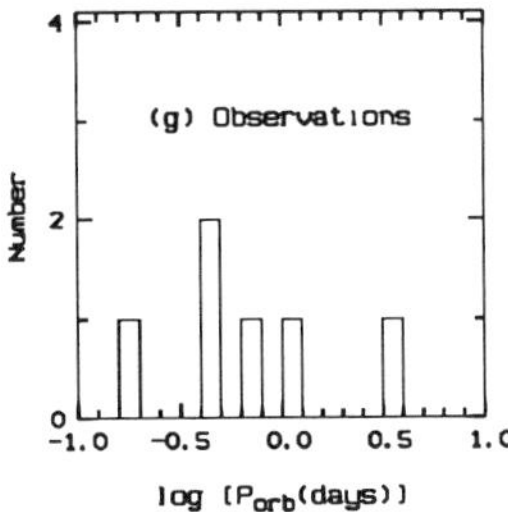

Figure 1. Orbital period distribution of supersoft sources.

X-RAY SPECTRA

Blackbody spectra are the most simple distributions which may be applied to SSS.[6] But it was found, that bolometric blackbody luminosities overestimate the true luminosities by orders of magnitude. In contrary WD atmosphere spectral models give bolometric luminosities in the range $10^{37} - 10^{38}$ $erg\ s^{-1}$ in full agreement with the nuclear burning WD model for all sources.[7,8] WD masses cannot be directly deduced from the observations but for Cal 87 it was shown, that a mass in excess of $1.2 M_\odot$ is needed.

RECURRENT SUPERSOFT SOURCES

SSS have been found to be transient, with turn-on times of days or decay times of years, recurrent with recurrence times of about 2 years and 40 years, eclipsing with the X-ray minimum coinciding with the optical minimum and persistent for ~10 years.[1,9,10,11,12,13] Transient and recurrent SSS can be understood in the model of recurrent SSS.[14] Such systems accrete at rates above the nova rate of $10^{-9} - 10^{-8}$ $M_\odot\ yr^{-1}$ and below the rate of steady-state nuclear burning of $\sim 1 - 4 \times 10^{-7}$ $M_\odot\ yr^{-1}$. Recurrent SSS are expected to have in the mean larger WD masses than steady-state nuclear burning WDs. It can be shown, that in the model of constant mass transfer two observables, the recurrence period t_{recurr} and the X-ray on or decay time t_{decay} together with the rate of steady-state nuclear burning $\dot{M}_{stable}$ determine the critical envelope mass of a WD M_{env}^{crit} and the effective mass accretion rate $\dot{M}_{accr}^{eff}$.[15]

$$M_{env}^{crit} = \dot{M}_{stable}(t_{decay} - \frac{(t_{decay})^2}{t_{recurr}})$$

$$\dot{M}_{accr}^{eff} = \dot{M}_{stable}(\frac{t_{decay}}{t_{recurr}})$$

The critical envelope mass M_{env}^{crit} and the stable accretion rate depend on the WD mass.[14,16] Applying to the transient and recurrent SSS one deduces accretion rates of $\sim 2 - 9 \times 10^{-8}$ $M_\odot\ yr^{-1}$ (lower bounds) and WD masses of $\sim 1.0 - 1.3$ $M_\odot$. These values fit well into the region of recurrent SSS.[14] Recurrent novae may be of a similar nature in case they are thermonuclear novae. Accretion rates $> 10^{-8}$ $M_\odot\ yr^{-1}$

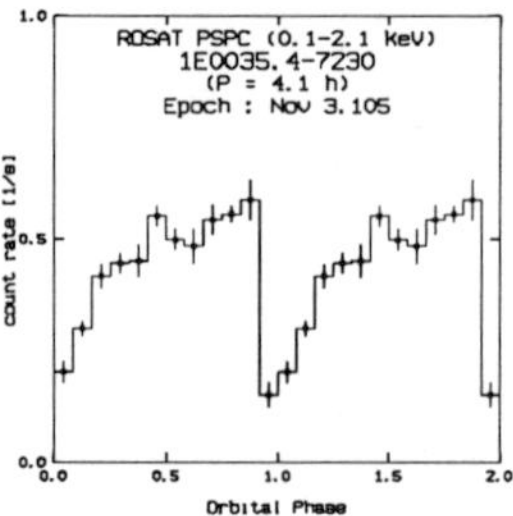

Figure 2. 4.1 hour orbital X-ray lightcurve of 1E0035.4-7230.

and WD masses $\geq 1.3\ M_\odot$ are deduced for these systems.[17] Such high-mass WD systems were considered to be candidates for supernovae type Ia.

ORBITAL MODULATION IN X-RAYS IN 1E0035.4-7230

1E0035.4-7230 has been discovered by the *Einstein* satellite.[18,19] *ROSAT* observations showed high and low intensity states in X-rays.[10] The source was identified in the optical with a variable blue star with strong UV excess and a 4.1 hour orbital period.[20,21] We report here about the discovery of X-ray modulation in 1E0035.4-7230 synchronous with the orbital modulation (c.f. Fig.2).

The orbital period found for 1E0035.4-7230 lies well outside the predicted range of orbital periods for steady-state hydrogen burning high-mass (0.7-1.2 $M_\odot$) WDs.[4] An alternative would be, that 1E0035.4-7230 is a low-mass (~0.5-0.7 $M_\odot$) WD and probably a recurrent supersoft source accreting at rates $\dot{M} \sim 10^{-9} - 10^{-8}\ M_\odot\ yr^{-1}$. Binary evolutionary models with magnetic breaking explain such high accretion rates.[22] Nuclear burning would then last for a few 100 years.[23] A puzzle remains the strong X-ray modulation with a pulsed fraction of ~70% as is known e.g. from magnetic polars (or intermediate polars).[24] If 1E0035.4-7230 would be a polar, synchronisation between the WD and the donor star affords a strong magnetic field of $\sim 7 \times 10^7 G$ (for a $0.7 M_\odot$ WD).[25] But a magnetic nature of the WD, which would have to be confirmed by optical polarimetry is probably not the only and even most straightforeward explanation. Orbital modulation in X-rays in phase with the optical modulation has also been found in Cal 87 which hardly can be synchronized with an orbital period of 10.6 hours. Models like partial X-ray eclipses make either an extended X-ray source (e.g. an accretion disk corona) necessary which seems difficult to achieve in case of a steady nuclear burning WD or a gas stream passing in front of the white dwarf preferently between orbital phases ~0.6 and ~0.9.[26]

CONCLUSIONS

From evolutionary calculations ~120 steady-state nuclear burning SSS are expected in the LMC, ~15 in the SMC, ~1000 in the Galaxy and ~3000 in M31.[4] This does not yet consider the recurrent SSS which add another significant number. The fact, that only about 10 systems are observed may be related to SSS being hidden from our view due to substantial absorption in the galaxies.[27] Supersoft sources have been discussed as progenitors of supernovae type Ia and as systems that eventually may undergo accretion induced collapse (AIC).[28] It is concluded, that supersoft systems having CO degenerates of mass 0.7-1.2$M_{\odot}$ will, if they grow due to accretion beyond the Chandrasekhar limit, experience carbon deflagration and henceforth a type Ia supernova event but are not a case for the AIC (initial WD masses in excess of 1.2$M_{\odot}$ are required). As type Ia supernovae comprise a rather inhomogenous class, the progenitors may be found among the SSS, recurrent novae but also among WD mergers. In late type galaxies CV like systems like the SSS and recurrent novae are favored and in early type galaxies double degenerates.[29] It is interesting to note, that SSS in which the WD does not grow beyond the Chandrasekhar limit, will become double degenerates.[30]

ACKNOWLEDGEMENTS

P. Kahabka is a Human Capital and Mobility fellow.

References

1. Hasinger G., 1994, Supersoft X-Ray Sources, in: AIP Conference Proceedings 308, The Evolution of X-Ray Binaries, eds. S.S. Holt, C.S. Day, 611
2. Kahabka P., Trümper J., 1995, Supersoft ROSAT Sources in the Galaxies, in: IAU Symposium 165, Compact Stars in Binaries
3. Cowley A.P., Schmidtke P.C., Crampton D., et al., 1995, Supersoft X-Ray Sources in the LMC, in: IAU Symposium 165, Compact Stars in Binaries
4. Rappaport S., Di Stephano R., Smith J.D., 1994, ApJ 270, L9
5. Mikolajewska J., 1995, Thermonuclear Runaway Events in Symbiotic Binaries, in: Inter-Relations between Cataclysmic Variables, Proc. of Padova Conference, ed. M. Della Valle, Kluwer
6. Kahabka P., 1995, Supersoft X-Ray Sources in the LMC and SMC, in: Inter-Relations between Cataclysmic Variables, Proc. of the Padova Conference, ed. M. Della Valle, Kluwer
7. Heise J., van Teeseling A., Kahabka P., 1994, A&A 288, L45
8. van Teeseling A., Heise J., Kahabka P., 1995, A&A (in prep.)
9. Schaeidt S., Hasinger G., Trümper J., 1993, A&A 270, L9
10. Kahabka P., Pietsch W., Hasinger G., 1994, A&A 288, 538
11. Greiner J., Wenzel W., 1995, A&A (subm.)
12. Schmidtke P.C., McGrath T.K., Cowley A.P., et al., 1993, PASP 105, 863
13. White N., Giommi P., Angelini L., et al., 1994, IAU Circ.
14. Fujimoto M.Y., 1982, ApJ 257, 767
15. Kahabka P., 1995, A&A (in prep.)
16. Van den Heuvel E.P.J., Bhattacharya D., Nomoto K., et al., 1992, A&A 262, 97
17. Livio M., 1994, Topics in the Theory of Cataclysmic Variables, in: Interactive Binaries, ed. H. Nussbaumer & A. Ott, 135
18. Seward F.D., Mitchell M., 1981, ApJ 243, 736
19. Wang Q., Wu X., 1992, ApJS 78, 391
20. Orio M., Della Valle M., Massone G., 1994, A&A 289, L11
21. Schmidtke P.C., Cowley A.P., McGrath T.K., et al., 1994, IAU Circ. No 6107
22. Pylyser E.H.P., Savonije G.J., 1988, A&A 208, 52
23. Shara M.M., Prialnik D., Koretz A., 1993, ApJ 406, 220
24. Mason K.O., 1985, Space Science Reviews 40, 99
25. Patterson J., 1994, PASP 108, 209
26. Lubov S.H., 1989, ApJ 340, 1064
27. Di Stephano R., Rappaport S., 1995, ApJ (subm.)
28. Livio M., 1994, On the Fate of Accreting White Dwarfs in Cataclysmic Variables and Related Systems, in: Millisecond Pulsars: A Decade of Surprise
29. Della Valle M., Livio M., 1994, ApJ 423, L31
30. Iben I., Tutukov A.V., 1984, ApJS 54, 335

ON THE NATURE, POPULATION, AND IMPLICATIONS OF LUMINOUS SUPERSOFT X-RAY SOURCES

R. DI STEFANO
Harvard-Smithsonian Center for Astrophysics
Cambridge, MA 02138

S. RAPPAPORT
Department of Physics and Center for Space Research, MIT
Cambridge, MA 02139

OVERVIEW

The discovery of a new class of object suggests three central questions: What is the physical nature of the objects? How many of them are there? What are the implications of the existence of this population?

THE NATURE OF THE SOURCES

An important clue to the possible nature of the sources is that the spectra measured by ROSAT seem to be thermal, so that the relationship between the inferred temperature (typically tens of eV) and the inferred bolometric luminosity (typically $10^{37} - 10^{38}$ ergs s^{-1}) leads to the derivation of a characteristic size on the order of that of a white dwarf. Indeed, soon after it became clear that ROSAT was on the trail of a new class of source, the elements of a white dwarf accretor model were outlined in a paper by van den Heuvel, Bhattacharya, Nomoto & Rappaport[1] (hereafter vdHBNR). The critical observation of vdHBNR was that for mass transfer rates in the range of $\sim 1-4 \times 10^{-7} M_\odot$ yr^{-1}, hydrogen accreted onto the surface of a C-O white dwarf can be steadily burned and can provide luminosities in the range observed for supersoft sources. The specific model proposed by vdHBNR is a close binary in which a white dwarf accretes mass from a more massive main sequence (or slightly evolved) star. A detailed evolutionary scenario was developed, and a population synthesis study was carried out by Rappaport, Di Stefano & Smith[2] (hereafter RDS). The result was that, even when a fairly strict set of criteria were used to identify supersoft sources, we computed that the Milky Way and M31 should each house on the order of a thousand active supersoft sources. Although there are uncertainties inherent in any such calculation, the bottom line was that the steady-nuclear-burning white dwarf accretor model passed one crucial test. That is, it seems to represents a phase in the evolution of a large enough fraction of primordial binaries, that a significant number of such systems should be actively involved in mass transfer today.

Perspective

The steady-nuclear-burning region, indicated in the figure below, can be physically realized in one of at least two ways. The first is in the close binaries considered by vdHBNR and by RDS; $4-5$ of the $\sim$ 15 well-studied supersoft sources have properties that seem compatible with this model. The second physical realization is in longer-period white-dwarf-containing binaries in which the donor star is a fairly well-developed giant; $3-4$ of the well-studied supersoft sources are apparently symbiotics of this variety. Systems that fall just below the steady-nuclear burning region will experience occasional nova outbursts. Novae which are still experiencing or which have recently experienced an episode of nuclear burning may also be expected to exhibit the properties of luminous supersoft sources, and indeed at least one of the well-studied sources was a recent nova. See Hasinger[3] and Kahabka[4] for reviews of the observations. It is also to be expected that systems for which $\dot{m}$ is just above the region associated with steady nuclear burning should pass through phases in which they exhibit supersoft behavior. Apart from the white dwarf models described above, it may be possible for accreting neutron stars or black holes to exhibit supersoft behavior.

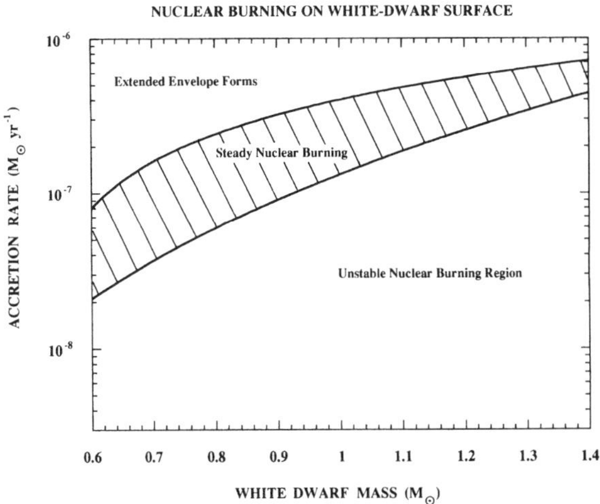

The first-principles population synthesis study of RDS leads to a population on the order of $\sim$ 1000 for the steady-nuclear-burning white dwarf accretors in close binaries alone. Thus, the total population of presently active sources may be larger.

OBSERVATION-BASED DERIVATION OF THE POPULATION

From the beginning it was recognized that, because radiation in the wavebands emitted by supersoft sources is strongly absorbed by the interstellar medium, the number of observed sources was likely to be only a fraction of the number of active emitters. A quantitative analysis to estimate the size of the active population, based on the sources already detected by ROSAT, was done by Di Stefano & Rappaport[5] for each galactic system in which supersoft sources have been observed. The approach we took is to "seed" each system with a set of supersoft sources. The ISM of each galaxy was also modeled, so that, for each individual source, we could compute the column density of neutral hydrogen between it and the Earth. By interfacing with the PIMMS software (Mukai[6]), we were able to compute the ROSAT PSPC count rate generated by each source and to thereby determine whether, given ROSAT's coverage of the relevant part of the sky, the source would have been detected. A Monte Carlo analysis led to the following ranges of values: for M31, 500 – 3000 sources; for the Magellanic Clouds, 20 – 100 sources; and for the Milky Way, 400 – 2000 sources. We also predict that M33 should house roughly 250 supersoft sources.

We note that our study was less sensitive to the lower temperature and lower luminosity sources, so that, should the total underlying population contain a larger fraction of such sources than we considered, our estimates would be underestimates. Furthermore, sources which might be (1) largely locally obscured by, for example, a thick accretion disk, or are (2) not presently luminous, were not covered by our analysis and would therefore supplement our estimate of the underlying population.

SUPERSOFT NEBULAE

On the one hand, the population estimates summarized above may be viewed as encouraging, since they indicate that supersoft sources may form a significant galactic population. On the other hand, these estimates may be viewed as discouraging, since they indicate that ROSAT and its near successors are not sensitive to the majority of the members of this population. We can "make the best of a bad situation", by utilizing the circumstances that make X-ray detection so difficult—the softness of the emitted radiation, and its consequent proclivity for interaction with the surrounding interstellar medium (ISM). The theory of the nebulae which may be associated with luminous supersoft sources has been studied by Rappaport, Chiang, Kallman, & Malina[7] (hereafter RCKM). Their observability depends on the properties of the source (luminosity and temperature) as well as the properties of the ISM. For source properties appropriate to supersoft sources, and for local ISM densities ~ 3 cm^{-3}, typical nebular radii are ~ 5 pc. During an optical survey of nine of the Magellanic Cloud supersoft sources (Remillard, Rappaport, & Macri[8]), one (CAL 83) was observed to have a nebula with properties compatible with the theoretical predictions of RCKM.

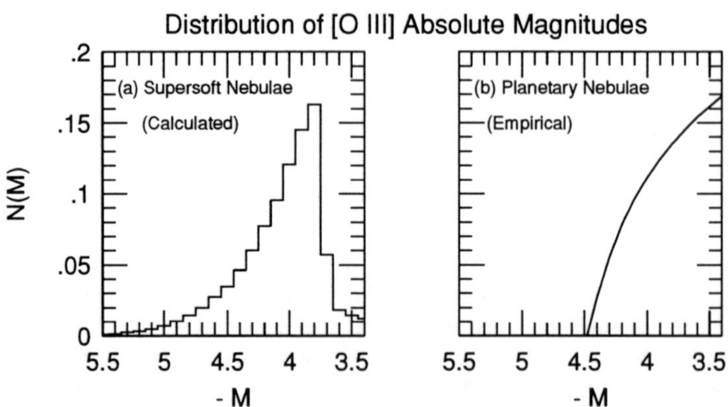

IMPLICATION OF SUPERSOFT NEBULAE

Should at least a subset of supersoft sources be associated with nebulae of the type described above, then possibly the most significant implication would be our ability to conduct systematic surveys for supersoft sources, and to study individual sources, even in distant galaxies. Those supersoft sources which illuminate the surrounding ISM may be discovered and identified though optical line emission of the nebulae. The intensity in individual lines can be large, with, for example 2 – 5% of the bolometric luminosity ultimately emerging in the 5007 line of [O III]. Thus, narrow band imaging alternating with off-band imaging can be used to identify such nebulae. This observational approach has been used to study another class of astrophysical nebula, planetary nebula. In fact, it was noted by Di Stefano, Paerels, & Rappaport[9] that the supersoft nebula luminosity function (SNLF) in the 5007 line of [O III] peaks near the cut of the planetary nebula luminosity function (PNLF). (See the figure above.) The PNLF, particularly near the cut–off has been well studied, since its apparently universal character allows it to be used to as measure of intergalactic distances.

TYPE Ia SUPERNOVAE

The subset of supersoft sources composed of accreting white dwarfs may include progenitors of type Ia supernovae. The rate at which such supersoft sources might undergo supernova explosions has been calculated for a variety of scenarios by RDS. For at least two of the scenarios, the supernova rate is comparable with the rate inferred from observations. A further piece of work (Di Stefano[1]1) utilizes the connection with supersoft nebulae, to compute potentially observable properties of possible type Ia progenitors.

IMPLICATIONS OF SUPERSOFT SOURCES

Relative to other X-ray binaries, supersoft sources are likely to exist in large numbers, with a galaxy like our own containing on the order of ~ 1000 sources, regardless of the fundamental nature of the sources. The radiation they emit is highly interactive with the ISM. Thus, in regions in which the local density of the ISM is high enough (on the order of a few atoms cm^{-3}), supersoft sources may be associated with distinctive nebulae. This opens the possibility that supersoft sources can be systematically studied, even in distant galaxies. Another interesting possible implication of the existence of supersoft sources is model-dependent—i.e., if some subset of them are accreting white dwarfs, then some may be progenitors of type Ia supernovae. In the oral presentation of this talk, an analogy between luminous supersoft X-ray sources and the tentacles of an octopus was drawn. The analogy seems apt, in that the implications associated with the existence of this class of object may reach into diverse areas of astrophysics.

REFERENCES

1. van den Heuvel, E.P.J., Bhattacharya, D., Nomoto, K., & Rappaport, S.A. 1992, *A&A*, **262**, 97-105.
2. Rappaport, S., Di Stefano, R., & Smith, J.D. 1994, *ApJ*, **426**, 692.
3. Hasinger, G. 1994, in '*The Evolution of X-ray Binaries*', Ed. S. S. Holt & C. Day (New York: AIP)
4. Kahabka, P. 1995 this volume.
5. Di Stefano, R., & Rappaport, S. 1994, *ApJ*, **437**, 733.
6. Mukai, K. 1993, *Legacy*, **3**, 21.
7. Rappaport, S., Chiang, E., Kallman, T., & Malina, R. 1994b, in press .
8. Remillard, R., Rappaport, S., & Macri, L. 1994, *ApJ*, in press.
9. Di Stefano, R., Paerels, F. & Rappaport, S. 1994, submitted to *ApJ*.
10. Di Stefano, R. 1995, submitted to *ApJL*.

Evidence for Particle Acceleration and Nuclear Reactions around Compact Relativistic Objects

Eduardo L. Martín

Instituto de Astrofísica de Canarias
38200, La Laguna, Tenerife, Spain

Abstract

In this paper I summarize recent observational evidence for processes of particle acceleration around compact objects and their interaction with the ambient medium, which leads to nucleosynthesis of light elements (D, Li, Be, B).

INTRODUCTION

The strong Li I resonance feature found by Martín et al. (1992) in the K0IV secondary of the X-ray binary V404 Cyg, which harbours the best known candidate for a stellar mass black hole (Casares et al. 1992), suggested the presence of an unanticipated phenomenon. Only about 1% of K-type evolved stars show a strong Li feature (e.g. Brown et al. 1989). This is due to the fragility of Li in stellar interiors, where it is rapidly destroyed by (p,α) reactions at temperatures above $\sim 2.5 \times 10^6$K. Depletion of the surface Li in the K-type secondary of V404 Cyg is expected to occur because of convective mixing, post-main sequence dilution and loss of the outer layers through the Lagrangian point. The Li detection in V404 Cyg implied a mechanism either producing Li or inhibiting its depletion. Searches for Li in other X-ray binaries has so far resulted in Li detections in A0620-00, also a black hole binary (McClintock & Remilliard 1986), and Cen X-4, a neutron star binary (McClintock & Remilliard 1990). These two X-ray binaries, together with V404 Cyg, are the brightest known representatives of the so-called soft X-ray transients (SXTs), sometimes referred to as X-ray novae. They are a subclass of low mass X-ray binaries (LMXBs), characterized by very energetic outbursts lasting several weeks, followed by long quiescent periods of several years. In outburst the SXTs become the brightest X-ray sources in the sky, reaching peak X-ray luminosities of $\sim 10^{38}$ ergs s^{-1}. On the other hand, while in quiescence, their quiescent optical luminosity is dominated by the low mass secondary star, thereby allowing a study of their properties. From the Li detections in these three SXTs, it seems that a high abundance of this element may be common among stars orbiting compact relativistic objects.
A search for Li in cataclysmic binaries (CVs) has very recently been carried out by Martín et al. (1995). They find no evidence for Li in seven late type CV secondaries. These binaries are very similar to the SXTs, but with a white dwarf as the accreting object, instead of a black hole or neutron star. In Table 1 we show the summary of observational searches for Li in SXTs and CVs.

Table 1: Li in binary systems with compact objects

Name	Sp.T.	P_{orb} days	ρ g cm^{-3}	W_{λ}(LiI) mÅ	$R_{Li/Ca}$	log N_{Li} LTE
			CVs			
AE Aqr	K5 V	0.41	1.14	<15	<0.04	<0.2
AH Her	K2-7 V	0.26	2.83	<100*	<0.59*	<1.7**
DX And	K1 V	0.44	0.99	<50*	<0.14*	<1.4
GK Per	K0 IV	1.99	0.05	<30	<0.27	<0.9
HS 1804	M2 V	0.21	4.34	<20	<0.15	<0.2
SS Cyg	K5 V	0.27	2.63	<20	<0.12	<0.4
SY CnC	G8-9 V	0.38	1.33	<30	<0.75	<2.3
Z Cam	K7 V	0.29	2.27	<40	<0.31	<0.7
			SXTs			
A0620-00	K3-5 V	0.32	1.81	235	1.23	2.0
V404 Cyg	K0 IV	6.47	0.005	285	1.36	2.6
Cen X-4	K5-7 IV	0.63	0.48	380	1.41	3.3

Notes: * Equivalent widths estimated from published spectra. ** Assuming a K5 spectral type. Measurement errors are less than 20%. The LiI equivalent widths listed have not been corrected for veiling. Li abundances are given in the customary scale with log N(H)=12.

DISCUSSION

From Table 1, it is evident that the spectral types and densities of the CVs and SXTs are similar. Therefore, the lack of Li in the CVs indicates that the Li found in the SXTs is not due to the intrinsic structure of the secondary star.

A satisfactory evolutionary framework for explaining the Li abundances in SXTs is not found in the literature (see Martín et al. 1994a for details), although the intriguing possibility that the secondaries could be extremely young cannot be ruled out with extant data. The most problematic star to be explained by the age hypothesis is Cen X-4, because late K stars deplete Li by about an order of magnitude in less than 10 Myr (see e.g. García López et al. 1994).

Martín et al. (1992, 1994a) proposed that the Li could come from $\alpha\alpha$ and/or spallation reactions during the outbursts. A small fraction of the fresh Li synthesized in the environment of the black hole or neutron star could be responsible for what we observe in the surface of the secondary. Two possible problems with this hypothesis are that the energetic requirements are high (Martín et al. 1994b), and that a significant fraction of the accreted matter has to be re-expelled. However, the observations of jets of relativistic particles in some X-ray binaries (e.g. Mirabel in this volume) indicate that ejection of matter can take place, and that more energy may be available in the outbursts than anticipated on the basis of X-ray luminosities. In this context it seems easier to understand the transient γ emission feature observed in the SXT Nova Musca (cf. Sunyaev et al. 1992) as partly due to Li production by $\alpha\alpha$ reactions.

PROSPECTS FOR THE FUTURE

Our understanding of particle acceleration and nuclear reactions around compact objects can greatly benefit from complementing the information that can be obtained in the whole electromagnetic spectrum. γ-ray spectroscopy is needed to observe emission lines from 0.4 MeV to 6.6 MeV arising from nuclear reactions due to the interaction of relativistic particles with the ambient medium. The light elements D, Li, Be and B are expected to be produced in such reactions, and the spectroscopic observations will tell us the ratios of their abundances, and the spectrum of the accelerated particles. Such information is essential for understanding the accelerating mechanism and the possible role of black holes and neutron stars in enriching the interstellar medium with light elements.
UV spectroscopy could detect absorption lines of Be and B in the secondaries of compact objects. The most feasible feature to observe with present-day technology is BeIIλ3131.
Optical spectroscopy of more secondaries to SXTs is required in order to confirm if all have high Li abundances, and to measure the surface Li abundance in each of them. However, the faintness of these stars implies that the Li search in most of them will only be possible with 8m class telescopes. Our group (Jorge Casares, Phil Charles, Rafael Rebolo and myself) is working now in trying to detect Li in the SXTs Nova Musca 1991 and Nova Per 1992 with 4m class telescopes, although it is very demanding. The search for Li should be extended to all secondaries of neutron stars and pulsars. We are currently working on Cyg X-2 and Her X-1.
A key observation to make is to re-observe the Li I λ6708 feature in the brightest SXTs (Cen X-4 or V404 Cyg) at very high spectral resolution (R$\geq$50000). The aim is to mesure the ^{7}Li/^{6}Li ratio, which should have a value around 2 if all the Li observed comes from direct $\alpha\alpha$ reactions.
IR spectroscopy is the only way of studying the nature of compact objects with large visual extinctions, for instance those near the Galactic center. In the central regions the black holes may be more massive, and they could be more efficient than stellar sized black holes in producing light-elements on a galactic scale (cf. Jin 1990).
High spatial resolution observations at radio frequencies have demonstrated the presence of jets of accelerated particles around 2 compact objects. Future radio observations of new outbursts will be of great importance for understanding the acceleration mechanism and some insight on the nuclear reactions taking place when the jet interacts with its surroundings will be gained. Fortunately, satellites like GRANAT and GRO will continue to provide us with new targets for multiwavelength studies.

REFERENCES

Brown, J.A., Sneden, C., Lambert, D.L. & Dutchover, E., Jr. 1989, ApJS, 71, 293

Casares, J., Charles, P.A., Naylor, T. 1992, Nature, 355, 614

García López, R.J., Rebolo, R. & Martín, E.L. 1994, AA, 282, 518

Jin, L. 1990, ApJ, 356, 501

Martín, E.L., Spruit, H. & van Paradijs, J. 1994b, AA, 291, L43

Martín, E.L., Rebolo, R., Casares, J. & Charles, P.A. 1992, Nature, 358, 129

Martín, E.L., Rebolo, R., Casares, J. & Charles, P.A. 1994a, ApJ, 435, 791

Martín, E.L., Rebolo, R., Casares, J. & Charles, P.A. 1995, AA, submitted

McClintock, J.E. & Remillard, R.A. 1986, ApJ, 308, 110

McClintock, J.E. & Remillard, R.A. 1990, ApJ, 350, 386

Mirabel, I.F. 1995, this volume

Sunyaev, R.A. et al. 1992, ApJ 389, L75

Evaporation of Companions in VLMXBs and Binary Millisecond Pulsars[a]

JACOB SHAHAM

Department of Physics
Columbia University
New York, NY 10027

Abstract. Irradiation-induced companion winds and a new model for the evaporation of the companion of the millisecond pulsar B1957+20 are discussed. The model may account for all companionless millisecond pulsars observerd today.

COMPANION WINDS

Evolution of Very Low Mass X-Ray Binaries (VLMXBs), binary systems consisting of a Neutron Star (NS) and a $\lesssim .1 - .2\ M_\odot$ companion, is not yet fully understood. It has, however, become of much interest recently, because some VLMXBs may well be the progenitors of isolated millisecond pulsars (mPSRs): It is possible that PSR B1957+20, where a wind is observed to come out of the companion[1], represents the stage of transition from a VLMXB to an isolated mPSR like PSR B1937+21, during which the companion is being evaporated by irradiation from the NS.

The self luminosity per unit area of a companion whose self surface temperature is $10^4 T_4$ °K is $\sigma T^4 \simeq (6 \times 10^{11}\ \mathrm{erg/cm^2 s}) T_4^4$. To upset this energy outflow by irradiation from the NS, assumed to be at a distance $a \equiv a_\odot R_\odot$ from the companion, the NS total luminosity L should be

$$L \gg 4 \times 10^{34} T_4^4 a_\odot^2\ \mathrm{erg/sec}. \tag{1}$$

Eq. (1) explains why companions in *tight* binaries are more likely to be seriously affected by a $L \sim 10^{38}$erg/sec NS than are companions in systems like Her X-1: when the inequality in (1) gets more pronounced, one can expect an expansion of the atmosphere that leads to extentive wind formation or, at high enough values of $L/a_\odot^2$, to a serious expansion of the star.

For an isolated star, an approximate expression for $\dot{\mu}$, the rate of wind outflow per unit area, is[2] (in *cgs* units) $\dot{\mu} \sim 10^{-17}\ell$, where ℓ is the illumination per unit area. One may expect that proximity to the Lagrangian $L1$ point may enhance significantly the irradiation-induced mass outflow. This is, actually, not completely true, because to take advantage of the "open gate" at $L1$ matter must first flow there from the rest of the stellar surface, and this can still not take place at a rate faster than the one determined by the speed of sound. Nevertheless, an enhancement factor of ~ 3 may be expected[2], so that for RL proximity

$$\dot{\mu} \sim 3 \times 10^{-17} \ell\ (cgs). \tag{2}$$

[a]This research has been supported, in part, by NASA grant NAGW-1618.

The above expressions may well not be related to the *last* stages of the companion evaporation, because then its mass m may well be small enough (i.e. of planetary order) that either the illumination exceeds the companion's Eddington luminosity or the resulting surface temperature exceeds its escape value, and an optically thick sonic point will ensue. In this situation, and when photons do not escape the flow (which is the situation when ℓ exceeds some critical value ℓ_0), the illumination essentially fuels the whole kinetic energy of the flow. Thus, for flow velocity u, $\ell \sim \dot{\mu} u^2$ and the absorption grammage $\tilde{\sigma}_{\rm in}$ for the incoming radiation satisfies $\dot{\mu}\frac{R}{u}\tilde{\sigma}_{\rm in}^{-1} \sim 1$ (R is the companion radius); hence[3]

$$u \sim \left(\tilde{\sigma}_{\rm in}^{-1}\ell R\right)^{1/3}, \quad \dot{\mu} \sim \left(\frac{\tilde{\sigma}_{\rm in}^{2}\ell}{R^2}\right)^{1/3} \sim 10^{-6}\ell^{1/3}(cgs). \tag{3}$$

When the evaporated wind, outflowing at a rate $\dot{m}_{\rm W}$, attempts to accrete on the NS, it encounters the pressure of the same radiation whose heat created it in the first place[4,5]. On comparing that pressure with the ram pressure of the wind, one finds two critical values for $\dot{m}_{\rm W}$, $\dot{m}_1$ and $\dot{m}_2$ ($> \dot{m}_1$), which have the following significance: when $\dot{m}_{\rm W} < \dot{m}_1$, expulsion of the wind from the binary will take place even if the system is in accretion; when $\dot{m}_{\rm W} > \dot{m}_2$ accretion will commence. The condition $\dot{m}_{\rm W} < \dot{m}_2$ suffices to expel the wind when there is no accretion, but if the wind already penetrates to inside of the pulsar's velocity-of-light cylinder, $\dot{m}_{\rm W}$ must be further reduced to below $\dot{m}_1$ in order to get expulsion. The values of $\dot{m}_1$ for B1957+20 and B1937+21 are $\sim$ (2.2 and 13) $\times 10^{-11}\, M_\odot\, {\rm yr}^{-1}$ correspondingly; this may be how these systems switched from VLMXBs to windy mPSRs, when the VLMXB Roche lobe (RL) overflow dropped below their respective $\dot{m}_1$ values.

Even when the wind does not accrete and $\dot{m}_{\rm W}/\dot{m}_2 \ll 1$, the outgoing flows are still "slow" (with velocities around the escape velocity from the *companion*, which is below that of the binary). This can have two effects. Firstly, that in the region close to the companion, wind particles can significantly torque the orbit to cause it to change and even to *reduce* its specific angular momentum and shrink[6]. And secondly, that as the wind proceeds to spiral further out it will settle into an "excretion" disk in which planets may, eventually, form[3].

THE WINDY BINARY PULSAR B1957+20 & ITS EVOLUTION

In PSR B1957+20, the observed orbital period seems to vary over timescales $\mathcal{T}$ of order $10^7 - 10^8$ yrs[1]. This variability could be due to either the direct action of the wind on the binary[6] or to quadrupolar (or other) structural changes in the companion[7]. Regardless of the cause, the change in orbital velocity of the binary over the above timescale must involve a rate of energy change (lost or gained) of $m v_{\rm orb}\dot{v}_{\rm orb} \sim \frac{GmM_{\rm NS}}{a\mathcal{T}} \sim 10^{31} - 10^{32}$ erg/sec. If a wind is the broker for these energy exchanges, then $\dot{m} \sim 10^{16} - 10^{17}$ g/sec (on comparing with Eq. (2), note that the RL surface in B1957+20 intercepts $\sim 6 \times 10^{32}$ erg/sec from the pulsar). To claim much smaller winds one must suggest an alternative way of converting the pulsar energy into orbital energy or else suggest an alternative energy broker.

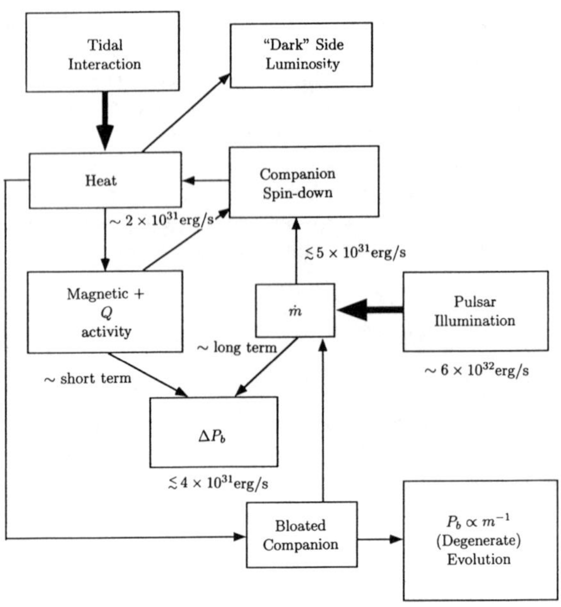

Fig. 1. Block diagram for the companion evaporation scenario in PSR B1957+20.

By combining self-consistently the idea of magnetic activity on the companion of B1957+20 with the idea of the intense irradiation-activated mass loss, Jim Applegate and I[8] were able to put together a scenario for the system that seems to be able to account for several features that have been mystifying us for a long time now. In our scenario (see Fig. 1), magnetic activity directs the wind to leave the surface on magnetic field lines essentially out to the eclipse shock front. As a result, the companion spin is being magnetically breaked continuously. Tidal torques from the NS act to restore corotation and deposit entropy in the companion via tidal friction. This heat is sufficient to support the companion at its bloated configuration as well as to power its "dark" side luminosity in steady state, thus making the companion of B1957+20 the *first discovered tidally-powered star.* The breaking torque is fairly independent of RL contact (it is driven by the irradiation-induced-wind) and, at steady state, it is equal to the tidal torque. But the heat generation rate is proportional to both that torque *and* to the offset between the angular rotations of the companion and the binary. When the companion tries to sink in its RL, that offset increases because the tidal synchronisation time *is* extremely sensitive to RL contact and it grows longer as contact is being lost. Thus, the tidal torques deposit now *more* heat in the companion, causing it to re-expand. This is how RL contact regulates itself. The heat generation, in turn, triggers the magnetic activity that created it, again in a self-regulatory manner. A glance into the future shows that the strong optically thick evaporative winds will set in at a companion mass of order a few Earth masses; that will bring about the final disappearance of the companion.

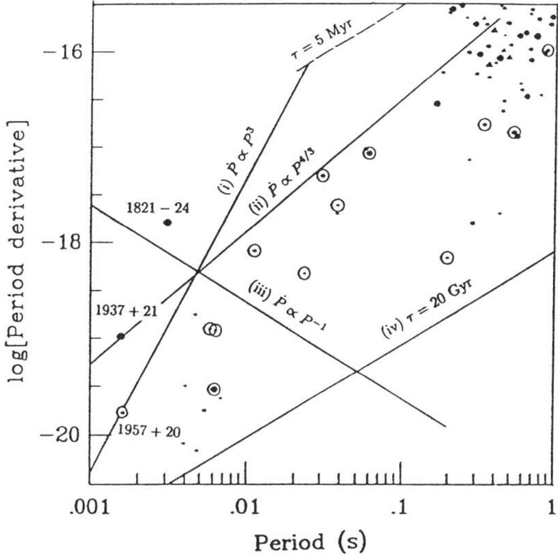

Fig. 2. The lower left part of Fig. 7 of the Taylor et al.[9] pulsar catalog. We added four straight lines: *(i)* the $\dot{P} \propto P^3$ line represents the boundary between pulsars more luminous (left) and less luminous (right) than B1957+20; *(ii)* the $\dot{P} \propto P^{\frac{4}{3}}$ line is the accretion-spin-up line, adjusted to go through B1937+21; *(iii)* the $\dot{P} \propto P^{-1}$ line is the aging curve for pulsars with constant magnetic moments (after spin-up, pulsars move down a line parallel to the one drawn); and *(iv)* the $\tau = 20Gyr$ line bounds from below pulsars younger than that age. All pulsars in the region bounded by *(i)*, *(ii)* and *(iv)* could have once been as luminous as B1957+20 and could have thus evaporated their companions by the Applegate-Shaham mechanism. The ones above line *(iii)* could not have been accretion-spun-up to luminosities as high as that of B1957+20. Except for 1821-24, whose measured period derivative may be contaminated by other effects, all single mPSRs are indeed in the "allowed" region, and could have thus evaporated their companions in the past.

Every single mPSR known today could have had in its past a B1957+20-like episode in which it fully evaporated its companion. We demonstrate this in Fig. 2.

REFERENCES

1. Arzoumanian, Z., Fruchter, A.S., and Taylor, J.H. 1994, ApJ, **426**, L85.
2. Ruderman, M., Shaham, J., Tavani, M., and Eichler, D. 1989a, ApJ, **343**, 292.
3. Banit, M., Ruderman, M.A., Shaham, J., and Applegate, J.H. 1993, ApJ, **415**, 779.
4. Ruderman, M., Shaham, J., and Tavani, M. 1989b, ApJ, **336**, 507.
5. Shaham, J., and Tavani, M. 1991, ApJ, **377**, 588.
6. Banit, M., and Shaham, J. 1992, ApJ, **388**, L19.
7. Applegate, J.H. 1992, ApJ, **385**, 621.
8. Applegate, J.H., and Shaham, J. 1994, ApJ, **436**, 312.
9. Taylor, J.H., Manchester, R.N. and Lyne, A.G. 1993, *ApJ (Suppl.)*, **88**, 529.

Accretion Discs Around Nonmagnetized Stars*

G.S.BISNOVATYI-KOGAN

Space Research Institute
Profsoyuznaya 84/32, Moscow 117810, Russia
(Permanent address)
and
Institute of Astronomy, Cambridge University,
Madingley Road, Cambridge CB3 0HA, U.K.

INTRODUCTION

Investigations of low-mass-X-ray binaries (LMXB) or cataclismic variables have lead to the conclusion that the accreting object can rotate with a Keplerian speed. The problem of accretion onto a rapidly rotating star was investigated in [1-4]. For self-consistency of the solution for a critically rotating star a flux of angular momentum of matter flowing inside must hold the equatorial speed at critical level[5]. The velocity profile in the transition zone to the accretion disc is smooth if star rotates rapidly.

The boundary layer (BL) is formed, when there is a disk accretion onto the slowly rotating nonmagnetized star [6]. The energy release inside BL can be comparable with that of the disk. Solution of BL structure is necessary also for completing the solution of the disk itself, because the integration constant must be found by matching BL solution. A successful approach to the problem was proposed in [7] basing on the method of matched asymptotic expansions (MAE)[8]. The problem of the disc structure together with BL was solved analytically in [9], using MAE method, for polytropic equation of state and α approximation[6] for viscosity.

ACCRETION INTO RAPIDLY ROTATING STAR

The demand of self-consistency during accretion onto a rapidly rotating star may be formulated as the condition that **the star absorb the accreted matter with a specific angular momentum such that the star remains in the state of critical rotation**[5]. If viscosity inside the star is high, the star rotates almost rigidly, and for polytropic equation of state $P = K\rho^{1+\frac{1}{n}}$,

*This work was supported in part RFFI grant 93-02-17106, Astronomy Programm of RMSTP topic 3-169 and Royal Society Kapitza Fellowship. DFG,DARA and MPE grant supported author's participation in the Symposium.

there is a self-similar solution for star-disc system not depending on mass. The value of the specific angular momentum of the accreted matter is determined by the structure of the star, described by nondimensional: function $\Theta(\xi,\mu)$, and angular velocity ω, so that[10]

$$\xi = r/r_*, \quad \Theta^n = \rho/\rho_c, \quad \omega = \Omega/\Omega_* \tag{1}$$

where ρ_c is the central density of the star and r_* and Ω_* are given by

$$r_* = \left(\frac{(n+1)K\rho_c^{\frac{1}{n}-1}}{4\pi G}\right)^{1/2}, \quad \Omega_* = \sqrt{2\pi G\rho_c} \tag{2}$$

Introduce nondimensional values of the mass $\mathcal{M}_n$ and of the total angular momentum $\mathcal{J}_n$, defined as

$$\mathcal{M}_n = \frac{1}{2}\int_{-1}^{1} d\mu \int_0^{\xi_{out}(\mu)} \Theta^n(\xi,\mu)\xi^2 d\xi, \tag{3}$$

$$\mathcal{J}_n = \frac{\omega}{2}\int_{-1}^{1} (1-\mu^2)d\mu \int_0^{\xi_{out}(\mu)} \Theta^n(\xi,\mu)\xi^4 d\xi, \tag{4}$$

During accretion along the critical states a star absorbs the matter with specific angulal momentum

$$j_a = \frac{dJ}{dM}_c = \frac{5-2n}{3-n}\frac{J}{M} = r_*^2\sqrt{2\pi G\rho_c}\frac{\mathcal{J}_n}{\mathcal{M}_n}\frac{5-2n}{3-n} \tag{5}$$

The specific angular momentum of matter j_e at the stellar equator r_e and the ratio j_a/j_e may be written as

$$j_e = \Omega r_e^2 = r_*^2\sqrt{2\pi G\rho_c}\xi_e^2\omega, \quad \frac{j_a}{j_e} = \frac{5-2n}{3-n}\frac{\mathcal{J}_n}{\mathcal{M}_n\xi_e^2\omega}. \tag{6}$$

This ratio is a function of n and ω. The nondimensional values of the critical ω_c, the equatorial radius $\xi_{out}(\theta=\pi/2)=\xi_e$, the mass $\mathcal{M}_n$, the momentum of inertia around the rotational axis $\mathcal{I}_n = \mathcal{J}_n/\omega_c$, the average specific angular momentum of the star ζ_s , the nondimensional derivative ζ_a, and ζ_e ($j_s = J/M$)

$$\zeta_s = j_s/(r_*^2\sqrt{2\pi G\rho_c}), \quad \zeta_a = \frac{5-2n}{3-n}\zeta_s, \quad \zeta_e = j_e/(r_*^2\sqrt{2\pi G\rho_c}) \tag{7}$$

for several polytropic indices n in the state of critical rotation are given in table 1, based on calculations [11–13].

SOLUTION FOR A DISC WITH BOUNDARY LAYER.

The equations for a thin accretion disk with α-viscosity $\mu = \alpha\rho u_{s0}z_0$, have an integral[6]

$$\frac{\dot{M}}{2\pi}(j - j_0) = \alpha\Sigma u_{s0}z_0 r^3\frac{d\Omega}{dr} \tag{8}$$

Table 1: Some parameters of critically rotating polytropic stars

n	ω_c	ξ_e	$\frac{\xi_e}{\xi_p}$	$\mathcal{M}_n$	$\mathcal{I}_n$	ζ_s	$\frac{\zeta_a}{\zeta_e}=\beta$
0.5	0.389 (0.367)		2.51				
0.808	0.326	4.7652	1.917	5.0248	24.43	1.585	0.3304
1.0	0.289	4.8265	1.792	4.289	18.73	1.262	0.2805
1.5	0.209	5.36	1.626	3.2138	12.18	0.792	0.176
2.0	0.147	6.307	1.555	2.6518	9.62	0.533	0.0894
2.5	0.09965	7.7623	1.522	2.30563	8.49	0.367	0.0
3.0	0.0639	10.123	1.535	2.0743	8.125	0.250	$-\infty$

where u_{s0} is a sound speed at $z = 0$ and z_0 is semi-thickness of the disk. The surface density Σ, mass flux over the disk $\dot{M}$, integrated pressure $\mathcal{P}$, angular velocity Ω and specific angular momentum j are determined as usual. The integration constant j_0 after multiplication by $\dot{M}$ gives the total (advective plus viscous) flux of angular momentum within the accretion disk.

Inside the BL variables change considerably over the small thickness of the layer $H_b \ll r_{in}$. The radial velocity term is negligible for small α, the pressure term must be left [7]; H_b is smaller then its vertical size z_0. In MAE we use stellar radius r_* as an inner boundary of the disk. The variable x

$$r = r_* + \delta x, \quad \delta = \frac{H_b}{r_*} \ll 1 \tag{9}$$

is used inside BL instead of r. The solution within BL is looked for in the region $0 < x < \infty$, while the outer solution is valid in $r_* < r < \infty$. According to MAE the solutions are matched so that values for the outer solution at $r = r_*$ are equal to the values of the inner ones at $x = \infty$. The equations inside BL

$$\frac{d\mathcal{P}}{dx} = -\Omega_{K*}^2 H_b \Sigma (1 - \omega^2), \quad \frac{d\omega}{dx} = -\frac{\dot{M} H_b}{2\pi \Sigma \nu_b r_*^2}(\beta_b - \omega) \tag{10}$$

with the viscosity coeffisient μ

$$\mu = \rho \nu_b, \qquad \nu_b = \alpha_b u_{s0} H_b \tag{11}$$

together with disc equations have a solution for an isothermal disk with $\mathcal{P} = K\Sigma$, using MAE

$$\Sigma = \left(-\frac{\dot{M}\sqrt{K}}{2\pi\alpha_b H_p}\right)\frac{1}{r_*^2 \Omega_{K*}^2}\left[(\omega - \omega_*)\left(1 + \frac{\omega + \omega_*}{2}\right)\right]^{-1} \tag{12}$$

$$\frac{H_b}{r_*} \approx \frac{2K}{r_*^2 \Omega_{K*}^2}\left[(1 - \omega_*)(3 + \omega_*)\right]^{-1}, \quad 1 - \beta = \frac{3}{\sqrt{2}}\frac{\sqrt{K}}{r_* \Omega_{K*}}\frac{\alpha}{\alpha_b} \tag{13}$$

$$(1-\omega)^{-2}(\omega-\omega_*)^{\frac{3+\omega_*}{1+\omega_*}}(2+\omega+\omega_*)^{-\frac{1-\omega_*}{1+\omega_*}} = \exp\left[2\frac{H_b}{K}\Omega_{K*}^2 x(1-\omega_*)(3+\omega_*)\right] \tag{14}$$

The (12) was obtained in [14], the solution for polytropic case see in [9]. The total luminosity of the accretion disc is equal to

$$L = \int_{r_i}^{\infty} 4\pi F r dr = (\frac{3}{2}-\beta)\dot{M}\frac{GM}{r_i} \tag{15}$$

When $\beta = 1$ half of the gravitational energy of accretion is radiated from the disc. For slowly rotating neutron star at $r_* > 3r_g$ the disc and BL luminosity each are equal to $GM\dot{M}/2r_s$. While the star absorbs angular momentum, the BL luminosity gradualy decrease to zero at critical rotation. A rapid increase of the disc luminosity (2-3 times) happens at that moment besause β decreases from 1 to j_a/j_e (see Table 1).

CONCLUSIONS

1. The self-consistent regime of accretion onto a rapidly rotating star implies that the specific angular momentum of matter accreting onto the star j_a is equal to the derivative along the equilibrium states $j_a = \frac{dJ}{dM}$

2.For slowly rotating star the integration constant β is less then 1 by a small value, of the order of the parameter z_0/r_*.

3.The luminosity of the disc with BL gradually decreases while the stellar rotation rate increases, and then rapidly increases by a factor ~2-3 , when the star becomes rotating critically.

REFERENCES

1. Narayan R., Popham R., 1989, Ap. J., **346**, L25.
2. Popham R., Narayan R., 1991, Ap. J., **370**, 604.
3. Paczynski B., 1991, Ap.J., **370**, 597.
4. Colpi M., Nanmurelli M., Calvani M., 1991, Mon.Not.R.A.S., **253**, 55.
5. Bisnovatyi-Kogan G.S., 1993, Astron.Astrophys. **274**, 796.
6. Shakura N.I., 1973, Sov. Astron., **16**, 756.
7. Regev O., 1983, Astron. Ap., **126**, 146.
8. Nayfeh A., 1973, Perturbation methods. John Wiley & Sons (New York).
9. Bisnovatyi-Kogan G.S., 1994, Mon.Not.R.A.S. **269**, 557.
10. Chandrasekhar S.,1989, Selected papers, vol.1, p.185; Chicago Univ press.
11. James R.A., 1964, Ap. J., **140**, 552.
12. Blinnikov S.I., 1975, Sov.Astron., **19**, 151.
13. Ipser J.R., Managan R.A., 1981, Ap. J. **250**, 362.
14. Shakura N.I., Sunyaev R.A., 1988, Adv. Space Res., **8**, (2)135.

Inclination effects in Z sources?[a]

ERIK KUULKERS AND MICHIEL VAN DER KLIS

Astronomical Institute "Anton Pannekoek", University of Amsterdam and Center for High Energy Astrophysics, Kruislaan 403, 1098 SJ Amsterdam, The Netherlands

INTRODUCTION

The brightest known low-mass X-ray binaries[18] are called Z sources, because most of them trace out a "Z"-like shape in the so-called X-ray colour-colour diagram[7,17] (CD). From the upper left to the lower right the branches of the "Z" are called horizontal branch (HB), normal branch (NB) and flaring branch (FB). It is thought that the mass accretion in these sources is at near-Eddington rates[12], and increases from the HB, through the NB, into the FB[6]. The properties of the fast timing behaviour of Z sources is correlated to their instantaneous position in the "Z" pattern[4]. On the basis of the behaviour of the Z sources it was suggested [4,9,10,15] that they can be divided into two subgroups. Here we briefly summarize the status of the research of Z sources based on EXOSAT observations.

COLOUR-COLOUR AND HARDNESS-INTENSITY DIAGRAMS

In this paper we used all available EXOSAT ME spectral data from 1984 to 1986 of the six Z sources. We divided the X-ray spectra into three or four energy bands and determined soft colour (hardness ratio of the count rates in the lowest energy bands), hard colour (hardness ratio of the count rates in the highest energy bands) and X-ray intensity (count rate of all energy bands added, corrected for background, dead time and collimator response).

In Figs. 1 and 2 we show a compilation of all hard colour vs. soft colour diagrams (CDs) and soft colour vs. intensity (HIDs)[b] of the six Z sources[3,8,9,10,11]. The energy bands used are: Cyg X–2: 0.9–3.1–6.4–19.7 keV, Sco X–1: 0.9–3.1–4.9–6.6–19.5 keV, GX 340+0: 1.2–4.7–6.6–19.9 keV, GX 17+2: 1.2–4.7–6.6–19.9 keV, GX 5–1: 1.4–3.6–6.0–17.1 keV and GX 349+2: 0.9–4.2–5.8–7.6–20.5 keV. All points are 200 s averages. The different symbols refer to different observation periods.

All Z sources exhibit all three of the branches (Fig. 1), except for GX 349+2, which has not been detected in the HB. Cyg X–2, GX 340+0 and GX 5–1 (hereafter "Cyg X–2 group") show shifts in the position of the Z in the CD between different observations periods. Sco X–1, GX 17+2 and maybe GX 349+2 (hereafter "Sco X–1 group"), however, show a relatively more stable Z in the CD. The HBs of the Cyg X–2 group is horizontally or diagonally oriented, while the HB of the Sco X–1 group is more vertically oriented. Also, the FB is very pronounced in the Sco X–1 group, whereas it is small in the Cyg X–2 group.

[a]This work was supported in part by the Netherlands Organisation for Scientific Research (NWO) under grant PGS 78-277.

[b]The HID of Sco X–1 has been kindly provided by Stefan Dieters.

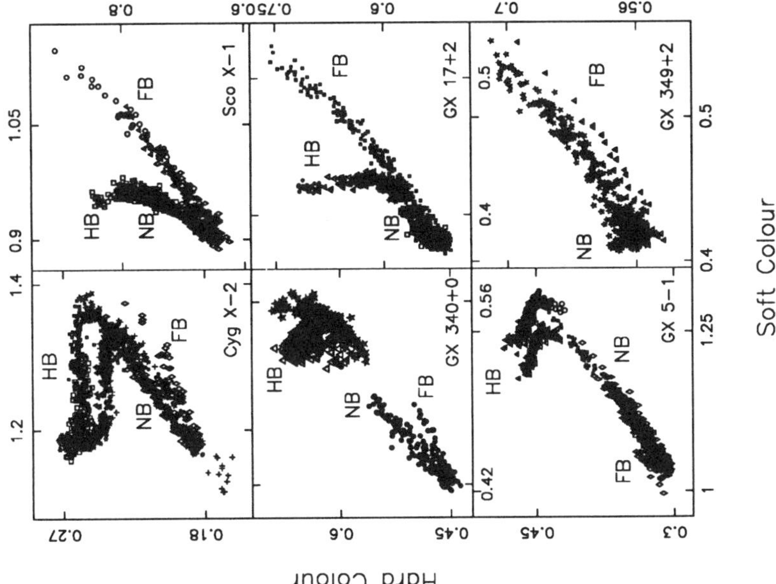

Figure 1: EXOSAT Colour-colour diagrams of the six Z sources

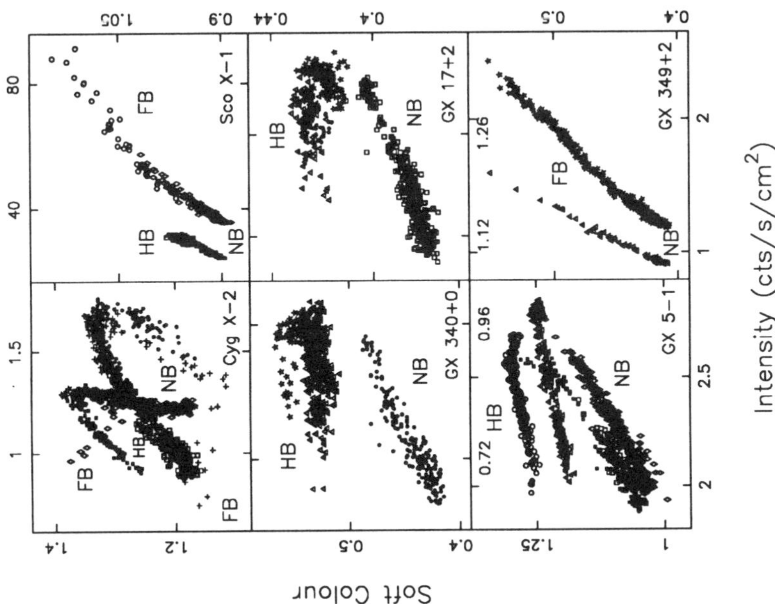

Figure 2: EXOSAT Hardness-intensity diagrams of the six Z sources

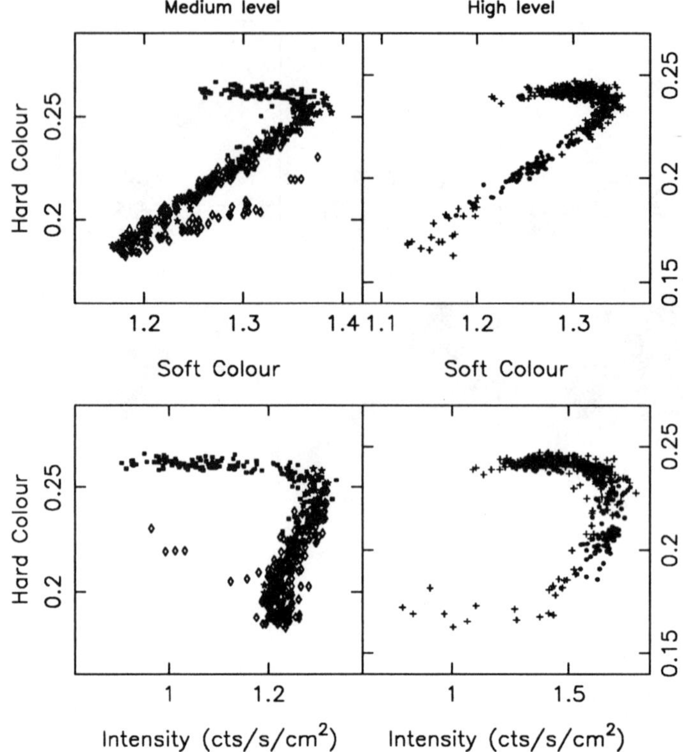

Figure 3: EXOSAT Cyg X-2 Colour-colour diagram (top) and Hardness-intensity diagram (bottom) during a medium level (left) and a high level (right).

The HIDs (Fig. 2) are more complex. We can see that although they are made up of 5 to 8 different observations, there are only two Z tracks in the HIDs of GX 5–1 and Cyg X–2. In the sources Sco X–1 and GX 349+2 we see also two shifted tracks, but here the HIDs have been compiled from only two observation periods. Clear is the FB in the latter sources, in which the intensity flares up (for GX 17+2 see Penninx et al.[14]). In Cyg X–2, however, the intensity drops, when the source moves up the FB[7,11]. This is also true for GX 5–1[9] and GX 340+0[15].

Because of the fact that the sources in the Cyg X–2 group shows secular variations in their Z pattern in the CD and HID and dip in intensity in the FB, whereas the sources in the Sco X–1 group show a more stable Z pattern in the CD and really flare up in the FB it was suggested[9] that this behaviour is governed by the inclination at which we view the sources[4,5]. the Cyg X–2 group is then viewed at a high inclination (more edge-on), while the Sco X–1 group is viewed at a lower inclination (more face-on). measurements of the inclination have only been reported for Sco X–1 (15–40°)[2] and for Cyg X–2 (65–75°)[1]. the new kind of QPO found during a dip in intensity in the FB of Cyg X–2[7] also favors the high inclination hypothesis.

CYG X–2

If we look at Cyg X–2 in more detail, we see that the same observations which are separated into two Z tracks in the HID are also slightly separated in the CD. in Fig. 3 we display both Z tracks separately in the CD (upper panel) and HID (lower panel; here the hard colour is used instead of, as in Fig 2, the soft colour.). Because of the differences in mean X-ray intensity in the NB of the two different Z tracks we refer to these observations as medium and high level[11]. We note that EXOSAT also observed Cyg X–2 at a so-called low level[11]; we will not dicuss this in the present paper.

In the medium level we see a clear FB in both CD and HID. The intensity decreases when the source moves into the upper FB, while both the soft colour and the hard colour increase. This has the effect of extending the FB in the CD to higher soft and hard colours, creating a true Z, whereas in the HID it extends to lower intensities and higher colours, so that the pattern is not like a Z. An FB-like dip was also observed during the high level, but no colour changes occurred in this dip so that no FB was found in the CD. In the HID we see an excursion to lower intensities while the colours stay approximately constant, which we interpret as a FB. Exactly the same FB/dip behaviour was seen in Ginga observations[6,13,19]. It is clear that the medium and high levels are recurrent[11].

In view of the interpretation of the Z track as a consequence of $\dot{M}$ changes, and since we observe the HB, NB and FB during both the medium and high levels we do not think that changes in the mass accretion rate are also responsible for the secular variations leading to the medium and high levels. A similar conclusion was derived from the study of the secular and timing behaviour in GX 5–1[9]. Therefore, the possibility was put forward[11] that in Cyg X–2 contains a precessing accretion disk, as has been suggested for other X-ray binaries[16] If the disk precesses in Cyg X–2, one would predict that the occurrence of the different levels is periodic[11].

References

1. Cowley A. P., et al., 1979, ApJ **231**, 539
2. Crampton D., et al., 1976, ApJ **207**, 907
3. Dieters S., van der Klis M., 1995, in preparation
4. Hasinger G., van der Klis M., 1989, A&A **225**, 79
5. Hasinger G., et al., 1989, ApJ **337**, 843
6. Hasinger G., et al., 1990, A&A **235**,131
7. Kuulkers E., van der Klis M., 1995a, A&A, in press
8. Kuulkers E., van der Klis M., 1995b, in preparation
9. Kuulkers E., et al., 1994, A&A **289**, 795
10. Kuulkers E., et al., 1995a, in preparation
11. Kuulkers E., et al., 1995b, in preparation
12. Lamb F. K., 1989, in: Two-Topics in X-ray, 23rd ESLAB Symp., J. Hunt, & B. Battrick (eds.), 1989, ESA SP-296, p. 215
13. Mitsuda K., Dotani T., 1989, PASJ **41**, 557
14. Penninx W., et al., 1990, MNRAS **243**, 114
15. Penninx W., et al., 1991, MNRAS **249**, 113
16. Priedhorsky, W. C., Holt S. S., 1987, Sp. Sc. Rev. **45**, 291
17. Schulz N. S., et al., 1989, A&A **225**, 48
18. van der Klis M., 1989, ARA&A **27**, 517
19. Vrtilek S. D., et al., 1986, ApJ **307**, 698

SN Ia: Light Curves, Spectra and H_o

P. Höflich[1], C. Dominik[2], A. Khokhlov[3], E. Müller[4] & J.C. Wheeler[3]
[1] *CfA, Harvard University, Cambridge, MA 02138, USA*
[2] *NASA Ames, Moffet Fields, CA 94035, USA*
[3] *Dept. of Astronomy, U Texas, Austin, TX 78712, USA*
[4] *MPA, D-8046 Garching, BRD*

MODELS FOR SNIa

Type Ia supernovae (SNe Ia) are thought to be thermonuclear explosions of carbon-oxygen white dwarfs. However, details of the explosion and of the progenitor structure are still under debate. A solution is relevant for our understanding of the explosion mechanism and both the chemical evolution of galaxies and the use of SNe Ia as distance indicators. To address these problems, theoretical B, V, R, and I light curves (LC) and NLTE-spectra are presented and compared to observations. The calculations are consistent with respect to the explosion mechanism, the optical and IR light curves, and the spectral evolution, leaving the description of the nuclear burning front and the structure of the white dwarf (WD) the only free parameters. The explosions are calculated using one-dimensional Lagrangian hydro [1] and radiation-hydro codes (PPM) including nuclear networks. Subsequently, light curves are constructed using our elaborated LC scheme including an implicit radiation transport, expansion opacities, MC-γ transport, and molecular and dust formation. Based on the previous results, detailed NLTE-spectra are constructed for several moments of time. For more details see [2-9].

A set of 33 SN Ia explosion models has been considered which encompasses all currently discussed explosion scenarios. The set consists of three deflagration models (DF1, DF1MIX, W7 [10]), two detonation models (DET1, DET2), seven delayed detonation models (N21, N32, M35-39), eight He-detonations of low mass white dwarfs (HeD2-HeD12, see Fig. 1), nine pulsating delayed detonation models (PDD1a-c,PDD3, PDD5-9), and three tamped detonation models (DET2env2-6), which may be regarded as a crude approximation to the merging scenario.

Different explosion models can be well discriminated by the slopes of the LCs and changes of spectral features (e.g. line shifts $\Rightarrow$ expansion velocities). The differences can be understood in terms of the expansion rate of the ejecta, the total energy release, the distribution of the radioactive matter, and the total mass and density structure of the envelope [2-5,7,8].

COMPARISON WITH OBSERVATIONS AND H_o

We found that fast rising LCs (e.g. SN1972e, SN1981b, 1994d) can be explained by "delayed detonation" or deflagration models. However, slow rising LCs (e.g. SN1990n) require models which have formed a compact envelope of typically 0.2 to 0.4 $M_\odot$ [6-8]. Such envelopes can be produced by a pulsation phase during the explosion or by merging white dwarfs. Our interpretation is backed up by the expansion velocities observed in the spectra [6,8]. In the framework of pulsating delayed detonation models, strongly subluminous supernovae (SN1991bg, SN1992k, SN1992bo) can be understood by models with a small ^{56}Ni production ($M_{Ni} \leq 0.4\ M_{\odot}$, Fig. 2, [7]). Consistent with the observations [11], the latter models show narrow maxima and fast postmaximum declines, are very red at maximum light and show no secondary IR-maximum. Note that all models with $M_{Ni} \geq 0.4 M_{\odot}$ are similar bright at maximum light ($-19.2^m \geq M_V \geq -19.6^m$). We failed to find any He detonation model within a range for the CO-core from 0.6 to 0.8 $M_{\odot}$ and with 0.12 to 0.22 $M_{\odot}$ of accreted He [12], which can reproduce strongly subluminous SN Ia. The main problems were connected to the heating of the outer layers by radioactive ^{56}Ni which causes both a very blue color at maximum light and a strong secondary IR-maximum. Further studies are under way to test whether spectra and LCs of normal bright SNe Ia can be understood by He detonation models. Note that this uncertainty does not affect the discussion on H_o (see below), since M_V is comparable with normal delayed detonation models if expansion opacities are used.

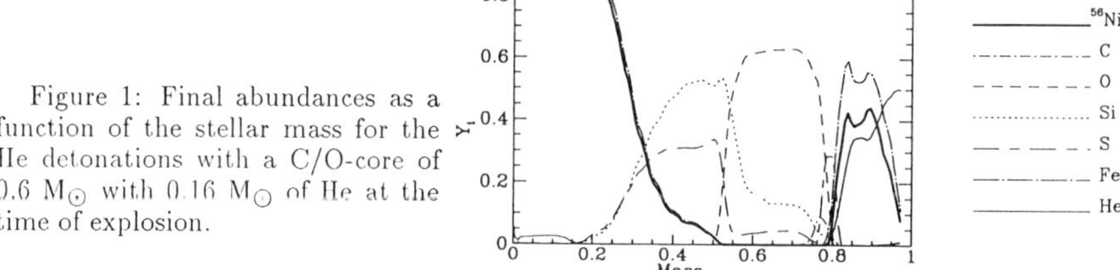

Figure 1: Final abundances as a function of the stellar mass for the He detonations with a C/O-core of 0.6 $M_{\odot}$ with 0.16 $M_{\odot}$ of He at the time of explosion.

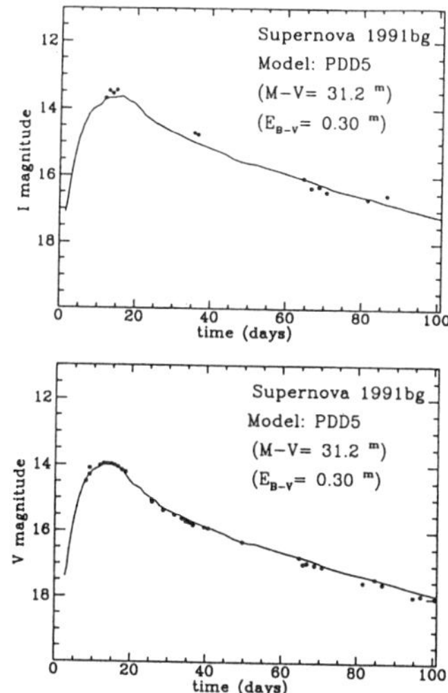

Figure 2: Observed V and I light curves for the subluminous SN 1991bg are compared with the calculated light curve of the pulsating delayed detonation model PDD5 which produces about 0.12 $M_{\odot}$ of ^{56}Ni. According to our calculations, strong molecular bands of CO should appear in the IR of CO starting about 2 weeks past maximum light [8]. Dust particles of about 300 atoms form during the first week after the initial explosion [9]. Further studies are required to decide whether the dust can survive and causes the extinction or whether a model with somewhat lesser ^{56}Ni production is needed for SN1991bg.

Despite the small number of SN Ia, it may be worthwhile to ask for the relation between the host galaxy and type of explosion. In spirals, about an equal number of single burning-mode models (delayed detonations and deflagrations), and pulsating and envelope models are successful whereas, in ellipticals, only the latter group seems to be present. This may give some hint on the progenitor evolution because ellipticals have no stars with masses $\geq 3M_\odot$.

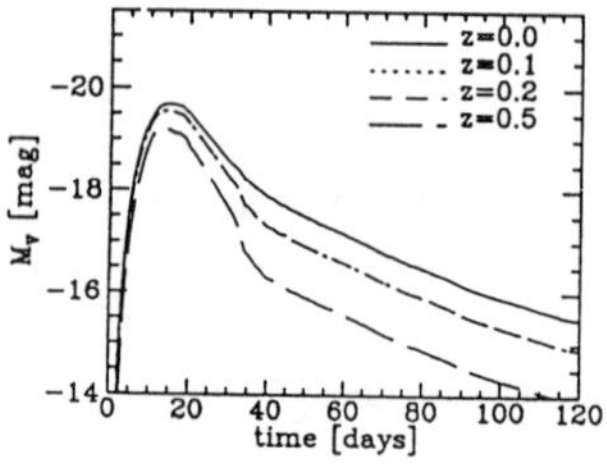

Figure 3: Change of the apparent shape of the V LC of W7 for different red shifts z. Models allow for a consistent determination of the k-correction.

Table 1: List of observed SNe Ia for which sufficient monochromatic light curve data are available to allow for a discrimination of theoretical models. Columns 2 to 5 give the parent galaxy, the distance D, the color excess according to our models, and the names of the models which can reproduce the observations within the error bars. The quoted errors in D include those of the measurements, reddening correction, filter functions and models [6].

Supernova	galaxy	Type	D[Mpc]	E(B − V)	acceptable models
SN 1937C	IC 4182	Sm	4.8±1	0.10	N32,W7, DET2
SN 1970J	NGC 7619	dE	66 ± 8	0.01	DET2ENV4/2, (PDD3/1c)
SN 1971G	NGC 4165	Sb	35 ± 9	0.0	N32, DET2, W7, M36
SN 1972E	NGC 5253	I	4.3±0:6	0.04	N21, M35
SN 1972J	NGC 7634	SBO	52 ± 8	0.01	N32, W7, DET2, DF1
SN 1973N	NGC 7495	Sc	70 ± 20	0.10	N32, W7
SN 1974G	NGC 4414	Sc	18.5 ±5	0.0	N32, W7, DET2
SN 1975N	NGC 7723	SBO	25 ± 7	0.32	PDD3/1c/5,DET2ENV2
SN 1981B	NGC 4536	Sb	23 ± 4	0.05	N21, M35
SN 1983G	NGC 4753	S	18 ± 4	0.30	N32, W7, M36
SN 1984A	NGC 4419	Ep(Sa)	17 ± 4	0.24	DET2ENV2, PDD3/1c/5
SN 1986G	NGC 5128	I	4:6 ± 1:2	0.90	N32, W7, M36, M37
SN 1988U	AC 118	-	1300^{+250}_{-150}	0.15	W7, M36, N32, DET2
SN 1989B	NGC 3627	Sb	8.3±3	0.60	N32, W7, M36
SN 1990N	NGC 4639	Sb	21 ± 5	0.01	DET2ENV2/4, PDD3/1c
SN 1990T	PGC 63925	Sa	195 ± 30	0.0	M37, M38
SN 1990Y	anonym.	E	210 ± 45	0.0	PDD6, W7, N32, M36, M37
SN 1990af	anonym.	-	290 ± 90	0.0	W7, N32, M36
SN 1991M	IC 1151	Sb	45 ± 10	0.12	PDD3,DET2ENV2
SN 1991T	NGC 4527	Sb	12 ± 2	0.01	PDDc3/1c/5, DET2ENV2
SN 1991bg	NGC 4374	dE	16 ± 5	0.30	PDD5/1a
SN 1992G	NGC 3294	Sc	26 ± 6	0.17	PDD3/1c, DET2ENV2, W7, N32
SN 1992K	ESO-269	SBb	43^{+15}_{-8}	0.28	PDD5/1c, M39
SN 1992bc	ESO-G9	S	91±	0.04	PDD6, PDD3/1c
SN 1992bo	ESO-G57	S	77 ± 10	0.00	PDD8
SN 1994D	NGC 4526	S0	16 ± 2	0.00	M36, (W7, N32)

SNe Ia should not be used as standard candles but the distances must be based on individual LC fits [6]. The time of the explosion is determined based on the horizontal shift. Because the intrinsic color is known from the models, we can correct for the interstellar reddening. We note that the determination of H_o does not depend on secondary distance indicators such as the surface brightness fluctuation for galaxies or δ-Cephii stars. From our fits of 26 SNe Ia, we find a value of $65 \pm 10\ km/(secMpc)$ [6] for H_o within a 2σ error up to a distance of about 1.3Gpc.

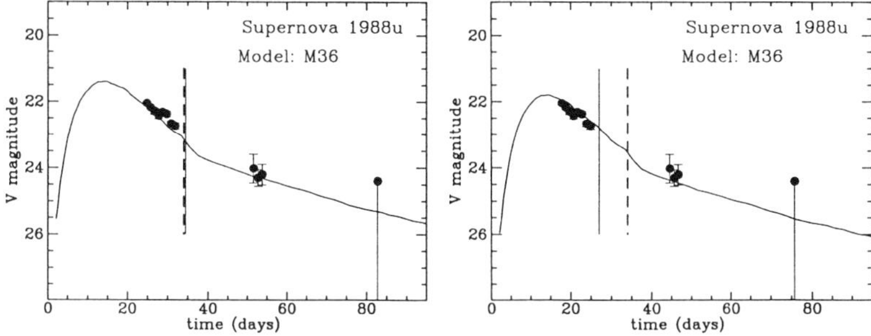

Figure 4: Comparison of observed (with error bars) and theoretical light curves of the delayed detonation model M36 [8] for SN 1988u at z=0.31 by Nordgard-Nielsen et al. [13]. The dashed line marks the time when the photospheric velocity of the model corresponds to the observed line shift (thin line). The advantage of a simultaneous analysis of light curves and spectra is evident for the distance determination of faint SNe Ia: from the light curve alone, both fits are equally good, but the time of the explosion is not well defined causing a 30% uncertainty in the distance alone. However, the right fit can be ruled out based on the photospheric velocity.

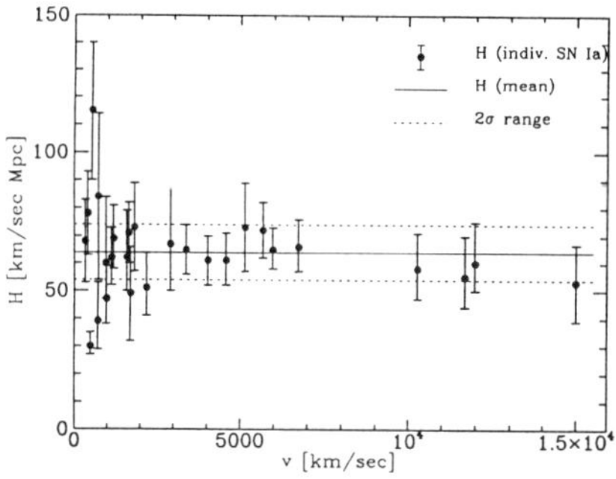

Figure 5: Hubble values H are shown as inferred by the distances derived for individual SNe Ia. Note that nearby SNe Ia ($v_o \leq 1000$ km/sec) give values inconsistent with a global H_o because these SNIa are not yet in the Hubble flow. SN1988u at v=91500 km/sec gives H= 70±10.

REFERENCES

1. Khokhlov, A., 1991, A&A 246, 383
2. Höflich, P., Khokhlov, A., Müller, E., 1992, A&A, 259, 549
3. Höflich, P., Müller, E.; Khokhlov, A., 1992, A&A 259, 243
4. Höflich, P., Müller, E,; Khokhlov, A., 1993, A&A, 268, 570
5. Khokhlov, A., Müller, E., Höflich, P., 1993, A&A 270, 23
6. Müller, E., & Höflich, P. 1994, A&A, 281, 51 and MPA 709/92
7. Höflich, P. 1995, ApJ, in press
8. Höflich, P., Khokhlov, A., Wheeler, C. 1995, ApJ in press
9. Dominik, C., Höflich, P., Khokhlov, A., Sedlmayr, E. 1995, ApJ, in preparation
10. Nomoto, K., Thielemann, F.-K., Yokoi, K., 1984, ApJ 286, 644
11. Hamuy, M. et al. 1995 ApJ, in press
12. Woosley, S.E., Weaver T.A. 1994, ApJ 423, 371
13. Norgaard, H., Nielsen, A.O., 1989, Nature 339, 522

Chandrasekhar Mass Models for Type Ia Supernovae

S. E. WOOSLEY[a], D. GARCIA-SENZ[a,b], J. NIEMEYER[a,c]
W. HILLEBRANDT[a,c], S. BLINNIKOV[a,d], and P. SASOROV[d]

[a]*Board of Studies in Astronomy and Astrophysics*
University of California, Santa Cruz

[b]*Department de Fisica i Enginyeria Nuclear*
UPC, Barcelona, Spain

[c]*Max Planck Institut für Astrophysik*
Garching, Germany

[d]*Institute for Theoretical and Experimental Physics*
Moscow, Russia

THE NEED FOR CHANDRASEKHAR MASS EXPLOSIONS

Given the difficulty arriving at a critical mass carbon-oxygen core, possible nucleosynthetic restrictions, and the existence of competing sub-Chandrasekhar mass models (see introduction, ref. 1), one may wonder if there are compelling reasons to think that *any* Type Ia supernovae must result from the explosion of what has been our "standard model" - a 1.39 $M_{\odot}$ white dwarf made mostly of carbon and oxygen. Ultimately the case must be made with spectra and light curves, but for now we begin by noting at least one compelling reason, namely the existence of unique nucleosynthesis - the abundances of ^{48}Ca, ^{50}Ti, ^{54}Cr, ^{58}Fe, and possibly ^{66}Zn that exist in the sun. For their production, these nuclei require material that has gone to nuclear statistical equilibrium (NSE) with a large neutron excess, $Y_e \approx 0.42$ and, moreover, that material must have low entropy so as to avoid an α-rich freeze out. Though neutron-rich zones may be found in Type II supernovae near the neutron star, they are always characterized by high entropy. Consequently ^{48}Ca and similar nuclei are never produced in appreciable quantities in massive stars (though ^{58}Fe and ^{66}Zn can be made by the s-process).

It has been shown[2], however, that the central 0.01 $M_{\odot}$ of carbon deflagration models provide ideal circumstances for the production of large quantities of these nuclei. Synthesis occurs during the initially slow propagation of a nearly laminar flame as a consequence of substantial electron capture which does indeed reduce Y_e at the center to about 0.41. Production is especially large in those explosions that ignite at high central density, say just short of the roughly 8×10^9 g cm^{-3} that leads to accretion induced collapse[2].

So this is a victory, and we should be grateful to find the origin of these nuclei, especially as they form some of the anomalous excesses found in the Allende meteorite (ref. 3 and references therein), suggesting that grain formation does indeed occur in Type Ia supernovae.

However, there is a problem - the large abundance of these nuclei made at the densities that characterize most modern Type Ia models, e.g., 3×10^9 g cm^{-3}. Production factors for a typical model[2] are 5, 85, 70, and 7 for ^{48}Ca, ^{50}Ti, ^{54}Cr, and ^{58}Fe respectively compared to ^{56}Fe (defined as one). In order not to overproduce ^{50}Ti, say, in the sun, such supernovae cannot have made more than 1% of the iron, yet chemical evolution[4] suggests that Ia's need to make about half of the solar iron. Hence this cannot be the common Ia event. What is wrong?

In order to avoid these overproductions, one must have less electron capture, which means either ignite the flame at a much lower density or else avoid an extended period of slow (~50 km/s) burning. The latter requires that the model not be ignited at a single point and burn as a laminar flame. Either the flame is born already moving fast or it ignites at many places concurrently (off center would be better) or both. This decreases the time required to burn enough fuel to significantly expand the white dwarf and that reduces the electron capture. Prompt carbon detonation models, for example, do not have this problem (though they have others[5]). Obviously burning faster also increases the probability of a prompt explosion.

THE IGNITION CONDITIONS

What does the interior of the star actually look like when the flame is first born? Garcia and Woosley[6] have examined this using a simple analytic approach that takes into account the fact that the runaway commences first as an extended period of stable convective burning. There comes a time, when the central temperature has reached 6 to 7×10^8 K, but well before matter burns in place on a hydrodynamic time ($\sim 1.5 \times 10^9$ K), when convection can no longer carry the flux. Specifically, the burning time becomes equal to the convective cycle time. Blobs rise and burn, growing more buoyant as they do, but they never come down again and they never go out. Starting with a background temperature, which is not critical, but was taken to be 6 to 7×10^8K, a blob is given a slight temperature excess which leads to accelerated burning. In response to this perturbation, the blob, which in this example is not allowed to grow in mass, becomes hotter, burns faster, becomes hotter still, decreases its density slightly, becomes buoyant, and starts to rise. As burning continues unabated the blob rises faster. Drag is included in the calculation in a simple way and the starting temperature and size of the blob are varied so as to find the blob that goes the farthest. In fact, the star will sample a continuum of temperatures as the runaway continues, but how finely it samples depends on how many blobs are launched. Since the distance and velocity a blob traverses before finally burning to NSE in place is very sensitive to its starting temperature, this will introduce aspects of chaos into the solution[6], by which we mean exponential sensitivity to initial conditions. Very similar starting conditions may ignite the burning at many locations whose radial extent varies from supernova to supernova.

Garcia and Woosley find that for starting temperatures around 7×10^8 K that blobs larger than 10 km will float to an altitude of about 200 - 300 km (more if the temperature is finely tuned) and achieve a speed of 100 to 200 km s^{-1} before running away in place. For smaller blobs, drag is more important and the speeds are not as great. This suggests that the initial burning may be ignited at many points inside a sphere of radius a few hundred km, and those points are already moving quite rapidly at birth.

Nor does the acceleration of the blobs end here. Once they burn to NSE, the transition from 1.5 billion to 9 billion K being essentially instantaneous, they become much *more* buoyant ($g_{eff} \sim$ few $\times 10^9$ cm s^{-2}) and accelerate faster. They are also endowed with negative specific heat because of the large abundance of free α-particles - expansion actually leads to energy generation which keeps the blob hot. In our simple study, the blobs rapidly accelerate to one to three thousand km s^{-1}, fast enough to seed the violent explosion of the star.

Why is this behavior not seen in multi-dimensional studies to date? Obviously the initial floatation of the blobs has not and cannot be followed using explicit hydrodynamics. The later acceleration after burning to NSE may be suppressed if the blob is not isolated, but partly surrounded with hot ashes. Or it may be that the calculations lack sufficient dimensionality and resolution to correctly follow the non-linear Rayleigh-Taylor (RT) instability, especially the detachment of small blobs from the burning pack.

THE LANDAU-DARRIEUS INSTABILITY

There have also been recent interesting developments in the microphysics of the flame, astrophysicists finally realizing what combustion physicists have known for 50 years. The kind of flames we are interested in is fundamentally unstable at all wavelengths. This instability, known as the Landau-Darrieus (LD) instability[7], arises because the decrease in density across the flame implies an acceleration in opposition to the density gradient, even in the absence of gravity. This instability has been analyzed in some detail recently by Blinnikov and Sasorov[8,9] and by Niemeyer and Hillebrandt[10]. The instability is damped by diffusion for wavelengths comparable to the flame thickness[11], but may exist at all larger scales, in particular much shorter scales than the RT-instability[5,11]. In the most detailed study to date, Blinnikov & Sasorov show that the effect of the LD-instability is to accelerate the flame by increasing its area by a factor given by a fractal dimension $D = 2 + 0.6\ \gamma^2$ where $\gamma = 1 - \rho_b/\rho_u$ and u and b indicate unburned and burned material. This expression is only valid for $\gamma << 1$, which certainly holds so long as the density is high. At densities of 10^9, 10^8, and 10^7 g cm^{-3}, γ is 0.19, 0.43, and 0.50 respectively[10], corresponding to $D =$ 2.022, 2.11, and 2.15. Taking the minimum unstable wavelength to be 10 flame thicknesses, 0.003, 0.3, and 40 cm at the given densities, and a maximum wavelength of 10^7 cm, one arrives at acceleration factors for the laminar flame of 1.6, 7, and 6. At the lower densities, it is unclear how to mesh the unstable wavelengths of the LD-instability with those of the RT-instability and turbulence. The LD-instability *per se*, may be important only on a smaller range of wavelengths. For the time being an over all acceleration factor of about two seems reasonable.

This will decrease the neutron-rich nucleosynthesis discussed in the first section. The interaction of the LD instability with others may also play an important role in triggering a delayed detonation, should such an event occur. Work is needed.

INDEPENDENCE OF THE MICROPHYSICS

Yet there is a growing consensus (eg., ref. 12) that the burning may propagate at a speed that is independent of the laminar flame and its associated microphysics. The burned material will rise and various instabilities will digest the encompassed material as needed. Interestingly, attempts to model Type Ia supernovae as resulting from fractal flames deformed by the RT-instability find that a fractal dimension very near 2.5 is needed[2]. It is precisely for $D = 2.5$ that the laminar speed drops out of the equation for the effective burning speed[5,11]. This happens because the minimum wavelength for RT instability depends on the square of the laminar speed. Thus the area factor $(R/\lambda_{min})^{D-2}$ cancels the laminar speed out front. The resulting speed is just $v = \text{const}(g_{eff} R)^{1/2}$, the same as one expects from simply watching a blob accelerate in response to gravity, g_{eff}, with no consideration of burning. A similar relation exists in mixing length convection theory.

More recently Niemeyer and Hillebrandt[12] have modelled the flame as a turbulent burning front using a subgrid extrapolation of Kolmogorov turbulence to calculate the flame speed. Assuming that the speed of turbulent eddies scales as $(\text{length})^{1/3}$, they use a prescription for the effective burning speed that depends only upon the strength of the subgrid turbulence. Again the solution can be characterized by a fractal with a minimum and maximum wavelength, here the minimum wavelength, known as the Gibson scale, being given by where the turbulent eddy speed equals the laminar flame speed. The fractal dimension of a turbulent flame that propagates independently of the laminar speed is[13] $D = 2.33$. This same value has been found for turbulent flames in laboratory experiments[14]. Flame propagation again depends only on macroscopic quantities, in this case the energy density being dumped into turbulence. Higher energy density decreases the Gibson scale and thus results in a faster moving flame.

There are several conclusions to be drawn. First, the subject of flames in white dwarfs is one of considerable current activity. One gets the impression that convergence may actually be occurring and in a few years we may understand, in a physical way, the complex physics of degenerate nuclear burning in the presence of turbulence, LD, and RT instabilities. The currently popular picture of a flame burn at point at the center of the star and burning slowly for an extended period of time is untenable and the ignition process needs to be considered carefully. Finally, from great complexity may eventually come simplicity. Chaos may be inherent in the ignition process, but the relevant physics of the ensuing explosion may occur on resolvable scales.

This work has been supported by the National Science Foundation (AST-91-15367) and by NASA (NAGW-2525)

REFERENCES

1. Woosley, S. E., & Weaver, T. A. 1994, ApJ, 423, 371
2. Woosley, S. E., & Eastman, R. G. 1995, Proceedings of Menorca Summer School on Supernovae (Sept., 1992), eds E. Bravo, R. Canal, J.M. Ibanez & J. Isern, World Scientific, in press
3. Völkening, J. & Papanastassiou, D. 1990, ApJL, 358, L29
4. Timmes, F. X., Woosley, S. E., & Weaver, T. A. 1995, ApJS, in press
5. Woosley, S. E. 1990, in *Supernovae*, ed. A. Petschek, (D. Reidel: Dordrecht), p. 182
6. Garcia-Senz, D., & Woosley, S. 1995, in preparation for ApJ
7. Landau, L. 1944, ZhETF 14, 240; Acta Physicochim, USSR, 19, 77
8. Blinnikov, S., & Sasorov, P. 1995, preprint submitted to Phys Rev E
9. Blinnikov, S., Sasorov, P., & Woosley, S. 1995, Proceedings of Gamow Symposium (St. Petersburg, Sept., 1994), Spac. Sci. Rev., in press
10. Niemeyer, J., & Hillebrandt, W. 1995a, preprint submitted to ApJ
11. Timmes, F., & Woosley, S. 1992, ApJ, 396, 649
12. Niemeyer, J., & Hillebrandt, W. 1995b, preprint submitted to ApJ
13. Kerstein, A. R. 1991, Phys. Rev. A, 44, 3633
14. North, G. L., & Santavicca, D. A. 1990, Combust. Sci. Tech., 72, 215

Type II Supernovae. The oldest new kid on the block

JASON SPYROMILIO AND BRUNO LEIBUNDGUT
European Southern Observatory
Karl-Schwarzschild-Str 2
Garching D-85748, Germany

INTRODUCTION

Type II Supernovae (SNe) were initially classified by Minkowski[1] as the SNe which differed from the group with homogeneous appearance namely the type Ia SNe. For these SNe, classified as type II, he states: *the individual differences in this group are large.* Following the identification of hydrogen in their spectra all SNe with with such lines have been classified as type II. It is worth noting that already in 1941 *the synthetic spectra agree better with spectra of type II* [SNe].

The Minkowski statements still hold and it could be argued that little progress has been made. This of course is not the case. We still group all SNe with hydrogen together. We now classify type I SNe in three subgroups (Ia, Ib and Ic) and every other SN exhibits hydrogen lines (Harkness & Wheeler[2]). The synthetic spectra seem to fit type II SNe reasonably well, at least for those objects that look 'normal'. Our understanding of all types of SNe has increased dramatically. In this short review we shall concentrate on the challenges ahead rather than the successes of the past.

With the increased understanding it has become possible to use type II SNe as distance indicators based on well understood and modeled physical properties. The expanding photosphere method (Schmidt, Kirshner & Eastman[3] and others) places the use of type II SNe as distance indicators on a firm physical basis.

FUTURE CHALLENGES

Chemistry of the ejecta

To date the only SN with detected molecular emission is SN 1987A. CO and SiO emission has been extensively modeled (see Spyromilio et al.[4], Roche et al.[5] and many others). Other species such as H_3^+ have been identified in the spectra of SN 1987A[6]. The lack of high quality infrared spectra of type II SNe at late times currently restricts the further development of this field. However their importance cannot be overstated. CO is the dominant coolant in the ejecta of SN 1987A a couple of hundred days after explosion as can be seen in Figure 1 where the fundamental band of CO dominates the flux at 62 THz (4.8μm). Derivation of the physical parameters of the SN without taking into account the chemistry could be compared to fitting the optical spectrum without including Hydrogen. The challenge is to the observers to obtain more late-time IR spectra and to the theorists to include molecules in their models.

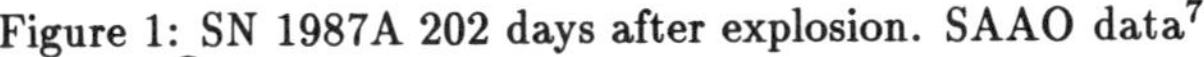

Figure 1: SN 1987A 202 days after explosion. SAAO data[7]

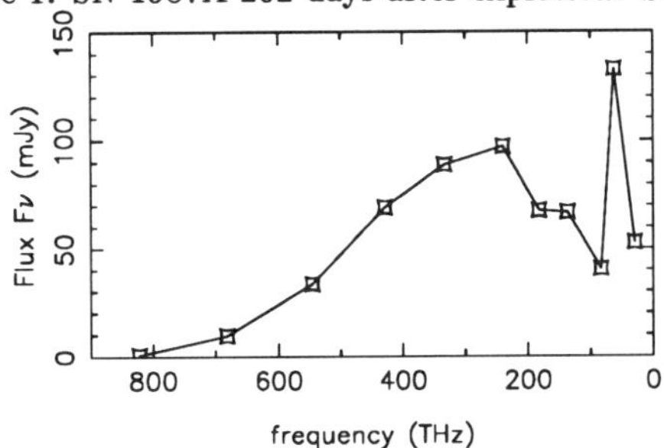

Clumping

The ejecta of type II SNe are clumped. Many studies of the line profiles in SN 1987A (eg Stathakis et al.[8], Spyromilio, Stathakis & Meurer [9]) have shown the characteristic signature of non-smooth distributions for the ejecta. The explosion models presented at this meeting (see contributions by Burrows and by Janka in these proceedings) seem to reproduce the observed non-uniform distribution of the ejecta. It is not clear how the clumping affects the radiative transfer.

Circumstellar and interstellar medium

Type II SNe are known to interact with their CSM. In some cases this interaction dominates the emission (see SN 1988Z; Stathakis & Sadler[10]). The output of a core-collapse SN in electromagnetic radiation is only a minor manifestation of the supernova event but is the most commonly observed one. The light from a SN is reprocessed by the surrounding gas and dust resulting in echoes often observed. Unfortunately for almost all observable SNe the angular extent of these echoes is outside the observable range. Extreme care is needed in the interpretation of spectra and photometric observations so as not to misinterpret the data. The interaction of the ejecta with the surrounding medium provides access to the enormous pool of kinetic energy. Even when tapped with small efficiency this can completely distort the view of a SN.

Asymmetry

To date every core-collapse SN with spectropolarimetric observations has shown some degree of polarization. The asymmetry in the ejecta required to produce the observed polarization is often large with axis ratios around 2. At face value this would imply large variations in the observed bolometric luminosity of these objects. It is difficult to reconcile the success of the EPM method, which assumes spherical symmetry, with the interpretation of the spectropolarimetric observations. Here again more data are needed as is the inclusion of asymmetric envelopes into radiative transfer codes (e.g. Jeffery[11]).

Figure 2: SN 1993J. [OI] residuals[12]

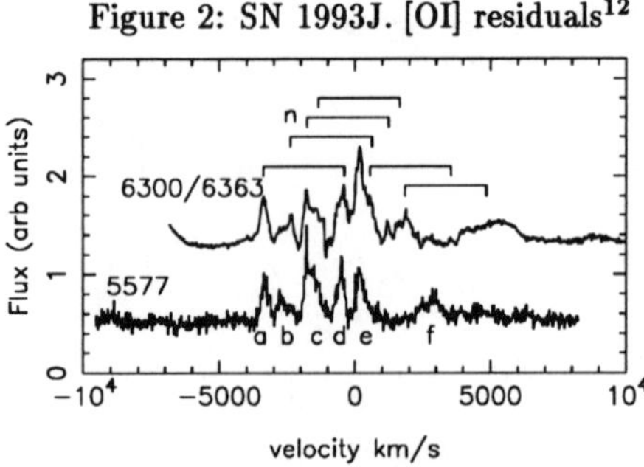

SN 1993J: A PECULIAR CASE STUDY

SN 1993J is amongst the best studied SNe. It exploded in M81 in mid 1993 and has provided a wealth of data in almost all wavelength bands. The spectrum exhibited hydrogen lines and the SN is classified as a type II. However the evolution of the spectrum is far from the canonical for this type. Its proximity has made possible studies which could not have been undertaken for more distant SNe. At the time of writing of this contribution no molecular emission from SN 1993J had been reported.

Clumping

Clumping has been detected from the analysis of the [OI] line profiles. In Figure 2 we show the residuals of the [OI] 5577Å and 6300/6363Å lines after the removal of a smooth component of the line profile[12]. The correspondence of features in 5577 and both the 6300 and 6363 lines unambiguously identifies these as being due to deviations from a smooth emissivity within the ejecta.

Interaction with the CSM

The strong radio flux from SN 1993J is attributed to interaction of the ejecta with the CSM[13]. We can now confidently attribute most of the current luminosity of the ejecta of SN 1993J to excitation by the interaction of the ejecta with the CSM[14,15]. Both the flux in the Hα line and the ratio of Hα to Hβ are in excellent agreement with the models of Chevalier & Fransson[16] and Fransson (these proceedings). These models invoke heating of the ejecta by the reverse shock associated with the outwards moving shock.

Asymmetry?

The double peaked profile exhibited by the hydrogen lines requires a lack of emitting material at low velocities with respect to the observer. In Figure 3 we show the observed profile and two models. The dashed line is based on a prolate spheroid with a power law emissivity viewed along its equator. The axis ratio of the spheroid is 3 to 2. The dotted line profile results from a torus. The inner and outer radii

Figure 3: SN 1993J. Hα profile and fit[15]

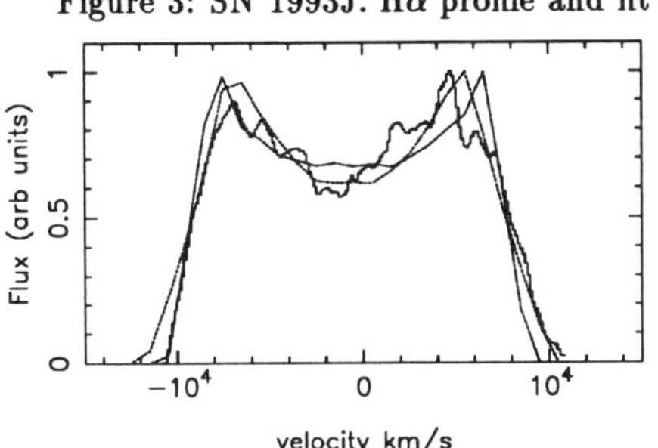

of both models are very similar at 7000 and 10500 km/s respectively. The prolate model is in good agreement with the spectropolarimetric observations[17] and could result from asymmetric expanding ejecta impacting a spherically symmetric CSM. The torus model could result from spherically symmetric ejecta impacting on an asymmetric CSM.

ACKOWLEDGEMENTS

Based in part on observations made with the William Herschel telescope operated on the island of La Palma by the Royal Greenwich Observatory in the Spanish Observatorio del Roque de los Muchachos of the Instituto de Astrofisica de Canarias. This work would not have been possible without the efforts of the LPO & RGO staff.

REFERENCES

[1] Minkowski R., 1941, PASP, 53, 224

[2] Harkness R.P., Wheeler J.C., 1990, in Petschek A.G., ed, Supernovae, Springer-Verlag, New York, p.1

[3] Schmidt B., Kirshner R.P., Eastman R., 1992, ApJ, 395, 366

[4] Spyromilio J., Meikle W.P.S., Learner R.C.M., Allen D.A., 1988, Nat. 334, 327

[5] Roche P.F., Aitken D.K., Smith C.H., 1991, MNRAS, 252, 39P

[6] Miller S., Tennyson J., Lepp S., Dalgarno A., 1992, Nat, 335, 420

[7] Catchpole R.M. et al., 1988, MNRAS, 231, 75P

[8] Stathakis R., Dopita M.A., Sadler E.M., Cannon R.D., 1991, in Woosley S.E., ed, Supernovae, Proc Xth Santa Cruz Summer Workshop. Springer-Verlag, New York, p.95

[9] Spyromilio J., Stathakis R., Meurer G.R., 1993, MNRAS, 263, 530

[10] Stathakis R., Sadler E.M., 1991, MNRAS, 250, 786

[11] Jeffery D.J., 1991, ApJ, 375, 265

[12] Spyromilio J., 1994, MNRAS, 266, L61

[13] Pooley G.G., Green D.A., 1993, MNRAS, 264, L17

[14] Patat F., Chugai N., Mazzali P.A., A&A, (in press)

[15] Spyromilio J., Leibundgut B., 1995, MNRAS (submitted)

[16] Chevalier R.A., Fransson C., 1994, ApJ, 420, 268

[17] Trammell S.R., Hines D.C., Wheeler J.C., 1993, ApJ, 414, L21

Type Ib/Ic/IIb/II-L Supernovae and Binary Star Evolution

K. NOMOTO[a], K. IWAMOTO[a], H. YAMAOKA[b], T. SUZUKI[a], O.R. POLS[c], E.P.J. VAN DEN HEUVEL[d], AND P. HÖFLICH[e]

[a]*Department of Astronomy, University of Tokyo*
[b]*Department of Physics, Kyushu University*
[c]*Institute of Astronomy, University of Cambridge*
[d]*Astronomical Institute, University of Amsterdam*
[e]*Department of Astronomy, Harvard University*

INTRODUCTION

Supernovae are classified as Type I and Type II from the absence and presence of hydrogen and further subdivided into Ia, Ib, Ic, II-P, II-L, IIb, and IIn (e.g., refs. 1, 2). Presence of strong Si lines and He lines defines Type Ia and Type Ib, respectively, while Type Ic is characterized by the lack or weakness of these lines. The light curves of Type II-P supernovae (SNe II-P) has a plateau, while those of Type II-L (SNe II-L) show linear decline.[3,4]

It has been suggested that the diversity of SNe II originates from progenitor's different main-sequence mass range, i.e., SNe II-L from 7–$10M_\odot$[5] and SNe II-P above $10M_\odot$, and that Type Ib/Ic supernovae (SNe Ib/Ic) originate from He stars of different mass range in binary systems.[6] However, the exact supernova–progenitor connection for these types has been a controversial issue.

Recent nearby supernovae, SN 1993J in M81 and SN 1994I in M51, have been identified as Type IIb (SN IIb: e.g., ref. 7) and Type Ic (e.g., ref. 8), respectively, adding more diversity of supernovae (i.e., IIb) but shedding new light on the classification and their evolutionary origins. Here we show that common envelope evolution in massive close binary stars can provide a plausible explanation of the observational diversity.

Here we propose for SN IIb 1993J that merging of two stars in a close binary is responsible for the formation of a thin H-rich envelope of the progenitor, as opposed to the conservative mass transfer scenario. For SN Ic 1994I, we present C+O star models and the mass dependence of the light curve. We then generalize the binary scenario to show that common envelope evolution in massive close binary stars leads to various degree of stripping of the envelope mass of massive star. This naturally explains the origin of supernova types, namely, II-L, IIb, Ib, and Ic, in a unified manner, depending on the mass ratio q of component stars and the initial separation R_0. Finally we briefly discuss circumstellar interactions of SN 1993J to compare with X-ray observations, since the extensive mass loss from the progenitor and the formation of dense circumstellar matter is predicted by the binary scenario.

MERGING AND THE PROGENITOR OF SN 1993J

SN 1993J has revealed important new features of supernovae (ref. 9 for reviews and references therein). It is identified as a Type II supernova (SN II) from hydrogen features. The light curve of SN 1993J shows a distinct difference from those of previously known SNe II. It was obvious that this peculiar light curve of SN 1993J cannot be accounted for by an explosion of an ordinary red supergiant with a massive hydrogen-rich envelope, which produces a light curve of a SN II-P.[10]

The light curve of SN 1993J can be understood as the explosion of a red-supergiant whose hydrogen-rich envelope is as small as $\lesssim 1 M_{\odot}$.[10,11] The thin envelope model has been confirmed by the spectral changes which shows growing features of helium and oxygen, so that SN 1993J can be classified as a SN IIb.[12,13]

The progenitor of SN 1993J is likely to have lost most of its H-rich envelope due to the interaction with its companion star in a binary system. The binary scenario raised the following questions: 1) what controls the mass of the remaining H-rich envelope of the progenitor, and 2) what is the relation of SN 1993J with other types of supernovae, such as SNe IIn, II-L and Ib/Ic.

Conservative Mass Transfer Scenario

As a possible evolutionary scenario leading to the progenitor of SN 1993J, a *Case C* binary evolution has been proposed.[11,14,15]. This scenario postulates that 1) the initial separation between the two component stars is so large that the mass transfer from the progenitor had started after helium was exhausted in the core, and 2) mass ratio $q = M_1/M_2$ between the progenitor 1 and the companion star 2 is close to unity, so that the mass transfer from the progenitor is more or less conservative.

In Case C mass transfer, however, the primary star has a convective envelope, thereby transferring mass to the companion star with a dynamical timescale at least at its early phase. Such an evolution is described by assuming the non-dimensional specific angular momentum loss from the system α and the ratio β between the mass accreted by star 2 and the mass lost by star 1 as arbitrary parameters.[16,17] It is seen that, even for $q < 1$ and $\beta \sim 1$, the radius of the mass donor exceeds the Roche lobe. Nevertheless the response of the companion star or the formation of the common envelope was not calculated, i.e., there has been no consistent set of calculations of both the dynamical mass loss from the progenitor and the response of mass receiving companion. Thus it is possible that even if q is initially smaller than 1 due to earlier wind mass loss from star 1, Case C mass transfer leads quickly to the formation of a common envelope. Although Case C scenario may be possible for a narrow parameter space, an alternative scenario is worth exploring.

Non-Conservative Mass Transfer Scenario

Evolutionary paths of close binaries depend significantly on the mass ratio q of component stars and the initial separation R_0. Here we consider binary systems consisting of star 1 and star 2 whose main-sequence mass ratio q is significantly smaller than unity. Star 1 evolves to form a He core and its H-rich envelope expands to fill its Roche lobe. Because of the extreme mass ratio, the mass transfer is highly non-conservative. This almost inevitably leads to the formation of a common envelope and the subsequent spiral-in of star 2 and the core of star 1.

The spiral-in deposits the orbital energy in the envelope due to viscous evolution. Subsequent common envelope evolution is so complicated that we take a simplified approach. From energy considerations, we assume that spiral-in yields the following outcome.

1. If deposited orbital energy E_g is larger than the binding energy of the common envelope E_b, almost all envelope materials are ejected before star 2 is dissolved. In other words, the binary system survives the spiral-in and then consists of helium star 1 of and main-sequence star 2 in a much closer orbit. Since E_g is larger for larger R_0, this case occurs when R_0 is larger than a certain limit. (Here the timescale is so short that radiation loss is negligible.)

2. If $E_g < E_b$, on the contrary, the two stars merge into a single star, i.e., star 2 is completely dissolved in the common envelope before all the envelope mass is ejected. The resulting single star 1 retains some envelope material. The envelope mass M_{env} after merging depends on the deposited energy E_g relative to E_b. Larger E_g/E_b induces larger amount of mass loss and forms a smaller mass envelope when the merging is completed. In terms of R_0, such a merging occurs when R_0 is smaller than a certain limit. The remaining envelope mass is smaller for larger R_0. Afterwards star 1 would expand to become a red-supergiant (RSG) but its M_{env} could be significantly smaller than, say, $5M_\odot$.

The non-conservative scenario predicts the formation of a single neutron star, in contrast to the binary neutron star in eccentric orbit as predicted by the conservative mass transfer scenario.

SUPERNOVA TYPES AND MERGING STARS

Based on the above spiral-in scenarios 1 and 2, we can present a new interpretation of the origin of various types of supernovae, in particular, IIb, II-L, and IIn.

1. **SN Ib**: In this case, the spiral-in forms a pair of helium star 1 and main-sequence star 2. If the helium star mass exceeds $\sim 2.5M_\odot$, it evolves through Fe core collapse and explodes as a SN Ib because of the presence of helium (e.g., refs. 18, 19).

   SN Ic: If the mass of helium star 1 is smaller than $\sim 5M_\odot$, its helium envelope expands possibly to exceed the Roche lobe.[20,21] After the loss of helium envelope, star 1 becomes a C+O star. The explosion of the C+O star is triggered by Fe core collapse and must be observed as a SN Ic.

2. In this case, the two stars merge to form a single core and lose a significant fraction of their common envelope due to frictional heating. Unless the envelope mass becomes as small as $10^{-2}\ M_\odot$, the envelope of the merged star should expand to a RSG size. If the helium core mass exceeds $\sim 2.5M_\odot$, a supernova explosion is triggered by Fe core collapse and observed as a Type II. We propose that the progenitors of IIb, II-L, and possibly IIn are these merged stars and the difference in the types originate from the difference in the mass of the H-rich envelope M_{env} as follows.

(a) **SN IIb**: If $M_{env} \lesssim 1M_{\odot}$, the presupernova configuration would be similar to that of SN 1993J. Their light curves around the second peak and tails must be similar to SNe Ib. SN 1993J continues to exhibit strong H_{α} emissions, while SN 1987K (also IIb[13]) did not show H-emission lines at late phase. Late time H_{α} emissions in SN 1993J is powered by X-rays from the reverse shock in the ejecta.[22] Further study is needed to understand whether such a difference stems from the difference in M_{env} (and thus density structure) or in the circumstellar matter density (or some other parameters).

(b) **SN II-L**: If $M_{env} \sim 2$–$3M_{\odot}$, the radius of RSG is as large as the ordinary RSG. However, because of small M_{env}, the expansion velocity at the bottom of the H-rich envelope higher. This leads to the shorter period of plateau, i.e., SN II-L (e.g., refs. 23, 5, 24).

(c) **SN IIn**: If the envelope mass remains to be such large as $M_{env} \gtrsim 5M_{\odot}$ possibly due to the initial large mass of star 1, the star becomes an ordinary RSG after merging. It eventually explodes as a SN II-P. However, its CSM would consist of the material ejected during spiral-in and the RSG wind material. The structure of CSM originated from spiral-in is likely to be asymmetric; the mass ejection may form a bipolar jet or disk like material. Mass loss rate during the RSG phase could be larger than the ordinary RSG because of the larger mass and extra heating due to merging. This case might correspond to SN IIn, like SN 1988Z (e.g., ref. 2).

LIGHT CURVE MODELS FOR TYPE Ic SUPERNOVA 1994I

We apply the above scenario to SN Ic 1994I. The simplest evolutionary path is as follows. The progenitor was a 13–18$M_{\odot}$ star on the main-sequence. Through Roche lobe overflow, the 13, 15, and 18$M_{\odot}$ stars became He stars of M_{α} = 3.3, 4.0, and 5.0$M_{\odot}$ and then lost their helium envelopes to become C+O stars of M_{C+O} = 1.8, 2.1, and 2.9$M_{\odot}$. Hereafter these models are called CO18, CO21, and CO29, respectively.[25]

The mass cut is chosen to produce 0.07$M_{\odot}$ of ^{56}Ni. The deposited energy is set to produce the kinetic energy of explosion $E = 1\times 10^{51}$ erg s^{-1} for CO18, CO21, and CO29, and $E = 6\times 10^{50}$ erg s^{-1} for the lower explosion energy model CO21L. It is noticeable that the ejecta masses of M_{ej} = 0.5 and 0.9$M_{\odot}$ for CO18 and CO21 are significantly smaller than the Chandrasekhar mass. Even for CO29, M_{ej} is still as small as 1.5$M_{\odot}$.[26]

Since the C+O star is compact with a radius of $\sim$ 0.2 $R_{\odot}$, the light curve is not due to shock heating but is powered by radioactive decay chain of ^{56}Ni $\rightarrow$ ^{56}Co $\rightarrow$ ^{56}Fe. For detailed comparison with observations, monochromatic light curves for the C+O star models are calculated and compare with observations[27] in FIGURE 1. The slopes of the calculated light curves are found to be sensitive to the models as follows:

CO18: Both the rise time and the decline are far too short compared with the observations. The main cause is that the diffusion time scales in the envelope are too short, suggesting the need for a more massive model to increase the diffusion time scale and, consequently, to produce a broader and flatter maximum.

CO21 : This model gives almost perfect agreement between the slopes of the theoretical and observed shapes of the light curves for the monochromatic B, V, R and I band

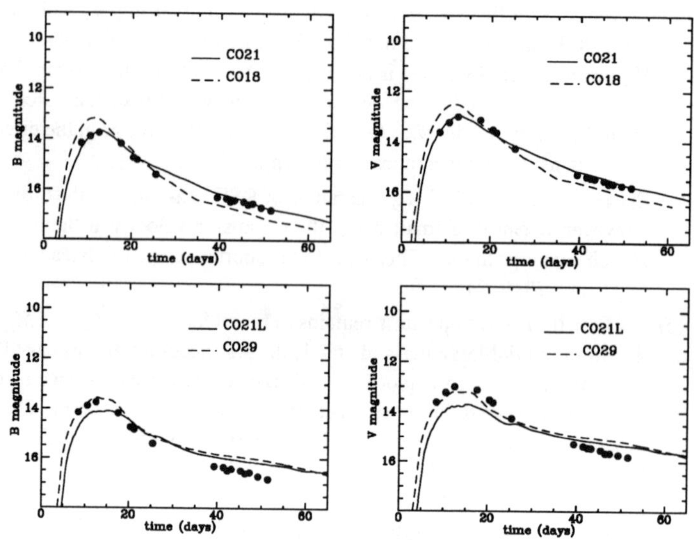

FIGURE 1. Theoretical monochromatic light curves of CO18, CO21, CO29, and CO21L (ref. 28) in comparison with observations of SN 1994I (ref. 27).

and the bolometric light curve as well.[28] This means that the diffusion time scales, the energy input and the temperature structure are about correct. From the light curve fits, a distance modulus of 29.2 ± 0.3 mag is derived for M51. Also a high interstellar reddening of $E(B-V) = 0.45^m$ ($A_V \approx 1.4$ mag) is required. The ejected ^{56}Ni mass is found to be $0.07\ ^{+0.035}_{-0.025} M_\odot$.

CO29 : The photosphere recedes nicely at maximum light. At later times, however, the light curve declines too slowly because the escape probability changes only little with time due to the lower expansion velocities than CO21.

CO21L : Due to the small expansion rate, the escape probability for gamma-rays is significantly higher compared to CO21. This keeps the photosphere hot, i.e., the opacity does hardly drop a maximum light and, consequently, the maximum is not well pronounced compared to CO21. At later times, the decline of the light curve is too slow for the same reasons as CO29.

Wolf-Rayet star models are clearly too massive to be consistent with the light curves of SN 1994I, because even the smallest WC star model has $2.7 M_\odot$ ejecta[29] which is significantly more massive than $1.5 M_\odot$ of CO29.

However, the late time oxygen line emissions of SN1994I suggest that models with somewhat slower expansion rate than CO21 are favored (Fransson, this workshop). Such slower models CO29 and CO21L are not consistent with the light curve, however. Mixing of ^{56}Ni would lead to a faster decline of the light curves for these models. Large scale mixing due to Rayleigh-Taylor instabilities may not be expected for these C+O star models because of the lack of both H and He layers and the associated density jumps.[30] If the expansion should certainly be slower than CO21, a new mechanism of mixing is required unless the star has a rather massive He envelope.

CIRCUMSTELLAR INTERACTION IN SN 1993J

The merging model for the progenitor of SN 1993J predicts that the presence of (probably) asymmetric outer circumstellar matter (CSM) formed through merging process and more symmetric inner CSM formed by the wind from the red-supergiant.

Suzuki *et al.*[31] have constructed a hydrodynamical model of interaction between the ejecta of SN 1993J and CSM to account for the basic features of X-ray emissions from SN 1993J as observed with OSSE, ASCA, and ROSAT for the first 570 days (also Fransson, this workshop). This model consists of a realistic ejecta model and clumpy CSM. The collision between the ejecta and CSM creates a reverse shock which is radiative to form a cooling dense shell in the ejecta. X-rays emitted from the reverse shock are mostly absorbed by this shell. Early hard X-rays are well modeled as thermal emissions from shocked CSM. The CSM density inferred from X-ray observations is as high as $\dot{M}/v_w$ = 3–4× 10^{-5} $M_\odot$ yr^{-1}/ 10 km s^{-1} which are rather high compared with the mass loss rate estimated for the 13–15$M_\odot$. In the merging progenitor model, this problem could be resolved because mass loss rate can be enhanced due to extra heating in the envelope produced by merging.

The above model indicates that CSM has a spatially variable density gradient. In the inner layer, the gradient is shallower than steady wind, while CSM is highly clumpy in the outer layers so that the density gradient of inter-clump matter is steeper. Soft X-rays at late times are mostly emitted from the shocked high density clumps.

The clumpy circumstellar structure can account for other important features of SN 1993J as well.[32] In particular, the steep density gradient of inter-clump matter leads to weak deceleration of the ejecta and shocked CSM, i.e., the expansion velocities of the shocked ejecta and CSM decrease only slowly. This is consistent with i) a roughly constant maximum velocity of hydrogen ($\sim$ 10,000 km s^{-1}) as observed in H_α features[33] (Fransson, this workshop) and ii) the average expansion velocity of the radio shell over 240 days, which is as high as 15,000 ± 2,000 km s^{-1} for the distance of 3.63 ± 0.34 Mpc.[34]

In the non-conservative binary evolution scenario for the progenitor of SN 1993J,[35,36] clumpy CSM could form as follows: Firstly the spiral-in of a companion star into the progenitor leads to the ejection of a common envelope. Later stellar wind from the red supergiant progenitor would collide with former common envelope matter. As a result of the deceleration of wind matter, the wind velocity would be slower at a larger distance, i.e., the density gradient would be shallower than steady wind, and CSM would be more clumpy as it is closer to the common envelope matter. This model predicts the collision of the blast shock wave with former common envelope matter, which would lead to some enhancement of soft X-ray fluxes and relatively low velocity H_α emissions; this might correspond to the latest ROSAT observations showing recent leveling off of the X-ray flux[37] and the HST observations of H_α features suggesting a collision with ring-like matter (Kirshner, this workshop).

Another future event would be an enhancement of X-ray fluxes when the reverse shock reaches the H/He interface where a density sharply increases.[38,14] If the enhancement is observed, its date will provide information on the thickness of the H-rich envelope and propagation speed of the reverse shock.

In this way, X-ray emissions from circumstellar interactions provide important information on the internal structure of the supernova ejecta (such as density gradient) as well as mass loss from the progenitor. Detailed study of circumstellar interaction would thus provide an important clue to the progenitor's evolution, especially its binary nature for SNe IIb, II-L, and IIn.

CONCLUDING REMARKS

We have proposed that except for *classical* SNe I and SNe II (SNe Ia and II-P), all other supernova subtypes (SNe Ib, Ic, IIb, II-L and IIn) can be explained as Fe core collapse of massive ($\gtrsim 10 M_\odot$) stars in close binary systems. In other words, the main-sequence mass range of the progenitor of SNe Ib, Ic, IIb, and II-L and thus their nucleosynthesis may be similar to those of SNe II-P except for the SNe II-P from the 8 - $10 M_\odot$ AGB stars.[39]

If the above proposal is correct, light curves and spectra of SNe Ib, Ic, IIb, II-L, and especially recent nearby ones, are particularly useful to obtain the ^{56}Ni and O masses as a function of the main-sequence mass. (The ^{56}Ni mass estimates is closely related with the neutron star mass estimate.[40]) From the light curve shape, we may infer the ejecta mass (i.e., the progenitor's mass) and the ^{56}Ni mass for the calculated extent of mixing. The produced ^{56}Ni mass (and the progenitor's main-sequence mass) is estimated as $0.09 \pm 0.02 M_\odot$ (13–14$M_\odot$) for SN 1993J ($A_V = 0.4$ mag) and $0.07\,^{+0.035}_{-0.025} M_\odot$ (14–15$M_\odot$) for SN 1994I. Combining with $0.075 \pm 0.01 M_\odot$ ^{56}Ni for SN 1987A ($\sim 20 M_\odot$[41,42]), the ejected ^{56}Ni mass seems to be not too sensitive to the progenitor's mass. (Earlier estimates of the $\sim 0.15 M_\odot$ ^{56}Ni for SNe Ib/Ic being $\sim$ 1/4 of SNe Ia[43] are higher than the above values, which probably suggests the necessity of reexamination of the distances and/or extinction for SNe Ib/Ic.)

The light curve tails of SNe II-L would provide a useful information on whether SNe II-L have massive progenitors. Recent observations of SNe II-L 1990K[44] have shown that its late light curve is similar to SN 1987A, suggesting the progenitors of *bright* II-L may be similarly massive. This is consistent with the binary merger model but not with the AGB model.[45] For normal II-L, more observations are needed.

To conclude, our hypothesis implies that the observational diversity of supernovae can be resulted from a single mechanism of explosion, i.e., a gravitational collapse of an Fe core but with merging of stars in close binary systems.

REFERENCES

1. Branch, D., Nomoto, K., & Filippenko, A.V. 1991. Comments on Astrophys. **15:** 221.
2. Filippenko, A.V. 1991. *In* Supernovae and Stellar Evolution. A. Ray & T. Velusamy, Eds. World Scientific, Singapore. 58.
3. Doggett, J. B., & Branch, D. 1985. Astron. J. **90:** 2303.
4. Young, T., & Branch, D. 1989. Astrophys. J. **342:** L79.
5. Swartz, D.A., Wheeler, J.C., & Harkness, R.P. 1991. Astrophys. J. **374:** 266.
6. Nomoto, K., Filippenko, A.V., & Shigeyama, T. 1990. Astron. Astrophys. **240:** L1.

7. Schmidt, B.P. *et al.* 1993. Nàture **364:** 600.
8. Wheeler, J.C. *et al.* 1994. Astrophys. J. **436:** L135.
9. Wheeler, J.C., & Filippenko, A.V. 1994. *In* IAU Colloq. 145, Supernovae and Supernova Remnants. R. McCray, Ed. Cambridge University Press, Cambridge. in press.
10. Nomoto, K., Suzuki, T., Shigeyama, T., Kumagai, S., Yamaoka, H., & Saio, H. 1993. Nature **364:** 507.
11. Podsiadlowski, Ph., Hsu, J.J.L., Joss, P.C., & Ross, R.R. 1993. Nature **364:** 509.
12. Woosley, S.E., Pinto, A., & Ensman, L.M. 1988. Astrophys. J. **324:** 466.
13. Filippenko, A.V. 1988. Astron. J. **96:** 1941.
14. Woosley, S.E., Eastman, R.G., Weaver, T.A., & Pinto, P.A. 1994. Astrophys. J. **429:** 300.
15. Ray, A., Singh, P., & Sutaria, F.K. 1993. J. Astrophys. Astron. **14:** 53.
16. Podsiadlowski, Ph., Joss, P.C., & Hsu, J.J.L. 1992. Astrophys. J. **391:** 246.
17. Rathnasree, N., & Ray, A. 1992. J. Astrophys. Astron. **13:** 3.
18. Ensman, L.M., & Woosley, S.E. 1988. Astrophys. J. **333:** 754.
19. Shigeyama, T., Nomoto, K., Tsujimoto, T., & Hashimoto, M. 1990. Astrophys. J. **361:** L23.
20. Habets, G.M.H.J. 1986. Astron. Astrophys. **187:** 209.
21. Nomoto, K. & Hashimoto, M. 1988. Phys. Rep. **163:** 13.
22. Clocchiatti, A., & Wheeler, J.C. 1994. IAU Circ. No. 6005.
23. Shigeyama, T., & Nomoto, K. 1990. Astrophys. J. **360:** 242.
24. Blinnikov, S.I., & Bartunov, O.S. 1993. Astron. Astrophys. **273:** 106.
25. Nomoto, K., Yamaoka, H., Pols, O.R., van den Heuvel, E.P.J., Iwamoto, K., Kumagai, S., & Shigeyama, T. 1994. Nature **371:** 227.
26. Hashimoto, M., Nomoto, K., Tsujimoto, T., & Thielemann, F.-K. 1993. *In* Nuclei in the Cosmos. F. Käppeler & K. Wisshak IOP Pub., Bristol. 587.
27. Schmidt, B., & Kirshner, R. 1994. e-mail circulation.
28. Iwamoto, K., Nomoto, K., Höflich, P., Yamaoka, H., Kumagai, S., & Shigeyama, T. 1994. Astrophys. J. **437:** L115.
29. Woosley, S.E., Langer, N., & Weaver, T.A. 1993. Astrophys. J. **411:** 823.
30. Hachisu, I., Matsuda, T., Nomoto, K., & Shigeyama, T. 1991. Astrophys. J. **368:** L27.
31. Suzuki, T., Nomoto, K., Iwamoto, K., Shigeyama, T., & Kumagai, S. 1995. Astrophys. J. submitted.
32. Van Dyk, S.D., Weiler, K.W., Sramek, R.A., Rupen, M.P., & Panagia, N. 1994. Astrophys. J. **432:** L115.
33. Patat, F., Chugai, N.N., & Mazzali, P.A. 1995. Astron. Astrophys. in press.
34. Marcaide, J. M. *et al.* 1995. Nature **373:** 44.
35. Nomoto, K., Suzuki, T., Pols, O. R., & van den Heuvel, E.P.J. 1994. *In* New Horizon of X-Ray Astronomy. F. Makino & T. Ohashi, Eds. Universal Academy Press, Tokyo. 139.
36. Nomoto, K., Iwamoto, K, & Suzuki, T. 1995. Phys. Rep. in press.
37. Zimmermann, H.-U. *et al.* 1994. IAU Circ. No. 6120.
38. Shigeyama, T., Suzuki, T., Kumagai, S., Nomoto, K., Saio, H., & Yamaoka, H. 1994. Astrophys. J. **420:** 341.
39. Hashimoto, M., Iwamoto, K., & Nomoto, K. 1993. Astrophys. J. **414:** L105.
40. Thielemann, F.-K., Nomoto, K., & Hashimoto, M. 1995. Astrophys. J. in press.
41. Arnett, W.D., Bahcall, J.N., Kirshner, R.P., & Woosley, S.E. 1989. Ann. Rev. Astron. Astrophys. **27:** 629.
42. Nomoto, K., Shigeyama, T., Kumagai, S., Yamaoka, H., & Suzuki, T. 1994. *In* Supernovae (Les Houche Session LIV). S.A. Bludman *et al.*, Eds. Elsever, New York. 489.
43. Panagia, N. 1987. *In* High Energy Phenomena Around Collapsed Stars. F. Pacini, Ed. Reidel, Dordrecht. 33.
44. Cappellaro, E., Danziger, I.J., Della Valle, M., Gouiffes, C., & Turatto, M. 1994. Astron. Astrophys. in press.
45. Swartz, D.A., & Wheeler, J.C. 1991. Astrophys. J. **379:** L13.

Convection in Type-II Supernovae: The First Second[a]

EWALD MÜLLER[b] AND H.-THOMAS JANKA[b,c]

[b]*Max-Planck-Institut für Astrophysik*
Karl-Schwarzschild-Str. 1
D-85740 Garching, Germany

INTRODUCTION

There is a large variety of observational hints that large-scale mixing processes occurred and played an important role in SN 1987A even during the early phase of the explosion. X-rays (e.g., [1,2]) and γ-rays (e.g., [3,4]) from the decay of ^{56}Ni and ^{56}Co and strongly Doppler-shifted, infrared emission lines of iron (e.g., [5,6]) were observed at a stage when Ni and Fe should have still been obscured by the overlying, expanding material of the progenitor star. This indicated that radioactive elements must have been mixed out with very high velocities from the place of their formation close to the center of the supernova far into the stellar mantle and envelope. The structure of the infrared emission lines was interpreted as an indication that the iron-peak elements were mixed outward macroscopically and inhomogeneously in form of about 60 to 100 similar-sized clumps[7].

The smoothness of the light curve of SN 1987A provided indirect evidence for the existence and strength of the mixing process. Mixing of hydrogen towards the center helps to explain the smooth and broad light curve maximum by the time-spread of the liberation of recombination energy[8]. Mixing of heavy elements into the hydrogen-rich envelope homogenizes the opacity and again smooths the light curve[9]. Moreover, the observation of a large number of fast-moving, young pulsars[10] might indicate the existence of violent, non-spherical processes in the early moments of the supernova explosion when the neutron star is formed. The large, dense explosion fragments seen on ROSAT X-ray images outside of the shock front in the Vela supernova remnant[11] might possibly also originate from such early instabilities.

Numerical simulations of Rayleigh-Taylor instabilities at the composition interfaces (metal-He, He-H) in the stellar mantle and envelope after shock passage could neither explain the extent of the required mixing nor could they account for the observed high velocities and the large scales of inhomogeneities and anisotropies[12–14]. This inspired supernova modellers

[a] This work was supported in part by the National Science Foundation under grant NSF AST 92-17969, by the National Aeronautics and Space Administration under grant NASA NAG 5-2081, and by an Otto Hahn Postdoctoral Scholarship of the Max-Planck-Society (H.-Th. J.).

[c] Present address: Department of Astronomy and Astrophysics, University of Chicago, 5640 S. Ellis Avenue, Chicago, Illinois 60637, U.S.A.

to perform multi-dimensional calculations of the early phases of the supernova explosion, i.e. the core-collapse, shock-formation, and neutrino-heating phases. Spherically symmetrical models had indeed shown that convectively unstable layers are present in the collapsed stellar core after formation and propagation of the prompt shock[15]. Theoretical considerations confirmed the possible importance of overturn and mixing in these layers for the evolution of the shock[16].

MULTI-DIMENSIONAL MODELS OF THE EXPLOSION

Herant *et al.*[17] first demonstrated by a hydrodynamical simulation that strong, turbulent overturn occurs in the neutrino-heated layer outside of the protoneutron star and that this helps the stalled shock front to start re-expansion as a result of energy deposition by neutrinos. Although the existence and fast growth of these instabilities was confirmed by Janka and Müller[18–20], the results of their simulations in 1D and in 2D indicated a very strong sensitivity to the conditions at the protoneutron star and to the details of the description of neutrino interactions and neutrino transport. Since the knowledge about the high-density equation of state in the nascent neutron star and about the neutrino opacities of dense matter is incomplete[21,22], the influence of a contraction of the neutron star and of the size of the neutrino fluxes on the evolution of the explosion has been tested by systematic studies.

In the following we shortly report the main conclusions that can be drawn from our set of 1D and 2D models with different core-neutrino luminosities and with varied temporal contraction of the inner boundary[23]. The latter was placed somewhat inside the neutrino sphere and was used instead of simulating the evolution of the very dense inner core of the nascent neutron star. This gave us the freedom to set the neutrino fluxes to chosen values at the inner boundary and also enabled us to follow the 2D simulations until about one second after core bounce with a reasonable number ($\mathcal{O}(10^5)$) of time steps and an acceptable computation time, i.e. several 100 h on one processor of a Cray-YMP with a grid of 400×90 zones and a highly efficient implementation of the microphysics. Note that doubling the angular resolution multiplies the computational load by a factor of about 4!

Our approach and aims differ from those chosen by other groups who recently performed 2D simulations of supernova explosions[24,25]. These groups attempted a self-consistent treatment that includes the central, high-density part of the neutron star up to a certain stage in the evolution, typically 100–200 ms after core bounce, but they did not reveal the dependence of their results on uncertain aspects of the input physics or its numerical description.

RESULTS

One-Dimensional Models

The evolution of the stalled, prompt supernova shock in 1D models turns out to be extremely sensitive to the size of the neutrino luminosities and to the corresponding strength of neutrino heating exterior to the gain radius[20,23].

Increasing the core-neutrino fluxes from low to higher values has the effect that the shock is pushed further and further out to reach a successively larger maximum radius during a short phase of slow expansion. Nevertheless, it finally recedes again to become a standing accretion shock at a much smaller radius. For a sufficiently high threshold luminosity, however, neutrino heating is strong enough to drive the shock front outwards and to cause a successful explosion. For even higher neutrino fluxes the explosion develops faster and gets more energetic. In case of our 15 $M_\odot$ star[26] with a 1.3 $M_\odot$ iron core we found that explosions occur for electron neutrino and antineutrino luminosities in excess of $2.2 \cdot 10^{52}$ erg/s in case of a contracting inner boundary (to mimic the shrinking protoneutron star) but of only $1.9 \cdot 10^{52}$ erg/s when the radius of the inner boundary is fixed.

The transition from failure to explosion requires the neutrino luminosities to be higher than some threshold value. Yet, this is not sufficient. High neutrino-energy deposition has to be maintained for a longer period of time to ensure a successful explosion. If the decay of the neutrino fluxes is too fast, e.g., if a significant fraction of the neutrino luminosity comes from neutrino emission by spherically accreted matter, being shut off when the shock starts to expand, then the outward propagation of the shock may break down again and the model fizzles. To drive a continuous shock expansion a sufficiently high push from the neutrino-heated matter must be maintained until the material behind the shock has achieved escape velocity and does not need pressure support to make its way out.

This contradicts a recent suggestion by Burrows and Goshy[27] that the explosion can be viewed at as a global instability of the star that, once excited, inevitably leads to an explosion. The analysis by Burrows and Goshy may allow one to estimate the radius of shock stagnation. The start-up phase of the explosion, however, can hardly be described by steady-state assumptions, because the times scales of shock expansion, of neutrino cooling and heating between protoneutron star and shock, and the corresponding time scales of temperature and density changes in the postshock region are all of the same order, although they are long compared to the sound crossing time scale and may be short compared with the characteristic times of luminosity changes or variations of the mass accretion rate into the shock. In particular, due to the high sound speed and rather slow shock expansion the propagation of the shock is very sensitive to changes of the conditions in the neutrino-heated layer. A contraction of the neutron star, enhanced cooling of the gas inside the gain radius, which is a process that also accelerates the advection of matter through the gain radius and reduces the time the matter behind the shock gets heated, or even a moderate decay of the neutrino fluxes can therefore be harmful to the outward motion of the shock.

Two-Dimensional Models

In spherical symmetry the expansion of the neutrino-heated matter and of the shock can occur only when also the overlying material is lifted in the gravitational field of the neutron star. In the multi-dimensional case this is different. Blobs and lumps of heated matter can rise by pushing colder material aside and cold material from the region behind the shock can get closer to the zone of strongest neutrino heating to readily absorb energy. Also, when buoyancy forces drive hot matter outward, the energy loss by re-emission of neutrinos is significantly reduced. This overturn of low-entropy and high-entropy gas results in an increase of the efficiency of neutrino-energy deposition external to the gain radius and leads to explosions in 2D already for lower neutrino fluxes than in the spherically symmetrical case. Our models, however, do not show the existence of a "convective cycle" or "convective engine"[24] that transports energy from the heating region into the shock. The matter between protoneutron star and shock is subject to strong neutrino heating and cooling and our high-resolution calculations reveal a turbulent, unordered, and dynamically changing pattern of rising and sinking lumps of material with very different thermodynamical conditions and with no clear indication of inflows of cool gas and outflows of hot gas at well-defined thermodynamical states.

2D models explode for core-neutrino luminosities which cannot produce explosions in 1D. There is a window of neutrino fluxes with a width of about 20% of the threshold luminosity for explosions in 1D, where convective overturn between the gain radius and the shock is a significant help for shock revival. For lower neutrino fluxes even convective overturn cannot ensure strong explosions but the explosion energy gets very low. We do not find a continuous "accumulation"[24] of energy in the convective shell until an explosion energy typical of a Type-II supernova is reached. For neutrino fluxes that cause powerful explosions already in 1D, turbulent overturn occurs but is not crucial for the explosion. In fact, in this case the effects associated with the fast rise of bubbles of heated material lead to a less vigorous start of the explosion and to the saturation of the explosion energy at a somewhat lower level. The explosion energy, defined as the *net energy of the expanding matter at infinity*, does not exceed 10^{50} erg earlier than after about 100 ms of neutrino heating. This is the characteristic time scale of neutrinos to transfer an amount of energy to the material that is roughly equal to its gravitational binding energy and it is also the time scale that the convective overturn between gain radius and shock needs to develop to its full strength. It is not possible to determine or predict the final explosion energy of the star from a short period of only 100–200 ms after shock formation. Typically, the increase of the explosion energy with time levels off not before 400–500 ms after core bounce, followed by only a very slow increase due to the much smaller contributions of the few $10^{-3}\,M_{\odot}$ of matter blown away from the protoneutron star in the neutrino wind. Since the wind material is heated slowly and can start expansion as soon as the internal energy per nucleon roughly equals its gravitational binding energy, the matter does not have a large kinetic energy at infinity.

Although the global evolution of powerful explosions in 2D, i.e., the increase of the explosion energy with time, the shock radius as a function of time, or even the amount of ^{56}Ni produced by explosive nucleosynthesis, is not much different than in energetic explosions of spherically symmetrical models, the structure of the shock and of the thick layer of expanding, dense matter behind the shock clearly show the effects of the turbulent activity. The shock is deformed on large scales and its expansion velocity into different directions varies by about 20–30%. The material behind the shock reveals large-scale inhomogeneities in density, temperature, entropy, and velocity, these quantities showing contrasts of up to a factor of 3. The typical angular scale of the largest structures is between about 30° and 45°. We do not find indications that the turbulent pattern tries to gain power on the largest possible scales and to evolve into the lowest possible mode, $l = 1$[24]. Turbulent motions are still going on in the extended, dense layer behind the shock when we stop our simulations at about 1 second after core bounce. We consider them as the origin of the anisotropies, inhomogeneities, and non-uniform distribution of radioactive elements which were observed in SN 1987A. The contrasts found behind the shock in our models are about an order of magnitude larger than the artificial perturbations that were used in the hydrodynamical simulations to trigger the growth of Rayleigh-Taylor instabilities in the stellar mantle and envelope.

From Core Bounce to 1 Second

Convective overturn outside of the protoneutron star develops within about 50–100 ms after shock formation. About 200–300 ms after bounce neutrinos have deposited a sizable amount of energy in the material below the shock front. The turbulent layer begins to move away from the region of strongest neutrino heating and to expand outward behind the accelerating shock. This is the time when we find turbulent activity around the protoneutron star to come to an end. Our models do not give an extended phase of convection and accretion outside of the protoneutron star. The inflows of low-entropy, proton-rich gas from the postshock region towards the neutrino-heated zone are not accreted onto the protoneutron star. Although the gas loses lepton number while falling in, it does not get as neutron-rich as the material inside the gain radius. In addition, neutrino heating and mixing with the surrounding, high-entropy gas increase the entropy in the downflows. Both high electron (proton) concentration and high entropy have a stabilizing effect and prevent the penetration of the gas through the gain radius into the cooler and more neutron-rich surface layer of the protoneutron star.

At about 400–500 ms the protoneutron star has become quite compact already and the density outside of it has dropped appreciably. This indicates the formation of the high-entropy, low-density "hot-bubble" region[28] and the phase of small mass loss from the nascent neutron star in the neutrino-driven wind, accompanied by slowly increasing entropies. As the outgoing supernova shock passes the entropy and composition discontinuity at the Si-O interface at about 6000 km, the density inversion between the dense shell behind the shock and the low-density hot-bubble region steepens into a strong reverse shock that forms a sharp discontinuity in the neutrino wind, slowing down the wind expansion from more than 10^9 cm/s to a few 10^8 cm/s. We do not know yet how this reverse shock develops on a longer time scale. Since the wind velocity decreases with time it is very likely that it will trigger the fallback of a significant fraction of the matter that was blown out in the neutrino wind. Once the infall of the outer wind material is initiated and the pressure support of the gas further out vanishes, the inward acceleration might even enforce the fallback of the more slowly moving parts of the dense shell behind the supernova shock.

Fallback of a significant amount of matter, between 0.1 and $0.2\,M_\odot$, has to be postulated to solve two major problems in the current supernova models. On the one hand the protoneutron star formed at the center of the explosion has quite a small (initial) baryonic mass, only about $1.2\,M_\odot$ in case of our $15\,M_\odot$ star with the $1.3\,M_\odot$ iron core. On the other hand the explosive nucleosynthesis yields of iron-peak elements are incompatible with observational constraints for Type-II supernovae as deduced from terrestrial abundances and galactic evolution arguments. In case of powerful explosions with explosion energies of $1\text{–}1.3\cdot 10^{51}$ erg about $0.2\,M_\odot$ of material are heated to temperatures above $4.5\cdot 10^9$ K and are ejected behind the shock during the early phase of the explosion. Only roughly half of this matter, 0.085–$0.1\,M_\odot$, has an electron fraction $Y_e > 0.49$ and will end up with ^{56}Ni as the dominant nucleosynthesis product. In that respect the models seem to match the observations quite well. Yet, only some part (about $0.05\,M_\odot$) of the matter that is shock-heated to $T > 4.5\cdot 10^9$ K has an electron fraction $Y_e \gtrsim 0.495$ and will end up with relative abundance yields in acceptable agreement with solar-system values. The amount of ^{56}Ni produced in neutrino-driven explosions turns out to be correlated with the explosion energy. In case of more energetic explosions the shock is able to heat a larger mass to sufficiently high temperatures.

SUMMARY AND CONCLUSIONS

Turbulent overturn between the zone of strongest neutrino heating and the supernova shock aids the re-expansion of the stalled shock and is able to cause powerful Type-II supernova explosions in a certain, although rather narrow, window of core neutrino fluxes where 1D models do not explode. The turbulent activity outside and close to the protoneutron star is transient and between 300 and 500 ms after core bounce the (essentially) spherically symmetrical neutrino-wind phase starts and the turbulent shell moves outward behind the expanding supernova shock. Our 2D simulations do not show a long-lasting period of convection and accretion after core bounce. Only very little of the cool, low-entropy matter that flows down from the shock front to the zone of neutrino-energy deposition is advected into the protoneutron star surface. Since the matter is proton-rich and its entropy increases quickly due to neutrino heating, it stays in the heated region to gain more energy by neutrino interactions and to start rising again. The strong, large-scale inhomogeneities and anisotropies in the expanding layer behind the outward propagating shock front will probably help to explain the effects of macroscopic mixing observed in SN 1987A and can account for moderately high recoil velocities of the neutron star[29,30].

Although the 2D models develop energetic explosions for sufficiently high neutrino luminosities and produce an amount of ^{56}Ni that is in good agreement with observational constraints, the initial mass of the protoneutron star is clearly on the low side of the spectrum of measured neutron star masses. Moreover, the models eject about $0.1\text{–}0.2\,M_\odot$ of material with

$Y_e < 0.495$, which implies an overproduction of certain elements in the iron peak by an appreciable factor compared with the nucleosynthetic composition in the solar system. The fallback of a significant fraction of this matter to the neutron star at a later stage of the evolution would ease these problems. It is possible that the reverse shock which develops in our models when the supernova shock passes the entropy discontinuity at the Si-O interface will trigger this fallback.

ACKNOWLEDGEMENTS

We are very grateful to S.W. Bruenn for providing us with the data of his core collapse calculations, which we used to construct the initial models for our simulations. The computations were performed on the CRAY-YMP 4/64 of the Rechenzentrum Garching.

REFERENCES

1. Dotani, T. *et al.* 1987. Nature **330:** 230.
2. Sunyaev, R.A. *et al.* 1987. Nature **330:** 227.
3. Matz, S.M. *et al.* 1988. Nature **331:** 416.
4. Mahoney, W.A. *et al.* 1988. Astrophys. J. **334:** L81.
5. Erickson, E.F. *et al.* 1988. Astrophys. J. **330:** L39.
6. Haas, M.R. *et al.* 1990. Astrophys. J. **360:** 257.
7. Li, H., R. McCray & R.A. Sunyayev. 1994. Astrophys. J. In press.
8. Shigeyama, T. & K. Nomoto. 1990. Astrophys. J. **360:** 242.
9. Arnett, W.D., J.N. Bahcall, R.P. Kirshner & S.E. Woosley. 1989. Ann. Rev. Astron. Astrophys. **27:** 629.
10. Lyne, A.G. & D.R. Lorimer. 1994. Nature **369:** 127.
11. Aschenbach, B., R. Egger & J. Trümper. 1994. Nature. In press.
12. Arnett, W.D., B.A. Fryxell & E. Müller. 1989. Astrophys. J. **341:** L63.
13. Den, M., T. Yoshida & Y. Yamada. 1990. Prog. Theor. Phys. **83:** 723.
14. Herant, M. & W. Benz. 1991. Astrophys. J. **345:** L412.
15. Burrows, A. & J.M. Lattimer. 1988. Phys. Rep. **163:** 51.
16. Bethe, H.A. 1990. Rev. Mod. Phys. **62:** 801.
17. Herant, M., W. Benz & S.A. Colgate. 1992. Astrophys. J. **395:** 642.
18. Janka, H.-Th. & E. Müller. 1993. *In* Proc. of the IAU Colloquium 145, Xian, China, May 24–29, 1993. Cambridge Univ. Press. Cambridge, New York.
19. Müller, E. & H.-Th. Janka. 1994, *In* Reviews in Modern Astronomy **7**. G. Klare, Ed.: p. 103. Astronomische Gesellschaft. Hamburg.
20. Janka, H.-Th. & E. Müller. 1995. Phys. Rep. In press.
21. Raffelt, G. & D. Seckel. 1994. Phys. Rev. D. To be published. HEP-PH/9312019.
22. Keil, W., H.-Th. Janka & G. Raffelt. 1994. Phys. Rev. D. Submitted.
23. Janka, H.-Th. & E. Müller. 1995. Astron. Astrophys. Submitted.
24. Herant, M., W. Benz, W.R. Hix, C.L. Fryer & S.A. Colgate. 1994. Astrophys. J. **435:** 339.
25. Burrows, A., J. Hayes & B.A. Fryxell. 1994. Astrophys. J. Submitted.
26. Woosley, S.E., P.A. Pinto & L. Ensman. 1988. Astrophys. J. **324:** 466.
27. Burrows, A. & J. Goshy. 1993. Astrophys. J. **416:** L75.
28. Bethe, H.A. & J.R. Wilson. 1985. Astrophys. J. **295:** 14.
29. Janka, H.-Th. & E. Müller. 1994. Astron. Astrophys. **290:** 496.
30. Janka, H.-Th. & E. Müller. 1995. Paper presented at this conference.

The Physics of Core-Collapse Supernova Explosions[a]

ADAM BURROWS AND JOHN HAYES

Departments of Physics and Astronomy
University of Arizona
Tucson, Arizona 85721

INTRODUCTION

Fresh insights and powerful numerical tools are revitalizing the theoretical exploration of the supernova mechanism. The realization that the protoneutron star is Rayleigh-Taylor unstable at various times and radii and, hence, that a multi-dimensional perspective is required is one agent of this revolution. However, a new physical understanding of the nature of explosions (even spherical explosions) that are driven by neutrino heating and that escape from deep within a gravitational potential well is also emerging.[1,2,3,4] This, together with the new multi-dimensional approach, promises to establish a new paradigm within which supernova explosions and their consequences can be studied in the future.

Supernova theory is in flux and a consistent model that fits the growing list of observational constraints does not yet exist. Nevertheless, the observed explosion energies, nickel yields, optical and IR line profiles, pulsar kicks, neutron star masses, nickel debris distributions, and nucleosynthesis are strengthening the connections between collapse theory and empirical astronomy. Two of the remaining embarrasments of theory concern the overproduction of neutron-rich species and the difficulty of achieving entropies sufficient to produce an r-process. However, recent simulations hint at how these problems can be solved.[2] In addition, the observation of high-speed pulsars suggests that asymmetries during and/or after core collapse might exist. Recent calculations by Bazan and Arnett[5] show that silicon- and oxygen-burning are hydrodynamic and that Mach-number and density variations in the core at collapse can be high. These asymmetries can amplify during infall and can result in asymmetric explosions that blow preferentially via the paths of least resistance. Burrows and Hayes[6] have recently demonstrated, via a 180° 2-D hydrodynamic simulation, such a "rocket" effect. The asymmetrically ejected matter causes the residual core to recoil with speeds of hundreds of kilometers per second. (This suggests that magnetic field effects are not necessary to achieve high

[a]The authors would like to acknowledge the NSF both for support under grant # AST92-17322 and for the use of NSF Supercomputer Centers where much of the heavy lifting was performed.

neutron star speeds.) In addition, there may be a correlation between the pulsar recoils and the distribution of the ejected ^{56}Ni.

There are far too many new and interesting questions, problems, and potential solutions to be comfortably contained in this short paper. Therefore, we will focus here on a discussion of the crucial ingredients of the explosion *mechanism* itself and the character of the blast after it starts. A more in-depth discussion of some of the constraints listed in Table 1 and the 1-D and 2-D hydrodynamic simulations that we have recently performed can be found in BHF[2]. Color figures from that paper and mpeg movies of one of its 2-D simulations can be acquired gratis via mosaic at URL address http://lepton.physics.arizona.edu:8000/.

NEUTRINO-DRIVEN EXPLOSIONS IN ONE AND TWO DIMENSIONS

a. The Quasi-Steady-State Phase Before Explosion and the Explosion Condition

After the "Chandrasekhar" core of a massive star becomes unstable to implosion, it evolves through various distinct hydrodynamic phases. These are infall, core bounce, shock formation, shock stagnation, the pre-explosion quasi-steady state, the onset of explosion, and the explosion proper. Twenty to one hundred milliseconds into the explosion, a distinct neutrino-driven wind emerges from the core, whatever the details of the mechanism. Since the direct hydrodynamic mechanism aborts for all progenitors (even the lightest massive stars), the nature and evolution of the quasi-steady state after the shock stalls takes on a new importance. How long does the steady-state phase last? What triggers the explosion? How does the explosion evolve? In what context does a black hole form? The answers to all these questions hinge on the proper understanding of the physics of the shock-bounded and accreting protoneutron star.

Burrows and Goshy (1993, BG)[1] have recently developed an approximate semi-analytic theory of such objects. Setting all partial derivatives with respect to time equal to zero, the equations of hydrodynamics and neutrino transfer become a set of coupled ordinary differential equations, subject to boundary conditions. Such a problem is an *eigenvalue* problem. With an equation of state, a prescription for the bounding shock jump conditions and the outer supersonic flow profiles, a given core mass (~ 1.1–$1.3 M_\odot$), and simple formulae for neutrino heating and cooling exterior to the neutrinospheres, BG[1] solved for the steady-state structure. The shock radius (R_s) was the eigenvalue and the electron neutrino luminosity (L_{ν_e}) and the mass accretion rate ($\dot{M}$) were the control parameters. Physically, the structure adjusts until the infall time from the shock to the core ($\frac{R_s}{u_1}$, where u_1 is the post-shock settling velocity) is "equal" to the cooling or heating timescale. This equality of timescales is similar to the equality of the free-fall and sound travel times in the context of hydrostatic equilibrium and to a similar condition in the context of AM Her stand-off shocks.[7,8] The quasi-steady assumption is proper as long as these adjustment timescales ($\sim$10 milliseconds) are shorter than the timescale for the decay of $\dot{M}$ (30-100 milliseconds). Note that the radius to which the bounce shock is initially thrown is almost unrelated to its later steady values.

Increasing L_{ν_e} for a given $\dot{M}$, increases R_s (roughly as $L^2_{\nu_e}/\dot{M}$). However, this behavior does not continue for arbitrarily high L_{ν_e}. BG[1] showed that for each $\dot{M}$ (and set of model assumptions), there is a *critical* L_{ν_e} above which there is no steady-state protoneutron star envelope (Figure 1). This is reached at a finite value of the eignevalue, R_s (generally less than 200 kilometers). BG[1] identified the implied *instability* with the onset of the supernova explosion. This onset is a critical phenomenon and is at the bifurcation between steady-state and wind solutions of the equations. If, for a given $\dot{M}$, L_{ν_e} could be increased by better neutrino transport or cross sections or by convective enhancement,[9,10] the 1-D models would explode more readily. The problem with previous 1-D calculations is that their L_{ν_e}–$\dot{M}$ trajectories passed below the critical curve, as Figure 1 depicts.

Recently, it was shown that the outer shocked envelopes of the protoneutron star are generically unstable to Rayleigh-Taylor overturn driven by neutrino heating from below.[11,12,13,2,14] This and other hydrodynamic instabilities before, during, and after explosion are redrawing our picture of the evolution and character of supernova blasts. An important question one may ask is: how do the multi-dimensional effects alter the explosion mechanism? On this there is much needless confusion that the next paragraphs may partially clear up.

Figure 2 depicts in cartoon form the shock-bounded protoneutron star before explosion in 1-D and ≥2-D. Relaxing spherical symmetry allows some parcels of matter that have just passed through the shock and that are being heated by neutrinos from the core to rise like balloons or cells in any "convectively" unstable region. This allows the matter to dwell longer in the gain region (where heating > cooling) and, hence, to achieve higher entropies than is possible in 1-D.[11,13]

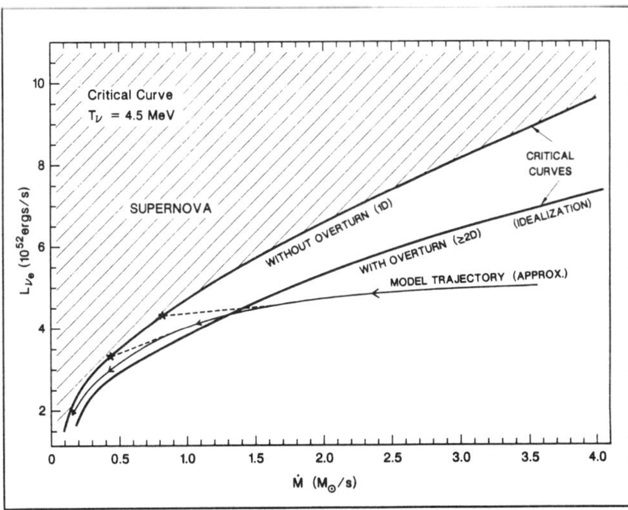

FIGURE 1. Approximate critical curves for explosion, with and without multi-dimensional effects, in L_{ν_e} versus $\dot{M}$ space. Superposed is a representative L_{ν_e} vs. $\dot{M}$ trajectory for a realistic calculation. Note that this curve intersects the lower ≥ 2-D critical curve at some time, whereas in 1-D it may never.

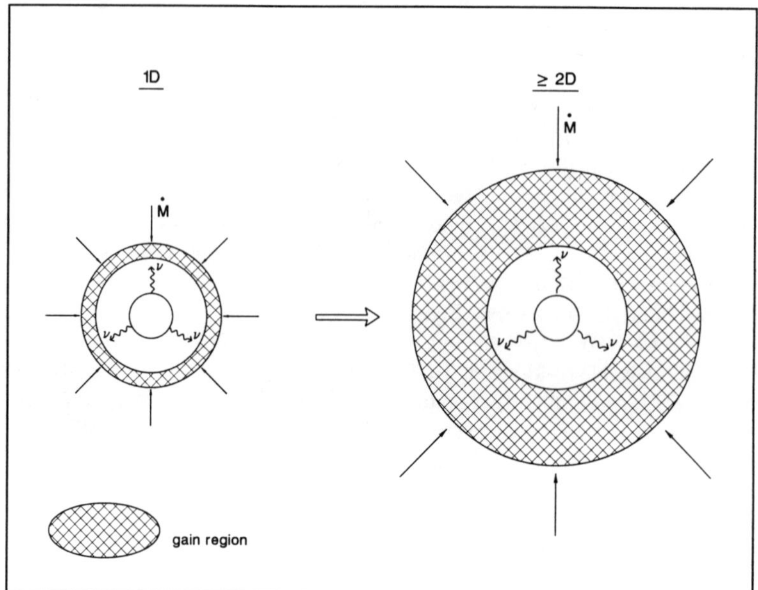

FIGURE 2. Schematics of the accreting protoneutron star in 1- and ≥2-dimensions. The gain region is indicated by the hatches. The main difference between the two is the larger steady-state radius that overturn allows.

In 1-D, the heated parcel would perforce fall directly into the cooling region interior to the gain radius (where cooling = heating) and lose its just recently gained energy. (Curiously, the cooling region lies closer to the neutrinospheres where the temperatures are higher.)

However, the rising balloons, upon encountering the shock, are immediately advected inward by the powerful mass accretion flux raining down. They do not dwell near the shock. In fact, the net mass flux through the shock is approximately equal to the mass flux onto the core and mass does not accumulate in the convective zone. The mass between the shock and the neutrinospheres *decreased* by about a factor of three in the calculations of BHF[2] during the pre-explosion boiling phase that lasted ~100 milliseconds (~30 convective turnover times). *All* the matter that participates in the "convection" before explosion eventually leaves the convection zone and settles onto the core. A given parcel of matter may "cycle" one or two times before settling inward (and a large fraction never rises), but more than three times is rare. The boiling zone is resupplied with mass by mass accretion through the shock and a secularly evolving steady-state is reached. This steady-state is similar to that achieved in 1-D, but due to the higher dwell time the average entropy in the envelope is larger and its entropy gradient is flatter. These effects, together with the dynamical pressure of the buoyant plumes, serve to increase the steady-state shock radius over its value in 1-D by 30%–100%. It is this effect of boiling that is central to its role in triggering the explosion, for it thereby lowers the critical luminosity threshold. As Figure 1 suggests, the lowering of the effective critical curve allows the actual model trajectory in L_{ν_e} vs. $\dot{M}$ space to intersect it. Even if in 1-D it can be shown that the two curves can intersect, they would intersect earlier and more assuredly with the multi-dimensional effects included. The physical reasons for the lowered threshold are straightforward: a large R_s enlarges the volume of the gain region, puts shocked matter lower in the gravitational potential well, and lowers the accretion ram pressure at the shock for a given $\dot{M}$. Since the "escape" temperature ($T_{\rm esc} \propto \frac{GM\mu}{kR}$) decreases with radius faster than the actual matter temperature (T) behind the shock, a larger R_s puts a larger fraction of the shocked material above its local escape temperature. $T > T_{\rm esc}$ is the condition for a thermally-driven corona to lift off of a star. In one, two, or three dimensions, since supernovae are driven by neutrino heating, they are coronal phenomena, akin to winds, though initially bounded by an accretion tamp. Neutrino radiation pressure is unimportant.

We conclude that the instability that leads to explosion in $\geq$2-D is of the same character as that which leads to explosion in 1-D. Since the explosion succeeds a quasi-steady-state phase, neither the total neutrino energy deposited during the boiling phase nor any putative coeval thermodynamic cycle is of relevance to the energy of the explosion or the trigger criterion. Energy does not accumulate in the overturning region before explosion (it in fact decreases) and the increasing vigor (speed) of convection is in response to the decay of $\dot{M}$. If $\dot{M}$ were held constant, the overturning would not grow more vigorous with every "cycle" and a simple, stable convective zone would be established. In fact, before explosion the average total energy fluxes $((\epsilon + P/\rho + \frac{1}{2}v^2 - \frac{GM}{r})\dot{M})$ due to the overturning motions are *inward*, not outward, since the net direction of the matter is onto the core. Figure 3 depicts such fluxes versus radius at various times for the 1-D and 2-D simulations conducted by BHF[2] of the core of a $15M_\odot$ star. The hump on the inside mirrors the corresponding plot for the total neutrino luminosities. During the boiling phase, the net fluxes in the gain region are negative, not positive, and they become positive only after the explosion commences. This interpretation of the role of 2-D and the nature of supernova explosions differs from that of HBC[11] and HBHFC[12].

b. The Explosion

Importantly, just after the explosion criterion is achieved, the explosion energy is still not determined. In fact, the matter that will eventually be ejected is often still *bound*, even correcting for the reassociation boost (the "after-burner"). This fact emphasizes that the explosion condition has nothing to do with the aggregate neutrino deposition before explosion and that this aggregate heating has nothing to do with the explosion energy itself. The onset of explosion causes the shocked matter to expand and to lower its temperature, thereby turning off cooling. The expansion runs away because heating continues unchecked by cooling, whose integral had usually dominated between the neutrinosphere and the shock during the quasi-state phase. The explosion and the shock are thereafter *driven* by neutrino energy deposition. This continuing source is necessary so that the ejecta can eventually achieve positive energies of supernova magnitude. Soon after the explosion commences, all convective cycling between the shock and neutrinospheres ceases permanently. Note that it is only after the shock has achieved many thousands of kilometers that the explosion has had a chance to "feel" the mantle binding energy that it will eventually have to overcome to succeed. If this binding energy is large, the explosion will be slow. This will allow the material of the explosion to be heated longer. This feedback effect partially compensates for the variety of envelope binding energies along the massive star continuum,[2] and may explain why supernova energies are all near 10^{51} ergs. Binding energies of massive star envelopes are of order 10^{51} ergs because this is approximately the binding energy of a Chandrasekhar white dwarf ($\sim m_e c^2 \mathrm{N_A M_{CH}}$). (Simple arguments show that the envelope and the core binding energies are comparable). Therefore, and very crudely, the envelope binding energies set the scale of the explosion energy (to within a factor of three?), though we still can't say whether it increases or decreases with ZAMS mass. It is suspected that the envelope binding energies can be too high and that the accretion tamp can be too oppressive and that above some ZAMS mass, the core will collapse to a black hole before or during explosion. It is not known whether, during the first seconds after bounce, a supernova and a black hole are mutually exclusive results. (We know from its neutrino signal that in SN1987A the neutron star lasted at least 12.5 seconds.)

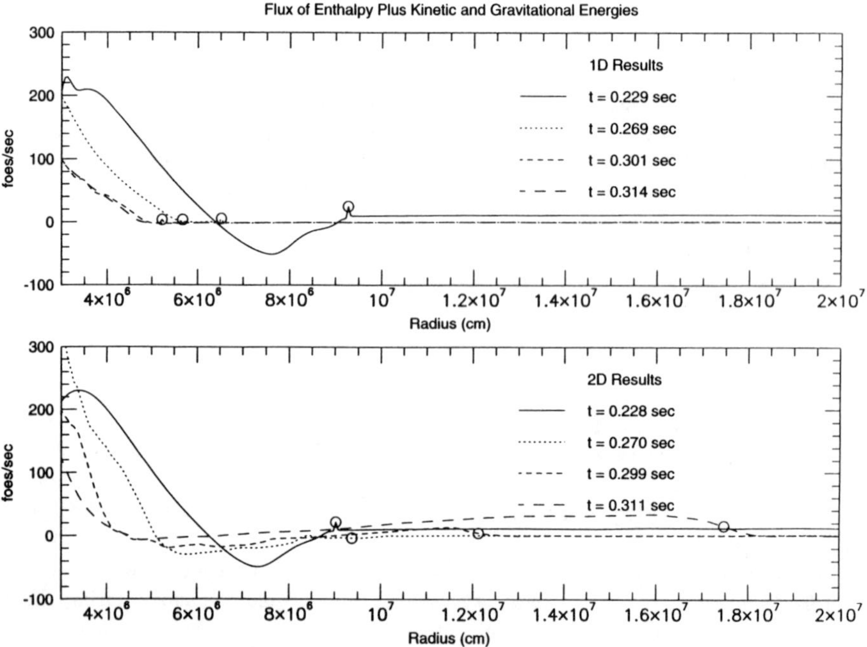

FIGURE 3. The angle-averaged total energy fluxes versus radius in the one- and two-dimensional calculations of BHF. The average position of the shock is identified with a circle. Note that the energy fluxes are not outward in 2-D, until the explosion begins.

DISCUSSION

Though we have made progress in understanding supernova explosions, we still don't know the interesting systematics as a function of progenitor mass. What are $E_{SN}(M_{\rm ZAMS})$, $M_{NS}(M_{\rm ZAMS})$, the critical mass for black hole formation, and the r-process and ^{56}Ni yields? In addition, all the 2-D explosion simulations to date eject too much neutron-rich matter. This problem may be solved if the delay to explosion (~50 milliseconds for HBHFC[12] and ~100 milliseconds for BHF[2]) were longer still, allowing the envelope mass to decrease and the density of the accreting matter to thin. (The latter is important because electron-capture rates are stiffly increasing functions of density.) Thus far, no multi-dimensional explosion simulation has involved multi-group transport, general relativity, or been done in 3-D. We are only at the beginning of a new theoretical assault on the supernova phenomenon that will provide scores of research projects for this and the next generation of modelers.

REFERENCES

1. BURROWS, A., J. GOSHY. 1993. Ap. J. Lett. **416**:L75 (BG).
2. BURROWS, A., J. HAYES, AND B. A. FRYXELL. 1994. submitted to Ap. J. (BHF).
3. JANKA, H.-T., E. MÜLLER. 1993. In the proceedings of the Internat'l Symposium on Neutrino Astrophysics, held in Takayama/Kamioka Japan Oct. 19–22, 1992, to appear in *Frontiers of Neutrino Astrophysics*, (Universal Academy Press Inc. Tokyo, Japan).
4. JANKA, H. T., E. MÜLLER. 1994. A&A in press.
5. BAZAN, G., AND D. ARNETT. 1994. Ap. J. Lett. **433**:L41.
6. BURROWS, A., AND J. HAYES. 1995. in preparation.
7. CHEVALIER, R. A., AND J. N. IMAMURA. 1982. Ap. J. **261**:543.
8. LANGER, S. H., G. CHANMUGAM, AND G. SHAVIV. 1981. Ap. J. **245**:L23.
9. BURROWS, A. 1987. Ap. J. Lett. **318**:L57.
10. MAYLE, R., J. R. WILSON. 1988. Ap. J. **334**:909.
11. HERANT, M., W. BENZ. S. A. COLGATE. 1992. Ap. J. **395**:642 (HBC).
12. HERANT, M., W. BENZ, J. HIX, C. FRYER, AND S. A. COLGATE. 1994. Ap. J. **435**: 339 (HBHFC).
13. BETHE, H. 1990. Rev. Mod. Phys. **62**:801.
14. JANKA, H. T., E. MÜLLER. This volume.

Introductory Remarks to the Gamma-Ray Line Astrophysics Symposium

V. SCHÖNFELDER
Max-Planck-Institut für extraterrestrische Physik
D-85740 Garching, Germany

Fine resolution spectroscopy has always been a powerful tool in astrophysics. Important discoveries have been made, which are based on line observations. Examples are the expansion of the Universe, the identification of quasars as extragalaactic distant objects, the chemical composition of stars, and the structure and dynamics of the Milky Way.

The field of gamma-ray line astronomy is still at its beginning. But, the fact that the organisers of the Texas Symposium decided to have this gamma-ray line symposium reflects that a real break-through has been achieved in this field. The progress is mainly due to results obtained by the COMPTEL and OSSE instruments aboard NASA's Compton Gamma-Ray Observatory.

Table 1 summarizes the gamma-ray line detections achieved so far.

TABLE 1. Overview of Astronomical Gamma Ray Line Observations

PROCESS	LINE ENERGY	SOURCE
Nucleosynthesis		
^{56}Co	847 keV and 1.238 keV	SN 1987a
^{57}Co	122 keV and 136 keV	SN 1987a
^{26}Al	1.809 MeV	Galactic disk and Vela SN-remnant
^{44}Ti	1.156 MeV	Cas A
$e^+ - e^-$ – annihilation	511 keV	Galactic disk
	511 keV	Galactic bulge
	480 ± 120 keV	1 E 1740 – 29
	479 ± 18 keV	Nova Muscae
Nuclear excitation lines	4.44 MeV and 6.13 MeV	^{12}C* and ^{16}O* in Orion complex
	511 keV	Solar flares
	1.24 MeV (^{56}Fe)	
	1.37 MeV (^{24}Mg)	
	1.63 MeV (^{20}Ne)	
	2.23 MeV (neutron capture)	
	4.4 MeV (^{12}C)	
	6.1 MeV (^{16}O)	

Most of these gamma-ray line measurements have been made with detectors of medium energy resolution (5 % to 10 % FWHM). Much more information can in principle be extracted from line measurements., if also their profiles are measured with high resolution. This will be one of the main objectives of ESA's INTEGRAL mission.

The Gamma-Ray Line Symposium is subdevided into three parts. The first one deals with observational results, the second one with the astrophysical interpretation of the results, and the third one with the prospects of future gamma-ray line observations. The aural presentations contained in this edition were complemented by a number of Poster contributions.

Gamma-Ray Line Observations with CGRO- COMPTEL

R.DIEHL

Max-Planck-Institut für extraterrestrische Physik, Garching,FRG

INTRODUCTION

The COMPTEL[1] γ-ray telescope aboard the NASA Compton Observatory enables γ-ray line spectroscopy, with a spectral resolution of ~8% (FWHM) and ~degree imaging resolution within its large (~1sr) field of view. Within its energy range (0.7-30 MeV), several gamma-ray lines from radioactive nuclei, as well as de-excitation lines can be imaged with a limiting sensitivity of ~10^{-5} ph $cm^{-2}s^{-1}$.

Radioactive nuclei are formed in nucleosynthesis processes as minor by-products of nuclear burning, and may be used as tracers to the nucleosynthesis processses themselves. The isotopes of interest to γ-ray instruments cover a range of decay times from ~100 days (^{56}Co) over years (^{22}Na, ^{44}Ti), to million years (^{26}Al, ^{60}Fe). Therefore the results from γ-ray line measurements cover a wide range of different interpretations: For the isotopes with short decay times, the relative abundances may reveal the evolution of the γ-ray transparency of the expanding envelope of the nucleosynthesis event, while for intermediate decay times (y) the total yield of a radioactive isotope is measured for an individual nucleosynthesis event. For the long decay times (million years), thousands of nucleosynthesis events are expected to contribute to the observed γ-ray line signal; the interpretation of the measurement in terms of nucleosynthesis requires assumptions or models of source event types, their yields, and their spatial distribution; also evolution of Galactic nucleosynthetic activity may play a role. Gamma-ray line observations constitute a direct measurement of freshly formed isotopes through their radioactive decay. In comparison, supernova light curves at other wavelengths only indirectly encode the radioactive energy input, and spectroscopic elemental composition measurements in optical/X show photospheric abundances, also only indirectly related to the nucleosynthesis process. Yet, the combination of those astronomical measurements with meteoritic studies provides proof of current nucleosynthesis processes in our Galaxy.

GALACTIC NUCLEOSYNTHESIS STUDY WITH ^{26}AL

COMPTEL measured the distribution of 1.809 MeV emission from radioactive ^{26}Al along the plane of the Galaxy[2]. Data from the sky survey and subsequent observations have confirmed the expectation of recent Galactic nucleosynthesis being traced by this isotope with its radioactive decay time of 1.04 10^6 years. It was commonly concluded that ~ 2-3 $M_\odot$ of ^{26}Al exist within the Galaxy[3,4]. ^{26}Al is expected to be produced mainly from two processes: hydrostatic nuclear burning in the core of very massive stars at the late stage of stellar evolution, and explosive nuclear burning in nova or supernova events[8,9]. Production of ^{26}Al depends critically on the balance of nuclear reaction rates in those environments, where convective mixing substantially affects abundances of seed nuclei and temperature. Main production reactions are proton capture reactions with intermediate β-decays, while destruction occurs through neutron capture; the β-decay time constants of ~10s introduce a critical dependence on the dynamics of the burning zone, particularly relevant for explosive burning in supernova shock waves or novae. Theoretical calculations have been performed for all candidate scenarios[9], although self-consistent models including stellar evolution still present a major challenge. Calculations for novae, supernovae, Wolf-Rayet and AGB stars all indicate that a Galactic yield in the range from a few tenths to several $M_\odot$ can be derived for each of the candidate sources. Thus yield considerations alone cannot decide among those: Correlation of the spatial distribution of the expected sources to 1.809 MeV imaging details is important.

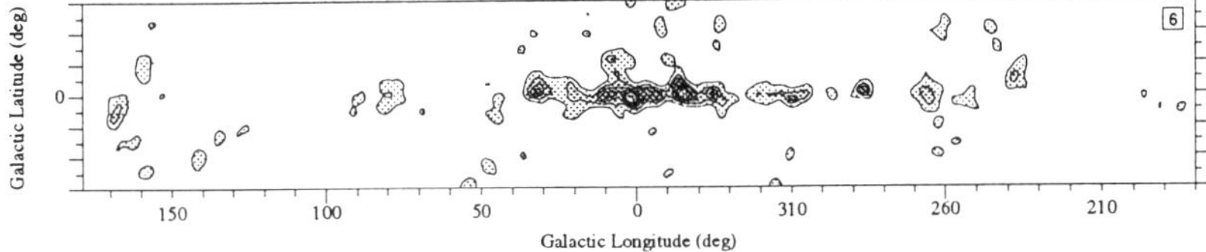

Fig. 1: *COMPTEL 1.8 MeV imag of the Galactic plane*

The COMPTEL *imaging data* at 1.809 MeV provide a new quality of 1.809 MeV measurements. Coarse software collimation analysis was applied[2], but also full imaging capability exploitation through maximum likelihood and maximum entropy analysis in the 3-dimensional imaging data space of COMPTEL spanned by the primary photon scatter angle and the secondary photon's scatter direction angles[5,6,7]. The overwhelming background (S/N~4%) was modelled from energy bands adjacent to but excluding the 1.8 MeV photopeak regime.

Overall 1.809 MeV emission from the plane of the Galaxy is clearly detected (>20σ), with individual features of the image ranging from insignificant fluctuations up to ~7σ peaks in the inner Galaxy. The observed distribution (see Figure 1), however, shows remarkable deviations from the expected smooth distributions previously attributed to the large number of candidate ^{26}Al sources[8,9]. In addition, now regions far off the inner Galaxy have been detected, namely the Vela, Carina, and Cygnus region.

The ambiguities inherent in Compton telescope data require a careful investigation of image detail significances. Source simulations confirmed that adjacent energy bands provide a representative background model free from image artifacts[10]. Those simulations indicate however that statistical background fluctuations may have an impact on image details near the sensitivity level, and some degree of irregularity arises for weak sources. Therefore a statistical bootstrap analysis[2,10] appears as the best method to investigate significances of image structures. This method assumes that a truly representative measurement has been taken, but does not make assumptions about instrumental response

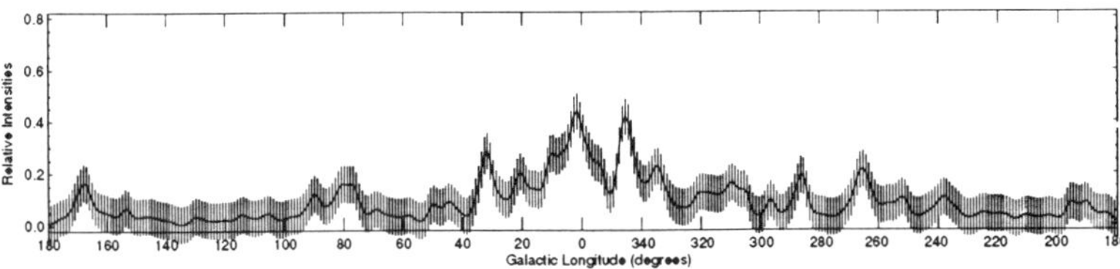

Fig. 2: *Estimate of the variability of the 1.8 MeV longitude profile due to statistical fluctuations, derived from bootstrap sample analysis*

and noise distributions. Figure 2 represents a profile along Galactic longitude, obtained from the flux image of Figure 1, integrated over a ±5° latitude region in 1° steps. In this profile, the distribution of flux values from 30 bootstrap data samples is included as a measure (±1σ) of the statistical variability of image details, outlining the band of confidence for the profile of 1.809 MeV emission. It is seen that a conservative smooth interpretation of the measured profile must ignore e.g. structures at l~20° or in the fourth Galactic quadrant, yet it confirms the general irregularity in the inner Galaxy with at least 3, possibly 5, distinct emission features and its steep falloff beyond l~35°, as well as the Vela, Carina, and Cygnus features. This is consistent with likelihood analysis for isolated features, which yields e.g.~5σ significance for the Vela emission[11].

Tracers of candidate sources have been compared to the observed 1.809 MeV image[2]. None of the classical candidate source distribution models for novae, supernovae, and Wolf Rayet stars has been found to satisfactory explain the data. We note the observational biases for radio supernova remnants and Wolf Rayet stars, and the generally small (~100) number of tracer objects that are supposed to represent the distribution of the few thousand or more ^{26}Al source objects. Yet, if Galactic spiral structure is included in the tracer model, suggestive correlations can be found at directions tangential to spiral arms[9]. A quantitative correlation is in progress. From present results, massive stars are favoured sources of the observed ^{26}Al, either from their hydrostatic burning phases or from their supernovae. Their ^{26}Al production may provide the best mapping of massive star Galactic distribution, through the 1.809 MeV γ-ray line.

INDIVIDUAL NUCLEOSYNTHESIS EVENTS

Individual supernovae or novae provide nucleosynthesis data free from source type and Galactic distribution uncertainties, if sufficiently close, or observed from an abundant and shortlived radioactive isotope. The COMPTEL sensitivity is adequate to study individual events in the ^{56}Co line up to ~15 Mpc, and in the ^{44}Ti line up to ~5kpc distance.

Even for ^{26}Al with its million year decay time, the detection of emission from the Vela region[11] indicates the possibility to assign the observed flux to a single object. Two relatively nearby candidate source objects, the Vela supernova remnant, and the Wolf Rayet binary system 'γ Velorum' are expected to dominate any 1.809 MeV signal from this region[12]. The γ Vel contribution (d~300-450pc) is probably small. The COMPTEL measurement indicates somewhat extended emission, compatible with the Vela supernova remnant[11]. The distance to the Vela supernova remnant is however uncertain within the wide range from ~150 to 600 pc. COMPTEL measurements combined with latest nucleosynthesis models[13] favour a distance of~300 pc or lower; the uncertainty in models from treatment of convection and overshooting is about one order of magnitude, the overall COMPTEL flux uncertainty ~30%. We must await the result of ongoing distance measurements in the radio band using pulsar proper motion, before the 1.809 MeV observation can be translated into a calibration of core collapse supernova yield for intermediate mass isotopes

The COMPTEL detection of the 1.157 MeV γ-ray line[14] (the first detection of nucleosynthesis in this γ-ray line) from the ~300 year-old Cas A supernova remnant presents another calibration mark for intermediate-mass nuclei synthesis. Yet, for this event the exact date, as well as its optical brightness and hence ^{56}Ni mass estimate, are uncertain. Therefore the apparent conflict of the COMPTEL-detected 1-3 10^{-4} $M_{\odot}$ of ^{44}Ti with recent supernova Ib nucleosynthesis models[13] still is not clear[15].

Recent re-analysis of the COMPTEL data on the supernova SN1991T in the Virgo cluster at ~13Mpc distance reveals an indication of γ-ray line emission from radioactive ^{56}Co at 847 and 1238 keV[16]. This confirms the Ia nature of this supernova with a high yield in the shortlived ^{56}Ni parent isotope, although the inferred ^{56}Ni mass is on the high side of (uncertain) supernova Ia model estimates[15]. Present data are still insufficient to decide among the models.

Nova nucleosynthesis is expected to produce significant amounts (~10^{-7} $M_{\odot}$) of ^{22}Na, a radioactive 1.275 MeV γ-ray line emitter with a decay time of 3.75 y. White dwarfs with enrichments of intermediate mass nuclei are the candidate objects for ^{22}Na production from proton capture reactions during explosive hydrogen burning of the nova. Within 1-2 kpc, these events should be observable with the instruments aboard the Compton observatory. Iyudin et al.[17] discuss the COMPTEL upper limits on ^{22}Na masses for the (more distant) Galactic novae of the last 3 years. They point out the inconsistency between optical/ultraviolet observations and nucleosynthesis calculations: the latter produce much lower ejected mass amounts from massive O-Ne-Mg white dwarfs than observed in neon novae. This may indicate that nova ^{22}Na production originates rather from low-mass white dwarfs enriched in intermediate mass seed nuclei from previous flashes, anti-correlated to the possible ^{26}Al contributing novae on massive white dwarfs with enrichments in O-Ne-Mg from hydrostatic burning in a massive star progenitor.

COSMIC RAY EXCITATION LINES

Interaction of cosmic ray nuclei with ambient matter is expected to result in characteristic de-excitation γ-ray line emission. Yet the COMPTEL discovery of excess MeV emission from the Orion region[18] indicating line contributions from $^{12}C^*$ and $^{16}O^*$ implies favourable circumstances for this process in this region. The stellar winds from OB star associations are plausible accelerators for heavy cosmic rays up to ~20 MeV/nucleon[19], the apparent absence of de-excitation lines from other nuclei suggests unusual matter compositions, possibly oxygen-enriched SN ejecta, or Wolf Rayet star wind material relatively depleted in hydrogen; also, pick-up ions from circumstellar dust may provide the cosmic ray seed[20]

REFERENCES

1. Schönfelder V. et al. 1993, Ap.J. Suppl.,86, 629-656
2. Diehl R., et al. 1995a, accepted for publication in A&A
3. Mahoney W.A., Ling J.C., Jacobson A.S., and Lingenfelter R.E. 1982, Ap.J., 262, 742
4. Share G., et al., 1985, Ap.J., 292, L61
5. Diehl R., et al. 1991, in: Data An. in Astr. IV, ed. V. diGesu et al., Plenum N.Y., 59, 201
6. de Boer H. et al. 1991, in: Data An. in Astr. IV, ed. V. diGesu et al., Plenum N.Y., 59, 241
7. Strong A.W. et al. 1991, in: Data An. in Astr. IV, Plenum N.Y., Vol 59, 251
8. Clayton D.D., and Leising M.D. 1987, Phys.Rep. 144, 1, 1-50
9. Prantzos N. and Diehl R., submitted to Phys.Rep.
10. Knödlseder J. 1994, Diploma Thesis, MPE Garching
11. Diehl R., et al. 1995b, accepted for publication in A&A
12. Oberlack U., et al., 1993, ApJ Suppl. 92, 2, 433
13. Hoffman R.D., et al, 1994, in 'The Gamma-Ray Sky with Compton GRO and Sigma', ASI
14. Iyudin A.F. et al., 1994, A&A, 284, L1-L4
15. Woosley S. E. et al., 1995, this conference
16. Morris D. et al., this conference
17. Iyudin A.F. et al., 1995, this conference; A&A in press
18. Bloemen H. et al., 1994, A&A, 281,L5-L8
19. Nath B.B. and Biermann P.L., 1994, MNRAS, 270, L33
20. Ramaty R, Kozlovsky B., Lingenfelter R.E., 1995, ApJ, 438, L21

Radioactivities Made in Supernovae

S. E. WOOSLEY[a,e], F. X. TIMMES[b], R. D. HOFFMAN[a],
D. H. HARTMANN[c], P. A. PINTO[d], and T. A. WEAVER[e]

[a]*Board of Studies in Astronomy and Astrophysics*
University of California, Santa Cruz

[b]*Laboratory for Astrophysics and Space Research*
Enrico Fermi Insititute, Chicago

[c]*Department of Physics and Astronomy*
Clemson University, Clemson

[d]*Steward Observatory*
University of Arizona, Tucson

[e]*Physics Department*
Lawrence Livermore National Laboratory

The results presented here are based chiefly upon three recent papers[1,2,3] recently submitted to the *Astrophysical Journal.* A grid of supernovae, 11 to 40 $M_\odot$, have been evolved and exploded[2] using a simple piston approximation to the explosion mechanism. The yields of 200 isotopes, including interesting radioactivities, were determined and incorporated into a Galactic chemical evolution calculation[1] that includes a reasonable representation of the stellar initial mass function, star formation, and mass infall into the Galaxy. The solar abundances of the isotopes lighter than zinc are well produced. This model gives a current rate of Type II and Ia supernova of three and 0.5 per century respectively and the Ia supernovae are responsible for 1/3 of the solar iron.

^{26}Al and ^{60}Fe

The model predicts[3] a production rate of ^{26}Al and ^{60}Fe of 2.0 ± 1.0 $M_\odot$ Myr^{-1} and 0.75 $\pm$ 0.4 $M_\odot$ Myr^{-1} respectively. This corresponds to 2.2 ± 1.1 $M_\odot$ of ^{26}Al and 1.7 ± 0.9 $M_\odot$ of ^{60}Fe in the present interstellar medium and implies that ^{60}Fe should have a signal 15% that of ^{26}Al, with a similar spatial distribution. The error bars include a range of supernova rates from 4.0 to 1.0 events per century. It is our expectation that the values given are more likely to be overestimates than underestimates. The error bars are derived from approximately 100 chemical evolution models that independently varied the total baryonic mass of the Galaxy, radial distribution of the baryonic mass, slope of the Salpeter IMF, the upper mass limit of the initial mass function and the efficiency of the Schmidt star formation rate. The most sensitive parameters are the total baryonic mass of the Galaxy, the rate of core collapse events, and the upper mass limit of the IMF. The amount of ^{26}Al and ^{60}Fe produced is linear with variations in the Galactic supernovae rate; a core collapse rate one half as large produces half as much ^{26}Al and ^{60}Fe.

The error bars above do *not* include any uncertainties associated with the stellar yields themselves which could be considerable. ^{26}Al and ^{60}Fe both have appreciable contributions from the neon-oxygen shell and reactions that occur prior to the explosion, especially during the last 1000 seconds of the star's life. Thus they are sensitive to the treatment of time-dependent convection and, ultimately, to multidimensional aspects of the fluid motion[4]. Still it is encouraging that, with no adjustment, the ^{26}Al comes out close to what is observed[5,6]. The predicted ^{60}Fe is close to current upper bounds. With 99.5 % confidence the upper flux limit from SMM on 1.17 MeV γ-ray line emission from the Galactic plane is[7] 8×10^{-5} cm^{-2} s^{-1}. This corresponds roughly to a steady state production of 0.8 M$_\odot$ Myr^{-1}, but the γ-ray fluxes depend, to some extent, on how the spatial distribution of the radioactive isotope is modeled.

A portion of ^{60}Fe may also be made in Type Ia supernovae[3,8], about 0.15 M$_\odot$ Myr^{-1} and this would have a different Galactic distribution.

^{44}Ti

The supernova models[2] also predict a yield of ^{44}Ti in the range 10^{-5} to 10^{-4} M$_\odot$ with a typical value of a few $\times 10^{-5}$ M$_\odot$. Values above 10^{-4} M$_\odot$ are difficult to achieve and getting any ^{44}Ti out at all is sensitive to the explosion mechanism and location of the mass cut. Unless the energy of the explosion increases with stellar mass (and it might), considerable fall back develops in the more massive stars, especially M $\gtrsim$ 25 M$_\odot$, and it becomes harder to eject ^{44}Ti. On the other hand, if the explosion is energetic enough to eject it, more gets made.

The COMPTEL team has reported[9] the detection of ^{44}Ti in Cas-A at a level that would imply a synthesis of $1.4 \pm 0.4 \times 10^{-4}$ to $3.2 \pm 0.8 \times 10^{-4}$ M$_\odot$ depending upon the uncertain half-life of ^{44}Ti and the explosion date of Cas-A. The larger values would be difficult to accommodate in the supernova models; the smaller ones might not. We have previously pointed out[10] that ^{44}Ti cannot be ejected unless a large quantity of ^{56}Ni is also ejected and this implies a bright supernova. How bright? An abundance of ^{44}Ti as large as reported would probably mean the ejection of at least 0.05 M$_\odot$ of ^{56}Ni and maybe 0.1 M$_\odot$. With or without a hydrogen envelope, the supernova would then have a luminosity brighter than 10^{42} erg s^{-1} at peak (e.g., SN 1987A). For a distance to Cas-A of 2.9 kpc this would imply an absolute magnitude at peak of -4! Cas-A was certainly not detected as such a bright object. However there may be a large reddening correction. Searle[11] estimates $A_v = 4.3$ magnitudes and if Cas-A was embedded in a dusty region, perhaps it could have been even more, though it is surprising that the supernova was not seen[12]. The ejection of ^{44}Ti from Cas-A would also imply that there was little or no fall back during the explosion. Without fine tuning, one might therefore expect a neutron star remnant. So far none has been found. We note that ^{44}Ti has not been detected from Cas-A by the OSSE experiment on CGRO[13].

SN 1987A should also have synthesized ^{44}Ti though the amount is uncertain. Guessing a value around 5×10^{-5} M$_\odot$ and using a half life of 58 years[14], one would have a flux, e.g., in the 1.157 MeV line, of $\sim 1-2 \times 10^{-6}$ cm^{-2} s^{-1}, too small for CGRO or INTEGRAL, but large enough that it might be detected in the next decade by post-INTEGRAL experiments.

^{60}Co

The isotope ^{60}Co ($\tau_{1/2}$ = 5.3 y) should also be produced in supernovae[15]. The mass ejected in 11, 15, 20, and 25 M$_\odot$ supernovae is, for example[2], 0.19, 1.5, 2.3, and 5.4×10^{-5} M$_\odot$ respectively. ^{60}Co is made chiefly in the oxygen and neon shells in the last stages of presupernova evolution, much like ^{60}Fe and ^{26}Al. Besides being an important tracer of uncertain nucleosynthesis, ^{60}Co is of interest because of its possible effect on the supernova energy budget. If each decay releases about 3 MeV, a five year old supernova with (initially) 3×10^{-5} $M_\odot$ of ^{60}Co would produce 6×10^{36} erg s^{-1} from ^{60}Co decay. The luminosity of SN 1987A at 1500 days[16] was $\sim 10^{37}$ erg s^{-1}, suggesting that ^{60}Co might have been contributing appreciably to the light curve. However, at 1500 days the γ-ray deposition efficiency has declined[17] to $\sim$0.13 (augmented slightly by the electron kinetic energy), so the fraction of the luminosity produced by ^{60}Co decay is correspondingly reduced.

The flux of γ-ray lines however, would be $\sim 4 \times 10^{-6}$ cm^{-2} s^{-1} (at the same age and at 50 kpc), larger than that of ^{44}Ti for a decade or so, a fact worth keeping in mind in the event of a Galactic supernova.

^{56}Ni

Of course the major radioactivity produced by supernovae, both in terms of abundance and consequence, is ^{56}Ni. Baring the occurrence of a Galactic supernova or the development of much more sensitive detectors, studies of this isotope and its daughter, ^{56}Co, are limited to Type I supernovae. Pinto[18] has recently computed the γ-ray light curves of a number of different types of Type Ia and Ib supernova models. Typical Type Ia models are given below.

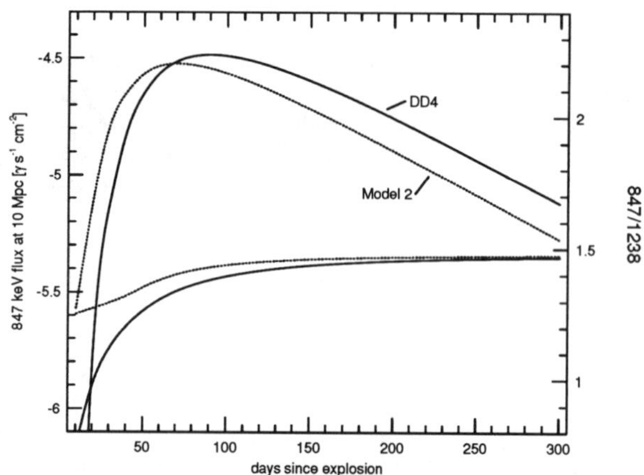

Fig. 1 The gamma-ray light curves[18] of two typical Type Ia supernova models at 10 Mpc. One[19] is a delayed detonation (DD4); the other a sub-Chandrasekhar mass model (Model 2[20]). Model DD4 made 0.62 $M_\odot$ of ^{56}Ni; Model 2 made 0.43 $M_\odot$. Also shown on the right hand axis is the ratio of the 847 keV line flux to that of the 1238 keV line (both from ^{56}Co decay). Early detection and measurements are critical in distinguishing two very different models.

At this meeting the COMPTEL team announced the detection of γ-ray lines from SN 1994 T. The distance to the host galaxy, NGC 4527, is uncertain, but adopting 13.5 Mpc, as does the COMPTEL team, gives a measured flux about 2 to 3 times greater than in Fig. 1.

However, the error bars on the COMPTEL results are large, SN 1991 T was an exceptionally bright supernova and may have made more ^{56}Ni, and the distance to the host galaxy is not certain.

Pinto also gives the light curves for Type Ib models[21]. They typically peak (for an assumed distance of 10 Mpc) at 6×10^{-6} cm^{-2} s^{-1} about 120 days after the explosion. This is a fainter display, by a factor of 5 to 10, and at a later time than shown for Ia's in Fig. 1. The Ib's make a smaller amount of ^{56}Ni ($\sim$ 0.1 $M_\odot$) and may have ejected more mass. This is below the threshold for experiments on CGRO, but observations might be attempted in the event of nearby Type Ib supernovae using INTEGRAL. Type Ib's occur more frequently than Ia's.

This work has been supported by the National Science Foundation (AST-91-15367) and by NASA (NAGW-2525, NAG 5-1578)

REFERENCES

1. Timmes, F. X., Woosley, S. E., & Weaver, T. A. 1995, ApJS, in press
2. Woosley, S. E., & Weaver, T. A. 1995, ApJS, submitted
3. Timmes, F. X., Woosley, S. E., Hartmann, D. H., Hoffman, R. D., Weaver, T. A., & Matteucci, F. 1995, ApJ, submitted
4. Bazan, G., & Arnett, D. 1994, ApJL, 433, L41
5. Diehl, R., et al. 1993, A&A Supp., 97, 181
6. Diehl, R. et al. 1995, A&A, in press
7. Leising, M. D. & Share, G. H. 1994, ApJ, 424, 200
8. Woosley, S. E., & Eastman, R. G. 1995, Proceedings of Menorca Summer School on Supernovae (Sept., 1992), eds E. Bravo, R. Canal, J.M. Ibanez & J. Isern, World Scientific, in press
9. Iyudin, A. F., et al. 1994, A&A, 284, L1
10. Hoffman, R. D., Woosley, S. E., Weaver, T. A., Timmes, F. X., Eastman, R. G., & Hartmann, D. H. 1994, Proceedings of the XXII NATO Advanced Studies School, eds. M. Signore, P. Salati, & G. Vedrenne, Kluwer publishers, in press
11. Searle, L. 1971, ApJ, 168, 41
12. Chevalier, R. A. 1976, ApJ, 208, 826
13. The, L.-S., Leising, M. D., Clayton, D. D., et al., 1995, ApJ, in press
14. Meissner, J. et al 1994, poster paper at "Nuclei in the Cosmos", Gran Sasso, Italy, July, 1994.
15. Clayton, D. D. 1982, in *Essays in Nuclear Astrophysics*, eds. C. Barnes, D. Clayton, and D. Schramm, (Cambridge Univ. Press), p. 401
16. Suntzeff, N. B., Phillips, M. M., Elias, J. H., DePoy, D. L., & Walker, A. R. 1992, ApJL, 384, L33
17. Woosley, S. E., Pinto, P. A., & Hartmann, D. H. 1989, ApJ, 346, 395
18. Pinto, P. A. 1995, in preparation for ApJ
19. Woosley, S. E. 1994 in *Supernovae: Les Houches 1990, Session LIV*, ed. S. Bludman, R. Mochkovitch, J. Zinn-Justin, (North Holland), p. 63
20. Woosley, S. E., & Weaver, T. A. 1994, ApJ, 423, 371
21. Woosley, S. E., Langer, N., & Weaver. T. A. 1995, ApJ, in press

Analysis and Implications of the Nuclear Line Emission from the Orion Complex

R. RAMATY[a], B. KOZLOVSKY[b] AND R.E. LINGENFELTER[c]

[a]Lab. for H.E. Astrophysics, NASA/GSFC, Greenbelt, MD 20771, USA
[b]Dept. of Physics and Astronomy, Tel Aviv University, Tel Aviv, Israel
[c]CASS, Univ. of California San Diego, LaJolla, CA 92903, USA

INTRODUCTION

We have studied the energy deposition, the rate of ionization and the production of light isotopes and ^{26}Al that accompany the production of deexcitation line emission by accelerated particle bombardment in Orion. The observations require a power of at least 4x10^{38} erg s^{-1} and a rate of ionization of 0.3 $M_{\odot}$ yr^{-1}. The Galactic inventory of ^{9}Be limits the currently active 'Orion-like' regions in the Galaxy to less than 200. The ^{26}Al to ^{9}Be production ratio precludes the formation of the protosolar ^{26}Al by accelerated particle bombardment.

This paper is an extension of our previously published[1] analysis of the COMPTEL Orion observations[2]. In our previous paper we showed that the observed gamma ray spectrum cannot distinguish between a pure broad line spectrum (i.e. a spectrum produced by only heavy projectiles) and a combined broad and narrow line spectrum (i.e. a spectrum produced by heavy projectiles as well as protons and α particles), but that the broad line spectrum is energetically more favorable. We also investigated the implications of the observed 1–3 MeV to 3–7 MeV flux ratio and proposed that the composition of the late phase winds of WR stars of spectral type WC produces a gamma ray spectrum which is consistent with the COMPTEL upper limit on the 1–3 MeV flux. We now also consider the composition that would result from the breakup of interstellar dust and the pickup of the resulting ions onto the fast winds of massive stars or supernova ejecta. We point out the gamma ray characteristics that distinguish these two compositions.

We also investigate the implications of the associated light isotope and ^{26}Al production. We perform our calculations using a more realistic accelerated particle spectrum resulting from acceleration by strong shocks.

ANALYSIS

We carry out thick target calculations of nuclear deexcitation line emission, and light isotope (^{6}Li,^{7}Li,^{9}Be,^{10}B,^{11}B) and ^{26}Al production. For the accelerated particles we take the compositions shown in Table 1 (SS – Solar System[3]; CRS – cosmic ray source[4]; SN35 – ejecta of a 35$M_{\odot}$ supernova[5]; WC – the late phase wind of a Wolf-Rayet star of spectral type WC[1,6]; GR – pick up ions resulting from the breakup of interstellar dust). For the ambient gas we assume the SS composition. Concerning the dust, in analogy with the anomalous component of the cosmic rays observed in interplanetary space[7], we consider the effects of the pick up of ions onto a magnetized high speed wind (e.g. the ejecta of supernovae or the winds of massive stars). As the ions acquire considerable energy during the pick up process, they form a seed population that is much more easily accelerated than the rest of the ambient plasma. In the solar system the pick up ions are interstellar neutrals that penetrate into the solar cavity where they are ionized. For Orion we propose that the incoming matter is essentially neutral dust that is broken up by evaporation, sputtering or other processes. The GR abundances in Table 1 are SS abundances modified by recently determined depletion factors[8]. The gamma ray and ^{26}Al calculations were described previously[1]. For the light isotopes we use the tabulated cross sections[9].

We assume that the accelerated particle source spectra are given by $N_i(E) = K_i E^{-1.5} e^{-E/E_0}$, where E is energy per nucleon, the K_i's are proportional to the abundances of Table 1, and the N_i's are measured in particles MeV^{-1} s^{-1}. The nonrelativistic spectral index of 1.5 is the consequence[10] of shock acceleration with a compression ratio at its maximal value of 4. The exponential turnover[10] characterizes the effects of a finite acceleration time or a finite shock size. This spectrum is different from the one that we used previously[1]. The differences in the predictions of these two spectral forms will be explored in a subsequent publication. Here we use the spectrum given above with values of E_0 ranging from 2 to 100 MeV/nucl.

TABLE 1. ASSUMED ACCELERATED PARTICLE COMPOSITIONS

	SS	CRS	SN35	WC	GR
^{1}H	1.18x10^{3}	2.19x10^{2}	2.72x10^{1}	1.00	2.00
^{4}He	1.18x10^{2}	2.19x10^{2}	7.60	9.20x10^{-1}	0.0
^{12}C	4.27x10^{-1}	8.86x10^{-1}	6.51x10^{-2}	6.90x10^{-1}	3.05x10^{-1}
^{13}C	4.82x10^{-3}	9.90x10^{-3}	3.05x10^{-4}	0.0	3.44x10^{-3}
^{14}N	1.32x10^{-1}	3.80x10^{-2}	2.28x10^{-2}	2.40x10^{-3}	2.82x10^{-2}
^{16}O	1.	1.	1.	1.	1.
^{20}Ne	1.28x10^{-1}	8.30x10^{-2}	8.50x10^{-2}	2.30x10^{-3}	0.0
^{22}Ne	1.56x10^{-2}	3.54x10^{-2}	1.27x10^{-2}	2.00x10^{-2}	0.0
^{23}Na	2.47x10^{-3}	1.71x10^{-2}	1.91x10^{-3}	4.41x10^{-5}	2.82x10^{-3}
^{24}Mg	3.53x10^{-2}	1.46x10^{-1}	2.66x10^{-2}	6.30x10^{-4}	4.03x10^{-2}
^{26}Mg	5.01x10^{-3}	3.32x10^{-2}	5.32x10^{-3}	4.00x10^{-3}	5.72x10^{-3}
^{27}Al	3.74x10^{-3}	2.70x10^{-2}	5.72x10^{-3}	6.70x10^{-5}	5.34x10^{-3}
^{28}Si	3.83x10^{-2}	1.66x10^{-1}	4.14x10^{-2}	6.90x10^{-4}	3.28x10^{-2}
^{32}S	1.81x10^{-2}	2.50x10^{-2}	1.50x10^{-2}	3.30x10^{-4}	0.0
^{40}Ca	2.68x10^{-3}	1.30x10^{-2}	1.63x10^{-3}	5.00x10^{-5}	3.83x10^{-3}
^{56}Fe	5.04x10^{-2}	1.61x10^{-1}	2.51x10^{-2}	9.10x10^{-4}	7.20x10^{-2}

The composition of the accelerated particles is strongly constrained by the COMPTEL limit on the 1–3 MeV emission. Calculations of $R = 2\int_{1\mathrm{MeV}}^{3\mathrm{MeV}} E_\gamma^2 Q(E_\gamma) dE_\gamma / \int_{3\mathrm{MeV}}^{7\mathrm{MeV}} E_\gamma^2 Q(E_\gamma) dE_\gamma$, together with the COMPTEL 2σ upper limit[2] $R < 0.13$, are shown in Fig. 1a. The discrepancy between this limit and the CRS values is greater than 3σ; and the SS and SN35 values are also inconsistent at greater than 2σ. On the other hand, both the GR and WC compositions yield values of R which are lower than the COMPTEL upper limit. However, while the WC composition predicts practically no emission in the 1–3 MeV region, the GR composition predicts significant broad line emission in this region, due to Mg, Si and Fe. The reduced value of R for the GR composition is due to the absence of Ne in grains. The value of E_0 cannot exceed about 30 MeV/nucl, because otherwise the accelerated particles would contribute to pion production and thus be in conflict with the high energy gamma rays observed from Orion with EGRET[11], since these data only require a cosmic ray flux equal to that observed locally near Earth.

For the observed[2] gamma ray line flux and the distance to Orion, the energy deposited by the accelerated particles, dW/dt, depends only on the composition and spectrum of the accelerated particles and the ionization and composition of the ambient gas. In Fig. 1b, we show the deposited power in a neutral gas. In a fully ionized plasma the rate is roughly twice. We see that the gamma ray line production is energetically most efficient for large values of E_0 and accelerated particle compositions that are poor in protons and α particles (i.e. the GR and WC compositions). But even for these compositions and $E_0 = 30$ MeV/nucl, the deposited power is between 4 to $5\text{x}10^{38}$ erg s^{-1}, which seriously limits the total duration of the irradiation. For example, if the accelerated particle bombardment were to last $\sim 10^5$ years, the total energy requirement would be $1.2\text{x}10^{51}$ ergs, equal to the total mechanical output a supernova. In a neutral gas this energy dissipation would correspond to a H ionization rate of $7\text{x}10^{48}$ s^{-1}, or 0.3 $M_\odot$ yr^{-1}.

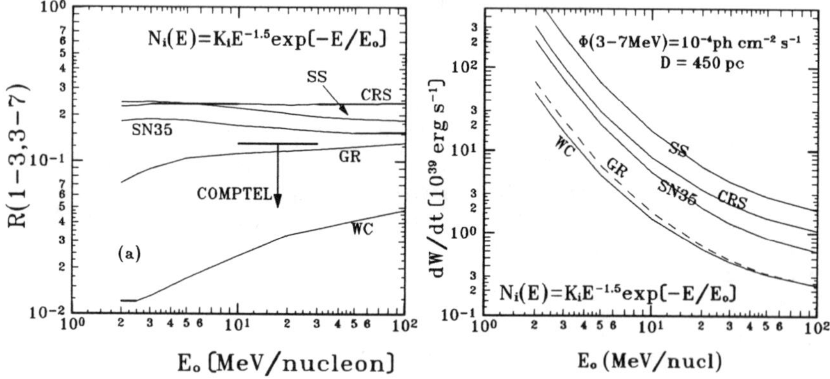

Fig. 1. (a) Ratio of 1–3 to 3–7 MeV emission for the various compositions; R is defined in the text. (b) Deposited power by the accelerated particles in a neutral medium.

It was shown[12] that cosmic rays, with flux equal to that observed locally, interacting with interstellar gas over the entire Galactic history can produce the observed abundances of ^{6}Li, ^{9}Be and ^{10}B. Since low energy cosmic rays with spectra similar to that of the accelerated particles in Orion can also produce these isotopes[13,14], the Galactic inventories of ^{6}Li, ^{9}Be and ^{10}B can set limits on the total Galactic irradiation by such low energy cosmic rays. In Fig. 2a we show the ratio of ^{9}Be to the 3–7 MeV nuclear deexcitation photon production. As both the 3–7 MeV photons and the ^{9}Be are produced predominantly in interactions involving C, N and O, this ratio is practically independent of composition. Adopting a total Galactic ^{9}Be inventory of 10^{57} atoms and assuming that currently there are N_{orr} regions in the Galaxy with the same level of low energy cosmic ray activity as Orion (i.e. producing 3–7 MeV nuclear gamma rays at a rate $Q_{orr} = 2.3\text{x}10^{39}$ s^{-1}), we find that $N_{orr} < 200$. This assumes that the average rate of irradiation is constant in time and ignores the destruction of ^{9}Be in stars. A comparable limit, which is also independent of composition, is obtained from ^{10}B.

The upper limit from light isotope production can be used to set limits on the 3–7 MeV nuclear line emission from the central radian of the Galaxy, $\Phi_{cr} = \xi 10^{-46} Q_G$; here Q_G is the total Galactic luminosity and ξ is a factor of order unity that depends on the distribution of the sources. Since $Q_G = N_{orr} Q_{orr}$, for $N_{orr} < 200$ we obtain that Φ_{cr}(3–7MeV) $< 5 \times 10^{-5}$ photons cm^{-2} s^{-1}. This limit is lower by about a factor of 6 than the upper limit obtained[15] from SMM data. We note, however, that the limit on N_{orr} obtained from ^{9}Be would become higher if a significant amount of ^{9}Be is destroyed by incorporation into stars. On the other hand, the limit would be lower if the rate of irradiation in the early Galaxy was much higher than the average. We clearly need more observations of nuclear gamma ray lines to map out the distribution of low energy cosmic rays in the Galaxy.

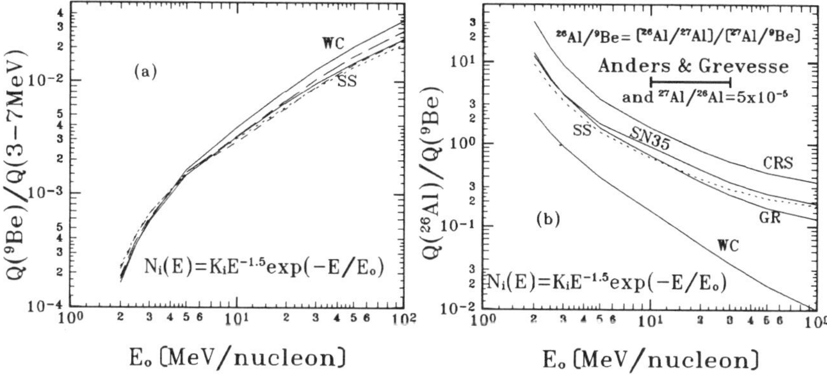

Fig. 2. (a) Production ratio of ^{9}Be to 3–7 MeV emission for the various compositions. (b) Production ratio of ^{26}Al to ^{9}Be for the various compositions; the horizontal bar represents the protosolar ^{26}Al abundance and the probable E_0 range.

We have shown[1] that the COMPTEL Orion observations, and in particular the upper limit on the 1–3 MeV emission, imply a significantly lower ^{26}Al production rate than the estimate[16] which was used to suggest that the protosolar ^{26}Al/^{27}Al of 5×10^{-5} could have resulted from accelerated particle bombardment. Our argument, however, depended on the total irradiated mass and on the level of protosolar low energy cosmic ray activity which could have been different from that in Orion. By combining our ^{9}Be and ^{26}Al calculations, we now provide an argument which is independent of these parameters. In Fig. 2b we show the ratio of ^{26}Al to ^{9}Be production. The solid bar is the ratio implied by the above value of ^{26}Al/^{27}Al and the SS ^{27}Al/^{9}Be. We see that for $E_0 > 10$ MeV/nucl (lower values are energetically very inefficient) the SS ^{9}Be abundance limits the contribution of particle bombardment to ^{26}Al/^{27}Al to less than 3 to 10% for compositions consistent with the 1–3 MeV flux limits. Moreover, if we take into account the facts that most of the ^{9}Be had already been produced prior to the formation of the protosolar nebula, and that some of the ^{26}Al must have decayed during irradiation, these limits become much lower. This shows that it is practically impossible to produce the protosolar ^{26}Al by accelerated particle bombardment.

CONCLUSIONS

We have considered the energy deposited in Orion by the accelerated particles responsible for the gamma ray line emission observed with COMPTEL. The gamma ray production is energetically more favorable for accelerated particle compositions poor in protons and α particles. For such compositions, the observations require at least $4x10^{38}$ erg s^{-1}, a power which is accompanied by an ionization rate of 0.3 $M_{\odot}$/yr. In addition to our previously suggested WC composition, we have proposed the acceleration, by shocks, of the pick up ions produced by the breakup of interstellar dust. Because of the absence of Ne in the dust particles, this mechanism will also yield a relatively low flux in the 1–3 MeV region, consistent with the COMPTEL upper limits. However, the broad Mg, Si and Fe lines at 0.85, 1.24, 1.64 and 1.78 MeV will distinguish the GR composition from WC composition which predicts essentially no emission at these energies.

The Galactic inventory of ^{9}Be limits the number of current 'Orion-like' active regions in the Galaxy to less than about 200. The implied upper limit on the 3–7 MeV nuclear line emission from the Galactic central radian is $5x10^{-5}$ photons cm^{-2} s^{-1}, about a factor of 6 lower than the limit established[15] from SMM data. Our limit, however, is subject to considerations of Galactic evolution.

Our calculated ^{26}Al to ^{9}Be production ratio sets strong limits on the ^{26}Al production in the protosolar nebula which make it practically impossible for accelerated particle bombardment to produce the protosolar ^{26}Al implied by meteoritic observations[17].

We wish to acknowledge V. Ptuskin for bringing up the possibility of producing pick up ions from the breakup of interstellar dust. We also acknowledge financial support from the US-Israel Binational Foundation and NASA under Grant NAG5–2811.

REFERENCES

1. Ramaty, R., B. Kozlovsky & R. E. Lingenfelter. 1995. ApJ, **438**: L21–L24.
2. Bloemen, H. et al. 1994. Astron. & Astrophys. **281**: L5–L8.
3. Anders, E. & N. Grevesse. 1989. Geochim. et Cosmochim. Acta, **53**: 197–214.
4. Mewaldt, R. A. 1983. Revs. Geophys. Space Phys., **21**: 295–305.
5. Weaver, T. A. & S. E. Woosley. 1993. Phys. Repts., **227**: 65–96.
6. Maeder, A. & G. Meynet. 1987. Astron. & Astrophys., **182**: 243–263.
7. Adams, J. H. 1991. ApJ, **375**: L45–L48.
8. Sofia, U. J., J. A. Cardelli & B. D. Savage. 1994. ApJ, **430**: 650-666.
9. Read, S. M. & V. E. Viola. 1984. Atomic Data & Nucl. Data Tabl., **31**: 359–397.
10. Ellison, D. C. & R. Ramaty. 1985. Astrophys. J., **298**: 400–408
11. Digel, S. W., S. D. Hunter & R. Mukherjee. 1995. ApJ, in press.
12. Reeves, H. 1994. Revs. Modern Physics, **66**: 193–216.
13. Reeves, H. & N. Prantzos. 1995. In Light Element Abundances, Ed. Ph. Crane, in press.
14. Casse, M., R. Lehoucq & E. Vangioni-Flam. 1995. Nature, in press.
15. Harris, M., G. H. Share & D. C. Messina. 1995. ApJ, in press.
16. Clayton, D. D. 1994. Nature, **368**: 222–224.
17. Lee, T., D. Papanastassiou & G. Wasserburg. 1977. ApJ, **211**: L107-L110.

Evidence for ^{56}Co Line Emission from the Type Ia Supernova 1991T using COMPTEL[a]

D.J. MORRIS[b,d], K. BENNETT[e], H. BLOEMEN[c], W. HERMSEN[c], G.G. LICHTI[b], M.L. McCONNELL[d], J.M. RYAN[d] AND V. SCHÖNFELDER[b]

[b] *Max-Planck-Institut für extraterrestriche Physik*
Postfach 1609
D-85740 Garching, Germany

[c] *SRON-Utrecht*
Sorbonnelaan 2
NL-3584 CA Utrecht, The Netherlands

[d] *Space Science Center*
University of New Hampshire
Durham, NH 03824-3525 USA

[e] *Astrophysics Division, ESTEC*
Keplerlaan 1
NL-2201 AZ Noordwijk, The Netherlands

Supernova 1991T was the brightest well-observed type Ia supernova to occur in recent years. It was discovered on April 13, 1991, just three days after the launch of the Compton Gamma-Ray Observatory[1] and was later observed by COMPTEL[2] during two two-week periods beginning 66 days (obs. 3) and 176 days (obs. 11) after the supernova explosion. The supernova occurred in NGC 4527, an Sb galaxy in the Virgo cluster, at an estimated distance of 13 Mpc[3]. The evolution of its optical spectrum showed a number of peculiarities[4,5], most notably the lack of absorption lines from intermediate mass elements before peak luminosity. It was also judged by some[5] to be brighter than most type Ia supernovae.

Here we present evidence for ^{56}Co emission lines from SN 1991T. While such lines have previously been seen[6] from the very nearby Type II supernova 1987A, this is the first indication of them from a type Ia supernova. Supernovae of type Ia are thought to result from the incineration and disruption of a white dwarf near the Chandrasekhar mass. They are expected to produce around ten times as much ^{56}Ni, the short-lived radioactive presursor of ^{56}Co, as other types of supernovae.

Upper limits to the ^{56}Co emission from SN 1991T have previously been reported by both COMPTEL[7] and OSSE[8]. The result reported here was obtained with a better understanding of the telescope response and the form of the background spectrum. Determination of the background is a general problem in COMPTEL data analysis, as it

[a] This work is funded in part by the Deutsches Agentur für Raumfahrtangelegenheiten (DARA grant 50 QV 90960), by NASA through contract NAS5-26645 and by the Netherlands Organization for Scientific Research (NWO).

varies with both time and position in the field of view. Background determination is in some ways even more problematic for spectral analysis in the energy range around the two principal ^{56}Co lines, at 847 keV and 1238 keV. The 847-keV line is near the COMPTEL energy threshold (720 keV) where the instrument response is a strong function of the energy. The 1238-keV line is near a large background feature, due principally to the emission of 1461-keV photons by ^{40}K in COMPTEL photomultiplier tubes, complicating determination of the underlying background. Adding to the instrumental background is a cosmic continuum source, the quasar 3C273, only 1.4° from SN 1991T (the angular resolution of COMPTEL is about 2° around these energies).

The most important innovation in this analysis is the use of a simple, smooth function,

$$N = b_o(1 - e^{-a_o(E-E_o)})e^{-\alpha E} + a_K e^{-0.5((E-E_K)/\sigma_K)^2} \quad \textbf{(1)}$$

to fit the background, where E is the γ-ray energy and all other parameters are determined in the fitting. The first term of (1) is the product of a threshold function and an exponential in energy, while the second is a Gaussian to fit the "^{40}K -line" feature. This approach avoids the problem of varying background and the increase in statistical uncertainty inherent in determination of the background from another point in the field of view or from another observation.

In the spectral fitting procedure a count spectrum over the energy range 720 to 1800 keV is first fit by the function (1) alone. The fitting is then repeated, adding a model of the expected ^{56}Co count spectrum derived from Monte Carlo simulations of the telescope response. A large reduction in χ^2 on adding the ^{56}Co model to the fit indicates possible ^{56}Co emission. These steps are repeated with different binnings of the energy spectrum (20, 25, 30, 40 and 50 keV) to check for consistency. This entire procedure is then repeated for other points in the field of view to assure that the emission comes from the suspected source. The ^{56}Co model includes weak lines at 1038, 1175 and 1360 keV in addition to the two principal lines, all weighted according to their production probabilities. The simulated emission model provides an accurate description of the telescope response, and fitting to the combined emission is a more sensitive test for a ^{56}Co source than fitting the individual lines.

Figure 1 shows the residual count spectrum at the position of SN 1991T, after subtraction of the fitted background. The dashed line shows the ^{56}Co emission model, normalized to the ^{56}Co strength found in the fitting. This spectrum was produced with combined data from observations 3 and 11. The individual observations each show ^{56}Co emission at a less significant level, such that the sum of the ^{56}Co counts in the individual observations is consistent with that found in the combined data. No significant emission is found at the same position during a series of observations totaling 56 days late in 1993, when 3C273 was also active but the ^{56}Co from SN 1991T had decayed away. Thus the positive signal seen in 1991 is unlikely to be associated with the quasar.

To better determine the location of the ^{56}Co source the spectral fitting was repeated for positions spaced by 1° along arcs of constant R.A. and declination through SN 1991T, and at the position of 3C273. The ^{56}Co response clearly peaks at the expected position, where it exceeds that seen at 3C273 by a bit over 1σ. Consistent results were found with maximum likelihood mapping[9] of combined data from the two observations and from two 2σ-wide energy windows around the principal ^{56}Co lines. Figure 2 shows the maximum likelihood map. Location contours for a point source at the 95% and 99% confidence levels are displayed together with the positions of SN 1991T (circle) and 3C373 (square).

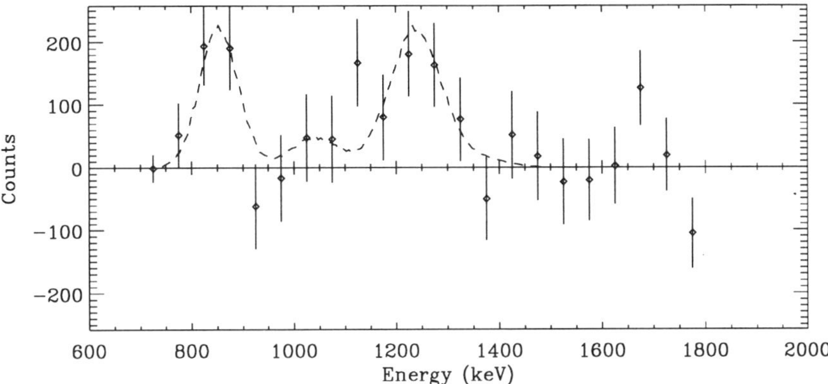

FIGURE 1. Residual count spectrum of SN 1991T for the combined observations 3 and 11, after subtracting the fitted background. The dashed line is the model of ^{56}Co emission, normalized to its fitted strength.

The combined flux in the two principal ^{56}Co lines determined from the spectral analysis is $(8.9 \pm 3.4 \pm 2.7) \times 10^{-5}$ cm^{-2} s^{-1} where the first uncertainty is statistical, and the second is a 30% systematic uncertainty common to all COMPTEL measurements. The maximum likelihood analysis gives a similar flux. Partitioning the flux between the lines by the expected production ratio gives $(5.3 \pm 2.0 \pm 1.6) \times 10^{-5}$ cm^{-2} s^{-1} for the 847-keV line and $(3.6 \pm 1.4 \pm 1.1) \times 10^{-5}$ cm^{-2} s^{-1} for the 1238-keV line. These fluxes are consistent with the published COMPTEL upper limits[7] at the 3σ level (after correction for known systematic errors), but exceed the OSSE 3σ upper limit[8] for the 847-keV line of 4×10^{-5} cm^{-2} s^{-1}. For a distance of 13 Mpc this flux implies the production of 1.3 ± 0.5 $M_\odot$ of ^{56}Ni in the supernova explosion. Although this is a very large mass of ^{56}Ni, it is consistent with what might be expected from an intrinsically bright type-Ia supernova. Given the uncertainty in the distance to SN 1991T, the flux is consistent with most type-Ia supernova models, in particular with three models (FD2H, FD2W[10] and W7DT[11]) which best reproduce the peculiar optical spectra[4,5] of SN 1991T. The flux is the same in the two individual observations to within the rather large uncertainties. The ratio of the strengths of the two principal lines in the ^{56}Co emission model can be varied by about a factor of two in either direction before the χ^2 of the fit is substantially increased.

It should be emphasized that these fluxes are *preliminary results*. Possible systematic errors in the conversion from spectral counts to flux have yet to be thoroughly investigated. They were obtained using the standard COMPTEL data selections and processing. While these standards optimize the data analysis overall, in this particular instance, of a spectral analysis near the energy threshold, there may be some advantage in using different data selections. For example, it may be possible to increase the significance of this result by using events with energy deposits in the two COMPTEL detector subsystems below the standard software thresholds or by placing a wider window on the time-of-flight between the two detector subsystems.

The primary significance of this result is the evidence that a very large mass of ^{56}Ni is produced in a type Ia supernova explosion. This evidence is of a direct sort, without the

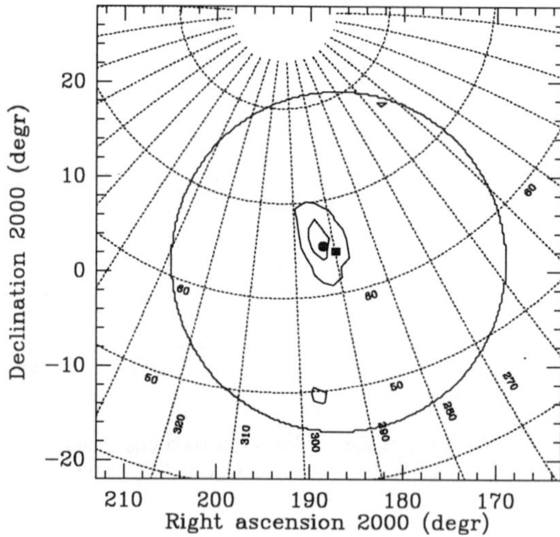

FIGURE 2. Likelihood map for the presence of a point source near SN 1991T based on combined data from observations 3 and 11 and from 2σ-wide energy windows around the two principal lines of ^{56}Co emission. The locations of SN 1991T (solid circle, to the left) and 3C273 (solid square, to the right) are indicated. Source location contours at the 95% and 99% confidence levels are shown. The large circle shows the field used in the maximum likelihood analysis.

inferences required in interpretation of optical observations. Unfortunately the uncertainty in the measurement is too large to distinguish between the various type Ia models. A nearer type Ia supernova in a galaxy at a well determined distance, such as might be obtained using Cepheid variables, will probably be needed for that. However this result does demonstrate COMPTEL's ability to detect type Ia supernovae out to the Virgo cluster. Using here techniques employed here it should be possible at the least, using the full COMPTEL data base, to obtain meaningful upper limits on rates of supernova explosions in a manner unaffected by absorption.

REFERENCES

1. GEHRELS, N., E. CHIPMAN & D. A. KNIFFEN. 1994. *In* AIP Conf. Proc. 304: The Second Compton Symposium (College Park, Md.), 1993. C.E. Fichtel, N. Gehrels & J.P. Norris, Eds.: 3-11. AIP Press. New York.
2. SCHÖNFELDER, V. *et al.* 1993. Astrophys. J. Suppl. **86**:657.
3. TULLY, R. B., E. J. SHAYA & M. J. PIERCE. 1992. Astrophys. J. Suppl. **80**:479.
4. PHILLIPS, M. M. *et al.* 1992, Astronom. J. **103**:1632.
5. FILIPPENKO, A.V. *et al.* 1992. Astrophys. J. **384**:L15.
6. MATZ, S. M. *et al.* 1988. Nature **331**:416.
7. LICHTI, G. G. *et al.* 1994. Astron. Astrophys. **292**:569.
8. LEISING, M. D. *et al.* 1993. *In* AIP Conf. Proc. 280: Compton Gamma-Ray Observatory (St. Louis), 1992. M. Friedlander, N. Gehrels & D.J. Macomb, Eds.: 137-146. AIP Press. New York.
9. DE BOER, H. *et al.* 1992. *In* Data Analysis in Astronomy IV, V. DiGesu *et al.*, Eds.: 241-249. Plenum, New York.
10. RUIZ-LAPUENTE, P. *et al.* 1993. *In* Origin and Evolution of the Elements. N. Prantzos, E. Vangiani-Flam & M. Cassé, Eds.: 326-330. Cambridge Univ. Press, Cambridge.
11. YAMAOKA, H. *et al.* 1992. Astrophys. J. **393**:L55.

Gamma-Ray Line Spectroscopy with INTEGRAL

PETER VON BALLMOOS[1], TONY DEAN[2], CHRISTOPH WINKLER[3]

1 *Centre d'Etude Spatiale des Rayonnements*
9, avenue du Colonel-Roche, 31029 Toulouse, FRANCE

2 *University of Southampton,*
Highfield, Southampton S09 5NH, United Kingdom

3 *ESA/ESTEC, Space Science Department*
Keplerlaan 1, 2001-AZ Nordwijk ZH, The Netherlands

INTRODUCTION

The substantial achievements of the Compton Observatory (CGRO) and GRANAT/SIGMA have revealed a wealth of fundamental astrophysical information available through the medium energy gamma-ray channel. Many new and exciting questions have been highlighted and require further attention. In this context, the future goals of gamma-ray spectroscopy have to be redefined : Where to go from here ? How could the better angular and spectral resolutions of future instruments be exploited ? How to obtain improved sensitivities as proposed for INTEGRAL and how to capitalize on them ?

The inventory of observed line features shows that emissions of a wide range of angular and spectral extent have to be handled : The INTEGRAL mission is expected to provide superior sensitivity while covering a wealth of scientific objectives spanning from compact sources as broad class annihilators, over long-lived galactic radioisotopes such as ^{26}Al with hotspots in the degree-range to the extremely extended galactic disk and bulge emission of the narrow e^+e^- line. As in most other wavebands, the full measurement capability of fine and large scale imaging together with fine and broad band spectroscopy is not technically possible within a single instrument. INTEGRAL will feature two separate and co-aligned instruments in order to obtain the broadest possible diagnostic information on imaging, positioning and spectroscopy. A SPECTROMETER and an IMAGER have been conceived and designed as a fully complementary pair : The SPECTROMETER will provide narrow line sensitivity and maintain an extended source imaging capability. The IMAGER will provide the complement of fine imaging, source identification and maintain the spectral sensitivity to broad lines and the continuum emission.

THE SPECTROMETER

The instrument presented below is a proposal in response to the ESA announcement (July 94) for INTEGRAL by a consortium between the CESR (Toulouse), MPE (Garching), CEA (Saclay), IFCTR (Milan), IEEC (Barcelona), Univ. of Louvain, Univ. of Valencia, Univ.of Birmingham and US Institutes.

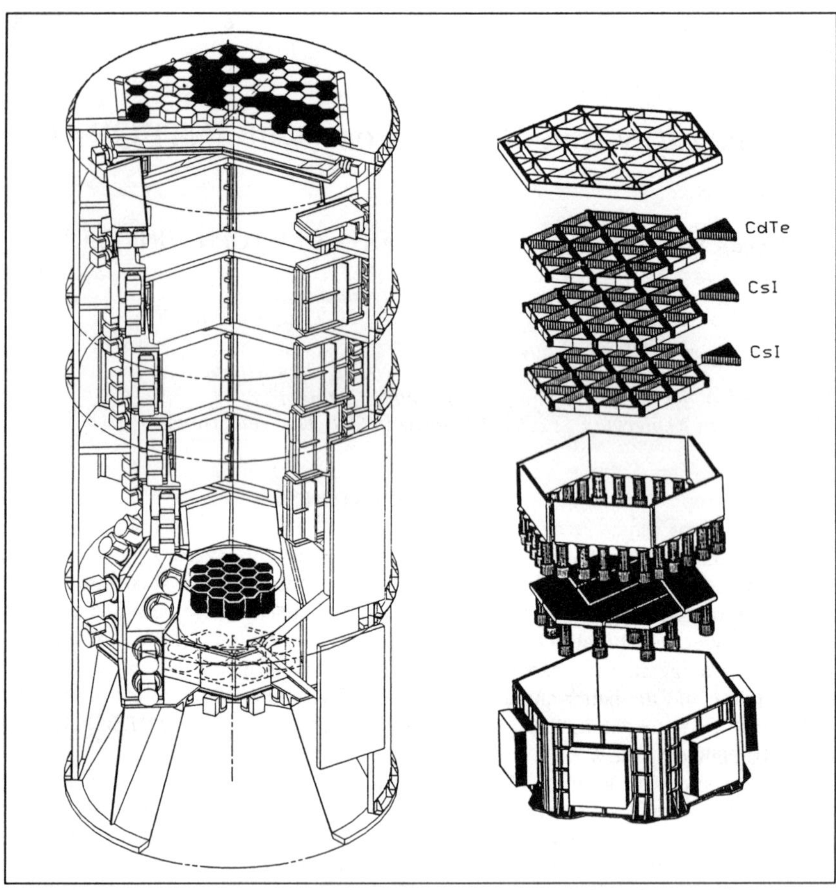

Fig. 1: Cutaway view of a) the SPECTROMETER and b) exploded view of the IMAGER detector

The SPECTROMETER features a compact array of high purity Germanium detectors entirely shielded by a BGO/CSI anticoincidence system. A coded aperture modulates the astrophysical signal. The anticipated performance for narrow line spectroscopy is characterized by an energy resolution in the parts per thousand range, an angular resolution of order 2.5°, and a sensitivity 2-5 10^{-6} ph $cm^{-2} \cdot s^{-1}$ in the energy range relevant for nuclear astrophysics.

The SPECTROMETER embodies three main hardware subsystems : the Germanium detector assembly, the anticoincidence shield system, and the aperture system. Each system includes its own electronics and software.

The Ge detector assembly consists of 19 high purity n-type germanium detectors cooled to 85K via pulsed tubes supplied by two cryogenic compressors. The detectors have hexagonal shape, 3.2 cm on a side, 7 cm thick, and a center-to-center distance of 6 cm. With a total geometric detection area of ≈ 500 cm^2 the detection plane is the largest one that could reasonably be expected to fit into the overall constraints of the INTEGRAL mission. The anticoincidence subsystem shields the detector assembly and defines an aperture of 25° FWHM. A massive bottom shield protects the lower 2π with 6 cm equivalent thickness of bismuth germanate (BGO). With the exception of the aperture, the upper 2π are completely shielded by a BGO/CSI collimator (equivalent thickness 3-4 cm) consisting of five hexagonal rings. In the standard option, the aperture subsystem features a tungsten coded mask (HURA with 127 elements) 171 cm above the detector. The coded mask together with the detector plane define an angular resolution of 2.5° within a fully coded field of view of 16° x 16°. The partially coded field of view is 34° x 34° and the collimator defines a hexagonal aperture of ~ 25° FWHM.

The narrow line sensitivity shown in Fig. 2 has been obtained by completely modeling the SPECTROMETER for the radiation environment conditions encountered outside the magnetosphere. The sensitivity is based on models for the instrument efficiency (GEANT) and background in ^{70}Ge detectors (Naya et al. 1995). Since the SPECTROMETER field-of-view has been significantly widened compared with the original phase A configuration (25° vs. 10° FWHM), the sensitivity for galactic plane sources will be improved due to multiple exposures of this region during the mission. During the galactic plane survey this "multiplex advantage" will produce an improvement in sensitivity of ~ 3/2. The effective observation time on any point in the inner Galaxy will be about $4 \cdot 10^6$ s. The sensitivity is therefore estimated for a) a point source on the optical axis of the instrument ($T_{obs}=10^6$ sec), and b) for a random point source in the inner Galaxy ($T_{obs}=4 \cdot 10^6$ sec). Fig.2 shows two curves for each exposure time case. The curve which is higher in the region above ~1 MeV in each case is obtained assuming a model β^+ background component (Naya et al. 1995b). The lower curve is that calculated without considering β^+production in the Ge detectors - a background component that has never been included in previous studies (Gehrels, 1992; INTEGRAL Phase-A report). These two cases represent optimistic and pessimistic extremes - the real case will lie in the band they define.

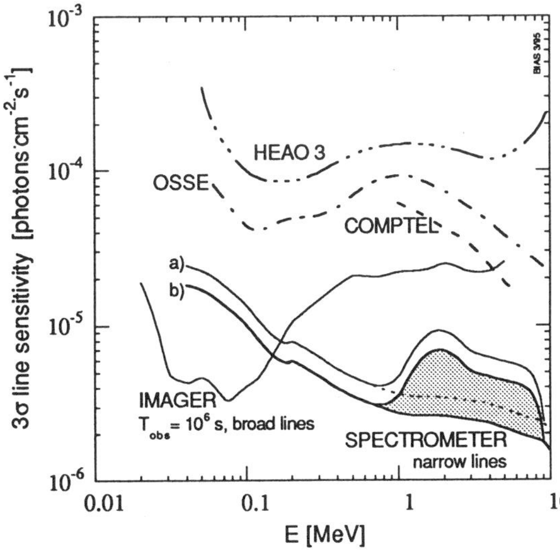

Figure 2 : narrow and broad line sensitivity of the SPECTROMETER and IMAGER respectively. Comparisons are for line sensitivities of OSSE (achieved sensitivity), GRO OSSE and COMPTEL ($T_{obs}=10^6$ sec). The SPECTROMETER sensitivity is estimated for a) a point source on the optical axis of the instrument ($T_{obs}=10^6$ sec), and b) for a random point source in the inner Galaxy during the survey. The curve which is higher in the region above ~1 MeV in each case is obtained assuming a model β^+ background component. The lower curve is that calculated without considering β^+production (see text).

For the baseline SPECTROMETER configuration (tungsten mask), the 511 keV line sensitivity for an on-axis point source $T_{obs}=10^6$ sec (case a) is $2.5 \cdot 10^{-5}$ $ph \cdot cm^{-2}s^{-1}$ but because of the high degree of multiplexing ***any*** point source brighter than $2.0 \cdot 10^{-5}$ $ph \cdot cm^{-2}s^{-1}$ in the inner galaxy would be detected at more than 3 standard deviations (during the first two years, case b). An active mask with the same signal modulation would achieve sensitivities of $1.3 \cdot 10^{-5}$ $ph \cdot cm^{-2}s^{-1}$ and $9.5 \cdot 10^{-6}$ $ph \cdot cm^{-2}s^{-1}$ respectively.

THE IMAGER

The IMAGER instrument described here was proposed in response to the ESA announcement of opportunity by a core team consortium between the University of Southampton, ITESRE (Bologna), CEA (Saclay), Rutherford Appleton laboratory, the University of Valencia, the University of Birmingham, and the University of Helsinki representing thirteen other research institutes.

The instrument concept is based upon the use of a multi-layer combination of typically 17,000 room temperature CdTe solid state detectors and 2880 CsI(Tl) scintillators arranged so as to provide a large area (3000 cm^2) multi- pixellated position sensitive array of discrete elements. The instrument will operate over the energy range 20 keV to about 8 MeV and provide a location for point source gamma-ray emitting objects at the 1' level. The CdTe plane provides the low energy response (20 - 200 keV), the CsI(Tl) second layer 3 cm thickness has been optimized for operation at ~500 keV, and the combined 6 cm of the two CsI(Tl) layers optimized for operation at MeV energies. The system has the capability to select both single site events as well as operate in Compton scatter mode. In the latter configuration the IMAGER operates as a sensitive (~10 mCrab) polarimeter. The spectral resolution is typically 10% FWHM or better over the entire operational range. The gamma-ray detection system (Fig. 1b) is enclosed within an active BGO shield/Tantalum collimator arrangement and operated with a 1663 order HURA rotating coded aperture mask, which is anti-symmetric on a 60° rotation, and is placed 3.2 m distant to provide a field of view of typically 10°x10°. The mask/antimask configuration will enable the deleterious effects of systematic errors to be effectively suppressed without recourse to the analysis of their origin.

The broadline sensitivity of the IMAGER, based upon a 10^6 s observation, is shown in figure 2 and has been obtained from a GEANT based Monte-Carlo model. Approximately 70,000 physical/mechanical elements have been included to describe the instrument as well as a simplified but representative approximation for the materials of the spacecraft and other instruments. This arrangement was exposed to the cosmic diffuse gamma-ray background and cosmic particle fluxes outside the earth's magnetosphere for both solar maximum and minimum conditions to give the on-axis spectral sensitivity.

THE INTEGRAL MISSION

INTEGRAL (with a total launch mass of 3.6 tons) will be launched early 2001 into a geosynchronous High Eccentric Orbit (HEO) with high perigee in order to provide long uninterrupted observation periods at nearly constant background and away from trapped radiation. ESA will be responsible for the overall spacecraft and mission design, procurement of bus and payload module, instrument integration into the payload module, integration of the spacecraft, system testing, spacecraft operations and acquisition and distribution of telemetry data.

The INTEGRAL payload will be completed by two monitor instruments covering the main instrument FOVs and providing concurrent monitoring of gamma-ray targets in the X-ray (2-100 keV) and optical domain. The spacecraft will consist of a service module (bus) containing all spacecraft subsystems and a payload module containing the scientific instruments. The bus will be identical for the two scientific missions INTEGRAL and XMM. Scientific instruments will be provided by the European science community through their national funding agencies and by the US science community through NASA. The INTEGRAL science operations center (ISOC) will be provided by ESA, the INTEGRAL science data center (ISDC) will be provided by the European science community through national funding.

INTEGRAL will be an observatory type mission. Most of the observing time, the "general program", will be awarded to the scientific community at large. Proposals will be selected on their scientific merit by a single review committee. A small fraction (i.e. 30% to 35%) of the total mission duration of nominal 2 years will be reserved for institutes which have developed and delivered instruments and the data center. The INTEGRAL science data center (ISDC) will monitor the instrument performance, perform data pre- processing and spot potential targets of opportunity. INTEGRAL will be able to respond to such targets within 24 hours. All scientific data will be assembled by the ISDC into an archive and will be made available (via ISDC or ESA) to the scientific community one year after they have been collected. This guarantees the use of the scientific data for different investigations beyond the aim of a single proposal.

CONCLUSION

The overall Integral payload has been designed to address the imaging, accurate positioning and spectroscopy of celestial gamma-ray emitting objects. The SPECTROMETER with an energy resolution in the per mille range and an angular resolving power on a scale of 2.5° - the IMAGER, with an energy resolution of a few percent and an angular resolution of 16' FWHM. This complementarity reflects our present understanding of the celestial gamma-ray emitters in the energy range of nuclear transitions : The fine spectroscopy/coarse imaging and fine imaging/coarse spectroscopy combinations are particularly appropriate since gamma-ray line emissions from violent compact objects are more likely to be extensively broadened, whereas phenomena such as the diffusion of radioactive isotopes into the interstellar medium will lead to narrow lines emitted on a broader angular scale.

The authors are grateful to the instrument teams of the SPECTROMETER and IMAGER consortia.

REFERENCES

Gehrels, N. 1992 Nucl. Instr. and Meth. **A313**, 513
Naya J., Jean P., Bockholt J., Matteson J., Vedrenne G., and v. Ballmoos P., submitted *NIM*
Naya J., Jean P., Bockholt J., Slassi S., Vedrenne G., and v. Ballmoos P., tbp *NIM*, 1995
INTEGRAL ESA SCI(93)1, ESA phase A study report, April 1993

Particle acceleration in Orion-like objects *

Andrei. M. BYKOV[a]

[a] *A.F.Ioffe Institute of Physics and Technology 194021, St. Petersburg, Russia*

INTRODUCTION

A study of regions of massive star formation and starburst galaxies attracts currently serious attention. A remarkable feature of these systems is the presence of powerful sources of kinetic energy due to colliding stellar winds and supernovae (SNe) confined in a relatively compact region. Important examples are the Galactic Center region[1], the Orion Complex[2], the open cluster Berkeley 87 [3] and many others.

An effective acceleration of non-thermal particles in the vicinity of an OB-association has been expected from the theory of particle acceleration[4].

The COMPTEL telescope[5] aboard the Compton Gamma Ray Observatory has recently detected the γ ray emission in the 3 -7 MeV range from the Orion complex - the nearest massive star formation region[6]. The γ lines have been identified with broad nucleus deexcitation lines of $^{12}C^*$ and $^{16}O^*$. The observations can be explained by the very efficient acceleration of energetic nuclei up to a few tens MeV/nucl due to collisions of stellar winds and supernova shocks in the OB association[7].

The Orion-like objects might represent a new class of galactic objects with greatly enhanced fluxes of non-thermal nuclei with non-standard composition. Nucleosynthesis and spallation reactions due to interactions of accelerated non-thermal nuclei with ambient medium would change drastically the isotope composition in the vicinity of such objects. These reactions can be responsible for variations of isotope ratios observed in ISM.

Below we will concentrate on the problem of particle acceleration in these sources of non-thermal nuclei. There are at least three main points to be highlighted.

(i) Composition of non-thermal nuclei and the injection problem .

(ii) Energy gain of non-thermal nuclei.

(iii) Interaction of accelerated nuclei with matter surrounding acceleration region taking into account of the transport processes and spatial inhomogeneity of distribution of accelerated nuclei.

ACCELERATION MODEL

To construct a model of non-thermal particle evolution in the Orion-like objects, let us adopt the following assumptions supported by the available observations[7]:

(1) Particle acceleration is produced by a powerful energy release in the form of the violent plasma motions which occur in a bubble created by stellar winds and SNe. The bubble is filled with hot rarefied plasma.

*This work was supported in part by the Russian Basic Research Foundation and the International Science Foundation (grant NU3000)

(2) Non-thermal particles accelerated within the bubble penetrate then into dense matter (shell or cloud) surrounding the bubble. The particles suffer Coulomb and nuclear interactions with dense matter which leads to nucleosynthesis and gamma ray production.

The elemental abundance inside the bubble filled with a rarefied hot gas can differ strongly from the standard cosmic abundance due to the ejection of matter enriched with some heavy elements from SNe and stellar winds of massive stars (WR and OB type). A bubble of 10 pc radius filled with the rarefied plasma of density 10^{-3} - 10^{-2} cm^{-3} contains, respectively, 0.1 or 1 $M_{\odot}$ of interstellar gas. Thus an ejection of about 1 $M_{\odot}$ of matter with SN or WR star abundance will fully determine the elemental composition. In particular, we might find a strong overabundance of He, O, C, Ne and Si nuclei within the hot bubble. For instance, strong O and Ne emission lines have been observed from the galactic SN remnant Pup A by the Einstein Observatory[8]. These x-ray lines indicate the presence of several $M_{\odot}$ of O and Ne in the hot plasma which fills Pupis A. Being at a distance of about 2 kpc, Pupis A can be considered as a candidate for search of emission γ-lines, at least with the INTEGRAL – the next post-CGRO major γ ray mission.

We can indicate two most important injection processes in the Orion-like systems. Firstly, a creation of suprathermal nuclei by collisionless shock waves within a hot bubble. According to numerical simulations[9] the number density of protons with energy well above, say, 60 times of the kinetic energy of the upstream protons (relative shock frame) exceeds $10^{-3} \times n_p$. Here n_p is the total number density of upstream protons in a collisionless shock of moderate strength. Since the injection produced by a collisionless shock depends on rigidity one may expect that the injection of He, O and C nuclei (as well as of other nuclei with $A/Z = 2$) in a chemically peculiar hot plasma of the Orion-like objects has the same high efficiency as the injection of protons in hydrogen plasma.

Secondly, the injection of superthermal nuclei might be associated with fast moving filaments or knots enriched very highly with oxygen and other star burning products observed in some SN remnants (due to explosions of massive stars[10]). These filaments move typically with velocities of about 1000 - 5000 km s^{-1}. A bow shock of a fast moving metal-rich filament should act as an efficient accelerator nuclei. Moreover, as noted by Ptuskin, any neutral atom which is evaporated from the filament and ionized then within the bubble will be picked up by moving magnetized plasma and injected into the acceleration.

(ii) Further energy gain of suprathermal nuclei injected within the bubble occurs due to large scale MHD motions of magnetized plasma.

A kinetic energy release within the bubble created by a stellar association may reach a few times of 10^{38} erg s^{-1} the for Orion-like objects at the stages of intense stellar winds and multiple SN explosions. The process is accompanied by formation of shocks, large scale flows and broad spectra of MHD fluctuations in a tenuous plasma with frozen-in magnetic fields. Vortex electric fields generated by the large scale motions of highly conductive plasma with shocks result in a non-equilibrium distribution of charged nuclei. The particle distribution within such a system is highly intermittent. A statistical description of intermittent systems differs from description of homogeneous systems[11].

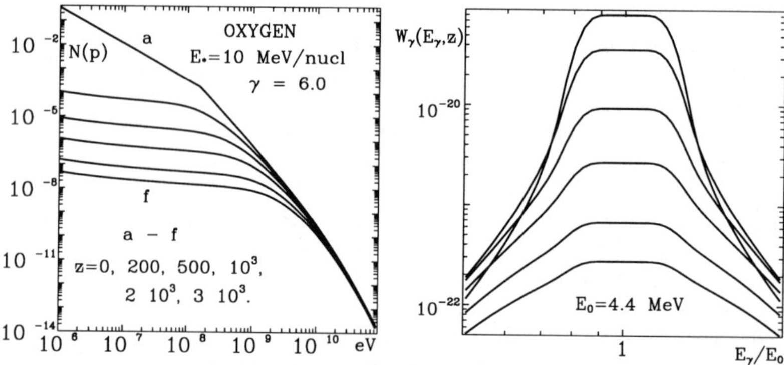

Figure 1: Left: Nonthermal oxygen distribution function (normalized) for several depths of penetration (a-f) into a massive cloud. Right: 4.4 MeV γ-line emissivity $W_\gamma(E_\gamma, z)$ (photons s^{-1} cm^{-3}) due to spallation of ^{16}O nuclei for some depths z (same as on the left).

A distribution function $N(\boldsymbol{r}, p, t)$ of non-thermal nuclei (with energies up to GeV range) averaged over ensemble of turbulent motions and shocks satisfies the transport equation[11]:

$$\frac{\partial N}{\partial t}+\frac{\partial}{\partial r_\alpha}\chi_{\alpha\beta}\frac{\partial N}{\partial r_\beta} = \left(\frac{1}{\tau_{sh}} + B\right)\hat{L}N+\frac{1}{p^2}\frac{\partial}{\partial p}p^4 D\frac{\partial N}{\partial p}+A\hat{L}^2N+2B\hat{L}\hat{P}N+F_j(p), \quad (1)$$

The integro-differential operators $\hat{L}$ and $\hat{P}$ as well as the kinetic coefficients A, B, D, τ_{sh}, and $\chi_{\alpha\beta}$ are expressed[11] in terms of spectral functions which describe correlations between large scale turbulent motions and shocks. The source term $F_j(p)$ is determined by the injection of nuclei of type j (see above).

The energy losses of accelerated nuclei in the bubble filled with very rarefied plasma are relatively unimportant and they are neglected in the Eq. (1). Eq. (1) takes into account particle acceleration by shocks and large scale plasma motions and also resonant particle acceleration by small scale MHD turbulence.

An approximate qusi-stationary solution to Eq. (1) for the spectrum of accelerated nuclei at the boundary of hot bubble is: $N_j(0, q) \approx \eta_j n_j p^{-3}$ for $p_{inj} \ll p \leq p_{*j}$, where η_j is the fraction of nuclei injected from thermal pool of number density n_j within the hot bubble. The high energy tail of distribution of nuclei ($p \gg p_{*j}$) may have power law ($\sim p^{-\gamma}$) or exponential asymptotic form depending on whether an ensemble of weak shocks is available or not. A model of nucleus acceleration by a single strong shock[13] yields the spectrum with single slope which is steeper in the range up to tens of MeV $nucl^{-1}$ than the spectra presented here.

Figure 1 displays calculated spectrum of oxygen nuclei and γ line emissivity for some depths in a dense cloud at the border of the hot cavity where the nuclei are accelerated. Diffusion propagation of the nuclei in the cloud is described by a diffusion coefficient $\chi(p) \sim vp^{0.3}$, with $\chi_0 = 10^{25}$ cm^2 s^{-1} at 1 MeV. Using the crossection[14] for the production of 4.4 MeV γ line due to spallation of^{16}O, we have calculated (in collaboration with S.V. Bozhokin) variations of the emissivity and width of broad γ-line with depth for the cloud of density $n = 10^3$ cm^{-3} (Fig.1, right). The emissivity $W_\gamma(E_\gamma, z)$ in Fig.1 corresponds to the model with ejection of about 1 $M_\odot$ of ^{16}O into the bubble and the injection efficiency $\eta_o \sim 10^{-3}$. The COMPTEL sensitivity allows one to detect at least the combined flux from the main carbon and oxygen lines in 3 - 7 MeV range[6].

In Fig. 1 we have used a dimensionless depth z (within the cloud). To obtain a real depth, one should multiply z by the factor

$$z_j = \left[\frac{p_0 \chi_0}{8\pi Z^2 a_0^2 \,\mathrm{Ry}\, n}\right]^{1/2}, \qquad (2)$$

where a_0 and Ry are the Bohr radius and Rydberg constant, respectively. For oxygen nuclei, we have $z_o \approx 10^{15}$ cm under assumed parameters. Thus one might conclude (see Fig.1) that the scale of the emitting region of oxygen lines is about 10^{18} cm (or larger for clouds with $n \ll 10^3$ cm^{-3}). It is important that the expected spatial resolution of the INTEGRAL would be sufficient to resolve the line emission region in Orion.

A comprehensive analisis of γ-line spectra in the thick target approximation[14,15] can be extended now to describe the actual structures of acceleration and interaction regions.

Acknowledgements

I would like to thank Prof. V.Schönfelder for his kind invitation and support of my participation in the Texas Symposium.

REFERENCES

1. Blitz, L. et al., 1993. Nature **361:** 417
2. Genzel, R. & Stutzki, J., 1989. Ann. Rev. Astron. Astrophys.**27:** 41.
3. Polcaro, V.F. et al., 1991. Astron. Astrophys. **252:** 590.
4. Bykov, A.M. 1992., In Evolution of Interstellar Matter and Dynamics of Galaxies. J.Palous, W.B.Burton, P.O.Lindblad Eds., Cambridge University Press, p.116
5. Schönfelder, V. et al., 1993. Astrophys. J. Suppl. **86:** 657.
6. Bloemen, H. et al., 1994. Astron. Astrophys. **281:** L5.
7. Bykov, A. & Bloemen, H., 1994. Astron. Astrophys. **283:** L1.
8. Canizares, C.R., & Winkler, P.F., 1981. Astrophys. J. **246:** L33.
9. Giacallone, J., et al., 1993. Astrophys. J. **402:** 550.
10. Winkler, P.F. & Kirshner, R.P., 1985 Astrophys. J. **299:** 981.
11. Bykov, A.M. & Toptygin, I.N., 1993. Physics-Uspekhi, **36:** 1020.
12. Bozhokin, S.V. & Bykov, A.M., 1994. Astron. Lett. **20:** 593.
13. Nath, B.B., & Biermann, P.L., 1994. MNRAS, **270:** L33.
14. Ramaty, R. et al., 1979. Astrophys. J. Suppl. **40:** 487.
15. Ramaty, R. et al., 1995. Astrophys. J. (in press).

Gamma-Ray Burst Mini-Symposium - Introductory Remarks

GERALD J. FISHMAN

Space Sciences Laboratory
NASA-Marshall Space Flight Center
Huntsville, AL 35812 USA

This mini-symposium is somewhat different than others at this Texas Symposium in that the papers are organized in a complementary manner so as to provide the reader with a comprehensive look at the present state of the field and to avoid redundant material. In this series of papers, one will also find a progression from straight-forward, non-controversial observational facts (e.g. time profiles of bursts) to the speculative models of their origin, as presented in the last two papers. Intermediate papers describe observational and interpretive subjects which are debatable and controversial (e.g. spectral features, gamma-ray burst repetition, and observations of a time dilation consistent with a cosmological red-shift).

It is interesting to note that at the Texas Symposium twenty years ago, Mal Ruderman gave an enlightening and humorous summary of the state of gamma-ray burst research at that time. His description of some of the exotic and wide variety of models then proposed makes interesting reading (Ruderman 1975). Now, twenty years later, there are over a hundred models published (Nemiroff 1994a,b) and we are apparently no closer to a resolution of the problem.

As you will see in these papers, the observational situation has vastly improved. Many of the new results presented here are from experiments on the Compton Gamma Ray Observatory, These observations follow on the very comprehensive data-sets obtained from primarily French, Russian, US and Japanese experiments over the past decade. Many of the most recent observations and theories are described in a conference proceedings published last year (Fishman, Brainerd and Hurley 1994). The field continues to be very active-there are three major international conferences in 1995 in which gamma-ray bursts are the primary topic. These are exciting times-- whatever their explanation, gamma-ray bursts are a unique phenomenon that may provide new insight into some heretofore unknown and very high-energy aspect of the Universe.

References

Fishman, G.J. Brainerd, J.J. and Hurley, K.C. eds. 1994 "Gamma-Ray Bursts", AIP Conf. Proc. # 307 (AIP Press:New York)

Nemiroff, R.J. 1994a Comm. Astrophys. v.17, p.189

Nemiroff, R.J. 1994b AIP Conf Proc. #307 p.730

Ruderman, M. 1975 Ann. N.Y Acad. Sci. v.262, p.164

TEMPORAL PROPERTIES OF GAMMA-RAY BURSTS

CHRYSSA KOUVELIOTOU*

NASA/Marshall Space Flight Center, ES-84
Huntsville, AL 35812 USA

INTRODUCTION

Time domain analysis of Gamma-Ray Bursts (GRBs) had been given appropriate attention by very few researchers in the past[1,2,3]. Strict morphological classification of temporal profiles was mostly unsuccessful and thus not conducive to GRB modeling. As a consequence, temporal analysis focused mainly on GRB duration calculations and periodicity searches. Burst durations were predominantly defined by looking for a predetermined deviation above the local background level before and after the event peak; it was found[4] that GRB durations span a range of <0.1 – 1000 s. On the other hand, taking into account all the pitfalls of confirming a periodic signal in a short, transient event, it is not surprising that results of periodicity searches were either negative or unconvincing (with one exception: the 1979, March 5 event which exhibited 20 cycles of an 8 s period)[5]. Thus, the general consensus in the field until about 1990 was that GRBs are each one of a kind, have a duration range of three orders of magnitude, and if anything they exhibit[6,7] quasi-periodic structures of $\sim$ 1 - 5 s.

On April 5, 1991, one of the NASA Great Observatories, the Compton Gamma-Ray Observatory (CGRO) was launched. One of the four instruments onboard CGRO is the Burst And Transient Source Experiment (BATSE), especially designed to detect GRBs with unprecedented sensitivity[8]. By the end of 1994 BATSE had detected over 1200 GRBs; it has accumulated sufficient statistics to redefine the observational constraints of the GRB problem. The multitude of the burst related data types collected with excellent temporal resolution (ranging from 2μs to 64 ms) and with different energy channel combinations, has enabled– among others– temporal analyses with breakthrough results. These analyses have concentrated mainly in the following areas: overall burst morphology, burst time scales and burst substructure. In the following I will review each of these areas separately.

BURST MORPHOLOGY

In the past 27 years there have been several attempts to classify GRB lightcurves into different types according to their common characteristics[1,2,3]. All have been unsuccessful, insofar that there is always the odd event: the event that will

* Universities Space Research Association

create its own unique class. Statistically speaking, when such "loner" cases dominate the sample, classification is useless. In Fig. 1 I try to provide an overview of the morphologies that are often encountered in GRBs. I have separated and examined the events according to duration into two subsets divided at 2 s; the choice of this separator is discussed in the next section. The diagram reveals that the same profile varieties are present at all time scales. GRBs can consist of single pulses or have many episodes, each of them either single or multi-pulsed. Each pulse can have a smooth overall envelope or it may contain rich substructure. The substructure can be entirely random (described as chaotic in the diagram) or it may exhibit quasi-periodicity. A particular pulse-form, exhibiting Fast Rise and Exponential-like Decay, has been termed FRED on the diagram; a word coined by the BATSE group to afford quick GRB description. The majority of bursts can be classified in the non-committal way described above; the remainder fall in the "other" box. Yet, there are light-curves that are unique; for instance, BATSE has detected only one symmetric GRB and only one anti-FRED (exponential-like rise and fast decay) burst so far.

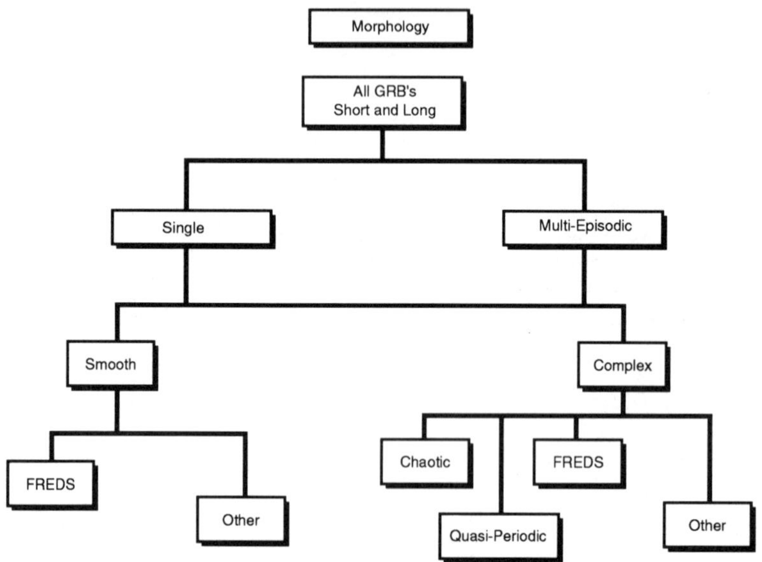

Fig. 1. Scheme of GRB morphological classification.

Although it is difficult to classify highly resolved GRB light-curves, when we look at an average GRB profile we can identify significant overall trends. Mitrofanov *et al.*[9] have created such a profile from a set of GRBs by first coarsely binning each event, then aligning all GRBs at their highest peak and adding the profiles after normalizing to the sample size; they found that the leading part of this average GRB (rise-to-peak) is shorter and has a harder spectrum than the trailing part (peak-to-decay). The asymmetry of the average GRB profile has also been quantitatively established by Nemiroff *et al.*[10], who have devised a simple test they call the Υ

statistic. This parameter is the ratio of the number of times that the counts in the previous time bin are higher to the number where the counts are lower. Their results indicate that GRBs show significant time-asymmetry, on all time scales from 64 ms to 4096 ms, between 50 – 300 keV.

TIME SCALES

Each GRB light-curve is a set of pulses that result from a random convolution of three distributions: individual pulse widths, separations between pulses and number of pulses per GRB. Given each of these distributions, one can reproduce an entire light-curve by assuming a few, simple functions for the pulse shapes, such as triangular, or FREDs. Determining the *total* duration of a transient event has always been a controversial issue. The criteria for "duration" have widely varied over the past and no consensus was reached for an algorithm. We introduced in the first BATSE catalog[11] an unbiased and reproducible way of estimating durations. We have used the sum of the counts above 25 keV in all triggered (GRB) detectors to reduce the effect of Earth scattering and the dependence of detector response on the burst arrival direction. We define T_{90} as the time during which the cumulative counts increase from 5% to 95% above background, thus encompassing 90% of the total GRB counts. The times thus defined are an intensity-independent measure of duration, unlike previous definitions[12]. Fig. 2a shows the distribution of T_{90} for the 3B catalog [13–14]. As reported previously, we find that there is a strong bimodality in the distribution; the two modes are separated at $\sim$ 2 s. The resulting two GRB subsets, with (logarithmic) mean T_{90} values of 0.57 ± 0.08 and 30.6 ± 1.9 s respectively, have been checked extensively for additional differences. We find that they are indistinguishable, except for their spectral characteristics.

We define as an integral Hardness Ratio (HR) the ratio of total burst counts above background in the 100 – 300 keV and 25 – 100 keV bands. The distribution of HRs is illustrated in Fig. 2b. We see that there is a strong trend for the short events to have relatively hard spectra while the spectra of the longer events are predominantly softer. Their corresponding average HRs are 1.44 ± 0.08 and 0.90 ± 0.03, respectively. This trend has been reported in the past[15], albeit not linked with a duration bimodality.

A very important test for the origin of GRBs would be the discovery of cosmological signatures (such as time dilation effects) in their light-curves. If GRBs are cosmological, and if they can be considered standard candles, the weaker events would originate in more distant sources and their durations should be time dilated, (*i.e.*, longer on the average), relative to these of the brighter events. Norris *et al.*[16] have searched for such an effect and reported that they found significant evidence of light-curve stretching by a factor of $\sim$ 2.3 between the extreme 10% dim and 10% bright GRBs. This amount of dilation would correspond to the dim events being at cosmological distances corresponding to $z \sim 1$. Their results have been disputed by Mitrofanov *et al.*[17] who claim that the measured stretching is a result of selection criteria for the two extreme (dim and bright) populations of GRBs. This is currently one of the hotly debated issues in the field.

In a recent search for precursor activity in GRBs, Koshut *et al.*[18] have found that only $\sim$3% of the bursts observed with BATSE can be associated with precursors; according to their definition a precursor is any case where the first episode has a

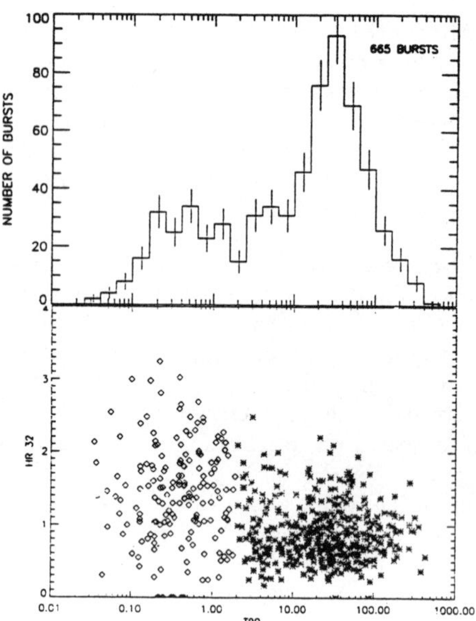

Fig. 2. *a:* Distribution of T_{90} for 665 GRBs from the 3rd BATSE catalog. *b:* Distribution of HRs *versus* T_{90} for the same set of 665 GRBs.

lower peak intensity than that of the main episode and is separated from it by a background interval that is at least as long as the main episode. The average precursor and main episode durations are 15 and 40 s, respectively. They find no significant correlations between or amongst any of the precursor or the main episode characteristics, except for a 3σ correlation between their durations. It seems that the characteristics of the main episode emission are independent of the existence of the precursor emission.

SUBSTRUCTURE

The subset of GRBs that are rich in substructure has been extensively searched for coherent pulsations. To prove the existence of coherent oscillations in short, transient phenomena has always been a challenging task. Before BATSE very few events have been identified to exhibit pulsations that are periodic and last for several cycles[5,6,7]. In a preliminary search using 22 out of the first 137 long (> 2 s) GRBs detected with BATSE, Kouveliotou *et al.*[19] reported that they found no significant coherent oscillations in any of the events. They also find that the power in the red noise component of the Fourier spectrum of GRBs does not extend above 1.5 Hz in 90% of the cases; this would mean that the characteristic GRB time scales are longer than ~ 0.7 s.

Bhat *et al.*[20] have studied the short (< 2 s) GRBs for evidence of variability. Their results from an analysis of 37 events show that more complex bursts (characterized by a Complexity Index, CI, which is a measure of the total number of peaks in a burst) have higher average HRs and sharper temporal features at higher energies. A search for photon bunching within the bursts had negative results.

Although no coherent oscillations have been unambiguously found in GRBs there is evidence in several cases for peaks in the GRB Fourier power spectrum around 2 - 7 s, reminiscent of the quasi-periodic oscillations (QPO) found in the intensity variations of X-ray binaries[21]. The interpretation of these peaks is currently unclear.

REFERENCES

1. Desai, U., 1981. Astrophys. Space Sci. **75:** 15.
2. Mazets, E., & Golenetskii, S., 1981. Astrophys. Space Sci. **75:** 47.
3. Barat, C., *et al.* 1981. Astrophys. Space Sci. **75:** 83.
4. Hurley, K., 1991. *In* AIP Conf. Proc. 265: Gamma-Ray Bursts. W. Paciesas & G. J. Fishman, Eds.: 3–12. American Institute of Physics, New York.
5. Mazets, E., *et al.* 1979. Nature. **282:** 587.
6. Barat, C., *et al.*, 1984. Ap. J. Letters. **286:** L5.
7. Kouveliotou, C., *et al.*, 1988. Ap. J. Letters. **330:** L101.
8. Fishman, G. J., *et al.*, 1989. *In* Proc. of the GRO Science Workshop. W. N. Johnson, Ed.: 39.
9. Mitrofanov, I., *et al.*, 1993. *In* AIP Conf. Proc. 307: Gamma-Ray Bursts. G. J. Fishman, J. Brainerd, & K. Hurley, Eds.: 187. American Institute of Physics, New York.
10. Nemiroff, R., *et al.*, 1993. *In* AIP Conf. Proc. 307: Gamma-Ray Bursts. G. J. Fishman, J. Brainerd, & K. Hurley, Eds.: 237. American Institute of Physics, New York.
11. Fishman, G. J., *et al.*, 1994. Ap. J. Suppl. Series. **92:** 229.
12. Kouveliotou, C., *et al.*, 1993. Ap. J. Letters. **413** (1993), L101.
13. Meegan, C. A., *et al.*, 1995. in preparation.
14. Kouveliotou, C., *et al.*, 1995. in preparation.
15. Dezalay, J–P., *et al.*, 1991. *In* AIP Conf. Proc. 265: Gamma-Ray Bursts. W. Paciesas & G. J. Fishman, Eds.: 304–309. American Institute of Physics, New York.
16. Norris, J. P., *et al.*, 1994. Ap. J. **424:** 540.
17. Mitrofanov, I., *et al.*, 1995. in preparation.
18. Koshut, T., *et al.*, 1995. Ap. J. in press.
19. Kouveliotou, C., *et al.*, 1991. *In* AIP Conf. Proc. 265: Gamma-Ray Bursts. W. Paciesas & G. J. Fishman, Eds.: 299–303. American Institute of Physics, New York.
20. Bhat, N., *et al.*, 1993. *In* AIP Conf. Proc. 307: Gamma-Ray Bursts. G. J. Fishman, J. Brainerd, & K. Hurley, Eds.: 197. American Institute of Physics, New York.
21. Hasinger, G. & van der Klis, M., 1989. A & A. **225:** 79.

BATSE OBSERVATIONS OF THE SPECTRA OF GAMMA-RAY BURSTS

MICHAEL S. BRIGGS

University of Alabama in Huntsville
Huntsville, AL 35899 USA

INTRODUCTION

Gamma-ray bursts (GRBs) acquired their name because their emission peaks in the gamma-ray domain. The Burst and Transient Source Experiment (BATSE) on the Compton Gamma-Ray Observatory has obtained spectral data on more than 1000 bursts, of which several hundred are sufficiently bright to permit detailed spectral studies. I report on the BATSE team's studies of GRB continuum emission and temporal evolution, and on our search, so far unsuccessful, for spectral features.

The BATSE instrument[1,2] consists of eight detector modules, each containing two scintillator detectors optimized for different purposes. The 2025 cm^2 area of the Large Area Detectors (LADs) enables the study of the continuum emission of bursts on short timescales in the energy range 20 to 1800 keV. The Spectroscopy Detectors (SDs), while having an area of only 125 cm^2, have 7.9% resolution at 511 keV and are thus useful for studies of spectral features. Depending on the gain setting of a SD, useful energy coverage can begin as low as 5 keV or end as high as a few tens of MeV.

CONTINUUM SPECTRA

Initially gamma-ray burst spectra were characterized with simple models such as exponentials or optically-thin thermal bremsstrahlung, but successive generations of instruments have revealed the spectra to be harder and more complex[2]. With the high signal-to-noise-ratio data obtained by BATSE, we have found that four-parameter continuum models are needed in order to acceptably fit the data of most bursts. One such model is the empirical "GRB" model[2], which blends a low-energy power law with a high-energy power-law via an intermediate exponential—in one parameterization the four parameters are the intensity, the low-energy index α, the high-energy index β, and the energy E_{peak} at which the energy luminosity is maximum (assuming $\beta < -2$, as is usually the case). This model has been extensively compared to the data collected with both the SDs and LADs, and with both time-resolved spectra and burst-integrated spectra, and has generally been found to acceptably match the observed counts[2,3].

Figures 1 and 2 show the spectrum of a 0.64 s interval of burst GRB 911118, both data and fitted "GRB" model. The fit was obtained using the standard forward-folding technique in which a photon model is assumed, folded through a model of the detector response and compared with the data, after which the photon model parameters are optimized so as to minimize χ^2. This particular spectrum is rather typical, with a peak luminosity at $E_{\rm peak} = 176$ keV. The value of $E_{\rm peak}$ for individual spectra within a burst ranges from below 100 keV to above 1 MeV. We have found that the peak energy $E_{\rm peak}$ of the spectra of GRBs *integrated over the burst* is typically between 80 and 800 keV[2,4]. EGRET has discovered an additional spectral component which peaks at much higher energies but only carries (based on certain assumptions) $\lesssim 2\%$ of the total energy emission [5,6].

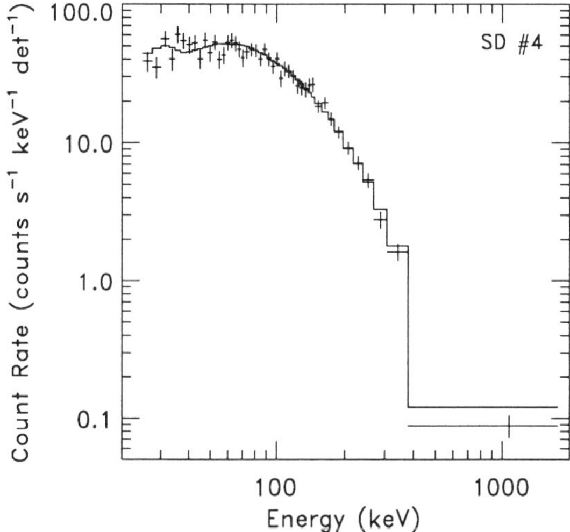

Fig. 1. An example spectrum, the 0.64 s interval starting 5.504 s after BATSE's trigger for burst GRB 911118. Some of the count rate data have been rebinned into larger bins for display purposes. The histogram shows the count rate model obtained by a forward-folding fit, using Band's GRB model as the photon flux model.

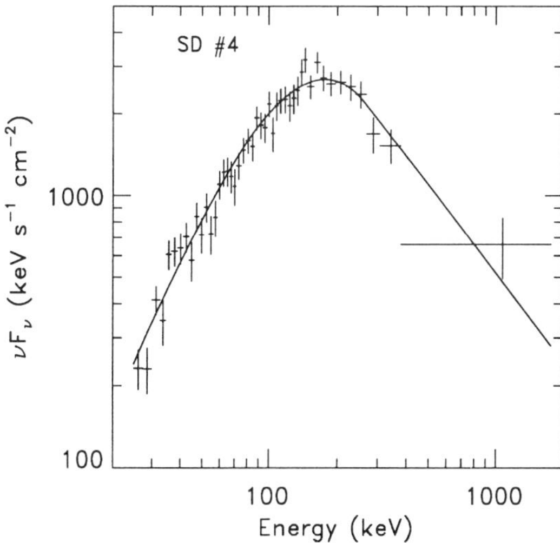

Fig. 2. This is the same data and fit as Fig. 1, here displayed as a deconvolved spectrum in νF_ν units. The index (in photon number flux) of the low-energy power law is $\alpha = 0.19 \pm 0.13$, and of the high-energy, $\beta = -3.1 \pm 0.2$. The spectrum peaks at $E_{\rm peak} = 176 \pm 8$ keV.

A theoretical five-parameter spectral model has been proposed, based upon Compton scattering and photoelectric absorption in the environs of a source located at a cosmological distance[7]. This model postulates a central object which produces a power law spectra which is modified by Compton scattering near the source and, in our frame, by the cosmological redshift of the source. This model makes predictions, e.g., it should be possible to fit the time-resolved spectra of a burst with a single value of redshift z, which have withstood preliminary testing.

SPECTRAL EVOLUTION

The spectral properties of gamma-ray bursts can change dramatically on short timescales, for example, in the burst GRB 930131, the hardness ratio changed significantly in $\lesssim 2$ ms[8]. The BATSE team has found the parameter E_{peak} obtained from fitting the GRB model to be useful to characterize the "hardness" and spectral evolution of GRBs[2,3]. When the index β of the high-energy power law is less than -2, as it almost always is, E_{peak} is the energy at which the energy flux is maximum. Figures 3 and 4 compare, for two GRBs, the temporal behavior of E_{peak} and the count rate history.

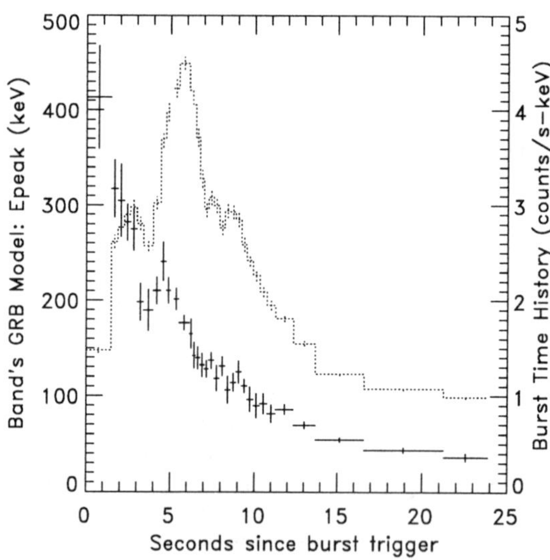

Fig. 3. Time history (histogram) and E_{peak} evolution (crosses indicating interval and $\pm 1\sigma$) for GRB 911118, SD 4. In order to obtain sufficient counts to perform spectral fitting, some of the time intervals are sums of the instrumentally provided spectra.

Before BATSE, there was a debate about whether hardness-intensity correlation[9] or hard-to-soft evolution[10] best characterized GRBs. We find that both behaviours are typically present[3], as can be seen in Figures 3 and 4. The GRB shown in Figure 3, GRB 911118, is fairly simple. It has an overall hard-to-soft evolution on which is imposed a hardening at the onset of the second peak at 6 seconds. This renewed hardening is probably best characterized as a hardness-intensity correlation, although it can also be characterized as hard-to-soft evolution of pulses. The burst GRB 920622 has a more complex time history. While there is still an overall hard-to-soft evolution, the increase in E_{peak} at the times of peaks in

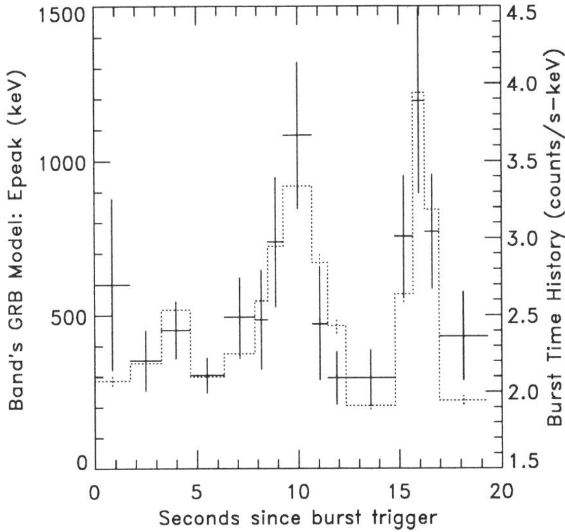

Fig. 4. The time history and spectral evolution of GRB 920622, SD 0.

the emission is more dominant. In some cases the peak in hardness leads the peak intensity (e.g., the emission peaks at 3 and 6 s in GRB 911118, the first peak in GRB 920622 at 4 s), while in other cases the peaks in hardness and intensity are coincident (e.g., the peaks at 10 and 16 s in GRB 920622). Every gamma-ray burst is unique!

SPECTRAL FEATURES?

Low-energy absorption features such as the 20 and 40 keV lines seen by Ginga[11,12] are commonly interpreted as resulting from cyclotron processes in the strong magnetic fields of Galactic neutron stars. BATSE's observations of the isotropy and inhomogeneity of GRBs are strong evidence for an origin of GRBs at cosmological distances [13,14,15]. The conflict between these interpretations of the distance scale of GRBs makes the existence and interpretation of spectral features of immense interest. We have visually searched the spectra of bright BATSE GRBs for low-energy spectral features— so far, no convincing line has been found[16].

Figure 5 shows the data of the best candidate: the interval 6.080 to 9.920 s after BATSE's trigger for GRB 930506 and a joint fit of the GRB model to 26 to 1300 keV for SD 2 and 46 to 4800 keV for SD 7. For eight channels the data of SD 2 lies well below the continuum model, yet the data of SD 7 is in excellent agreement with the continuum model. If the data of SD 2 alone are fit, the improvement in χ^2 from adding a line at 55 keV indicates a chance probability for the feature of about 10^{-4}. [16,17] A joint continuum fit to both detectors has a reasonable χ^2 of 435.9 for 425 degrees of freedom—adding a line reduces χ^2 by only 6. The likely resolution of the apparent contradiction between the data of SDs 2 and 7 is that we have examined enough spectra to make a feature like that of SD 2 probable by chance in the ensemble. The presence of lines in the Ginga data and their absence to date in the BATSE data is consistent with chance[16,18]. We are continuing our

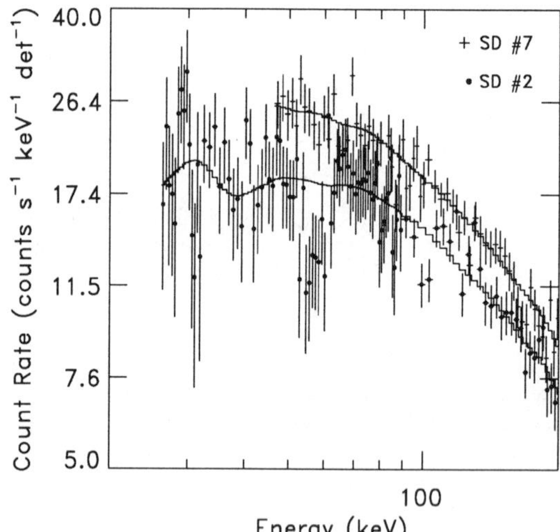

Fig. 5. The 55 keV line candidate in GRB 930506. The count rate data of SDs 2 & 7 and the best-fit count rate models obtained from forward-folding optimization of the GRB model are shown. The count models are different for a single photon model because of the differing burst-to-detector angles. The feature (data & model) at 33 keV is due to the K-shell energy of Iodine in the NaI detectors.

search and are improving it by implementing an automatic computer search.

REFERENCES

1. Fishman, G. J., C. A. Meegan, R. B. Wilson, *et al.* 1994. Astrophys. J. Supp. **92:** 229–283.
2. Band, D., J. Matteson, L. Ford, *et al.* 1993. Astrophys. J. **413:** 281–292.
3. Ford, L. A., D. L. Band, J. L. Matteson, *et al.* 1995. Astrophys. J. **439:** 307–321.
4. Mallozzi, R. S., W. S. Paciesas, G. N. Pendleton, *et al.* 1995. Astrophys. J. Submitted.
5. Hurley, K., B. L. Dingus, R. Mukherjee, *et al.* 1994. Nature. **372:** 652–654.
6. Schneid, E., 1995. *In* These proceedings.
7. Brainerd, J. J. 1994. Astrophys. J. **428:** 21–27.
8. Kouveliotou, C., R. Preece, N. Bhat, *et al.* 1994. Astrophys. J. **422:** L59–L62.
9. Golenetskii, S. V., E. P. Mazets, R. L. Aptekar & V. N. Ilyinskii. 1983. Nature. **306:** 451–453.
10. Norris, J. P., G. H. Share, *et al.* 1986. Astrophys. J. **301:** 213–219.
11. Fenimore, E. E., J. P. Conner, R. I. Epstein, *et al.* 1988. Astrophys. J. **335:** L71-L74.
12. Murakami, T., M. Fujii, K. Hayashida, *et al.* 1988. Nature. **335:** 234–235.
13. Hakkila, J., C. A. Meegan, G. J. Fishman, *et al.* 1994. Astrophys. J. **422:** 659–670.
14. Briggs, M. S., W. S. Paciesas, G. N. Pendleton, *et al.* 1994. Astrophys. J. submitted.
15. Hartmman, D. H., M. S. Briggs & G. N. Pendleton. 1995. *In* These Proceedings.
16. Palmer, D. M., B. J. Teegarden, B. E. Schaefer, *et al.* 1994. Astrophys. J. **433:** L77–L80.
17. Ford, L., D. Band, J. Matteson, *et al.* 1994. *In* AIP Conf. Proc. 307: Gamma-Ray Bursts. G. J. Fishman, J. J. Brainerd & K. Hurley, Eds.: 261–265. American Institute of Physics, New York.
18. Band, D. L., L. A. Ford, J. L. Matteson, *et al.* 1994. Astrophys. J. **434:** 560–569.

EGRET Observations of Gamma-Ray Bursts

[1]E J SCHNEID, [2]D L BERTSCH, [2]C E FICHTEL, [2]R C HARTMAN, [2]S D HUNTER, [2]D J THOMPSON , [3]G KANBACH, [3]H A MAYER-HASSELWANDER, [4]Y C LIN, [4]P F MICHELSON, [4]P L NOLAN, [5]B L DINGUS, [5]C VON MONTIGNY, [5]P SREEKUMAR, AND [6]D A KNIFFEN

[1]Northrop Grumman
Bethpage,New York 11714

[2]NASA Goddard Space Flight Center
Greenbelt, Maryland 20771

[3]Max-Planck-Institut fur
extraterrestrische Physik
D-85740 Garching, Germany

[4]Stanford University
Stanford, California 94305

[5]Universities Space Research Association
NASA Goddard Space Flight Center
Greenbelt, Maryland 20771

[6]Hampden-Sydney College
Hampden-Sydney, Virginia 23943

ABSTRACT

The Energetic Gamma-Ray Experiment Telescope (EGRET) has observed gamma-rays bursts with the highest energy gamma-rays and the longest high energy emission to date. EGRET measures the high energy gamma-rays with its large NaI scintillator (1 to 200 MeV) and its spark chamber (30 MeV to 30 GeV). The spark chamber also measures time and arrival directions of individual photons allowing locations for the energetic bursts to be determined. Since the Compton Gamma Ray Observatory launch in 1991, EGRET has observed five bursts in the spark chamber with several having gamma-ray energies grater than 1 GeV. The recording breaking burst, GRB940217, had gamma-rays up to 18 GeV and lasted over 5000 seconds. The results for the energetic bursts are presented. The high energies observed from these gamma-ray bursts set constraints for the burst distances .

INTRODUCTION

The launch of the Compton Gamma-Ray Observatory (CGRO) in 1991 has provided a unique suite of four instruments that can detect and measure gamma-ray bursts from the keV region with the Burst And Transient Source Experiment (BATSE)[1] to the GeV region with EGRET. Intermediate energy regions are measured by the Oriented Scintillation Spectrometer Experiment (OSSE)[2] and the Compton Telescope experiment (COMPTEL)[3]. Prior to the launch of CGRO, the Gamma-Ray Spectrometer on the Solar Maximum has measured gamma-ray bursts spectra up to 10 MeV (Matz,et al.[4]) and one burst up to almost 100 MeV (Share, et al.[5]). Shortly after the CGRO launch, the first burst seen by EGRET occuring on 1991 May 3, gamma-rays with energies over 100 MeV were observed (Schneid, et al.[6],Dingus, et al.[7]).

Five energetic gamma-ray bursts with energies exceeding 100 MeV have been been detected and measured by EGRET. Several of these bursts have energies exceeding 1 GeV. The emission for the high energies gamma-rays occur over a longer time interval when compared to the low energy emissions.

THE EGRET INSTRUMENT

EGRET has the standard elements of a high energy gamma-ray telescope, specifically an anticoincidence scintillator dome to discriminate against charged particles, a particle track detector consisting of spark chambers with interspersed high z material to convert the gamma-rays into electron pairs, and a energy measurement device which is a large NaI crystal. A description of the instrument and its general capabilities is given by Kanbach[8].The results of the instrument calibration, both before and after launch, are given by Thompson[9]. The spark chamber telescope covers the energy range from about 30 MeV to 30 GeV. The telescope is designed to be free of internal background, and the calibration tests have verified that is at least an order of magnitude below the extragalactic gamma radiation.

High energy gamma-rays from a burst may be studied in two ways with EGRET. First, if the energy of the gamma-ray is greater than 30 MeV and in telescope field of view, it may pair produce in the spark chamber and the electron and positron can be imaged. In this mode, directions and energy of individual gamma-rays can be measured. Second, the energy spectrum may be measured by the large NaI scintillator. The NaI has an independent, self triggered low energy mode of analysis which spans the energy range from 1 to 200 MeV.

OBSERVATIONAL RESULTS

For all BATSE gamma-ray bursts within the EGRET telescope field of view (~30 degrees), a careful search of the spark chamber events was conducted for burst related gamma rays. Gamma-rays from five bursts were detected and they correspond to some of the most intense bursts measured by BATSE. Table 1 lists, for each of these bursts, the key parameters for the high energy gamma-ray emissions : maximum energy, duration of emission, indication of high energy emission continues after the low energy emission, and the power law fit function to the spectra. All five bursts have energies greater than 100

TABLE 1. EGRET Energetic Gamma-Ray Burst Observations

Burst ID.	Max. Energy (GeV)	Duration Emission	Delayed Emission	Spectral Function	Refs.
GRB910503	10	84 s	X	$E^{-2.2}$	6,7
GRB910601	0.314	200 s	X	$E^{-3.7}$	10,7
GRB930131	1.2	100 s	X	$E^{-2.0}$	11,7
GRB940217	18	1.5 h	X	$E^{-2.6}$	12
GRB940301	0.16	30 s	-	$E^{-2.5}$	13

MeV and three of the bursts have emissions over 1 GeV. The duration of the high energy emissions, all except for the March 1, 1994 burst, extend beyond the duration of the low energy emissions. For the GRB910601 burst, all high energy spark chamber events occured after the gamma rays seen in the NaI.

An impressive example of the spark chamber data is that for GRB940217. Figure 1 presents energy and the recorded time for the 28 high energy gamma-rays seen in the spark chamber. Ten photons were recorded while the low-energy burst was in progress during the first 180 secs. Following this, an additional 18 photons were recorded for ~5400 s including an 18 GeV photons ~4500 s after the low energy emission has ended. For this burst, the probaliity of the delayed emission being a background anomaly has been studied[12] and is less than 2×10^{-6}. Interruption of data collection due to Earth-occultation and lack of telemetry are indicated by the solid lines in this figure.

The gamma-rays observed in the spark chamber can be used to determine the burst location[6,7,10,11,12].The 95% confidence contours, obtained by maximum-likelihood analysis of the photons overlap the arc from the Interplanetary localization, resulting from the Ulysses and BATSE observations and the 2 sigma COMPTEL burst location. The EGRET data with the data from other experiments can be used to help refined the location of these bursts.

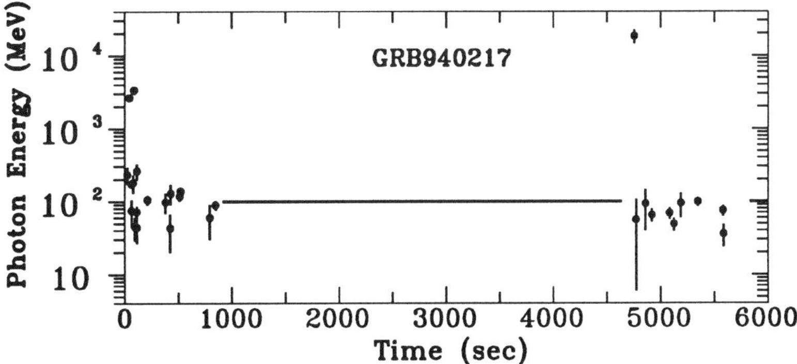

FIGURE 1. EGRET spark chamber energies as a function of time; horizontal lines indicate periods when no data were available. Low energy emission is complete in 180 sec.

DISCUSSION

The burst related gamma-rays detected by EGRET have increased the highest energy by two orders of magnitude. For these bursts, the power-law spectra , extending over three decades in energy, rule out any model which predicts black body or thermal spectra. The models must be able to predict the production of relativistic particles required to produce these gamma-rays. The observed power-law spectra are consistent with radiation from shock-accelerated particles.

Delayed , high-energy emission from a gamma-ray burst is now a well established fact, but there is still not enough evidence to state this for all bursts because EGRET detections corresponded to only the strongest BATSE bursts. The acceleration of the relativistic particles could persist over the emission time or the particle lifetimes are comparable to the emission time with the particles interacting with burst debris[14]. However , the delayed emissions can increase hopes for detection of bursts by other means that could lead to identification of these objects.

The presence of high energy gamma-rays apparently unattenuated by pair-production[15] with the microwave background photons[14] put an upper limit of $z<200$ for ~ GeV photons and $z\sim55$ for ~18 GeV photons. Both distances rule out cosmic strings at $z=1000$ [16,17] for these bursts.

The authors gratefully acknowledge support from the following : Bundesministerium fur Forschung and Technologie, grant 50 QV 9095 (MPE); NASA grant NAG 5-1742 (HSC); NASA grant NAG 5-1605 (SU), and NASA contract NAS 5-31210 (NG).

REFERENCES

1. Meegan, C., et al. 1992, Nature, **335** : 143
2.Johnson,W.N.,et al. 1993,Astrophys J. Suppl.,**86** :693.
3.Schonfelder, V. et al. 1993, Astrophys J. Suppl., **86** :657.
4. Matz, S.M., et al. 1992, Astrophys J., **228** : L 37
5. Share, G.H., et al. 1986, Adv. Space Res., **5** :15
6. Schneid, E.J., et al. 1992, Astr. Astrophys., **225** : L13-L16
7. Dingus, B.L.,et al. 1994 in Gamma-Ray Bursts. G.Fishman,G. Brainard, & K. Hurley, Eds.:22-26. AIP 307 Am. Inst. Phys.,New York
8. Kanbach, G., et al. 1988, Space Sci. Rev., **46** : 69
9. Thompson, D.J., etal. 1993,Astrophys J. Suppl., **86** : 629-656
10.Kwok,P.,et al. in Compton Gamma-Ray Observatory. M. Friedlander,N. Gehrels,&D. Macomb, Eds. : 855-859 .AIP 280, Am Inst Phys., New York
11.Sommer, M. et al.1994,Astrophys. J. **422** :L63-L66
12. Hurley,K. et al. 1994, Nature **372** : 652-654.
13. Schneid, E.J. et al. ,manuscript in prep.
14 Meszaros, P. and Rees, M. 1994, Mon.Not.R.Astr.Soc. **269:** L41-L43.
15. Fazio,G.G,& Stecker,F.W.,1970, Nature **226** :135.
16 Babul,A.,Paczynski,B.,&Spergel,D. 1987, Astrophys J. **316** : L49.
17. Paczynski,B. 1988, Astrophys J. **335** :525.

COMPTEL Observations of Gamma-ray Bursts[a]

R. M. KIPPEN,[b] J. RYAN,[b] A. CONNORS,[b] M. MCCONNELL,[b]
V. SCHÖNFELDER,[c] J. GREINER,[c] M. VARENDORFF,[c]
W. COLLMAR,[c] W. HERMSEN,[d] L. KUIPER,[d] C. WINKLER,[e]
L. O. HANLON[e] AND K. S. O'FLAHERTY[e]

[b]*Space Science Center, University of New Hampshire*
Durham, NH 03824

[c]*Max-Planck Institut für Extraterrestrische Physik*
D-85748 Garching, FRG

[d]*SRON-Utrecht*
Sorbonnelaan 2, 3584 CA Utrecht, NL

[e]*Astrophysics Division, ESTEC, ESA*
NL-2200 AG Noordwijk, NL

INTRODUCTION

The origin of cosmic γ-ray bursts is as mysterious today as it was when they were discovered more than 25 years ago. Despite a wealth of new observational data obtained with the BATSE instrument on board the *Compton* Gamma Ray Observatory, many of the fundamental questions remain unanswered. For instance, although BATSE has provided a tremendous statistical advantage (allowing the most accurate measurement of the spatial isotropy and inhomogeneity of burst sources[1]), its limited angular resolution and spectral range have given us an incomplete picture of the small-scale angular source distribution and high energy emission properties.[2,3] Furthermore, the limited angular resolution has also made it difficult to search for burst counterparts at other wavelengths.

The COMPTEL instrument on board *Compton* measures the locations and spectra (0.75-30 MeV) of several strong γ-ray bursts per year which occur within the ~1 sr field-of-view of the main ("telescope") instrument. A secondary ("burst") observing mode provides supplementary spectral data at lower energies (0.3-10 MeV). The statistics of COMPTEL burst observations are modest, but good location accuracy and spectral coverage at higher energies make them a valuable source of information which may help to understand the nature of bursts in those areas where BATSE is most limited. Many of the COMPTEL burst observations have been reported elsewhere.[4] Here, we present an overview of results from recent spatial and spectral analyses and report on the ongoing campaign using COMPTEL burst localizations to quickly search for fading low-energy counterparts.

[a]The COMPTEL project is supported by NASA under contract NAS5-26645, by the German government through DARA grant 50 QB 90968 and by the Netherlands Organization for Scientific Research (NWO).

ANGULAR DISTRIBUTION AND BURST RECURRENCE

In its first three years of operation from April 1991 through April 1994, COMPTEL detected 18 significant ($\gtrsim 4\sigma$) γ-ray bursts in the main instrument. These bursts are localized through direct imaging using a maximum-likelihood technique.[4] Statistical location accuracy (1σ) ranges from better than 0.5° for strong bursts to ~2° for weak detections, with a mean of ~1°. Systematic location errors are estimated to be $\lesssim 0.5°$.

The COMPTEL burst locations (FIGURE 1) are consistent with an isotropic distribution of sources within the large statistical uncertainty produced by the small sample size (errors caused by location uncertainty are negligible). The Galactic dipole and quadrupole moment statistics[5] corrected for non-uniform COMPTEL sky exposure are $\langle \cos \Theta \rangle = (-0.13 \pm 0.14)$ and $\langle \sin^2 b - \frac{1}{3} \rangle = (+0.01 \pm 0.07)$, respectively. Because all COMPTEL bursts are also observed by BATSE, these results provide an important, independent confirmation that a subset of some of the strongest BATSE bursts are consistent with isotropy. The full BATSE sample, which now incorporates more than 1000 bursts, clearly provides the most constraining measure of the large-scale angular distribution by merit of its overwhelming statistical advantage.[6]

In the COMPTEL sample of source locations, two bursts coincide spatially within instrumental uncertainties.[7] GRB 930704 and GRB 940301 occurred ~8 months apart and are completely consistent with the same position on the sky within the statistical location errors (see FIGURE 1). The probability of such a coincidence occurring by chance in an isotropic distribution has been estimated using Monte Carlo simulations to be $\lesssim 3\%$ — suggesting the *possibility* that these two events are related to a single source.[8] Independent localizations determined by BATSE, EGRET and the Interplanetary Network (using BATSE, *Ulysses* and *Mars Observer* arrival time analysis) cannot rule out this possibility. Gravitational lensing has been excluded since the lightcurves and

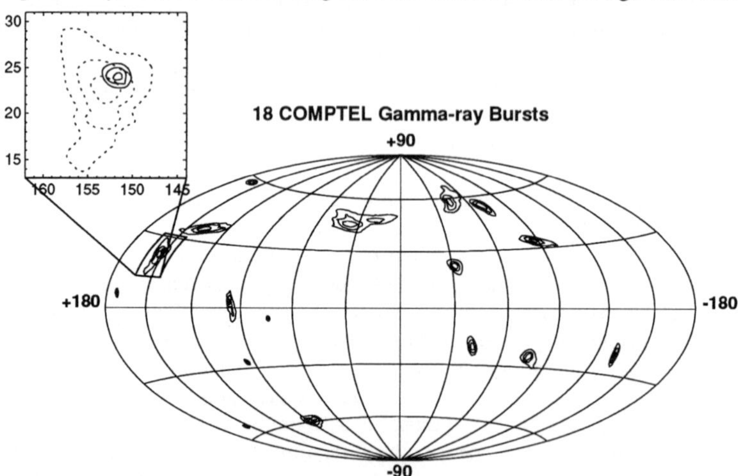

FIGURE 1. The Galactic distribution of 18 γ-ray bursts localized by COMPTEL. For each burst, 1σ, 2σ and 3σ statistical location uncertainty contours are shown. The enlarged region at left shows localizations of the spatially coincident bursts GRB 930704 (dashed contours) and GRB 940301 (solid contours).

spectra of the two bursts were shown to differ significantly using *Compton*-OSSE and BATSE data.[9] The remaining explanation is that the coincidence is a result of two distinct bursts from a single source. However, without arc-minute localization, the only way to confirm that this is a repeating burst source (i.e., $\gtrsim 4\sigma$ detection) and not an accidental coincidence would be the observation of a third outburst within the next few years.

SPECTRAL RESULTS & GRB 940217

The time-averaged energy spectra of most of the bursts observed by COMPTEL are well described by a single power-law model with spectral index in the range 1.5 to 3.5 — consistent with earlier results obtained with *SMM*.[10] However, at least two bursts show significant evidence of a "break" or transition in their spectra at MeV energies. This behavior agrees with BATSE findings that indicate most bursts have a smooth break in their spectra with turnover energies ranging from below 100 keV to above 1 MeV.[3] Attempts to supplement the keV-range BATSE data with the MeV spectra of COMPTEL have shown that spectral modeling is very sensitive to the high energy end of the spectrum and that simple models are insufficient to describe the wide-band spectra.[11]

The extraordinary γ-ray burst of 17 February 1994 was the strongest event yet observed by COMPTEL with fluence $S(> 0.3\text{ MeV}) = 2 \times 10^{-4}\text{ erg/cm}^2$. It exhibited complex, multi-peaked emission throughout the COMPTEL energy range lasting for more than 160 s. Rapid intensity fluctuations and spectral variations were observed throughout this event and during several of the individual pulses — including hard-to-soft evolution both within and between the pulses. Preliminary integrated photon spectra from the independent COMPTEL burst and telescope modes indicate a "break" in the spectral slope from ~2.1 to ~2.6 near 1 MeV.[12]

During the most intense pulse, measurement of significant (3σ) power-law emission up to 4.4 MeV coincident with rapid (100 ms) variability indicates that the source must be closer than 1.3 kpc in order to avoid the opacity of γ–γ interactions (assuming the emission is isotropic at the source).[13] If the source of this burst lies at cosmological distances (> 1 Gpc) significant relativistic beaming is required. This makes γ–γ opacity an unlikely cause of the observed break at ~1 MeV because the required relativistic motion would naturally force the break to be above the COMPTEL energy range.[14]

The *Compton*-EGRET instrument observed high energy photons (up to 18 GeV) from this burst for a period of ~120 minutes following the main 160 s interval — suggesting that high energy emission is extended in time beyond that at lower energies.[15] During the same post-burst interval, COMPTEL did not measure (due to its lower sensitivity) any significant emission from the direction of GRB 940217. The COMPTEL upper limits are consistent with the flux level observed by EGRET.

RAPID BURST RESPONSE

Several of the COMPTEL bursts have been localized within *hours* of their occurrence following timely notification from BATSE. The locations of such rapidly localized events are promptly distributed to a world-wide network of multiwavelength observatories. This "Rapid Burst Response" campaign[16] was initiated in July 1992 to take advantage of COMPTEL's good angular resolution, which allows relatively deep searches to be performed. Since then, continuous improvements have allowed deep searches for fading GRB counterpart emission in record time.

No *obvious* fading counterparts at any wavelength have yet been identified. The most constraining measurements were for GRB 940301,[17] where the absence of an optical counterpart in deep searches ($m_V \approx 16$) seven hours after the burst places an upper limit on the ratio of optical to γ-ray flux of 2.5×10^{-6}. This upper limit is more than three orders-of-magnitude lower than the best previous measurements. The lack of radio counterparts in a comprehensive follow-up study places significant constraints on currently favored cosmological models, which predict transient radio emission.[18]

COMPTEL observes new bursts at a rate of ~1/month and localization response-times continue to improve using automated burst notification via BACODINE.[19] In the near future COMPTEL locations of strong bursts (location accuracy ~0.5°) will be distributed to observers automatically in as little as ~10-15 min. For future bursts like GRB 940217, this capability will allow observations in optical and radio while GeV γ-rays are still being detected (e.g., by EGRET).

CONCLUSIONS

We have shown that COMPTEL γ-ray burst observations are providing information in areas where there are fundamental gaps in our knowledge. Unique measurements of individual bursts like GRB 940217 and GRB 940301 have given us a glimpse of the MeV emission process as well as providing important limits on low energy fading counterparts. Continued observations insure that in the near future COMPTEL measurements of the small-scale distribution of burst sources will provide constraining limits on recurrence.

REFERENCES

1. MEEGAN, C., *et al.* 1992. Nature **355:** 143-145.
2. STROHMAYER, T. E., E. E. FENIMORE & J. A. MIRALLES. 1994. Astrophys. J. **432:** 665-671.
3. BAND, D., *et al.* 1993. Astrophys. J. **413:** 281-292.
4. HANLON, L. O., *et al.* 1994. Astron. Astrophys. **285:** 161-178.
5. PACZYNSKI, B. 1990. Astrophys. J. **348:** 485-494.
6. FISHMAN, G. J. 1994. These proceedings.
7. RYAN, J. M., *et al.* 1994. Bull. Am. Astr. Soc. Vol. 26, **No. 2:** 881.
8. KIPPEN, R.M., *et al.* 1995. Astron. Astrophys. **293:** L5-L8.
9. HANLON, L. O., *et al.* 1995, Astron. Astrophys. Submitted.
10. MATZ, S. M., *et al.* 1985. Astrophys. J. **288:** L37-L40.
11. GREINER, J., *et al.* 1995. Astron. Astrophys. In press.
12. WINKLER, C., *et al.* 1995. Astron. Astrophys. In preparation.
13. SCHMIDT, W. K. H. 1978. Nature. **271:** 525-527.
14. BARING, M. 1993. Astrophys. J. **418:** 391-394.
15. HURLEY, K., *et al.* 1994. Nature **372:** 652-654.
16. KIPPEN, R. M., *et al.* 1994. *In* AIP Conf. Proc. No. 307, Gamma-Ray Bursts. G. J. Fishman, J. J. Brainerd & K. Hurley, Eds.: 418-422. AIP. New York, New York.
17. HARRISON, T.E., *et al.* 1995. Astron. Astrophys. In press.
18. FRAIL, D. A., *et al.* 1994. Astrophys. J. **437:** L43-L48.
19. BARTHELMY, S.D., *et al.* 1994. *In* AIP Conf. Proc. No. 307, Gamma-Ray Bursts. G. J. Fishman, J. J. Brainerd & K. Hurley, Eds.: 643-647. AIP. New York, New York.

Search for GRB counterparts

J. GREINER

Max-Planck-Institute for Extraterrestrial Physics, 85740 Garching, Germany

INTRODUCTION – SEARCH STRATEGIES

Historically, the search for γ-ray burst (GRB) counterparts began with the check of photographic exposures taken contemporaneously with the burst event[1]. Due to the lack of obvious associations of the GRB positions with well-known transients such as supernovae or flare stars, the search for correlations rapidly spread to very different energy ranges and detection methods. One example is the search for absorption events in the VLF (164 kHz) propagation in the Earth ionosphere caused by the transit of (ionising) X-ray emission from the bursts[2], which earlier proved successful for the detection of Sco X-1.

After these early (mid-seventies) attempts of simultaneous observations later on (eighties) the counterpart search was dominated by searches for non-correlated flaring as well as quiescent sources in different wavelength bands. This kind of investigation always used the implicit assumption that bursts repeat on timescales less than 50 years. Searches for archival optical transients in small GRB error boxes using large photographic plate collections were initiated at Harvard Observatory[3] and then also performed at other observatories[4,5]. In the same manner, data of the *Einstein* , *EXOSAT* and *ROSAT* satellites were used to search for quiescent soft X-ray sources[6–10]. In parallel, very few of the well-localised GRB error boxes of the second interplanetary network were examined with larger telescopes to search for faint quiescent optical counterparts[11,12]. No reliable counterpart candidate emerged from all these studies.

With the launch of the *Compton* Gamma-Ray Observatory (GRO) and the BATSE (Burst and Transient Source Experiment) measurements of about one burst per day improved possibilities emerged for the correlation with other (simultaneously performed) observing programmes. This includes the correlation of GRBs with (1) regularly exposed wide-field photographs of most major observatories[13], (2) a CCD-based multi-camera monitoring of 0.75 ster of the sky[14,15], (3) atmospheric airshower events due to TeV emission [16–19], (4) neutrino or muon detection rates[20–22]. Due to an unexpected incident (the failure of the GRO tape recorders) and the advent of fast data transmission, rapid follow-up observations of GRB error boxes became possible with the development of BACODINE[23] and the COMPTEL/BATSE/NMSU network[24]. In the following I will concentrate on the new developments connected with these rapid follow-up observations as well as the availability of times and locations of many bursts for correlated investigations (see also other recent reviews[25,26]).

SIMULTANEOUS OBSERVATIONS

Monitoring with Dedicated Automated Telescopes

The Explosive Transient Camera (ETC) is a wide (0.75 ster) field-of-view (FOV) CCD array consisting of 8 camera pairs which operates entirely by computer control on Kitt Peak National Observatory[27]. During its more than 4 years of operation there were five cases in which a BATSE GRB occurred during an ETC observation and within or near an ETC FOV[14,15]. No optical transients were detected during these observations, resulting in upper limits for the **fluence** ratio $L_\gamma/L_{opt} \geq$ 2–120[14,15]. Unfortunately, in all cases only a part (20%–80%) of the rather large BATSE error box was covered. For those GRBs also seen by other satellites, triangulation will reduce the error box size and may increase the actual coverage.

Photographic Sky Patrol

For the correlation of BATSE GRBs and photographic wide-field plates a logistic network of 11 observatories with regular photographic observations has been established, including Sonneberg, Tautenburg and Hamburg (all FRG), Calar Alto (Spain), Ondřejov (CR), Odessa, Crimea (Ukraine), Dushanbe (Tadshikistan), Kiso (Japan), Australian UK and ESO (La Silla, Chile). For nearly 60 GRBs detected by BATSE since the launch of GRO simultaneous plates with typically $m_{lim} \approx$2–3 mag for a 1 sec duration flash have been identified[13,28,29]. No optical transient was found, resulting in limits for the **flux** ratio $F_\gamma/F_{opt} \geq$ 1–20[13].

Neutrino and Muon Fluxes

Searches for neutrinos, anti-neutrinos and muons during times of BATSE GRBs have been performed using the data of the Irvine-Michigan-Brookhaven (IMB) detector[21], the Soudan 2 detector[20] and the Mont Blanc Neutrino Telescope[22]. Both samples are believed to contain primarily atmospheric events, i.e. neutrinos produced by the decay of secondary particles from cosmic-ray interactions in the Earths atmosphere. The IMB correlation with 183 Venera and Ginga GRBs revealed no neutrino coincidences. The resulting flux limit of 1.7×10^7 neutrinos cm^{-2} corresponds to 2.7×10^4 erg/cm^2 for neutrinos in excess of 200 MeV[21]. Also, no muon coincidences were found. A flux limit of 5.8×10^{-7} muons cm^{-2} from neutrino interactions during GRBs is derived[21].

A correlation of 180 BATSE GRBs with the Soudan 2 muon detector events revealed 18 trigger coincidences, near to the number expected by chance[20]. A correlation of the first 40 BATSE GRBs with events of the Mont Blanc telescope revealed no excess interactions. Depending on the neutrino flavour, the upper limits range between 2.5×10^{10} cm^{-2} s^{-1} ($\tilde{\nu}_e$ at $20\leq E_\nu \leq 50$ MeV) and 7.9×10^{12} cm^{-2} s^{-1} ($\tilde{\nu}_{\mu+\tau}$ at $20\leq E_\nu \leq 100$ MeV)[22].

TeV Emission

A correlation of the 1991/1992 season 0.4–4 TeV events observed with the 10m the Whipple Observatory Reflector with BATSE bursts has resulted in no positive detection which allows to set limits on the density of primordial black holes and the local density of cosmic strings[19].

The CYGNUS-I air shower array detects ultra-high-energy (UHE) radiation at about 3.5 triggers/sec for E>50 TeV from zenith and E>100 TeV for $\theta>30°$. For a total of 52 (out of 260) GRBs and 6 (out of 18) IPN GRBs having been in the CYGNUS FOV ($\theta<60°$) no evidence for UHE emission was found[16] The limits for 3 GRBs are inconsistent with an extrapolation of the BATSE spectra, indicating either a softening of the production spectrum at high energies or the presence of UHE γ-ray absorption. If the latter possibility is true this would imply cosmological GRB distances[16].

All GRBs of the BATSE 2B catalog with a zenith angle <60° and an location error <15° have been correlated with the HEGRA triggers. There are simultaneous data for 85 bursts but no significant excess could be found around the burst trigger times resulting in an upper limit between $10^{-7}...10^{-10}$ cm^{-2} s^{-1} for energies in the range of 40–500 TeV[17].

NEAR-SIMULTANEOUS OBSERVATIONS

Monitoring Projects

Due to the nature of the monitoring observations it is natural that the above discussed projects also allow searches for near-simultaneous coverage of burst positions. While simultaneous photographic exposures[28] (12 hours before or after the burst) are available for dozens of GRBs with typical limiting magnitudes of $m_{lim}\approx 7$ mag for 1 sec flash (and as short as about 1 hour after the burst), a recent ETC observation of GB 941014 happened to be only 150 sec after the burst onset. The 5 sec exposure with no optical transient sets a limit of $m_{lim}=10.2$ mag[15].

A HEGRA search in a time window of 12 hours after the bursts also revealed no significant excess rate[17]. This excludes strong extended TeV emission, as one could expect from the observation of the extended GeV emission from GB 940217[30]

Rapid Follow-up Observations

BACODINE[23] (BATSE Coordinates Distribution Network) automatically evaluates the real-time telemetry stream from GRO, calculates approximate coordinates for a BATSE burst and distributes this position to interested observers. The COMPTEL/BATSE/NMSU network[24] derives improved positions for those bursts occurring in the COMPTEL FOV using its imaging capabilities. The characteristics of both these systems are given in the Table.

	BACODINE	COMPTEL/BATSE/NMSU	HETE
Time delay	5 sec	16 min	5 sec
Location error	±10°	±1-2°	±10′
Rate	20–50 yr^{-1}	2-5 yr^{-1}	5–20 yr^{-1}

From the wealth of observational results I want to pick out only three examples, which should demonstrate the progress possible with the rapid burst notification: **(1)** The Gamma Ray Optical Counterpart Search Experiment (GROCSE) located at Lawrence Livermore National Laboratory adapted a wide-FOV telescope with a total of 23 cameras with $7^\circ\!.7\times11^\circ\!.5$ FOV each for optical burst follow-up observations[31]. During 0.5 sec integration time a limiting magnitude of $m_{lim}\approx8$ mag is reached, and images are taken at a repetition rate of every 5 sec. During the first year of operation 8 BACODINE triggers were received and observations of the burst locations started for two of these events while the GRB was still bursting. For event 1 (2) the first image was taken 24 (17) sec after the onset of the burst which lasted 40 (90) sec[32]. While the analysis of these events still is in progress, the limits achieved are considerably deeper than any previous observations.
(2) Within the observing campaign of GB 940301 by the BATSE/COMPTEL/NMSU rapid response network the COMPTEL error box was observed seven hours after the burst with the 1m Schmidt telescope at Socorro reaching a limiting magnitude of $m_V \approx 16$ mag[33]. Still, no optical transient was found. Radio observations at the same time at 8.42 GHz with the 34 m Goldstone antenna also found no new radio object[33].
(3) The bright gamma-ray burst GB 940301 was monitored at 1.4 GHz ($2^\circ\!.6$ FOV) and 0.4 GHz ($8^\circ\!.1$ FOV) at the Dominion Radio Astrophysical Observatory (DRAO) Synthesis Telescope starting starting 3 days after the burst[34]. A total of 245 radio sources were identified in the summed image in an intensity range between 0.8–110 mJy, but none was identified as a candidate for a flaring/fading radio counterpart based on variability analyses. The daily upper limits on the nondetection are $\approx$ 3.5 mJy at 1.4 GHz and 55 mJy at 0.4 GHz and constrain some fireball parameters for cosmological GRB models[34].

SUMMARY AND PROSPECTS

Despite the increasing efforts in the search for GRB counterparts no convincing candidate has been identified yet. While the limits for simultaneous emission are still not yet deep enough, those for up to a few hours after the bursts are getting constraining for models which predict 1) predict strong and/or long-duration afterglows, or 2) simple extrapolations of the measured X-ray to γ-ray spectra to lower and higher energies.

Observational improvements in the very near future can be expected mainly from (1) the next generation device of GROCSE which will allow observations with a smaller FOV down to $m_{lim}\approx$ 12–13 mag for a 1 sec flash[32], and (2) the launch of HETE[35] which will measure burst positions with better than $10'$ accuracy (see the Table) and after subsequent data transmission to ground observers will allow observations with smaller FOV telescopes (and thus fainter limiting magnitudes) than feasible up to now. This for the first time will allow optical observations with a sensitivity which is comparable to the flux estimates deduced by extrapolating burst spectra down from the X-ray/γ-ray range.

ACKNOWLEDGEMENTS

I'm grateful to H.-S. Park, S. Barthelmy, T. Harrison and R. Vanderspek for communicating their results prior to publication. JG is supported by the Deutsche Agentur für Raumfahrtangelegenheiten (DARA) GmbH under contract FKZ 50 OR 9201.

REFERENCES

1. Grindlay J.E., Wright E.L., McCrosky R.E., 1974. ApJ **192**, L113
2. Kasturirangan K., Rao, U., Sharma, D. et al. 1974. Nat. **252**, 113
3. Schaefer B.E., Bradt, H., Barat, C., et al. 1984. ApJ **286**, L1
4. Hudec R., Borovicka, J., Wenzel, et al. 1987. A&A **175**, 71
5. Greiner J., Flohrer J., Wenzel W., Lehmann T., 1987. ApSS **138**, 155-171
6. Pizzichini G., Gottardi, M., Atteia, J., et al. 1986. ApJ **301**, 641
7. Boër M., Hurley, K., Pizzichini, G., et al. 1991. A&A **249**, 118
8. Greiner J., Boër M., Motch C., et al. 1991. 22nd ICRC Dublin, vol. **1**: 53-56
9. Greiner J., et al. 1995. NATO ASI C450, eds. M.A. Alpar et al. , Kluwer, p. 519-522
10. Boër M., Greiner J., Kahabka P., et al. , 1993. A&A Suppl. **97**, 69
11. Vrba F., Hartmann D.H., Jennings M.C., 1995. ApJ (in press)
12. Sokolov V., et al. 1995. Flares and Flashes, eds. J. Greiner et al. , LNP (in press)
13. Greiner J., et al. 1994. *Gamma-Ray Bursts*, eds. G.J. Fishman et al. , AIP 307: 408-412.
14. Krimm H. et al. 1994. *Gamma-Ray Bursts*, eds. G.J. Fishman et al. , AIP 307: 423-427.
15. Vanderspek R. et al. 1995. Flares and Flashes, eds. J. Greiner et al. , LNP (in press)
16. Alexandreas D.E., Allen G.E., Berley D., et al. (CYGNUS Collab.), 1994. ApJ **426**, L1
17. Matheis V., 1995. Ph.D. thesis, Heidelberg University
18. Aglietta M. et al. (EASTOP Collab.) 1993. 23rd ICRC, Calgary 1993, vol. **1**: 61
19. Connaughton V. et al. 1994. *Gamma-Ray Bursts*, eds. G.J. Fishman et al. , AIP 307: 470
20. DeMuth D.M. et al. 1994. *Gamma-Ray Bursts*, eds. G.J. Fishman et al. , AIP 307: 475
21. LoSecco J.M., 1994. ApJ **425**, 217
22. Aglietta M., Antonioli P., Badino G., et al. 1993, 23rd ICRC, Calgary 1993. vol. **1**: 69
23. Barthelmy S. et al. 1994. *Gamma-Ray Bursts*, eds. G.J. Fishman et al. , AIP 307: 643
24. Kippen R.M., Ryan J., Connors A. et al. , 1995. this volume
25. Schaefer B.E., 1994. *Gamma-Ray Bursts*, eds. G.J. Fishman et al. , AIP 307: 382-391.
26. Hartmann D.H., 1995. *Flares and Flashes*, eds. J. Greiner et al. , LNP (in press)
27. Vanderspek R. et al. 1994. *Gamma-Ray Bursts*, eds. G.J. Fishman et al. , AIP 307: 438
28. Greiner J., et al. 1995. *Flares and Flashes*, eds. J. Greiner et al. , LNP (in press)
29. Hudec R., 1995. *Flares and Flashes*, eds. J. Greiner et al. , LNP (in press)
30. Hurley K., Dingus B.L., Mukherjee R., et al. 1994. Nat. **372**, 652-654
31. Akerlof C. et al. 1994. *Gamma-Ray Bursts*, eds. G.J. Fishman et al. , AIP 307: 633-637.
32. Park H.-S., 1994. AAS HEAD meeting, Napa Valley, Nov. 1994
33. Harrison T.E., McNamara B.J., Pedersen H., et al. 1995. A&A (in press)
34. Frail D.A., Kulkarni S.R., Hurley K.C., et al. 1995. ApJ **437**, L43-L46
35. Ricker G. et al. 1988. *Nuclear Spectroscopy of Astrophysical Sources*, eds. N. Gehrels et al. , AIP 170: 407

Global Distribution Constraints on GRB Models

D. H. HARTMANN[a], M. S. BRIGGS[b], & G. N. PENDLETON[b]

[a] *Department of Physics and Astronomy*
Clemson University, Clemson, SC 29634

[b] *Department of Physics*
University of Alabama, Huntsville, AL 35899

INTRODUCTION

The mystery of the nature and origin of gamma-ray bursts (GRBs) has persisted over two decades of intensive research, observational and theoretical. The basic problem is the absence of quiescent or transient counterparts at other wavelengths, and the absence of an established distance scale. The recent transition from the previous paradigm of Galactic neutron stars to the new paradigm of cosmological sources was generated by the global statistical properties of GRBs. The non-uniform brightness distribution combined with the angular isotropy argues against Galactic source distributions and favors cosmological origins. However, the situation is not as straightforward as it appears. Large-scale isotropy does not exclude clustering on small angular scales, and clustering could reflect a spatial or temporal property of bursts. Burst repetition is hard to explain in some of the currently favored cosmological scenarios, but does not rule out a cosmological origin. A population of sources residing in an extended Galactic halo has been considered as an alternative explanation. BATSE data continue to tighten the constraints at an approximate rate of one burst per day.

This paper summarises the observational constraints on GRB models derived from their global statistical properties. Other papers in this workshop discuss the observational details of light curves (Kouveliotou *et al.*) and spectra (Briggs *et al.*). The recent discovery of extremely delayed emission (90 minutes) of photons with energies extending up to ~ 20 GeV by the EGRET Team (Hurley *et al.* 1994) places tight constraints on photon generation and transport (near the source as well as through the intergalactic medium), and impacts the question of source beaming. The EGRET constraints are discussed in this session by Schneid *et al.*, and the art of modeling GRBs given current observational constraints is discussed in the two papers by Meszaros and Woosley. Searches to identify "optical" counterparts (sofar unsuccessful) are reviewed by Greiner.

THE BRIGHTNESS DISTRIBUTION

The observed brightness distribution of GRBs constrains spatial distribution models, although the lack of an established GRB luminosity function complicates the interpretation. Progress in analyses of the brightness data was due to the introduction of the so-called V/Vmax test (Schmidt, Higdon, & Hueter 1988). The GRB intensity statistic obtained with this tool removes selection effects due to uncertainties in the spectral deconvolution process and allows an individual treatment of events that takes variations in detector backgrounds into account. The V/Vmax test has been applied to data sets from HEAO-A1 (Schmidt, Higdon, & Hueter 1988), KONUS on Venera 11/12 (Higdon & Schmidt 1990), SMM (Matz, *et al.* 1992), SIGNE on Venera 11/12 (Atteia, *et al.* 1991), the Lilas/Apex experiments on the Soviet Phobos mission (Atteia, *et al.* 1991), the PVO burst detector

(Hartmann, *et al.* 1992; Chuang *et al.* 1992), PHEBUS-GRANAT (Terekhov *et al.* 1994), the Ginga detector (Ogasaka *et al.* 1991), and the BATSE experiment (Meegan *et al.* 1992). If the observed bursts are uniformly distributed in Euclidean space, then a uniform V/Vmax distribution with mean value 1/2 should be obtained. The data clearly imply that the γ-ray burst space distribution is inhomogeneous and that faint bursts are systematically deficient.

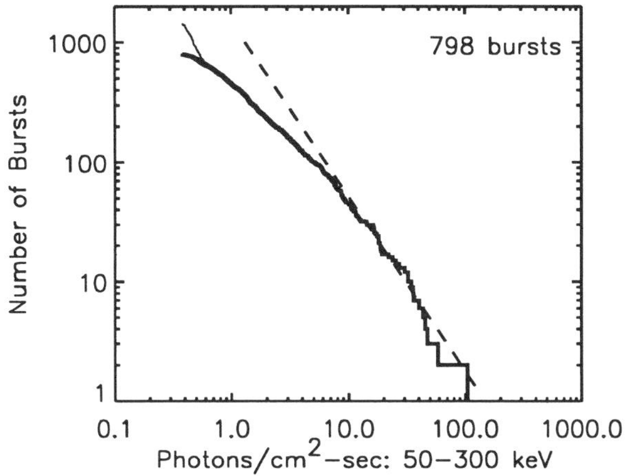

FIGURE 1. Cumulative distribution of burst brightness (apparent peak energy flux on the 256 ms timescale in the energy interval from 50 keV to 300 keV) for 798 bursts observed with BATSE (Pendleton *et al.* 1995). A spatially uniform distribution of sources would yield a power law with index $-3/2$ (dashed line), which obviously only holds for the brighter bursts. Selection effects cause some uncertainties near the detector threshold, but the break in the distribution occurs well above that flux level.

While the V/Vmax statistic is sufficient to establish the overall fact that we are observing the "edge" of the GRB distribution in space, it is not appropriate for direct comparisons to physical models predicting energy- or photon fluxes. The BATSE Team has provided peak fluxes (photons $\mathrm{cm}^{-2}\ \mathrm{s}^{-1}$) for $\sim$ 800 bursts (Pendleton *et al.* 1995), whose cumulative distribution function is shown in Figure 1. The brightness distribution derived from BATSE observations was combined with data from PVO (Fenimore *et al.* 1992) to extend the observed range to higher fluxes (PVO had more exposure than BATSE, due to a longer lifetime and less sky blockage). The combination of these two datasets implies that BATSE samples to a redshift of order unity. Furthermore, the observed distribution is consistent with the assumption of negligible source evolution in density and/or luminosity for a standard Friedman universe with zero cosmological constant (e.g., Wickramashinge *et al.* 1992; Fenimore *et al.* 1992). Emslie & Horack (1994) relaxed the standard candle assumption and showed that the luminosity function must be fairly narrow in that case (see also Horack *et al.* 1994). On the other hand, it seems likely that source evolution would occur in nature (it does in many of the proposed models) and that the observed brightness distribution in that case would suggest sampling redshifts that could be significantly higher than unity, and a cosmological constant that may be non-zero (Horack, Emslie, & Hartmann 1995). Thus, whether GRBs reside in an extended Galactic halo or associate with the large scale structure of the universe, the peak flux distribution provides significant constraints that theorists must not ignore.

THE ANGULAR DISTRIBUTION

Lacking reliable counterpart identifications, one turns to global properties of γ-ray bursts to derive a rough distance scale (Epstein & Hurley 1988). The apparent distribution of sources on the sky provides clues to the distance scale. Any correlation of source positions with Galactic features (plane, poles, center, globular clusters, etc.) or extragalactic features (nearby galaxies, galaxy clusters, super-galactic plane, etc.) would be sufficient to deduce at least a rough distance scale, and thus indirectly suggest the most likely energy sources. The angular distribution of 1005 GRBs observed with the BATSE instrument is shown in Figure 2. Since the eye does not discern any deviations from isotropy, no obvious hint of the GRB distance scale is contained in these data.

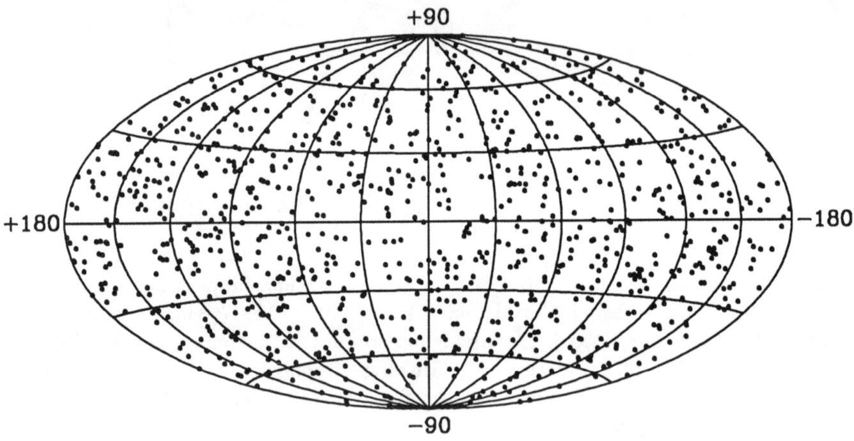

FIGURE 2. The sky distribution (in Galactic coordinates) of the first 1005 gamma ray bursts observed by BATSE shows no evidence for deviations from isotropy exceeding the 1 sigma level (Briggs *et al.* 1995).

The angular distribution of bursts on the sky has been analyzed via a low order multipole expansion (Hartmann & Epstein 1989; Briggs 1993). The first 1005 GRBs observed by BATSE have been examined for dipole and quadrupole moments in a coordinate frame dependent as well as frame-independent manner (Briggs *et al.* 1995). The only marginal detection of a moment was due to the known uneven sky exposure of the detectors (confirming that the statistical techniques would find an underlying anisotropy at a $\sim$ 10% level), but no concentration towards the Galactic plane or center were found. Briggs *et al.* (1995) compared the observational data against Galactic models that were developed after the launch of BATSE in 1991. The majority of models (often just geometric constructions of assumed density profiles) did not fit the data within the two sigma significance level, and the best extended halo model (see discussion below) predicted a dipole moment to the Galactic center 1.4 sigma larger than the observed value. These authors (see also Hakkila *et al.* 1994) find that although a spherical dark matter halo with Hubble profile could in principle meet the constraints, the required core radius was much larger than commonly inferred from other tracers of the Galactic mass distribution. Of course, GRBs may not correlate with either the luminous or the dark matter component of the halo, but constitute a dynamically unimportant component whose density profile is very different from the profiles of halo field stars, globular clusters, or the rotation curve of the Galactic disk.

To save the galactic hypothesis under the tight constraints of isotropy one is driven to consider an extended halo population with a sampling distance of at least 200 kpc. On the other hand the sampling distance should not be much larger than 200 kpc, because otherwise BATSE should have observed an excess of bursts from the direction of M31 (e.g., Hakkila *et al.* 1994; Briggs *et al.* 1995). Creating an extended halo of such dimensions could occur in several distinct ways. Either the GRB source objects were created only during the relatively short birth period of the Galaxy or in a continuous process lasting perhaps to the present time. In the latter case formation of the progenitors could occur through injection of high velocity objects from the disk or from the halo. Obviously, the disk option requires somewhat higher injection velocities. Injection of very high velocity neutron stars from the disk was favored by Li & Dermer (1992) and Podsiadlowski, Rees, & Ruderman (1995). Births of neutron stars with somewhat lower velocities in the halo was favored by Eichler & Silk (1992) and Hartmann (1992). The disk-injection picture faces severe constraints. To avoid a large number of bursts in the disk (which would generate an unacceptable quadrupole moment) one requires that bursts should not occur immediately after birth of the star, since at that time they are still very much concentrated towards the Galactic plane, and that the great majority of burst sources have velocities comparable or greater than Galactic escape velocity ($\sim$ 600 km s^{-1}). Only a small fraction of all neutron stars born in the disk appear to have such high velocities (Lyne & Lorimer 1994).

Isotropy requires a minimum size of the expanding neutron star cloud of a few 100 kpc, but to avoid an excess of bursts from the direction of M31 the detector sampling distance is restricted to less than a few 100 kpc. Apparently fine-tuning of the parameters is required to match the data. The Galactic dipole moment of the best-fit model (Podsiadlowski *et al.* 1994) indeed agrees with the BATSE data to within 1.4 sigma significance (Briggs *et al.* 1995). However, a velocity of order 10^3 km s^{-1} implies that all stars older than a few 10^8 yrs will have left the observed volume. Given a birth rate of one pulsar every 100 yrs only about 1 million stars were created during that time interval. Since only a few percent of these stars have the required velocities we only have about 10^4 neutron stars generating the observed rate of 10^3 bursts per year. Each burst source should thus repeat, on average, every 10 yrs. Unless the burst clock is very regular this recurrence rate appears to be in conflict with the constraints on small-scale clustering which would result from burst recurrence (Meegan *et al.* 1995; Hartmann *et al.* 1995). If beaming is significant the problem of small source numbers becomes worse. However, if future pulsar motion studies demonstrate that in fact $\sim$ 10% of all pulsar are born with velocities above the Galactic escape velocity, recurrence times of $\sim$ 100 yrs would not be a problem.

Since isotropy argues so strongly against a Galactic origin, the now most frequently adopted option is to place the sources at cosmological distances. If γ-ray bursts indeed trace the large scale matter distribution in the universe, then the observed lack of small-scale angular correlations can be used to derive a minimum sampling depth for cosmological models (Hartmann & Blumenthal 1989; Blumenthal *et al.* 1994). If the underlying spatial clustering resembles that of galaxies or galaxy clusters, we must observe events to distances in excess of $\sim$ 200 Mpc, corresponding to a redshift of $\sim$ 0.05. Fits of the brightness distribution discussed in the previous section as well as observational evidence of time dilation (Norris *et al.* 1994) suggests that BATSE samples bursts to redshifts of z $\sim$ 1, consistent with the angular correlation limit.

However, these arguments may not apply if bursts trace cosmological dark matter or perhaps a new component of the intergalactic material for which we do not even have a hint about its clustering properties. The most straightforward assumption is to associate GRBs with phenomena occurring inside (or near) galaxies. Accreting black holes, supernovae of some kind, or merging compact objects are popular building blocks of current theoretical models (e.g., Hartmann & Woosley 1995). If redshift limits are of order unity than the total number of spiral galaxies within the sampled volume is of order 10^9 (e.g., Hartmann 1994). The inferred BATSE rate of $\sim 10^3$ bursts per year (full sky average) thus implies one event per host galaxy every 10^6 yrs. While supernova models yield much larger rates, the estimates of compact object merger rates are of the same order. Thus geometric beaming into a small fraction of the sky (as opposed to beaming of the radiation due to relativistic expansion) becomes an important issue. Beaming would allow more photons of higher energy to escape the source region and would also reduce the overall energy requirements. However, it also requires a larger number of sources, which may be hard to accomplish with some models. An optical constraint on extragalactic burst models (Schaefer 1991; see also Fenimore *et al.* 1992), based on the absence of bright galaxies inside γ-ray burst error boxes, appears to imply even larger distances of $\sim$ 1 Gpc, but is confined to models in which bursts would occur in M31-like galaxies. However, the majority of galaxies is much fainter than M31. Inventories of several small GRB error boxes down to 24th V magnitude (e.g., Vrba, Hartmann, & Jennings 1995) do not rule out galactic counterparts, and even suggest that QSOs might be linked to GRBs.

Small-scale clustering consistent with the instrumental point-spread function could be evidence for repeating sources of GRBs. Evidence for repetition in the first BATSE catalog was found by Quashnock & Lamb (1993) and by Wang & Lingenfelter (1993). However, these discoveries are not confirmed in the second BATSE catalog (Meegan et al. 1995, Brainerd et al. 1995, Hartmann et al. 1995). Since in almost all currently popular models of GRB origin at cosmological distances a burst destroys its source, further evidence of the third BATSE catalog is eagerly awaited.

From large-scale isotropy constraints we conclude that bursts most likely reside in either a very extended halo (200 kpc) or are truly cosmological. While a Galactic origin can not be ruled out, it is clear that well established source populations associated with the Galactic disk are not likely to be responsible for the GRB phenomenon. The debate about the distance scale of GRBs is not yet settled, but it is clear that neither the local (Oort cloud), or Galactic (disk, halo, extended halo), or cosmological scale yield solutions to the GRB puzzle that are in concord with all of the observational data. The Shapley-Curtis debate (1920) on nebulae was eventually settled, and the Lamb-Paczynski debate (1995) on GRBs hopefully has a similar fate in the not too far future. The constraints derived from an increasing database obtained with BATSE and other instruments will be essential for this debate, and probably for its solution, although the holy grail of GRB studies remains the unambiguous identification of even a single counterpart.

This work was supported in part by NASA grant NAG 5-1578.

REFERENCES

Atteia, J.-L., *et al.* 1991, Nature, 351, 296

Blumenthal, G. R., Hartmann, D. H., & Linder, E. V. 1994, in *Gamma Ray Bursts*, AIP Proceedings **307**, ed. G. J. Fishman, J. J. Brainerd, & K. Hurley, p.117

Brainerd, J. J., *et al.* 1995, ApJL, in press

Briggs, M. S. 1993, ApJ 407, 126

Briggs, M. S., *et al.* 1995, ApJ, submitted
Chuang, K., White, R., Klebesadel, R., & Laros, J. 1992, ApJ, 391, 242
Eichler, D., & Silk, J. 1992, Science, 257, 937
Emslie, A. G., & Horack, J. M. 1994, ApJ, 435, 16
Epstein, R. I., & Hurley, K. 1988, *Astr. Lett. Comm.*, 27, 229
Fenimore, E. E., *et al.* 1992, Nature, 366, 40
Hakkila, J., *et al.* 1994, ApJ, 422, 659
Hartmann, D. H. 1992, Comm. Astrophys., 16, 231
Hartmann, D. H. 1994, in *The gamma-ray sky with COMPTON and SIGMA*, (Kluwer Acad. Publ.), eds., M. Signori, P. Salati, and G. Vedrenne, in press
Hartmann, D. H., & Epstein, R. I. 1989, ApJ, 346, 960
Hartmann, D. H. & Blumenthal, G. R. 1989, ApJ, 342, 521.
Hartmann, D. H., *et al.* 1992, in *Gamma Ray Bursts*, (Cambridge Univ. Press), eds. C. Ho, R. I. Epstein, and E. E. Fenimore, p. 45
Hartmann, D. H., *et al.* 1995, ApJ, in press
Hartmann, D. H., & Woosley, S. E. 1995, Proc. of 30th COSPAR Sci. Assembly, (Pergamon Press), eds. K. Pollock & N. Gehrels, in press
Higdon, J. C., & Schmidt, M. 1990, ApJ, ApJL, 355, L13
Horack, J. M., Emslie, A. G., & Meegan, C. A. 1994, ApJL, 426, L5
Horack, J. M., Emslie, A. G., & Hartmann, D. H. 1995, ApJ, in press
Hurley, K., *et al.* 1994, Nature, 372, 652
Li, H., & Dermer, C. 1992, Nature, 359, 514
Lyne, A. G., & Lorimer, D. R. 1994, Nature, 369, 127
Matz, S. M., *et al.* 1992, in *Gamma Ray Bursts*, ed. C. Ho, R. Epstein, & E. Fenimore, (Cambridge Univ. Press), 175
Meegan, C. A., *et al.* 1992, Nature, 355, 143
Meegan, C. A., *et al.* 1995, ApJL, in press
Norris, J. P., *et al.* 1994, ApJ, 424, 540
Ogasaka, Y., *et al.* 1991, ApJL, 383, L61
Pendleton, G., *et al.* 1995, ApJ, in preparation
Podsiadlowski, P., Rees, M. J., & Ruderman, M. 1995, MNRAS, in press
Quashnock, J. M., & Lamb, D. Q. 1993, MNRAS, 265, L59
Schaefer, B. E. 1991, in *Gamma Ray Bursts*, ed. C. Ho, Cambridge Univ. Press, p. 107
Schmidt, M., Higdon, J. C., & Hueter, G. 1988, ApJL, 329, L85
Terekhov, O. V., Denisenko, D. V., Lobachev, V. A., Sunyaev, R. A., & Kovtun, A. V. 1994, Astron. Letters, 20, 265
Vrba, F. J., Hartmann, D. H., & Jennings, M. C. 1995, ApJ, in press
Wang, V. C., & Lingenfelter, R. E. 1993, ApJ, 416, L13
Wickramasinghe, W. *et al.* 1993, ApJL, 411, L55

Gamma-ray Burst Models: General Requirements and Predictions

P. MÉSZÁROS

Pennsylvania State University, 525 Davey Laboratory, University Park, PA 16803, USA

ABSTRACT

Whatever the ultimate energy source of gamma-ray bursts turns out to be, the resulting sequence of physical events is likely to lead to a fairly generic, almost unavoidable scenario: a relativistic fireball that dissipates its energy after it has become optically thin. This is expected both for cosmological and halo distances. Here we explore the observational motivation of this scenario, and the consequences of the resulting models for the photon production in different wavebands, the energetics and the time structure of classical gamma-ray bursters.

1. OBSERVATIONAL REQUIREMENTS

A main requirement of almost any model for gamma-ray burst sources (GRB), which has been long appreciated, is that the "working surface" must be expanding due to radiation pressure. This is expected either for an extended galactic halo ($D \lesssim 10^{24}$ cm) or for a cosmological ($D \sim 10^{28}$ cm) spatial distribution (distances much smaller than 100 kpc are not favored by current observational limits on the isotropy of the location distribution, e.g. Fishman, 1994). For a characteristic observed fluence $F = 10^{-6} F_{-6}$ ergs cm^{-2} the energy in a solid angle θ^2 is

$$E = 10^{43} F_{-6} D_{24}^2 \theta^2 = 10^{51} F_{-6} D_{28}^2 \theta^2 \text{ ergs,} \tag{1}$$

so the luminosity is highly super-Eddington, $E/t_b \gg 10^{38}(M/M_\odot)$ erg s^{-1} for the characteristic burst durations $t_b \gtrsim$ s and masses $M \sim M_\odot$ expected for most progenitors. Coupled with the fact that gamma-ray photons above $e^\pm$ pair threshold constitute most of the observed flux, this led to the concept of an expanding pair-photon fireball (Cavallo and Rees, 1978, Paczyński , 1986, Goodman, 1986). However, from the above information alone it is not possible to determine whether the expansion will be slow or fast. If the baryon load of the radiation-pressure ejected shells were large ($E/Mc^2 \lesssim 1$, the expansion would be subrelativistic, and the flow would remain optically thick, leading to a degradation of the gamma-rays (Paczyński , 1990).

A second element which must enter any successful GRB scenario is implied by the detection in many GRB of photons with energies $\epsilon_\gamma \gtrsim 1$ GeV. Pair formation sets in for

$$\epsilon_\gamma > 2(m_e c^2)^2 [E_t(1 - \cos\alpha)]^{-1} \simeq 4(m_e c^2)^2 \epsilon_t^{-1} \alpha^{-2} \tag{2}$$

in the laboratory frame, where ϵ_t is the lab frame target photon energy and α is the relative angle of the two photons. However in the same frame, causality implies

$\alpha \lesssim \Gamma^{-1}$ where Γ is the bulk Lorentz factor. Therefore $\Gamma^{-1} \gtrsim \alpha \gtrsim 2m_e c^2/\sqrt{\epsilon_t \epsilon_\gamma}$, or since $\epsilon_t \sim 1$ MeV is where much of the lab-frame GRB photons are,

$$\epsilon_{\gamma,\ \mathrm{MeV}} \lesssim 10^4 \epsilon_{t,\ \mathrm{MeV}}^{-1} \Gamma_2^2 \ \mathrm{MeV}\ . \tag{3}$$

This (e.g. Fenimore, Ho and Epstein, 1993, Harding and Baring, 1994) is sometimes referred to as the need for a "beaming" characterized by a factor Γ. However, it must be stressed that this refers only to relative photon angles – the GRB could perfectly well be emitting isotropically. (The word "beaming" can be misleading, and is better reserved for actual channeling of the total emission into a restricted solid angle). In any case, one infers a highly relativistic expansion, $\Gamma \gg 1$, and this in turn implies that, somehow, the GRB deposits much of its radiation energy in a low density region where it significantly exceeds the baryon rest mass energy density. This "low baryon loading" is *required* by the observation of high energy photons $\epsilon_\gamma \gtrsim 0.1 - 1$ GeV.

A third requirement of a satisfactory GRB model is that it must account for the generally non-thermal appearance of the spectra, strongly suggesting an optically thin spectrum from power-law relativistic electrons. In principle a thermal electron distribution in a scattering-thick medium could produce a power law up to an energy $\sim 3kT_e$, but in order to have the photon power law extend to GeV energies the "temperature" would have to be of the same order, and at these energies nonthermal distributions are more likely. The requirement is therefore very likely to be that the emitting plasma is optically thin and nonthermal. The comoving electron density is $n_e' = L/4\pi r^2 m_p c^3 \eta \Gamma$, where primes denote comoving-frame (CF) quantities, and the scattering depth $\tau_e \sim n_e' \sigma_T r/\Gamma$ must satisfy

$$\tau_e \simeq L\sigma_T/(4\pi r m_p c^3 \eta \Gamma^2) \sim 1\ r_{12}^{-1} L_{51} \eta_2^{-3} \lesssim 1\ , \tag{4}$$

requiring the observed radiation to be produced at radii $r_{rad} \gtrsim 10^{12} r_{12} L_{51} \eta_2^{-3}$ cm (cosmological) or $r_{rad} \gtrsim 10^7 r_7 L_{43} \eta_1^{-3}$ cm (halo).

An additional model requirement is that, for photons observed above pair threshold, the "compactness parameter" (or photon-photon optical depth) must be $\tau_{\gamma\gamma} < 1$ at the radius r_{rad}. In the lab-frame (LF) we have $\tau_{\gamma\gamma} \simeq n_\gamma' \sigma_T r/\Gamma$, where the CF photon density in the frame moving with Γ is $n_\gamma' = L'/4\pi r^2 c\epsilon_\gamma' = L/4\pi r^2 c\epsilon_\gamma \Gamma$. Thus

$$\tau_{\gamma\gamma} \simeq L\sigma_T/(4\pi r c\epsilon_\gamma \Gamma^2) \lesssim 1 \tag{5}$$

for photons observed above the threshold for that Γ. Otherwise, the above equation implies a cutoff of the photon spectrum above ϵ_γ. For the much smaller radii $r_o \sim 10^6 r_6$ cm expected as the initial size of the "primary" GRB event from a neutron star progenitor, the compactness parameter is large at MeV energies, which implies that the initial stages of an impulsive fireball (or for a continuous input, the lower portion of the wind) is an optically thick $e^\pm\gamma$ flow, which eventually becomes thin at larger radii.

2. THE GENERIC GRB MODEL

The observational requirements discussed above provide very strong evidence for a relativistically expanding fireball scenario. This has some straigthforward consequences. The bulk Lorentz factor initially grows linearly with the radius, $\Gamma \propto r$, until the plasma becomes subrelativistic in its own rest frame. After this Γ saturates to the average value of the initial radiation to rest-mass ratio, $\eta = E/Mc^2 = L/\dot{M}c^2$,

$\Gamma \to \eta$. The saturation radius $r_s \sim \eta r_o$ is usually much smaller than the radiation radii r_{rad} required by the (optically thin spectrum) observations.

One problem faced by simple ($\lesssim$ 1992) fireball models, caused by the saturation phenomenon, is that beyond the radius r_s most of the initial energy has been converted to kinetic energy of the baryons, the radiation energy content decreasing adiabaticaly. This would raise enormously the initial energy required to explain the observed photon luminosity (Paczyński , 1990), and in addition it leads to photon energy degradation. Another major problem with simple fireball models (e.g. Paczyński , 1986, Shemi and Piran, 1990) is that the only radiation observed would be from the fireball photons that escape at the optical thinness radius, and these would have a quasi-thermal spectrum. In addition, if the initial event has a timescale comparable to a neutron star collapse dynamic time $r_o/c \sim 10^{-3}$ s, the observed time over which the fireball becomes thin is also comparable (Goodman, 1986). Both the spectrum and the timescale would be unacceptable.

There are, however, at least two ways in which a relativistic fireball can produce a longer ($10^{-1} - 10^3$ s) duration, nonthermal γ-ray burst with reasonable energy and spectrum. One of them results from the fact that the baryons entrained in the ejecta will eventually have to run into an external medium, and there they will reconvert their kinetic energy into radiation (e.g. Mészáros , Laguna and Rees, 1993; Rees and Mészáros , 1992; Katz, 1994). The external medium may either be a pre-ejected wind from the progenitor, or the interstellar medium. If its density is n_o cm^{-3}, deceleration occurs at a radius

$$r_{dec} \sim 10^{17}(E_{51}/n_0)^{1/3}(\theta\eta_2)^{-2/3} \text{ cm} \tag{6}$$

and the time-delayed LF duration of the burst is

$$t_{dec} \sim\sim r_{dec}/c\Gamma^2 \sim 5 \times 10^2 (E_{51}/n_0)^{1/3}\theta^{-2/3}\eta_2^{-8/3} \text{ s.} \tag{7}$$

Here the total (initial) energy $E \sim 10^{51}E_{51}$ in a solid angle θ^2 is assumed to be released during an intrinsic time shorter than t_{dec} (otherwise, the intrinsic timescale t_w would be the observed duration). The total photon energy produced in the deceleration, for very modest subequipartition magnetic fields which ensure high radiative efficiency, is the entire baryon kinetic energy, comparable to the initial burst radiative energy. The large radius $r_{rad} \sim r_{dec}$ ensures optical thiness, and the strong deceleration and reverse shocks ensure ideal conditions for relativistic particle acceleration leading to synchrotron and inverse Compton (IC) nonthermal radiation.

In addition, shocks may also arise internally in the ejecta itself, before any deceleration by the external medium occurs. Such internal shocks could arise for various reasons. For instance, the energy or matter input may be time-variable, so that higher Γ shells overtake lower Γ shells (e.g Paczyński and Xu, 1994, Rees and Mészáros , 1994). If the energy or matter input at the base of the wind occurs during an intrinsic time t_w but is modulated on some timescale t_v (shorter than t_w) with $\Delta\Gamma \sim \Delta\eta \sim \eta$ around an average final Lorentz factor $\Gamma \sim \eta$, an overtaking collision (internal dissipation shock) occurs at

$$r_{dis} \sim ct_v\eta^2 \sim 3 \times 10^{14} t_v \eta_2^2 \text{ cm.} \tag{8}$$

This occurs beyond the wind Thomson photosphere

$$r_{ph} \sim \dot{M}\sigma_T/4\pi m_p c^2 \eta^2 \sim 10^{12} L_{51}\eta_2^{-3} \text{ cm} \tag{9}$$

(larger than the saturation radius) for $0.3(L_{51}/t_v)^{1/5} \lesssim \eta_2 \lesssim 10^2(L_{51}/t_v)^{1/4}$. Also, in general $r_{dis} < r_{dec}$. For the magnetic fields turbulently generated in the shocks, or for a young pulsar wind, the radiative efficiency of the internal shocks is sufficient to radiate an appreciable fraction of the wind bulk kinetic energy. Other mechanisms for randomizing the wind energy might be the dissipation of Alfvén turbulence beyond the photosphere (Thompson, 1994; see also Duncan and Thompson, 1992) or the conversion of Poynting flux into photons after the charge density becomes insufficient to sustain the frozen-in magnetic field of a young pulsar type wind (Usov, 1994, 1992). Convective Rayleigh-Taylor instabilities may also arise and become nonlinear beyond the saturation radius (Waxmann and Piran, 1994). Below the photosphere this would become pressure that reaccelerates the flow, while above the photosphere it would be expected to result in freely coasting shells of different η as discussed above.

3. SOME OBSERVATIONAL CONSEQUENCES

What are some of the predictions of the dissipative relativistic fireball scenario?

One consequence of models based on this scenario is that emission is expected at energies other than in gamma-rays. A simultaneous burst of the same duration and low but significant fluence is predicted at X-ray and optical energies (Mészáros and Rees, 1993b). Breaks in the photon power-law spectrum are also predicted (Mészáros , Rees and Papathanassiou, 1994), in qualitative agreement with γ-ray spectra (Band, *et al.* , 1994), and detailed comparison of such calculations with observations could provide useful insights into the source physics. For a typical MeV fluence $F_\gamma \sim 10^{-6}$ ergs cm^{-2} the optical and X-ray fluences predicted for cosmological models are compatible with the few detected X-ray flashes as well as the lack of widespread X-ray identifications (X-ray paucity), and with the lack (so far) of optical detections. Typically , however, they are above the expected HETE threshold (Ricker, *et al.* , 1994) of $\sim 10^{-10}$ ergs cm^{-2}. As an example, for some cosmological models in the optical U-band, $m_U \sim 11 - 2.5\log(F_u/10^{-9}\ \text{ergs cm}^{-2}) + 2.5\log(t_b/\ \text{s})$, e.g. Mészáros , Rees and Papathanassiou, 1994. Other satellite or ground-based observations triggered by BATSE via systems such as BACODYNE could, for appropriate sensitivities and pointing times less than the burst duration, also detect the GRB at other wavelenghts. Previous attempts at simultaneous optical detection (e.g. Vanderspek, *et al.* , 1994, Krimm, *et al.* , 1994) did not yet have the sensitivity needed for a meaningful comparison, while most other searches were not simultaneous.

A much delayed (days to weeks) radio outburst at the mJy level could also become observable when the ejecta has expanded sufficiently for the radio opacity to become negligible (Paczyński and Rhoades, 1993).

Simultaneous (and delayed) GeV emission is also a fairly straightforward consequence of this scenario, for electron power-law indices not too steep and moderate shock magnetic field strengths (which can be inferred from MeV spectra). A sustained or delayed GeV emission (as observed, e.g. by Hurley, *et al.* , 1994) can be understood in terms of a wind with internal dissipation *and* an external deceleration shock (Mészáros and Rees, 1994). The wind, with an $\eta \lesssim 10^2$ and duration t_w produces via internal shocks an MeV burst extending up to $\lesssim$ few GeV, and also produces a longer event ($t \sim t_{dec}$, which may be up to hours, see eq[7]) as the wind is decelerated in the external medium. At these η the external burst is below BATSE threshold at MeV, but is prominent at GeV and due to the low compactness it extends easily above 30 GeV. An alternative possibility for delayed GeV bursts is discussed by Katz, 1994.

4. DISCUSSION AND PROSPECTS

A dissipative relativistic fireball scenario is motivated (and practically required) by the key observations discussed in §1. It is to a large degree generic, and is expected whether the "ultimate source" is, e.g., a young high-field pulsar, a failed supernova Ib, a neutron star binary merger, a halo neutron star quake, or comets crashing into magnetospheres. The basic assumption common to all such sources is that they deliver the required energy in a small initial volume ($r_o \lesssim 10^6 - 10^7$ cm) in a short time. Whatever the ultimate source is, it should in any case remain hidden behind an opaque $e^{\pm}\gamma$ veil, and the subsequent evolution of the fireball (during which the object manifests itself observationally) is independent of the birth details. This may be phrased as a *GRB "No-Hair" theorem*: the only thing that characterizes observationally a GRB is the initial energy, the initial volume and (possibly) the initial energy deposition timescale. Knowing the exact nature of the primary GRB source would, of course, greatly help in estimating expected rates per galaxy per year and details of the spatial distribution. However, this should be (to a good first approximation) irrelevant for understanding the physics of the observable intrinsic GRB properties.

A fireball model is, in principle, expected also if GRB arise in the galactic halo. In this case the nature, the rate of events and the spatial distribution might be understood (see, e.g. Podsiadlowski, Rees and Ruderman, 1995; Colgate and Leonard, 1994; Wasserman and Salpeter, 1994) if one makes some new assumptions about the source physics. In a cosmological setting, on the other hand, there are plausible astrophysical sources with relatively uncomplicated physics that could supply the observed rate and distribution (e.g. Eichler, *et al.* , 1989, Narayan, Paczyński and Piran, 1992, Mészáros and Rees, 1992, Woosley, 1993). The dissipative fireball scenario described above explains then in a straightforward manner (Mészáros and Rees, 1993a, Mészáros , Laguna and Rees, 1993) the GRB energies, the typical overall durations and the non-thermal spectra. What is not specifiable in detail in such a model is how the energy gets deposited in a low density region to provide a high Γ, which is puzzling either for halo or cosmological sources. However, the conclusion that this *does* occur seems unavoidable, since the observations demand such conditions (see §1).

Several other questions can be addressed within the context of such models. One of them is the difference in spectral and temporal properties of halo and cosmological models. The former have smaller total energy and lower magnetic field strengths in the wind or at the shocks. If the minimum particle energies accelerated in the shocks were very high this could still yield halo GRB spectra which satisfy the X-ray paucity and have acceptable MeV breaks, but the time-delayed durations are all less than a second (Mészáros and Rees, 1993b). This may be helped if the durations are explained with a (so far unspecifiable) internal time t_w, but they would still tend to overproduce X-rays. In cosmological models, on the other hand, either the calculated dynamic time t_{dec} or an internal time t_w can give acceptable spectra. Details of the spectrum at several wavelenghts, if observed simultaneously, could provide discriminants between these cases.

Another question that may be addressed in these models is the reported bimodal duration distribution (Kouveliotou, *et al.* , 1993 and these proceedings). One possibility is that the durations are given by the deceleration time (7) and bursts occur in two main types of external environment. One may speculate, e.g., that the progenitor's random spatial velocity causes a fraction of the bursts to occur in the galactic halo, while others occur in the disk. The density contrast would be at least 10^{-3} and the duration difference at least a factor $\Delta n_o^{-1/3} \sim 10$. Alternatively, there may be a fraction of bursts for which deceleration occurs in a pre-ejected denser wind, while in others the latter is unimportant compared to the ISM.

Acknowledgements: I am grateful to Martin Rees for many discussions and insights into these problems. The research is partially supported through NASA NAGW-1522 and NAG5-2362.

References

Band, D., *et al.* , 1993, Ap.J., 413, 281

Cavallo, G. and Rees, M.J., 1978, M.N.R.A.S., 183, 359

Colgate, S.A. and Leonard, P.J.T., 1994, in *Gamma-ray Bursts*, ed. G. Fishman, *et al.* , p. 518 (AIP 307, NY)

Duncan, R.C. and Thompson, C., 1992, Ap.J.(Letters), 392, L9

Eichler, D., Livio, M., Piran, T. and Schramm, D., 1989, Nature, 340, 126

Fenimore, E.E., Epstein, R.I. and Ho, C., 1993, Astr.Ap.Suppl, 97, 59.

Fishman, G., these proceedings

Goodman, J., 1986, Ap.J., 308, L47

Harding, A.K. and Baring, M.G., 1994, in *Gamma-ray Bursts*, ed. G. Fishman, *et al.* , p. 520 (AIP 307, NY)

Hurley, K., *et al.* , 1994, Nature, 372,652

Katz, J.I., 1994, Ap.J., 422, 248

Katz, J.I., 1994, Ap.J.(Lett.), 432, L27

Kouveliotou, C., *et al.* , 1993, Ap.J., 413, L101

Mészáros , P. and Rees, M.J., 1992, Ap.J., 397, 570

Mészáros , P. and Rees, M.J., 1993a, Ap.J., 405, 278

Mészáros , P. and Rees, M.J., 1993b, Ap.J. (Letters), 418, L59.

Mészáros , P., Laguna, P. and Rees, M.J., 1993, Ap.J., 415, 181.

Mészáros , P. and Rees, M.J., 1994, M.N.R.A.S., 269, L41

Mészáros , P., Rees, M.J. and Papathanassiou, H., 1994, Ap.J., 430, L93

Narayan, R., Paczynski, B. and Piran, T., 1992, Ap.J.(Letters), 395, L83

Paczyński, B., 1986, Ap.J.(Lett.), 308, L43

Paczyński, B., 1990, Ap.J., 363, 218.

Paczyński, B. and Rhoads, 1993, Ap.J., 418, L5

Paczyński, B. and Xu, G., 1994, Ap.J., 427, 708

Podsiadlowski, P., Rees, M.J. and Ruderman, M., 1995, M.N.R.A.S., in press

Ricker, G., *et al.* , 1992, in *Gamma-ray Bursts*, C. Ho, R. Epstein and E. Fenimore, eds. (Cambridge U.P.), p. 288

Rees, M.J. and Mészáros , P., 1992, M.N.R.A.S., 258, 41P

Rees, M.J. and Mészáros , P., 1994, Ap.J.(Letters), 430, L93-L96

Shemi, A. and Piran, T., 1990, Ap.J.(Lett.), 365, L55

Thompson, C., 1994, M.N.R.A.S., 270, 480

Usov, V.V., 1992, Nature, 357, 472

Usov, V.V., 1994, M.N.R.A.S., 267, 1035

Wasserman, I. and Salpeter, E.E., 1994, Ap.J., 433, 670

Waxmann, E. and Piran, T., 1994, Ap.J., 433, L85

Woosley, S., 1993, Ap.J., 405, 273

Gamma-Ray Bursts - What Are They?

S. E. WOOSLEY

Board of Studies in Astronomy and Astrophysics
University of California, Santa Cruz
Santa Cruz, CA 95064

INTRODUCTION

The observational data show conclusively that the Earth is situated at or very near the center of the gamma-ray burst universe. Yet, seemingly we are seeing to its edge. Three possibilities arise then for the burst locations - 1) a local Oort cloud of comets, though it is doubtful that this could provide sufficient spherical symmetry[1], let alone credible models; 2) an extended Galactic halo, presumably populated by neutron stars; and 3) cosmological sources at red shifts z $\sim$ 1 - 2. We consider here, in part because of limited space, only a few representative models from each of these last two classes.

HALO MODELS

Halo models have received recent encouragement from the high velocity observed for pulsars[2] and the association of at least two soft gamma-ray repeaters with supernova remnants and high velocity neutron stars with slow rotation rates (refs. 3-8). New results were presented at the meeting[9], however, suggesting a possible stellar counterpart for SGR 1806-20, weakening the evidence for its high velocity and drawing into question the assumption that soft repeaters are solitary neutron stars. Many, though not all of those favoring halo models claim an association between SGR's (youth stage) and the regular bursts. In these models the regular bursters are solitary neutron stars, in part because of the lack of any bright optical couterpart. Several recent studies[10−12] show what is necessary for halo models to be concordant with the BATSE data. First, not all pulsars can participate. The low energy tail of the velocity distribution, and especially that group which is kicked in a direction opposing Galactic rotation, would imply a strong disk association which is not seen. Bursters would need to be born with kicks of at least 700 km/s (the *average* pulsar is under 500 km/s). Moreover, their bursting activity would need to be restricted for first 10 million years while they moved away from the disk, and it would help if they turned off after a billion years or so. Finally, it helps if the Galactic gravitational potential is non-spherical[11]. None of these things is impossible, though, in total, the models seem somewhat contrived. Models also exist in which the neutron stars are not born in the disk, but in the halo[13] or during Galaxy formation[14]. These more naturally have a spherical distribution. However, such models must explain why neutron stars from Pop III should become bursters and not their modern counterparts. The model by Woosley[14] suggests that blue supergiants, which were more numerous in the early Galaxy, would more readily become bursters. The one by Eichler & Silk[13], requires that the halo be populated with a large number of white dwarfs, some of which merge to form neutron stars.

Finally, Li, Duncan, & Thompson[15] have pointed out that if the bursts are preferentially beamed along the velocity vector of the neutron star, the BATSE limits can be more easily accommodated. Interestingly, this would also lead to the strong suppression of bursts from Andromeda. Since the spherical symmetry must be broken anyway in order to give the neutron star a "kick" at birth, it might not be too surprising if the gamma-ray burst occurred along the same axis (i.e., the rotational or magnetic pole). Lacking a compelling model for either neutron star kicks or gamma-ray bursts, it is hard to say.

The production of the burst itself in halo models is often attributed to magnetic instabilities[16,11] or accretion[17], but what is there for a 700 km/s neutron star to accrete? Marc Herant and I have proposed that, during the supernova, a neutron star may capture a portion of its own ejecta. Especially in the more massive stars, a neutron star of this speed (or more) will find itself moving faster than $\sim$1 $M_\odot$ of ejecta. Calculations show the material through which the neutron star moves is clumpy[18], with clump masses on the order of Jupiter and distance scales $\sim$1 AU. If captured, this material could form an accretion disk, and ultimately planetesimals[19]. Accretion of these, at 10^{22} gm per event, could in principle power the burst. Colgate and Leonard[17] have proposed a similar model where the material is captured from a companion star rather than the supernova itself. A major problem with all planetesimal accretion models, however, is how to move in from the tidal radius ($\sim 10^{10}$ cm) to the Alfven radius ($\sim 10^8$ cm) without experiencing complete tidal disruption[20] and spreading of the fragments. The efficiency for this will be small compared to all the planetesimals either ground to dust or accreted by the planets (W. Benz, private communication). If successful in finding its way to the neutron star, the planetesimal might make the burst by torquing the magnetic field of during its impact.

Those who would invoke a strictly magnetic explanation of the bursts (eg. refs. 11, 16) must preserve a minimum of 10^{46} erg in the field, more if not all (10^9) neutron stars participate. This implies field strengths of $\gtrsim 10^{15}$ gauss. Whether such fields can be generated and preserved for $\gtrsim 10^9$ years is unknown. Such fields would provide a natural explanation for the slow rotation rate of SGR's[16] and possibly regular bursters as well. But it is not clear why the duration of SGR activity would be the same as the lifetime of a supernova remnant, what would cause the delay in onset of regular burst production, or why the ordinary pulsars (presumably with much weaker fields) would have, to a factor of two, the same kick velocity as the bursters. The model does make a prediction that bright, slowly rotating pulsars should occasionally be found.

COSMOLOGICAL MODELS

If the halo models appear contrived, current cosmological models do not fare much better. The most popular is merging neutron stars[21–23]. In its simplest version, tidal heating between the two stars leads to neutrino emission. Neutrino meets antineutrino along the rotational axis and annihilates, leading to pair production[23]. The pairs energize a jet which makes the gamma-burst either directly, as a fireball, or indirectly by producing a relativistic jet of baryons. As Meszaros & Rees have shown elsewhere[24,25] and in this volume, bulk relativistic $\Gamma's \sim 100$ to 1000 are required. The model has the merit of tapping a large energy supply (the neutron star binding energy) and thus requiring only moderate efficiency, about 1%, to make a 10^{51} erg event.

However, recent studies[26,27] and a poster paper by Ruffert et al.[28] show that this efficiency is not likely to be realized, at least by neutrino processes. The calculations of Ruffert et al. are three dimensional and follow the last $\sim$10ms of the merger. Estimates of the gamma-ray emission prior to 10 ms are even smaller. While quite high temperatures are reached in the neutron star interior, and the neutrino emission rate rises, near the end, to $\sim 10^{53}$ erg s^{-1}, the neutrino annihilation power is only $\sim 2 \times 10^{50}$ erg s^{-1}. During the 0.01 sec during which it exists, this provides about 1000 times too little energy to power an isotropic burst. Yet highly beamed bursts, would require too many mergers, much more than the expected rate of $\sim 10^{-5}$ per year per Galaxy[22]. While this calculation and others might be criticized from leaving out magnetic fields or use of a particular representation of viscosity, a successful model would have to overcome this very large factor.

Yet the story is not over for merging neutron stars. After the merger, and in a time of a few ms, a black hole will form ($M \approx 2.7$ $M_\odot$) surrounded by an accretion disk (~ 0.1 $M_\odot$).

This disk still has more than enough energy to power a burst[22], about 10^{53} erg. The viscous time scale is very uncertain, but might be $\sim$1 minute[29]. Energy release could occur by a variety of mechanisms like those discussed for quasars[30] - magnetic dissipation, Blanford-Zajnek extraction, etc. A jet will presumably form and its interaction with surrounding material makes the burst.

But if the burst source is to be a minature quasar, there is more than one way to form it. Obviously a merging neutron star and stellar mass black hole could produce the same configuration, but with a more massive disk. Even more efficient, both in terms of its estimated frequency (about 10^{-3}/yr) and the mass to be accreted, is the "failed supernova"[29]. This comes from a rotating Wolf-Rayer star whose iron core collapses without producing a successful outgoing shock. Ten $M_\odot$ or more may then be accreted into a black hole that starts at around 2 $M_\odot$ and grows with time.

The chief problem with all these models is "baryonic contamination", by which one means too much matter and not enough energy, in other words, too little entropy. To produce $\Gamma \gtrsim 100$, the specific energy must be $\gtrsim 10^{23}$ erg g^{-1}, i.e., 10^{51} erg in 10^{-4} $M_\odot$, 10^{52} erg in 10^{-3} $M_\odot$, etc. During the last stages of merger, the neutron star pair will blow a wind[31,32], $\dot{M} \approx 0.01$ $M_\odot$ s^{-1}. With great uncertainty, the estimated mass surrounding the merged pair is $\gtrsim 10^{-4}$ $M_\odot$. Dissipation in the disk of the order required to make a burst in a few seconds will probably generate much more. In the failed supernova, one still has the residual of the collapsed mantle, about 0.01 Ω $M_\odot$ where Ω is the solid angle into which the jet is beamed. Clearly beaming can reduce the energy requirements on these models and beaming into a solid angle as small as 0.01 is allowed by the possible large event rate of failed supernovae. Beaming into even $\Omega \lesssim 0.1$ would pose problems for the merging neutron star event rate.

In addition, the failed supernova model might also be surrounded by $\sim$10 $M_\odot$ of precollapse stellar wind extending out many parsecs. This wind, which has Thomson depth of order unity, could both provide the anvil which the jet strikes and a Comptonizing layer for scattering soft radiation out of the beam[33]. Additional time structure may be imposed on the beam if the black hole an disk are not coplanar in which case the black hole will precess with a time scale comparable to that of the burst duration[34].

CONCLUSIONS

Currently, there are no good theoretical models for gamma-ray bursts. All invoke miracles, either minor or major, but they form the speculative basis from which real models may someday emerge. The current controversy regarding distance scale will probably be resolved on a time scale of say five years by a variety of observational techniques - precise localization of a common couterpart, association of faint bursts with other galaxies of the local group, low energy absorption features, high spectral energy cut off, or other means. However, that resolution will not necessarily lead to the prompt realization of a physical model. For twenty years, "knowing" that gamma-ray bursts were neutron stars in our galaxy did not lead to unanimity regarding models. My own view is that it will be a long, and enjoyable voyage for observers and theorists alike before this mystery is solved, even to the level of understanding which now typifies, say, pulsars and quasars.

This work has been supported by NASA under MIT subcontract SC-A-292701 and NAGW 2525.

REFERENCES

1. Clarke, T. E., Blaes, O., & Tremaine, S. 1994, AJ, 107, 1873
2. Lyne, A. G., & Lorimer, D. R. 1994, Nature, 369, 127
3. Cline, T., et al. 1982, ApJL, 255, L45

4. Kulkarni, S., et al. 1994, Nature, 368, 129
5. Murakami, T., et al. 1994, Nature, 368, 127
6. Rothschild, R., et al. 1994, Nature, 368, 432
7. Ulmer, A., et al., 1993, ApJ, 395, 235
8. Hurley, K., et al., 1994, ApJL, 431, L31
9. Kulkarni, S., et al. 1995, preprint "Optical and Infrared Observations of SGR-1806-20", submitted to ApJ.
10. Hartmann, D., et al. 1994, ApJS, 90, 893
11. Podsiadlowski, P., Ruderman, M., & Rees, M. 1994, MNRAS, in press
12. Bulik, T., & Lamb, D. 1994, BAAS, 26, No. 4, 1508
13. Eichler, D., & Silk, J. 1992, Science, 257, 937
14. Woosley, S. 1993, in Planets Around Pulsars, ed. J. Phillips, S. Thorsett, and S. Kulkarni, ASP Conf. Series Vol. 36, 355
15. Li, H., Duncan, R., & Thompson, C. 1994, in Gamma-Ray Bursts, AIP Conf. Proc. 307, eds. G. Fishman, J. Bainerd, and K. Hurley, p. 600
16. Duncan, R. C., & Thompson, C. 1992, ApJL, 392, L9
17. Colgate, S. & Leonard, P. 1994, in Gamma-Ray Bursts, AIP Conf. Proc. 307, p. 581
18. Herant, M., & Woosley, S. 1994, ApJ, 425, 814
19. Lin, D., Woosley, S., & Bodenheimer, P. 1991, Nature, 353, 827
20. Tremaine, S., & Zytkov, A. 1986, ApJ, 301, 155
21. Paczynski, B. 1986, ApJL, 308, L43
22. Narayan, R., Paczynski, B., & Piran, T. 1992, ApJL, 395, L83
23. Meszaros, P., & Rees, M. J., 1992, ApJ, 397, 570
24. Rees, M., & Meszaros, P. 1992, MNRAS, 258, 41P
25. Meszaros, P., Rees, M., & Papathanassiou, H. 1994, ApJ, 431, 181
26. Wilson, J. R., & Mathews, G. J. 1994, Proc. 7th Marcel Grossman Meeting on General Relativity, LLNL preprint 118330
27. Davies, M., Benz, W., Piran, T., & Thielemann, F. 1994, ApJ, 431, 742
28. Ruffert, M., Janka, H.-T., Keil, W., & Schäfer, G. 1994, poster paper "Coallescing Neutron Stars as Gamma-Ray Bursters", this meeting.
29. Woosley, S. E. 1993, ApJ, 405, 273
30. Blandford, R. D. 1989, in Theory of Accretion Disks, ed. F. Meyer et al., (Kluwer Academic), p. 35
31. Woosley, S. E. 1993, A&A Supp, 97, 205
32. Woosley, S. E., & Baron, E. 1992, ApJ, 391, 228
33. Brainerd, J. 1994, ApJ, 428, 21
34. Hartmann, D., & Woosley, S. E. 1995, Proceedings of COSPAR meeting, August, 1994

The Ultra-High-Energy Cosmic Rays : An Astrophysical Puzzle

MURAT BORATAV

Laboratoire de Physique Nucléaire et de Hautes Energies
Université Paris 6
4 Place Jussieu - 75252 Paris Cedex 05 - France

INTRODUCTION

The "all particle" cosmic ray spectrum presents several interesting features, especially above 10^{14} eV. Some of these can be seen on Figure 1. This differential energy spectrum falls very rapidly with energy : it follows a power law decreasing roughly as $E^{-2.7}$ up to a region called the "knee" (around 10^{15} eV) and somewhat steeper above. This first structure is followed by a second one which occurs at above 10^{18} eV (or 1 EeV) where the spectrum seems to flatten (a structure usually called the "ankle"). There the data become very scarce to finally completely (or, as we shall see, almost completely) vanish above 100 EeV.

A special attention should be given to the rates with which the cosmic rays arrive in the vicinity of the earth's atmosphere, in terms of unit aperture and unit time, since those figures have dramatic consequences on the experimental techniques and methods. Up to the "knee" region, the rates are sufficient to use more or less small sized detectors shipped aboard satellites or balloons. Between the "knee" and the "ankle", the incident intensities become very small so that only large aperture ground detectors (including optical devices with their small -around 10%- duty cycles) can be used to reach reasonable statistics. Above the ankle, and especially above the so-called Greisen-Zatsepin-Kuzmin (GZK) cutoff, the fluxes are so infinitesimal that only giant ground arrays can be expected to help accumulating data.

Many questions remain partially or totally open in the understanding of the cosmic ray phenomena in this energy region.

- Although the existence of the "knee" structure in the spectrum can be understood to some extent (e.g. by a reasonable limit to the energies that can be produced by charged particle acceleration in supernova blast-waves[2]), no natural explanation can be found for the sharpness of this structure.

- Since the study of the chemical composition of the cosmic rays by ground detectors is delicate and model dependent, it is important that a future generation of experiments make the link between composition measurements by direct and indirect methods around the "knee" region.

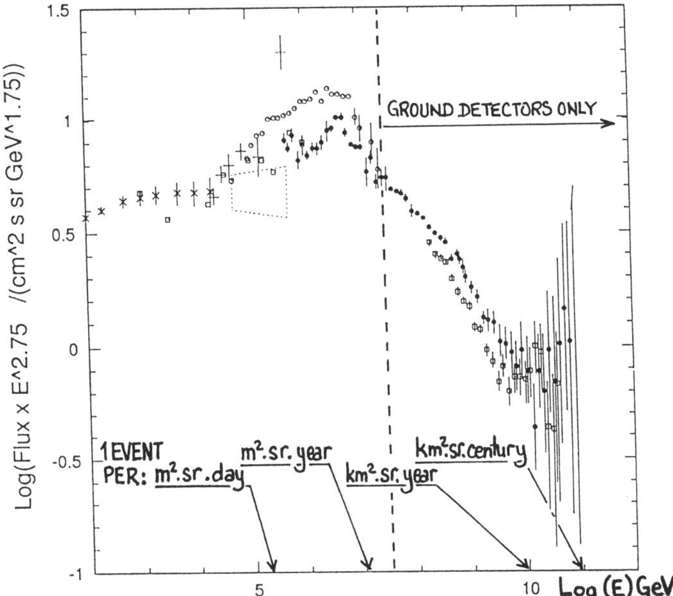

Figure 1: The "all-particle" cosmic ray spectrum[1] above 10^{11} eV. The data points are those from the JACEE, Aoyama-Hirosaki, BASJE and the Proton satellite experiments below the "knee", the Akeno and Fly's Eye experiments above.

- There is a general agreement on the fact that the conventional acceleration mechanisms have difficulties in explaining the observations made of UHE cosmic rays up to the end of the spectrum. Although many authors propose models which, by using extreme values for the free parameters, can reach very high energy acceleration values,[3] to our knowledge no satisfactory theory has been proposed up to now which can give the answers to the whole set of experimental observations available beyond the GZK cutoff.

A comprehensive review of the observations near the "knee" region is given elsewhere.[4] In the following, we'll concentrate on the questions left open by the data available above the GZK cutoff and on a possible way to provide answers to them.

EXPERIMENTAL SITUATION AROUND AND ABOVE THE GZK CUTOFF

For more than 30 years, ground based detectors have been observing cosmic rays with energies equal to or larger than 100 EeV. Let us call such particles super cosmic rays (SCR) for brevity. Although, because of about 20% uncertainty in the values of the reconstructed energies, such a distinction is somewhat arbitrary, about 10 SCR were detected during this period.[5] One must emphasize the fact that the detectors which reported these observations are designed with various techniques : Volcano Ranch (scintillators), Haverah Park (water Cerenkov), Akeno (scintillators and muon detection), Yakutsk (scintillators and Cerenkov), Fly's Eye (atmospheric fluorescence). So there is no doubt that the observations are reliable and not artefacts due to some misleading modelisation of the air showers. Many articles have been written about the SCR, and especially the most energetic among them, a 320 EeV event observed on 15 October 1991 by the Fly's Eye at Dugway (USA).[6] The difficulties in interpreting such events can be summarized in a few words :

- There are very few astrophysical mechanisms able to accelerate particles to energies exceeding 100 EeV. If the mechanism is an electromagnetic process, the energy of the particle is then proportional to the magnetic rigidity in the region where the acceleration takes place. This puts very strong constraints on the site where the mechanism, whatever it is, can take place. Hillas[7] proposed an often used plot showing the necessary magnetic fields and corresponding site dimensions to produce particles accelerated up to 100 EeV. The sites passing the threshold values are scarce : pulsars with extreme magnetic field values, active galactic nuclei, radiogalaxy lobes. Alternative ways of producing the SCR exist, such as the disintegration of grand-unification gauge bosons of masses as large as 10^{15} GeV and resulting from the collapse of cosmic strings or other topological defects.[8] Such hypotheses are of course highly speculative and have been criticized by some authors.[9]

- Once produced, the particles have to escape from the production sites. The above mentioned conventional mechanisms all present a very hostile environment where a large part of the initial energy should be lost, either by synchrotron radiation or through interactions with very dense radiation.

- If one ignores the preceding limitations, the current particles (except neutrinos) interact with the 3K microwave background radiation if their energies are above photoproduction thresholds (typically 200 TeV for photons, 70 EeV for protons). The corresponding attenuation lengths are less than a few tens of Mpc : this is the origin of the GZK cutoff. Also, at energies around 100 EeV or beyond, the bending of the charged particle trajectories by the extragalactic magnetic fields is weak enough so that the direction of the incident SCR must point within a few degrees at the source. It is a fact that no known conventional high-energy source is to be found within, say, 100 Mpc of our galaxy and in the directions of the incident SCR.

There have been many attempts to interpret the experimental data with the constraints listed above.[10] The main conclusions one can reach in going through them is the following. There is no satisfactory way to explain the bulk of existing data with conventional physics. All the other possible explanations either lead to new interpretations of what is thought to be well known (conventional mechanisms but not yet observed or discovered, complete revision of our present knowledge about the extragalactic magnetic fields etc...) or to new physics (diffuse sources such as cosmic strings, neutrino interactions inconsistent with the Standard Model etc...).

THE AUGER PROJECT

There is no doubt that the observation of the SCR raises some of the most exciting problems to be solved in the field of high-energy astrophysics. The answers can only come from a dedicated experiment which should provide enough statistics of high-quality data so that we have a better understanding of the chemical composition of these objects, of their incoming direction and of the shape of their energy spectrum beyond the GZK cutoff. The main constraint on such an experiment is the expected incident flux of these cosmic rays. A back-of-the-envelope formula on this flux is :

$$I(E > E_0) \approx \frac{100}{E_0^2}$$

which gives their intensity in $km^{-2}sr^{-1}yr^{-1}$ above the energy E_0 (in EeV). Above the 70 EeV GZK cutoff the rate is roughly 2 events per *century* and per square kilometer for a standard ground detector. Thus a 100 events per year experiment needs a 5000 km^2 ground array efficient 100% of the time. This is the starting point of the international giant array project presently known as the "Auger Observatory".[11]

The design of the proposed detector is under way and it is the aim of a six-month (January-July 1995) working group centered on Fermilab to prepare a final technical proposal. The present status of the project is that of a hybrid detector which combines the statistical power of a giant array (5000 km^2 with about 3000 scintillator detectors spread over the site) with the observational quality of a High-Resolution Fly's Eye. The detector should use unconventional techniques such as solar power, data transfer and acquisition with cellular telephony or some other telecommunications standard, synchronisation through GPS satellites etc... It is designed in such a way that the identification of the incident particles should be possible, at least to some extent (distinction between photons, protons and heavy nuclei). It is hoped that in a very few years, the Auger Observatory will help to disentangle the threads of this astrophysical enigma.

ACKNOWLEDGMENTS

I am very grateful to the organizers of the mini-symposium on "Ultra-High-Energy Cosmic Rays" for their invitation and partial financial support.

REFERENCES

1. HAYASHIDA, N., *et al.* 1993. *In* Proceedings of the Tokyo Workshop on Techniques for the Study of Extremely High-Energy Cosmic Rays : 63-73. Tokyo, Japan, September 27-30 1993. Institute of Cosmic Ray Research. University of Tokyo.
2. GAISSER, T.K., *et al.* (Report of group R3b) 1994. *In* Proceedings of the Snowmass Summer Conference on Particle and Nuclear Astrophysics in the next Millenium : in press. Snowmass-Aspen, USA, June 29-July 14, 1994.
3. See e.g. J.R.JOKIPII and V.S.PTUSKIN, these proceedings.
4. MULLER, D., these proceedings.
5. For a detailed review of the experimental situation on the highest energy cosmic rays, see A.A.WATSON. *In* Proceedings of the Snowmass Conference.
6. A description of the Fly's Eye can be found in G.L.CASSIDAY, 1985. Ann. Rev. Nucl. Part. Sci. **35**: 321-349. A summary of the most recent observations made by this detector is in D.J.BIRD *et al.*, 1993. Phys. Rev. Lett. **71**: 3401-3404.
7. HILLAS, A.M., 1984. Ann. Rev. Astron. Astrophys. **22**: 425-444.
8. SIGL, G., D.N.SCHRAMM, P.BHATTARCHARJEE. 1994. Astroparticle Physics **2**: 401-414.
9. GILL, A.J., T.W.B.KIBBLE. 1994. Phys. Rev. **D50**: 3660-3665.
10. DAWSON, B. These proceedings. See also P.SOMMERS. Proceedings of the Snowmass Conference.
11. CRONIN, J.W. 1992. *In* Proceedings of the International Workshop on Techniques to Study Cosmic Rays with Energies Greater than 10^{19} eV. M. Boratav, J.W.Cronin and A.A.Watson Eds. Nuclear Physics (Proc. Suppl.) **28B**: 213-226.

The Cosmic Ray Composition Towards The Knee: Experimental Status And Implications

DIETRICH MÜLLER AND SIMON SWORDY

Enrico Fermi Institute and Department of Physics, University of Chicago, Chicago, Illinois 60637, USA

INTRODUCTION

Sources of the cosmic particle radiation cannot directly be observed. Rather, their properties must be inferred from measurements of the ambient cosmic ray flux near Earth. This requires an assessment of the changes in composition and energy spectra that occur while the cosmic rays propagate from the source regions to the observer. Among the relevant processes are ionization energy losses, nuclear spallation reactions, diffusive escape from the galaxy, and, locally, solar modulation. At high energies, above ~10 GeV/nucleon, solar modulation and ionization energy losses are insignificant, and nuclear spallation cross sections become independent of energy. Therefore, interstellar propagation effects are much easier to quantify at high energies than in the low energy region. Unfortunately however, because of the decreasing flux the quality of available data on cosmic ray composition decreases rapidly with increasing energy: much detail is known in the GeV-region, but only the spectra of some of the major components are measured up to TeV energies, and virtually no information on cosmic ray composition exists beyond $\sim 10^{14}$ eV per particle.

At present, all detailed information on the cosmic ray composition is derived from *direct* measurements with detectors above the atmosphere, flown either on balloons or spacecraft. *Indirect* measurements with air shower observations have determined the "all-particle" spectrum from $\sim 10^{14}$ eV to $\sim 10^{20}$ eV and have detected a conspicuous steepening of the spectrum between 10^{15} and 10^{16} eV (usually referred to as "the knee") but have encountered great difficulty in providing more detail about the primary flux than an average classification such as "light" (proton-like), "heavy" (iron-like), or "mixed". Even if new installations improve this situation, it would be highly desirable if the energy ranges between future direct and indirect measurements overlap to permit a cross calibration of the techniques.

OVERVIEW OF EXPERIMENTAL DATA

Direct measurements on the cosmic ray composition at high energy have been performed with detectors on board the HEAO-3 satellite (1), the Spacelab-2 mission (2), the Russian Sokol mission (3), and with several balloon payloads (4-9) including the JACEE program (10). In figures 1 and 2 we show a compilation of data for the more abundant primary nuclei. The elements C, N, O and also Ne to S are grouped together although individual spectra have been reported up to $\sim 10^{12}$ eV/nucleon. For higher energies, data with individual element resolution are not available for the heavier nuclei, and the statistical accuracy of the results deteriorates with increasing energy. Except for the most abundant elements, H and He, no significant data exist above $\sim 2 \times 10^{13}$ eV/nucleon. A most conspicuous feature in figure 1 is the apparent steepening of the proton spectrum above 10^{13} eV. The dashed lines in figure 2 refer to a model fit to the data which will be described below. Key information on the progagation of the primary cosmic ray nuclei through the galaxy comes from measurements of the abundances and spec-

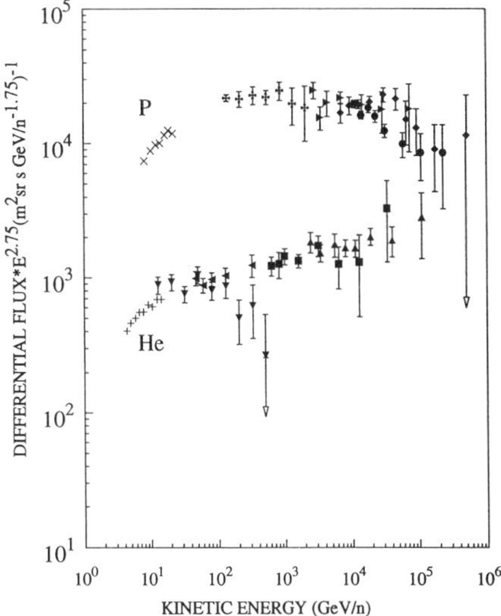

FIGURE 1. Compilation of data of the differential energy spectra of protons and helium. Note that here and in the following figures, the flux is multiplied with $E^{2.75}$.

FIGURE 2. Differential energy spectra of several primary elements. The lines refer to the model described in the text: dotted line-model A, dashed line-model B, dash-dot line-model C.

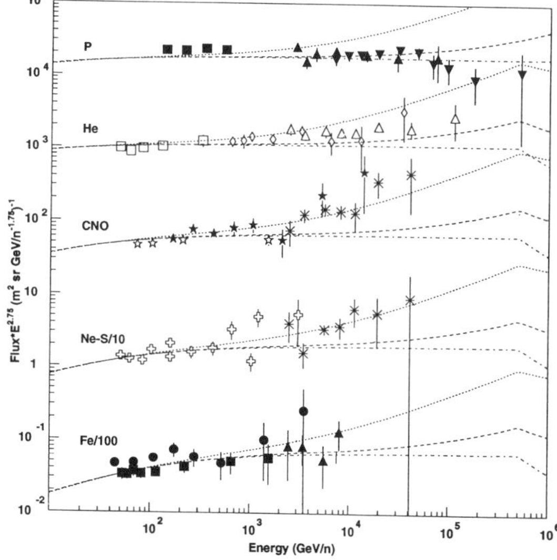

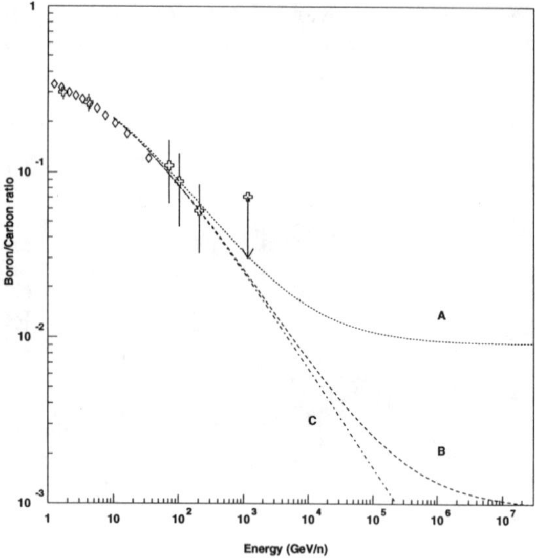

FIGURE 3. Boron/Carbon flux ratio versus energy. The lines refers to models with different residual pathlength $\wedge_r$ described in the text.

tra of the secondary nuclei, such as Li, Be, and B. In figure 3, we show the abundance ratio B/C as function of energy, as observed on HEAO-3 and Spacelab-2. The significant decline of the B/C ratio shows that the interstellar propagation path length $\wedge$ encountered by the primary cosmic rays decreases with rigidity R in the following way:

$$\wedge(R) = \wedge_\circ(\frac{R}{20GV})^{-0.6}, \quad \wedge_\circ \approx 6g/cm^2, \quad R > 20GV \tag{1}$$

This dependence of the containment of cosmic rays to the galaxy on energy or rigidity appears to be valid at least up to energies of a few 100 GeV/nucleon (or total energies $\approx$ 1 TeV/particle) (11) [note that for highly relativisitic particles rigidity and energy are proportional to each other].

COMPOSITION OF THE COSMIC RAY SOURCE

The "leaky box" model of the galaxy is the simplest propagation model which is consistent with the measured data just described. This model assumes equilibrium between the production of cosmic rays at their acceleration sites and their subsequent loss from the galaxy. The most significant loss processes for high energy nuclei are spallation interactions in the interstellar gas and diffusive loss through escape from the galaxy. We may parametrize these two processes by the propagation path length $\wedge(E)$ (eq. 1) and a spallation path length $\wedge_s$, respectively. The spallation cross sections are assumed to be energy-independent, therefore $\wedge_s$ does not depend on energy but becomes smaller with increasing mass of nucleus. For instance, $\wedge_s$ is about 5.5 g/cm^2 for oxygen, but only 2.3 g/cm^2 for iron. At sufficiently high energy (typically E $\gtrsim$ 100 GeV/nucleon), we will always have $\wedge(E) << \wedge_s$, and the fate of the cosmic rays is then dominantly determined by escape. It appears that all primary cosmic ray nuclei asymptotically reach a differential energy spectrum

$\propto E^{-2.75}$. The model then predicts a much harder spectrum at the source, approximately $\propto E^{-2.15}$. At least up to energies around 1 TeV/nucleon, this seems to be very well in agreement with measured data (12), perhaps with

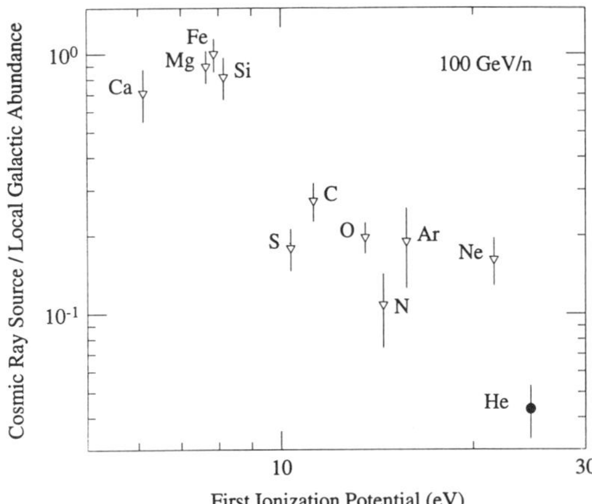

FIGURE 4. Cosmic ray source abundances relative to the local galactic element abundances vs. the first ionization potential of the elements(from ref. 8).

the exception of the element silicon which may have a steeper spectrum (2,9). This power-law spectrum for the source region agrees well with predictions from supernova-shock acceleration models and, in fact, may be taken as strong evidence for the validity of such models.

We may use the leaky box model to deduce the relative abundances of the cosmic ray nuclei at the sources, taking both escape and spallation losses quantitatively into account (2,8). The result of such a propagation calculation is shown in figure 4 where we compare the source abundances relative to the "local interstellar" abundance scale of the elements, with the first ionization potential (FIP) of the elements. This figure demonstrates that the relative underabundance of cosmic ray elements with high FIP wich is well known at lower energies, seems to persist to the highest energies for which measurements are available. A similar FIP depletion is also known to exist locally for energetic particles ejected from the Sun. The origin of this effect for solar particles is not fully understood, but must have to do with fractionation processes in the solar atmosphere. This similarity may indicate that the material leading to cosmic rays at all energies, originates in stellar atmospheres and winds.

We should note that propagation models that are more complicated than the leaky box have been proposed, for instance models that include secondary acceleration during propagation (13). However, in order to be forced to adopt such a model, we need more detailed constraints from measured data than are presently available.

EXTRAPOLATION TO HIGHER ENERGIES

To understand cosmic ray phenomena at energies beyond the TeV region, we must rely on extrapolations from what is known at lower energy. While, as mentioned, some measurements on the primary composition extend to higher energy, nothing is known about the secondary abundances, i.e. about the propagation path length Λ. We may assume that Λ continues decline for all energies as a power law in rigidity as described in equation (1). Alternatively, it could also be that Λ reaches a residual value Λ_r at high rigidity which could, for instance, reflect a minimum amount of matter that cosmic rays traverse near their sources ("nested leaky box model"):

$$\Lambda(R) = \Lambda_\circ(\frac{R}{20GV})^{-0.6} + \Lambda_r \quad (2)$$

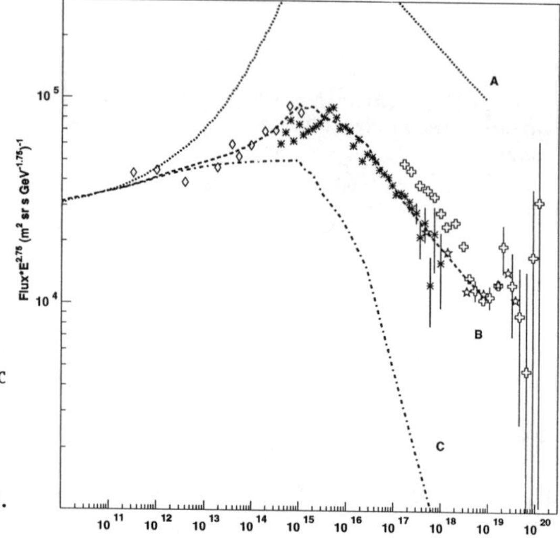

FIGURE 5. The all-particle spectrum of cosmic rays. The lines labeled A, B, C, respectively, are the predicted all-particle spectra, according to the model described in the text.

Another complicaton in attempts to extrapolate the data to higher energies comes from the fact that supernova shock acceleration is expected to be efficient only up to particle rigidities of a few 10^{14}V/c. Let us assume that the "knee" in the all-particle spectrum is due to this fact i.e. that the cosmic ray source, with relative elemental abundances as described above, and with a source energy spectrum $\propto E^{-2.15}$, extends to a rigidity of 10^{15}V/c, but steepens to $\propto E^{-3.0}$ beyond that rigidity. Clearly, this assumption is ad hoc, but it does have illustrative value: we now calculate the cosmic ray flux and energy spectra that we would expect to measure near Earth for such a source if, in addition, the propagation path length is given by equation (2). We consider three different values for Λ_r: (A) $\Lambda_r = 0.13g/cm^2$, (B) $\Lambda_r = 0.013g/cm^2$, (C) $\Lambda_r = 0$. These three model fits are shown as dashed lines in figures 2 and 3. Figure 2 shows that the measured individual spectra are not very sensitive to the choice of parameters in our model, except for the fact that the model cannot describe the drop in the proton spectrum at high energy.

We now sum up the individual intensities predicted by the model and construct the expected all particle spectrum. Figure 5 shows our extrapolation compared with the measured data. It is obvious that the choice $\Lambda_r = 0.013g/cm^2$ leads to excellent agreement with the measured all particle spectrum. While this agreement may not be proof for the validity of the model, it is striking that the other two values of $\wedge_r$ give grossly divergent fits. Thus, seemingly small details in the propagation mode of cosmic rays which are very difficult to measure, can have a strong effect on the overall spectral shape.

CONCLUSION

We have attempted to describe the presently known cosmic ray data at high energies in a self-consistent model based on the leaky box of the galaxy. This model is quite successful below TeV energies, although a few experimental details, for instance a very steep spectrum of silicon nuclei, cannot be explained. Extrapolation to higher energies becomes increasingly uncertain. A major source of uncertainties is the lack of knowledge of the modes of galactic propagation of high energy nuclei. A striking and unexplained feature in the measured data is the apparent steepening of the proton spectrum around 10^{14} eV. As all cosmic ray phenomena, such as acceleration or galactic containment, are expected to scale by magnetic rigidity, one should expect a corresponding steepening in the spectra of helium and heavier nuclei at the same rigidity. This has not been observed. Therefore, if future measurements confirm this contrasting behavior of proton and heavier nuclei spectra, a more radical explanation outside conventional cosmic ray models must be found.

Clearly, future experimental work should be directed towards extending the measurements to higher energy and should provide higher systematic and statistical accuracy. This requires large detectors and long esposure times: exposure factors of the order of several 100 m^2sr days are needed for measurements up to energies around 10^{15} eV/particles. Several techniques to accomplish these, even with balloon borne instruments, do exist (14). These might include repeated long duration flights of passive instruments such as JACEE, or flights of new transition radiation detectors of very large area. One should hope that new data from these detectors become available within a few years.

REFERENCES

1. ENGELMANN, J.J., et al. 1990. Astron. & Astrophys. **233:** 96.
2. MÜLLER, D., et al. 1991. Ap.J. **374:** 356.
3. IVANENKO, I.P., et al. 1993. Proc. 23rd ICRC (Calgary).**2:** 17.
4. RYAN, M.J., et al. 1972. Phys. Rev. Letters.**28:** 15.
5. SMITH, L.H., et al. 1973. Ap.J. **180:** 987.
6. SIMON, M., et al. 1980. Ap.J. **239:** 712.
7. SEO, E.S., et al. 1991. Ap.J. **378:** 763.
8. BUCKLEY, J., et al. 1994. Ap.J. **429:** 736.
9. ICHIMURA, M., et al. 1993, Phys. Rev. **D48**, 1949.
10. ASAKIMORI, K., et al. 1993. Proc. 23rd ICRC (Calgary). **2:** 21.
11. SWORDY, S., et al. 1990. Ap.J. **349:** 625.
12. SWORDY, S., et al. 1993. Ap.J. **403:** 658.
13. SEO, E.S., PTUSKIN, V.S. 1994. Ap.J. **431:** 705.
14. WADDINGTON, J., et al. GOAL Report of NASA Cosmic Ray Working Group, Washington 1992.

The Highest Energy Cosmic Rays

BRUCE R. DAWSON

Department of Physics and Mathematical Physics
University of Adelaide, Adelaide 5005, AUSTRALIA

INTRODUCTION

Recent reports of observations of three cosmic rays with energies greater than 10^{20}eV have re-focussed the attention of many to the central question of the origin of these extraordinary particles.[1,2,3] Provided that they have low or zero charge, the paths of these energetic particles are little affected by the galactic magnetic field and the assumed extragalactic fields, and so some sort of cosmic ray positional astronomy may be possible for the first time. Upper limits can be placed on the source distance based on the expected interactions of the particles with radiation fields, resulting in a belief that the sources have distances less than 50 Mpc. Unfortunately, no obvious point sources have yet been associated with the three particles in question. We begin with a very brief review of the current experimental situation at cosmic ray energies above 10^{17}eV, before turning to the very energetic events.

OBSERVATIONAL TECHNIQUES

At these high energies, the low cosmic ray flux precludes any direct measurement above the Earth's atmosphere. The integral flux of approximately 1 particle km^{2}yr^{-1} above 10^{19}eV requires large-area detectors which take advantage of interactions of the cosmic ray with the atmosphere. The interactions produce huge particle cascades (called extensive air showers, or EAS) which, at these energies, are still very large at sea-level. The EAS have large area footprints on the ground as a result of finite particle emission angles and Coulomb scattering of the shower electrons. A typical 10^{19}eV shower would contain 3×10^9 particles at sea level spread over an area of roughly 2 square kilometres.

Large arrays of particle detectors have been set up over the past 30 years to detect these most energetic EAS, starting with Volcano Ranch in New Mexico and followed by Haverah Park (UK), Sydney (Australia), Yakutsk (Russia) and Akeno (Japan). The new array at Akeno, AGASA, represents the state of the art for this detection method, with almost 300 detectors spread over an area of 100 km^2 to measure the various components of the EAS as they hit the ground. Apart from measuring the arrival direction of the EAS (and hence that of the primary cosmic ray), arrays estimate the energy of the primary and attempt to characterise the primary chemical composition in a coarse way (e.g. proton, CNO group, or iron nucleus).[4]

A quite different technique was first proposed in Japan and the US in the late 1950's and was first successfully implemented in Utah in 1979 in the form of the Fly's Eye. Here, detectors composed of large mirrors and photomultipliers observe fluorescent light from atmospheric nitrogen induced by the passage of the EAS. In this way it is possible to map out the entire development of the EAS, and therefore achieve a calorimetric energy measurement and a somewhat more direct determination of the primary particle composition.[4] While the light source is weak (only 5 photons/m are emitted for every EAS electron), the light is emitted isotropically allowing a very large collecting area provided the showers are energetic enough. In fact the threshold energy for the Fly's Eye detector was 10^{17}eV with an acceptance above 10^{20}eV of 1000 km^{2}sr. Unfortunately, this large area is effectively reduced by a factor of ten because of the requirement that the detector be run only on clear, moonless nights. The Fly's Eye detectors have now been superseded in Utah by HiRes, a system with larger mirrors and smaller photomultiplier aperture, which promises a collecting area in excess of 6000 km^{2}sr when in full operation in 1997.[4]

ENERGY SPECTRUM, COMPOSITION AND ANISOTROPY

There is very good agreement between experiments on the form of the energy spectrum at the highest energies. This is especially heartening, given the different detection and analysis techniques used. There is agreement on the features of the spectrum, and the normalisation is in agreement to approximately 20%.[4] In summary, the spectrum above 10^{17}eV falls as a power law with index very close to 3.0, steepens at $10^{17.6}$eV to an index of about 3.25 and then flattens to an index of about 2.7 above $10^{18.5}$eV. This behaviour is best seen in the Fly's Eye spectrum for showers seen by two Fly's Eye installations in "stereo". The stereo data has superior energy resolution and shows the features very clearly.[5]

There is a question about whether the spectrum cuts off at the highest energies. The Griesen-Zatsepin mechanism predicts a cut-off beyond about 6×10^{19}eV if the cosmic rays are protons from distant extragalactic sources, because of interactions with photons of the 2.7K CMB. Unfortunately, the event statistics are too poor at this stage to be conclusive about the extent of this effect, especially given the handful of events recorded with energies in excess of 10^{20}eV.[6]

Despite an expectation many years ago that a clear anisotropy might be seen in the very energetic cosmic rays, no such clear signal has been seen.[4] Of course, the galactic magnetic field of 2 or 3 μG can be blamed for this situation for galactic sources at moderate energies. If, however, the galaxy produces protons with energies above 10^{19}eV, it is generally believed that the current experiments should have seen a strong galactic plane excess. Thus, if protons make up the flux at the highest energies, it might be necessary to consider a large number of isotropically distributed extragalactic sources.

Recent Fly's Eye results on chemical composition contribute to this hypothesis.[5] The Fly's Eye measures the depth in the atmosphere (X_{max}) at which the EAS reaches its maximum size. For a given primary energy this is a measure of the mass of the primary particle, with iron initiated showers developing about 100 g cm^{-2} shallower than proton showers. The measurements of X_{max} as a function of energy show a steeper slope than expected on the basis of a constant chemical composition at energies above 10^{17}eV. Comparing with shower development models, it appears that the data are consistent with a "heavy" composition at 10^{17}eV which becomes lighter at higher energies, and perhaps purely protonic above 10^{19}eV. Of course, these predictions are model dependent (relying on theory and extrapolations of accelerator data), but a number of plausible models provide similar conclusions.

The behaviour of the energy spectrum, anisotropy and composition is consistent with a model which "heavy" cosmic rays are produced in our galaxy with a steeply falling spectrum, which is overtaken with a flatter spectrum of protons from extragalactic sources at the highest energies.[5]

THREE VERY ENERGETIC EVENTS

A total of eight events have been assigned energies above 10^{20}eV during thirty years of observations. However three events have recently been reported with energies well in excess of this threshold. One of them, seen with the Yakutsk array could have an energy as high as 2×10^{20}eV, but the EAS was highly inclined and muon-rich, and as such was unlike other events in the Yakutsk catalog.[1] As such the energy is uncertain.

The "record" is held by the Fly's Eye with an event observed on October 15, 1991 with an energy of $(3.2 \pm 0.9) \times 10^{20}$eV, the error derived from estimates of statistical and systematic errors.[2] While the event was not seen in "stereo", the view provided by a single Fly's Eye allowed a reasonable energy estimate and an error box of $\pm 0.5°$ in RA and $\pm 6°$ in declination. The particle arrived from the galactic anticentre direction, just north of the plane ($l = 163°, b = 9.6°$). A thorough investigation of reconstruction uncertainties was made. For example it was shown that if the reconstruction was forced to give a more "normal" energy of 10^{20}eV, the shower would need to be initiated at an extreme depth in the atmosphere, and the shower profile would then be unphysically narrow. The best estimate of the primary cosmic ray mass is around 20 amu, but because of intrinsic EAS development fluctuations and detector resolution, the primary could easily have been a proton, an iron nucleus or a gamma ray. (It would be extremely difficult to identify the composition of a single particle even with a perfect detector, using this necessarily indirect approach).

The third extreme event has been reported by the AGASA experiment.[3] It was detected on December 3, 1993 and has an energy estimate of $(1.7-2.6) \times 10^{20}$eV, and an arrival direction of $l = 131°, b = -41°$. The event appeared normal in every way except for the extreme numbers of particles detected by the array of scintillators, with densities ranging from 0.3 m^{-2} at the edge of the shower to 24,000 m^{-2} closer to the shower core. No estimate of the mass of the primary particle was made.

A galactic plot showing the arrival directions of the Yakutsk and Fly's Eye events is shown in Figure 2 of a recent report by Sigl, Schramm and Bhattacharjee.[7] Apart from the apparent coincidence in arrival directions of the two events (1% chance probability given the size of the error boxes and the exposure of the experiments)[8] there is nothing striking about the directions. An investigation of possible sources requires a discussion of galactic and extragalactic magnetic fields, and interactions of the cosmic rays in intergalactic space. Two excellent summaries have recently been prepared.[7,9]

Constraints on objects capable of accelerating particles to these energies seem to rule out galactic objects, although this is not at all certain.[10] If the particles are extragalactic, then it is likely that the sources are quite local, with distances less than 50 Mpc. This results from interaction mechanisms that limit the propagation distance for all the likely candidate particle types. These include interactions of nucleons with the 2.7 K CMB (pion photoproduction, mean free path about 5 Mpc, with the leading particle losing about 20% of its energy per interaction), photodisintegration of heavy nuclei by photons of the CMB (loss rate about 4 nucleons/Mpc at 3×10^{20}eV, severe), and pair production interactions for primary gamma-rays (with CMB and radio photons, together giving a mean free path of 7 Mpc).[9] A primary neutrino would avoid all of these interactions, but would have a 10^{-5} probability of interacting in the Earth's atmosphere. (The EAS profile measured by the Fly's Eye looks very normal, and does not require an exotic initiator). The 50 Mpc source distance limit can be stretched if, for example, a source produces an extremely energetic proton ($> 10^{23}$eV) at a larger distance (say 125 Mpc).[7] However it is clear that for unexotic particles and processes the distance limit is quite severe.

Very little is known about the strength and orientation of extragalactic magnetic fields, but it is likely that a 3×10^{20}eV proton would be deviated by less than 10° from its source direction.[7,9] (The 10° maximum angle results from a 10^{-9}G field perpendicular to the particle's path and coherent over a scale of 100 Mpc). Based on this assumption, and assuming that the particle was indeed a proton, Elbert and Sommers searched for a source of the Fly's Eye cosmic ray.[9] No obvious source (e.g a nearby radio galaxy with radio lobe hotspots) was detected within the 10° search region. Sources like M87 and Cen A are 87° and 136° respectively away from the nominal source direction, and may be plausible sources of heavier nuclei (especially the closer Cen A) which could be deviated in the magnetic fields, especially if stronger than expected extragalactic fields exist.

One viable but exotic possibility for the source of these particles is the decay of topological defects (e.g. cosmic strings).[7] Such a decay would result in the production of GUT mass particles ($mc^2 \sim 10^{24}$eV) which would decay to super-high energy hadrons.

It is presently impossible to decide between the competing possibilities for the origins of these three events, especially as no outstanding candidate sources are visible in the nominal event directions. However we are fortunate that two large detectors are either operational (AGASA) or under construction (HiRes). In the next several years we can hope for a further handful of extreme events. On the horizon is the Auger Project, which may become operational at the turn of the next century, and which promises up to 50 events a year above 10^{20}eV.[11]

REFERENCES

1. Efimov,N.N. *et al.* 1990. *In* Proc. Astrophysical Aspects of the Most Energetic Cosmic Rays. M.Nagano, F.Takahara Eds.: 434. World Scientific. Singapore.
2. Bird,D.J. *et al.* 1995. Ap.J. In press (March 1).
3. Hayashida,N. *et al.* 1995. Phys. Rev. Lett. In press.
4. Sokolsky, P., P. Sommers & B.R. Dawson. 1992. Phys. Rep. **317:** 225.
5. Bird,D.J. *et al.* 1993. Phys. Rev. Lett. **71:** 3401.
6. Bird,D.J. *et al.* 1994. Ap.J. **424:** 491.
7. Sigl, G., D.N. Schramm & P. Bhattacharjee 1994. Astropart. Phys **2:** 401.
8. Sommers, P. 1993. *In* Proc. Workshop on Tech. for the Study of Extremely High Energy Cosmic Rays. M.Nagano Ed.: 23. University of Tokyo. Tokyo.
9. Sommers, P. & J.W. Elbert. 1995. Ap.J. In press (March 1).
10. Hillas, A.M. 1984. Ann. Rev. Astron. Astrophys. **22:** 425.
11. Boratav,M. 1995. *In* Proc. this conference.

Propagation and Acceleration of Ultra-High Energy Cosmic Rays

VLADIMIR S. PTUSKIN

Institute of Terrestrial Magnetism, Ionosphere and Radio Wave Propagation
Russian Academy of Sciences
Troitsk, Moscow District 142092, Russia

INTRODUCTION

Present-day data on the most energetic cosmic rays with energies not less than 1 10^{20} eV include 5 events at the Haverah Park array, single events at Volcano Ranch, Yakutsk, Fly's Eye (with energy 3 10^{20} eV), and AGASA installations, and 8 events at the Sydney array. The extension of cosmic ray spectrum beyond 10^{20} eV imposes a strong restriction on the particle age $T < 3\ 10^8$ yr (and $T < 10^8$ yr for 3 10^{20} eV) when particle interaction with the universal microwave radiation is taken into account. Hence, these particles are accelerated within the Local Supercluster or even within the Galaxy.

On the other hand, the interpretation of cosmic ray, gamma ray, and radioastronomical data shows that the main fraction of cosmic rays in the interstellar medium is of galactic origin [1]. Extragalactic component may dominate only at very high energies above 10^{17}-10^{19} eV.

DIFFUSIVE SHOCK ACCELERATION AND THE PROBLEM OF PERPENDICULAR DIFFUSION

Diffusive shock acceleration is the most generally employed in the investigations of cosmic ray origin. The acceleration is due to repeated crossing of the shock front in a random particle walk and to an energy gain in the head on collisions with scattering centers embedded in the background plasma. A distinguishing feature of these acceleration is a power-law form of the particle momentum spectrum [2]. The maximum energy (momentum) of accelerated particles is determined from the condition $D(p_{max}) = Ru$ (a numerical factor is omitted). Here D is the particle diffusion coefficient increasing with momentum, R and u are the radius and the velocity of spherical shock wave, $u \ll c$. Using the Bohm value $D_B = r_g v/3$ for diffusion in strongly turbulent magnetic field (here v is the particle velocity, and r_g is the Larmor radius), one has the following estimate for the particle Larmor radius at the maximum energy

$$r_{g,max} = uR/c. \quad (1)$$

The use of Eq.(1) gives not extremely high energy of accelerated particles. The spherical shock produced in the interstellar medium by a supernova outburst with initial energy $3\ 10^{50}$ erg may provide cosmic ray acceleration up to 10^{14}Z eV, here Z is the charge of accelerated particle [3,4]. The supernova blast wave moving in the stellar wind of a presupernova star (the red giant star or the Wolf-Rayet star) with high magnetic field may give the maximum particle energy about 10^{16}Z eV [5,6]. The acceleration at the termination shock of a hypothetical galactic wind [7] allows to reach $3\ 10^{16}$Z eV.

In a shock moving strictly perpendicular to the regular magnetic field, the maximum energy of accelerated particles could be higher than given by Eq.(1) since perpendicular diffusion coefficient is smaller than the Bohm value [8]. At its best, the maximum Larmor radius of accelerated particles could reach $r_{g,max} = R$ which would rise the maximium energy by 2-3 orders of magnitude compared with the Bohm limit. However the shock front can not be perfectly parallel to the magnetic field if for no other reason than that the magnetic field has a random component which is necessary for particle scattering and spatial diffusion.

We consider a simple model to illustrate the problem. The length over which particle acceleration takes place is approximately equal to the diffusion length D/u, where D is the diffusion coefficient perpendicular to the shock front. Let us assume that the average magnetic field B_0 is parallel to the shock front. There is also weak random magnetic field B_1 with extended spectrum of inhomogeneities. The dimensionless amplitude of random field with the scales larger than l is $A_l = B_1(>l)/B_0 << 1$. The scattering on inhomogeneities with $l = r_g$ provides the spatial particle diffusion coefficient $k_{ll} = r_g v/(3A_g^2)$ along the local magnetic field and $k_l = r_g v A_g^2/3$ in the perpendicular direction (here $A_g = B_1(>r_g)/B_0$). The long-wave wandering of magnetic field lines leads to the following averaged diffusion coefficient perpendicular to the field B_0:

$$D = k_l + A_L^2 k_{ll} = \left(A_g^2 + A_L^2 / A_g^2\right) r_g v/3 > r_g v/3 = D_B, \quad (2)$$

where the level of turbulence in the principal scale L is assumed to be larger than at the scale r_g. Hence the value of the average diffusion coefficient D which determines the diffusion length is not smaller than the Bohm value. Our consideration is correct when one has a resonant scattering on the inhomogeneities of the order of r_g and the principle scale of the random field is larger than the parallel mean free path, i.e. $L >> k_{ll}/v$. More general and rigorous treatment of the problem is needed. It is worth to point out that the process of cosmic ray transport across the magnetic field remains not completely understood by itself [9].

COSMIC RAYS WITH THE HIGHEST ENERGIES

The maximum energy which particles can reach in a course of diffusive shock acceleration at the supernova shock is hardly sufficient to explain the extended cosmic ray spectrum beyond the knee at $3\ 10^{15}$ eV. Additional acceleration to energies in the range 10^{16}-10^{19} eV by the regular electric field in pulsar-driven SNR was considered [10]. A small fraction of energetic particles entering the envelope from the outer shock at the pole and exiting near the equator will gain the energy $E = ZecB_0(R_s/R)^2/9$, here B_0 is the magnetic field at the outer radius of the remnant R, and $R_s << R$ is the radius of internal cavity filled with relativistic pulsar wind. The particle motion is controlled by grad-B drift. The energy gain is about $E = 4\ 10^{15}Z$ eV in the Crab Nebular, but the energy gain can reach $10^{19}Z$ eV for a few msec pulsar. The potential available for acceleration is the same whether the charged particles orbit the neutron star near the light cylinder or near the outer edge of the remnant.

In principle, the maximum particle energy gained at an individual shock may be increased by additional acceleration at different numerous expanding shells [11-13]. In particular, collective acceleration may operate in OB star associations and probably increase particle energy up to about $10^{17}Z$ eV.

The difficulties with acceleration up to the highest energies in galactic sources and the observed high isotropy of cosmic rays stimulate the search of potential sources outside the Galaxy. As was mentioned in the Introduction, the detection of event at $3\ 10^{20}$ eV imposes a strong restriction on particle age $T < 10^8$ yr and hence the sources lie within the Local Cluster.

There are few models under the consideration. One of them consists of the diffusive shock acceleration in the hot spots of extended jets throwing away from the active galactic nuclei with velocities $u > 0.3c$ [14,15]. The maximum particle energy according to Eq.(1) reaches $10^{19}Z$ eV for the jet velocity $u = 0.3c$, the magnetic field $B = 10^{-5}$G, and the shock curvature radius R = 3 Kpc. The first order Fermi acceleration between approaching galactic magnetic coronas or between galactic winds could work in the case of colliding galaxies [16]. The maximum Larmor radius $r_{g,max} = R(u/c)^{1/2}$ which is larger than the Bohm limit (1) may be achieved in this model. The corresponding maximum particle energy is estimated as $10^{20}Z$ eV for the galactic wind velocities $u = 10^3$ cm/s, and the field $B=10^{-5}$ G in the magnetic coronas with the sizes R = 100 Kpc. A couple of galaxies M81 and M82 (the last has a star burst nuclei) might serve as a close source of ultra high energy cosmic rays. Another not well developed model employs the collective cosmic ray acceleration by the ensemble of galaxies within the Virgo cluster.

The study of cosmic ray propagation after their exit from the sources makes an integral part of the solution of cosmic ray origin problem. The modelling [17] of particle transport in galactic magnetic fields at energies more than 10^{18} eV has shown that two different models are possible. The two component model [18] includes a heavy galactic component with the steepening of the spectrum at $3\ 10^{17}$ eV plus light extragalactic component with flat spectrum. The extragalactic component dominates above $3\ 10^{18}$ eV. Pure galactic origin of all observed cosmic rays is assumed in the one component model [17]. The variations of particle energy spectrum and composition are explained then as a result of rigidity dependent escape of cosmic rays from the Galaxy.

Future measurements of cosmic ray anisotropy above 10^{18} eV together with the development of the theory of particle acceleration in the violent astronomical objects will play a decisive role in the selection of the proper model.

ACKNOWLEDGEMENT

The author thanks Frau W. Butsmann for her help in the preparation of this paper.

REFERENCES

1. Berezinskii V. S., S. V. Bulanov, V. A, Dogiel, V. L. Ginzburg & V.S. Ptuskin. 1990. Astrophysics of Cosmic Rays. North Holland.
2. Krymsky G. F. 1977. Sov. Phys.-Dokl. **23**: 327.
3. Prishchep V. L. & V. S. Ptuskin. 1981.Sov. Astron. **25**: 446.
4. Lagage P. O. & C. J. Cesarsky. 1983. Astron. Astrophys. **118**: 223.
5. Völk H. J. & P. L. Biermann. 1988. Astrophys. J. **333**: L65.
6. Berezinsky V. S. & V. S. Ptuskin. 1989. Astron. Astrophys. **215**: 399.
7. Jokipii J. R. & G. E. Morfill. 1985. Astrophys. J. **290**: L1.
8. Jokipii J. R. 1987 Astrophys. J. **313**: 842.
9. Chuvilgin L. G. & V. S. Ptuskin 1993. Astron. Astrophys. **279**: 278.
10. Bell A. R. 1992. Mon. Not. R. Astron. Soc. **257**: 500.
11. Berezhko E. G. & G. F. Krymsky. 1988. Sov. Phys. Uspekhi. **31**: 27.
12. Axford W. I. 1991. *In* Astrophysical Aspects of the Most Energetic Cosmic Rays. M. Nagano, F. Takahara. Eds.: 406. World Scientific.
13. Bykov A. M. & I. N. Toptygin. 1993 Sov. Phys. Uspekhi. **163**: 19.
14. Quenby J. J. & R. Lieu. 1989. Nature. **342**: 654.
15. Ip W.-H. & W. I. Axford. 1991. *In* Astrophysical Aspects of the Most Energetic Cosmic Rays. M. Nagano, F. Takahara. Eds.: 273. World Scientific.
16. Cesarsky C. J. & V. S. Ptuskin 1993. 23th Intern. Cosmic Ray Conf. Calgary. **1**: 341.
17. Pochepkin D. N., V. S. Ptuskin, S. I. Rogovaya & V. N. Zirakashvili. 1995. 24th Intern. Cosmic Ray Conf. Rome (submitted).
18. Bird D. J. & S. C. Corbato, H. Y. Dai et al. 1994. Astrophys. J. **424**: 491.

Possible Extragalactic Sources of the Highest Energy Cosmic Rays[a]

JÖRG P. RACHEN

Max-Planck-Institut für Radioastronomie
53010 Bonn, Germany

INTRODUCTION

The observation of cosmic ray events clearly above 100 EeV by the Fly's Eye[1] and AGASA[2] experiments on the one hand, and further evidence for an extragalactic origin of cosmic rays above about 3 EeV on the other hand (see below), create a dilemma for cosmic ray physics: because of the inevitable existence of the Greisen-Zatsepin-Kuzmin (GZK) cutoff at 60 EeV for protons and about 100 EeV for nuclei sources of such "super-GZK" cosmic rays have to be nearby, if not inside the Galaxy. Because the bending of particle paths in galactic and extragalactic magnetic fields over short distances is expected to be small, the sources should be observable somewhere close to the event directions, and it has been claimed that this is not the case.[3,4] However, it depends on the particles considered what "nearby" and "close" means, and it is not so easy to make a search for extragalactic sources complete in the galactic disk direction, where the Fly's Eye event was observed.

WHY EXTRAGALACTIC?

Obviously, the problem we discuss may be constructed by the claim that cosmic rays above 3 EeV are of extragalactic origin. We therefore want to give some arguments that support this claim.

First, the statistically significant observation of a simultaneous change in spectrum and composition of cosmic rays at the "ankle" at about 3 EeV by the Fly's Eye[5] strongly suggests a change of components of different origin. The light and flat component shows a spectral slope of -2.7 *above* the ankle; but if we deduce its spectrum *below* the ankle from composition data, we find that here the slope is near to -2.0.[6] Even though this steepening of the light component is less significant, it is interesting to note that this is exactly what is expected for extragalactic cosmic rays due to Bethe-Heitler losses.

Second, in spite of the observation of the highest energy events the existence of a GZK "break" is not in contrast to the data: most experiments show a lack of particles above 60 EeV, compared with the expectation for the case of a continuing $E^{-2.7}$ spectrum.[2,5] Some experimental groups claim the existence of a "gap" between 60 and 200 EeV, because they did not observe events in this region. However,

[a]This work was developed in part at the Bartol Research Institute, University of Delaware, Newark DE. Work of the author at the Bartol Research Institute was supported by a DAAD Doktorandenstipendium HSP-II/AUFE.

the Volcano Ranch, Haverah Park and Yakutsk arrays did observe in total 6 events slightly above 100 EeV.[7] In any case, the rather large energy errors allow a continuous distribution of the highest energy events as well, and, within the statistical flux errors, the "world data set" seems to be consistent with a spectral steepening at 60 EeV, which is expected for a universal source distribution with at least one local source to provide the highest energy events.[8]

Third, even if a gap exists (which would also suggest that the sub-GZK cosmic rays are extragalactic), the new component arising at 200 EeV is unlikely to be galactic, because all presently known acceleration models for compact objects fail to reach these energies by orders of magnitude, and an acceleration on galactic scales is unable to produce the extremely flat spectrum required for a rapid recovery.

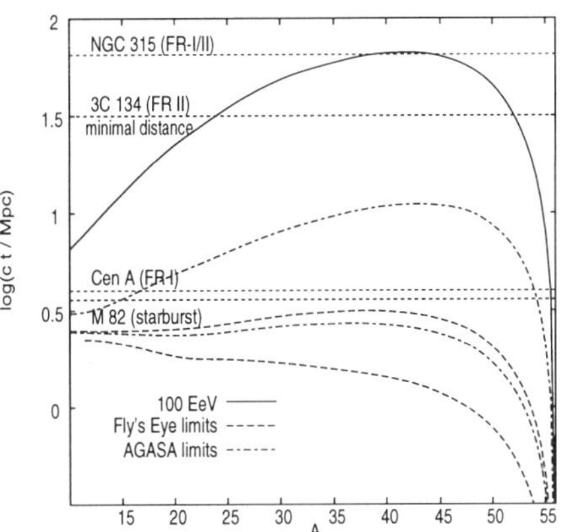

FIGURE 1: Range of iron nuclei, observed with mass A, for energies corresponding to the nominal GZK value for nuclei and the 1-σ statistical limits of the highest energy events.

RANGE AND DEFLECTION OF COSMIC RAY PARTICLES

At ultra high energies any particle may produce hadronic showers, which are more or less consistent with the data. Only neutrinos are expected to interact, if at all, much deeper in the atmosphere; however, particle physics at the interaction energy under discussion is far from being explored. We therefore are free to express the heuristic hypothesis of the existence of background immune particles (BIPs) with large hadronic cross section at extremely high energies. We now can distinguish the following possibilities:

Protons, with an average range of about 50 Mpc per decade of energy loss (above 300 EeV), and a 3-σ upper limit allowed by fluctuations about three times larger. The maximum deflection is less than 10°.

Heavy nuclei, which allow for large deflection, but a smaller range at high energies following FIGURE 1, and only small fluctuations.

Photons, propagating on straight paths and limited by interactions with the radio background to an average range of about 20 Mpc at 300 EeV (3-σ upper limit: 100 Mpc), strongly decreasing for smaller energies.

BIPs, which have an infinite range by definition. To be immune to photon interactions, they obviously have to be neutral and therefore point to their sources.

COSMIC RAY SOURCES AND THEIR ENERGY LIMITS

Basically we may distinguish between two kinds of sources for highest energy cosmic rays: *accelerators* and *quantum processes.*

Accelerators are usually easy to observe by the radio emission of accelerated electrons. Radio luminosity may thus be used as a measure for acceleration efficiency, but it depends on the proton-to-electron ratio k_p. FR-II galaxies, the strongest extragalactic radio sources, can not only accelerate cosmic ray protons and nuclei up to a few 100 EeV in their hot spots, but also give the correct flux and spectral slope to explain the sub-GZK cosmic rays,[9] assuming a value k_p consistent with the hadronic explanation of other observations.[10–13] Some of the weaker FR-I radio galaxies and compact jets in AGN may also be able to reach the energies,[14] but charged particles leaving the galaxy suffer adiabatic losses. Secondary neutral particles, however, can escape, thus FR-I galaxies may contribute energetic neutrons, and beamed jets in flat spectrum radio quasars (FSRQs) may supply neutrinos up to some 100 EeV.[15] Starburst galaxies, due to their high supernova rate, are expected to be sources of energetic heavy nuclei, which may reach highest energies.

For accelerating sources the cosmic ray flux can be estimated from observations. In contrast, the injection of energetic particles by decay of quantum-cosmological relics from the early universe, i.e. topological defects (TD), is much less observationally constrained.[4,16] These sources should appear as gamma pulses, observable above 100 EeV and maybe simultaneous in the TeV regime, but no systematic observations exist yet. Moreover, the energetics of the model has been questioned.[17]

CORRELATION AND CONSEQUENCES

The assumptions about particles *and* their sources fall into two classes, depending on the continuation of the cosmic ray spectrum beyond the GZK break: if it just steepens, we can assume classical particles (protons and nuclei) originating in classical sources (radio or maybe starburst galaxies); if it cuts off and resumes at about 200 EeV, we probably have to assume a new particle component (photons or BIPs), arising from a new source population.

If the spectrum steepens we have the following choices: For the Fly's Eye event, there is the FR-II radio galaxy 3C 134 close to it, of which the redshift is unknown because no optical counterpart has been found due to galactic obscuration. From the size of the radio structure, we can estimate its distance between 30 and 300 Mpc, making it a possible proton source candidate. Close to the AGASA event we find the FR-I galaxies 3C 31 and NGC 315, which are inside the 3-σ upper limit range of protons originating with >500 EeV. Considering heavy nuclei, we are less restricted in direction, and nearby starburst galaxies, as M 82 and NGC 253, can be discussed.

If there is a recovery of the spectrum, the new component may consist of photons from topological defects. However, the speculative BIP hypothesis opens a new possibility, based on an interesting observational coincidence: *both* the Fly's Eye and the AGASA event directions include a bright quasar in their error boxes; the FSRQ 3C 147 at $z \simeq 0.5$, and the radio weak PG 0117+213 at $z \simeq 1.5$, respectively. Since only a few 100 quasars with comparable brightness exist in the northern hemisphere, the probabilty to find this correlation by chance is less than 5%. While 3C 147 could be a source of extremely energetic neutrinos, this is unlikely for radio weak quasars.[15] However, both populations are connected by the expectation that these are *the only* objects where we have free view to the "surface" of a black hole, which is probably one of the most likely places for unexpected physics to happen.

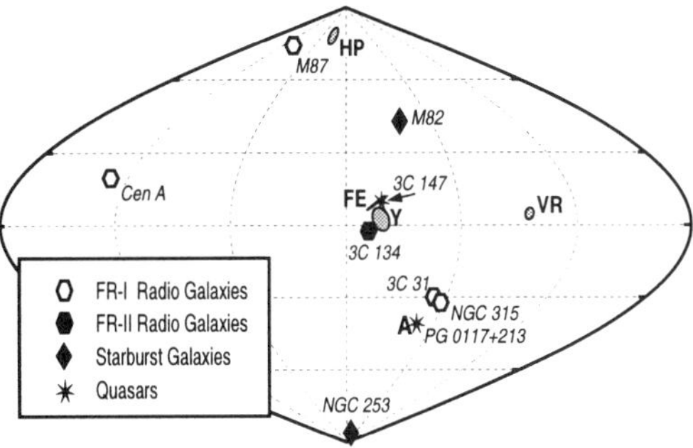

FIGURE 2: Distribution of the highest energy events of the experiments **A**GASA, **F**ly's **E**ye, **H**averah **P**ark, **V**olcano **R**anch and **Y**akutsk, and their possible sources discussed in the text, in galactic coordinates.

In conclusion, to constrain the speculations about the origin of the highest energy cosmic rays, we first need to get better event statistics above 100 EeV. If the high event energies and the high flux in this region can be confirmed, it would be interesting to check whether the correlation to quasar positions holds. Otherwise, if the events appear to be exceptional, they may be explained as the energetic tail of "normal" extragalactic cosmic rays.

ACKNOWLEDGEMENTS

This work was developed in collaboration with Drs. T. Stanev and P.L. Biermann.[18]

REFERENCES

1. BIRD, D.J. *et al.* 1995. Astrophys. J. **441:** in print.
2. HAYASHIDA, N. *et al.* 1994. Phys. Rev. Lett. **73:** 3491.
3. ELBERT, J.W. & P. SOMMERS. 1995. Astrophys. J. **441:** in print.
4. SIGL, G., D.N. SCHRAMM & P. BHATTACHARJEE. 1994. Astropart. Phys. **2:** 401.
5. BIRD, D.J. *et al.* 1994. Astrophys. J. **424:** 491.
6. RACHEN, J.P., T. STANEV & P.L. BIERMANN. 1993. Astron. Astrophys. **273:** 377.
7. WATSON, A.A. 1994. Paper pres. at the APS meeting in Snowmass, CO, July 1-14.
8. BEREZINSKIĬ, V.S. & S.I. GRIGOR'EVA. 1988. Astron. Astrophys. **199:** 1.
9. RACHEN, J.P. & P.L. BIERMANN. 1993. Astron. Astrophys. **272:** 161.
10. MANNHEIM, K. 1993. Astron. Astrophys. **269:** 67.
11. HARRIS, D.E., C.L. CARELLI & R.A. PERELY 1994. Nature **367:** 712
12. BÖHRINGER, H. *et al.* 1993. Mon. Not. R. Astron. Soc. **264:** L25.
13. FALCKE, H. & P.L. BIERMANN. 1995. Astron. Astrophys. **293:** 665.
14. BIERMANN, P.L. & P.A. STRITTMATTER. 1987. Astrophys. J. **322:** 643.
15. MANNHEIM, K. 1995. Astropart. Phys. in print, and personal communication.
16. HINDMARSH, M.B. & T.W.B. KIBBLE. 1995. Submitted to Rep. Prog. Phys.
17. GILL, A.J. & T.W.B. KIBBLE. 1994. Phys. Rev. **D50:** 3660.
18. BIERMANN, P.L. 1994. Paper presented at "Trends in Astroparticle Physics", Stockholm, Sweden, September 21-25. To be published in Nucl. Phys. B, Proc. Suppl.

Search for Cosmic Gamma-Rays above 24 TeV with the HEGRA Detector

E. LORENZ
HEGRA Collaboration
Max Planck Institut für Physik,
Werner Heisenberg Institut,
Föhringer Ring 6, D 80805 Munich, FRG

INTRODUCTION

Cosmic ray (CR) physics has recently entered a new phase of interest. The research is both driven by new questions in high energy Astrophysics, particle physics (above current accelerator energies) and Cosmology and on the other hand by considerable progress in detector performance allowing for a quantum jump in sensitivity. Many open questions are linked to the detection of energetic gamma rays (γ) as only these can be extrapolated back to their origin that should be also the origin of the charged CRs. The fraction of very high energy or ultra high energy (VHE/UHE) cosmic γ's is extremely low and experimenters have only modest tools at hand to discriminate them against the hadronic CRs. For the search of the origin of cosmic radiation the recent satellite experiments (COS B, EGRET) allow us to make extrapolation to the VHE/UHE range of what sensitivity is at least needed. Physics studies in this energy range have to be carried out by large, ground based detectors. Currently the most successful detectors are air Cerenkov telescopes (ACTs) and first high significance observations in the VHE range of γ sources, the Crab nebula, Mrk 421 and PSR 1706, have been reported while the majority of the detectors in the UHE range, the wide angle scintillator arrays, have still to report the first high significance observations. Their main drawback is the high energy threshold close to 50-200 TeV, a modest angular resolution and a very poor gamma hadron (γ/h) separation power. ACTs have on the one hand a much lower detection threshold compared to extensive air shower scintillator arrays, thus they can see much deeper into the universe (above $\approx$ 100 TeV the universe is not more transparent over cosmological distances due to the γ-2.7° background interaction), but on the other hand they have only very limited angular acceptance and can only be used for the study of preselected astronomical objects. A possible way out might be the construction of wide angle Cerenkov detectors that sample the Cerenkov light disc at many positions.

The HEGRA collaboration has recently built the first full sized wide angle Cerenkov detector, dubbed AIROBICC, and added it to the HEGRA cosmic ray detector complex on the Canarian island La Palma. Here we report first results of a search for γ emission of potential sources and a first direct search for isotropic γ radiation predicted by some models of topological defects.

THE HEGRA DETECTOR

The HEGRA detector complex is located on the site of the Instituto Astrophysico Canarias at the Roque de Los Muchachos on La Palma (28.8° N, 17.8° W, 2220 m asl). The main element is a 13 by 13 matrix of scintillation counters spaced 15 m apart. This matrix is complemented by 52 additional scintillation counters in the central region to lower the detection threshold and to achieve a higher precision in shower core position. Interspersed to the scintillation counters are 49 air Cerenkov detector stations in a 7x7 matrix of 30 m grid spacing. These detectors, hemispherical photomultipliers augmented by Winston cones, view the night sky directly. From the timing information of the various fired detector stations one can deduce the incident direction with high precision while from the analog signals one can determine the incident energy and some air shower parameters like the maximum of the shower development. Details of this novel detector can be found in ref.[1]. The main performance parameters of both arrays are summarised in table 1. Besides these two matrices the detector complex comprises 17 large area muon detectors (not used in this analysis) and two air Cerenkov telescopes. The combination of many different detector elements at one site allow us, besides the normal task to record many shower parameters at once, to cross check performance data of the various elements from simultaneous observations.

THE DATA

The data for this analysis were recorded during the 'run-in' phase of AIROBICC between May 1992 and March 1993. We recorded the data from all stations whenever at least 14 stations in the scintillator array or at least 6 stations in AIROBICC were fired. We used for this analysis only events seen at least in AIROBICC because this allowed us to determine the energy with high precision and, in case of the sample above 60 TeV energy, γ/hadron (γ/h) separation with high efficiency. In total a sample of $2.4 \cdot 10^7$ showers was recorded, corresponding to 821 hours observation time. The data were processed by our standard reconstruction program yielding the core position, the primary particle direction, the incident energy (ambiguous with particle type), the shower size N_e and from the analysis of the radial Cerenkov light

TABLE 1. Main performance data of the HEGRA arrays.

	Scintillator array	AIROBICC array
No of detector stations	223	49
gross detection area	36000 m^2	36000 m^2
γ-threshold	40 TeV	16 TeV
p-threshold	40 TeV	40 TeV
Fe threshold	90 TeV	90 TeV
mean angular resolution	0.63° (for 60 TeV γ's)	0.2° (for 30 TeV γ's)
trigger rate	12 Hz (≥ 14 hits)	14 Hz (≥ 6 hits)
Up-time	continuous	≈ 15%
γ/h enrichment	-	typically 3
γ/h separation	none	≈ 100 at 60 TeV
angular acceptance	≈ 1 sterad	≈ 1 sterad

distribution a measure of the height of the shower maximum. We demanded that at least 12 AIROBICC counters had fired for the fit of the Cerenkov data. Finally a series of quality cut was applied to the data set.

The threshold for various particles was calculated by Monte Carlo (MC) simulations from detector and atmospheric parameters and cross checked with measurements of the cosmic radiation flux. Good agreement has been found.

SEARCH FOR γ POINT SOURCES

The data set recorded with AIROBICC allows to lower considerably the energy threshold for γ observation from potential sources and to achieve a higher sensitivity compared to the scintillator array data although the mean 'up-time' of AIROBICC is only 15% per day. This is due to the better angular resolution and the nearly a factor two larger Cerenkov light yield in γ induced showers compared to hadron induced showers of identical energy. After all cuts the final data set used for this analysis contained $8.475 \cdot 10^6$ showers that were then searched for possible γ emission from a catalogue of about 60 potential sources, mostly of galactic origin. About half of the potential sources were SNRs that are considered to be the main sources of VHE/UHE cosmic radiation.We treated SNRs as point sources.

The search region for a given source candidate is defined as a circular area around its nominal position on the sky. In the normal case of the background being Gaussian distributed the optimal source radius for maximal significance is $\vartheta_S = 1.1 \cdot \sigma_{63}$ with σ_{63} being the radius of a cone that contains 63% of all events from a point source. Besides the statistical error, σ_{63}, we have to consider the contribution from the pointing uncertainty of 0.1°. In this analysis we used 0.39° as the radius of the source region that contains then at least ≥ 73% of all possible point source events The number of background events expected in the source region is extrapolated from the environment just outside it. We defined the background region as the ring centred on the source with an inner radius of $\vartheta_i = 0.5°$ and an outer one of $\vartheta_O = 2.4°$. The statistical error in the number of extrapolated background events is then almost negligible in comparison to the error on the number of source events. The systematic over- or underestimate of the background is less than 0.5% in our analysis, e.g., the systematic errors are negligible when compared to the purely statistical one of around 6% resulting from the number of events in the source region. By following the prescription of Li and Ma[2], one can calculate the significance of a possible signal from the difference of the number of observed events N_S in the source region and the number of background events N_{BG} extrapolated from the surrounding region. This method takes the fluctuation of the extrapolated background properly into account. If no significant excess is observed the results can be used to calculate an upper limit $N_{UL}(CL)$ at a certain confidence level (CL) which we have chosen to be 90%. With the measured cosmic ray flux around the source region one can convert $N_{UL}(90\%)$ to the flux limit of the chosen source.

No significant excess from any of the 60 potential sources has been found. The relevant event statistics, the energy thresholds and the derived flux limits for previously unpublished objects are listed in table 2 while the limits from 36 sources can be found in ref.[3] The threshold energy for possible γ-radiation ranges from 24 to 52 TeV with an average value of 28 TeV. The upper limits to the integral flux of on average $2.6 \cdot 10^{-13}$ events/cm^{2}s for steady radiation have been determined from the rather short observation time of typically 200 hours per object and are similar to or better than those from data recorded over several years with large EAS scintillator arrays when scaling their flux limits by an $E^{-1.7}$ dependence to our energy.

In case of the Crab and Mrk 421 our limits are in the range or below the flux values extrapolated from HE and VHE observations. For the object G 78.2 -2.1 Aharonian et al [4] predict a large γ rate originating from the shock wave of the SNR hitting a dense molecular cloud, DR4, while our flux limit is at least a factor 10 below the expectations.

SEARCH FOR ISOTROPIC γ's BETWEEN 65 AND 200 TEV

The search for isotropic γ's relies entirely on the ability of a detector to suppress efficiently the large hadronic background in the same energy range. The combined information from the HEGRA scintillator array and AIROBICC allow for a hadron suppression of about a factor 100 for γ energies above 60 TeV. From models[5] of topological defects one expects in the energy range around 60-200 TeV a fraction of γ's to the total cosmic ray flux of around 5 %. This is for masses of about 10^{16} eV for the X particles released by the defects. The underlying process for the large γ fraction is the interaction of extreme high γ's, from the X particle decay into hadrons, with the 2.7° background radiation resulting in a large multiplication and accumulation of secondary γ's around 10^{14} eV.

The efficient γ/h separation of the HEGRA detector relies on the fact that γ and hadron induced showers have a different longitudinal development. On one hand the scintillator array allows for a measurement of the shower size N_e at the shower tail while the AIROBICC measurements give a (calorimetric) measurement of the shower energy and the position of the shower maximum. The ratio of N_e over the energy, corrected by the shower maximum position, differs nearly by a factor two for γ's and hadrons and can be used for efficient separation. The limits of the method are given by the inevitable fluctuations of the shower process. Details of the method can be found in ref.[6].

The data for the search were selected from a sample taken under excellent optical conditions and restricted to zenith angles below 25° where our angular acceptance is nearly flat in energy. We have chosen an energy range between 80-200 or 65-160 TeV where according to the model prediction the γ fraction should be maximal. Above about 100-150 TeV cosmological γ's should be completely suppressed. No excess signal has been found above the tail of hadronic events extending into domain expected to be populated by γ's. Treating the hadronic tail as potential γ's we estimated 90% confidence upper limits for these γ's of

$f_\gamma (E_\gamma > 80 \text{ TeV}) < 7.8 \cdot 10^{-3}$ of the charged cosmic ray flux above 80 TeV and

$f_\gamma (E_\gamma > 65 \text{ TeV}) < 10.3 \cdot 10^{-3}$ of the charged cosmic ray flux above 65 TeV.

These results rule out the predictions of Aharonian et al.[5], but it should be mentioned that in case of a large infrared background the predicted γ flux can be seriously altered, e. g., the maximum ratio would be expected at lower energies. Additional details can be found in ref.[7].

Table 2.: Energy threshold, γ flux limits of galactic SNR's and significances

Source	γ threshold in TeV	flux limit 10^{-13}/cm^2.sec	significance in units of σ
G 78.2 DR4 position	25.1	2.25	-0.55
G 78.2, EGRET pos.	25.0	1.93	-0.94
G 78.2, COS B pos.	25.0	2.45	-0.35
G 78.2 SNR center	25.1	1.99	-0.78
G 180.0 -1.7	21.0	3.76	0.37
G 189.1 +3.0	20.9	1.75	-1.63
G 205.5 +0.2	27.9	4.00	1.13
G 30.7 +0.1	53.2	1.70	0.30
G 34.6 +0.0	40.9	2.66	0.88
G 40.5 -0.5	32.6	1.32	-1.34
G 41.1 +0.3	31.7	1.26	-1.51
G 156.4 -1.2	24.6	2.57	-0.27
G 33.0 +0.1	45.8	1.58	-0.17
G 35.6 -0.0	39.0	1.60	-0.46
G 37.6 -0.1	35.8	1.78	-0.38
G 39.2 -0.3	33.9	4.17	1.17
G 43.3 -0.2	30.4	2.28	-0.11
G 45.5 +0.1	29.0	2.64	0.15
G 46.8 -0.3	28.3	3.90	1.10
G 49.0 -0.3	26.9	2.82	0.16
G 53.7 -2.2	26.0	2.34	-0.38
G 74.8 +0.6	24.7	3.67	0.65
G 78.3 +2.5	25.0	3.34	0.43
G 78.6 +1.0	25.1	2.38	-0.41
G 78.9 +3.7	25.4	1.78	-1.12
G 79.8 +1.2	25.2	1.67	-1.31
G 111.7 -2.1	42.3	1.83	0.02
G 47.0 +10.6	33.3	2.59	0.38
G 69.9 +1.5	33.5	2.47	0.28

ACKNOWLEDGMENT

I would like to thank my colleagues from the HEGRA collaboration for the provision of data from the current AIROBICC analysis.

REFERENCES

1. Karle A. et al.: Design and performance of the angle integrating Cerenkov array AIROBICC. 1994, to be published in Astroparticle Physics.
2. T. Li T. and Ma Y., Ap. J., **272**, 317 (1983)
3. Karle A. et al. : Search for astronomical point sources of γ radiation above 25 TeV energy. to be published in Astroparticle Physics
4. Aharonian F., Drury L. and Völk H., Astron. Astrophys, **285** (1994) 645
5. Aharonian F. et al.; Phys. Rev. D, Vol **46**, No 10 (1992) 4188
6. Arqueros F. et al.: Separation of gamma and hadron initiated air showers with energies between 20 and 500 TeV. 1994, submitted to Nuc. Inst. Meth.
7. Karle A. et al., : Search for isotropic gamma radiation of cosmological origin between 65 and 100 TeV. 1994, MPI Preprint MPI-PhE 94-28

Necessity and reality of experimental investigation of ultrahigh energy cosmic rays ($10^{19} \div 10^{21}$ eV)

YU.A.FOMIN, G.B.KHRISTIANSEN, G.V.KULIKOV

Institute of Nuclear Physics, Moscow State University, Moscow, 119899, Russia

Many sharp problems such as black hole identification, clarification of barster nature and so on were arised at this symposium. The problem of superhigh energy cosmic ray sources is also one of them. The main part of superhigh energy cosmic rays is evidently consists of protons and nuclei which diffuse effectively from a source to an observer. However at energy about 10^{20} eV the proton Larmour radius appears to be more of the Galactic dimension and the diffusion is ended [1]. At the same time in principle the possibility of determination of cosmic ray source direction (proton astronomy) appears [2, a]. Taking into account the rudimentary state of the neutrino astronomy and difficulties of the high energy gamma-astronomy in PeV-region the conviction was recently confirmed that the problem of origin of particles with energies $10^{15} \div 10^{20}$ eV may be solved probably by the investigation of upper limit of mentioned range but not the low limit. This statement turns into the confidence if we take into account also the existence of the cosmic proton generation problem in known astrophysical objects at energies greater 10^{20} eV. At present any astrophysical objects with parameters (object dimension and magnetic field) wich sufficient for particle acceleration upto mentioned above ultrahigh energies are not known [3]

The last decade investigations with help of Haverah Park array (10 km^2), Yakutsk array (15 km^2), Akeno (20 km^2) and Fly's Eye show that aperture of these arrays and the accuracy of primary particle parameters (energy and arrival direction) are not sufficient for the decision of problems mentioned above.

The suggestion of the construction of array with total area of 1000 km^2 and with great sensitive area of scintillation detectors was published yet in 1987 [4]. This great sensitive area gives the possibility to improve essentially the main parameter accuracy. In 1989 the project of such array called EAS-1000 [b] was published in Texas Symposium Proceedings in the same year [5], and then this project was discussed and published repeatedly in the

[a] At some lower energies (~ 10^{18} eV) the neutron decay life time in observer coordinate system becames large enough to pass the Galaxy and the neutron astronomy may be developed.

[b] The primary particle main parameters are determinated by the registration of extensive air showers (EAS) of secondary particles generated by primary particle in the atmosphere of the Earth.

Cosmic ray International Conferencies (Adelaide, 1990, Dublin, 1991, Calgary 1993) and also in the International and European Symposia (Kofu, 1990, Paris, 1992, Ann-Arbor, 1992, CERN, 1992, Tokyo, 1994, Balatonfuered 1994, Moscow, 1994).

The main feature of the EAS-1000 project is the use of the EAS electron component for shower registration and the use of optical cables for the synchronization all detectors and for the collection of experimental data.

The EAS registration with electron detectors has the evident preference to another used methods, for example optical one, as it gives the possibility to register showers non-stop at a day, at a night and at any weather. It is important especially for investigation of cosmic ray anisotropy[6, 7, c]. The stability of registration is guaranteed technically by use of optical cables and this communication system is undoubtedly reliabler than high frequency radiocommunication (~ 1 Ghz) which is necessary for synchronization with given accuracy. Moreover the optical cable construction may be used for high electric power feed to scintillation detectors and so this system does not propose the use of solar batteries with accumulators.

According to the EAS-1000 project the array consists of about 4000 scintillation detectors of the area 1 m^2 each. with the distance 500 m. between them. These detectors cover uniformly the total area. Moreover the array has 10 muon detectors of 2000 m^2 total area and about 100 undirectional Cherenkov detectors[8].

The scintillation detectors are used for the determination of such EAS parameters as ρ_{600} (the primary energy measure) and the orientation and position of the axis, the Cherenkov detectors are used for absolute energy calibration and muon detectors together with Cherenkov detectors - for study of primary particle nature (chemical compositon). Before the EAS-1000 array construction the prototypes for electron component registration (MSU and AS RK)[9] and for muon registration (MSU and Yakutsk)[10] were worked out and constructed. For the acquisition system the Xilinx multifunctional programmable matrix logics is used[11].

The prototype for the electron component registeration (16 scintillation detectors) was operated during 1000 hours in the real conditions. As a result the necessary technical data on the electronic system and the optical cable communication were received. In particular it was established that the temporal stability of laser transmitters and receivers is about 1 ns if the temperature changes in the wide range from -10 C to +20 C.

The prototype of the 200 m^2 muon detector is included in present into the Yakutsk EAS array for the study of the primary cosmic ray mass composition.

The amplitude measurements of scintillation detectors are made in the range of $1 \div 5.10^4$ relativistic particles with the accuracy of 10%, the integration time changes from 5 ns to 10 μs. The temporal measurements are made with accuracy in the range of 5ns ÷ 5 μs.

The muon detectors with area of 100 m^2 have the more narrow dynamic range for amplitude measurements, namely from 1 to 3.10^3 particles per detector. There is the possibility to change for muon detectors the temporal gate width and to displace it relatively the front of the EAS electron component.

[c] It is not excluded the using of optical detectors in content of the EAS-1000 array for EAS absolute energetic calibration and for determination of such important EAS parameter as is the shower maximum position in the atmosphere.

The Cherenkov detectors allow to measure the density of the photon flux in range of 20 ÷ 20000 photons for wavelengths 300 ÷ 600 nm.

There is also a possibility to measure the arrival time of the photon front with accuracy 5 ns and the halfwidth of optical pulse. All these technical characteristics of detectors allow to determine such main parameters of the primary particle as the arrival direction (with 1 deg.accuracy), the position of the point of transection of the observation level by the primary particle trajectory (with ~10 m. accuracy), the primary particle energy (with 10% accuracy) and the shower maximum depth. It should be noted that the scintillation and Cherenkov detector data gives the opportunity to realize the mutual control of these data.

Above mentioned accuracies of the primary particle parameter determination allow to approach the solving of the main astrophysical problem - the problem of ultrahigh energy cosmic ray origin. The more special problems of the cosmic ray physics will be also studied, namely, 1/ the existance and position of the irregularity of the primary cosmic ray spectrum at energies $3.10^{18} \div 10^{19}$ eV, 2/ the study of the relic cut off effect and the choice of Galactic or Metagalactic cosmic ray origin model, 3/ the measurement of cosmic ray anisotropy at energies $10^{18} \div 10^{20}$ eV, 4/ the estimation of the cosmic ray mass composition at energies $10^{18} \div 10^{19}$ eV [12], 5/ the measurement of the temporal-lateral distribution functions (TLDF) of electrons, muons and photons at large (more than 1.5 km) distances from the shower axis. The latter problem is essentially important for the construction of a new generation of arrays (with greater aperture) and for quantitative separation of the Cherenkov radiation and luminescence.

The detail description of the array acquisition system is given in [13] (see also [8] and [9]). Here we notice that the total information flux from all detectors is about 10^5 B/s. This flux includes the experimental data together with detector background, control measurements of detector parameters, data for the amplitude and temporal calibrations, data on the atmosphere conditions and so on. Before data transferring to computer the selection of correlated in time and space detectors is made that corresponds to the selection of four-fold coincidences with resolution time of 10 μs. This allows to reduce the information flux $\sim 10^3$ times comperatively with the initial one and to use only 10 usual IBM PC/AT 386. During 1 year of the operation of the array such as the EAS-1000 array the information of 5 ÷ 10 GB will be stored up. Strimmers of HP35-48 with memory of 8 GB will be used for keeping of this information.

Really the creation of the EAS-1000 array in FSU will be began with the construction of the EAS-100 array having 10 times smaller aperture. The technology of the EAS-1000 array will be tested with the EAS-100 array and the creation of the EAS-1000 array will be equivalent to the construction of 10 copies of the operating the EAS-100 arrays. The construction of the EAS-100 array including 400 scintillation detectors, 100 Cherenkov detectors and 200 m^2 muon detector (see [9]) is of great interest inspite of operation of the AGASA array (Japan) with effective area of 80 km^2 It is connected with the following reasons: 1/ The EAS-100 array gives the possibility to realize the energetic calibration of showers, 2/ The energetic threshold of the EAS-100 array is four times smaller that the AGASA array and the upper energetic limit of detected showers is approximately the same. It is important for anisotropy measurements at ultra-high energies ($\sim 10^{18} \div 10^{19}$ eV). 3/ The double temp of the world data statistics increase after the EAS-100 array putting into operation, 4/ The mutual control of both array data.

At present the realisation of EAS-100 array depends on the financial support only. The technical project of the EAS-100 array is worked out. The prototypes of scintillation and muon detectors and even the technological samples of scintillation, Cherenkov and muon detectores are elaborated. Scientific group of Moscow State University has the great expierence in cosmic ray physics and known scientific results.

The EAS-1000 project uses so called "classical" approved communication with the optical cable for data collection, synchronization and also for power supply[d].

For the analysis of the EAS-1000 array experimental data it is necessary to use the EAS-100 experimental results on the TLDF of electrons, muons and Cherenkov photons for the great distancies from the shower axis.

At present the possibility of construction the of EAS array with area of 5000 m^2 (EAS-5000 or AUGER-project) is discussed [14]. This project proposes the distribution of detectors of 10 m^2 area with the distance of 1.5 km between them, the use of solar batteries as power sources and the satellite communication between detectors.

Do not considering a technical difficulties of the AUGER-project it should be noted that the experimental data on TLDF up to distances at least several kilometers from the shower axis are needed for array projects with such large distances between detectors.

These data can be got with the EAS-1000 array. Using real detected showers with this array it is possible to simulate the patterns of the registered showers with the EAS-5000 array and in this way to estimate the accuracies of shower parameter determination (primary energy, axis position, axis orientation etc) which are expected on this array.

So the creation of the arrays of a new generation is needed to realize gradually step by step the transition from the EAS-100 array to the EAS-1000 array and then to the arrays of greater areas.

[d] This method was approved on the Yakutsk array, Akeno and Prototype of EAS-1000.

REFERENCIES

1. Zirakashvili V.N. et.al. Izvestia RAN, Ser.Fiz. 1995, v.59: in press
2. Pogorely V.G., Khristiansen G.B. Izvestia RAN, Ser.Fiz. 1993, v.57, p.82-85
3. Hillas A.M. Ann.Rev.Astron.Astrophys. 1984, v.22, p.425-444
4. Khristiansen G.B. Uspekhi Fiz.Nauk, 1987, v.152, p.341-344
5. Khristiansen G.B. et.al. Proc. 14th Texas Symp. on Relativistic Astrophys. Annals N-Y Acad. Sci. 1989, v.571, p.640-644
6. Khristiansen G.B. et.al. Proc. 21st ICRC, Adelaide, 1990, v.10, p.282-285
7. Khristiansen G.B. et.al. Nucl.Phys.B (Proc.Suppl.) 1992, v.28B, p.40-48
8. Khristiansen G.B. et.al. Proc.14th ECRS, Balatonfured, 1994: in press
9. Khristiansen G.B. et.al. Proc. 22nd ICRC, Dublin, 1991,v.4, p.484-487
10. Efimov N.N. et.al. ICPRA, Yakutsk, Otchet, 1991
11. Khristiansen G.B. et.al. Izvestia RAN, Ser.Fiz., 1994, v.58, p.63-66
12. Atrashkevich V.B. et.al. Pis'ma, ZETF, 1981, v.33, p.236-239
13. Fomin Yu.A. et.al. Proc 8th Intern.Symp. VHECRI, Tokyo, 1994, p.536-551
14. Cronin J.W. Proc. Tokyo Workshop Techniques for Study Extremely HE CR, 1993, p.87-98

LISA

Laser Interferometer Space Antenna for gravitational wave measurements

KARSTEN DANZMANN for the LISA Study Team

Max-Planck-Institut für Quantenoptik, D–85748 Garching and Universität Hannover, D–30167 Hannover, Germany

Abstract

LISA (Laser Interferometer Space Antenna) is designed to observe gravitational waves from violent events in the Universe in a frequency range from 10^{-4} to 10^{-1} Hz which is totally inaccessible to ground based experiments. It uses highly stabilised laser light (Nd:YAG, $\lambda = 1.064\,\mu$m) in a Michelson-type interferometer arrangement.

A cluster of six spacecraft with two at each vertex of an equilateral triangle is placed in an Earth-like orbit at a distance of 1 AU from the Sun, and 20° behind the Earth. Three subsets of four adjacent spacecraft each form an interferometer comprising a central station, consisting of two relatively adjacent spacecraft (200 km apart), and two spacecraft placed at a distance of 5×10^6 km from the centre to form arms which make an angle of 60° with each other. Each spacecraft is equipped with a laser.

A descoped LISA with only four spacecraft has undergone an ESA assessment study in the M3 cycle, and the full 6-spacecraft LISA mission has now been selected as a cornerstone in the ESA Horizon 2000-plus programme.

The LISA Assessment Report is available as ESA document SCI(94)6, May 1994.

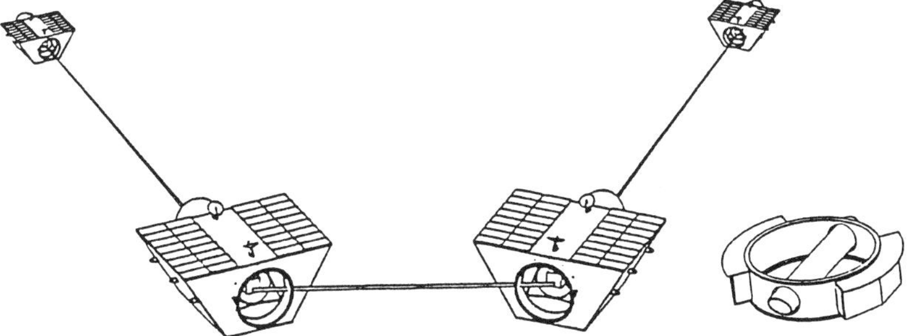

Fig. 1. *Configuration of a single LISA interferometer with four spacecraft. This earlier design of trapezoidal boxes was later changed into flat circular disks, their axes normal to the interferometer plane, as indicated for a single spacecraft (with top lid removed) on the righthand side.*

Overview

The goal of LISA (Laser Interferometer Space Antenna) is to detect and study low-frequency astrophysical gravitational radiation. The data will be used for research in astrophysics, cosmology, and fundamental physics. LISA is designed to detect the gravitational radiation from regions of the Universe that are strongly relativistic, e.g. in the vicinity of black holes. Such regions are difficult to study by conventional astronomy. The types of astrophysical sources potentially visible to LISA include galactic binaries of black holes, extragalactic supermassive black hole binaries and coalescences, and background radiation from the Big Bang. LISA will also observe galactic binary systems which are theoretically well-understood and observationally known to exist. Observation of these will provide strong verification of the instrument performance and a direct test of General Relativity.

LISA, with an array of six spacecraft, will measure such gravitational waves interferometrically. A single two-arm Michelson-type interferometer is formed from a vertex (actually consisting of two closely-spaced 'central' spacecraft), and two remote spacecraft defining the end-points of the arms, as indicated in Fig. 1. The full six-spacecraft configuration, with two spacecraft at *each* vertex of an equilateral triangle, thus consists of three separate, but not fully independent, interferometers. This configuration provides redundancy against component failure, gives better detection probability, and it allows the determination of polarisation.

When a gravity wave passes by it causes a strain distortion of space. LISA will detect these strains down to a level of order 10^{-23} in one year of observation time, by measuring the fluctuations in separation between shielded proof masses located 5×10^6 km apart. The measurement is performed by optical interferometry which determines the phase shift of laser light transmitted between the proof masses. Each interferometer has two symmetric arms in order to cancel out effects due to laser frequency noise. All spacecraft have a laser on board. The lasers in the two central spacecraft (which are 200 km apart) are phase-locked together, so they effectively behave as a single laser. The lasers in the end spacecraft are phase-locked to the incoming light, and thus act as amplifying mirrors.

Each proof mass is shielded from extraneous disturbances (e.g. solar radiation pressure) by the spacecraft in which it is accommodated. Drag-free control servos enable the spacecraft to precisely follow the proof masses. The relative displacement between the spacecraft and proof mass is measured electrostatically and the drag compensation is effected using proportional electric thrusters. Careful thermal design ensures the required mechanical stability.

The need for space-based detectors

LISA will complement the next-generation ground-based detectors (VIRGO, LIGO) by accessing the important low-frequency regime (10^{-4} to 10^{-1} Hz) which will *never* be observable from the Earth because of terrestrial disturbances. This low-frequency window allows access to the most exciting signals, those generated by massive black hole formation and coalescences, as well as the most certain signals, such as from galactic binaries. Ground-based detectors, on the other hand, are most likely to observe the rapid bursts accompanying the final stages of a compact binary coalescence. LISA would observe the 'low-frequency' epoch where the binary systems spend most of their life.

Cosmic background gravitational radiation, which spans a wide frequency range, may be detectable. With comparable energy sensitivites, LISA and the ground-based detectors will, in combination, provide much extended spectral coverage, essential to test cosmological models.

Extragalactic astronomy. The ideas that many galaxies (including our own) contain massive black holes, and that mergers of galaxies were common in the past, are gaining widespread acceptance. There is even evidence of binary black hole systems; an example is 3C66B which shows a precessing jet. Mergers of galaxies should produce *mergers of their supermassive black holes,* and their gravitational waves would be detected wherever in the Universe the event occurred. Recent calculations suggest that the event rate might even be as frequent as once per month.

The signal-to-noise ratio is typically several thousand for $10^6 M_\odot$ black holes. Waves this strong might not only be useful in testing gravity, as remarked above, but may make an important contribution to fundamental cosmology. By monitoring the amplitude and phase of the merger waves while the detector rotates, both the direction and total amplitude of the waves may be determined. Then, if the direction can be used to identify the source of the waves within a known cluster of galaxies, the amplitude will give an independent distance measurement to the source. A single redshift measurement would then determine the *deceleration parameter* q_0, and hence the mean density of the Universe, and thus measure the total density of *dark matter*. Merging galaxies may also trigger the *formation of massive black holes,* since they may replicate conditions at the time of galaxy formation. These formation events would also be detectable and identifiable. They may also be common, even the dwarf elliptical M32 seems to have a black hole.

Experiment description

A *single* LISA interferometer, as shown in Fig. 2, consists of a V-formation of four proof masses each shielded by a drag-free spacecraft. The vertex of the antenna's V-formation is formed by the two central spacecraft. In principle, one central spacecraft would be sufficient, but the optical system and attitude control requirements would be prohibitive. The four (6) spacecraft are in heliocentric orbits. They lie in a plane which is 60° to the ecliptic such that their relative orbit is a stable circular rotation with a period of 1 year. The 'constellation' should be located as far behind the Earth as possible (but limited to about 20° due to launch vehicle constraints) to minimise Earth-induced relative velocities of the spacecraft which would lead to excessive Doppler-shifts of the transponded light. The two central spacecraft are 200 km apart, and the distance to the remote spacecraft, defining the interferometer arm length, is $5{\times}10^6$ km.

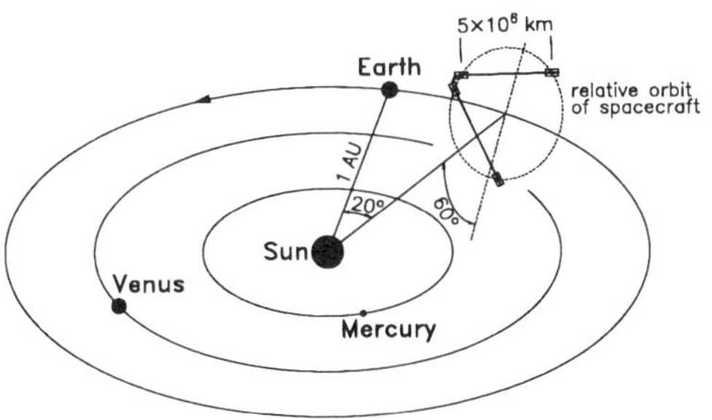

Fig. 2. *Proposed LISA orbit.*

The payload module consists of the inner structural carbon-epoxy cylinder with four stiffening rings, surrounded by a carbon-epoxy payload thermal shield cylinder. The thermal shield, cut at a 30° angle at both sides, keeps sunlight from the thermally stable payload interior throughout the heliocentric orbit. The payload cylinder houses four major assemblies: the telescope assembly, the optical bench, the preamplifier disk and the radiator disk.

The telescope assembly contains a 38 cm diameter f/1 Cassegrain telescope. The primary mirror is a double-arch light-weight ultra-low expansion (ULE) design. The secondary is supported by a three-leg carbon-epoxy spider. The final quality of the plane wavefront leaving the telescope is $\lambda/30$.

The optical bench contains the laser beam injection, detection, and beam shaping optics, and the drag-free sensor (or "accelerometer"). The proof mass of the drag-free sensor acts as the mirror at the end of the interferometer arm. The bench consists of a solid ULE plate to which all components are rigidly attached. Light from the laser is delivered to the optical bench by a single-mode fibre. About 1 mW is split off the 1 W main beam to serve as the local reference for the heterodyne measurement of the phase of the incoming beam from the far spacecraft. Also, about 1 mW is split off and directed towards a triangular cavity which is used as a frequency reference.

Recent Developments

A descoped LISA with only four spacecraft has been proposed for the M3 selection cycle of medium-sized missions. Of the 150 proposed missions, 7 were selected and underwent an assessment study in 1993/1994, and LISA was one of them.

A LISA mission with four spacecraft would be lost if one of the spacecraft is lost. This is also true for any mission with only one spacecraft. For small failure probabilities, the risk for LISA is in first order higher by only a factor of four. In any case, the baseline for the Cornerstone proposal assumes six spacecraft, whereby the loss of up to two spacecraft (not at the same vertex) could be tolerated without loss of the entire mission. In fact, this makes the mission more reliable than even a single spacecraft mission. Incidentally, with CLUSTER, there is a precedent for ESA to fly a multi-spacecraft misson with no backup, where the primary scientific goal would not be achieved in the event of a single-spacecraft failure.

The LISA M3 Assessment Study assumed four spacecraft in order to keep costs down. Using conventional technology, that mission was costed at 694 MAU. Preliminary assessments at ESTEC reveal that six spacecraft based on conventional technolgy can be accommodated in a single Ariane 5. Only three propulsion modules are required (two spacecraft per module) – not six. In combination, these factors suggest that a six-spacecraft mission need not be much more expensive than a four spacecraft mission.

More significantly, one can easily envisage that technological advances in the next 15 years will substantially reduce the mass and cost of various spacecraft and payload elements.

The full 6-spacecraft LISA misson has now been selected as a cornerstone in the ESA Horizon 2000+ program, to be launched in the 2017 timeframe, and it has been recommended for immediate commencement of the technology development program.

The implementation of this program is subject to a small increase in the ESA science budget starting in 2000. In the more immediate future, funding should be available for technical research and development of the mission concept.

Wide-band Spherical Gravitational Wave Detector

M. KARIM*, M. BOCKO†, L. E. MARCHESE†, G. ZHANG†

*Department of Physics, St. John Fisher College
Rochester, NY 14618

†Department of Electrical Engineering, University of Rochester
Rochester, NY 14627

INTRODUCTION

No events of significance have been found in the data gathered over several years from the continuous operation of a number of cryogenic gravitational wave antennae. Several groups are now looking to the future for improvements in antenna design. One of the promising designs to emerge is an antenna of spherical shape, more precisely a truncated icosahedron (acronym TIGA). There are several advantages in this design, particularly the enlarged cross-section to gravitational wave impulses, an omni-directional radiation pattern etc.[1] Furthermore, the new design is a step up from proven technologies such as high Q aluminum alloys, cryogenics on a large scale, vibration isolation, sensitive transducers and amplifiers that have been developed for bar antennas.

DESIGN

TIGA Geometry

The design of the TIGA that is taking shape has a diameter of 2.5 meters and weighs 2.6 tonnnes. With these parameters the fundamental mode will be around 1 kHz.

We propose to suspend the TIGA from its barycenter using three thin steel wires in a pyramid geometry; a possible configuration is shown in the FIG.1. Attached to the wires will be vibration isolation springs inside the dewar, and alternating lead/rubber stacks outside.

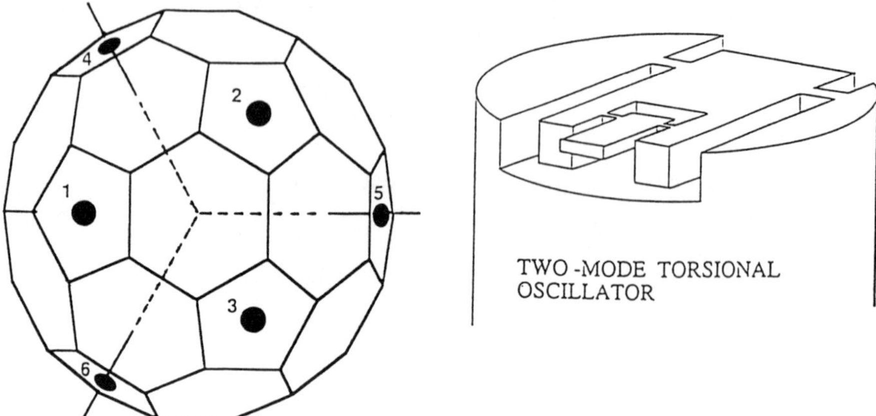

FIG.1 Top view of TIGA with locations of transducers and suspension wires

FIG.2 Geometry of two-mode oscillator

Multi-mode Transducer

For rapid transfer of vibrational energy from the antenna to the electromechanical transducer we propose a two-mode resonant mechanical transducer of a design which follows our earlier work on nested transducers[2]. A sketch of the design is shown in FIG. 2.

In order to make truly degenerate resonant transducers we have developed a method of fabrication which in summary is as follows:

Using the noise characteristics of the electromechanical transducer, one can calculate the optimum values for the two masses (for unity signal/noise ratio). The required dimensions of these masses are then calculated and used in the generic design of the two-mode transducer. The smaller mass is first machined out of a heavy 5056 aluminum alloy base (in order to minimize any effect on the Q due to attachment to the antenna). The spring elements are cut while the base is mounted on a milling table, and the vibration frequency of the small mass is monitored with a spectrum analyzer, until the desired frequency (antenna fundamental frequency) is reached. Next the larger mass is machined. As the second set of springs is cut successively thinner the two coupled modes, between the two masses, are monitored, until their spacing is minimum. The frequencies of the individual resonators are now equal. We have been able to tune the frequencies to within ±1

Hz of the desired frequency. Measured and calculated tuning curves shown in FIG. 3 confirm the success of our design method.

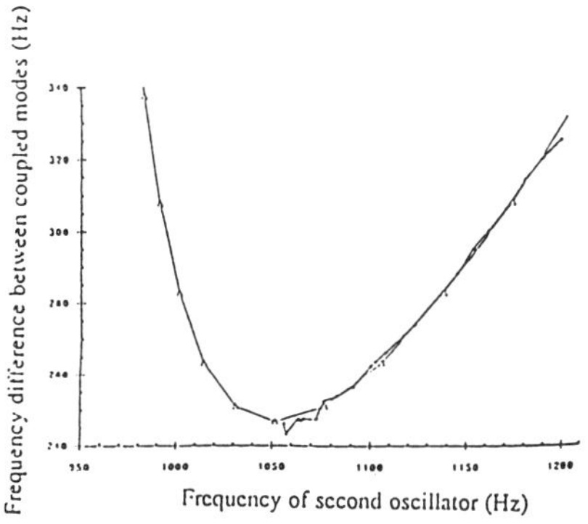

FIG.3 Open triangles denote calculated values, solid points are measurements.

Results

The experimental and theoretical performance characteristics were compared by mounting a two-mode mechanical transducer to a dumb-bell test antenna. (Although the transducer can be machined into the face of the antenna itself). The three coupled modes appeared at 770, 899 and 1029 Hz, which are within ±2 Hz of the predicted frequencies. The time evolution of the displacements of the three modes to an impulse was calculated and compared with the measured response. These are shown in FIG. 4. There is remarkable agreement between the calculated and measured curves.

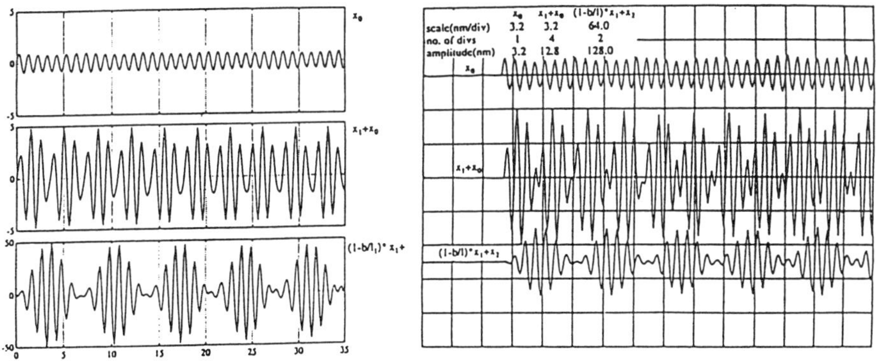

FIG.4 Theoretical (left) and measured (right) responses of a multimode transducer.

The band-width of 259 Hz has several consequences. If one follows Shannon's theorem then the signal from the antenna can be sampled at a maximum frequency which is twice the band-width. The sampling interval comes out to be 2 msec, which implies that one can perform time-delay coincidence between a pair of antennas with a minimum separation of 600 km. In other words if a pulse of gravitational wave was to excite a pair of antennas which are separated by 600 km it would be possible to distinguish between the responses of each antenna. With a triangular array of antennae one would be able to locate the direction of the source of the pulse, complementing the information from the response of the six double-resonant transducers mounted on each TIGA.

Furthermore one can attempt to compare the sensitivity of a wide-band spherical antenna to that of a 4 km long laser antenna. This is shown in FIG. 5. Some complementarity between the two types of antennae is evident.

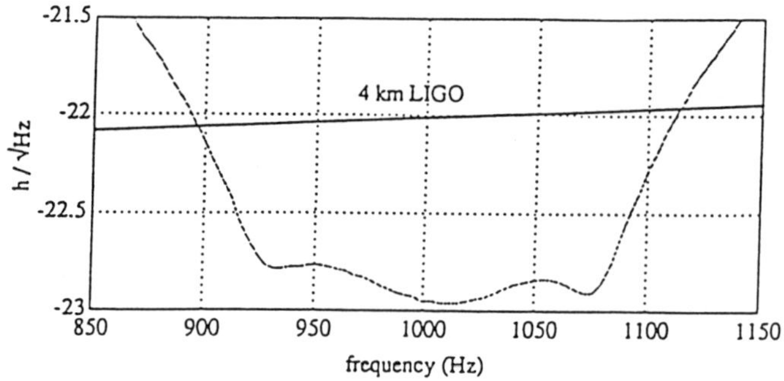

FIG.5 Strain sensitivity of one mode channel of a 2.6 meter diameter TIGA detector with a three mode transducer. The ratio of the mass from one stage to the next is 100. The temperature is 0.05 K; all mechanical Q's are assumed to be 40 million and the transducer noise number is unity.

REFERENCES

1. Johnson, W.W., Merkowitz, S.M. 1993. Phys. Rev. Lett. 70: 2367- 2370
2. Marchese, L., Bocko, M., Zhang, G. & Karim, M.. 1994. Rev. Sci. Instrum. 65(8): 2627-2634.

Binary neutron star inspiral, LIGO, and cosmology*

Lee Samuel Finn
Department of Physics and Astronomy,
Northwestern University, Evanston, Illinois 60208-3112

I. INTRODUCTION

The most promising anticipated source for the United States Laser Interferometer Gravitational-wave Observatory (LIGO)[1] or its French/Italian counterpart VIRGO[2] is the radiation emitted during the final moments of inspiral before the coalescence of a neutron star - neutron star (ns-ns) binary system. The instruments that will operate in the LIGO facility will evolve over time, and the often-discussed "advanced detectors" will be sensitive to inspirals at luminosity distances greater than[3] 1 Gpc. At such great distances the rate and character of the inspiral signal depends on the Hubble expansion, and several authors have investigated how LIGO and/or VIRGO binary inspiral observations can be used to measure the Hubble constant and other cosmological parameters[4–6]. This brief report summarizes a more lengthy investigation into estimates of the rate and character of the inspiral signals that LIGO and VIRGO should observe. A complete presentation, which also discusses black hole binaries, is in preparation for submission to Physical Review D.

II. DETECTION BY LIGO AND VIRGO

To speak of detection rates implies the choice of a detection criterion. Here I adopt a particularly simple criterion: an inspiraling binary is said to have been detected if the anticipated signal-to-noise ratio ρ in the detector is greater than a fixed threshold ρ_0.

Detector noise may conspire to appear as an inspiraling binary, leading to a false alarm; or, noise may mask the signal from an inspiraling binary, leading to false dismissal. The importance of false alarms increases as ρ_0 is decreased, while the importance of false dismissals increases as ρ_0 is increased. I will not consider either false alarms or false dismissals here. Preliminary estimates indicate that a threshold of $\rho_0 = 8$ corresponds to a "false alarm" rate of less than $1\,\mathrm{yr}^{-1}$ for binary ns inspiral[7].

---

*This work was supported grants from the Alfred P. Sloan Foundation and the National Science Foundation (PHY 9308728).

The ns-ns binary inspiral detection rate will be proportional to the ns-ns binary coalescence rate density. The best estimate of this rate density at the current epoch is[8,9] $\dot{n}_0 = 1.1 \times 10^{-7} h\,\mathrm{Mpc}^{-3}\,\mathrm{yr}^{-1}$, where h is the Hubble constant in units of $100\,\mathrm{Km\,s^{-1}\,Mpc^{-1}}$. Estimates of $\dot{n}_0$ rely on the 3 observed binary pulsar systems that will coalesce in less than a Hubble time. Phinney has estimated that, while unlikely, the actual rate could be two orders of magnitude higher or lower without coming into serious conflict with current observations[9].

Given the detection criterion $\rho > \rho_0$, not all binaries within a fixed distance will be detected:

1. *More massive binaries are visible at greater distances than less massive one.* The signal-to-noise ratio depends directly on the intrinsic chirp mass $\mathcal{M}_0$, where

$$\mathcal{M}_0 \equiv \mu^{3/5} M^{2/5} \tag{2.1}$$

and μ and M are the reduced and total mass of the binary.

2. *Binary systems are not isotropic radiators.* The intensity of the radiation in the each polarization mode depends on the orientation of the binary system with respect to the line of sight to the detector.

3. *The detector antenna pattern is not isotropic.* LIGO and VIRGO are more sensitive to binaries in one part of the sky than binaries in another.

When all these effects are taken into account, but cosmological ones are ignored, the expected rate of detection of ns-ns binary inspiral in the advanced LIGO detector is found to be[3]

$$\frac{dN}{dt d\mathcal{M}_0}(\rho > \rho_0) = 69.\,\mathrm{yr}^{-1}\,\frac{\dot{n}_0}{8 \times 10^{-8}\,\mathrm{Mpc}^{-3}\,\mathrm{yr}^{-1}} \left(\frac{8}{\rho_0}\right)^3 \left(\frac{\mathcal{M}_0}{1.2\,\mathrm{M}_\odot}\right)^{5/2} P(\mathcal{M}_0), \tag{2.2}$$

where $P(\mathcal{M}_0)$ is the probability that a ns-ns binary has *intrinsic chirp mass* $\mathcal{M}_0$. Observational[10] and theoretical[11] evidence suggests that the ns distribution is sharply peaked about $1.4\,\mathrm{M}_\odot$, corresponding to an intrinsic chirp mass of $1.2\,\mathrm{M}_\odot$. If the mass of all neutron stars is assumed to be $1.4\,\mathrm{M}_\odot$, then $P(\mathcal{M}_0)$ is a delta-function.

The Hubble expansion modifies these results in several ways:

1. *Distant sources are more luminous.* The signal-to-noise ratio of an inspiral event depends directly on the binary system's chirp mass

$$\mathcal{M} \equiv \mathcal{M}_0(1+z), \tag{2.3}$$

where z is the binary system's cosmological redshift; thus, the Hubble expansion leads to an *increase* in the visibility of moderately distant sources relative to what would be expected without accounting for the redshift. Note that this is in the opposite sense to the usual cosmological K-correction: ns-ns binary inspiral is more luminous at higher frequencies than at lower ones.

2. *Distant sources are less frequent.* The cosmological expansion redshifts the rate of distant coalescence events, which reduces their contribution to the *observed* inspiral rate.

3. *Cosmological volumes are not proportional to the cube of the luminosity distance.* The volume of spheres of constant luminosity distance d_L increases more slowly than d_L^3; consequently, the number of sources at great distances is not as large as would be expected if cosmological effects could be neglected.

4. *Distant sources are older.* The number density of coalescing binaries and the mass distribution of their components changes as the universe evolves. Estimates of the density and distribution in $\mathcal{M}_0$ at the present cosmological epoch do not necessarily apply at earlier times.

5. $\mathcal{M}$ — *not* $\mathcal{M}_0$ — *is observed.* When a binary system is observed, $\mathcal{M}$ is measured. Since $\mathcal{M}$ depends on $1+z$ the observed binaries will be distributed across a range of $\mathcal{M}$. The nature of the distribution depends on the Hubble constant, which permits the observed distribution to be used to measure H_0[5,12].

Assuming an Einstein-deSitter cosmology, negligible evolution over the observed sample of neutron stars, and that the intrinsic chirp mass of all ns binaries is a constant $\mathcal{M}_0$, the anticipated rate of detected ns-ns inspiral can be written as an expansion in large ρ_0:

$$\begin{aligned}\frac{dN}{dt}(\rho > \rho_0) &= 69.\frac{\dot{n}_0}{8\times 10^{-8}\,\mathrm{Mpc}^{-3}\,\mathrm{yr}^{-1}}\left(\frac{8}{\rho_0}\right)^3\left(\frac{\mathcal{M}_0}{1.2\mathrm{M}_\odot}\right)^{5/2}\\ &\quad\times\left[1-0.32x+0.05x^2+\mathcal{O}\left(x^3\right)\right]\end{aligned}\tag{2.4}$$

where

$$x \equiv \frac{8h}{\rho_0}\left(\frac{\mathcal{M}_0}{1.2\,\mathrm{M}_\odot}\right)^{5/6}. \tag{2.5}$$

Assuming the "best guess"[9] $\dot{n}_0$ and $\mathcal{M}_0$ corresponding to ns-ns binaries composed of $1.4\,\mathrm{M}_\odot$ neutron stars, dN/dt ranges between 29 and $43\,\mathrm{yr}^{-1}$ as h ranges from 0.5 and 0.8.

There is a maximum distance (or redshift) beyond which a ns-ns binary of fixed $\mathcal{M}_0$ will always have a signal-to-noise ratio less than ρ_0; for ns-ns binaries this limiting redshift is given approximately by

$$z_0 \simeq 0.473x\left[1+0.276x+0.056x^2+\mathcal{O}(x^3)\right]. \tag{2.6}$$

Assuming $\mathcal{M}_0$ corresponding to ns-ns binaries composed of $1.4\,\mathrm{M}_\odot$ neutron stars, z_0 ranges between 0.27 and 0.48 as ranges h from 0.5 and 0.8.

Since $\mathcal{M}$ depends on z, the distribution of observed chirp masses will reflect how the spatial volume increases with z. The signal-to-noise ratio ρ depends on the luminosity distance d_L; consequently, the number of binaries observed with $\rho > \rho_0$ depends on how the volume of space increases with luminosity distance. Thus, the observed distribution of $\mathcal{M}$ in binaries with $\rho > \rho_0$ depends on the relationship between redshift and luminosity distance, and can be used to measure the Hubble constant. This is the crux of the cosmology test first proposed by Chernoff and Finn[5]. Preliminary estimates show that observation of 200 ns-ns binaries with $\rho > 8$ are sufficient to determine h to within 10%.

III. CONCLUSIONS

Given present estimates of the ns-ns binary coalescence rate and the ns mass distribution the advanced LIGO detectors can expect to observe between 30 (if $h = 0.5$) and $40\,\text{yr}^{-1}$ (if $h = 0.8$) ns-ns binary inspirals with signal-to-noise ratio ρ greater than $\rho_0 = 8$. Present estimates of the coalescence rate are uncertain to several orders of magnitude, and the actual rate could be two orders of magnitude higher. The observed binaries will be at redshifts less than 0.48 in a $h = 0.8$ universe, or 0.27 in a $h = 0.5$ universe. The observed distribution of binary chirp masses reflects h; under these conditions, 200 observations are sufficient to determine the Hubble constant to 10%.

I am glad to thank Kip Thorne for helpful conversations.

REFERENCES

1. A. Abramovici *et al.*, Science **256**, 325 (1992).
2. C. Bradaschia *et al.*, Nucl. Instrum. Methods Phys. Research **A289**, 518 (1990).
3. L. S. Finn and D. F. Chernoff, Phys. Rev. D **47**, 2198 (1993).
4. B. F. Schutz, Nature (London) **323**, 310 (1986).
5. D. F. Chernoff and L. S. Finn, Astrophys. J. Lett. **411**, L5 (1993).
6. D. Marković, Phys. Rev. D **48**, 4738 (1993).
7. B. F. Schutz, in *The detection of gravitational waves*, edited by D. G. Blair (Cambridge University Press, Cambridge, 1991), pp. 406–452.
8. R. Narayan, T. Piran, and A. Shemi, Astrophys. J. Lett. **379**, L17 (1991).
9. E. S. Phinney, Astrophys. J. **380**, L17 (1991).
10. L. S. Finn, Phys. Rev. Lett. **73**, 1878 (1994).
11. S. E. Woosley and T. A. Weaver, in *The Structure and Evolution of Neutron Stars*, edited by D. Pines, R. Tamagaki, and S. Tsuruta (Addison-Wesley, Redwood City, California, 1992), pp. 235–249.
12. L. S. Finn, 1995, in preparation.

On the detectability of post-Newtonian effects in gravitational-wave emission of a coalescing binary[1]

ANDRZEJ KRÓLAK[a] KOSTAS D. KOKKOTAS[b] GERHARD SCHÄFER[c]

[a]Institute of Mathematics
Polish Academy of Sciences
Śniadeckich 8, 00-950 Warsaw, Poland

[b]Department of Physics
Aristotle University of Thessaloniki
540 06 Thessaloniki, Macedonia, Greece

[c]Max-Planck-Research-Group Gravitational Theory
at the Friedrich-Schiller-University
07743 Jena, Germany

INTRODUCTION

It is currently believed that the gravitational waves that come from the final stages of the evolution of compact binaries just before their coalescence are very likely signals to be detected by long-arm laser interferometers [?]. The two projects to build such detectors - LIGO and VIRGO - are already approved and they are rapidly progressing, and the third one - GEO600 - is likely to be funded soon. A standard optimal method to detect the signal from a coalescing binary in a noisy data set and to estimate its parameters is to correlate the data with the filter matched to the signal and vary the parameters of the filter until the correlation is maximal. It has recently been realized[?] that the correlation is very sensitive even to very small variations of the phase of the filter because of the large number of cycles in the signal. Consequently, the addition of small corrections to the phase of the signal due to the post-Newtonian effects decreases the correlation considerably. Thus the post-Newtonian effects in the coalescing binary waveform can be detected and estimated to a much higher accuracy than it was thought before[?]. We present an analysis of the estimation of parameters of the post-Newtonian signal. We also examine the detectability of the post-Newtonian signal and estimation of its parameters using the Newtonian waveform as a filter. This filter can be used as the simplest search template.

GRAVITATIONAL-WAVE SIGNAL FROM A BINARY

Let us first give the formula for the gravitational waveform of a binary with the currently known post-Newtonian corrections. In this paper we work within the so called "restricted" post-Newtonian approximation, i.e., we only include the

[1]This work was supported by KBN Grant No. 2 P302 076 04

post-Newtonian corrections to the phase of the signal keeping the amplitude in its Newtonian form; this is because the effect of the phase on the correlation is dominant. The inclusion of post-Newtonian effects in amplitudes will not qualitatively change our results. Due to radiation reaction the orbit of the binary is rapidly circularized, nevertheless, we include the first order correction due to eccentricity for completeness[2]. The tidal effects are known to be very small and we neglect them. We also include the contributions due to dipole radiation predicted by the Jordan-Fierz-Brans-Dicke (JFBD) theory (assuming that they are small) to investigate the possibility of testing GR against alternative theories of gravity.

The analysis of the signal is best performed in the Fourier domain. The expression for the Fourier transform of our signal in the stationary phase approximation is given by

$$\tilde{s} = \frac{1}{(30)^{1/2}} \frac{1}{\pi^{2/3}} \frac{\mu^{1/2} m^{1/3}}{R} \tilde{h} \tag{1}$$

where

$$\begin{aligned} \tilde{h} = f^{-7/6} \exp i[2\pi f t_c - \phi_c - \pi/4 - \\ \frac{7065}{187136} \frac{k_e}{(\pi f)^{34/9}} + \frac{3}{128} \frac{k}{(\pi f)^{5/3}} + \frac{5}{96} \frac{k_1}{(\pi f)^1} - \frac{3}{32} \frac{k_{3/2}}{(\pi f)^{2/3}} + \frac{3}{128} \frac{k_2}{(\pi f)^{1/3}} \\ - \frac{5}{14336} \frac{k_D}{(\pi f)^{7/3}}], \end{aligned} \tag{2}$$

(for $f > 0$, and by the complex conjugate of the above expression, for $f < 0$) where

$$\begin{aligned} k &= \frac{1}{\mu m^{2/3}}(1 - \frac{F(C_1, C_2, m_1, m_2)}{2+\omega}), \; k_e = \frac{1}{\mu m^{2/3}} e_o^2 (\pi f_o)^{19/9}, \\ k_1 &= \frac{1}{\mu}(\frac{743}{336} + \frac{11}{4}\frac{\mu}{m}), \; k_{3/2} = \frac{m^{1/3}}{\mu}(4\pi - s_o), \\ k_2 &= \frac{m^{4/3}}{\mu}(3.0 + c_1 \frac{\mu}{m} + c_2 (\frac{\mu}{m})^2 + s_s \frac{m}{\mu}), \\ k_D &= \frac{1}{\mu m^{4/3}} \frac{(C_1 - C_2)^2}{2+\omega} \end{aligned} \tag{3}$$

where e_o is the eccentricity of the binary at gravitational frequency f_o (k_e is constant to order e_o^2), s_o and s_s are spin-orbit and spin-spin parameters, respectively[?]. The constants c_1 and c_2 in the 2nd post-Newtonian contribution are only recently being calculated [?] and calculation of further corrections is in progress. C_i, i=(1,2), is the sensitivity of the body i to changes of the scalar field, $F(C_1, C_2, m_1, m_2)$ is a complicated function of sensitivities and masses of the bodies and ω is the parameter of the JFBD theory. Current observational tests restrict $\omega > 600$ and observations of the Hulse-Taylor binary pulsar restrict $\omega > 200$.

ESTIMATION OF THE POST-NEWTONIAN PARAMETERS

A basic method proposed to detect the gravitational-wave signal of a coalescing binary and to estimate its parameters is matched-filtering and maximum likelihood

[2]This correction was derived by N. Wex

(ML) estimation. The performance of the method is determined by signal-to-noise ratio d and the Fisher information matrix Γ. Signal-to-noise ratio determines the probability of detection of the signal and the diagonal elements of the inverse of the Fisher matrix are approximately the variances of the maximum likelihood estimators of the parameters for large d.

We have considered three representative binary systems with parameters summarized in Table I.

Table I. Parameters of Neutron Star (NS) and Black-Hole (BH) Binary Systems

Binary	$m_1 M_\odot$	$m_2 M_\odot$	$\mathcal{M} M_\odot$	s	$k M_\odot^{-5/3}$	$k_1 M_\odot^{-1}$	$k_{3/2} M_\odot^{-2/3}$
NS-NS	1.4	1.4	1.2	2.4×10^{-2}	0.72	4.1	25
NS-BH	1.4	10	3.0	4.0	0.16	2.0	16
BH-BH	10	10	8.7	3.9	2.7×10^{-2}	0.58	4.7

Assuming that the noise is Gaussian, neglecting the effects of eccentricity, 2nd post-Newtonian effects, dipole radiation, and assumining that binaries are located at a distance of 200Mpc we have calculated the signal-to-noise ratios and approximate values of the variances of the ML estimators. The results are summarized in Table II (cf. [?] Table II).

Table II. Signal-to-Noise Ratio and RMS Errors for Parameters of the 3 Binaries

Binary	S/N	$\sqrt{n}$	Δt_ams	$\Delta\mathcal{M}/\mathcal{M}$	$\Delta\mu/\mu$	$\Delta m/m$	$\Delta s/s$
NS-NS	15	32	0.75	0.023%	6.4%	9.6%	33×10^2%
NS-BH	32	15	0.65	0.061%	4.8%	7.0%	8.2%
BH-BH	77	6	0.46	0.19%	16%	23%	38%

The number $\sqrt{n}$ gives the improvement of the signal-to-noise ratio due to matched filtering.

We have also examined the effects of eccentricity and dipole radiation We have added their contributions to the phase of the 3/2post-Newtonian signal separately and we have calculated the Fisher matrix and its inverse. For a NS-NS binary the relative rms error in eccentricity is given by $\frac{\Delta e}{e} = \frac{0.63\times10^{-6}}{a_e}$. For the Hulse - Taylor binary pulsar we have $a_e = e_o^2 f_o^{19/9} = 1.8\times10^{-13}$. Thus the effects of eccentricity would be practically undetectable for such a binary. For the NS-BH binary the relative rms error in the dipole radiation coefficient k_D in JFBD theory is given by $\frac{\Delta k_D}{k_D} = 0.75\,(\frac{\omega}{100})$. Thus, the effects of the dipole radiation could be determined to about the same accuracy as from timing of the binary pulsar in our Galaxy.

It is well known that the rms errors of the estimators of the parameters increase with the number of parameters. We have investigated this effect with the increasing number of post-Newtonian parameters. We have considered a reference binary of $\mathcal{M} = 1M_\odot$ located at the distance of 100Mpc.

Table III RMS Errors for Parameters at Various post-Newtonian Orders

Δt_c msec	$\Delta\phi_c$	$\Delta k M_\odot^{-5/3}$	$\Delta k_1 M_\odot^{-1}$	$\Delta k_{3/2} M_\odot^{-2/3}$	$\Delta k_2 M_\odot^{-1/3}$	$\Delta k_e M_\odot^{-19/9}$
0.17	0.10	8.3×10^{-6}	-	-	-	-
0.27	0.33	4.0×10^{-5}	5.8×10^{-3}	-	-	-
0.54	1.9	1.7×10^{-4}	0.70×10^{-1}	0.52	-	-
1.6	24	6.6×10^{-4}	0.50	7.2	28	-
2.3	45	2.3×10^{-3}	1.3	17	59	1.2×10^{-6}

We were able to include the 2nd post-Newtonian correction because the Fourier transform of the signal is linear in the mass parameters k_i and consequently, the Fisher matrix for these parameters does not depend on their numerical values. In Table IV we show the degradation of accuracy of estimation of the chirp mass, the reduced mass, and the total mass with the increasing number of parameters in the template for the NS-NS binary at a distance of 200Mpc.

Table IV. RMS Errors for Masses at Various post-Newtonian Orders

pN order	$\Delta\mathcal{M}/\mathcal{M}$	$\Delta\mu/\mu$	$\Delta m/m$
1 pN	0.0054%	0.55%	0.81%
3/2 pN	0.023%	6.4%	9.6%
2 pN	0.080%	42%	63%

The rms errors in μ and m do not depend on the value of 2nd post-Newtonian mass paramter k_2.

THE NEWTONIAN FILTER

On the one hand the correlation of the signal with the template is very sensitive to small corrections in the phase of the signal on the other hand the accuracy of estimation of the parameters is significantly degraded with increasing number of corrections even though a correction may be small. Moreover we cannot entirely exclude unpredicted small effects in the gravitational-wave emmission (e.g. corrections to general theory of gravity) that we present cannot model. Thus, there is a need for simple filters or *search templates* that will unable to scan the data effectively and isolate stretches of data where the signal is most likely to be[?]. The simplest such filter is just a Newtonian waveform h_N the Fourier transform of which, in stationary phase approximation, is given by $\tilde{h}_N = f^{-7/6} \exp i[2\pi f t_c - \phi_c - \pi/4 + k\frac{3}{128}(\pi f)^{-5/3}]$. Using such a suboptimal filter decreases the correlation and, consequently, decreases the signal-to-noise ratio. Also the estimate of the k parameter is shifted by a definite amount depending on the noise of the detector and the parameters of the binary. In Table V we have given the drop in signal-to-noise given by $FF = (h|h_N)/(h|h)$ where $(h_1|h_2)$ is the correlation of functions h_1 and h_2 (see Ref.[?] Eq.(2.3) for definition) and the parameter $l = \sqrt{FF}$. In the case of the Newtonian filter probability of detection is determined by two parameters: optimal signal-to-noise ratio $d = \sqrt{(h|h)}$ and $d_o = \sqrt{(h|h_o)} = l \times d$. We have also calculated the shift in the estimate of the k parameter and the accuracy of the determination of k. We have included the 1st and the 3/2 post-Newtonian corrections.

Table V. Performance of the Newtonian Filter

Binary	l	FF	Shift $\delta k M_\odot^{-5/3}$	Accuracy $\Delta k M_\odot^{-5/3}$
NS-NS	0.90	0.81	0.01560	0.13×10^{-3}
NS-BH	0.87	0.76	0.004898	0.026×10^{-3}
BH-BH	0.98	0.96	0.001209	0.011×10^{-3}

When the amplitude and phase modulations are taken into account it was shown[?] that in the worst case FF = 0.39 (l = 0.63).

Using the Newtonian filter we would not like to loose any signals. We can achieve this by suitably lowering the detection threshold when filtering the data with the Newtonian filter. By this procedure we would isolate stretches of data with all the signals that would be detected with optimal filter and also an increased number of false alarms. The next step would be to analyse the reduced set of data with accurate templates and the initial threshold to make the final detection and estimate the parameters of the signal.

In Table VI we have given examples of the performance of the above procedure. We assume the detection threshold T = 5 and we assume that we have one signal for the optimal signal-to-noise ratio d. N is the expected number of detected signals with the optimal filter, N_F is number of false alarms, N_N is the number of detected signals with the Newtonian filter, T_N is the lowered threshold, N_L is number of signals with the lowered threshold and N_{FL} is number of false alarms with the lowered threshold.

Table VI

d	FF	N	N_F	N_N	T_N	N_L	N_{FL}
15	.81	27	0.055	20	4.5	28	0.16
15	.36	27	0.055	5.6	3.225	28	2.1
30	.81	225	1.1	165	4.5	230	2.2
30	.25	225	1.1	31	2.875	229	32

A different search template consisting of the post-Newtonian waveform with spin effects stripped off has been analysed in Ref.[?].

ACKNOWLEDGEMENTS

A.K. thanks the Max-Planck-Gesellschaft for support and the Arbeitsgruppe Gravitationstheorie an der Friedrich Schiller Universität in Jena for hospitality during the time this work was done. We would like to thank T. Apostolatos and K.S. Thorne for helpful discussions.

References

[1] K.S. Thorne, in *300 Years of Gravitation*, edited by S.W. Hawking and W. Israel (Cambridge University Press, Cambridge, 1987), pp. 330-458.

[2] C. Cutler *et al.*, Phys. Rev. Lett. **70**, 2984 (1993).

[3] A. Królak, in *Gravitational Wave Data Analysis*, edited by B.F. Schutz (Kluwer, Dordrecht, 1989), pp. 59-69.

[4] L.E. Kidder, C.W. Will, and A.G. Wiseman, Phys. Rev. D **47**, R4183 (1993).

[5] See contributions of L. Blanchet and A.G. Wiseman in this volume.

[6] C. Cutler and E. Flanagan, Phys. Rev. D **49**, 2658 (1994).

[7] T. Apostolatos, PhD Thesis, Caltech 1994.

Gravitational Wave Signals from Collapsing Rotating Polytropes

EWALD MÜLLER AND THOMAS ZWERGER

Max-Planck-Institut für Astrophysik, Karl-Schwarzschild-Str. 1, D-85740 Garching, Germany

INTRODUCTION

Among potential sources which emit gravitational waves core collapse supernova are thought to be one of the most promising sources to be detected on earth by gravitational wave detectors presently under construction (for a review, see Thorne, this volume). Thus, it is very important to provide gravitational wave experimentalists with reliable theoretical predictions of the signal strength and signal form of gravitational waves produced during a core collapse supernova event.

Such theoretical predictions have already been made in the past being based on 2-d hydrodynamical core collapse calculations with a more or less sophisticated treatment of the complicated microphysics and transport processes involved in the event[1–4]. However, all attempts up to now were restricted to a few specific core collapse models, by which only a small subset of the parameter space spanned by possible initial (pre-collapse) models was explored.

For this reason we have performed a comprehensive set of axially symmetric hydrodynamical simulations of collapsing rotating stellar cores to determine the strength, form and frequency of the gravitational wave signal produced in such events.

MODEL DESCRIPTION

In order to examine a large set of initial conditions, obviously some simplifications had to be made. Therefore, we have approximated the collapsing core by a rotating polytrope and used a simplified analytical equation of state of the form $p = K\rho^\Gamma$, where p, ρ and K are the pressure, the density and the polytropic constant, respectively. Γ is the variable adiabatic index which mimics the microphysics of core collapse[5]. In our study we have computed the evolution of models with $\Gamma = 1.325$, 1.3225, 1.32, 1.315, 1.31, 1.30, 1.29 and 1.28, respectively.

Contrary to all previous studies we have used rotating equilibrium configurations as initial models. All initial models have been calculated with the method of Eriguchi & Müller[6], and are rotating $\Gamma = 4/3$ polytropes. Collapse was induced by suddenly reducing Γ to one of the above specified values. Varying the initial amount and distribution of angular momentum

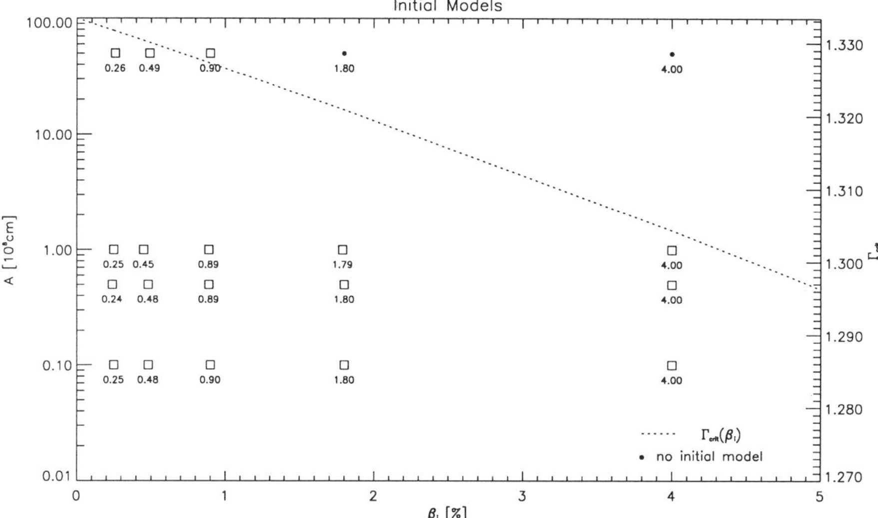

Fig. 1. Overview of initial models: The labels of the axis on the right only refer to the dotted line. This line separates models which can collapse for a given β_i (below the line) from those which cannot (above the line) because of their amount of rotation. The numbers at the symbols give the corresponding value of β_i.

we have computed a total of 81 initial models (see Fig.1) and followed their evolution well beyond core bounce.

The amount of rotation can be conveniently characterized by the ratio $\beta \equiv T/|W|$, where T and W are the rotational energy and the gravitational binding energy of the configuration. In our study we have considered initial models with $\beta_i = 0.0025$, 0.0045, 0.009, 0.018 and 0.04, respectively (see Fig.1). All models initially rotated according to the rotation law[6]

$$\Omega(\widetilde{\omega}) = \frac{\Omega_0}{1 + (\widetilde{\omega}/A)^2} \tag{1}$$

where $\widetilde{\omega}$ is the distance from the rotation axis. The constant Ω_0 is uniquely related to the parameter β_i, while A is a free parameter determining the initial angular momentum distribution. For large values of A (compared to the size of the initial model) one obtains almost rigidly rotating configurations, while small values of A correspond to strongly differentially rotating ones. Specifically, we have examined models with $A = 5\ 10^9$ cm, 10^8 cm, $5\ 10^7$ cm and 10^7 cm, respectively (see Fig.1). Note that in all models the angular velocity is constant on cylinders initially.

The evolution of the models has been computed with a modified version of the multidimensional hydrodynamical code PROMETHEUS developed by Fryxell & Müller[7,8], which is based on the piecewise parabolic method (PPM) of Colella & Woodward[9]. The Poisson equation was solved with the method of Müller & Steinmetz[10], and the gravitational wave signal was computed using a post-Newtonian approach, where numerically troublesome higher order time derivatives of the quadrupole moment are transformed into much better tractable spatial derivatives[11,3]. More details can be found in Zwerger[12], and in Zwerger & Müller[13].

RESULTS

A subset of models shows a bounce caused by centrifugal forces at sub-nuclear densities. This possibility was already pointed out by Shapiro & Lightman[14] and discussed in more detail by Tohline[15]. Previous numerical examples of such a low-density bounce have been given, too[16,17,3]. For a given value of β_i the bounce density decreased with increasing Γ and decreasing A, i.e., strongly differentially rotating models with Γ close to 4/3 showed the lowest bounce densities ρ_b. Models with $A = 10^7\,\mathrm{cm}$ and $\beta_i \geq 0.009$ bounced with $6.2\,10^{12}\mathrm{g/cm}^3 \leq \rho_b \leq 7.0\,10^{13}\mathrm{g/cm}^3$ the model with the lowest ρ_b having $\beta_i = 0.04$ and $\Gamma = 1.30$. But even for quite rigidly rotating configurations with $A = 10^8\,\mathrm{cm}$ the collapse was stopped in some cases well below nuclear matter density, e.g., $\rho_b = 3.6\,10^{13}\mathrm{g/cm}^3$ for $\beta_i = 0.018$ and $\Gamma = 1.32$, and $\rho_b = 7.0\,10^{13}\mathrm{g/cm}^3$ for $\beta_i = 0.009$ and $\Gamma = 1.325$.

Models suffering a bounce due to centrifugal forces show large amplitude oscillations of the inner core during which the central density can vary by up to a factor of ten. For example, in the model with $A = 5\,10^7\,\mathrm{cm}$, $\beta_i = 0.018$ and $\Gamma = 1.31$ the bounce density is $\rho_b = 4.6\,10^{13}\mathrm{g/cm}^3$. Then this model re-expands the central density dropping down to $\rho = 7.5\,10^{12}\mathrm{g/cm}^3$. Eventually the model achieves a new equilibrium state with a central density of $\rho = 1.3\,10^{13}\mathrm{g/cm}^3$.

Because of angular momentum conservation the effects of rotation become more important during collapse, which is reflected by an increase of the value of β. For several models β exceeds the critical value $\beta = 0.1375$, where MacLaurin spheroids become secularly unstable against triaxial perturbations[18]. Two of the most differentially ($A = 10^7$ cm) and rapidly ($\beta_i = 0.04$) rotating models reached ($\Gamma = 1.30$) and even exceeded ($\Gamma = 1.28$) the critical value for dynamical stability ($\beta = 0.2738$). However, such large values of β only prevailed for only one millisecond. Most of the secularly unstable models with $A \leq 10^8\,\mathrm{cm}$, $\beta_i \geq 0.009$ and $\Gamma \leq 1.32$, remained in the unstable regime for several milliseconds, and for $\beta_i = 0.04$ even for several ten milliseconds. Moreover, all potentially unstable models are located in a region of the parameter space, which is not very likely realized in nature.

The analysis of the gravitational wave signals shows that the amount of energy radiated in form of gravitational waves during axisymmetric core collapse lies in the range $2\,10^{-10}M_\odot c^2 \lesssim E_{GW} \lesssim 3\,10^{-7}M_\odot c^2$. The corresponding dimensionless wave amplitudes are in the range $8\,10^{-24} \lesssim h \lesssim 9\,10^{-23}$

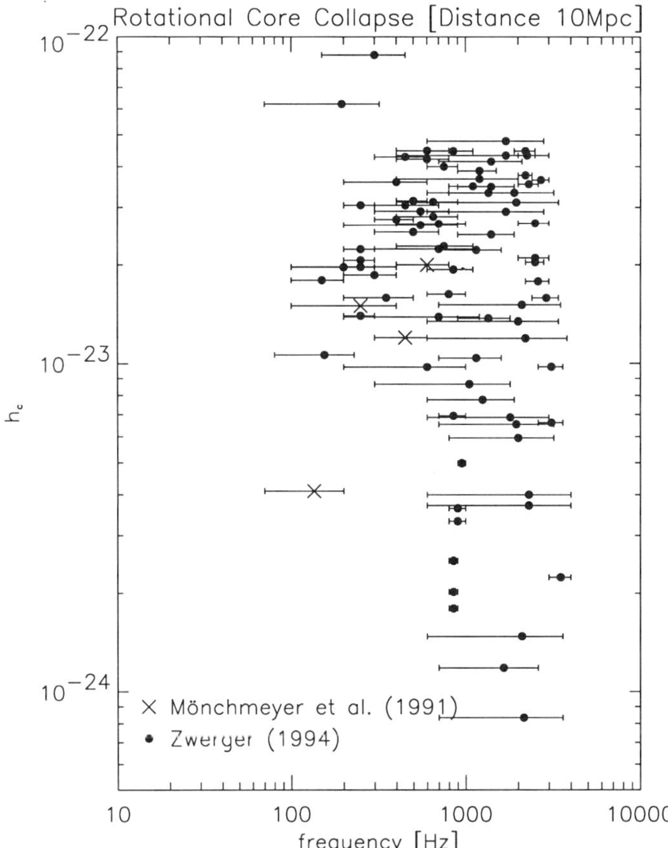

Fig. 2. The dimensionless gravitational wave amplitude of all models is plotted versus frequency. It is assumed that the source is located at a distance of 10 Mpc. The amplitude of the four models calculated by Mönchmeyer et.al. [3] are also shown and are marked by crosses.

for a source at a distance of 10 Mpc (see Fig.2). The largest signals either are produced by models which are slowly ($\beta_i \leq 0.009$) rotating and have an adiabatic index $\Gamma \geq 1.32$, or which are rapidly ($\beta_i \geq 0.018$), strongly differentially ($A = 10^7\,$cm) rotating and have a relatively small adiabatic index ($\Gamma \leq 1.30$). The first class of models experiences a bounce at densities above nuclear matter density with a fast deceleration of the collapsing core. This gives rise to a rapid change of the quadrupole moment of the core and consequently to a strong gravitational wave signal. In the other models the quadrupole moment is larger due to fast rotation, but changes less rapidly.

We find both type I and type II gravitational wave signals[3]. Which type occurs is solely determined by the adiabatic index, and in particular does not dependent on whether the central density of the core reaches nuclear matter density or not. Type I signals, which are characterized by a large amplitude spike at the time of bounce followed by a "ring-down", are produced by models with a "soft" equation of state ($\Gamma \lesssim 1.31$), while type II signals require a "stiff" equation of state ($\Gamma \gtrsim 1.32$). The latter form shows several distinct amplitude spikes, which nicely correlate with the maxima of the central density, and are caused by a volume oscillation of the core. We have also found models with intermediate signal forms.

The spectra are characterized by two maxima at 400 Hz to 1 kHz and at 1 kHz to 4 kHz, respectively. In some models both maxima strongly overlap and are difficult to separate. The high frequency maximum is caused by the bounce. Type II signals show oscillations, which are superimposed on the smooth maxima. The spectra are not very sensitive to β_i and the degree of differential rotation.

REFERENCES

1. Müller, E. 1982. *Astron. Astrophys.* **114**. 53
2. Finn, L.S. & C.R. Evans. 1990. *Astrophys. J.* **351**. 588
3. Mönchmeyer, R., G. Schäfer, E. Müller & R. Kates. 1991. *Astron. Astrophys.* **246**. 417
4. Yamada, S., T. Shimizu & Sato, K. 1993. *Prog. Theor. Phys.* **89**. 1175
5. Janka, H.-Th., T. Zwerger & R. Mönchmeyer. 1993. *Astron. Astrophys.* **268**. 360
6. Eriguchi, E. & E. Müller. 1984. *Astron. Astrophys.* **147**. 161
7. Fryxell, B.A. E. Müller & W.D. Arnett. 1989. *Max-Planck-Institut für Astrophysik, Preprint* **449**.
8. Müller E., B. Fryxell & W.D. Arnett. 1991. in *ESO/EIPC Workshop on SN1987A and other Supernovae*, I.J. Danziger and K. Kjär (Eds.), ESO Workshop and Conference Proceedings No. **37**, Garching, FRG, p. 99
9. Colella, P. & P.R. Woodward. 1984. *J. Comp. Phys.* **54**. 174
10. Müller, E. & M. Steinmetz. 1995. *Comp. Phys. Comm.* in press
11. Blanchet, L., T. Damour & G. Schäfer. 1990. *Mon. Not. Roy. Astron. Soc.* **242**. 289
12. Zwerger, T. 1995, *Ph.D thesis.* Technical Univ. München, unpublished
13. Zwerger, T. & E. Müller. 1995, *Astron. Astrophys.* in preparation
14. Shapiro, S.L. & A.P. Lightman. 1976, *Astrophys. J.* **207**. 263
15. Tohline, J.E. 1984, *Astrophys. J.* **285**. 721
16. Müller, E., W. Hillebrandt & M. Ròżyczka 1980, *Astron. Astrophys.* **81**. 358
17. Symbalisty, E.M.D. 1984, *Astrophys. J.* **285**. 729
18. Tassoul, J.-L. 1978, *Theory of Rotating Stars*, Princeton Univ. Press, New Jersey

Gravitational Waves from Coalescing Neutron Stars

XING ZHUGE, JOAN CENTRELLA, AND STEPHEN MCMILLAN

Dept. of Physics & Atmospheric Science
Drexel University
Philadelphia, PA 19104

Coalescing binary neutron stars are among the most promising sources of gravitational waves.[1] The inspiral waveforms can be calculated quite accurately using post-Newtonian expansions in the point-mass limit,[2] and it is expected that they will give the masses and spins of the neutron stars, as well as the orbital parameters of the binary systems.[3] When the separation of the binary is comparable to the neutron star radius, hydrodynamic effects dominate and coalescence takes place within a few orbits. The coalescence regime probably lies at or beyond the upper end of the frequency range accessible to broad-band detectors such as LIGO and VIRGO, but it may be observed using specially designed narrow band interferometers[4] or resonant detectors.[5] Observations of the waveforms and spectra in this range may yield valuable information about neutron star radii, and thereby the nuclear equation of state.[6]

We have carried out 3-D numerical simulations of binary neutron star coalescence[7] using the smooth particle hydrodynamics code TREESPH.[8] The gravitational field is Newtonian, and the gravitational radiation is calculated using the quadrupole approximation. We use nonrotating neutron stars of mass $1.4M_\odot$ modeled as polytropes with equation of state $P = K\rho^{1+1/n}$, where n is the polytropic index. We vary both n and the neutron star radius R. In this paper we briefly describe our calculations and summarize our results, referring the interested reader to reference 7 for further details and a more complete set of references.

Our initial conditions consist of identical spherical polytropes on (nearly) circular orbits with separations large enough that tidal effects are negligible. The stars are thus initially in the point-mass regime, and the inspiral is caused by the loss of orbital energy due to gravitational radiation reaction. Once the stars get close enough to produce significant tidal distortions, Newtonian effects dominate and lead to rapid merger and coalescence[9]. Since the gravitational field is purely Newtonian, we must explicitly include the radiation reaction losses to cause the inspiral. This is accomplished by adding a frictional term to the particle acceleration equations that removes energy at the rate given by the point-mass expression. This frictional term is applied until tidal effects dominate; the actual merger and coalescence then proceed by purely Newtonian hydrodynamics. See reference 10 for a similar approach.

We have focused on calculating the spectrum dE/df. For point-mass inspiral, $dE/df \sim f^{-1/3}$. To see a cutoff frequency, due to the expected shutoff of radiation after coalescence, in our simulation data requires a fairly long region of point-mass inspiral in the frequency domain. Although our models begin in the point-mass regime, the binaries actually merge and coalesce within just a few orbits. We therefore match the waveforms h_+ and $h_\times$ onto point-mass waveforms that extend back to much larger separations and thus lower frequencies. The resulting spectra show suitably long regions of point-mass inspiral at low frequencies, followed by the effects of merger and coalescence.

Run 1 used $N = 4096$ particles per star and has $R = 10$km (so $GM/Rc^2 = 0.21$), $n = 1$, and initial separation $a = 4R$. The gravitational waveform rh_+ for an observer on the axis at $\theta = 0$ and $\varphi = 0$ is shown for this run in Figure 1a. Here, time is measured in units of the dynamical time for a single star, $t_D = (R^3/GM)^{1/2}$. The waveform initially matches the point-mass prediction, then increases in both frequency and amplitude as tidal bulges form and cause the stars to spiral in faster than the point-mass case. The waves reach their maximum amplitude during the merger at $t \sim 105 - 110t_D$, then decrease as the stars coalesce. The gravitational waves have shut off and the system is essentially axisymmetric by $t \sim 180t_D$.

The gravitational wave energy spectrum for Run 1 is shown in Figure 1b, where the solid line gives the results of the simulation matched onto the point-mass waveforms and the dashed line shows the point-mass result. The initial dip in the spectrum below the point-mass value near $f_{dyn} \sim 1500$Hz is due to the dynamical instability[9] that causes the stars to spiral together faster than they would on point-mass trajectories. The broad maximum at $f_{peak} \sim 2500$Hz is due to a transient, rotating, bar-like structure that forms immediately after coalescence, during the time $t \sim 120 - 150t_D$. The secondary peak at $f_{sec} \sim 3200$Hz also appears during this time; we attribute it to transient oscillations induced in the coalsecing stars during the merger process.

Run 2 is the same as Run 1, except that we chose $R = 15$km (so that $GM/Rc^2 = 0.14$) and used $N = 1024$ particles per star. The gravitational waveform rh_+ for this run is shown in Figure 2a and the spectrum dE/df is shown in Figure 2b. Here, dynamical instability causes the spectrum to dip below the point-mass result near $f_{dyn} \sim 850$Hz. It then drops sharply just beyond $f_{peak} \sim 1500$Hz and rises again to a secondary peak at $f_{sec} \sim 1750$Hz. The lack of a strong peak at f_{peak} and the weaker maximum at f_{sec} result from

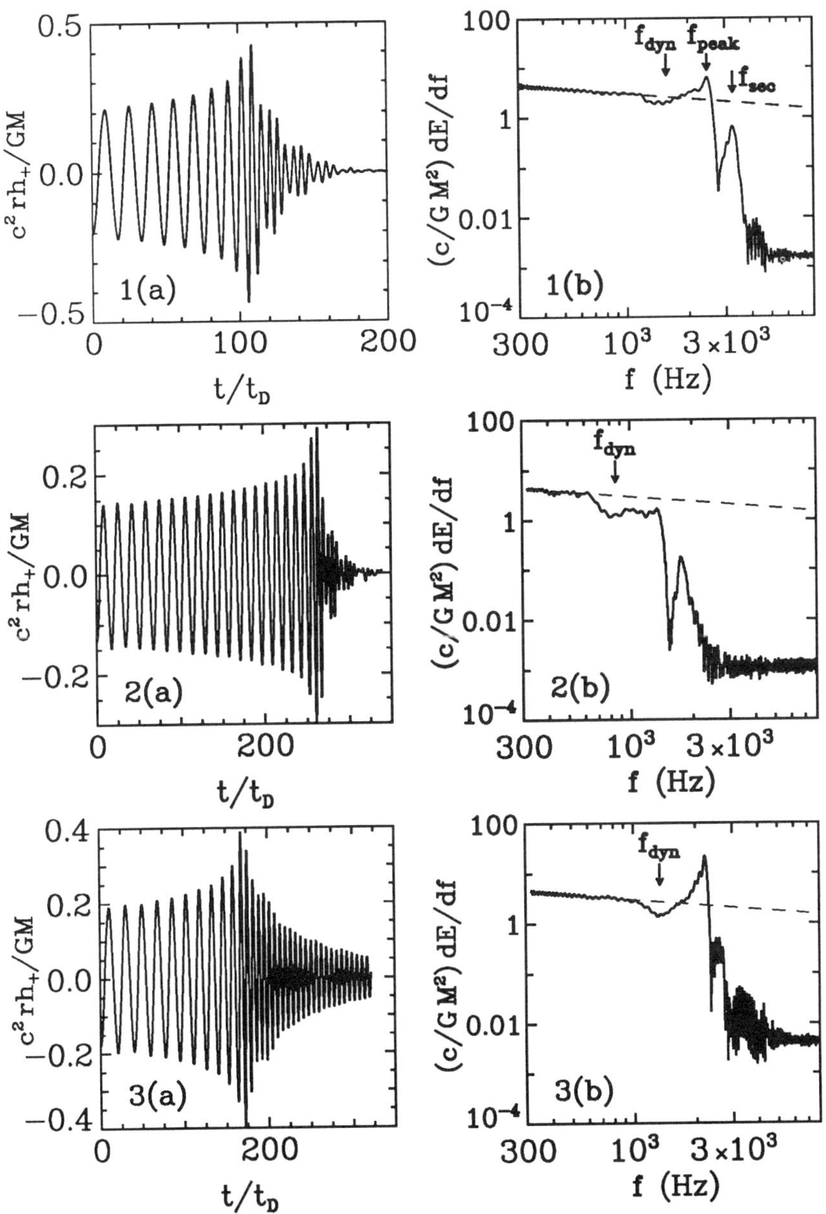
1(a)
2(a)
3(a)
1(b)
2(b)
3(b)
$c^2 rh_+/GM$
t/t_D
$(c/GM^2)\, dE/df$
f (Hz)
f_{dyn}
f_{peak}
f_{sec}

the weaker tidal forces at the point of dynamical instability that produce a less prominent and shorter-lived bar. The gravitational wave frequency for point-mass inspiral $f \sim a^{-3/2} \sim R^{-3/2}$, so we expect that the spectral features in Run 2 should occur at lower frequencies than in Run 1, roughly in the ratio $f_1/f_2 \sim (R_1/R_2)^{-3/2} \sim 1.8$. Our numerical results confirm this behavior.

Finally, Run 3 is the same as Run 1 except that we chose the polytropic index $n = 0.5$ and used $N = 1024$ particles per star and initial separation $a = 4.5R$. In this case, the merged object is slightly nonaxisymmetric, and the waveform rh_+ in Figure 3a shows the late-time ringing due to the rotating, bar-like merger product.[11] The spectrum in Figure 3b drops below the point-mass value near $f_{dyn} \sim 1340$Hz, and then rises to a sharp peak at $f_{peak} \sim 2200$Hz. It then drops sharply and rises again to a secondary peak at $f_{sec} \sim 2600$Hz. The gravitational radiation due to the rotating bar is in the frequency range of the first peak. By comparing the separations at which dynamical instability occurs[9] in Runs 1 and 3 we estimate that the spectral features should have roughly the ratio $f_1/f_3 \sim 1.2$. Our simulations do approximate this behavior.

The gravitational waveforms and energy spectra from these models are rich in information about the hydrodynamical merger and coalescence of binary neutron stars. Our results demonstrate that detailed information about both GM/Rc^2 and the equation of state can in principle be extracted from the spectra.

ACKNOWLEDGMENTS

We thank M. Davies, D. Kennefick, D. Laurence, and K. Thorne for interesting and helpful conversations. This work was supported in part by NSF grants PHY-9208914 and AST-9308005, and by NASA grant NAGW-2559. The simulations were carried out at the Pittsburgh Supercomputing Center.

REFERENCES

1. Abramovici, A., et al. 1992. Science 256: 325.
2. Will, C. 1994. In Relativistic Cosmology, Proceedings of the 8th Nishinomiya-Yukawa Memorial Symposium, M. Sasaki, Ed.: 83-98. Universal Academy Press, Tokyo.
3. Cutler, C., et al. 1993. Phys. Rev. Lett. 70: 2984.
4. Krolak, A., J. Lobo, & B. Meers. 1993. Phys. Rev. D 47: 2184.
5. Johnson, W. & S. Merkowitz. 1993. Phys. Rev. Lett. 70: 2367.
6. Kennefick, D., D. Laurence, & K. Thorne. 1994. Unpublished.
7. Zhuge, X., J. Centrella, & S. McMillan. 1994. Phys. Rev. D 50: 6247.
8. Hernquist, L. & N. Katz. 1989. Ap. J. Suppl. 70: 419.
9. Lai, D., F. Rasio, & S. Shapiro. 1994. Ap. J. 420: 811.
10. Davies, M., W. Benz, T. Piran, & F. Thielemann. 1994. Ap. J. 431: 742.
11. Rasio, F. & S. Shapiro. 1994. Ap. J. 432: 242.

Coalescing compact binaries to second–post-Newtonian order

Luc Blanchet
Département d'Astrophysique Relativiste et de Cosmologie
Centre National de la Recherche Scientifique (UPR176),
Observatoire de Paris, 92195 Meudon Cedex, France

I. INTRODUCTION

The problem of the emission of gravitational radiation by astrophysical sources is relevant to the future LIGO/VIRGO/GEO network of gravity-wave detectors. The most promising targets for this network are the gravitational waves emitted by coalescing compact binaries, i.e. binary systems of neutron stars or black holes in their late radiation-reaction-driven stage of evolution[1]. Observed coalescences will occur typically at cosmological distances. Once a binary signal has been identified, further analysis of data using the technique of matched filtering will provide accurate measurements of the masses and angular momenta of the two bodies[2]. Furthermore one will determine the cosmological distance of the coalescence (independently of any assumption concerning the nature of the compact objects[3]). Tests of the nonlinear structure of general relativity[4] and tests of the viability of alternative gravitational theories[5] will also be possible. The technique of matched filtering consists of correlating the detector's output with a filter whose Fourier transform is equal to that of the expected signal itself, and divided by the power spectral density of the noise. This technique evidently necessitates the knowledge of the signal, and as a matter of fact an extremely accurate (in the post-Newtonian sense) signal deduced from general relativity is needed for the matched filtering technique to be successful – at least in the case of coalescing compact binaries.

This contribution reports the result of a recent computation of the gravitational radiation field generated by a coalescing compact binary to the so-called second-post-Newtonian (2PN) order of general relativity, which takes into account the relativistic corrections in the radiation field up to the order $(v/c)^4$ beyond the usual quadrupole, where v is the orbital velocity of the objects and c the speed of light. This work, which is done in collaboration with Thibault Damour and Bala R. Iyer[6], is based on a previously derived 2PN-accurate generation formalism valid for arbitrary slowly moving isolated systems[7]. An independent computation based on a different method yields confirmation of the result[8].

II. THE 2PN-ACCURATE WAVE GENERATION FORMALISM

This formalism rests on several steps, the first of which consists of assuming that the gravitational field in the *exterior* and *weak-field* region of an isolated gravitating system admits an asymptotic expansion in powers of Newton's parameter G,

$$h^{\mu\nu} = Gh_1^{\mu\nu}[M_L, S_L] + G^2 h_2^{\mu\nu} + \cdots + G^n h_n^{\mu\nu} + \cdots , \tag{2.1}$$

where the first term, linear in G, is given as an infinite decomposition of multipolar waves parametrized by two sets of multipole moments representing the source, namely mass-type moments (M_L) and current-type ones (S_L). The index L on the moments refers to a multi-index composed of ℓ indices, where ℓ is the order of multipolarity. With no specialization of these moments, the first term in (2.1) represents the most general solution of the *linearized* Einstein equations outside the source[9]. By substitution of (2.1) into the fully *nonlinear* (vacuum) equations, and iterative resolution of these equations, one obtains in principle a complete control of the solution in all the exterior-weak-field region. This includes notably the region far from the source where the detector is located, and also, when the source is slowly moving, the exterior part of the so-called *near-zone* of the source. In the latter zone, one can expand formally the solution when $c \to +\infty$. Terms of relative order $(v/c)^7$ in this expansion are seen a posteriori to correspond to an order higher than 2PN in the radiation field. Neglecting these terms (which we symbolize by $O(7)$), we find

$$h^{\mu\nu} = Gh_1^{\mu\nu}[M_L, S_L] + \text{ Finite Part } \Box_R^{-1}\overline{\Lambda}^{\mu\nu}[M_L, S_L] + O(7) , \tag{2.2}$$

where the second term reads as the integral operator of the retarded potentials (denoted by $\Box_R^{-1}$) acting on a gravitational source term $\overline{\Lambda}^{\mu\nu}$ which incorporates the nonlinearities of Einstein's equations up to the considered order. This term (the overbar of which indicates that it is valid only up to $O(7)$) is given as some explicit functional of the linear coefficient $h_1^{\mu\nu}$, and therefore is given similarly to it as an infinite multipole decomposition parametrized by M_L and S_L. The "Finite Part" in front of the retarded integral denotes a certain operation of regularization made indispensable by the fact that multipole expansions are valid solely in the exterior of the source, and present a singularity at the origin of the coordinates in the source (when considered formally inside the source).

The second step consists of matching the near-zone-expanded external solution (2.2) of the Einstein vacuum equations to a solution of the *non-vacuum* equations, which is generated by the actual material source and is valid up to the same precision. This is in order to obtain the explicit expressions of the multipole moments M_L and S_L parametrizing the external solution in terms of source's physical quantities. One must take into account with due care the fact that in general the interior and exterior solutions differ by a coordinate transformation. As a result, we end up with the matching equation[7]

$$h_1^{\mu\nu}[M_L, S_L] = -\frac{4}{c^4}\sum_{\ell=0}^{\infty}\frac{(-)^\ell}{\ell!}\partial_L\left[\frac{1}{r}\overline{F}_L^{\mu\nu}(t-r/c)\right] + \partial\varphi^{\mu\nu} + O(7) \ , \qquad (2.3)$$

giving explicitly, modulo a gauge transformation $\partial\varphi^{\mu\nu}$, the linear term in (2.1)-(2.2) as a function of the (reducible) multipole moments

$$\overline{F}_L^{\mu\nu}(t) = \text{ Finite Part}\int d^3\mathbf{x}\hat{x}_L\int_{-1}^{1}dz\delta_\ell(z)\overline{\tau}^{\mu\nu}(\mathbf{x}, t+z|\mathbf{x}|/c) \ , \qquad (2.4)$$

in which we have introduced the effective total stress-energy tensor

$$\overline{\tau}^{\mu\nu} = \mathcal{T}^{\mu\nu} + \frac{c^4}{16\pi G}\overline{\Lambda}^{\mu\nu} \ . \qquad (2.5)$$

This tensor describes the non-gravitational or matter fields in the source (first term $\mathcal{T}^{\mu\nu}$) *and* the gravitational field (second term $\overline{\Lambda}^{\mu\nu}$ now expressed in terms of source potentials). In (2.4) we denote by $\hat{x}_L$ the tracefree part of a product of ℓ spatial coordinate vectors, and by $\delta_\ell(z) = [(2\ell+1)!!/(2^{\ell+1}\ell!)](1-z^2)^\ell$ a weighting function whose appearance is linked to the fact that the waves propagate inside the source at the finite velocity c. It is trivial to deduce from (2.3)-(2.5) the looked-for expressions of the (irreducible) multipole moments M_L and S_L.

Finally the last step consists of expanding at large distance from the source the external field (2.1), now endowed thanks to (2.3)-(2.5) with full physical content determined by the actual gravitating source. This entails introducing a set of "observable" multipole moments U_L and V_L parametrizing the far-away gravitational field and directly "observed" in the detector. These moments are given by some nonlinear functionals of the corresponding moments M_L and S_L, which one can symbolize by

$$U_L = \mathcal{U}[M_L, S_L] \ , \qquad (2.6a)$$

$$V_L = \mathcal{V}[M_L, S_L] \ . \qquad (2.6b)$$

The functionals $\mathcal{U}$ and $\mathcal{V}$ have the property of being *non-local* in time, in the sense that U_L and V_L depend at some given (retarded) time on the values of M_L and S_L at all times equal or anterior to this retarded time. The dominant non-local time dependence in (2.6) has been shown to arise at the 1.5PN order and to be associated with "tails" in the wave zone[10].

III. APPLICATION TO COALESCING COMPACT BINARIES

The implemention of the previous 2PN-accurate wave generation formalism to a specific source composed of two point-masses moving on a circular orbit (this is the adequate model for a coalescing compact binary) is in principle straightforward except for getting a certain cubically nonlinear term in the mass-type quadrupole moment of the binary. This computation yields the 2PN gravitational field generated by the binary and, most importantly, the total energy in the field or power of the gravitational wave emission. The latter power is found[6] to be

$$\mathcal{P} = \frac{32c^5}{5G}\nu^2\gamma^5 \left\{1 - \left(\frac{2927}{336} + \frac{5}{4}\nu\right)\gamma + 4\pi\gamma^{3/2} + \left(\frac{293383}{9072} + \frac{380}{9}\nu\right)\gamma^2\right\} . \quad (3.1)$$

Some comments on the various terms entering this expression are in order. The dominant term is the Newtonian one, which is issued from the usual Einstein quadrupole formula (ν denotes the mass ratio μ/M where μ is the reduced mass and M the total mass of the binary). This term is followed by a post-Newtonian expansion ordered by the parameter $\gamma = GM/c^2r$, where r is the separation between the two bodies (in harmonic coordinates). The 1PN correction, proportional to γ, was known[11] long before this work. The 1.5PN correction ("4π" term proportional to $\gamma^{3/2}$) is the direct consequence of the existence of tails in the wave zone[10,12–14], as contained in (2.6). The 2PN correction investigated here is the term proportional to γ^2. It involves besides a term constant in ν a term linear in ν. The constant term was known previously thanks to a method of perturbation of the Schwarzschild geometry, dealing with the test-body $\nu \to 0$ limit[15,16]. Interestingly, the 2PN correction in (3.1) does not involve a term quadratic in ν, although all the separate pieces making up this correction do contain such terms. The final coefficient of ν^2 in (3.1) turns out to be zero. The 2PN correction term is shown[17] to make a significant contribution to the accumulated phase of theoretical filters to be used in the future LIGO/VIRGO/GEO network of gravity-wave detectors. The next-order correction term, namely 2.5PN (proportional to $\gamma^{5/2}$), makes also (probably) a non-negligible contribution to the accumulated phase; its computation is in preparation[18].

To conclude, we emphasize that the result (3.1) is derived in a fully consistent and mathematically well-defined way. In particular the multipole moments used in the computation are given by well-defined integrals such as (2.4) (in contrast with ill-defined integrals present in other wave generation formalisms). Also the computation makes a clear separation between the effects arising from relativistic corrections in the moments of the source (1PN and 2PN terms in (3.1)), and the effects due to nonlinear (and, in general, non-local) interactions of these moments in the wave-zone (e.g. 1.5PN term).

REFERENCES

[1] See the contribution of K. S. Thorne in this volume.
[2] C. Cutler *et al*, Phys. Rev. Lett. **70**, 2984 (1993).
[3] B.F. Schutz, Nature **323**, 310 (1986).
[4] L. Blanchet and B.S. Sathyaprakash, Phys. Rev. Lett. **74**, 1067 (1995).
[5] C. M. Will, Phys. Rev. D **50**, 6058 (1994).
[6] L. Blanchet, T. Damour and B.R. Iyer, Phys. Rev. D, in press.
[7] L. Blanchet, Phys. Rev. D, in press (and references therein).
[8] See the contribution of A.G. Wiseman in this volume.
[9] K.S. Thorne, Rev. Mod. Phys. **52**, 299 (1980).
[10] L. Blanchet and T. Damour, Phys. Rev. D**46**, 4304 (1992).
[11] R.V. Wagoner and C.M. Will, Astrophys. J. **210**, 764 (1976).
[12] E. Poisson, Phys. Rev. D**47**, 1497 (1993).
[13] A.G. Wiseman, Phys. Rev. D**48**, 4757 (1993).
[14] L. Blanchet and G. Schäfer, Class. Quantum Grav. **10**, 2699 (1993).
[15] H. Tagoshi and T. Nakamura, Phys. Rev. D**49**, 4016 (1994).
[16] H. Tagoshi and M. Sasaki, Prog. Theor. Phys. **92**, 745 (1994).
[17] L. Blanchet, T. Damour, B.R. Iyer, C.M. Will and A.G. Wiseman, Phys. Rev. Lett., in press.
[18] L. Blanchet, in preparation.

Gravitational Waves From a Particle Orbiting around a Rotating Black Hole: Post-Newtonian Expansion

Masaru SHIBATA, Misao SASAKI

Department of Earth and Space Science, Faculty of Science, Osaka University, Toyonaka, Osaka 560, Japan

Hideyuki TAGOSHI and Takahiro TANAKA

Department of Physics, Kyoto University, Kyoto 606-01, Japan

Using the Teukolsky and Sasaki-Nakamura equations for the gravitational perturbation of the Kerr spacetime, we calculate the post-Newtonian expansion of the energy and angular momentum luminosities of gravitational waves from a test particle of mass μ in a circular orbit around a rotating black hole of mass $M \gg \mu$ up through $P^{5/2}N$ order beyond the quadrupole formula.

The last stage of an inspiraling compact binary such as binary neutron stars is one of the promising sources of gravitational waves for the near-future laser-interferometric detectors such as LIGO[1] and VIRGO.[2] Such a binary is not only a strong source of gravitational waves, but also has a possibility to become a treasury of physics of neutron stars,[3] cosmology,[4] and so on, provided we obtain the data about binaries such as masses, spins, distance to the earth, and so on. To obtain those data with a sufficient accuracy, it is necessary to construct theoretical template waveforms whose phasing has a fractional accuracy of less than 10^{-4}.[3] This work is one of such efforts.

To calculate gravitational waves from inspiraling compact binaries, the standard method employed is the post-Newtonian expansion (PNE) of the Einstein equations,[5,6] in which the equations are expanded in terms of a small parameter $v \sim (M/r)^{1/2}$, where M and r are the total mass and orbital length-scale of the system, respectively. Despite much efforts, however, calculations have been successful to only a few orders in v beyond the leading (Newtonian) order so far. More fundamentally, the nature of PNE has not been clarified due to its complexity; nobody knows the convergence property of PNE nor the validity of the polynomial expansion in v. Given this situation, it is highly desirable to have a method which is complementary to the standard PNE. The perturbative study of a black hole spacetime is one of such, in which we consider gravitational waves radiated by a particle of mass $\mu \ll M$ orbiting around a black hole. This method, though restricted to the case of $\mu \ll M$, is very powerful because we can calculate fully general relativistic corrections of gravitational waves by means of relatively simple analyses. It is then fairly straightforward to evaluate the post-Newtonian corrections which are to be calculated in the standard PNE[7]. In this work, we extend the analysis by Tagoshi and Sasaki[7] in the case of a non-rotating black hole to the case of a rotating black hole to see the effect of spin.

Details of the calculation are shown in Shibata *et al.*[8], so we here only describe the outline of the method. To calculate gravitational radiation at infinity from a particle orbiting around a Kerr black hole, we start with the Teukolsky equation[9] for the fourth Newman-Penrose quantity, ψ_4, which may be expressed as

$$\psi_4 = (r - ia\cos\theta)^{-4} \int d\omega e^{-i\omega t} \sum_{\ell,m} \frac{e^{im\varphi}}{\sqrt{2\pi}} {}_{-2}S^{a\omega}_{\ell m}(\theta) R_{\ell m\omega}(r), \tag{1}$$

where ${}_{-2}S^{a\omega}_{\ell m}$ is the spheroidal harmonic function of spin weight $s = -2$ and λ is the eigenvalue. The radial function $R_{\ell m\omega}(r)$ obeys the Teukolsky equation with spin weight $s = -2$,

$$\Delta^2 \frac{d}{dr}\Big(\frac{1}{\Delta}\frac{dR_{\ell m\omega}}{dr}\Big) - V(r) R_{\ell m\omega} = T_{\ell m\omega}(r), \tag{2}$$

where $T_{\ell m\omega}(r)$ is the source term. The solution of the Teukolsky equation at infinity ($r \to \infty$) is expressed as

$$R_{\ell m\omega}(r) \to \frac{r^3 e^{i\omega r^*}}{2i\omega B^{in}_{\ell m\omega}} \int_{r_+}^{\infty} dr' \frac{T_{\ell m\omega}(r') R^{in}_{\ell m\omega}(r')}{\Delta^2(r')} \equiv \tilde{Z}_{\ell m\omega} r^3 e^{i\omega r^*}, \tag{3}$$

where $r_+ = M + \sqrt{M^2 - a^2}$ and $R^{in}_{\ell m\omega}$ is the homogeneous solution which satisfies the ingoing-wave boundary condition at horizon,

$$R^{in}_{\ell m\omega} \to \begin{cases} \Delta^2 e^{-ikr^*}; & r^* \to -\infty, \\ r^3 B^{out}_{\ell m\omega} e^{i\omega r^*} + r^{-1} B^{in}_{\ell m\omega} e^{-i\omega r^*}; & r^* \to +\infty, \end{cases} \tag{4}$$

where $k = \omega - ma/2Mr_+$ and r^* is the tortoise coordinate.

Thus in order to calculate gravitational waves emitted to infinity from a particle in circular orbits, we need to know 1)the explicit form of the source term $T_{\ell m\omega}(r)$, which has support only at $r = r_0$ where r_0 is the orbital radius in the Boyer-Lindquist coordinate, 2)the ingoing-wave Teukolsky function $R^{in}_{\ell m\omega}(r)$ at $r = r_0$, and 3)its incident amplitude $B^{in}_{\ell m\omega}$ at infinity. We consider the expansion of these quantities in terms of a small parameter, $v^2 \equiv M/r_0$. Note that v is approximately equal to the orbital velocity, but not strictly equal to it in the case of $a \neq 0$.

In addition to these, we need to expand the spheroidal harmonics and their eigenvalues in powers of $a\omega$. Since $\omega = O(\Omega)$ where Ω is the orbital angular velocity of the particle, we have $a\omega = O(M\omega) = O(v^3)$. Thus the expansion in powers of $a\omega$ is also a part of PNE. Note also that the spin parameter of the black hole a does not have to be small but can be of order M. Since the angular eigenvalues λ comes into play in the radial equation, we must also expand it as $\lambda = \lambda_0 + a\omega\lambda_1 + a^2\omega^2\lambda_2 + O(v^9)$. These expansions are performed by extending the method shown in ref.[10].

To perform the PNE of the Teukolsky equation for a rotating black hole we use the Sasaki-Nakamura (SN) equation, which is obtained by a certain transformation of the Teukolsky equation. The reason to use it is that the SN equation is a generalization of the Regge-Wheeler (RW) equation for $a = 0$ to the case of $a \neq 0$ and hence has a more tractable structure than the Teukolsky equation. In addition, we can make use of algebraic formulas for the PNE of the RW equation developed previously.[7]

To obtain the source term, we need to specify the motion of a test particle. We consider the case in which a particle moves along a constant radius $r = r_0$, but precesses around the symmetric axis(i.e., $\theta \neq \pi/2$). The degree of precession is determined by the value of Carter constant C. If r_0 and C are given, the energy E and the z-component of the angular momentum l_z are obtained. In this work, we only consider the case in which a particle is in the orbit with small inclination because in general case, we can not solve the geodesic equation analytically. To be specific, we introduce a dimensionless parameter y defined by

$$y = \frac{C}{Q^2}; \quad Q^2 = l_z^2 + a^2(1 - E^2), \tag{5}$$

and regard it as small. Since $Q^2 \sim l_z^2$ and $C \sim l_x^2 + l_y^2$, this is equivalent to assuming $l_x^2 + l_y^2 \ll l_z^2$. Gathering these PNE, we obtain the luminosities up to $O(v^5)$ and $O(y)$ as follows;

$$\left\langle \frac{dE}{dt} \right\rangle = \left(\frac{dE}{dt} \right)_N \left[1 - \frac{1247}{336} v^2 + 4\pi v^3 - \frac{73}{12} q v^3 \left(1 - \frac{y}{2} \right) - \frac{44711}{9072} v^4 + \frac{33}{16} q^2 v^4 - \frac{527}{96} q^2 v^4 y - \frac{8191\pi}{672} v^5 + \frac{3749}{336} q v^5 \left(1 - \frac{y}{2} \right) \right], \tag{6}$$

$$\left\langle \frac{dJ_z}{dt} \right\rangle = \left(\frac{dJ_z}{dt} \right)_N \left[\left(1 - \frac{1247}{336} v^2 + 4\pi v^3 - \frac{44711}{9072} v^4 - \frac{8191}{672} \pi v^5 \right) \left(1 - \frac{y}{2} \right) - \frac{61}{12} q v^3 \left(1 - \frac{3y}{2} \right) + q^2 v^4 \left(\frac{33}{16} - \frac{229}{32} y \right) + q v^5 \left(\frac{417}{56} - \frac{4301}{224} y \right) \right], \tag{7}$$

where $(dE/dt)_N$ and $(dJ_z/dt)_N$ denote the quadrupole luminosity. From the results, we find the following features:

1) The coefficients of qv^3 and qv^5 in dE/dt are proportional to $1 - y/2$. Since $1 - y/2 \sim \cos\theta_i$, these terms may be regarded as proportional to the inner product $\vec{S} \cdot \vec{L}$ of the spin angular momentum $\vec{S}$ and the orbital angular momentum $\vec{L}$.

2) Contrary to the feature 1), the coefficient of the q^2v^4 term does not seem to be expressible as a simple function of $\cos\theta_i$. We suspect that a major part of it is attributable to the quadrupolar gravitational field around the Kerr black hole which modifies the particle orbit because for $y = 0$, the $2q^2v^4$ term can be explained in terms of the Newtonian quadrupole formula as the contribution from the quadrupole moment of the Kerr black hole. However, the $1/16\, q^2v^4$ term in $\eta_{2\pm10}$ cannot be explained in this way. So that, our result suggests the existence of a new type of spin-dependent terms in the energy flux when a PN analysis beyond the present level is carried out.

3) In the limit $q \to 0$, $dE/dt = dJ_z/dt(1 - y/2)v^3M$ holds, but in the case $q \neq 0$, any simple relations between dE/dt and dJ_z/dt do not hold. This shows that we need to calculate the change rate of the Carter constant to

calculate the back reaction of the test particle in general orbits around a rotating black hole.

Using the above formula, we also analyse the phase error of coalescing binaries. Summarizing the analyses, we obtain the following conclusions:

1) The spin-orbit coupling term at the $P^{5/2}N$ order is important for the evolution of NS-NS binaries with $P_{ms} \lesssim 2$ and BH-NS binaries with $q \gtrsim 0.2$. Since the rotation of $P_{ms} \sim 1$ would be the fastest possible period that a NS could have, inclusion of the phase corrections up through the $P^{5/2}N$ order seems to be enough for NS-NS binaries. On the other hand, the terms higher than the $P^{5/2}N$ order are likely to be important for BH-NS binaries since the rotation of $q \gtrsim 0.2$ for a black hole seems quite possible.

2) The q^2 terms at the P^2N order becomes important for BHs with $q \gtrsim 0.2$ or NSs with $P_{ms} \lesssim 2$. However, the latter value is based on our formula which is valid only for a rotating BH. An estimate base on a realistic NS model gives $P_{ms} \lesssim 5$. Thus it will be possible to distinguish a small mass BH from a NS if the phase corrections to P^2N order can be detected by matched filtering.

3) At any order of PN corrections, the effect of a finite inclination angle to the number of the phase cycles must be taken into account whenever the spin terms become important.

REFERENCES

1. K. S. Thorne, in *Proceedings of the 8th Nishinomiya-Yukawa Memorial Symposium: Relativistic Cosmology*, ed. M. Sasaki (Universal Academy Press, Tokyo, 1994), p67.
2. C. Bradaschia et al., Nucl. Instrum. and Method **A289**, 518 (1990).
3. C. Culter et al., Phys. Rev. Lett. **70**, 2984 (1993).
4. D. Markovic, Phys. Rev. **D48**, 4738 (1993).
5. C. M. Will, ref.[1], p83, and references therein.
6. As for a recent development, see L. Blanchet, this volume.
7. H. Tagoshi and M. Sasaki, Prog. Theor. Phys. **92**, 745(1994) and references therein.
8. M. Shibata, M. Sasaki, H. Tagoshi and T. Tanaka, Phys. Rev. **D51**, No.4(1995).
9. S. A. Teukolsky, Ap. J. **185**, 635 (1973).
10. W. H. Press and S. A. Teukolsky, Ap. J. **185**, 649 (1973).

Active Galactic Nuclei across the Electro-magnetic Spectrum

THIERRY J.-L. COURVOISIER

Geneva Observatory
CH-1290 Sauverny
Switzerland

Introduction

The emission of Active Galactic Nuclei spans the electro-magnetic spectrum from the far infrared domain (or radio domain for radio loud objects) to the gamma rays. This emission is made of a number of components, each described by a number of parameters which, at present, cannot be deduced from a comprehensive and self consistent underlying model. Observations spanning large parts of the electro-magnetic spectrum are therefore crucial to make our understanding of these objects progress. Data can now in some cases span 15 decades in energy in near simultaneous observations, from the radio domain to high energy gamma rays. Such spectral energy distributions are, however, seldom, the reason being the difficulty of organising coordinated sets of observations with a wide variety of instruments on the ground and in orbit.

AGN are also variable sources, the time scales observed span many decades, from minutes to years. This characteristic has caused additional problems to the observers, as it imposes that observations in different spectral regions are coordinated in time. This variability is, however, a powerful diagnostic tool to probe the emission mechanisms and the geometrical arrangement of the emission components. It is by using variability that one can hope to understand which components are primary and which re-processed radiation.

Some of the questions studied using multi-wavelength techniques are:

- What is the nature of the emission mechanisms in the different spectral domains?
- What are the relations between the components? What is the geometrical arrangement of the sources?
- One would also like to be able to understand the relation between the radiation we observe and the (most probably) gravitational origin of the energy. Radiation is always a loss of energy by a region. These regions have been energetised prior to the radiation (probably often in shocks) but the energy ultimately comes from the fall of matter in the deep gravitational well at the centre of the AGN. The latter link is still largely unexplored.

- The relationship between the nucleus and the host galaxy is also a subject of high interest.

In this review I will briefly touch on three results, one related to BL Lac objects, the second to Seyfert galaxies and the last one to quasars. These results will be based on studies of single objects rather than on samples. This is also an illustration of the difficulty of obtaining high quality repeated sets of data on many objects.

MULTI-WAVELENGTH OBSERVATIONS OF THE BL LAC OBJECT PKS 2155-304

The idea behind the BL Lac observation campaigns is that in these objects relativistic beaming enhances one of the emission components, the synchrotron emission produced by the jet, and not the others. It is therefore expected that these objects should provide relatively clean laboratories for the study of the component singled out by the beaming effect.

With this in mind a large international collaboration has set up a campaign to observe PKS 2155-304 in November 1991 from the radio domain to the X-rays. The results have been published in a series of papers recently[1,2,3,4]. In summary the results show that the light curves in the X-ray , the UV and the infrared domains have been very similar, the amplitude of the variations have been a factor of approximately 2. The variations were achromatic. This means that the spectral slopes observed during the campaign were not correlated with the flux (very little variations in the slopes were indeed observed). The X-ray flux most probably lead the UV flux by a few hours. No lags have been observed between the bands in the visible. The amplitude of the polarised flux variations have been larger than those of the total flux and have shown little correlations with the latter.

We can conclude from this set of observations that the variability properties do not match those expected from a synchrotron emitting cloud. Were this true, decreases in the flux would have occurred faster at the higher frequencies than at the lower frequencies. We also cannot decompose the flux in a constant unpolarised component and a variable polarised component. We can also rule out that the ROSAT X-rays are due to an inverse Compton process on the optical or UV photons. Such a process would imply that the X-rays lag behind the soft photons, contrary to what is observed.

A second campaign of observations was organised on the same object in May 1994[5]. The results of this second campaign are drastically different from those observed in 1991. The X-rays varied by a larger factor than the UV flux. The shape of the light curves differ, showing, therefore, spectral evolution in time. There were some very fast variations of about 40% in the UV flux on a time scale of a few hours. Work in progress possibly suggests that these results might be accommodated by a synchrotron model.

The most surprising result from these two campaigns is the large difference in the qualitative behaviour of the multi-wavelength flux variations in one and the same object observed at 2.5 years interval.

IS THE BLUE BUMP OF NGC 5548 RE-PROCESSED X-RAY RADIATION?

The X-ray emission of AGN has complex properties Mushotzky, Done and Pounds[6] give a review of some of these properties. In particular it is shown that in several objects (not all) there is mounting evidence for the existence of a cold medium that re-processes a large fraction of the "primary" X-rays.

Clavel et al.[7] have shown that at least at certain epochs the UV flux of NGC 5548 is well correlated with the X-ray flux with no measurable lag. The correlation is such that the UV flux is proportional to the X-ray flux. They interpret this result with a model in which most of the UV flux is due to the surface of an accretion disk heated by the strong X-ray flux of the source. There is then little room for additional heat sources in the disc. This type of data and interpretation are compatible with the presence of a significant amount of cold matter in the vicinity of the X-ray source.

A campaign of UV and X-ray observations of NGC 5548 took place in early 1993 with the IUE and ROSAT satellites. The results of this campaign[8] are, however, difficult to reconcile with this model. The point is that the hard X-rays (in ROSAT terms) have barely varied while the UV and the soft X-ray fluxes changed by a factor of about 2. It is difficult to understand how the primary energy source should remained constant while the re-processed component varied very significantly.

MULTI WAVE-LENGTH CORRELATIONS IN THE BRIGHT QUASAR 3C 273

The bright Quasar 3C 273 has been the object of a long set of observations in all spectral domains (ref.[9,10] and work in progress). This set of observations is sufficient to allow us to find a number of relationships between the emission components.

The first relationship was already found in the references given above. It shows that the UV and the radio light curves are well correlated, the UV leading the radio emission by a few months. This result has firmed up as time elapsed and the number of observations increased. This result is confirmed by a similar correlation found between the optical light curves and the radio light curves. This confirmation is independent, because the UV and the optical observations have very different time samplings. The UV and the optical light curves are, furthermore, very well correlated with at most a very short lag, considerably less than a year. This last result is very difficult to interpret in terms of standard accretion disk models[11].

The X-ray spectral index as measured in the 2-10keV region is anti-correlated to the logarithm of the X-Ray to UV fluxes in 3C 273. This result can be well explained if the X-Ray emission is due to inverse Compton boosting of the UV photons by a thermal electron population of few 100 keV[12].

A further result links the observed spatial structure of 3C 273 as it is observed by the VLBI technique with powerful IR and optical flares. The IR flux of 3C 273 is mostly stable, this emission is probably due to warm dust. However, in 1988 the IR (and the optical) flux increased very rapidly (time-scales of few days) by more than 50%. This very unusual behaviour is best understood if the flares are due to synchrotron emission[13]. VLBI observations in the months following this flare have shown that a new component of the VLBI jet had appeared at an epoch which coincides with the synchrotron flare[14]. This is the first evidence that structures in VLBI jets of AGN are related to the synchrotron activity of the core.

CONCLUSIONS

Up to now we knew that AGN are variable sources, now we also know that the way in which they vary also varies with time. This is certainly (yet) another piece of complexity in these objects.

Multi-wavelength studies of at least few AGN over long periods of time (many years) is proving to be a powerful tool to un tangle the emission components and will certainly show which of the emission mechanisms is primary and which are re-processed. It is interesting to also note that the features of the jets can be related to activity in the core of the objects.

REFERENCES

1. Urry C.M. et.al. 1993. Ap.J. **411**:614
2. Brinkmann W. et.al. 1994. A&A in press
3. Courvoisier T.J.-L. et.al. 1995. Ap.J. in press
4. Edelson R. et.al. 1995. Ap.J. in press
5. Urry C.M. et.al. 1995. B.A.A.S. **26**:1467
6. Mushotzky R., Done C. & Pounds K. 1995. ARAA. **31**:717
7. Clavel J. et.al. 1992. Ap.J. **393**:92
8. Walter R. et al. This conference
9. Courvoisier T.J.-L. et.al. 1987. A&A **176**:197
10. Courvoisier T.J.-L. et.al. 1990. A&A **234**:73
11. Courvoisier T.J.-L. and Clavel J. 1991 A&A **248**:349
12. Walter R. and Courvoisier T.J.-L. 1992 A&A **258**:255
13. Courvoisier T.J.-L. et.al. 1988. Nature **335**:330
14. Krichbaum T.P. et.al. 1990. A&A **237**:3

RECENT X-RAY SPECTRAL RESULTS FROM SEYFERT 1 GALAXIES

A.C. FABIAN[a], R.F. MUSHOTZKY[b], K. NANDRA[a], C.S. REYNOLDS[a]

[a]Institute of Astronomy
Madingley Road
Cambridge CB3 0HA, UK

[b]Code 666, NASA/GSFC
Greenbelt, MD 20771

INTRODUCTION

Recent results from GINGA, ROSAT and, in particular, ASCA show that the X-ray spectra of Seyfert 1 galaxies are complex. The primary power-law continuum is commonly modified by oxygen absorption edges indicative of highly ionized gas along the line of sight (the warm absorber) and fluorescent iron emission features, with an associated hard continuum, indicative of reflection from relatively cold material. The excellent sensitivity and spectral resolution of ASCA now enables us to resolve the spectral features and so probe and map out the innermost structures of AGN.

WARM ABSORBERS

Halpern[1] first introduced warm absorbers to explain a spectral complexity in MR 2251-178. Oxygen absorption was clearly seen[2] in the ROSAT spectra of MCG-6-30-15 and many other Seyfert 1 galaxies[3,4,5] and then clearly resolved with ASCA spectra[6,7,8,9]. The principal absorber tends to be OVII or OVIII and the (equivalent) hydrogen column density (assuming cosmic abundances) $\sim 10^{22}\,\mathrm{cm}^{-2}$. The edge is sharp and consistent with gas moving at less than a few per cent of the speed of light in the rest frame of the object. Some theoretical discussions of warm absorbers can be found in refs. 10–12.

The 1993 ASCA data for MCG-6-30-15 show that the warm absorber is variable on timescales of 3 weeks[6] to 3 hours[13]. The variation in the absorber is principally an increase in column density, by about a factor of two. The data show that although the luminosity, L, of the source varied considerably during the observations, the ionization state of the gas seems to be independent of L. This is curious since the ionization state, measured by fitting absorption models obtained from Ferland's photionization code CLOUDY and parametrized by the ionization parameter, $\xi = L/nR^2$, where n and R are the gas density and its radius from the source respectively, is proportional to L.

The photoionization timescale $\sim 1R_{16}^2L_{43}^{-1}$ s and the recombination time for oxygen $\sim 500n_9^{-1}T_5^{1/2}$ s, where the subscripts are the log of the parameter value and T is the gas temperature. The lack of a luminosity correlation can

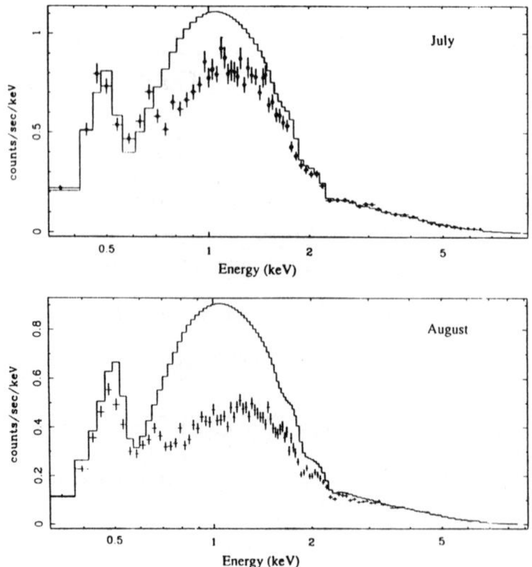

Fig 1. ASCA SIS spectra of MCG-6-30-15 taken[6] in 1993. The solid line is the best fitting power-law to the data above 3.2 keV, extrapolated to lower energies with Galactic absorption. Note the large deficit of counts between 0.7 and 2 keV, which is due to oxygen absorption in warm gas, and the change in the depth of absorption between the two observations.

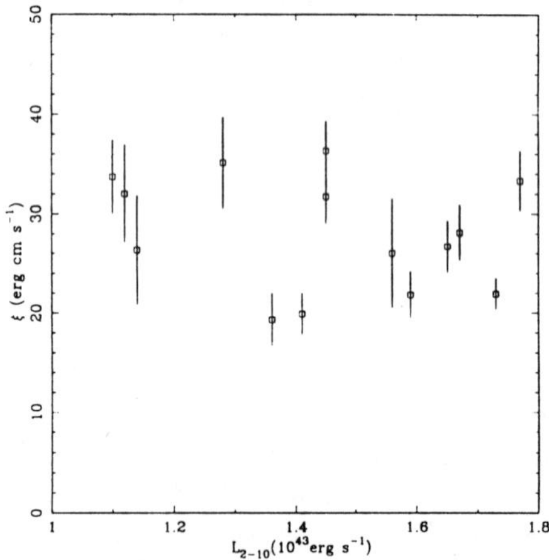

Fig 2. Ionization parameter, ξ, plotted[13] against luminosity in the 2–10 keV band during the July 1993 observation of MCG-6-30-15. If the ionization state of the warm absorber is determined through photoionization by the X-radiation and lies close enough to the source that the photoionization and recombination times are less than a few hours then we would expect $\xi \propto L$.

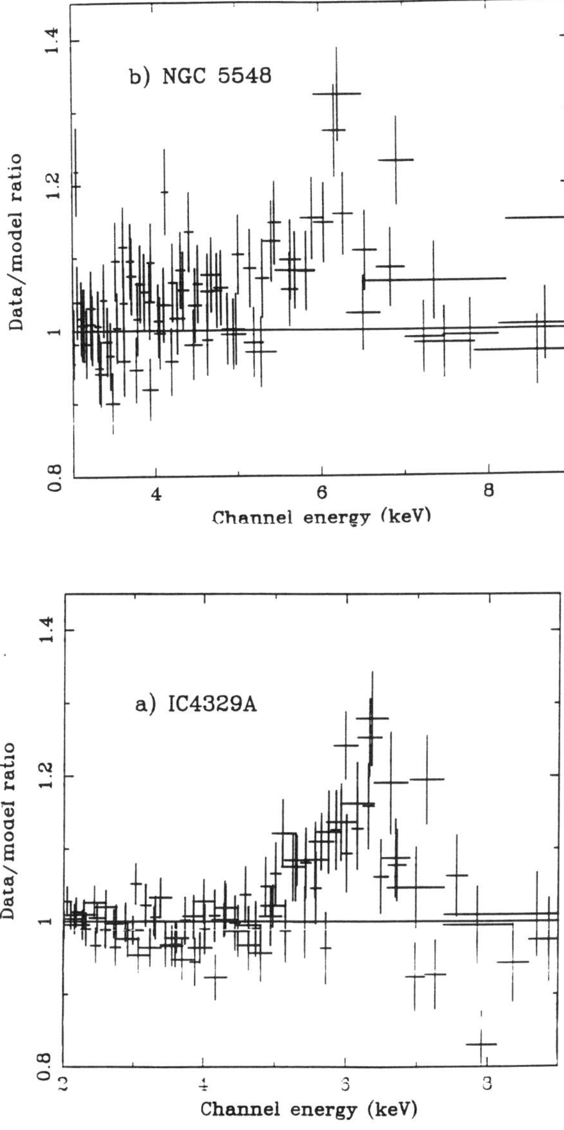

Fig 3. Iron lines from NGC 5548 and IC 4329A from the ASCA SIS data[21]. The plots are made by fitting the best power-law and gaussian line to the data, then setting the normalization of the line to zero and forming the ratio of data to (power-law) model.

then be due to the gas being either at large radii, where a 3 hr variation is implausible, or at small radii in a dense gas where collisional, or other, processes determine the ionization state. A simple photoionized shell of gas appears not to be a correct model for the warm absorber. Note that this is one instance where both the absorption edge and the ionizing continuum are directly observed in an AGN and the behaviour is found to be complicated. This may have ramifications for optical/UV reverberation mapping of these objects since there the relevant edges and ionizing continuum are not directly observed.

THE IRON LINE AND REFLECTION

Spectral complexities around the iron-K energy and reflection were predicted to occur in AGN[14,15] and discovered in GINGA data[16,17]. The observations are consistent[18] with the primary power-law continuum being seen both directly and after 'reflection' by Thomson-thick, cold matter (i.e. most metals relatively neutral although hydrogen and helium may be ionized) subtending about 2π sr at the source. The cold matter could either be the accretion disk or the outer torus postulated[19,20] in unified models for AGN.

The iron emission line is resolved in the ASCA data[6,21]. The FWHM of the line in NGC 5548, for example, corresponds to about $54,000\,\mathrm{km\,s^{-1}}$. Broad lines are resolved too in MCG-6-30-15 and IC4329A. The large breadth of the line immediately rules out the torus as the main source of line emission and indicates the inner accretion disk. Good fits are obtained with models predicted for a relativistic disk[22].

A recent 4 day observation of MCG-6-30-15 has produced the best iron line profile yet obtained from a Seyfert 1 galaxy. The line is clearly skewed to the red, suggesting the effects of gravitational redshift and transverse doppler shift. The line cannot be broadened by Comptonization since in order to produce the large observed shift, the continuum would also be significantly down-scattered, contrary to observation.

Further studies of variability of the line should enable future observations to measure the mass of the black hole. The immediate surroundings of accreting black holes are now open for detailed observational study.

ACKNOWLEDGEMENTS

We are very grateful to Professor Y Tanaka and the ASCA PV team for the opportunity to work with the ASCA data and Professor H Inoue for further collaboration on MCG-6-30-15.

REFERENCES

1. Halpern J.P., 1984, ApJ, 281, 90
2. Nandra K., Pounds K.A., 1992, Nature, 359, 215
3. Nandra K., et al., 1993, MNRAS, 260, 504
4. Fiore F., et al., 1993, ApJ, 415, 129
5. Turner T.J., et al., 1993, ApJ, 419, 127
6. Fabian A.C., et al., 1994, PASJ, 46, L59
7. Mihara T., et al., 1995, PASJ, 46, L137
8. Ptak A., et al., 1995, ApJ, in press
9. George I.M., Turner T.J., Netzer H., 1995, ApJ, 438, L67
10. Netzer H., 1993, ApJ, 411, 594
11. Krolik J.H., Kriss G., 1995, preprint
12. Reynolds C.S., Fabian A.C., 1995, MNRAS, in press
13. Reynolds C.S., et al., 1995, MNRAS, in press
14. Guilbert P.W., Rees M.J., 1988, MNRAS, 233, 475
15. Lightman A.P., White T.R., 1988, MNRAS, ApJ, 335, 57
16. Pounds K.A., et al., 1990, Nature, 344, 132
17. Matsuoka M., et al., 1990, ApJ, 361, 440
18. Nandra K., Pounds K.A., MNRAS, 268, 405
19. Krolik J.H., Madau P., Zycki P.T., 1994, ApJ, 420, L57
20. Ghisellini G., Haardt F., Matt G., 1994, MNRAS, 267, 743
21. Mushotzky R.F., et al., 1995, MNRAS, 272, L9
22. Fabian A.C., Rees M.J., Stella L., White N.E., 1989, MNRAS, 238, 729

Correlated Optical and Gamma–Ray Variability in Blazars[a]

STEFAN WAGNER

Landessternwarte Heidelberg
Königstuhl
69117 Heidelberg
Germany

INTRODUCTION

One of the exciting results from the EGRET instrument on board the CGRO spacecraft is the discovery of a large number (> 40 by 1994) of gamma–bright AGN.[1] Almost all of them fall into the group of Blazars, i.e. AGN which are radio-loud, have flat radio spectra and show strongly variable and highly polarized optical emission. The properties in the low-energy part of the electromagnetic spectrum ($\nu < 10^{16}$ Hz) indicate that the emission is dominated by synchrotron radiation.

In particular, it is interesting to note that many of the sources with large gamma–ray fluxes are 'Intra-day variable' sources (IDV), i.e. objects which vary significantly within one day or even faster. Several such objects have been identified whose variations on time scales of a few hours imply diameters of a few 10^{15} cm, and hence brightness temperatures at 5 GHz of up to 10^{18} K.[2] Simultaneous optical and radio observations found correlated variations, illustrating that the fast radio variability cannot be explained by interstellar scintillation.[3] Intrinsic variations can only be reconciled with the Compton limit of 10^{12} K of incoherent synchrotron radiation, if the emission is relativistically beamed with Doppler factors as high as $\mathcal{D} = (T_B/10^{12}\,K)^{1/3} \sim 100$.[4] High Doppler factors are also required to explain high gamma–ray fluxes, since photon-photon pair production strongly absorbs the gamma flux of an isotropic emitter.[5] Although the constraints on the Doppler factor derived from the gamma–ray luminosity are not as severe as suggested by IDV, high gamma–ray fluxes may be expected from sources with ultra-relativistic jets.

Indeed, a large fraction of IDV sources was detected by EGRET at photon energies above 100 MeV.[6] Since both the groups of gamma–bright objects and IDV sources are small in comparison with the total number of AGN known, the high degree of coincidence of the two phenomena suggests that a common explanation is needed.

The mechanisms which have been discussed in the context of the high gamma–ray luminosities include hadronic cascades[7] as well as leptonic models.[8] The latter can be subdivided in pair plasma scenarios[9,10] and comptonisation of either generic synchrotron photons[11] or photons which are scattered into the jet. Such photons may be scattered from the vicinity of the central engine[12] or – in order to gain additional energy from the blueshift in head-on collisions – from a more extended medium.[13] All of those suggestions have successfully fitted the broad-band energy distribution, and spectral signatures alone may be insufficient to distinguish the dominant processes.

[a]This work was supported by the Deutsche Forschungsgemeinschaft through SFB 328

The different models predict different variability properties, however. Many of the stronger gamma–bright AGN have shown changes in gamma–ray flux from one observation to another and several cases have been reported which exhibited significant variation within a single week of observations.[14] Together with the outbursts of many of these sources at lower frequencies mentioned above, this illustrates the necessity of carrying out simultaneous campaigns over the entire electromagnetic spectrum to reveal the actual broad-band energy distribution and to study the relationship of variations in different frequency regimes.

Correlations between major radio outbursts and detections in gamma–ray emission have been reported for several sources.[15,16] Usually the epoch of gamma–ray detection (or a state of high gamma–ray emission) is followed (with a delay of several months) by an outburst at centimeter frequencies. Although the delay can be understood in terms of optical thickness in the radio-regime, a principal problem with such correlations is the sparse sampling in the gamma–ray range and the high rate of flaring at radio-frequencies. Well-sampled radio-lightcurves of Blazars show flares at rates of about one per year, implying that correlations with delays of several months are difficult to establish. This problem may be less severe in the optical domain where the nonthermal radiation is generally assumed to be optically thin. In order to investigate possible correlations of the gamma–ray fluxes with the synchrotron emission we therefore monitored several candidates at optical wavelengths.

LONG-TERM VARIABILITY

Many of the sources discussed by von Montigny et al.[14] exhibit variations of their gamma–ray flux-density by a factor of two or three on time-scales of several months. One of them is the superluminal quasar PKS 0420-014, which is discussed in detail by Radecke et al.[17] This source has been well observed in various ground-based monitoring campaigns in the radio and optical regimes, resulting in lightcurves which span a period of 3 decades.[18] During the lifetime of the EGRET mission the source produced the strongest historical outburst at optical wavelengths in February 1991, at a time when the source was also observed with the EGRET instrument. During

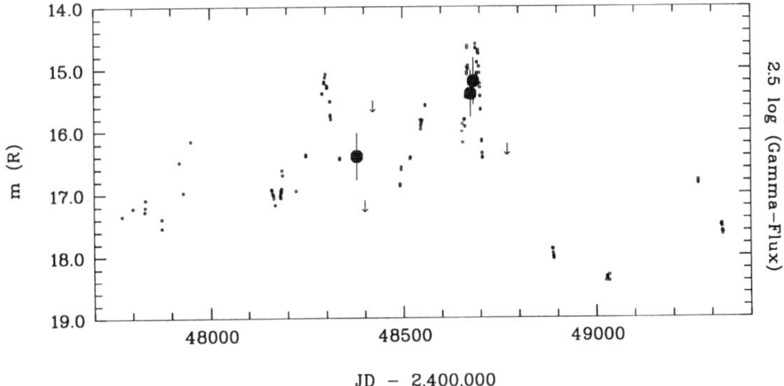

Figure 1. Lightcurve of PKS 0420-014. The third of the three major outbursts markes the highest optical flux level of this source in 30 years. It coincided with the detection of the highest state of gamma–ray emission.[18] EGRET measurements are indicated by open boxes.

this strongest optical flare (which also shows a delayed radio counterpart) the gamma–ray flux was particularly high – larger then in any other epoch. Figure 1 illustrates the correlation of optical and gamma–ray fluxes which have comparable amplitudes during changes on comparable time-scales. Although the sampling is still poor in both lightcurves, the simultaneity of the high states in both observations is noteworthy. We also draw attention to the flaring rate of about 13 months, which is reminiscent of a similar behaviour in 3C345.[18,19] In this particular case, however, no enhanced gamma–ray emission had been detected immediately prior to the optical flare.[20] This is not in disagreement with the assumption of correlated optical and gamma–ray emission, since very fast variation has also been observed in both bands.

RAPID VARIATIONS

Already during the first discovery of strong gamma emission from a blazar (3C 279), this source showed an increase of a factor of 2 within 2 d and a subsequent decay which was even faster (factor of two within 24 h).[21] A similar behaviour was found in PKS 0528+134[22], 1633+382[23], and PKS 2251+158[24].

During observations in January 1993 EGRET found PKS 1406-076 in a state of high gamma–ray luminosity.[25] More detailed analysis revealed fast variations (Figure 2),[26] i.e. a well-defined flare, lasting for about one day in the frame of the source (which has a redshift of $z = 1.5$). Simultaneous monitoring at optical frequencies found the source to be unusually bright and also revealed an outburst. The flares recorded in the optical and gamma–ray regimes have similar shapes (within the limited statistics of the gamma–ray data), but are separated in time. The optical flare precedes the gamma–ray maximum by about 22 hours.[26]

This object has not been studied in detail prior to its detection as a gamma–ray source. Our optical monitoring since the observation in January 1993 did not record any similar optical flare. This implies that such bursts are rare events and strongly supports the assumption that the nearly simultaneous occurrence of the outbursts shown in Figure 2 are not due to a chance coincidence.

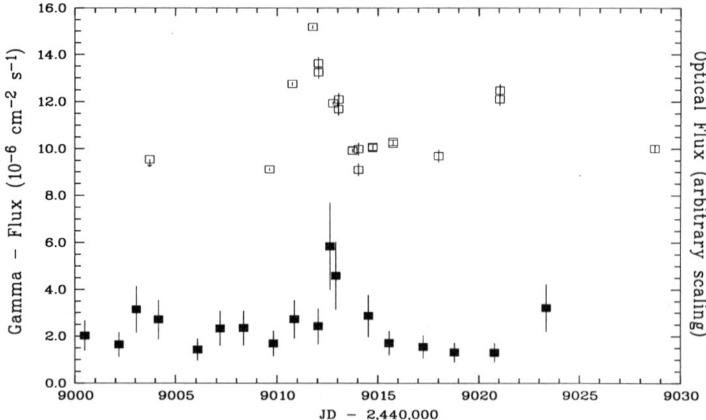

Figure 2. A three-week period of PKS 1406-076 during a high state of gamma–ray flux shows a pronounced flare lasting for one day in the source frame. An optical outburst of similar shape preceded the gamma–ray flare by 22 hours.[26]

Similar observations of the BL Lac object 0716+714, which is also a well-known IDV source, also suggests correlated variations (Mattox et al., in preparation).

IMPLICATION FOR GAMMA–RAY EMISSION MECHANISMS

In case the optical and gamma–ray flare are indeed correlated, the observed delay between the two bands suggests that the gamma–rays are reprocessed. If the optical burst traces a flare of synchrotron radiation which is later Compton-scattered into the gamma–ray range, the delay implies that the scattering cannot be strictly due to the very same electron population causing the optical synchrotron flare. The close similarity in shape (and width) of the flares constrains the amount of geometrical dilution in any model which involves back-scattering of photons into the jet as a source for the comptonised photons. The optical photons clearly cannot be decay-products of cascades starting at much higher energies. It is well possible, however, that the optical flare marks the trigger of the very acceleration process which injects the highly relativistic particles which later cause the high-energy flare.[4,26]
The variability properties also show that truely simultaneous monitoring is required to determine the broad band energy distributions which have to be explained by a realistic model.

REFERENCES

1. Fichtel, C.E, et al. 1994. *Ap.J. Suppl.* 94, 551
2. Wagner, S.J., & Witzel A. 1992. In *Extragalactic Radio Sources - From Beams to Jets.* Eds. J. Roland, H. Sol, G. Pelletier. Cambridge University Press. p.59.
3. Quirrenbach, A., et al. 1991. *Ap.J. Lett.* 372, L71
4. Wagner, S.J., & Witzel, A. 1995. *Annu. Rev. Astron. Astrophys.* Vol. 35, in press.
5. McBreen, B. 1979. *Astron. Astrophys.* 71, L19
6. Wagner, S.J. 1995. *Astron. Astrophys.*, submitted
7. Mannheim, K., & Biermann, P.L. 1992, *Astron. Astrophys.* 253, L21
8. Schlickeiser, R. 1995. In: Proceedings of the Heidelberg Workshop on *TeV Astronomy, Sp. Sci. Rev.*, in press
9. Bednarek, W., & Kirk, J.G. 1995. *Astron. Astrophys.*, in press
10. Henri, G., Pelletier, G., & Roland, J. 1993. *Ap.J. Lett.* 404, L41
11. Maraschi, L., Ghisellini, G., & Celotti, A. 1992, *Ap.J. Lett.* 397, L5
12. Dermer, C.D., Schlickeiser, R., & Mastichiadis, A. 1992, *Astron. Astrophys.* 256, L27
13. Sikora, M., Begelman, M.C., & Rees, M.J. 1994. *Ap.J.* 421, 153
14. von Montigny, C., et al. 1995. *Ap. J.*, in press
15. Reich, W., et al. 1993. *Astron. Astrophys.* 273, 65
16. Zhang, Y.C., et al. 1994. *Ap.J.* 432, 91
17. Radecke, H.-D., et al. 1995. *Ap.J.* 438, 659
18. Wagner, S.J., et al. 1995. *Astron. Astrophys.*, in press
19. Schramm, K.J., et al. 1993. *Astron. Astrophys.* 278, 391
20. Hartmann, R.C., et al. 1994. In *The second Compton Symposium.* Eds. C.E. Fichtel, N. Gehrels, J.P. Norris. AIP. p.563
21. Kniffen, D.A., et al. 1993. *Ap.J.* 411, 133
22. Hunter, S.D., et al. 1993. *Ap.J.* 409, 134
23. Mattox, J.R., et al. 1993. *Ap.J.* 410, 609
24. Hartman, R.C., et al. 1993. *Ap.J. Lett.* 407, L41
25. Dingus, B.L., et al. 1993. IAUC 5690
26. Wagner, S.J., et al. 1995. *Ap.J. Lett.*, submitted

DECELERATING RELATIVISTIC JETS, BL-LAC OBJECTS AND THE FANAROFF-RILEY CLASSIFICATION

GEOFFREY V. BICKNELL

Mt. Stromlo and Siding Spring Observatories, Australian National University and JILA, University of Colorado.

INTRODUCTION

There are good arguments (*e.g.* [4],[9]) as to why FRI radio galaxies are the unbeamed parent population of relativistically beamed BL-Lac objects. One caveat has been the commonly held view that FRI jets are subrelativistic on the kpc scale. Thus if unification based upon the FRI–BL-Lac connection is to hold, deceleration of relativistically moving jets is implicated. In this talk I summarize detailed work [2] which *does* support the notion of a relativistic velocity for the core jets in NGC 315 and NGC 6251. Moreover, the notion of deceleration of relativistic jets plays an important rôle in explaining the remarkable division between FRI and FRII sources in the radio - optical plane, recently discovered by Owen and Ledlow [8].

CONSERVATION LAWS FOR DECELERATION OF A RELATIVISTIC JET

Deceleration of a relativistic jet may occur through turbulent entrainment [1, 5] or mass injection from stellar mass-loss [10, 6, 7]. In either case, momentum and energy are approximately conserved as expressed in the following equations:

$$\begin{aligned}(1+\mathcal{R}_2)\,\gamma_2^2\beta_2^2 &= \left[(1+\mathcal{R}_1)\,\gamma_1^2\,\beta_1^2+\frac{\Delta\,p_1}{4\,p_1}\right]\left(\frac{p_1A_1}{p_2A_2}\right)\\ \left[(\gamma_2-1)\mathcal{R}_2+\gamma_2\right]\gamma_2\,\beta_2 &= \left[(\gamma_1-1)\mathcal{R}_1+\gamma_1\right]\gamma_1\,\beta_1\left(\frac{p_1A_1}{p_2A_2}\right)\end{aligned}$$

Subscripts 1 and 2 refer to the pc and kpc scales respectively; the parameter $\mathcal{R} = \rho c^2/4p$ is the ratio of cold matter energy density to enthalpy; the relativistic Mach number, $\mathcal{M} = (2 + 3\mathcal{R})^{1/2}\gamma\beta$; Δp is the excess of pressure over the ambient value; $\beta = v_{\rm jet}/c$; γ is the Lorentz factor and A is the cross-sectional area. These equations may be used in two ways: (1) To calculate velocity and Mach number on kpc scales given the Lorentz factor and pressure on the pc scale and the pressure on the kpc scale; (2) To calculate Mach number as a function of velocity through elimination of the factor $(p_1 A_1)/(p_2 A_2)$ from the energy and momentum equations.

A result of the latter calculation is that when the Mach number of an initially relativistic jet decreases to a transonic value ($\mathcal{M} \approx 2$) the velocity is about $0.6c$. This relation between velocity and Mach number is due to the fact that the ratio of energy flux to momentum flux is c.

MACH NUMBER AND PC-SCALE LORENTZ FACTOR FOR NGC 315 and NGC 6251

For NGC 315, there is enough information from the observational data [12, 11] to show whether it is plausible that NGC 315 has slowed to a transonic Mach number from relativistic initial conditions. The solution of the energy and momentum equations with input data derived from the observations shows that a transonic kpc-scale Mach number is indeed consistent with a pc-scale Lorentz factor. A similar result has been obtained for NGC 6251.

RELATIVISTIC JETS AND THE FANAROFF-RILEY CLASSIFICATION

Whilst the NGC 315/NGC 6251 analysis could be carried out for other sources another statistical approach gives substantial justification to the idea that all jets are launched at relativistic speeds. We begin with the premise that sources at the borderline between class I and II become transonic at approximately the core radius of the parent galaxy, where the medium is most dense and where the effects of turbulence are likely to be the greatest. The energy flux of a relativistic jet can be expressed as:

$$\begin{aligned}\log F_E \;&=\; [\log(4\pi c)] + \left[\log\left\{\left(1 + \tfrac{\gamma-1}{\gamma}\mathcal{R}\right)\gamma_{\rm jet}^2\beta_{\rm jet}\right\}\right] \\ &\quad + \left[\log(p_{\rm c} r_{\rm c}^2) + 2\log\left(\tfrac{r_{\rm jet}}{r_{\rm t}}\right) + \log\left(\tfrac{p_{\rm t} r_{\rm t}^2}{p_{\rm c} r_{\rm c}^2}\right)\right]\end{aligned}$$

where $p_{\rm jet}$, $r_{\rm jet}$, are the jet pressure and radius. Subscripts t and c refer to the transition radius (to turbulent flow) and the core radius respectively. Since the jet velocity for transonic flow is more or less unique ($\sim 0.6c$) and $r_{\rm jet}/r_{\rm t} \sim 0.1$, the major dependence of this expression is on the core pressure ($p_{\rm c}$) and the core radius

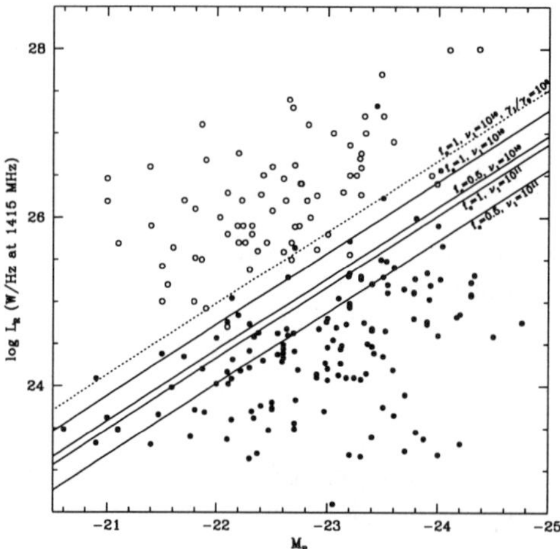

Figure 1: Theoretical FRI/II dividing lines overlaid on the data of Owen and Ledlow [8]. Filled circles are FRI galaxies, open circles FRII's

itself (r_c). These parameters are in turn related to the absolute magnitude via a sequence of empirical relations: The X-ray luminosity, L_X is related to p_c, r_c and velocity dispersion, σ and L_X, p_c, r_c and σ are related to the absolute magnitude M_B. Forsaking the details (see [3]) the expression for the critical energy flux is:

$$\log F_{\rm E} = 24.1 - 0.85\, M_{\rm B}$$
$$+ \left[2 \log \left(\frac{r_{\rm jet}/r_{\rm t}}{0.1} \right) + \log \left(\frac{p_{\rm t}\, r_{\rm t}^2}{p_{\rm c}\, r_{\rm c}^2} \right) + \log \left\{ \left(1 + \frac{\gamma - 1}{\gamma} \mathcal{R} \right) \gamma_{\rm jet}^2 \, \beta_{\rm jet} \right\} - \frac{1}{2} \log \left(\frac{I_{\rm X}}{5} \right) \right]$$

implying $F_E \approx 2 \times 10^{42} {\rm ergs\,s}^{-1}$ for $M_B = -21$. I_X parameterizes the X-ray emission.

The final step is to relate energy flux and monochromatic radio luminosity: The synchrotron luminosity of a lobe fed by a jet is

$$\mathcal{L}_S \approx \frac{3}{4} \frac{t}{t_{\rm rad}} F_E \tag{1}$$

where $t_{\rm rad}$ is the radiative time scale ($\propto B^2$) and t is the age of the source. Frequently, t is estimated from the synchrotron cooling time scale $t_c \propto B^2$ deduced from spectral steepening. The ratio of the two is independent of the magnetic field and can be estimated, for typical parameters, giving a ratio ~ 0.05 for the ratio of radio luminosity to jet energy flux. A similar argument relates the ratio of monochromatic luminosity to jet energy flux.

Putting all of this together, gives an estimate of the slope and intercept of the FRI/II dividing line in the radio-optical plane. This is plotted in figure 1 overlaid on the data points of Owen and Ledlow [8]. The slope of the line is theoretically well determined and is very close to the slope of the division in the data. The intercept is not as well determined. However, one can see that for reasonable parameters, (principally the electron/positron fraction of the internal energy and the upper cut-off frequency) it lies within an order of magnitude of the value implied by the data.

This characterization of the FRI/II dividing line depends upon fairly simple physics and well-established empirical relations and therefore provides strong support for the notion that all jets begin at relativistic velocities.

This paper was prepared and presented during the time I was recipient of a JILA Visiting Fellowship and I would like to acknowledge the generous support of that program.

References

[1] G. V. Bicknell. *ApJ*, 305:109–130, 1986.

[2] G. V. Bicknell. *Ap. J.*, 422:542–561, 1994.

[3] G. V. Bicknell. *ApJ (submitted)*, 1994.

[4] R. D. Blandford and M. J. Rees. In A. M. Wolfe, editor, *Pittsburgh Conference on BL Lac Objects*, page 328, Dordrecht, 1978. Kluwer.

[5] D. S. DeYoung. *ApJ*, 405:L13–L16, 1993.

[6] S. S. Komissarov. In R. J. Davis and R. S. Booth, editors, *Sub-Arcsecond Radio Astronomy*, pages 349 – 351, Cambridge, UK, 1994. Cambridge University Press.

[7] S. S. Komissarov. *MNRAS*, 269:394, 1994.

[8] F. N. Owen and M. J. Ledlow. In G. V. Bicknell, P. J. Quinn, and M. A. Dopita, editors, *The First Stromlo Symposium: The Physics of Active Galaxies*, PASP Conference Series, pages 319–324. Astronomical Society of the Pacific, 1994.

[9] P. Padovani and C. M. Urry. *ApJ*, 368:373–379, 1990.

[10] E. S. Phinney. PhD thesis, University of Cambridge, 1983.

[11] T. Venturi, G. Giovannini, and L. Ferretti.

[12] A. G. Willis, R. G. Strom, A. H. Bridle, and E. B. Fomalont. *A&A*, 95:250, 1981.

PAIR CASCADE MODELS OF GAMMA-RAY BLAZARS

Amir Levinson & Roger Blandford[a]
Theoretical Astrophysics, 130-33 Caltech, Pasadena CA 91125.

1. INTRODUCTION

The detection of bright, variable γ-ray emission from about 40 blazars by the EGRET instrument on CGRO [1,2] and the nondetection of radio quite AGN or extended radio sources, has lent additional support to the unified model of AGN. It is now widely believed that the γ-ray emission seen originates from a relativistic jet pointing along our line of sight, and several models involving X-and γ-ray production in jets have been proposed recently [3-9]. Below we consider a specific class of models involving pair-cascade processes, and discuss some of the observational consequences of these models.

2. INHOMOGENEOUS PAIR-CASCADE MODELS

a) X-and γ-Ray Production

In a simple version of the inhomogeneous pair-cascade model developed recently [9-12], the energetic X-and γ-rays observed are produced through inverse Compton scattering of ambient X-ray photons by nonthermal pairs accelerated locally in an outflowing relativistic jet pointed towards us. The jet is approximated as 1D relativistic flow with varying cross sectional area and constant bulk Lorentz factor, Γ. The UV/soft X-ray background also provides an opacity to pair production, which can largely exceed unity at sufficiently small jet radii and, therefore, the observed γ-rays at a given energy must originate from near the associated γ-sphere - the radius at which the opacity to pair production on the soft X-ray background is unity. The injection of pairs, at a given radius, to energies well above the corresponding threshold energy above which pair production becomes important, results in the development of a cascade down to energies at which the γ-rays can escape freely to infinity. In general, we find that the γ-spheric radius increases with increasing γ-ray energy. The slope of the emitted γ-ray spectrum reflects the soft-radiation spectrum and the radial variations of soft-photon intensity and pair injection rate, but is insensitive to the injected pair spectrum. The beamed X-ray spectrum is produced at the base of the γ-ray jet, near the annihilation radius within which the pair density is limited by annihilation.

b) The Background X-Ray Spectrum

Several sources of soft-radiation have been discussed in the literature [3,5,8,9]. The one which is believed to play a dominant role at least in the powerful sources, and considered here in detail is Thomson scattering of nuclear radiation by accreting gas far from the black hole. The isotropic UV/X-ray spectrum is supposed to be similar to those observed in radio quite sources which exhibit prominent "blue bumps", flattening at ~ 0.5 keV [13], and high energy cutoffs at around 100 keV.

[a] This work was supported by NASA grants NAGW 2816, 2372, and NAG5 2504

The observed hard X-ray emission, however, is believed to be dominated by the beamed component produced at the base of the γ-ray emitting jet. The detection of uncorrelated flux variability in the ROSAT hard (1.5-2.4 keV) and soft (0.1-0.3 keV) bands in 3C273 [14] seems to support this view, namely that the soft X-ray spectrum in this source is the extension of the "blue bump", whereas the hard component is produced in the jet. However, in some other blazars, in particular BL Lac objects, there is no evidence for a "blue bump" and it could be that in these sources the overall observed spectrum is dominated by beamed emission.

c) γ-Ray Breaks and the Cutoff of the Background X-Ray Spectrum

There are now six EGRET AGN sources for which the combined COMPTEL and EGRET data indicate spectral breaks (or peaks) at energies between a few to 30 MeV [15,16], with softening of the spectrum above the break. Although more (simultaneous) data at lower X-ray energies is required to determine the location of the break and the change in the spectral slope accurately, the existence of such a break in the COMPTEL band appears to be an important diagnostic of the powerful EGRET AGN sources. In our model, the γ-ray break is associated with the cutoff of the scattered X-ray spectrum which is not observed directly in the compact radio sources, but is assumed to be similar to those observed in radio quite sources, as explained above. Specifically, the break energy, E_{break}, is related to the cutoff energy E_{cutoff} through, $E_{break} = (m_e c^2)^2 / E_{cutoff}$. The portion of the spectrum below the break is produced mostly at the base of the jet, near the annihilation radius, and is determined by the distribution function of cascading pairs there. Above the break the γ-rays are subject to pair production, as discussed in §2a, resulting in a steepening of the spectrum (Fig. 1). It could be that in the weaker BL Lac objects the dominant source of soft photons is the beamed emission from the jet. The observational consequence is then that powerful sources should exhibit γ-ray breaks whereas in Low power BL Lac the spectrum should not steepen in the EGRET band.

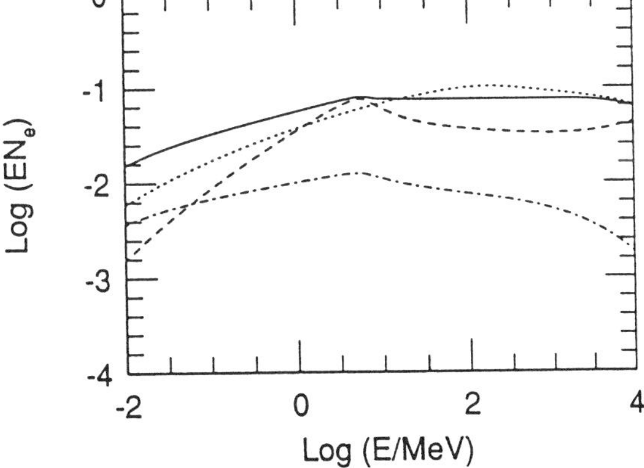

Fig. 1. Emergent photon spectra normalized to the total energy injected, for different pair injection rate and soft-photon intensity.

d) Synchrotron Emission by the Cascading Pairs

In recent work [12] the pair-cascade model has been extended to incorporate magnetic fields and synchrotron emission. Preliminary results show, as expected from simple analysis, that typically, the γ-spheres at EGRET energies are located well within the GHz radio cores. Consequently, variations of the radio flux are expected to be slower or later than variations of the γ-ray flux, consistent with the temporal behavior observed in PKS 0528+134 [17]. The emission at higher frequencies (mm to UV) is produced from radii similar to the X-and γ-ray emission. The spectra of the unresolved cores depend on geometry, gradient of magnetic field and particle acceleration rate in the jet, and soft-photon intensity. Typically, the spectra peak at mm to optical wavelengths. The turnover frequency equals the local peak frequency of the self-absorbed spectrum near the annihilation radius below which particle acceleration is strongly suppressed, but could also reflect the maximum injection energy in the case in which the product of pair injection rate and magnetic field declines sufficiently slowly with radius. Some examples are shown in Fig. 2.

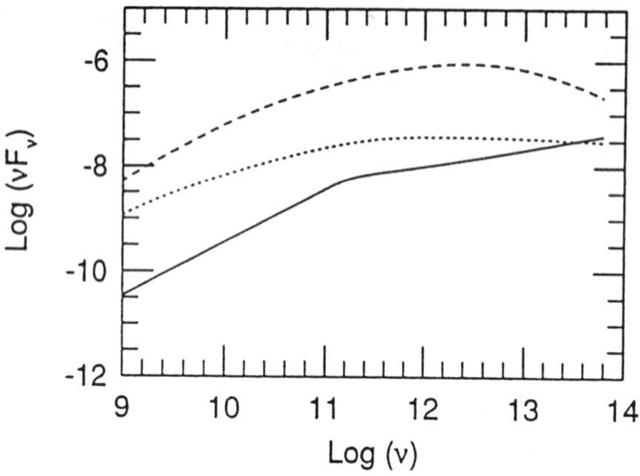

Fig. 2. Emergent synchrotron spectra produced by cascading pairs in a relativistic jet. In all cases shown the magnetic field $B = 10^{-3}(r/1pc)^{-1}$G. The spectrum represented by the solid line was computed using $S \propto r^{-2}$, where S is the pair injection rate. In this case the synchrotron emissivity declines sufficiently rapidly with radius so that most of the flux at a given frequency originates from near the corresponding photosphere. The turnover frequency is associated with the annihilation radius (see text). The other two cases were computed for different soft-photon intensities and $S \propto r^{-1}$. Here, the flux is produced mostly at radii beyond the associated cores, where the source is optically thin, and the spectrum reflects essentially the energy distribution of the pairs there. The peak frequency in these cases is related to the maximum pair injection energy.

3. DYNAMICAL CONSIDERATIONS

In the following we briefly outline a physical model of jet formation and collimation [9,10], and consider yet another implication of the assumed ambient X-ray spectrum. In this model the energy is extracted from a spinning black hole by externally supported electromagnetic field and collimated by a surrounding hydromagnetic wind. Initially, the energy and momentum flux in the jet are carried by electromagnetic fields as a Poynting flux. However, as the jet propagates away from the central black hole, there is a steady conversion of electromagnetic energy into energy carried by electrons and positrons. The transition from electromagnetic to particle dominance was envisage to occur in the vicinity of the annihilation radius at the end of the inner jet. At small radii, within r_{ann}, the bulk flow is supposed to be steadily accelerated, so that Γ steadily increases. When Γ exceeds $\Gamma_c \sim (m_e c^2/E_{cutoff})$, background X-rays can be scattered as γ-rays with energies above the threshold energy to pair production on the ambient X-rays, thereby giving rise to copious pair production which will increase the inertia of the jet, create a greater radiative drag on the outflow, and will eventually limit the jet Lorentz factor at a modest value $\Gamma \sim \Gamma_c$. The details of this limiting process depend upon the prescription for particle acceleration and the spectrum of the external radiation field. It is interesting to note that for typical X-ray spectra observed in radio quite sources Γ_c is of order that inferred from VLBI observations of superluminal sources [18]. If SL motions observed with VLBI reflects the speed of the bulk flow rather than that of a shock propagating down the jet, as appears to be the case in the SL galactic source GRS 1915+105 [19], then correlations between γ-ray breaks and SL speeds are expected.

ACKNOWLEDGEMENTS

We thank M. Sikora, M. Rees, M. Begelman, P. Coppi, A. Laor, J. Mattox, and P. Michelson for useful discussions

REFERENCES

1. Fichtel, C.E., et al. 1994. Astrophys. J. Suppl., **94**, 551.
2. Thompson, D. J., et al. 1994. IAU Symposium 159, Active Galactic Nuclei across the Electromagnetic Spectrum, ed. A. Blecha & T. Courvoisier, Dordrecht: Kluwer, in press.
3. Dermer, C. & Schlickheiser, R. 1993. Astrophys. J. Suppl., **416**, 458.
4. Ghisellini G. & Maraschi L. 1989. Astrophys. J., **340**, 181.
5. Sikora, M., Begelman, M. & Rees, M. 1994. Astrophys. J., **421**, 153.
6. Melia, F., & Königl A. 1989. Astrophys. J., **340**, 162.
7. Mannheim, K. 1992. Astron. Astrophys., **269**, 67.
8. Bloom, S.D. & Marscher, A.P. 1993. Compton Gamma Ray Observatory ed. N. Gehrels & M. Friedlander New York: American Institute of Physics, p.578.
9. Blandford, R. D. 1993. Compton Gamma Ray Observatory ed. N. Gehrels & M. Friedlander New York: American Institute of Physics, p.533.
10. Blandford, R. D., & Levinson, A. 1995. Astrophys. J., in press.
11. Levinson, A., & Blandford, R.D. 1995. Astrophys. J., submitted
12. Levinson, A. 1995. In preparation
13. Laor, A., et al. 1994. Astrophys. J., **435**, 611.
14. Leach, C.M., McHardy, I.M., & Papadakis, I.E. 1995. MNRAS, in press.
15. Collmar, W., et al. 1994. Second Compton Symposium, New York: American Institute of Physics, in press.
16. Lichti, G., et al. 1994. Second Compton Symposium, New York: American Institute of Physics, in press.
17. Zhang, Y.F., et al. 1994. Astrophys. J., **432**, 91.
18. Vermeulen, R. C. & Cohen, M. H. 1994. Astrophys. J., **430**, 467.
19. Mirabel, F.I. 1995. These proceedings.

Soft X-Ray Excesses as a Probe of the Conditions at the Innermost Part of Accretion Flow in AGN[a]

B. CZERNY, P.T. ŻYCKI AND Z. LOSKA

Nicolaus Copernicus Astronomical Center
Bartycka 18
00-716 Warsaw, Poland

UNIVERSAL SHAPE OF BIG BLUE BUMP ?

It is widely accepted that AGN are powered by accretion onto a massive black hole. Observations prove that this flow consists of at least two phases: hot optically thin phase responsible for X-ray emission and cool optically thick phase responsible for the big blue bump. However, it is still far from being clear in which one of the two phases most of the energy is liberated. Comparison of the bolometric luminosities of the two components is a key test to any model. Unfortunately, the big bump emission peaks in the unobserved frequency range between the far-UV and soft X-rays and any estimate of its bolometric luminosity heavily relies on extrapolations which may differ by an order of magnitude.

In the early eighties when the very idea of the big blue bump started to emerge there was no good X-ray data. Therefore the big bump was thought to extend only to far-UV. The idea was supported by a turn-off observed in the UV data. After the first paper containing the analysis of three objects[1] more papers came and a claim was made[2] that the big blue bump has a *universal* shape corresponding to a temperature of 27 000 K. There was no good explanation of this peculiarity.

Fortunately, good X-ray data came and in a few objects the soft X-ray emission was seen well above the hard power law component typical for all objects with intensity rising towards longer wavelengths[3,4]. A new claim was made that this emission is a high frequency tail of the big bump. The previous problem of its universal shape vanished.

Unfortunately, more data were collected and soft X-ray excesses are now seen in about 90% of all quasars and Seyfert galaxies[5,6]. The idea of the universal shape of the big bump came back[6] although this time the corresponding temperature is higher, about 10^6 K. Again, there is no explanation of this peculiarity. So perhaps we should be more careful this time before jumping into such a conclusion and consider alternative explanations for the universally seen soft X-ray emission.

MECHANISMS OF SOFT X-RAY EMISSION

As most of the soft X-ray excesses extend only up to ~0.5 keV their coverage by data points is relatively poor so the shape of this feature is not well constrained.

[a] This work was supported in part by grant No. 2P30401004 financed by the State Committee for Scientific Research.

In particular, it is difficult to tell whether this extra emission rises steeply towards longer wavelengths or shows signs of flattening[7]. However, we can change approach and look for natural physical mechanisms which give extra emission just at the required energy band. A number of such mechanisms were suggested.

(1) High frequency tail of accretion disk spectrum[3,4]. This is an attractive possibility if not required to be present in all sources as the extension of the spectrum depends on the mass of black hole and accretion rate. The predicted shape is very steep, as the spectrum is characterized by exponential cut-off.

(2) Accretion disk spectrum comptonized by thin (extended?) corona[4]. The presence of the corona with low or moderate value of the Compton parameter y changes the spectrum from exponential cut-off to a power law with the slope determined by y. Again, this model will never explain a universal shape as it depends on a number of parameters.

(3) Comptonization of soft photons by cool pair plasma in non-thermal models[8]. This model differs from the previous one since the pair plasma does not necessarily fully cover the entire disk.

(4) Hot spots on accretion disk below X-ray emitting plasma blobs[9]. Such an effect is predicted if hard X-rays are emitted not by a relatively continuous medium but by large regions of magnetic field reconnections.

(5) Warm absorbers. As absorption by partially ionized gas produces a depression around ~0.8 keV it may also be mistaken for an excess emission below 0.5 keV. Warm absorbers unquestionably modify spectra of a number of Seyfert galaxies (but not necessarily all of them) and contribute to the soft X-ray excess but may not fully explain it, like for NGC 4051[10].

(6) Reflection features, i.e. enhanced reflection below the carbon edge supplemented by emission lines and recombination continua[11]. Such a model fitted to the combined ROSAT & EXOSAT data was able to explain the soft X-ray excesses in 5 out of six Seyfert galaxies. Soft X-ray emission of Mkn 335 was far too profound to be explained through atomic features.

(7) Absorption of X-rays by dust[12]. Contribution to the total absorption from the dust is similar to absorption by a neutral gas above 0.4 keV but much lower below that energy (Fig. 1) so the drop in absorption may be mistaken for an excess emission in intrinsically absorbed sources.

(8) Free-free emission. It may come from upper layers of accretion disk[13], optically thin clouds close to the nucleus[14] or extended medium at much larger scale. Plasma at the required temperature (of order of 10^6 K) may exist in thermal equilibrium under strong irradiation[15] and the predicted narrow range of temperatures corresponds to the universal position of the soft X-ray contribution. However, it should be stressed that the free-free emission cannot account for the entire big bump emission from optical to soft X-ray band[15].

The conclusion from this list is that a number of mechanisms is available, both producing a universally positioned spectral feature ((5) to (8)) as well as giving a varying spectral shapes ((1) to (4)). It is not necessarily true that only one of the suggested explanations is correct as different mechanisms may be responsible for the soft X-ray excesses in different objects. Excellent data are really required in order to distinguish between the possibilities.

The information about the variability of the entire spectrum is particularly valuable as the predictions based on various models are significantly different. If a bare accretion disk is responsible for both UV and soft X-rays we expect that the emission in the two bands vary simultaneously. If the disk is embedded in an extended corona then soft X-rays will lag UV radiation. Similar lags are expected in the case of reflection and perhaps free-free reprocessing although the time delays should be different. Any fast variability rules out the influence of dust. There are already interesting constraints available for a few sources[16,17].

THEORETICAL CONSTRAINTS ON THE EXTENSION OF THE BIG BLUE BUMP

Why are the details concerning soft X-ray excesses important? The answer is simple: they decide on the extension of the big blue bump into unobserved EUV range. If the soft X-ray emission of an object is due to atomic features then the bolometric luminosity of the big bump is of the same order or perhaps even smaller than that of X-rays. On the other hand if the big bump tail extends into soft X-rays it may dominate even by an order of magnitude. This information is absolutely crucial if we want to understand the physics of accretion. Analogy with X-ray novae suggests that this ratio depends critically on the current accretion rate switching to the domination by hard X-ray emission below some 0.01 of the Eddington luminosity[18].

We have to determine the ratios of the bolometric luminosities of big blue bump and hard X-rays from observations because the theory does not provide much of support so far.

The earliest studies of disk coronae heated by acoustic flux or electromagnetic waves suggested that only a small fraction of energy, $\sim 10\%$, can be liberated in the corona (in the form of hard X-rays). Models introduced later simply incorporated the fraction of energy liberated in the corona as a free parameter. An attempt to predict the corona strength based on the α-viscosity description of accretion flow both in the disk and in the corona poses more problems than offers solutions[19]. For low accretion rates we predict that a weak corona forms above an optically thick disk and it carries an increasing fraction of energy as the accretion rate increases (Fig. 2). For high accretion rates there are no solutions in hydrostatic equilibrium. However, if the solutions with outflow can exists (the problem is still under investigation) they will most probably predict that the fraction of energy liberated in the corona increases with the accretion rate, contrary to the trends observed in the galactic X-ray novae.

CONCLUSIONS

The multiphase accretion flow onto a massive black hole is so complex that the only chance to understand it is through observations, classifications of objects according to the strength of the big blue bump with respect to the hard X-ray emission and correlations of such classes with the ratio of the luminosity to the Eddington luminosity. This may finally help to reveal the nature of the coupling between the cool-disk and hot-corona phases of the flow.

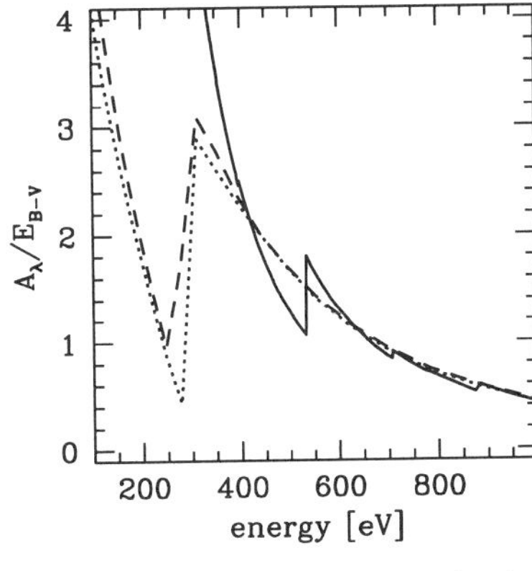

Fig. 1 Extinction curves in X-ray band: neutral gas (continuous line) and carbon curves multiplied by 3 (dashed curves)

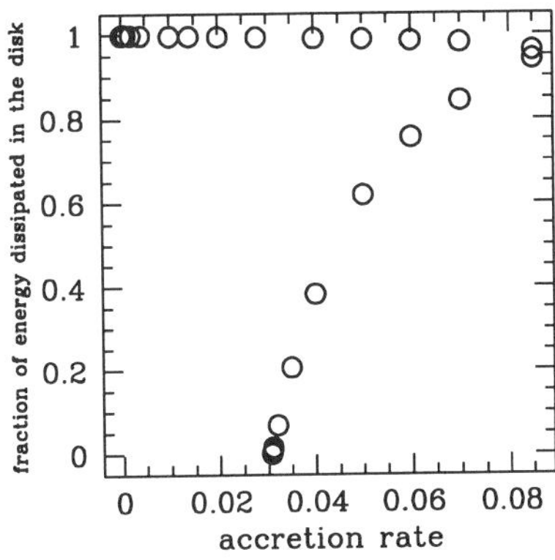

Fig. 2 The fraction of energy liberated in the optically thick disk below the hot thin corona as predicted by α-viscosity model

REFERENCES

1. MALKAN, M.A. 1983, Astrophys. J. **268**:582
2. EDELSON, R.A. & M.A. MALKAN. 1986, Astrophys. J. **308**:59
3. PRAVDO, S.H. & F.E. MARSHALL. 1984, Astrophys. J. **281**:570
4. CZERNY, B. & M. ELVIS. 1987, Astrophys. J. **321**:305
5. WALTER, R. & H.H. FINK. 1993, Astr. Astrophys. **274**:105
6. MASNOU, J.-P., B.J. WILKES, M. ELVIS, J.C. MCDOWELL & K.A. ARNAUD. 1992, Astr. Astrophys. **253**:35
7. YAQOOB et al. 1994, Publ. Astron. Soc. Japan **46**:L49
8. ZDZIARSKI, A.A. & P.S. COPPI. 1991, Astrophys. J. **376**:480
9. HAARDT, F., L. MARASCHI & G. GHISELLINI. 1994, Astrophys. J. Letters **432**:L95
10. MIHARA, T. et al. 1994, Publ. Astron. Soc. Japan **46**:L137
11. CZERNY, B. & P.T. ŻYCKI. 1994, Astrophys. J. Leters **431**, L5
12. CZERNY, B., Z. LOSKA, R. SZCZERBA, J. CUKIERSKA & G. MADEJSKI. 1995, submitted to Astrophys. J.
13. ROSS, R.R. & A.C. FABIAN. 1993, Mon. Not. R. Astr. Soc. **261**:74
14. FERLAND, G.J, K.T. KORISTA & B.M. PETERSON. 1990, Astrophys. J. Letters **363**:L21
15. COLLIN-SOUFFRIN, S., A.-M. DUMONT, B. CZERNY & P.T. ŻYCKI. 1995 (in preparation)
16. GONDALEKHAR, P. & B.J. KELLET. Poster
17. WALTER R., T. COURVOISIERE, C. DONE, L. MARASCHI, K.A. POUNDS & C.M. URRY. Poster
18. WHEELER, J.C., S. MINESHIGE, M. KUSUNOSE, S.-W. KIM & M. MOSCOSO. This volume.
19. ŻYCKI, P.T., S. COLLIN-SOUFFRIN & B. CZERNY. 1995 (submitted to Mon. Not. R. Astr. Soc.)

Activity of Rotating Magnetospheres in AGNs: Collimated Propagation of MHD Waves near a Black Hole

Kouichi HIROTANI
Department of Astronomy, Kyoto University
Sakyo-ku, Kyoto 606-01, JAPAN
E-mail: hirotani@kusastro.kyoto-u.ac.jp
and
Akira TOMIMATSU
Department of Physics, Nagoya University
Chikusa-ku, Nagoya 464-01, JAPAN

Abstract

We have studied non-stationary and axisymmetric perturbations of a magnetohydrodynamic accretion onto a rotating (Kerr) black hole. Assuming that the magnetic field dominates the plasma accretion, we find that the meridional current becomes large near the fast-magnetosonic surface located close to the horizon. This is due to the effects of the particle's inertia and the strong gravity near the horizon. As a consequence, the fluid suffers a large radial acceleration resulting from the Lorentz force, and becomes highly variable compared with the electromagnetic field there. In fact, we further find an interesting perturbed structure of the plasma velocity with a large peak in some narrow region just beyond the fast-magnetosonic surface. If the critical surface has an oblate shape for radial poloidal field lines, this perturbation of plasma motion is induced by outgoing waves propagating from the equatorial super-fast-magnetosonic region to the polar sub-fast-magnetosonic one. Thus, the effective acceleration of particles due to the outgoing fast-magnetosonic waves will work mainly in the polar direction as a mechanism of trigger of jet production in active galactic nuclei.

1 Introduction

It is generally accepted that the jets are powered by the associated active galactic nuclei (AGNs), where accreting mass is converted into energy with several percent efficiency. We must explain how the energy goes into a small fraction of the gas in the gravitational field of the prime mover, probably a supermassive black hole. For example, the magnetohydrodynamic (MHD) disturbances produced at a galactic nucleus with a compact nuclear disk would be strongly collimated polewards resulting in jets, if the Alfvén velocity in the disk is much higher than its surroundings (Sofue 1980).

The purpose here is to explore a little further the problem whether MHD waves can convey energy from the equatorial region to the poles in a close vicinity of the central black hole of an AGN. Adopting the magnetically dominated limit in which the rest-mass energy density of particles is negligible compared with the magnetic energy density, we will show that plasma accretion is intrinsically highly variable and that the energy of this large amplitude fluid disturbance will be collimated by propagating polewards.

2 Highly Variable Accretion

We will begin by considering basic equations describing a magnetosphere around a non-rotating black hole. Since the self-gravity of the electromagnetic field and plasma around the black hole is very weak, the background geometry of the magnetosphere surrounding a rotating black hole is described by the Kerr metric. We assume the ideal MHD conditions in which the electric field vanishes in the fluid rest frame. Furthermore, the motion of the fluid is described in the cold limit. The proper number density obeys the continuity equation.

In order to look closely at the problem of how the existence of the horizon affects the MHD interactions, we would like to focus attention on an analysis near to the horizon ($\Delta \equiv r^2 - 2Mr + a^2 \ll 1$), where r is the radial coordinate in Boyer-Lindquist coordinate, M the hole's mass, and a the hole's angular momentum per unit mass. Since the poloidal magnetic field near to the horizon tends to be bent into a radial shape, due to fluid inertia, even in the magnetically dominated limit (Hirotani et al. 1992), it may be safely assumed that the unperturbed field is radial. Because particles are frozen to the radial magnetic field lines, flow lines also become radial. Under this condition, the unperturbed electric field becomes meridional. Furthermore, we should notice that the fast-magnetosonic point ($r = r_{\rm F}$) is located very close to the horizon ($r = r_{\rm H} \equiv M + \sqrt{M^2 - a^2}$) in the magnetically dominated limit.

Let us next consider a non-stationary and axisymmetric perturbation superposed on the stationary and axisymmetric magnetically dominated magnetosphere. We consider the Fourier components of purturbed quantities and replace temporal derivatives into frequency (i.e., $\partial_t \to i\omega$) and adopt a local approximation in the meridional direction (i.e., $\partial_\theta \to -ik_\theta$). Solving linear perturbation equations, we obtain relations describing relative amplitude among perturbed quantities. To see the relative amplitude, it is convenient to take the short-wavelength limit not only in the meridional direction but also in the radial direction. Then we can see that the equipartition of energy between the electromagnetic field and the fluid is achieved near the fast-surface, although the perturbation energy is supplied mainly in the form of electromagnetic disturbances far from the horizon. In other words, a lot of perturbation energy is supplied to the fluid from the electromagnetic field, as a result of the large amplitude radial Lorentz force resulting from large meridional current (J_θ). This effect is due to a redshift effect near the horizon, and disappears if the fast-magnetosonic point is located far from it (Hirotani et al. 1993).

3 Structure of Fluid Disturbance

We see in the last section that the magnetically dominated accretion is intrinsically highly variable near the fast-surface located very close to the horizon. Let us next emamine the spatial structure of fluid disturbance. Combining linear perturbation equations, we obtain a single wave equation which fluid's radial velocity, energy, and angular momentum obey. Assuming that the perturbed quantities are proportional to $\exp(i\omega - ik_\theta)$, we can regard the wave equation as a 2nd order differential equation with respect to r. The solution can be written as

$$u^r = C_1 \frac{x^{i\sigma}}{[x - x_F(\theta) + (dx_F/d\theta)^2 + i\epsilon]^{1+2i\sigma}}, \tag{1}$$

where x is a non-dimensional radial coordinate defined by $x \equiv \Delta/\Sigma_{\rm H} = 2(r_{\rm H} - M)(r - r_{\rm H})/({r_{\rm H}}^2 + a^2\cos^2\theta)$; the imaginary part $i\epsilon$ is much smaller than the real part $(dx_{\rm F}/d\theta)^2$. A sharp peak of the amplitude appears at the surface $x = x_{\rm F} - (dx_{\rm F}/d\theta)^2$ just beyond the fast-surface. The infalling motion of the plasma in the sub-fast region is mainly disturbed slightly outside the fast-surface, because the outgoing waves disperse and the amplitude decays like $\mid u^r \mid \sim x^{-1}$ for $x \gg x_{\rm F}$. The steady supply of outgoing waves is due to the meridional propagation from the super-fast equatorial region to the sub-fast polar one, provided that the fast-surface is oblate $(dx_F/d\theta > 0)$. A large part of the outgoing waves can escape from the narrow super-fast region in the range $x_{\rm F} - (dx_{\rm F}/d\theta)^2 < x < x_{\rm F}$, before reaching the polar region. Hence, the peak amplitude of u^r at $x = x_{\rm F} - (dx_{\rm F}/d\theta)^2$ exponentially decreases as θ decreases. In the next section, we confirm the importance of meridional propagation in detail.

4 Collimated Propagation of MHD Waves

To consider the meridional propagation, it is useful to calculate the characteistics of the wave equation. The characteristics of the wave equation in the super-fast region are presented by the non-linear differential equation

$$dx/d\theta = \mp\sqrt{x_{\rm F}(\theta) - x}. \tag{2}$$

The minus sign corresponds to the wave propagating to lower lattitudes, while the plus sign to a polewardly propagating wave. Here, we must notice that $dx < 0$ holds near the horizon. Both waves are depicted in Figure 1. We are considering a situation in which accretion is mainly supplied from the equatorial disk. In this case, accretion rate η, and hence Alfvénic Mach number at the horizon $\mathcal{M}_{\rm H}(\theta) = (4\pi m_{\rm p}\eta^2/n)^{1/2}$ becomes large in low lattitudes. Because $x_{\rm F}$ is essentially proportional to ${\mathcal{M}_{\rm H}}^2$ (Hirotani et al. 1992), the magnetosphere has an oblate critical surface as described in Figure 1.

If the fast-surface is oblate, only the waves propagating polewards can escape into a sub-fast region. This could be clearly seen if we solved the equation (2) (Hirotani et al. 1994), but can also be understood intuitively from the figure. If the fast-surface were prolate, on the contrary, only the waves propagating into the equatorial region could escape.

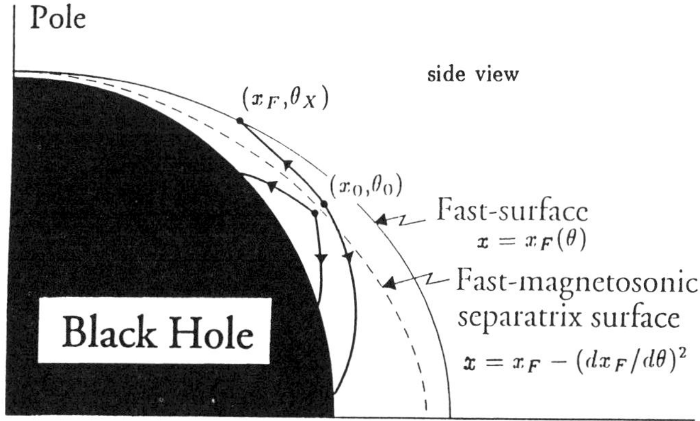

Fig. 1. Schematic figure of a black hole magnetosphere near to the horizon. A pair of thick solid curves denotes a pair of characteristics of equation (2). The maximum-amplitude surface $[x = x_F - (dx_F/d\theta)^2]$, which is indicated by the dashed line, coincides with the fast-magnetosonic separatrix surface.

5 Summary and Discussion

Let us summarize the main points that have been made in this paper. We have studied non-stationary and axisymmetric perturbations of magnetically dominated accretion onto a non-rotating black hole. A slight fluctuation of the electromagnetic field is accompanied by fluid motion which is highly disturbed in the super-fast region just beyond the fast-surface. Some portion of the energy of this fluid disturbance can escape to a sub-fast region by virtue of the meridional propagation and concentrate toward the polar sub-fast region in the form of outgoing fast-magnetosonic waves.

It would be possible to argue that the poleward collimation of waves may bring about the acceleration of particles in jets. To understand how the wave energy is converted into particle's kinetic energy is an important problem for developing the MHD scenario of jet formation in the very central region of AGNs. For example, in addition to heating of the surrounding plasma due to the dissipation of wave energy into thermal energy, the wave stress exerted on the expanding atmosphere may result in a direct transfer of momentum and energy from the waves to the outgoing winds (Jacques 1977). This will be the subject of future investigations.

References

Bogovalov S. V. 1992, in Proceed. of 4th International Toki Conf. on Plasma Physics and Controlled Nuclear Fusion, ed Guyenne T. D., Hunt, J. J., (Europian Space Agency, Netherland) p. 317.

Hirotani K., Takahashi M., Nitta S., and Tomimatsu A., 1992, ApJ 386, 455.

Hirotani K., Tomimatsu A., and M. Takahashi 1993, PASJ 45, 431.

Hirotani K., and Tomimatsu A. 1994, PASJ 46, 643.

Jacques S. A. 1977, ApJ 215, 942

Sofue Y. 1980, PASJ 32, 79.

Accretion Disks in Active Galaxies: The Sub-Keplerian Paradigm[a]

SANDIP K. CHAKRABARTI[b]
NASA/GSFC, Greenbelt, MD 20771

In the 1970s, the standard accretion disk models were constructed[1,2] to largely explain observations from the binary systems. In these systems, angular momentum supplied at the outer edge is necessarily Keplerian. Viscosity drives the inflow by removing angular momentum outwards and keeping the entire disk Keplerian in the process. Since then these binary disk models are being used also to explain big blue bump observed in the continuum of active galaxies and quasars and with some success[3]. Occasionally, one invokes additional components along with the standard thin disks, such as corona, warm absorbers, etc. in order to explain continuum as well as variable components of X-rays and γ-rays. There are several recent observations of almost zero time-lag correlated variabilities between X-rays and optical[4] which cannot be explained by simple Keplerian disk models. Temporal variation of line profiles from objects, such as, ARP102B and 3C390.3 is impossible to explain by using axisymmetric disk models[5]. Recent HST observation of M87 shows clear evidence of non-axisymmetry in the ionized disk around the black hole and the association of the spiral shock in NGC4258 with the accretion disk cannot be ruled out. These examples suggest the existence of large scale spiral shock waves in both M87 and NGC 4258[6]. It is also generally believed that the radiation from our own galactic center is most likely coming from a low efficiency, quasi-spherical, accretion flow.

That the disks need not be of 'standard' type was sensed by theoreticians even in late '70s and throughout the '80s. Thick accretion disk, transonic accretion disk and slim accretion disk models (For a general discussion and references, see Chakrabarti[7]) came about. In thick disks, angular momentum is assumed to be almost constant but the radial motion is ignored. In transonic disks, the flow is thin but the radial motion is included. In slim disks, matter is allowed to pass through the inner sonic point, thus improving on the standard disk model. However, in all these models, unsuccessful attempts were made to match the flow with *Keplerian disks* at some distance and the disk structure depended strongly upon the matching radius and other parameters invoked.

This intrinsic problem with these theoretical models, as well as problems in explaining a large number observations with Keplerian disks, particularly when applied to active galaxies, disappear with the realization that the accretion disks in active galaxies need not be Keplerian anywhere including the outer boundary! The outer boundary condition here is completely different from that of the binary systems, since matter is largely supplied by winds from

[a]Supported in part by NRC Senior Research Associateship of National Academy of Science

[b]On leave from Tata Institute of Fundamental Research, Bombay, 400005, INDIA

mass lossing and colliding stars in random motion. Matter could loss most of its angular momentum before it brings itself together to form an accretion disk.

After the matter starts with highly sub-Keplerian angular momentum, its subsequent behaviour depends strongly upon the accretion rate and the viscosity in the flow. If the accretion rate is small enough, the flow passes through the outer sonic point (just as a Bondi flow) and remains supersonic before falling onto a black hole. If the accretion rate is high, matter would pass through the inner sonic point[8–9]. This description is valid if the entropy remains almost constant, i.e., the viscosity is small enough. Even when the flow starts with small entropy, some entropy could be generated at a shock or in the flow by viscosity which then allows the flow to pass through the inner sonic point as well[10–12]. If the entropy is higher (for a given accretion rate) it is likely that strong winds may be formed from flows with positive energy[9,11]. Except when the viscosity is high, the flow radiates with a very low efficiency as in a Bondi flow as discussed in these works.

Contrary to a Newtonian star, a black hole has no hard surface. If the flow is unable to loss angular momentum efficiently, the centrifugal barrier causes the flow to have a shock close to the black hole[7–12]. The postshock flow is the boundary layer equivalent of a black hole accretion. This is clearly the case when the viscosity is small enough. The postshock flow has all the features of a thick accretion disk only more self-consistent since the radial motion is also included[11]. The preshock flow is mainly advected towards the black hole, just as a Bondi flow, but the immediate post-shock flow is rotationally dominated as in a Keplerian disk. Further on, the flow picks up radial motion and supersonically enters into the black hole. If the viscosity is high (typically, if the α parameter[1] is larger than about 10^{-2}) the *stable* shock disappears and the disk eventually becomes Keplerian[10] except close to the boundaries.

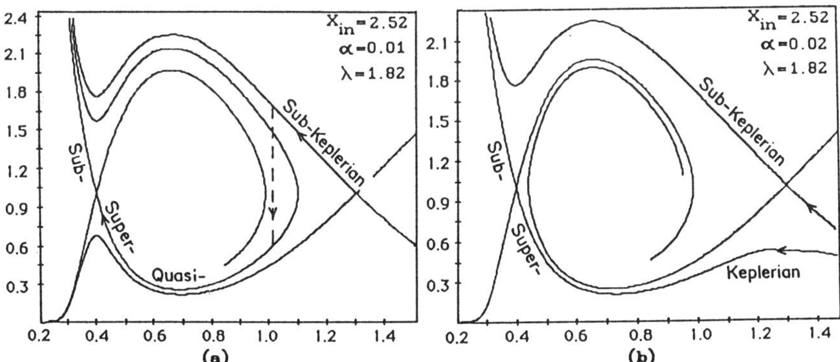

Fig. 1(a-b): Mach Number (Y-axis) as a function of logarithmic radial distance (X-axis) for viscous isothermal flows. In (a), a weak shock is present ($\alpha = 0.01$). In (b) with $\alpha = 0.02$, the shock disappears. Out of two choices (arrowed curves), the one passing through the inner sonic point is preferred due to higher dissipation.

Fig. 1(a-b), adapted from Chakrabarti (1990) which solves the fully viscous, isothermal flows shows this behaviour. This prediction is verified by numerical simulation of viscous flows[12]. Fig. 2(a-b) shows the time variation (curves drawn at intervals of $1000GM/c^3$) of the Mach number and the angular momentum distribution in a viscous ($\alpha = 0.1$) flow which is sub-Keplerian at the outer boundary. Here the angular momentum is transported more efficiently in the post-shock flow and increasingly higher centrifugal barrier pushes the shock further out making it weaker in the process (2a) and making the disk shock free and quasi-Keplerian (2b). These highly viscous, shock-free, quasi-Keplerian solutions self-consistently pass through the inner sonic point. If the viscosity is very small, a weak shock can survive[10,12] and the angular momentum distribution remains sub-Keplerian everywhere except close to the marginally stable orbit where it is super-Keplerian. The dotted curve in Fig. 2a is the shock in the inviscid flow.

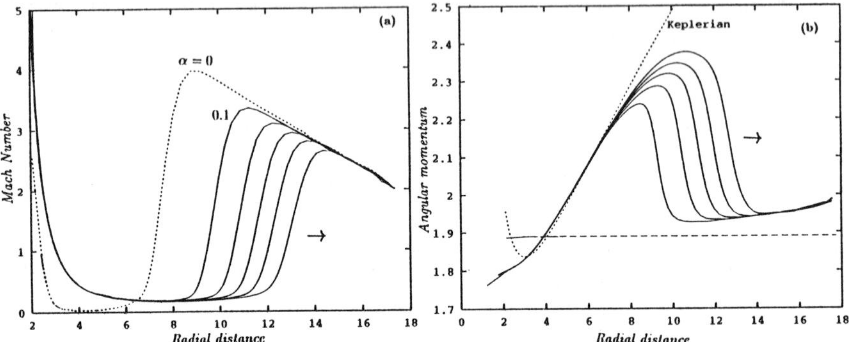

Fig. 2(a-b) Time evolution of Mach number (a) and angular momentum (b) variation as a function of logarithmic radial distance in a viscous flow.

Fig. 3: Composite disk model around a black hole with sub-Keplerian outer boundary condition and height dependent viscosity.

We thus notice that one may have a thick, thin, slim or transonic (with or without shocks) disk from an initially sub-Keplerian inflow at large distance depending upon accretion rate and viscosity. Based on this experience, in order to explain observations across the electromagnetic spectrum, a generic model for the accretion disk could be built (Fig. 3). A disk of this kind would form if there is a significant variation of viscosity in the vertical direction inside the disk. Higher viscosity on the equatorial plane produces an optically thick standard Keplerian disk, which is vertically flanked by warm, optically thin halo of low angular momentum gas. The halo forms a standing shock close to the black hole ($\sim 10\ R_g$). The post-shock flow ($T_p \sim 10^{11}\ K$) heats up soft-photons coming from the disk to produce observed γ-rays. Such a model has the potential to explain most of the steady state as well as time dependent behaviour of the continuum and line emissions and is under active study[13].

In active galaxies, jets are also seen. It is generally assumed that in the absence of a binary companion (which is a sink of angular momentum in a binary system), jets has to carry away angular momentum of the disk. Various jet models are constructed in order to achieve this goal. It is to be remembered that the 'angular momentum' problem is present only in models which start with Keplerian flows at the outer boundary. In the sub-Keplerian paradigm such problems are not present. Even in the presence of high viscosity, the flow may have just enough (not excess!) *total* angular momentum to redistribute itself to form a quasi-Keplerian disk. In the case of low angular momentum flow, shear is strong only close to the black hole. Strong shear produces very strong toroidal field. In the gas pressure dominated hot disks (with $T_p > 4 \times 10^{10}\ K$) magnetic tension catastrophically brings the flux tubes close to the black hole axis[14], ejecting matter from the inner accretion disk in the form of jets. Thus the jets would be blobby, rather than continuous. This general feature may have been observed in detail in GRS 1915 | 105[15].

REFERENCES

1. Shakura, N.I. & R.A. Sunyaev. 1973. Astron. Astrophys. **24:** 337.
2. Novikov, I. & K.S. Thorne. 1973. in: Black Holes, eds. C. DeWitt and B. DeWitt (Gordon and Breach, New York).
3. Sun, W.H. & M.A. Malkan. 1989. Astrophys. J. **346:** 68.
4. Clavel, J. et al. 1990. Mon. Not. R. Astron. Soc. **246:** 668.
5. Chakrabarti, S.K. & P.J. Wiita. 1994. Astrophys. J. **434:** 518.
6. Chakrabarti, S.K. 1995. Astrophys. J. (March 10th, in press)
7. Chakrabarti, S.K. 1995. Physics Reports (to appear).
8. Chakrabarti, S.K. 1990. Theory of Transonic Astrophysical Flows (World Scientific, Singapore, 1990)
9. Chakrabarti, S.K. 1989. Astrophys. J. **347:** 365.
10. Chakrabarti, S.K. 1990. Mon. Not. R. Astron. Soc. **243:** 610.
11. Molteni, D., G. Gerardi & S.K. Chakrabarti. 1994. Astrophys. J. **436:** 249.
12. Chakrabarti, S.K. & D. Molteni. 1995. Mon. Not. R. Astron. Soc. **272:** 80.
13. Chakrabarti, S.K., L. Titarchuk, & D. Kazanas. 1995. (in preparation)
14. Chakrabarti, S.K. & S. D'Silva. 1994. Astrophys. J. **424:** 138.
15. Mirabel, I.F. & Rodriguez, L.F. 1994. Nature **371:** 46.

On the Hot Spot near a Kerr Black Hole

ALEXANDER ZAKHAROV

Institute of Theoretical and Experimental Physics,
B.Cheremushkinskaya, 25, 117259, Moscow, Russia,

INTRODUCTION

The connection of black holes with activity of quasars and active galactic nuclei was considered by several authors, e.g. Blandford & Znajek[1]; Begelman, Blandford & Rees[2]; Bičak, Semerák & Hadrava[3]. Recently Abramowicz et al[4,5] proposed a model for the variability of X -ray sources as a result of hot spot motion near a black hole. Abramowicz et al[4]; Karas & Bao[6]; Bao[7]; Rauch & Blandford[8] analyzed this model for the Schwarzschild black hole. Cunningham & Bardeen[9]; Zakharov & Polnarev[10]; Karas et al[11] obtained the light curves of hot spots for a Kerr black hole. In the last paper the authors used interpolation by Chebyshev polynomials but we used Monte Carlo simulations in our papers since a strong dependence on problem parameters was found[12] and therefore an interpolation is not accurate enough in this case. We also used the direct solution of the differential equations for the photon motion since the result of numerical calculations of the elliptical integrals depends simetimes strongly on problem parameters (e.g. small changes of constants of motion may cause qualitative changes of results[12,13,14]).

DESCRIPTION OF NUMERICAL METHOD

We will calculate, similarly to Karas et al[11], the observational effects in the frame of a distant observer using Boyer - Lindquist coordinates (r, θ, ϕ, t)[15].

We realized the simulations of the hot spot emission using a Monte Carlo approach. We will describe briefly our method. In random time moments the hot spot emits the photons which have an isotropic distribution in the hot spot frame. Below we will obtain the expression of the photon's 4-pulse in the local Lorentz frame of the hot spot. We suppose that we know the photon's energy and angles θ, ϕ which define the direction of photon emission.

We integrate the system of ordinary differential equations for the hot spot motion and we obtain the characteristic time of the hot spot - τ_p (for example, τ_p is the time of a change of the ϕ coordinate by the 2π). We suppose that the photons are emitted by hot spot in random time moments on segment $[0, N_2\tau_p]$. We consider $N_3 = N_1 * N_2$ random numbers which have a uniform distribution on segment $[0, N_2\tau_p]$ (approximately N_1 numbers for a segment with the lengh τ_p). It is clear that a random variable τ has a uniform distribution on segment $[0, N_2\tau_p]$ if $\tau = \gamma_3 N_2 \tau_p$, where γ_3 is a random variable with uniform distribution on the segment [0,1]. Thus we get the array of random numbers $\tau_i \in [0, N_2\tau_p], i = 1, ...N_3$ after we made a sorting. As the result we have the array $\tau'_1, ..., \tau'_{N_3}$ and $(\tau'_i \leq \tau'_{i+1}, \forall i)$.

Thus in the time moment τ'_i using the independent random numbers γ_1, γ_2 (which have uniform distribution on segment [0,1]) we obtain the angles θ_0 and ϕ_0, find the photon's 4-pulse in local Lorentz coordinates, make the boost, and after that we obtain motion constants.

We partition the sphere angle 4π into the equal elements and for any element we store the time moment when the following inequality for the r-coordinate is valid $r \geq r_\infty$. We store also the value p_t that is characteristic for the Doppler effect and the gravitational red shift affecting the photon's energy in the local Lorentz frame of the hot spot. Thus we may find the histogram that characterizes the light curve for a fixed element of the sphere angle, i.e. we have a bolometric light curve when we add the values p_t for photons in the element of the sphere angle if we have $t_i < t < t_i + \Delta t = t_{i+1}$ for $r > r_\infty$. We partition the sphere angle into N_4 equal elements by N_5 meridians and by $N_6 - 1$ paralles and therefore $N_4 = N_5 * N_6$.

NUMERICAL RESULTS AND CONCLUSIONS

We consider our model with the following values of parameters[16]. The hot spot moves in a circular equatorial orbit in the gravitational field of the Kerr black hole with a rotation parameter value $a = 0.998$ and radius $r = 3 \quad (\sim \frac{M}{M_\odot} * 4.44 * 10^5 cm)$. The hot spot period is

$T = 20.2(\sim \frac{M}{M_{\odot}} * 9.95 * 10^{-5}s)$. We consider approximately $5 * 10^3$ photons for one period of the hot spot, in total the emission of 17321 photons, including 7321 photons that were captured by the black hole. We partition the sphere angle into 40 elements. We found that 838 photons lie in the sphere angle element that is closest to the equatorial plane. If we have a uniform distribution of photons over sphere angle then there will be 500 photons in the element. The light curve of the hot spot for an observer, which is near to the equatorial plane, is given[16]. We conclude that there is the strong variability of emission (similarly, Karas et al[11]). of the black hole. There are 306 photons in the sphere angle element near the pole[16]. Change of intensity is the result of statistical fluctuations (on average the intensity must be constant in this case). There is the light curve for an observer position near the angle $\theta = \pi/4$. There are 395 photons in the sphere angle element[16]. There is also strong variability shown in the case (similar results were obtained by Karas et al[11]).

Thus we described a new Monte Carlo model for simulations of short-time variability. A similar model for the variability of X -ray AGN was discussed[4,5,7]. If we consider the X-ray variability of the Seyfert galaxy NGC 6814 with a period of 10^4 s as the result of a hot spot motion[5] then we may apply our model with a black hole mass $10^8 M_{\odot}$ in the object.

The structure of the optical caustics in a Kerr metric and the origin of rapid X - ray variability in AGN were cosidered by Rauch & Blandford[8]. They obtained the natural result that the light curve shows strong variability near the caustics. We will get similar results with an increasing amount of random numbers and a decrease of lengths of segments on the t - axis for the histograms.

The Monte Carlo approach is universal and useful for this problem since it is possible to consider different numbers of hot spots, different types of hot spot motion etc, whereas for example in the approach of Karas et al[11] it is impossible to consider a photon's motion which does not cross the equatorial plane. But it will be necessary to consider such photon trajectories if the hot spot does not move in the plane. If we compare their algorithm with our's we see that their relationship between photon's trajectories and their grid points is not one to one mapping. Our algorithm uses the method of direct solution of the system of differential equations for finding the photon trajectories, a qualitative analysis of the photon geodesics[13] and a careful consideration of the treatment of different types of trajectories in numerical calculations[12].

Using similar numerical model we considered the stellar explosion near

Kerr black hole[18].

ACKNOWLEDGMENTS

I am grateful to Organizing Committee of the 17 th Texas Symposium on Relativistic Astrophysics, especcially to prof. J. Trümper for the excellent Symposium.

REFERENCES

1. BLANDFORD R.D. & ZNAJEK R.L. 1977. Mon.Not.R.Astron.Soc. **179**: 433.
2. BEGELMAN M.C., BLANDFORD R.D. & REES M.J. 1984. Rev. Mod. Phys. **56**: 255.
3. BIČAK J, SEMERÁK O. & HADRAVA P. 1993. Mon. Not. R. Astron. Soc. **263**: 545.
4. ABRAMOWICZ M.A., BAO G., LANZA A. & ZHANG X.-H. 1991. Astron. & Astrophys. **245**: 454.
5. ABRAMOWICZ M.A., LANZA A., SPIEGEL E.A.& SZUSKIEWICZ E. 1992. Nature. **356**: 41.
6. KARAS V.& BAO G. 1992. Astron. & Astrophys. **257**: 531.
7. BAO G. 1992. Astron. & Astrophys. **257**: 594.
8. RAUCH K.P.& BLANDFORD R.D. 1993. Preprint Caltech. GRP-334.
9. CUNNINGHAM C.T.& BARDEEN J.M. 1973. Astrophys. J. **183**. 237.
10. ZAKHAROV A.F.& POLNAREV A.G. 1988. Preprint ITEP 181.
11. KARAS V., VOKROUHLICKY D.& POLNAREV A.G. 1992. Mon. Not. R. Astron. Soc. **259**: 569.
12. ZAKHAROV A.F. 1991. Astron. Zhurn. **35**. 147.
13. ZAKHAROV A.F. 1986. Sov. Phys. JETP. **64**. 1.
14. ZAKHAROV A.F. 1989. Sov. Phys. JETP. **68**. 217.
15. MISNER C.W., THORNE K.S.& WHEELER J.A. 1973. Gravitation. W.H.Freeman. San Francisco.
16. ZAKHAROV A. 1994. Mon. Not. R. Astron. Soc. **269**: 283.
17. MÄHÖNEN P.& ZAKHAROV A. 1994. Preprint University of Oulu. UO-TP-117/94.

Non-stationary Accretion With Ordered Magnetic Fields and Outflows in AGNs and Quasars*

M.M.ROMANOVA[a] AND R.V.E.LOVELACE[b]

[a] *Space Research Institute, Profsoyuznaya 84/32, Moscow 117810, Russia*

[b] *Department of Applied Physics, Cornell University, Ithaca, NY 14853*

INTRODUCTION

The processes of matter accretion onto a compact object and matter outflow to jets are closely related. Outflows from the disk carry matter, angular momentum, and energy and this influences the physical state of the accretion disk. From the other side, the physical state of the outflowing matter and the mechanism of outflow depend on the conditions in the accretion disk- its temperature, magnetic field, etc. Recent theoretical work on jet formation has focused on magnetohydrodynamic (MHD) models of jet formation from accretion disks. This mechanism can be pertinent to matter outflow from cold disks owing to the driving action of the magnetic pressure connected with the twisting of magnetic field lines[1,2]. However, other mechanisms of outflow connected for example with the high temperature of the disk or high radiation pressure are possible. The jets can carry an essential part of the angular momentum from the disk thus increasing the rate of accretion dramatically[3,4,5]. Here, we first present results of 1.5D MHD simulations of disk accretion with magnetically driven outflows. Later we describe 2.5D MHD simulations of outflow from a hot accretion disk where both thermal and magnetic pressures are important. Applications to AGNs and Quasars as well as to X-ray burst Novae are discussed.

NON-STATIONARY ACCRETION WITH OUTFLOWS

A model and simulation code have been developed for time-dependent axisymmetric disk accretion onto a compact object including the influence of an ordered magnetic field and outflows. The magnetic field of the wind was treated in a phenomenological way suggested by self-consistent wind solutions[1,2]. The magnetic diffusivity of the disk was assumed to be of the order of the

*This work was supported in part by RFFI grant 93-02-17106, Astronomy Program of RMSTP topic 3-169, by National Science Foundation AST-93-20068 and Soros Foundation grants. DFG,DARA and MPE grant supported MMR's participation in the Symposium.

turbulent viscosity[6]. The non-stationary MHD equations for surface density, radial velocity, azimuthal velocity, temperature, and $B_z(r, z = 0, t)$ were integrated numerically[5], including the radiation pressure. In the test case without magnetic field, we found that any non-equilibrium disk settles into the equilibrium solution of Shakura and Sunyaev[7]. However, when the ordered magnetic field is not zero, the behaviour of the accretion is markedly different. We took the superposition of a background magnetic field and an enhancement or "bump" in the magnetic field distribution at some radius $r = r_0$ from the compact object. We discovered that at relatively low amplitudes of the magnetic field the bump evolves moving inward and diffusively spreading. For a larger amplitude of the magnetic field in the bump and/or a larger background field, the bump moves inward more rapidly, it narrows in radial width, and forms a soliton-like disturbance. The disturbance propagates inward while growing in strength up to the time when it is "swallowed" by the compact object, that is, when it passes within inner radius of the disk r_i (Figure 1a). The density and temperature in the wave also grow with time, so that the bump evolution gives a burst in thermal radiation of the disk. At the same time the enhancement of the magnetic field in the region of the disturbance leads to enhancement of magnetically driven outflow. Thus, the disturbance produces almost simultaneous bursts in the power output in the winds and in the disk radiation.

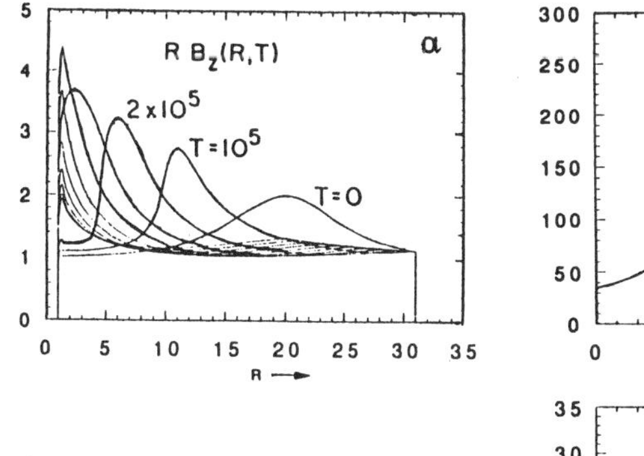

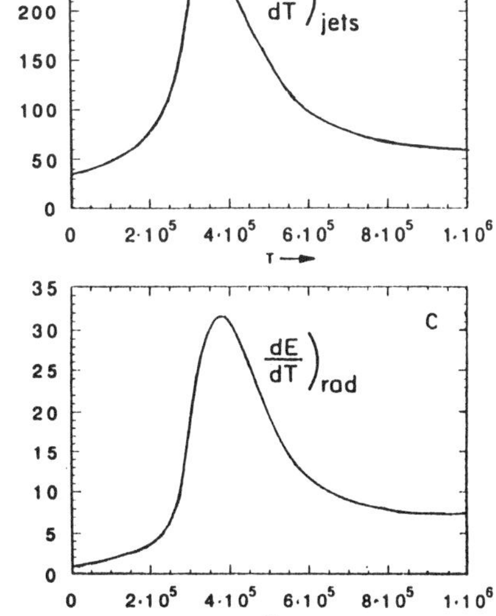

Figure 1. The figure shows the behavior of the magnetic field (a), and energy outbursts from the disk (b), and from the jet (c).

The inward radial accretion speed $u(r, t)$ of the disk matter is shown to be the sum of the usual viscous contribution and a magnetic contribution $\propto r^{\frac{3}{2}} B_p^2/\sigma$, where $B_p(r,t)$ is the poloidal field threading the disk and $\sigma(r,t)$ is the disk's surface mass density. For comparison, for typical parameters in the disk[5] the viscous ($B=0$) accretion time from $r=20r_i$ to $r=r_i$ is ~6000 yr, whereas the "magnetic" accretion time is ~1-10 yrs.

This model is pertinent to any system involving accretion and magnetically driven outflows. High resolution VLBI radio observations of the some quasars has shown that the jets (0.1 - 10pc) typically consist of bright moving components and a core. The extrapolated zero-point time of formation of the moving components often coincide with the brightness amplification of the continuum radiation in the optical, and X-ray wavebands[8,9]. The data appears to confirm the earlier hypothesis of Kinman[10] that both observed features, jet formation and flux-density outburst occurs simultaneously and reflect the same physical process. Recently discovered jets from X-ray novae[11] also show this kind of correlation. The prediction of this theory is that transient jets of small scale may also be discovered in other low luminsoity X-ray novae.

NUMERICAL SIMULATION OF OUTFLOWS

Here, we present 2-D results of axisymmetric ideal MHD simulations of outflows from a **hot** magnetized accretion disk. Simulation of magnetically driven outflows from a **cold** accretion disk was performed earlier[3], but for cases where the disk was far from Keplerian and initially close to free-fall. Here, the disk is considered as a lower boundary condition at $z=0$ with a hot homoentropic equilibrium corona above it. Both the disk and corona are considered to be initially threaded by an ordered magnetic field represented as superposition of monopoles located on the z axis below the disk plane. At some moment say $t=0$ we start to push matter from the disk with speed less than slow magnetosonic velocity and at the same time start to rotate the disk with Keplerian velocity. Disk matter starts to flow and accelerate along the z axis and forms a collimated jet (Figure 2a). Poloidal magnetic field lines which diverged initially become pinched towards the z axis by the toroidal component of magnetic field[12]. The flow accelerates and passes through the slow magnetosonic surface located at an axial distance $z_{sm} \sim r_i$ (the point dashed line in Figure 2a), then through the Alfvén surface (the dashed line), then through the fast magnetosonic speed in the upper part of the considered region (the long dashed line). At this time the v_z velocity on the top of the region is larger than about twice the escape speed for a single particle from this location. The velocity distribution has a maximum on a cylinder of radius $\sim (3-5)r_i$ parallel to the z-axis. This region coincides with the maximum of the toroidal magnetic field distribution, thus confirming that magnetic pressure connected with twisting of magnetic field lines is important. Analysis of forces and fluxes has shown that here both thermal and magnetic pressures are important in driving matter to the jet, and the magnetic force is very important in collimation of the flow. Figure 2b shows the twisting of the magnetic field line as a result of disk rotation. The magnetic energy density at the inner radius of the disk r_i is much less than gravitational energy (0.03), whereas the thermal energy-density is comparable with gravitational (0.9). Thus, even though the energy-density of the magnetic field at the disk is relatively small, it strongly influences to the flow, especially the collimation.

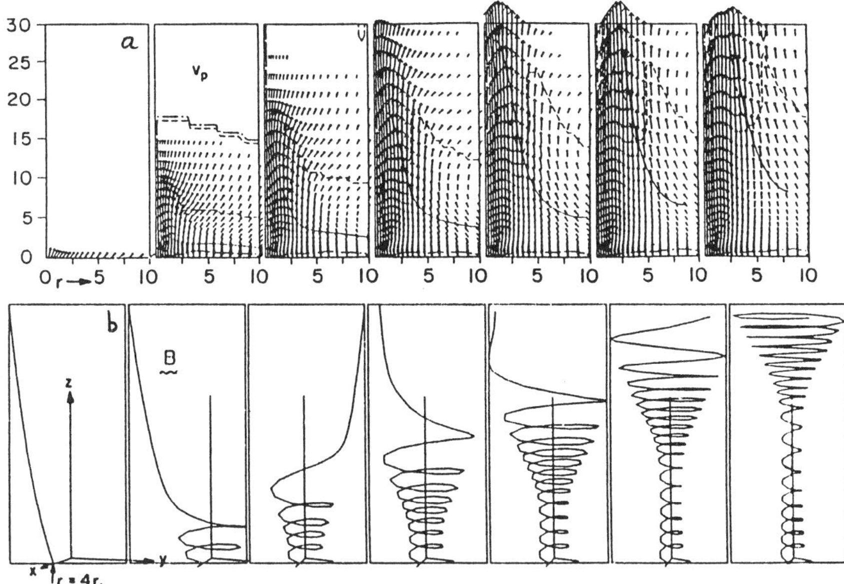

Figure 2. a- The poloidal flow velocity $\mathbf{v}_p(r, z)$ at a sequence of times, $t = 0, 4, 8, .., 24$, where t is measured in units of the rotation period of the inner edge of the disk, t_i. The lengths of the arrows are proportional to $|\mathbf{v}_p|$. Boundary of the matter coming from the disk (solid line). b- the 3D projection of magnetic field line started at $r = 4r_i$.

This model is pertinent to jet formation from AGNs and quasars and also to any objects where the accretion disk is hot, for example, SS433, X-ray novae, etc. If the central object is a black hole, then matter reaches a mildly relativistic velocity $\sim$ 0.3 c which is of the order of that observed in SS433.

REFERENCES

1. BLANDFORD, R.D., & D.G. PAYNE. 1982. MNRAS, **199**, 883.
2. LOVELACE, R.V.E., H.L. BERK & J. CONTOPOULOS. 1991. ApJ, **379**, 696.
3. SHIBATA, K., & Y. UCHIDA. 1986. Pub. Astron. Soc. Japan, **38**, 631.
4. STONE, J.M., & M.L. NORMAN. 1994. ApJ, **433**, 746.
5. LOVELACE, R.V.E., M.M. ROMANOVA & W.I. NEWMAN. 1994. ApJ, **437**, 136.
6. BISNOVATYI-KOGAN, G.S. & A.A. RUZMAIKIN. 1976. Ap&SS, **42**, 401.
7. SHAKURA, N.I., & R.A. SUNYAEV. 1973. Astron. & Astrophys., **24**, 337.
8. BELOKON, E.T. 1987. Astrophysics, **27**, 588.
9. KRICHBAUM, T.P., et al. 1990. Astron. & Astrophys., **237**, 3.
10. KINMAN, T.D. 1977. Nature, **267**, 798.
11. MIRABEL, I.F. & L.F. RODRIGUEZ. 1994. Nature, **371**, 46.
12. USTYUGOVA, G.V., A.V. KOLDOBA, M.M. ROMANOVA, V.M. CHECHETKIN, & R.V.E. LOVELACE. 1995. ApJ (Letters), in press.

Cosmological Origin of Quasars

ABRAHAM LOEB

Astronomy Department, Harvard University
60 Garden St., Cambridge, MA 02138

INTRODUCTION

Observations of high-redshift quasars and absorption systems provide a rich data set on the early formation of structure in the universe. Theoretical and observational investigation of the physics leading to the formation of quasars and their environments can shed some light on early structure formation. In this contribution, I summarize briefly a few aspects of quasar physics that are of cosmological interest.

ORIGIN OF QUASAR BLACK HOLES

The most recent evidence that massive black holes exist in the centers of galaxies comes from the existence of compact gaseous disks with high rotation velocities. Examples include the HST imaging and spectroscopic observations[1] of M87 that revealed a ~ 20 pc disk with a rotation velocity of ~ 500 km s^{-1} and implied the presence of a $2 \times 10^9 M_\odot$ black hole, and VLBA observations[2] of the powerful water maser emission from a ~ 0.1 pc disk rotating at $\sim 10^3$ km s^{-1} at the center of NGC 4258. In the latter case, the central mass density exceeds $4 \times 10^9 M_\odot$ pc^{-3} and is unlikely to be associated with anything other than a central black hole with a mass of $4 \times 10^7 M_\odot$. Independent observational constraints on the compactness of the energy source in active galactic nuclei come from unresolved lensed images (indicating[3] a source size $\lesssim 1$ pc from HST imaging of the quasar 2237+0305), gravitational microlensing (indicating[4] a continuum source of size $\lesssim 2 \times 10^{15}$ cm in 2237+0305), milliarcsecond jets (showing an unresolved core $\lesssim 10^{17}$ cm in some objects[5]), and rapid X-ray variability[6].

The integrated light of quasars can be used to find the minimum mass density of black hole remnants today. This calculation[7] implies that a fraction $\gtrsim 3 \times 10^{-5}$ of all the baryons in the universe have ended inside black holes. Formation of black holes must therefore be a non-negligible consequence of gravitational instability in the early universe, yet the origin of these $\sim 10^{6-10} M_\odot$ black holes in standard cosmologies is enigmatic[8]. In fact, most of the luminous mass in the universe is locked up in gas and stars that are prevented from condensing in the centers of galaxies by their angular momentum. A galaxy as a whole is a highly non-relativistic system which has a typical size that is $(v/c)^{-2} \sim 10^6$ times bigger than its Schwarzschild radius. Loeb & Rasio[9] have performed hydrodynamic simulations of the collapse of protogalactic gas clouds and concluded that baryons are unlikely to reach a relativistic state by merely sinking spontaneously to the center of a galaxy, unless a massive

($\gtrsim 10^6 M_\odot$) seed black hole is already in existence there. Without a massive seed, the rotationally-supported cold gas is strongly unstable to fragmentation due to its self-gravity, and is converted to stars long before it approaches relativistic scales. However, if a central massive seed is artificially added to the system, it could dominate gravity near the center and stabilize a smooth accretion disk of gas around it. The minimum seed mass necessary for that purpose, $\sim 10^6 M_\odot$, is consistent with the existence of a lower bound on the empirically determined black hole masses in active galactic nuclei. Various such determinations[10] all yield black hole masses $\gtrsim 10^6 M_\odot$, with the lowest mass objects not necessarily being close to the detection threshold. Nevertheless, it is puzzling as to why a small fraction of the mass in the universe ended up in relativistic seeds while the rest was strongly prevented from doing so by its angular momentum. To answer this question, we must first consider the origin of angular momentum of collapsing systems in cosmology.

An initially overdense region in the universe that eventually forms a virialized object acquires angular momentum about its center of mass through tidal torques from its environment[11]. It is therefore possible to imagine that different environments could result in different amounts of rotation for the final virialized object. As the initial conditions can be well-specified in terms of a Gaussian random field of density perturbations with some power-spectrum, one can calculate the distribution function of angular momenta for collapsed objects in the universe either numerically[12] or analytically[11,13]. This distribution has a tail of low-spin objects which by chance happened to reside in an environment with a low tidal shear. When the analytical calculation is extended into the far low-spin tail of this distribution one finds an astrophysically interesting abundance of low-spin objects[13]. The cosmological collapse of low-spin systems is found to be close to spherical because of their spherical initial shape, and the low shear in their cosmological environment. In addition, because of the unusually low amplitude of the external shear, the gas in these systems is unlikely to be disrupted by external torques before it forms a compact disk. In more than $\sim 10^{-4}$ of the objects on the $10^{6-7} M_\odot$ mass scale, the baryons can settle after the initial collapse and cooling phases to a compact disk of an initial size $\sim 10^{17}$ cm and a rotation velocity $\gtrsim 500$ km s^{-1}. Because of its small initial size, such a disk has a viscous evolution time $\lesssim 10^6$ yr, shorter than the characteristic time it takes a star or a supernova to form in it. The compact disk is therefore expected to evolve into a seed black hole[13]. Most of these seed black holes form just above the cosmological Jeans mass and have a mass $\sim 10^6 M_\odot$.

Each galactic bulge contains about $\sim 10^4$ subunits on the $10^6 M_\odot$ mass scale. Among these subunits there is a class of rare objects that are high density peaks which acquire low angular momentum during their cosmological collapse. These high peaks collapse early ($z \gtrsim 20$), long before any other object in their nearby environment starts to form. Because of its low angular momentum, the gas in these rare peaks forms a deep potential well as it cools to a compact disk. The initial disk can then evolve to a massive black hole on a short viscous timescale ($\lesssim 10^6$ years), well before star formation or supernovae could act to disrupt it. After the bulge of the surrounding galaxy virializes, the already formed seed sinks to the center of the potential-well by dynamical friction. This process provides just the initial $10^6 M_\odot$ seed necessary to stabilize later accretion of gas around the center[9]. The later accretion allows the further growth of the black hole there. Qualitatively, the above sequence of events must take place at some level in the universe. The only open question is quantitative: *for a given power spectrum of initial density perturbations, what fraction of the* 10^4 *subunits belongs to this low-spin class?*

An analytical calculation of the distribution function of angular momenta[13] shows that there is of order one low-spin subunit per bright galaxy, which is a $>2.5\sigma$ peak with 10^{6-7} solar masses in gas and is capable of forming a black hole seed shortly after its initial cosmological collapse. In principle, it is also possible to get black hole binaries in galactic centers by the formation and sinking of more than one seed per galaxy. If more than two seeds sink to the center, slingshot ejection of black holes from the bulge becomes important.

It can be shown mathematically[11] that a high density peak on the $10^{6-7} M_\odot$ mass scale is very likely to be surrounded by a high density region on the $\sim 10^{10} M_\odot$ mass scale. Therefore, the seed black holes that form out of high peaks are very likely to be surrounded by galactic mass systems that collapse later and feed them with additional gas, thus resulting in the bright quasar activity[13]. The observed maximum in the comoving quasar density at $z \approx 2$ may just reflect the epoch of galaxy formation when considerable infall feeds these seed black holes[14]. The subsequent decline in the abundance of bright quasars at low redshifts would then result from the dilution of their gas supply. The lack of starlight around some nearby quasars[15] may indicate that star formation does not necessarily precede the accretion process.

There are various observational ways to probe the above sequence of events. Searches for the progenitors of quasars at very high redshifts ($z \gtrsim 10$) may be best undertaken in the infrared or millimeter regimes; optical surveys are limited by intrinsic dust extinction or intergalactic absorption. The emission of fine-structure lines from the host systems of quasars can be detected by millimeter telescopes and provide information about the velocity dispersion and gas content of the hosts[16]. For example, the [C II] 158 μm line flux from a bulge surrounding a bright quasar at a redshift of 10 can reach ~ 2 mJy, and be detected at the 3σ level with a 1" beam and a velocity resolution of 150 km s^{-1} after 40 minutes of integration by the future Millimeter Array telescope.

A novel method to set a lower limit on individual quasar lifetimes makes use of the Lyα forest[17]. It is well-known that the ionizing radiation from a bright quasar can dilute the population of Lyα clouds in its vicinity out to a characteristic distance of $\sim 10^{7-8}$ light years[18]. When lines of sight separated by $\sim 1°$ from the line of sight to the quasar are used to probe this "proximity effect", they are sensitive to radiation that left the quasar $\sim 10^{7-8}$ years earlier than the radiation arriving from the quasar today. Therefore, two lines of sight can be used to set a lower limit on the quasar lifetime. By coincidence, this limit happens to be just in the regime of interest for the expected duty cycle of quasars[14].

PROBING CLUSTERING AT HIGH REDSHIFTS THROUGH QUASAR ABSORPTION LINES

If quasars form in high density regions then they are likely to be surrounded by concentrations of galaxies. Groups and clusters of galaxies hosting a quasar can be found through the detection of Lyα absorption lines beyond the quasar redshift. The effect occurs whenever the peculiar velocities of the quasar and the Lyα clouds combine to lower the quasar redshift below that of its nearest Lyα cloud. For this to be observable, the distortion to the redshift distribution of Lyα clouds induced by the cluster potential must extend beyond the proximity effect of the quasar. For any specific cosmological model, it is possible to predict the probability for finding lines beyond the quasar redshift ($z_{\rm abs} > z_Q$) under the assumption that the physical properties of Lyα clouds are not affected by flows on large scales ($\gtrsim$ Mpc) in the quasi-linear regime. If quasars randomly sample the underlying galaxy distribution, the expected number of lines with $z_{\rm abs} > z_Q$ per quasar can be as high as $\sim 0.25 \times [(dN/dz)/350]$ at $z = 2$ for Cold Dark Matter cosmologies, where dN/dz is the number of Lyα lines per unit redshift far from the quasar[19]. The probability is enhanced if quasars typically reside in small groups of galaxies. In addition, a statistical excess of Lyα lines is expected near very dim quasars or around metal absorption systems. The expected magnitude of these clustering effects should be detectable by forthcoming observations with the Keck telescope. Finally, it can be shown[19] that the standard approach to the proximity effect overestimates the ionizing background flux at high redshifts by up to a factor of ~ 3, as it ignores clustering. This result weakens the existing discrepancy between the deduced background flux and the contribution from the known population of quasars.

ACKNOWLEDGEMENTS

I thank Daniel Eisenstein, Fred Rasio, and Ed Turner for many fruitful discussions.

REFERENCES

1. Ford, H. C., et al. 1994, ApJL, 435, 27; Harms, R. J., et al. 1994, ApJL, 435, L35

2. Miyoshi, M. et al. 1995, Nature, Jan. 12th issue

3. Rix, H. W., Schneider, D. P., & Bahcall, J. N. 1992, AJ, 104, 959

4. Rauch, K. P., & Blandford, R. 1991, ApJ, 381, L39

5. Baath, L. B., et al. 1992, A& A, 257, 31

6. Remillard, R. A., et al., 1991, Nature, 350, 589

7. Sołtan, A. 1982, MNRAS, 200, 115; Chokshi, A., & Turner, E. L. 1992, MNRAS, 259, 421

8. Turner, E. L., 1991, AJ, 101, 5

9. Loeb, A., & Rasio, F.A. 1994, ApJ, 432, 52

10. Peterson, B. M. 1993, PSAP, 105, 247; Wandel, A., & Mushotzki, R. F. 1986, ApJ, 306, 61; Netzer, H. 1990, in Active Galactic Nuclei (Berlin: Springer); Wandel, A., & Yahil, A. 1985, ApJL, 295, 61; Padovani, P., Burg, R., & Edelson, R. A. 1990, ApJ, 353, 438

11. Eisenstein, D., & Loeb, A. 1995, "An Analytical Model For The Triaxial Collapse of Cosmological Perturbations", ApJ, in press

12. Warren, M. S., Quinn, P. J., Salmon, J. K., & Zurek, W. H. 1992, ApJ, 399, 405

13. Eisenstein, D., & Loeb, A. 1995, "Origin of Quasar Progenitors From The Collapse of Low-Spin Cosmological Perturbations", ApJ, in press

14. Haehnelt, M. G., & Rees, M. J. 1993, MNRAS, 263, 168

15. Bahcall, J. N., Kirhakos, S., & Schneider, D. P. 1994, ApJL, 435, 11

16. Loeb, A. 1993, ApJL, 404, 37

17. Loeb, A., & Maoz, E. 1995, in preparation

18. Bechtold, J. 1994, ApJS, 91, 1

19. Loeb, A., & Eisenstein, D.J. 1995, ApJ, in press

Testing Cosmogonic Models with Gravitational Lensing

JOACHIM WAMBSGANSS[a,b], RENYUE CEN[b]
JEREMIAH P. OSTRIKER[b], and EDWIN L. TURNER[b]

[a] *Astrophysikalisches Institut Potsdam
An der Sternwarte 16, 14482 Potsdam, Germany*

[b] *Princeton University Observatory
Peyton Hall, Princeton, NJ 08544*

INTRODUCTION

Conventional tools for comparing cosmogonic theories with observations rely on either galaxy density or galaxy velocity information, both of which unavoidably suffer from the uncertainties with regard to density or velocity bias of galaxies over the underlying mass distribution, hampering our attempts to understand the more "fundamental" questions concerning the mass evolution and distribution. In contrast, gravitational lensing *directly* measures fluctuations in the gravitational potential along lines of sight to distant objects. Thus, gravitational lensing provides a powerful independent test of cosmogonic models. Narayan and White[1] were the first to study this using the Press-Schechter formalism, Cen *et al.*[2] (subsequently paper I) used surface mass density distributions from realistic n-body simulations.

Each model for the development of cosmogonic structure (e.g. the HDM [hot dark matter] or CDM [cold dark matter] scenario) has at least one free parameter, the amplitude of the density (or potential) power spectrum. But now in the light of COBE observations[3], that parameter is fixed by the ($\pm 15\%$) determination on the $5^o - 10^o$ scale in the linear regime. With its amplitude fixed, a secure determination of the potential fluctuation on any scale provides a test; any single conflict between the theory and reality can falsify the former. The most leverage is obtained for tests made on scales as far as possible from the COBE measurements. The reason is that all models have an assumed power spectrum that passes through the COBE normalization point at the very large comoving scales ($\lambda \approx 1000$Mpc) fixed by that measurement. Since the slope of the power spectrum is a primary model dependent feature, the maximum variations amongst models occur typically at the smallest scales. Thus one looks for tests at scales as small as possible, but they should not be so small as to be greatly influenced by the difficulty in modelling the physics of the gaseous, baryonic components ($\leq$ 10kpc). Thus critical tests are best made on scales 0.01Mpc $< r <$ 1Mpc. Here we will present first results from using gravitational lensing from matter distributions on these scales to test the standard CDM scenario.

METHOD

The model used is the "standard" CDM scenario with $\Omega = 1$, $\lambda = 0$ and $H_0 = 100h = 50$km/s/Mpc. The normalization was done according to the COBE first year results[4], it corresponds to $\sigma_8 = 1.05$. In order to have detailed small scale information (which is crucial for effects of "strong lensing") we ran a total of 10 independent simulations with $L = 5h^{-1}$Mpc, having $500^3 = 10^{8.1}$ cells and $250^3 = 10^{7.2}$ particles. Because this box size does not account for power on large scales, we also ran an $L = 400h^{-1}$Mpc size box with $500^3 = 10^{8.1}$ cells and $250^3 = 10^{7.2}$ particles.

Knowing the distribution of overdensities on the $5h^{-1}$Mpc scale from the large simulation box, we statistically convolve the small and large scale runs to produce simulated sheets or screens of matter spaced $5h^{-1}$Mpc apart between the observer at $z = 0$ and a putative galaxy or quasar in the source plane at $z = z_S$. A large number of independent runs (ten runs were done) is required so that identical structures do not repeat along a line of sight. The details of the convolution method and tests of it using a high resolution P^3M simulation provided by Bertschinger and Gelb[5] will be presented in a subsequent paper. The method is statistically reliable for describing structures in the range 30kpc $< \Delta Lh <$ 1.2Mpc, which corresponds roughly to splitting angles $5'' < \theta < 200''$. On these scales we expect that dark matter dominates over baryons so that a dark matter only simulation is approximately valid.

A very preliminary attack on this problem has been presented in Paper I. In that work no ray tracing was done. Rather it was simply checked whether or not surface mass densities were greater than the critical surface mass density[6] at which multiple imaging will occur. In addition to the much better method of treating the lensing aspect used here, the convolution algorithm has been modified and improved over the one adopted in Paper I.

We follow light rays in a cone of fixed angular size (about 6 arcmin) through realistic matter distributions in many lens planes, determine their deflection angles in each plane and collect them at source planes of different redshifts. We use the multi-plane lens equation for the light rays[7]. The deflection angles are calculated for 500^2 rays in each plane, according to all the matter inside this plane. Subsequently we study the lensing properties of the matter in these lens planes.

RESULTS AND DISCUSSION

We performed 100 independent lensing runs of a square field with about 6 arcmin at a side. In Figure 1 we present the probability of multiple imaging as a function of quasar redshift. We considered only multiple images with separations greater than $5''$ and magnitude difference less than 1.5 mag. In fact "amplification bias"[6] will increase the probabilities over those shown in Figure 1 by a significant amount. From this figure it can be seen that splittings

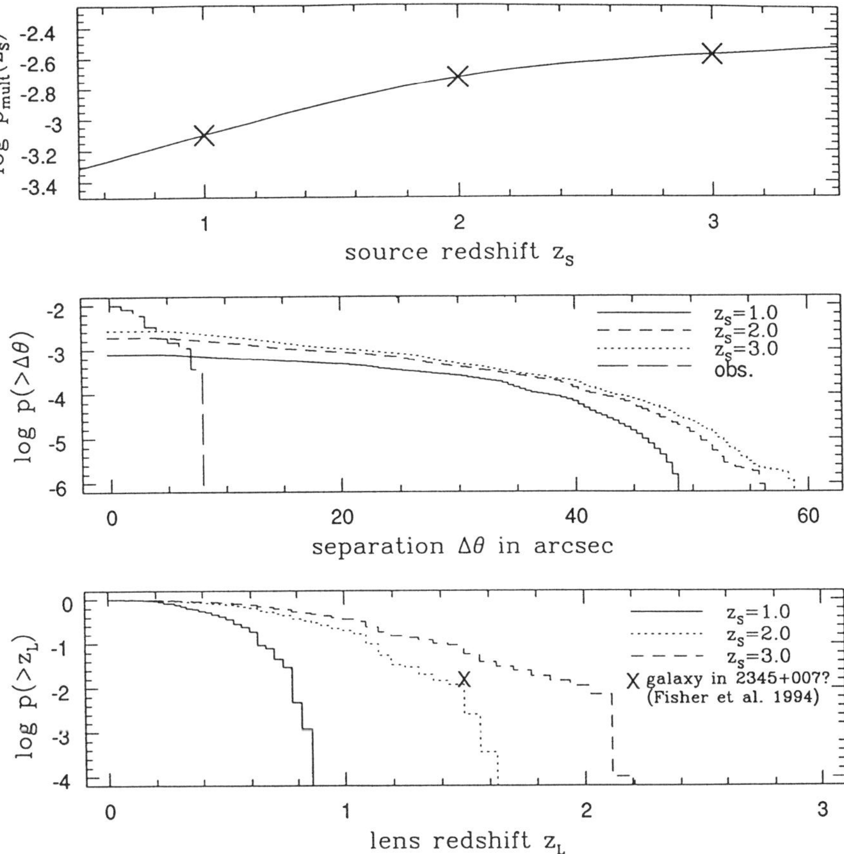

Fig. 1 (top) Probability of multiple lensing (with image splitting $\Delta\theta \geq 5''$ and magnitude difference $\Delta m \leq 1.5$mag as a function of source redshift z_S.

Fig. 2 (middle) Multiple-lensing probability distribution as a function of image separations for sources at $z_s = 1$, 2 and 3. Also shown as long dashed curve is the observed distribution[9].

Fig. 3 (bottom) Integrated lensing probability distribution as a function of expected lens redshift. The symbol X indicates the recent observation[8] of a lens candidate for the double quasar QSO 2345+007.

larger than 5" should be common (when several thousand quasars have been examined), if this cosmogonic model were correct.

A very revealing statistic is the distribution of image separations expected for multiple images as shown in Figure 2 for quasar redshifts of $z_S = 1$, 2 and 3. Large and very large splittings should be the rule. Also revealing is the distribution of expected lens redshifts as shown in Figure 3. Most lenses in the standard CDM scenario should be relatively nearby, close enough to be seen in almost all known gravitational lens cases. On this issue the recent observation[8] of a lens candidate for the double quasar QSO2345+007 is quite relevant. The separation of the two images is 7".06, the quasar redshift is $z_S = 2.15$ and the putative lens is at $z_L = 1.49$. We see from Figure 3 that, although 7 arcsecs separation can be produced in the CDM model, the probability that the lens is as far away as $z = 1.49$ is very small (2%) due to the relatively late formation of structure in this model. In open models structure formation occurs earlier.

It appears that all three of these results (the total frequency of large separation multiple quasars, the distribution with splitting angle, and the distribution of lens redshifts) do not agree with the existing observations, the first two are in serious conflict. Surveys and occasional serendipitous discoveries have revealed 27 confirmed or possible multiply-imaged QSO's according to a recent compilation[9]. Detailed analysis of the most statistically useful of these surveys yields a lensing rate in the vicinity of a few tenth to one percent[10,11], consistent with the CDM predictions quoted above making allowance for plausible magnification biases. However, as shown in Figure 3, all observed QSO lens systems have image splittings of less than 10", and the large majority, less than 5".

This sharply contradicts and thus falsifies the model. Since the large splitting, modest brightness ratio systems predicted by the model would be typically much easier to detect and recognize than those 27 which have actually been found, no escape by appeal to observational selection seems possible.

This failing of the model is not presented as an entirely new result, but only as a new and more robust manifestation of a previously recognized problem, namely the excessively deep potential wells produced by the dark matter component in COBE normalized standard CDM. These excessively deep potential wells lead both to excess galaxy pair-wise velocity dispersions and to the predicted excessive rate of large splitting lensing events. The virtue of the lensing test is that it is independent of other tests and is not subject to the same caveats concerning "bias" of galaxies with respect to dark matter.

Are there variant models that would not fail these tests? A lower Ω is clearly useful, but it would be premature to argue that the present results by themselves indicate $\Omega < 1$, as many other properties of the scenario ("temperature" of the dark matter, shape of the power spectrum etc) contribute to lensing properties. However the *directness* of gravitational lensing as a test for the growth of inhomogeneities, coupled with the rapidly increasing power of computers and numerical algorithms, makes one optimistic that calculations of the type reported on here should become a major tool for testing and discriminating among competing cosmological scenarios.

ACKNOWLEDGEMENTS

We thank J.R. Gott, C.S. Kochanek and P. Schneider for useful discussions, NCSA for use of the Convex-3880, and NASA grants NAGW-2448 and NAGW-2173, NSF grants AST91-08103 and HPCC ASC-9318185 for financial support.

REFERENCES

1. Narayan, R. & White, S.D.M. 1988, MNRAS, 231, 97p

2. Cen, R., Gott, J.R., Ostriker, J.P., & Turner, E.L. 1994, ApJ, 423, 1

3. Smoot, G.F., *et al.* 1992, ApJ(Letters), 396, L1

4. Efstathiou, G., Bond, J.R, & White, S.D.M 1992, MNRAS, 258, 1p

5. Bertschinger, E., & Gelb, J. 1994, ApJ, in press

6. Turner, E.L., Ostriker, J.P., & Gott, J.R. III 1984, ApJ, 284, 1

7. Schneider, P., Ehlers, J., & Falco, E.E., *Gravitational Lensing* (Springer Verlag, Berlin, 1992)

8. Fischer, P., Tyson, J.A., Bernstein, G., & Guhathakurta, P. 1994, ApJL, 431,L71

9. Surdej, J., & Soucail, G. 1993, in "Gravitational Lenses in the Universe", Proceedings of the 31st Liege International Astrophysical Colloquium, p205

10. Kochanek, C.S. 1993, ApJ, 417, 438; 419, 12

11. Maoz, D., & Rix, H.-W. 1993, ApJ, 416, 425

GIANT LUMINOUS ARCS

François Hammer[a], Isabella Gioia[b], Olivier Le Fèvre[a], Gerry Luppino[b]

[a]DAEC, Observatoire de Meudon, France

[b]Institute of Astronomy, University of Hawaii, USA

1 Introduction.

Since 80s, clusters of galaxies were not believed to be enough massive and compact to reach their lensing critical density, i.e. to provide strong lensing events by their own. In 1981, a filament structure in A370 was reported[1], and has been identified 5 years latter to the so-called giant luminous arc[2], simultaneously with a similar structure in Cl2244-02[3]. The first lensing models [4] were quickly made and reproduced well the arcs, providing a strong support that arcs are images of background galaxies gravitationally distorted by cluster core masses. The definitive proof came from the redshift measurement of the A370 arc, which yielded a value almost twice the cluster redshift [5]. Arc studies are now very numerous, most of them intending to estimate the amount and the distribution of the dark matter component in clusters, and more than 40 giant luminous arcs have been already discovered. In this short paper, we summarize the main results concerning the giant luminous arcs (defined as having $R<22.5$ and axis ratio $l/w > 10$) including the new constraints derived on the dark matter density profile within the cluster cores, and the new insights well below 1 kpc scale of the morphology of faint distant galaxies.

2 What can be derived from arc observations.

Mass estimation of the lensing cluster core can be fairly accurate when it is based on the observation of a large arc ($l/w > 10$) with a known redshift, although some estimations can be affected by the presence of cluster substructures. They lead to M/L values ranging from 100 to 300 with uncertainties as low as 10% for the best configurations [6].

For images having a large axis ratio ($l/w >> 1$) one[7] can establish the relationship between the observed arc width (w) and the source diameter (d) as $w = 1/2\ d\ (1 - K(arc))^{-1}$, where K(arc) is the lensing matter term at the arc location, which is proportional to the local mass surface density. For a very compact density profile (point mass), K(arc)=0, providing $w = 1/2$ d, for a singular isothermal sphere, K(arc)=0.5 and w=d while for an isothermal profile with a core (with radius $\approx$ arc impact parameter), K(arc) $\approx$ 0.75, w=2d. Since most of the giant luminous arcs are generally unresolved from the ground (w < 0"5), either the arc sources are very compact, or the matter term K(arc) should be not very larger than 0.5. From the observations of faint galaxy sizes, it seems rather conservative to estimate that the core radius of the most of the arc clusters cannot be significantly higher than the observed arc radius, which range from 50 to 150 kpc.

The former relationship provides a fundamental uncertainty to our knowledge of the magnification factor (ratio of the arc luminosity to the source luminosity), which can be expressed as

$$Amp = 1/4(l/w)(1 - K(arc))^{-2}, \quad (1)$$

and also strongly depends on the lens density profile. Distant rich clusters of galaxies were often claimed to be gravitational "telescopes" allowing redshift measurement of galaxies otherwise too faint, but we still have a poor knowledge of the accurate powers of such "telescopes". It prevents us to draw an Hubble diagram for the arc sources, in order, for example, to go beyond the magnitude limit of current spectroscopic surveys.

3 Statistics of luminous arcs.

The previous relationships show that modelling of arcs in clusters can be generally done assuming a large variety of cluster density profiles. An interesting possibility to constrain the cluster mass density profile from arc modelling has been done in Cl 2137-23 [8], assuming that several images of the same source are lying at different impact parameters (including the so-called radial arc). Another alternative would be to measure redshifts of several arcs at different impact parameters in the same cluster.

Since the magnification probabilities depend so strongly on the lens density profile, solid constraints on the latter can be derived through arc statistic studies. Wu and Hammer [9] have developped a statistical model, assuming that lensing clusters follow the X-ray luminosity function and that arc sources follow a Schechter luminosity function scaled on the density of faint galaxies from deep spectroscopic surveys. In order to compare the results to the observations, the generated arcs have been limited by their magnitude and by their axis ratio. The predicted distribution for arc cluster and source redshifts is in good agreement with the available data as well as for the X-ray luminosity distribution of arc clusters. Recall that most, if not all, of the arc clusters are X-ray luminous. Several mass density profiles have been tested through that model, resulting to the expectation of one giant luminous arc per three ($L_X >$ 410^{44} erg/s) clusters, if latter have density profile following singular isothermal sphere, and ten times less giant luminous arcs, if lensing clusters have $\approx$ 200 kpc core radii. Indeed, in the last case, the predicted number of giant luminous arcs over the whole sky would drop to 12, much less than the number of already known arcs!

In order to have an observational data basis directly comparable to the theoretical expectations, a systematic survey of a complete sample of clusters have been done [10]. Target clusters (39 clusters with $L_X >2\ 10^{44}$ erg/s and z >0.15) have been selected from the Einstein Medium Survey Catalog, and have been imaged in two colors (V and I or B and R) at a moderate depth at C.F.H.T. and U.H.22' (20mn exposure time in each filter at C.F.H.T., $\mu(V) < 25.3$ and $\mu(I) < 25.2$) with good seeing conditions (0".5 < FWHM < 1"). Recall that the luminosity function assumed for lensing clusters in the Wu and Hammer's work have been derived from the EMSS catalog. 14 luminous arcs (R < 22.5 and l/w > 4) have been detected including 8 giant luminous arcs (l/w> 10). The arcs have been identified by eye by two

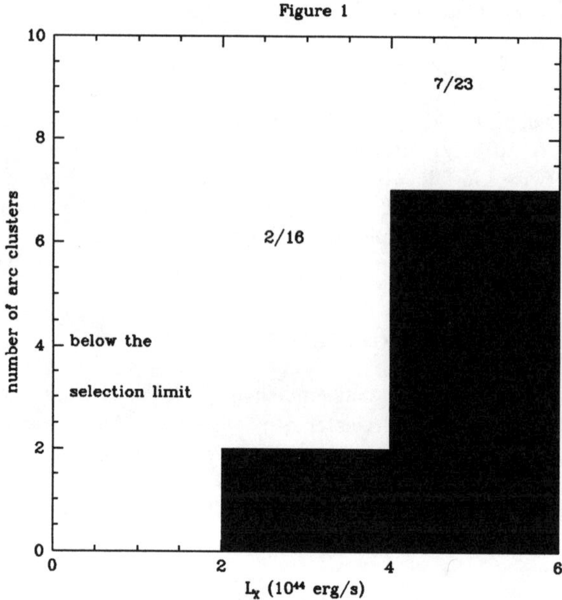

independent groups, and one can not expect a large contamination by cluster edge-on spiral galaxies [11].In such a double limited sample (in flux and in axis ratio), no bias should be expected against arcs having intermediate axis ratio ($4 < l/w < 10$) since at a given magnitude, the smaller arcs should have brighter surface brightnesses. Figure 1 shows the number of arc clusters in function of the cluster X-ray luminosity. 7 of the 23 very luminous clusters ($L_X > 4\ 10^{44}$ erg/s) have an arc or more in their core, which is consistent with the prediction of Wu and Hammer, if cluster mass density profiles have very small or null core radii. Conversely, the 16 moderately luminous X-ray clusters ($2\ 10^{44} < L_X < 4\ 10^{44}$ erg/s) only show 2 arcs in their cores. One can interpret Figure 1 as revealing that only very powerful X-ray clusters have reached the lensing critical density, i.e. they have much more compact mass density profiles (or small core radii) than the other ones.

Another test of the cluster density profile can be provided by examining the distribution of their axis ratio, which also depends on the density profile, since the fraction of arcs having large arc axis ratio increases for decreasing cluster core radii[6]. 8 of the 12 arcs detected in the sample of the 23 very luminous clusters have an axis ratio larger than 10, which is marginally consistent with a singular isothermal sphere density profile for lensing clusters, but highly inconsistent with the presence of core radii larger than ≈ 50 kpc (probability of rejection ≈ 0.9996 [6]). It is also interesting to notice that the 2 arcs detected in the 16 intermediately powerful X-ray clusters have small axis ratio ($l/w < 10$).

4 Giant arcs as magnified galaxies and new HST observations.

Magnification factor along the arc length can easily reach values larger than 10. It results that one can resolve details ten times smaller than if the source was not magnified. For example, it has been suggested [12] that the patchy appearance of the Cl2244-02 arc is the result of intrinsic source structures (arms and bulge of a nearly face-on galaxy), convolved by the spatial resolution (FWHM $\approx$ 0"6, the arc width is not resolved from the ground).
The cluster MS0440+0204 has been observed by the HST in October 1994, with a total integration time of 6 hours in a broad band filter centered on 7000A. Several arcs in this cluster were already identified from the ground[13], and the HST image strikingly reveals the circular symmetry of their distribution around the cluster core. MS0440+0204 deserves further studies but already appears to be an almost ideal lens, for which it would be easy, from a minimum of assumptions, to derive the density profile (indeed the core radius appears to be very small). On this image, the 14 arcs all have their widths spatially resolved (pixel size of WFPC2 = 0".1), and several of them show structures and compact or unresolved sources. One should notice that when a spatially unresolved source is found in an arc, it implies an intrinsic size smaller than 0".1/(Amp(1-K)), which can be as small as 0".01 in most cases. Such very compact sources have then intrinsic sizes smaller than 120 pc if the surrounding galaxy lies at z=1, and could be compact HII regions for example. Figure 2 shows a tentative of source reconstruction for one arc detected around the MS0440+0204 core, assuming a singular isothermal density profile for the cluster, and a magnification factor of 9 for the arc. The arc length is 6", the reconstructed pixel size is 0".011, and the reconstructed source size 0".75. Most of the reconstructed sources have irregular morphologies, although one could identify arm-like structures in some of them. More definitive results are in progress.

5 Summary.

There are several tens of giant luminous arcs which are now discovered, and more than one hundred are expected. Their modelling can provide fairly good estimation of the projected masses in rich cluster cores, and the derived M/L ratio are ranging from 100 to 300. Their statistical properties (number occurence, axis ratio distribution) lead to a dark matter component much more concentrated to the cluster center than the X-ray gas or the galaxy distribution. The first survey of a complete sample of X-ray clusters has led to the discovery of 14 luminous arcs, most of them lying in the brightest X-ray clusters. It results that only Lx > 4 10^{44} erg/s clusters have reached the critical lensing density and have core radii likely smaller than 50 kpc, while less luminous clusters are either less massive or less compact.
First results from a new HST image of MS0440+0204 are presented. Not less than 14 arcs are found in the core of this very rich cluster, which appears by many aspects, to be an ideal lens. First source reconstruction technics are experienced from this high quality data, and reveals the internal structure of background sources, i.e. on faint distant galaxies. These galaxies, otherwise barely resolved from the ground,

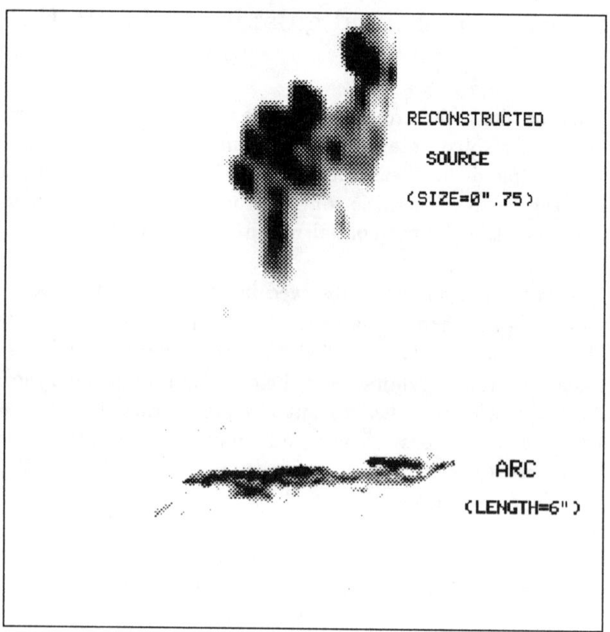

show arm-like structures and compact sources (HII regions ?) as small as 120 kpc, if the source is lying at z=1.

References.

[1] Hoag, A., 1981, B.A.A.S., 13, 799

[2] Soucail, G.,*et al.* A& A, 172, L14

[3] Lynds, R., Petrosian, V., 1986, B.A.A.S., 18, 1014

[4] Hammer, F., 1987, in Bergeron et al, eds., 3rd IAP Astrophys. Meeting on High Redshift Objects, Edts Frontières, Gif sur Yvette, P. 467

[5] Soucail, G.,*et al.* 1988, A& A, 191, L19

[6] Hammer, F.,*et al.* 1993, in the 31st Liege Int. Astrophys. Coll. on Gravitational Lenses, Institut d'Astrophysique, Eds J. Surdej et al, P. 609

[7] Hammer, F., 1991, ApJ, 383, 66

[8] Mellier, Y., Fort, B., Kneib, J.P., 1993, ApJ, 407, 33

[9] Wu, X.P., Hammer, F., 1993, MNRAS, 262, 187

[10] Luppino, G.,*et al.* 1995, ApJ, submitted

[11] Le Fèvre, O.,*et al.* 1994, ApJ, 422, L5

[12] Hammer, F., Rigaut, F., 1989, A& A, 226, 45

[13] Luppino, G.,*et al.* 1993, ApJ, 416, 444

The investigation of cluster mass distribution by gravitational lenses

G. SOUCAIL

Observatoire Midi-Pyrénées, URA 285
14 Avenue E. Belin
Toulouse, France

INTRODUCTION

Gravitational lensing in clusters of galaxies is a recent topic which brought a large number of original results in the understanding of clusters of galaxies, as well as of distant galaxies. Some of the results on the determination of mass distribution in clusters of galaxies are presented. In the most favorable cases, the cluster center mass distribution can be explored in great details, as well as some individual galaxies considered as local deflectors. Moreover, the recent detection of weak shear in the outerparts of a few clusters is complementary in these determinations. So the dark matter distribution may be inferred from gravitational optics on scales of 10 to 1000 kpc. Moreover the faint field population of distant galaxies can also be observed with such gravitational telescopes. Some clues about their distribution and properties will be presented in the context of the lens modeling of the best studied cluster-lenses.

OBSERVATIONS OF CLUSTER LENSES

Up to now, in more than 30 clusters containing arcs or arclets have been detected [1], so we can now understand what are the optimal conditions for such detections. From the observational point of view, two conditions are required: high resolution (*i.e.* sub-arcsecond) imaging for an analysis of the morphology of giant arcs, the identification of intensity breaks and the detection of multiple images, and ultra-deep imaging for the detection and the photometry of arclets as faint as possible, for a better statistics.

The criteria of selection of the cluster-lenses correspond to rich clusters, as the number of arclets varies roughly as the square of the Einstein radius R_E^2, or correspondingly to σ^4 if σ is the velocity dispersion of the cluster galaxies, for an isothermal distribution. Initially chosen from the richest clusters of the Abell catalogue, the best cluster-lenses can also be selected from other samples of X-ray clusters [2], provided the redshift range varies typically from 0.15 to 0.5 or 0.6. A recent compilation of the known lenses can be found in Fort and Mellier [1], although the list is probably not any more complete!

MASS DISTRIBUTION IN CLUSTERS OF GALAXIES

The mass distribution in clusters of galaxies can be determined from the modeling of arcs and arclets. The main ingredients of any model are the choice of the analytic form of the gravitational potentiel, parametrized with 6 or 7 variables such as: the center $(x_0; y_0)$, the ellipticity and the orientation of the distribution $(\varepsilon; \theta)$, the central mass, the total mass and the core radius $(\Phi_0; \sigma_{1D}; r_c)$ which caracterize the profile of the distribution. In addition, the source has also to be modeled as an extended entity, with its position $(x_S; y_S)$, its shape ($\varepsilon_S; \theta_S$ for an ellipse, or more detailed morphology), and its redshift. Finally, the cosmology has to be fixed, as the scaling factor D_{LS}/D_{OS} depends on q_0, Λ, Ω. In order to adjust these parameters, one has to add as many constraints as possible on the position of the arcs, their elongation. Stronger contraints may come from the identification of multiple images which represent the splitting of one point in the source plane in several images and which trace the gradient of the potential in two or three directions. Also intensity breaks are important to identify as they trace the location of the critical lines, where the inverse of the magnification matrix cancels.

Gravitational lensing allows the scanning of different mass scales: the main deflector has typically M $\sim 10^{14} M_\odot$ and $r_c \sim 100$ kpc, but small scale structures along the arcs may be due to "mini-lenses" with masses ranging from 10^6 to 10^{11} $M_\odot$ and typical size of 10 kpc. Finally lensing effects outside the central mass distribution are detected through the statistical weak shear pattern: in the case of the rich cluster Cl0024+1654, weak shear effects are detected up to 1.5 h_{100}^{-1} Mpc [3]. This may open a new window to search for distant clusters of galaxies and to map the large scale structures, as shown below.

MS2137–23: THE OPTIMAL CASE FOR LENS MODELING

The argumentation developped above has been successfully applied in the cluster MS2137–23, a cluster at a redshift of 0.315 dominated by a cD galaxy with an extended enveloppe and also a strong X-ray emitter. In this cluster, two large arcs are detected [4], namely a tangential one, 15" long, and a radial one, 5" long. For both of them no redshift are available yet.

In order to reconstruct the dark matter distribution in the cluster, the main assumption underlying the proposed modeling of the lens is that the light of the extended enveloppe of the cD traces the dark matter. In other words, this means that the DM distribution has the same orientation and ellipticity than the cD enveloppe. With such a model, it has been possible to reconstruct the position and the shape of the two main arcs, and to predict the position of each of their counter images, then identified on the CCD images and used to refine the lens parameters. The main results of the model show that only a small core radius of about 40kpc is compatible with the image configuration, and that the velocity dispersion of the cluster is in the range from 900 to 1200 km/s, the main uncertainty depending on the unknown redshifts of the sources [5]. The strength of this proposed model is that it is overconstrained, contrary to other previous attempts in other clusters. This means that a certain predictive powercan be used, as shown on the way the counter images were identified.

A MORE COMPLEX LENS: ABELL 370

Many attempts to model this well-known lens have been proposed since the beginning of the "arc saga" ([6,7,8,9]). Most of them presented strong anisotropies or ellipticities to reproduce the shape of the giant arc. More recent high quality images from CFHT (seeing 0.5") allowed to identifiy near the cluster center two ellongated small images which gave strong indications for a bi-modal potential. A new model was proposed in that sense, again using the assumption that the DM distribution followed the light distribution of the enveloppes of the two giant elliptical galaxies dominating the cluster center [10]. Again, successful reconstruction of the giant arc was possible and the predictive power of the method was used to reconstruct the image pair so-called B2–B3. In particular, its redshift is predicted to range at 0.865 ± 0.01.

Morevoer the regular shape of the giant arc means that the galaxies located near it cannot distort it too much. In other words, some upper limits on the mass of individual galaxies were proposed with a M/L ratio ranging from 4 to 20. The most successfull confirmation of this lens reconstruction came from a deep ROSAT HRI image centered on A370, which also clearly shows a double peaked distribution of the X-rays, definitely proving its bi-modal nature [1].

EXTENSION OF THE ANALYSIS OUTSIDE THE CLUSTER CENTERS

The arclet regime

The so-called "arclets" correspond to the distorted but single images of background sources and their use in lens reconstruction is more difficult for an individual analysis. But they can be usefull for statistical approach to probe the matter distribution outside the cluster center, and also the redshift distribution of the faint background sources. In particular in a case such as Abell 370 for which a well defined model of the matter distribution exists, such an approach was used on a sample of 50 initially detected arclets [11]. For a more restricted sub-sample, multicolor photometry allowed a possible identification of a photometric redshift. The evolution of the source ellipticity versus its redshift also allowed a redshift interval for which the source ellipticity was compatible with the (partly unknown) ellipticity distribution of faint galaxies. The final sub-sample of well studied arclets was only 19 objects, but after correction of their magnification, their B-magnitude ranges from 25 to 2, far below the limit of present-day spectrographs! Their redshift distribution is mainly centered around a redshift of 0.9, and only 4 of them are at $z > 1$! This results is important in the framework of galaxy evolution ([12] and referefences therein) as it seems that even at such faint magnitudes, the high redshift tail of the galaxy distribution is quite small, an assumption in favor of a non-evolving scenario of galaxies, including merging processes of sub-units, to explain the high number counts at faint magnitudes. But we must stress that the analysis is still preliminary, with probalby an incomplete and biased sample of arclets and it has to be extended to other well-studied clusters.

The weak shear regime

This concerns a regime in which individual distorsion is not possible to identify, but a statistical average over many similar objects can lead to the measurement of the local shear. Preliminary results were proposed by Tyson et al. [13] but the method was greatly emphasized by Kayser and Squires [14] and more recently by Schneider and Seitz [15,16]. More recently, measurements in several clusters were proposed by different groups ([3,17]). For example the weak shear effect outside the cluster Cl0024+17 was detected up to 1.5 h_{100}^{-1} Mpc at a shear level of only 5% and the shear profile is compatible with either an isothermal profile or a de Vaucouleurs' law. The total mass necessary to create this distorsion profile is of 1 to 2 $10^{15} M_{\odot}$, a huge value implying a M/L value of about 1000 at the periphery of the cluster [3]! If this result is confirmed in further studies, this may have strong cosmological implications.

Another expectation of the weak shear method is to detect new mass structures through their gravitatioal effects on the background sources. This has partly been done in the field of the double QSO 2345+007. Although already identified with a gravitational lens, the lensing agent of this multiple QSO was not known, and it was considered as a "dark lens". Using deep images of this field, Bonnet et al. [18] were able to detect a shear field around the QSO, not centered on it but slightly shifted towards a concentration of faint blue objects. This might be the signature of a distant cluster, although we do not know yet its redshift, and wether the mass distribution is sufficient to split the images of the QSO.

CONCLUSIONS AND FUTURE PROSPECTS

In summary, for most of the well modeled and understood clusters, it appears that they are dominated by one (or two) gE or cD galaxies. Their mass distribution indicates that the DM is peaked towards the cluster center, with very small core radii, and that the ellipticity and orientation of the cluster potential is similar to the ones of the cD. The implications of these results on the general question of the DM distribution in clusters are not clear yet. The main difficulty is that with giant arcs we mainly probe the central mass distribution, but the extension is not well constrained. In particular the slope of the DM needs to be checked through a weak shear approach. More generally, as we need to separate the cluster effects from the cD effects, we also plan to study in details some examples of non-cD clusters, such as Cl2244–02, for which deep HST images are expected soon. In fact, any improvements of the models will come from the addition of optical data from the HST (identification of multiple images, distorted arclets, etc ...) and with ROSAT X-ray images which may constrain the shape of the potential of the mass distributions in the clusters. This will be the case for the clusters Abell 2218 [19] in which spectacular HST images have revealed several couples of multiple and thin arcs, or for Abell 2390 in which the ROSAT X-ray image confirms the strong ellipticity of the cluster [20]. Finally the more difficult but exciting task of measuring the weak shear at the periphery of the clusters will strongly benefit from the improvement of the deep and wide field capabilities now available on some of the large ground-based telescopes.

REFERENCES

[1] Fort B., Mellier Y., 1994, A&AR 5, 239
[2] Gioia I.M., Luppino G.A., 1994, ApJS 94, 583
[3] Bonnet H., Mellier Y., Fort B., 1994, ApJ 427, L83
[4] Fort B., Le Fèvre O., Hammer F., Cailloux M., 1992, ApJ 399, L125
[5] Mellier Y., Fort B., Kneib J.P., 1993, ApJ 407, 33
[6] Soucail G., Mellier Y., Fort B., Hammer F., Mathez G., 1987, A&A 184, L7
[7] Grossman S., Narayan R., 1989, ApJ 324, L37
[8] Narasimha D., CHitre S.M., 1988, ApJ 344, 51
[9] Mellier Y., Soucail G., Fort B., Le Borgne J.-F., Pello R., 1990, *Gravitational Lensing*, p. 261, Mellier Y., Fort B., Soucail G. (eds), Springer, 1990.
[10] Kneib J.P., Mellier Y., Fort B., Mathez G., 1993, A&A 273, 367
[11] Kneib J.P., Mathez G., Fort B., Mellier Y., Soucail G., Longaretti P.Y., 1994, A&A 286, 701
[12] Koo D.C., Kron R.G., ARAA 30, 613
[13] Tyson J. A., Valdes F., and Wenk R., 1990, ApJ 349, 381
[14] Kaiser N., Squires G., 1993, ApJ 404, 441
[15] Schneider P., Seitz C., 1995, A&A 294, 411
[16] Seitz C., Schneider P., 1995, A&A in press
[17] Fahlmann G.G., Kaiser N., Squires G., Woods, D., 1994, ApJ 437, 56
[18] Bonnet H., Fort B., Kneib J.P., Mellier Y., Soucail G., 1994, A&A
[19] Kneib J.P., Mellier Y., Pelló R., Miralda-Escudé J., Le Borgne J.F., Böhringer H., Picat J.P., 1995, A&A in press
[20] Pierre M., Le Borgne J.F., Kneib J.P., Soucail G., 1995 submitted

DETERMINING THE MASS DISTRIBUTION OF CLUSTERS FROM GRAVITATIONAL LENSED BACKGROUND GALAXIES

CAROLIN SEITZ
Max Planck Institut für Astrophysik
85740 Garching, Germany

INTRODUCTION

The weak distortion of high redshift galaxies caused by gravitational light deflection through clusters of galaxies can be be used to reconstruct the projected (two-dimensional) surface mass density of low-redshift clusters. This technique, pioneered by Tyson, Valdes & Wenk[1], and Kaiser & Squires[2], is generalized to critical clusters, i.e., to such clusters which are capable of producing multiple images and giant luminous arcs. The resulting integral equation for the surface mass density has a kernel which can not be fully determined through local observations of lensed background galaxies if the cluster is critical, because for a critical lens there exists a degeneracy between the shape of lensed background galaxies and the local lensing strength. However, with an iterative prodedure, we obtain a fairly good reconstruction of the mass distribution of a critical cluster. In particular, the mass within the inner few arcminutes of the cluster can in principle be determined with an error of only a few percent, thus demonstrating the potential applicability and accuracy of this method for cluster mass determinations.

1. IMAGING OF EXTENDED SOURCES

One can describe an extended source by the quadrupole matrix Q_{ij} of its surface brightness distribution, and we define a complex ellipticity of the source as

$$\chi^{(s)} = \frac{Q_{11}^{(s)} - Q_{22}^{(s)} + 2iQ_{12}^{(s)}}{Q_{11}^{(s)} + Q_{22}^{(s)}} \tag{1.1}$$

The imaging of a small source can be described by the linearized lens mapping, provided the cluster mass distribution is smooth on scales of galaxy images (typically $\lesssim 2$ arcsec). If $\boldsymbol{\beta}$ and $\boldsymbol{\theta}$ are the corresponding position vectors of a light ray in the source and image plane respectively, we obtain $d\boldsymbol{\beta} = A(\boldsymbol{\theta})d\boldsymbol{\theta}$, where

$$A(\boldsymbol{\theta}) = (1-\kappa)I + \begin{pmatrix} \gamma_1 & \gamma_2 \\ \gamma_2 & -\gamma_1 \end{pmatrix} , \tag{1.2}$$

is the Jacobian matrix of the lens equation. Here, κ is the dimensionless surface mass density defined as usually in gravitational lens theory * (see Schneider,

*Throughout this contribution we assume that the sources are far behind the lens, thus that the lensing strength is the same for all sources, independent of their individual redshifts. If the galaxy sample is sufficiently faint, which it must be anyway to observe a sufficiently large number density of sources, this assumption is satisfied reasonably well for a low-redshift cluster with $z \lesssim 0.2$.

Ehlers & Falco[3]) and $\gamma = \gamma_1 + i\gamma_2$ is the complex shear calculated from

$$\gamma(\boldsymbol{\theta}) = \frac{1}{\pi} \int_{\mathbb{R}^2} \mathrm{d}^2\theta' \, \mathcal{D}(\boldsymbol{\theta} - \boldsymbol{\theta}') \, \kappa(\boldsymbol{\theta}') \,, \quad \text{with} \quad \mathcal{D}(\boldsymbol{\theta}) = \frac{\theta_1^2 - \theta_2^2 + 2\mathrm{i}\theta_1\theta_2}{|\boldsymbol{\theta}|^4} \,. \tag{1.3}$$

We define in analogy to the quadrupole matrix $Q_{ij}^{(s)}$ of the source the quadrupole matrix Q_{ij} of the image. Since the surface brightness is conserved in gravitational light deflection one can easily derive the transformation between both and obtains $Q^{(s)} = AQA^T = AQA$. Defining in analogy to Eq. (1.1) the complex ellipticity χ of the image, we obtain for the transformations between $\chi^{(s)}$ and χ

$$\chi^{(s)} = \frac{\chi + 2g + g^2\chi^*}{1 + |g|^2 + 2\mathcal{R}\mathrm{e}(g\chi^*)} \quad , \quad \chi = \frac{\chi^{(s)} - 2g + g^2\chi^{(s)*}}{1 + |g|^2 - 2\mathcal{R}\mathrm{e}(g\chi^{(s)*})} \tag{1.4}$$

Here, an asterisk denotes complex conjugation and $g = \gamma/(1-\kappa)$ is the lens parameter, which depends, for fixed redshifts of the lens and the sources, only on the mass distribution of the lens.

2. LOCAL OBSERVABLES

For the reconstruction of the surface mass density of a lens from images of gravitationally lensed background galaxies one has to assume that the sources are intrinsically randomly oriented on the sky, so that the net orientation of images is due to the tidal effects of the deflecting mass distribution. This basic assumption transforms to

$$\left\langle \chi^{(s)} \right\rangle = 0 \quad , \tag{2.1}$$

where the average is performed over an ensemble of sources.Inserting Eq. (1.4) into Eq. (2.1) and solving for g we find that the solution is invariant under the transformation $g \to 1/g^*$. Hence, only a combination of g and g^*, the (complex) *distortion*

$$\delta = \frac{2g}{1 + |g|^2} = \frac{2\gamma(1-\kappa)}{(1-\kappa)^2 + |\gamma|^2} \tag{2.2}$$

can be determined locally from images of background galaxies. Thus, we can not derive γ nor κ from lensed background galaxies, but only the combination δ of both. Detailed discussions concerning the rms error in determining the distortion from galaxy images can be found in Sect. 3.2 of Schneider& Seitz[4] and in Seitz & Schneider[5], hereafter Paper I and Paper II.

3. THE INVERSION EQUATION

Eq.(1.3) can be inverted to obtain the surface mass density as an integral over the shear,

$$\kappa(\boldsymbol{\theta}) = \frac{1}{\pi} \int_{\mathbb{R}^2} \mathrm{d}^2\theta' \, \mathcal{R}\mathrm{e}\left[\mathcal{D}(\boldsymbol{\theta} - \boldsymbol{\theta}') \, \gamma^*(\boldsymbol{\theta}')\right] \quad , \tag{3.1}$$

which shows that we could reconstruct the surface mass density $\kappa(\boldsymbol{\theta})$ of the lens, if the shear $\gamma(\boldsymbol{\theta})$ caused by the deflector could be measured locally as a function of angular position $\boldsymbol{\theta}$. Unfortunately, however, the shear is not observable in general, but a combination δ of surface mass density and shear, as was pointed out in Sect.2. Solving Eq. (2.2) for the shear

$$\gamma = \frac{(1-\kappa)}{\delta*}\left[1 \pm \sqrt{1-|\delta|^2}\right] \quad , \tag{3.2}$$

we see that for given values of δ and κ two solutions γ exist. The (local) degeneracy caused by the two signs in (3.2) cannot be resolved from a local measurement of image distortions. The correct sign is opposite to that of $\det A(\boldsymbol{\theta})$; however, the sign of $\det A$, i.e., the parity of the local lens mapping, cannot be observed locally. Inserting (3.2) into (3.1) yields

$$\kappa(\boldsymbol{\theta}) = \frac{1}{\pi}\int_{\mathbb{R}^2} \mathrm{d}^2\theta' \, \mathcal{R}\mathrm{e}\left[\mathcal{D}(\boldsymbol{\theta}-\boldsymbol{\theta}')\left(\frac{1 \pm \sqrt{1-|\delta(\boldsymbol{\theta}')|^2}}{\delta(\boldsymbol{\theta}')}\right)(1-\kappa(\boldsymbol{\theta}'))\right] \quad . \tag{3.3}$$

We like to mention three problems arising from calculating the mass density with Eq. (3.3) and also make suggestions to deal with them:

a.) Since images are observed only at discrete points, and since the sources have an intrinsic ellipticity distribution, noise is introduced in the local estimate for δ. We therefore recommend to use a grid, determine the distortion at each gridpoint from neighbouring galaxy images observed within a smoothing radius. The smoothing radius is adapted to the lensing strength, such that the rms error in determining δ is almost the same on the whole data field (for detailed discussions see Papers I& II).

b.) We do not know which sign to choose locally. Thus, the kernel of the integral equation is not fully determined from observations. We suggest to solve Eq. (3.3) iteratively: we start with $\kappa = 0$ and the minus sign on the whole lens plane (i.e. the cluster is first assumed to be uncritical). We obtain a first estimate for κ, calculate from this the shear with Eq. (1.3), and from both the sign of $\det(A)$ on the whole lens plane. Then we insert the present solution for κ and the local signs in (3.3) and repeat the last three steps, until convergence is reached. (For details see Paper II.)

c.) Data (images) are restricted to a finite field (CCD), whereas in Eq. (3.3) the integral is performed over the whole lens plane. This introduces boundary artefacts. One way to reduce these effects is to extrapolate the deduced distortion field to the outside of the data field, and use the increased distortion field for the reconstruction in (3.3).

4. RECONSTRUCTION OF A MODEL CLUSTER

In the following we try to reconstruct a critical cluster obtained from a simulation of structure formation in a CDM universe (Bartelmann, Steinmetz & Weiss[6]). The surface mass density is shown in Fig. 1a. We use sources with random positions and with random ellipticities drawn from a Gaussian distribution with mean $\sqrt{\langle\chi\rangle^{(s)2}} \approx 0.3$. The number density of the sources is 20 galaxies/arcmin^2. Using Eq.(1.4) we calculate the image ellipticities.

We apply the method derived by Kaiser & Squires[2] (herafter KS) on these galaxy images (for specialist: the size of the window function is $s = 0.2'$).

The result is shown in Fig. 1b: first, the central peak and also the secondary maxima are badly reconstructed. This is because the KS method was designed for the weak lensing regime, whereas in the center of our cluster strong lensing takes place. Second, the reconstructed mass increases towards the edges and has a (negative) minimum between those at the boundaries. This boundary artefacts come in because the inversion of Eqs. (3.1&1.3) is not exact on the finite data field ($10' \times 10'$ in our example). Third, the reconstruction shows noise especially in regions of small mass densities (see contourline with $\kappa = 0$). There is a kind of 'shot noise' because in the KS method one has to sum up over the ellipticities of individual randomly distributed galaxies. In addition, the ellipticities of the sources introduce a noise, which is small if the observed image ellipticities are dominated by the lensing effect, and large if they are dominated by the intrinsic ellipticities.

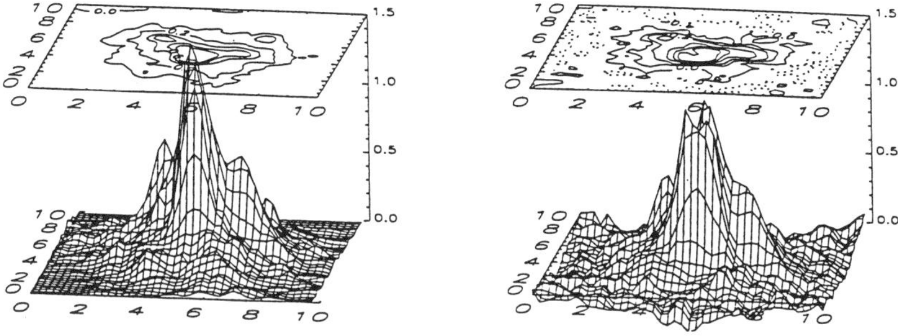

Fig. 1. a) The projected dimensionless surface mass density κ of a numerically modelled cluster with a sidelength of about 10'. The levels of the contourlines are 0.1,0.2,0.4,0.6,0.8 (left). b) The reconstructed mass density obtained with the method derived by Kaiser & Squires[2] (right)

Next, we use the synthetic galaxy images, derive the distortion δ and use Eq. (3.3) for the reconstruction. The result is shown in Fig. 2a.: first, the central peak and the secondary maxima are much better reconstructed. This demonstrates that our method works in the weak and strong lensing regime. Second, from comparing the contourlines, say for $\kappa = 0$ we find that Fig. 2a shows less noise than Fig. 1b. This is because we do not have any shot noise and we use an elaborated smoothing procedure. Third, boundary artefacts are still in. To reduce them we extrapolate the deduced distortion field outwards, such that the extrapolated distortion field has twice the sidelength of the data field. With the increased field we do the reconstruction again and show the results in Fig. 2b. Now, no boundary artefacts can be detected per eye.

We mention that recently Kaiser[7] developed a different method for a non-linear lens reconstruction, which can be used for constructing an inversion formula for κ which in contrast to Eq. (3.1) uses only data in a finite field (Schneider[8]).

From the mass distribution shown in Fig. 2b, we calculate the mass within a distance θ_g from the cluster 'center' and compare it with the mass found within the same distance of the model cluster (Fig. 1a). For $\theta_g = 1', 2', 4'$ we obtain the values 95%, 97%, 90% respectively, which we find are surprisingly accurate.

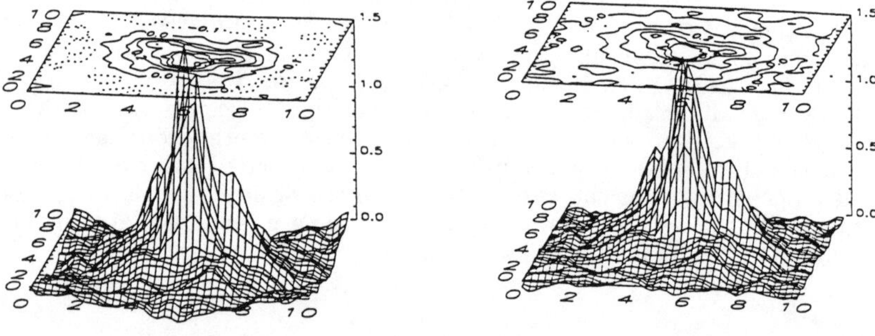

Fig. 2. a) The reconstructed surface mass density obtained with Eq.(3.3) (left) and b) the surface mass density obtained with Eq.(3.3) using an enhanced distortion field (right, see text)

CONCLUSION

We have developed a method to derive the mass distribution of a lens from the images of lensed background galaxies. We have demonstrated that this method simultaneously works in the weak and strong lensing regime. After correcting for the finite data field we find that the mass distribution and the total mass is reconstructed fairly accurate. We point out that we need no information on the ellipticity distribution of the sources, besides that the orientation of the sources is random on the sky.

The method as presented here, can be applied only on low-redshift cluster ($z_d \lesssim 0.2$), because we treated all sources as to be at the same effective distance, i.e. $D_{ds}/D_s \approx$ const. We currently generalize the method, such that the redshift distribution of the sources is accounted for. Then, the reconstruction of higher-redshift clusters is possible, if one has a resonable model for the redshift distribution of the sources.

REFERENCES

1. J.A. Tyson, F. Valdes & R.A. Wenk, 1990, *ApJ*, **349**, L1.
2. N. Kaiser & G. Squires, 1993, *ApJ*, **404**, 441 (KS).
3. P. Schneider, J. Ehlers & E. Falco, 1992, *Gravitational Lenses*, Springer-Verlag, Heidelberg.
4. P. Schneider & C. Seitz, 1994, *A&A*, in press. (Paper I)
5. C. Seitz & P. Schneider, 1994, *A&A*, in press. (Paper II)
6. M. Bartelmann, M. Steinmetz & A. Weiss, 1994, *A&A*, in press.
7. N. Kaiser, 1994, preprint.
8. P. Schneider, 1994, preprint.

Ray Propagation in the Cheese Slice Universe[a]

SYLVIE LANDRY AND CHARLES C. DYER

Department of Astronomy, Scarborough College
University of Toronto, 1265 Military Trail
Scarborough, Ontario, Canada, M1C 1A4

INTRODUCTION

Most studies of gravitational lensing and the impact on observations concentrate on lensing structures which are bounded, that is, of some finite size in an otherwise reasonably smooth background universe. Here we consider a model of the universe, the "cheese slice" universe where the lensing is caused by very large scale structures, in this case large slabs of alternating Friedmann-Lemaître-Robertson-Walker (FLRW) dust and pure vacuum.

The matching of FLRW space-times onto a Kasner vacuum region has been investigated by Dyer, Landry and Shaver[1]. We have shown that only a spatially flat Einstein-de Sitter region can be joined smoothly to a special case of the Kasner space-time. The metrics involved, in cylindrical coordinates, are:

$$ds^2 = dt^2 - a^2 t^{4/3}(dr^2 + r^2 d\phi^2 + dz^2) \tag{1}$$

in the Einstein-de Sitter region, and

$$ds^2 = dt^2 - a^2 t^{4/3}(dr^2 + r^2 d\phi^2) - b^2 t^{-2/3} dZ^2 \tag{2}$$

in the Kasner region. The two metrics are identical for surfaces with z and Z constant. Constants a and b are introduced to ensure proper unit bookkeeping. For the interface to be represented in a consistent way as seen from both regions, we must require that $a^2 t^{4/3} dz^2 = b^2 t^{-2/3} dZ^2$ at the interface. This matching suggested the "cheese slices" cosmological model, which is constructed by alternating layers of FLRW and Kasner regions. This model evolves from being a universe which is initially almost all vacuum with only thin dense slices of matter to eventually becoming almost entirely filled with cosmological dust with only thin slices of vacuum. In what follows, we investigate the changes in a few key properties of the beam as it propagates through this universe.

RAY PROPAGATION

From now on, we will use the symbols F and K to designate indexed quantities or scalars associated with the FLRW and Kasner regions respectively[b]. The null

[a]This work was supported in part by Kappa Kappa Gamma Foundation and NSERC, Canada.

[b]Throughout this paper Greek indices take the range 1-3 and Latin indices the range 0-3. The indices (0,1,2,3) correspond to (t,r,ϕ,z) or (t,r,ϕ,Z) depending on the metric used.

tangent vector, k^a, can be obtained directly from the Euler-Lagrange equations and was found to be a function of t only (note that symmetry considerations enable us to choose $k^2 \equiv k^\phi = 0$ in both regions). We need to know how k^a crosses the boundary.

We first consider a point comoving with the boundary surface separating the two regions. The observer's 4-velocity and the null tangent vector to the geodesic must satisfy the following conditions: $(u_a k^a)_{\rm F} = (u_a k^a)_{\rm K}$ and $(u_a u^a)_{\rm F} = (u_a u^a)_{\rm K} = 1$. The first relation simply means that an observer fixed to a point would measure the same frequency whether it is measured in the Einstein-de Sitter or Kasner space-times. Since the boundary is a comoving surface in both regions, the conditions imply that *at the surface* $(k^0)_{\rm F} = (k^0)_{\rm K}$.

Consider now an observer moving along the radial coordinate. Once again, since the metric elements involved are the same on both sides of the boundary, $(u_a)_{\rm F} = (u_a)_{\rm K}$. The condition that the frequency be continuous and $(k^0)_{\rm F} = (k^0)_{\rm K}$ requires that *at the surface* $(k^1)_{\rm F} = (k^1)_{\rm K}$.

Finally, the physical symmetries of the model and the fact that k^a is a null vector imply that k^3 is not continuous across the boundary but, as mentioned earlier, $g_{33}k^3$ is.

Bending Angle

We define Ψ as the angle between a 3-vector normal to the boundary surface, $\vec{V}$, and the tangent 3-vector to the geodesic, $\vec{k}$. Ψ can easily be calculated from the familiar equation for scalar products $\vec{V} \cdot \vec{k} = |\vec{V}||\vec{k}| \cos \Psi$. Since k^a is a null vector, it follows that $\cos \Psi = -|g_{33}|^{1/2} (k^3/k^0)$. From the simple trigonometric relation $\cos^2 \Psi + \sin^2 \Psi = 1$ we can write $\sin \Psi = |g_{11}|^{1/2}(k^1/k^0)$ and $\tan \Psi = -\sqrt{|g_{11}|/|g_{33}|} \times (k^1/k^3)$. Finally, the following results are obtained:

$$\tan \Psi_{\rm F} = \text{constant} \quad \text{and} \quad \tan \Psi_{\rm K} = \frac{\tan \Psi_{\rm F}}{(t/t_{\rm in})}, \tag{3}$$

where $t_{\rm in}$ is the time calculated when the light ray reaches the boundary and $\Psi_{\rm F}$ was evaluated in the *previous Einstein-de Sitter slice.* The first result was of course expected since the Einstein-de Sitter space-time is homogeneous and isotropic. When crossing the boundary into the Kasner region, no bending will occur at the surface. As the beam propagates forward (backward) in time through the Kasner slice, it will be forced to move towards (away from) the normal to the surface separating the two regions (see figure 1).

Redshift Factor

The contribution of a given slice to the overall redshift factor is $x \equiv 1+$ redshift $= (u_a k^a)_{\rm in}/(u_a k^a)_{\rm out} = \dot{t}_{\rm in}/\dot{t}_{\rm out}$, where $\dot{}$ denotes total differentiation with respect to the affine parameter (the redshift factor for the entire model is the product of each contribution). The following results are obtained

$$x_{\rm F} = \left(\frac{t_{\rm out}}{t_{\rm in}}\right)^{2/3} \quad \text{and} \quad x_{\rm K} = \frac{(t_{\rm out}/t_{\rm in})^{2/3}}{\cos \Psi_{\rm in} \sqrt{\tan^2 \Psi_{\rm in} + (t_{\rm out}/t_{\rm in})^2}} \tag{4}$$

for the Einstein-de Sitter and Kasner regions respectively. Future pointing rays will experience the usual redshift effect each time a FLRW slice is crossed. The contribution of each Kasner region, for different $\Psi_{\rm in}$ values, is qualitatively illustrated in figure 2a. When travelling through the slice, future pointing rays will be blueshifted as long as the tilt angle $\Psi_K <\sim 35°$. If $\Psi_K >\sim 35°$ (i.e. the expansion along the radial coordinate dominates the contraction along z), then they will be redshifted. Since Ψ is a function of the thickness of the slice and its value at entry, the net contribution of a Kasner slice may be a redshift or a blueshift effect.

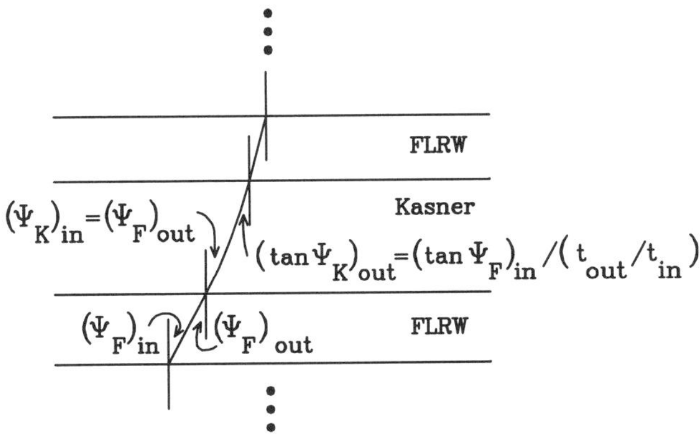

FIGURE 1: Light ray propagating forward in time through the Einstein-de Sitter-Kasner universe.

Shape and Size of the Beam

The geometric-optics approximation for a test electromagnetic field introduces a sequence of null hypersurfaces whose normals are null, and tangent to an irrotational congruence of null geodesics[2]. From the Ehlers-Sachs theorem, the cross-section of such a congruence of geodesics is expanded and sheared at the respective rates[3]

$$\Theta = \frac{1}{2}k^a_{||a} \qquad \text{and} \qquad |\sigma| = \left(\frac{1}{2}k_{(a||b)}k^{a||b} - \Theta^2\right)^{1/2} \tag{5}$$

with respect to the affine parameter τ along the beam. $||a$ denotes covariant differentiation with respect to x^a and () denotes symmetrization over the enclosed indices. We define the real quantities $C_\mp$ and $\alpha_\mp$ in terms of the two principal curvatures of the wave front $\Theta \pm \sigma = [\ln(C_\mp + i\alpha_\mp)]^\cdot$, where Θ is expected to be real. The area of the beam is $\propto C_+C_-$, while the distortion factor is $\propto C_+/C_-$. Because of the physical symmetries of the model, we do not expect the direction of the principal axis of curvature to change. The shearing rate σ is known up to a constant phase factor. It then follows that $\Theta \pm |\sigma| = \dot{C}_\mp/C_\mp$, which can be solved analytically. After a convenient normalization with respect to output values from the previous slice, the dimensions of the beam are:

$$(C_-)_{\rm F} = (C_-)_{\rm in}\left(\frac{t}{t_{\rm in}}\right)^{2/3}\left(\frac{r}{r_{\rm in}}\right) \, , (C_+)_{\rm F} = (C_+)_{\rm in}\left(\frac{t}{t_{\rm in}}\right)^{2/3} \tag{6}$$

for Einstein-de Sitter slices and

$$(C_-)_K = (C_-)_{in} \left(\frac{t}{t_{in}}\right)^{2/3} \left(\frac{r}{r_{in}}\right) , (C_+)_K = (C_+)_{in} \cos \Psi_{in} \frac{\sqrt{\tan^2 \Psi_{in} + (t/t_{in})^2}}{(t/t_{in})^{1/3}} \quad (7)$$

for Kasner slices. One must keep in mind when evaluating these equations that future pointing rays are incoming rays (i.e. $dt > 0$ and $dr < 0$). The effect of the Kasner space-time on the dimensions of the beam are qualitatively illustrated in figure 2b. It is worth mentioning that all physical properties of a beam of light travelling through the cheese slice model are known analytically. The only exception is: t =function(τ), which can be quite easily integrated numerically.

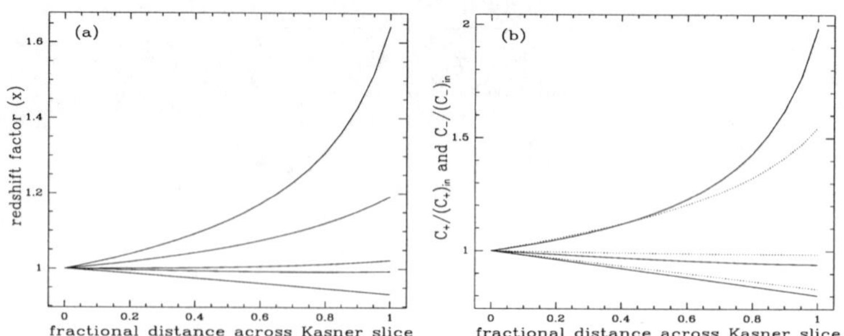

FIGURE 2: Kasner slice: (a) x evaluated using $\Psi_{in} = 10°$ (bottom curve), 30°, 35°, 50° and 60° (top curve). (b) Normalized C_+ (full lines) and C_- (dotted lines) evaluated using $\Psi_{in} = 10°$ (bottom curve), 35° and 60° (top curve).

CONCLUSION

We have investigated a beam of light as it propagates through the cheese slice cosmological model. The ray tracing problem has been solved and shows that only the Kasner regions will introduce a bending in the beam as it propagates. We have demonstrated that the Kasner slices introduce anisotropic redshift effects. The optical scalars were used as a tool to obtain the cross-sectional area and shape of the beam. Complete multi-slice models are being studied further using a computer code.

The cheese slice universe could be used, in a restricted way, to model the observational relations expected in a sheet-like universe. Particular emphasis should be given to anisotropy effects in the observations and any apparent periodicities in the data[4].

REFERENCES

1. Dyer, C. C., S. Landry & E. G. Shaver. 1993. Phys. Rev. **D47:** 1404.
2. Trautman, A. 1964. *In* Brandeis lectures on general relativity. Prentice-Hall, Inc., Englewood Cliffs, New Jersey, .
3. Sachs, R. K. 1961. Proc. Roy. Soc. London **A264:** 309.
4. Broadhurst, T. J., R. S. Ellis, D. C. Koo & A. S. Szalay. 1990. Nature **343:** 726.

Gravitational Lensing by Cosmic Strings

F. MARLEAU, C.C. DYER and J.H. PALMER
Department of Astronomy, University of Toronto
60 St. George St., Toronto, Ontario, Canada

INTRODUCTION

The cosmic string scenario has been recently developed and proposed as a model for galaxy formation. The model predicts density fluctuations and microwave background anisotropies compatible with the detection limit of the present observations. These calculations rely on the metric of the string and are only correct if the spacetime of the cosmic string can be represented by a conical spacetime. The Minkowski conical spacetime is one possible exterior solution for a non-gravitating string[1] and we discuss in detail the physical properties of this particular spacetime.

The conical spacetime was also derived by Vilenkin in his earlier attempt to study cosmic strings in the context of the linear approximation of general relativity. The original form of the metric derived by Vilenkin is derived from the energy-momentum tensor of the string Lagrangian and has curvature. Vilenkin applies a coordinate transformation and obtains the conical representation of spacetime. We find that this transformation is incorrect and that the interior spacetime of the non-gravitating string given by Vilenkin is not the conical Minkowski spacetime. This result is in agreement with the model we have developed of a self-gravitating cosmic string which, we find, doesn't have a conformally flat metric but has a metric with curvature. With this metric describing the interior spacetime of the gravitating string, we have been able to compute interesting properties of the string, including true gravitational lensing effects. These effects are truly gravitational lens effects since the lensed images show distortion and amplification. This is in fact quite different from previous analysis of lensing due to the conical nature of the spacetime which creates images by purely geometrical means, as arises with optical prisms. We present these new results and discuss the implications for the observational evidence for strings.

The Conical Spacetime

The simplicity of the conical spacetime has led to a non-relativistic treatment of the lensing and redshift properties of cosmic strings. In particular, the effects of a moving string have been modelled by moving a wedge representing the deficit angle with respect to otherwise "stationary" observers. It is not known how to move a string and calculate the back-reaction of the rest of the universe; moving the string is not equivalent to moving the source and observer. Kaiser and Stebbins[2] use this picture of a conical spacetime as Minkowski minus a wedge to calculate the effect of a moving string on temperature fluctuations of the microwave background.

A relativistic treatment of lensing on a conical spacetime requires the solution of the null geodesic equations. The metric describing a conical spacetime has the form:

$$ds^2 = -dt^2 + dz^2 + dr^2 + \alpha^2 r^2 d\phi^2 \tag{1}$$

The conical nature of this spacetime is due to the deficit angle α. A value of $\alpha = 1$ gives the Minkowski spacetime. We will consider only rays which travel perpendicular to the straight string. The null orbit is straight forward to derive and results in the following equation joining an initial point i and a final point f through a closest approach radius r_c:

$$\alpha(\phi_f - \phi_i) = cos^{-1}(\frac{r_c}{r_i}) + cos^{-1}(\frac{r_c}{r_f}) \tag{2}$$

The radius of curvature is given by:

$$R = r_c \frac{(\alpha^2 sin^2(\alpha(\phi - \phi_c)) + cos^2(\alpha(\phi - \phi_c)))^{3/2}}{cos^4(\alpha(\phi - \phi_c))(1 - \alpha^2)} \tag{3}$$

If $\alpha = 1$, $R \to \infty$ and we have a straight line. If $\alpha = 0$, $R = r_c$ and we have a circle. From equation (3) we see that the radius of curvature of the orbit increases with r_c increasing. Thus orbits passing closer to the string are more highly curved than distant orbits. The radius of curvature also varies with ϕ along a given orbit, contrary to the picture used by Kaiser and Stebbins of Minkowski space minus a wedge.

The total bending can be computed from equation (2). For the cases $0 < \alpha < 1.0$, as $r \to \infty$, $\phi \to \pi/2\alpha$. This implies that the total bending on the orbit is $\beta = \pi(1/\alpha - 1)$. While the curvature at any point along the orbit is a function of r_c, the total bending is independent of r_c. Examination of these relations shows that every ray is just a scaled copy of any other ray, as would be generated by moving every point of a ray out along the generators of the cone in the ratio of the impact parameters.

Double Imaging

Because of the conical nature of the spacetime, we expect that there will be regions where two null orbits will connect the same initial and final points. The total bending will define the region of double imaging. For a given observer, a source within a region bounded by the two null orbits with r_c approaching zero from either side of the string will produce two images. For these two null rays to leave from the same point (ϕ_i, r_i) and meet on the opposite side of the string at the same point (ϕ_f, r_f), the equation $\Delta\phi_1 + \Delta\phi_2 = 2\pi$ must be satisfied:

$$cos^{-1}(\frac{r_{c2}}{r_i}) + cos^{-1}(\frac{r_{c2}}{r_f}) = \alpha 2\pi - cos^{-1}(\frac{r_{c1}}{r_i}) - cos^{-1}(\frac{r_{c1}}{r_f}) \tag{4}$$

It follows from this equation that we can also solve for the closest approach radius of each of the two null orbits.

We can now examine what an observer would see in the double image region as a source moves across the sky behind the string. We can compute angular separation of the two images, θ, observed at r_f. If θ_i is the angle between the tangent vector to orbit "i" at point r_f and the radial vector, then we have:

$$\theta = \theta_1 + \theta_2 = tan^{-1}\left(\frac{1}{\alpha\sqrt{(\frac{r_f}{r_{c1}})^2 - 1}}\right) + tan^{-1}\left(\frac{1}{\alpha\sqrt{(\frac{r_f}{r_{c2}})^2 - 1}}\right) \tag{5}$$

We find that the angular separation is constant throughout the double image region for a given r_f and r_i.

In addition, we can compute the point amplification as a function of r_c:

$$A(r_c) = \frac{1}{1 + \frac{1-\alpha^2}{\alpha^2}\left(\frac{r_c}{r_f}\right)^2} \tag{6}$$

For a moving point source, the observer sees a single image until the double image region is entered. Then the second image turns on until the source leaves the double image region. At this point the first image turns off. In the case of an extended object, such as a disk, there will be in addition shearing in the direction perpendicular to the string and the line of sight and amplification of the images as they approach the string. In addition, the appearance of the secondary image will be incremental as first the near limb passes into the double image region, followed by the rest of the disk. The original primary image will vanish in a similar manner.

The Self-gravitating Cosmic String

The spacetime associated with a self-gravitating cosmic string was introduced in 1981 by Vilenkin[3]. In the weak-field limit, he found that the interior spacetime of a non-gravitating string was conical Minkowski. By recalculating Vilenkin's result, we find that an error has been made in his derivation of a purely conical Minkowski spacetime in the weak-field limit.

In the linear approximation he derives the following vacuum string metric (Vilenkin's equation (27)):

$$ds^2 = dt^2 - dz^2 - (1-\lambda)(dr^2 + r^2 d\phi^2) \tag{7}$$

where $\lambda = 8G\mu ln(r/r_0)$ and, according to Vilenkin, r_0 is a constant that can be set equal (approximately) to the radius of the string. This metric has curvature since its Riemann curvature tensor is non-zero. However, the conical metric has a null curvature tensor. Since the vanishing of a tensor is coordinate independent, the coordinate transformation must be invalid, and inspection shows that his coordinate transformation is not possible.

In order to verify the origin of Vilenkin's spacetime, it is necessary to find a complete, self-consistent solution to the field equations. This involves solving simultaneously the Euler-Lagrange equations and the Einstein equations[4]. Using the following form of the metric (static case),

$$ds^2 = -X^2(r)(dt^2 - dr^2) + \frac{W^2(r)}{X^2(r)}d\phi^2 + X^2(r)dz^2 \tag{8}$$

where X is an exponential function, we solve such a set of coupled differential equations and obtain numerical solutions using a Taylor-series method. As shown in Figure 1, the energy-momentum tensor of the interior of the string is not constant but a function of r. Although the T_ϕ^ϕ and T_r^r components of the energy-momentum tensor can be zero, we find that for other valid string solutions these components are not necessarily null. In all solutions, the energy-momentum tensor follows the

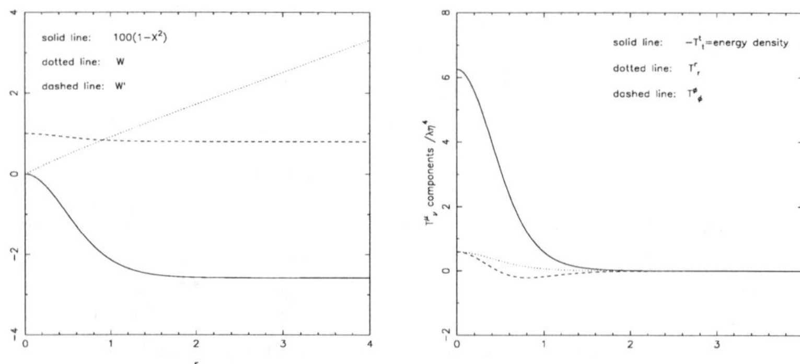

Figure 1: Metric coefficients and energy-momentum tensor components of a representative string solution of the Einstein-Euler-Lagrange equations.

behaviour of the gauge and scalar fields which define the interior of the string and when it becomes zero, the exterior metric of the string is the conical Minskowski spacetime. These solutions show that the interior string spacetime is curved and that there is real gravitational lensing for light rays propagating through such a spacetime. The fact that the Weyl tensor has non-zero components implies that there will be shearing of images formed by light rays going through the string. We expect to see as well amplification of far away objects, located behind the string. In terms of observational consequences, the new results suggest that the detection of string could be based on a completely different set of visible effects.

Acknowledgements

This work has been supported by the Natural Sciences and Engineering Research Council of Canada through a Postgraduate Scholarship (F. R. M.), a Postdoctoral Fellowship (J. H. P.) and an operating grant (C. C. D.).

Bibliography

1. Gott III, J.R., 1985, *Astrop. J.* **288**, 422.
2. Kaiser, N. and Stebbins, A., 1984, *Nature* **310**, 391-393; Stebbins, A., 1988, *Astrop. J.* **327**, 584.
3. Vilenkin, A., 1981, *Phys. Rev. D* **23**, 852.
4. Dyer, C.C. and Marleau, F., "Complete Model of a Self-gravitating Cosmic String I. A New Class of Exact Solutions and Gravitational Lensing", paper submitted to *Phys. Rev. D*, October 25, 1994.

THE CORRELATION OF 1-JANSKY SOURCES TO ZWICKY CLUSTERS

STELLA SEITZ
Max-Planck Institut für Astrophysik
Karl-Schwarzschild-Str.-1, 85740 Garching, Germany

INTRODUCTION

Investigations of the distribution of galaxies or clusters of galaxies near to the line of sight to QSOs are very instructive for several reasons: The correlation of AGNs with galaxies or clusters of galaxies at the same redshift and the dependency of this correlation on the redshift of the AGN reflect common formation conditions in terms of the density fluctuations in the early universe, relate QSO activity to the richness of the environment and thus give hints for QSO-feeding and triggering mechanisms (see Ellingson, Yee & Green[1] and references therein). Correlations with foreground objects, i.e. objects with a smaller cosmological redshift, yield statistical properties of the interviening media along the lines of sight to the QSOs. A positive correlation of QSOs from a flux-limited sample with galaxies has been found and interpreted as gravitational-lens induced magnification of the background population, where the galaxies themselves act as gravitational lenses (Drinkwater et al.[2] and references therein) or trace larger lensing structures like cluster of galaxies or the Large Scale Structure of the universe (Bartelmann & Schneider[3] and references therein). Also an anti-correlation of different QSO samples with foreground clusters of galaxies was found (Boyle et al.[4], Romani & Maoz[5]) and absorbing dust in these foreground objects was hold responsible for it. Recently, Rodrigues-Williams & Hogan[6] (hereafter RH94) have reported on a statistically significant overdensity of high-redshift LBQS-QSOs around Zwicky clusters. The overdensity claimed amounts to a factor of about 1.7 and differs from unity at a formal 4.7σ level. Such a high overdensity is much larger than expected from the lensing action of clusters, as noted by RH94, and also cannot be attributed to other effects like foreground obscuration by patchy dust in our galaxy; in addition, this result is in apparent conflict with previous results of Boyle et al. which indicate an anticorrelation of high-redshift UVX QSOs with cluster galaxies. Furthermore, RH94 briefly noted an underdensity of low-redshift QSOs around Zwicky clusters.

In these proceedings we reconsider the association of QSOs with Zwicky clusters, using a different and better selected sample of QSOs (for a more extensive discussion see Seitz & Schneider[7], (hereafter SS95). For multiple reasons, i.e. minimized dust bias, a possible strong magnification bias and a large coverage of the sky (stated in more detail in SS95), we use the 1-Jy catalog of radio sources (Kühr et al.[8]).

CLUSTER AND QUASAR SAMPLES

As quasar sample we use the 1-Jansky catalog (Kühr et al.[8]) in the version of May 1992 provided by M. Stickel and supplemented by additional optical identifications obtained by Stickel & Kühr[9,10]. This catalog contains all radio sources brighter than 1 Jansky at 5 GHz which lie outside the galactic plane ($|b| \geq 10°$) and not towards the Magellanic Clouds. 424 of these sources are optically identified and have spectroscopically determined redshifts. Most high redshift sources ($z \geq 0.5$) are flat spectrum QSOs or BL Lacs, the low-redshift ones are predominatly radio galaxies. For simplicity we refer to all of them as 'quasars' in the next two Sections. Our quasar sample contains all those 236 optically identified sources with determined redshift which have a nonnegative declination, $\delta \geq 0°$; the selection area of these quasars ($|b| \geq 10°$, $\delta \geq 0°$) then covers about 40 percent of the sky. Several subsamples in redshift space were investigated, see Fig. 1(a); in these proceedings we present only results from those no explicit optical flux limit. As in RH94,

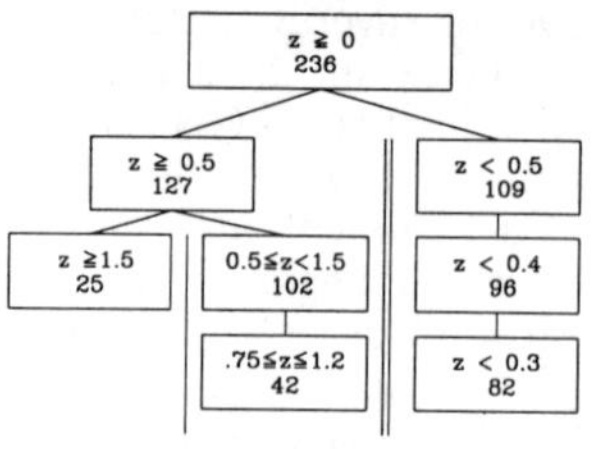

Fig. 1. a) The quasar samples investigated: each box represents one sample containing quasars within the indicated redshift range

Fig. 1. b) The fraction f of the total selection area covered by the association area for clusters with $\delta \geq -3°$ and $|b| \geq 10°$ (solid line) and for clusters with $20° \leq \delta \leq 40°$ and $|b| \geq 20°$ (dashed line)

the clusters of galaxies were taken from the Zwicky Catalog of Galaxies and Clusters of Galaxies[11]; they have been selected in those regions of the sky with $\delta \geq -3°$, $|b| \geq 10°$; hence their selection area slightly exceeds that of the quasar sample; this is advantageous, see SS95. In order to reduce the contribution of nearby clusters, we only admit clusters with a Zwicky radius less than or equal to 15 arcminutes. Thus we obtain 6342 clusters with a *mean value for the Zwicky radius of 8.5 arcminutes*, and a distribution in distance classes shifted to Very Distant and Extremly Distant clusters. In agreement with RH94 we estimate the mean redshift of the cluster sample to be 0.2. For details concerning the correspondence of Zwicky clusters and Abell clusters, distance class-redshift relation or redshift-Zwicky radius relation, see RH94, Sect. 2 and 3, and references therein.

STATISTICAL METHOD AND RESULTS

For the analysis of the correlation we use a similar method as RH94. We define an area associated to the clusters by a disc centered on each cluster with an angular radius θ proportional to the Zwicky radius θ_{ZW} of the cluster. The union of all such discs in $|b| \geq 10°$ and $\delta \geq 0°$ we call 'association area'. If the quasars are not correlated with the clusters, the number density of quasars in this 'cluster area' or 'association area' is expected to be the same as the average number density of quasars. Choosing the radius of the disc proportional to the Zwicky radius of the cluster ensures that for similar clusters at different distances one examines approximately the same physical scale. With Monte-

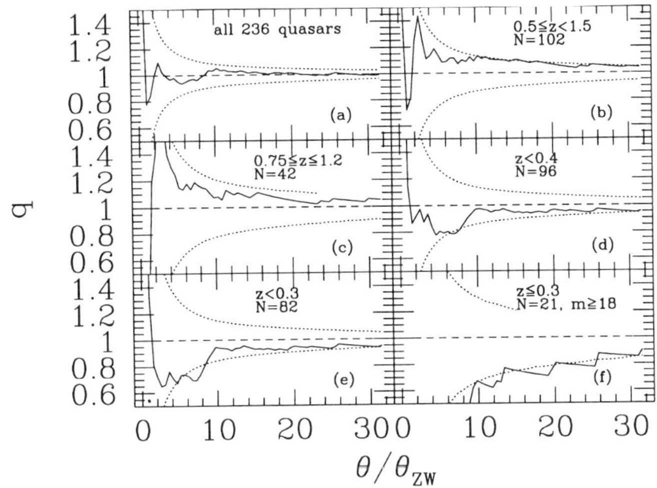

Fig. 2. The density enhancement q (solid curve) is plotted as a function of $x = \theta/\theta_{ZW}$ for several quasar samples: (a): all 236 quasars; (b): all quasars with $0.5 \leq z < 1.5$ (102 quasars); (c): all quasars with $0.75 \leq z \leq 1.2$ (42 quasars); (d): all quasars with $z < 0.4$ (96 quasars). (e): all quasars with $z < 0.3$ (82 quasars). (f): all quasars with $z < 0.4$ and $m \geq 18$ (21 quasars). The dotted curves in each panel correspond roughly to a 98% significance level for a density enhancement or depletion on a given angular scale (or better, given value of x). To obtain these 'significance curves', we have used the binomial distribution which yields the probability that one finds, given a fixed total number N of quasars, $N_a(f)$ quasars in a fraction f of the total area. Since the binomial distribution is discrete, we obtained our smooth significance curves by a linear interpolation of the cumulative distributions $p(\leq n)$ and $p(\geq n)$. If the number N of quasars in a sample is smaller than 78 (57), there exist scales above (below) which there exist no density enhancement (reduction) events which have a probability of two percent or less. Then the upper (lower) enhancement curve is no longer defined on these scales, see c and f. The dashed curve shows the expectation value of q for a random distribution of points. For further details, see the text

Carlo integration we have calculated the fractional association area $f(x = \theta/\theta_{ZW})$, i.e., the ratio of the area covered by circles of angular radius $x\theta_{ZW}$ centered on each Zwicky cluster to the *total selection area of the quasars*, and we plot the result as a function of x in the solid curve shown in Fig. 1b. For small scales, it increases quadratically since the individual cluster discs do not overlap and it saturates only slowy to 1, due to the large incompletness of the Zwicky catalog at low galactic latitudes and a moderate incompleteness at very low and very high declinations; this fact and the clustering of clusters causes the distribution of Zwicky clusters on the sky to be highly non-Poissonian. This improves remarkable for clusters in the area of $20° \leq \delta \leq 40°$ and $|b| \geq 20°$ as can be seen in the dashed curve in Fig. 1(b).

Let N be the number of quasars in a particular subsample, Ω the solid angle of the sky in which the 1-Jy quasars are selected, and $N_a(f)$ the number of quasars within the association area. Then the density of quasars is $n = N/\Omega$, and the densities of associated and non-associated quasars are $n_a = N_a/[f\Omega]$ and $\bar{n}_a = (N - N_a)/[(1-f)\Omega]$, respectively. We define the *density enhancement* q to be

$$q = \frac{n_a}{n} = \frac{N_a}{f\,N} \quad , \tag{1}$$

i.e., the ratio of the QSO density on the association area to the mean QSO density.

In Figs. 2a) to e) we have plotted the density enhancements (solid curves) for the quasar sample and different subsamples for 61 equidistant values of x in the interval $0.5 \leq x \leq 31.5$; all curves are drawn on the same scale to allow for easy comparison. To assess by eye what kinds of events are rare for a random distribution we have included two significance lines: for each value of $x = \theta/\theta_{ZW}$ (or association area), the probability to find a density enhancement equal to or above (below) the upper (lower) dotted curve is less than or equal to 2 percent for a random point distribution. The dashed curve in each panel indicates the expectation value (1) of the density enhancement for a random distribution of points. Since the significance lines are only indicators of rare density distributions we obtained the significance for a deviation of a density distribution from random by Monte Carlo simulations: for a subsample with N quasars and an enhancement curve which is on or above (on or below) the upper (lower) 98% significance curve at j of the 61 grid points, we have calculated the probability, $P_N(\leq j-1)$ that a random distribution with N quasars falls on or above (on or below) this curve at $j-1$ or fewer gridpoints, by using M realizations of N randomly positioned points.

The results are the following:

(A) *Quasars with a redshift of about unity are overdense near to the line of sight to clusters*: the $0.5 \leq z < 1.5$ ($0.75 \leq z < 1.2$) (see Fig. 2b) and c)) quasar sample is above the upper singficance curve on 10 (1) of the 61 gridpoints; with 10000 (1000) MC-simulation this yields a significance for a density enhancement of 97.67% (89.1%). The decrease of the significance for the $0.75 \leq z < 1.2$ sample is due to the strong decrease of quasar number in this sample.

(B) *Quasars with a redshift below 0.4 are avoiding clusters*: The enhancement curve for the $z < 0.4$ ($z < 0.3$) (see Fig. 2d) and e)) sample is below the lower significance line on 3 (7) of the 61 gridpoints; The significance for a density reduction is 93.7% (96.3%). Faint quasars with $m_V \geq 18^{th}$ are extremely underdense near to the line of sight to clusters; this is demonstrated in Fig. 2f): the enhancement curve is below the lower significance curve on 39 of the 61 gridpoints; this translates to a significance of 99.89% for a density depletion.

(C) *The density distribution of quasars with* $z > 1.5$ *is compatible with a*

uniform distribution, (not shown here).
(A) and (C) can be understood in the *lensing picture*: the foreground mass associated with the clusters at $z_{mean} = 0.2$ enhances the number counts of the background quasar population via gravitational lensing; this is most efficient for a source redshift of order unity. However, for $z_{source} \gtrsim 1.5$ the most effective lenses are at higher redshift than the mass distributions in the Zwicky catalog and hence these sources are not expected to be strongly correlated to very low-z mass distributions.
A possible explanation of (B) is an *environmental effect*. For very low redshift, the 'quasars' (which are actually mostly radio galaxies then) are likely to be at the same redshift as the Zwicky clusters. We suggest that the very strong depletion of radio galaxies in clusters on a scale of roughly one degree reflects low fueling efficiency of AGNs in clusters (stripping off the gas of AGN host galaxies due to ram-pressure?). *Other interpretations, such as patchy dust obscuration in the Galaxy or dust in the Clusters of galaxies cannot explain the observed effect and its tendency with redshift.* For a more detailed discussion and for results about the optical flux-limited quasar samples, see SS95.

REFERENCES

1. Ellingson, E., Yee, H.K.C., 1991, ApJ, **371**, 49.
2. Drinkwater, M.J., Webster, R.L., Thomas, P.A. 1991, Crampton, D. (ed.), *The Space Distribution of Quasars*, ASP Conference Series, Vol. 21.
3. Bartelmann, M., Schneider, P. 1994, A&A, **284**, 1.
4. Boyle, B.J., Fong, R., Shanks, T 1988, MNRAS, **231**, 897.
5. Romani, R.W., Maoz, D. 1992, ApJ, **386**, 1992.
6. Rodrigues-Williams, L.L., Hogan, C.J., 1994, AJ, **107**,(2) 451.
7. Seitz, S. & Schneider, P., submitted to A&A (SS95).
8. Kühr, H., Witzel, A., Pauliny-Toth, I. I. K. & Nauber, U. 1981, A&AS, **45**, 367.
9. Stickel, M., Kühr, H., 1993, A&AS,**100**, 395.
10. Stickel, M., Kühr, H., 1993, A&AS, **101**, 521.
11. Zwicky, F., Herzog, E., Wild, P., Karpowicz, M. & Kowal, C. 1961 -68, *Catalogue of Galaxies and Clusters of Galaxies*, California Institute of Technology, Pasadena.

Dark Matter from Quasar Microlensing

M.R.S. HAWKINS

Royal Observatory
Blackford Hill
Edinburgh EH9 3HJ

INTRODUCTION

This paper argues that almost all quasars are being microlensed, so that in addition to short timescale intrinsic variations quasars also vary on a timescale of several years due to microlensing. This hypothesis makes a number of specific predictions related to the incidence and nature of quasar variability, which are tested by analysis of a large sample of quasars which has been monitored in a homogeneous way over 17 years, and by examination of the light curves of individual quasars from the literature. The results of these tests suggest that nearly all quasars are being microlensed by a large population of small compact bodies, and the consequences of this are discussed.

QUASAR LIGHT CURVES

Since 1975 a large scale quasar survey and monitoring programme has been undertaken at the Royal Observatory, Edinburgh using measures from the COSMOS fast measuring machine of a large set of plates from the UK Schmidt telescope at Siding Spring, Australia. Some 200 plates were obtained covering timescales from 1 hour to 17 years in ESO/SERC field 287 at 21h 28m, $-45°$ and galactic latitude $-47°$. The monitoring was done primarily in blue and red passbands, but extensive coverage was also obtained in the U, V and I bands. The survey covered an area of 19 square degrees in the centre of the Schmidt field, containing over 1000 quasars candidates, of which 400 now have redshifts. Complete samples according to well defined criteria were selected using variability, ultraviolet excess and other colour discriminators, giving redshift coverage in the range $0.1 < z < 3.5$.

Light curves were obtained for all quasars over a variety of timescales, in red and blue passbands. A 17 year run of data sampled essentially every year was divided into subsamples on the basis of luminosity and redshift, and a time varying autocorrelation function (ACF) evaluated to examine the dependence of these quantities on timescale. A 10 year run of data in two colours was used to look for any change of colour with luminosity, and a closely monitored six month period was used to look for short term variability.

TESTS FOR MICROLENSING

Examination of the 400 quasar light curves shows variations dominated by undulations on a time scale of several years, with little evidence for flares or eclipses. The overall appearance of the light curves bears a strong resemblance to simulations of variability due to microlensing[1]. The geometrical nature of microlensing results in a number of definite predictions with regard to the nature of the variation, and these can be tested using the quasar sample described above, and obsrvations from the literature.

Time dilation If quasar variability is intrinsic, then the timescale of variability will be subject to a time dilation effect, increasing by a factor of $(1+z)$. Variability from microlensing on the other hand will be subject to time dilation at the redshift of the lens. Turner *et al.*[2] have shown that there is a high probability that the lens will be at a redshft of about 0.5, with only small dependence on the redshift of the source, and so any time dilation effect should be small. FIGURE 1 shows timescale evaluated from an ACF as a function of redshift, in three luminosity bins denoted by different symbols. Also shown as a solid line is the expected effect of time dilation.

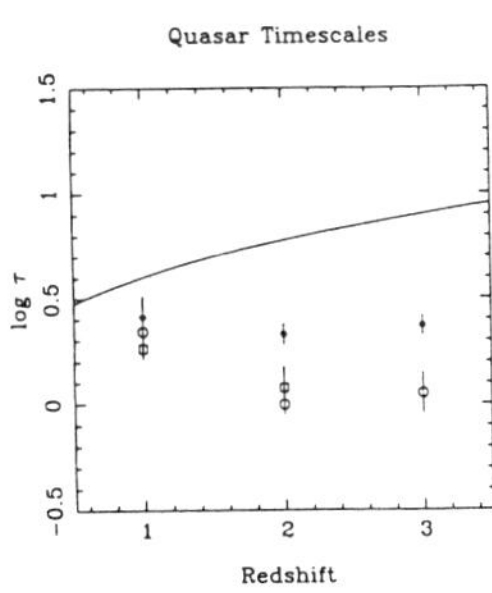

FIGURE 1. Timescale τ as a function of redshift. Squares, open circles and filled circles are for the luminosity ranges $-24 < M_B < -23$, $-25 < M_B < -24$ and $-26 < M_B < -25$ respectively. The solid line is the expected increase of timescale with redshift due to time dilation.

It will be seen that there is no indication of the time dilation effect which would be expected from intrinsic variation. If anything there is some evidence for a decrease in timescale with redshift, which can be accounted for in the microlensing picture. There is also some evidenc for an increase of timescale with luminosity. This is not surprising, as most models of quasar variability would predict this. In the case of microlensing it is due to the increased time taken by the larger disk of a more luminous quasar to cross the line of sight.

Nearby quasars If quasar microlensing is a general phenomenon then sufficiently nearby quasars should not be affected. This is observed in the present sample where the lowest redshift quasars do not show the characteristsic undulating variations, but only the occasional isolated short term flare. Microlensing only becomes dominant above a redshift of about 0.6, in agreement with models of optical depth to gravitational lensing[3]. Independent support for this point comes from the light curve of the luminous nearby quasar 3C273, which has been monitored for many years, and only exhibits the occasional brief low amplitude flare which is presumably intrinsic. There is no sign of the wave like variations which dominate at higher redshifts.

Statistical symmetry of variation Although the observed quasar light curves would not be expected to be symmetrical due to the asymmetric distribution of the lenses along the line of sight, statistically the variation should be symmetric, and it has been shown[4] that for the sample as a whole the rising parts of the light curves mirror the falling parts. In most models of intrinsic variation this would not be the case.

Achromatic light curves Inspection of the light curves for the quasar sample strongly suggests that the variation is achromatic. This can be checked in a more quanitative way by examination of FIGURE 2 which shows the change in B magnitude plotted against tha change in R. It will be seen that the relation is well fitted by a line of unit slope, implying achromatic variation. This is consistent with the predictions of the microlensing model, but not with most models of quasar variability.

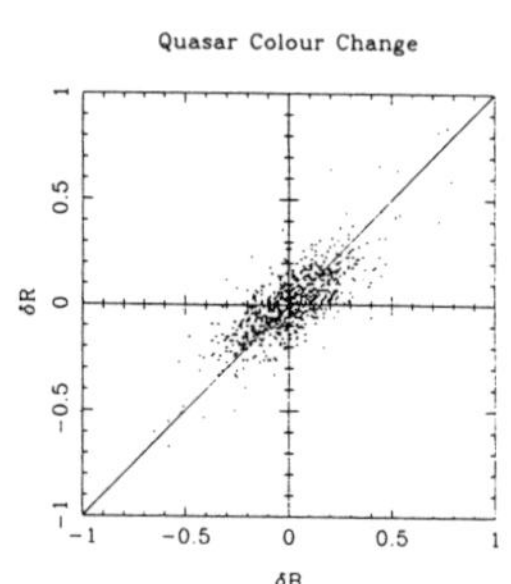

FIGURE 2. Plot of yearly magnitude changes in B versus the corresponding change in R for quasars in the sample. Also shown is a line of unit slope.

Microlensing of macrolensed quasars Quasars which have been split into two images by the gravitational lensing effect of a massive compact object such as a galaxy along the line of sight, form a useful laboratory for the effects of microlensing. For example the quasar Q0957+561 has been closely monitored to detect the time lag for intrinsic variations due to differences in the light travel time to the two images. When these variations are removed there remains a long term change in magnitude for which the only plausible explanation is microlensing[5].

Changes in emission line strength Spectrophotometric monitoring of Seyfert galaxies and quasars over a timescale of weeks shows a change in emission line strength associated with intrinsic changes in the continuum region. However, monitoring of emission line strength versus continuum on a timescale of several years shows essentially no change in line strength for large changes in continuum[6]. This is the effect expected from microlensing, where the continuum region would be small enough to be lensed but the emission line region would not.

NATURE OF THE DARK MATTER

If the hypothesis that all quasars are being microlensed is correct, it is possible to estimate the mass and cosmological density of the population of lensing objects. The mass can be calculated from the timescale of variation which gives the size of the Einstein radius, and from the distribution of amplitudes[7]. These two methods are independent, but both give masses in the range $10^{-3} - 10^{-4} M_{\odot}$. The mass density comes from a result of Press & Gunn[8] which states that if every line of sight is being microlensed then the mass density of the lensing objects (Ω_{λ}) is about equal to the cosmological critical density i.e. $\Omega_{\lambda} = 1$. This is in agreement with the result of Schneider[7] based on the distribution of amplitudes. Constraints on baryon synthesis imply that such a large density of material cannot be baryonic. The most plausible explanation is that the lenses are Jupiter mass primordial black holes formed at the quark hadron transition[9]. These objects would behave like Cold Dark Matter and would be sufficient to make up the dark matter needed for a flat universe.

REFERENCES

1. LEWIS, G.F., J. MIRALDA-ESCUDÉ, D.C. RICHARDSON & J. WAMBSGANSS. 1993. Mon. Not. R. Astron. Soc. **281**: 647.
2. TURNER, E.L., J.P. OSTRIKER & J.R. GOTT. 1984. Astrophys. J. **284**: 1.
3. TURNER, E.L. 1990. Astrophys. J. **365**: L43.
4. HAWKINS, M.R.S. 1993. Nature. **366**: 242.
5. SCHILD, R.E. & R.C. SMITH. 1991. Astron. J. **101**: 813.
6. GONDHALEKAR, P.M. 1990. Mon. Not. R. Astron. Soc. **243**: 443.
7. SCHNEIDER, P. 1993. Astron. Astrophys. **279**: 1.
8. PRESS, W.H. & J.E. GUNN. 1973. Astrophys. J. **185**: 397.
9. CRAWFORD M. & D.N. SCHRAMM. 1982. Nature. **298**: 538.

Cosmological Parameters and Gravitational Lensing Statistics

PHILLIP HELBIG

Hamburger Sternwarte · Gojenbergsweg 112
D-21029 Hamburg · Germany

INTRODUCTION

Lensing statistics gives a method of testing the cosmological constant at intermediate redshifts—at low and high redshifts the possibilities for measuring λ_0 are limited and even at intermediate redshifts λ_0 often cancels out in observable relations—using well-understood lensing theory and standard astrophysical assumptions.

It has recently been suggested by many authors (see, for example, Fukugita et al.[1] and references therein) that gravitational lensing statistics can provide a means of distinguishing between different cosmological models, most effectively concerning the value of the cosmological constant. Kochanek[2] has suggested a method based not on the total number of lens systems but rather on the redshift distribution of known lens systems characterised by observables such as redshift and image separation. Looking at a few different models, he concludes that flat, λ-dominated models are five to ten times less probable than more 'standard' models. The advantage of this method is that it is not plagued by normalisation difficulties as are most schemes involving the total number of lenses.

Since Kochanek was apparently able to get some interesting results using statistics based on only four gravitational lens systems, I wanted to exlpore this more thoroughly by looking at not just a few but all models characterised by λ_0 and Ω_0 as well as varying degrees of homogeneity. I also looked at selection effects and did some simulations to get a handle on what the results mean.

THEORY

I make the 'standard assumptions' that the Universe can be described by the Robertson-Walker metric and that lens galaxies can be modelled as non-evolving singular isothermal spheres (SIS). This leads an equation for the relative differential optical depth[3]

$$\frac{\mathrm{d}\tau}{\mathrm{d}z_{\mathrm{d}}} = (1+z_{\mathrm{d}})^2 \frac{a}{a*}\frac{\gamma}{2}\left(\frac{a}{a*}\frac{D_{\mathrm{s}}}{D_{\mathrm{ds}}}\right)^{\frac{\gamma}{2}(1+\alpha)} \qquad \times$$

$$D_{\rm d}^2 \frac{1}{\sqrt{Q(z_{\rm d})}} \exp\left(-\left(\frac{a}{a*}\frac{D_{\rm s}}{D_{\rm ds}}\right)^{\frac{\gamma}{2}}\right) \tag{1}$$

where $a* := 4\pi\left(\frac{v*}{c}\right)$ ($v* := v$ of an $L*$ galaxy), γ is the Faber-Jackson/Tully-Fisher exponent, α the Schechter exponent, $D_{\rm d}$ the angular size distance between the observer and the lens and

$$Q(z_{\rm d}) := (1+z_{\rm d})^2(\Omega_0 z_{\rm d} + 1 - \lambda_0) + \lambda_0. \tag{2}$$

The optical depth depends on the cosmological model through $Q(z_{\rm d})$ as well as through the angular size distances, because of the fact that $D_{ij} = D_{ij}(z_i, z_j; \lambda_0, \Omega_0, \eta)$. The influence of η, which gives the fraction of homogeneously distributed, as opposed to compact, matter is felt only in the calculation of the angular size distances, whereas the cosmological model in the narrower sense makes its influence felt here as well as through $Q(z_{\rm d})$. The angular size distances can be calculated for arbitrary cosmological models by the procedure given in Kayser & Helbig.[4]

CALCULATIONS

The following gravitational lens systems meet my selection criteria: 0142-100 (UM 673), 0218+357, 1115+080 (Triple Quasar), 1131+0456, 1654+1346 and 3C324. (For more details on these systems see Refsdal & Surdej.[5])

I considered the following ranges of values for the cosmological parameters:

$$\begin{array}{rcccl} -10 & < & \lambda_0 & < & +10 \\ 0 & < & \Omega_0 & < & 10 \end{array}$$

In order to measure the relative probability of a given cosmological model, I defined the quantity f as follows: p

$$0 < f := \frac{\int_0^{z_l} d\tau}{\int_0^{z_s} d\tau} < 1, \tag{3}$$

where z_l is the *observed* lens redshift for a particular system. ($z_{\rm d}$ is used to denote the variable corresponding to lens redshift as opposed to the measured value for a particular lens.) The distribution of the different f values (one for each lens system in the sample) in b equally-sized bins in the interval]0,1[gives the relative probability p of a given cosmological model, with

$$p = \prod_{i=1}^{b} \frac{1}{n_i!} \tag{4}$$

where n_i is the number of systems in the i-th bin. (This definition allows only a few discrete values, of course.) The variable b is a free parameter; since the most information is obtained when b is equal to the number of systems, I adopt this value for b.

RESULTS AND DISCUSSION

My results are in Fig. 1, plot **b**. (This is for $\eta = 0.5$; the results do not depend strongly on η.[3] See Kayser & Helbig[4] for a discussion of this parameter.) One can see basically that areas of equal probability occur in some fashion which is not stochastic. Although there are only a few discreet values for the probability as defined in Eq. (4), nevertheless one sees a degeneracy—there is a wide range of cosmological models for a given probability. Plot **c** shows the result of neglecting the observational bias, *e.g.* assuming that the lens could have its redshift measured whatever this redshift were. As a comparison with plot **b** shows, this leads to a bias against models with a high median expected lens redshift—those near the de Sitter model.

For comparison, I have also tested the method on the systems used by Kochanek,[2] using $m_{\rm lim} = \infty$ und $\eta = 1$, both of which he implicitly assumes. (Of course, when one considers finite values for $m_{\rm lim}$, one cannot include systems with lens redshifts which have been determined by means other than measured emission redshifts, such as absorption lines (which assumes that the lens is also the absorber).) The results are in plot **d** where the relative probabilities are 0, $\frac{1}{6}$, $\frac{1}{2}$ and 1 and comparing the various models examined by Kochanek confirm his conclusions. For instance, the relative probabilities of the Einstein-de Sitter and de Sitter model are 1 and $\frac{1}{6}$, confirming his result that flat, λ-dominated models are 5–10 times less probable than standard ones. (However, taking $m_{\rm lim}$ into account and/or using only directly measured lens redshifts would produce quite different results, as discussed above.) This plot artificially indicates a low probability for models near the de Sitter model for the same reasons as those discussed in connection with plot **c**.

NUMERICAL SIMULATIONS

For the numerical simulations, the observables θ'' (the radius of the Einstein ring or *half* the image separation corresponding to the diameter of the Einstein ring), z_s and galaxy type were chosen randomly from an interval roughly corresponding to the observed range of values in order to produce synthetic data comparable to real observations. For a given cosmological model, the corresponding lens redshift z_l for each system was calculated from the observables and a randomly generated f through (numerical) inversion of Eq. (3). This catalog was then used to determine a relative probability for each of the points in the λ_0-Ω_0 plane in the same manner as for the real systems.

The Kolmogorov-Smirnov (K-S) test is of course a well understood method for testing if two distributions are statistically significantly different. (See, e.g., Press et.al.[6] for a general discussion and definition of the K-S probability.) However, this test can only be used for distributions with more than ≈ 20 data points. Therefore I plot in Fig. 1 in plots **b**, **c**, and **d** the probability given by Eq. (4) and in plot **e** the K-S probability.

I have done simulations for a variety of world models and also for numbers of systems between 20 and 50. In the interest of brevity, I present only one plot. Plot **e** in Fig. 1 shows the results derived from a catalogue of simulated gravitational lens systems. Since, even with 50 systems, no area can be excluded based on the K-S

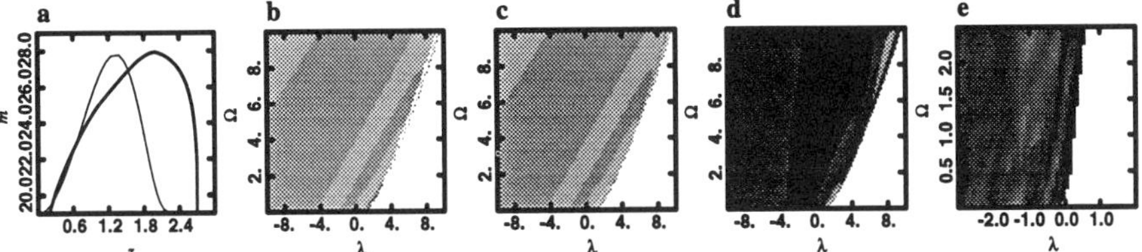

Figure 1: **a.** The relative differential optical depth (thin curve) and the calculated lens brightness m (thick curve) as functions of z_d. The world model is the de Sitter model ($\lambda_0 = 1.0, \Omega_0 = 0.0$) and the observables are those for the gravitational lens system 0142−100. The ordinate gives the magnitude in Johnson R. For realistic limiting spectroscopic magnitudes ($\approx 24^{\mathrm{m}}$) it is clear that one cannot sample the probability distribution without a strong bias. **b.** Relative probability for the systems mentioned in the text, with a realistic value for m_{lim}. Here and in the other halftone plots, the relative probability increases linearly from white to black. **c.** The same as b., but *neglecting* m_{lim}, which, as in plot d., makes the models near the de Sitter model appear more probable than they are. **d.** Relative probability for the systems used by Kochanek. **e.** Results based on a catalogue of 50 simulated systems. (Note the different scale on the axes.) The cosmological model used to generate the lens redshifts is the homogeneous Einstein-de Sitter model ($\eta = 1.0, \lambda_0 = 0.0, \Omega_0 = 1.0$).

probability—the white area has $p = 0$ due to the fact that at least one lens would be fainter than m_{lim} in these world models, as discussed in Helbig & Kayser[3]—I conclude that, although one can qualitatively understand the physics which at least in part is responsible for the results presented in Fig. 1, the actual relative probabilities are more indicative of intrinsic scatter in the redshifts of the lenses than a hint of the correct cosmological model.

SUMMARY AND CONCLUSIONS

For known gravitational lens systems the redshift distribution of the lenses was compared with theoretical expectations for 10^4 Friedmann-Lemaître cosmological models, which more than cover the range of possible cases. The comparison was used for assigning a relative probability to each of the models. However, my simulations indicate that a reasonable number of observed systems cannot deliver interesting constraints on the cosmological parameters using this method. Therefore, it seems that lensing statistics can tell us something about the cosmological model only if one makes use of all information, which means coming to grips with normalisation difficulties.

REFERENCES

1. Fukugita, M., K. Futamase, M. Kasai & E.L. Turner. 1992. ApJ **393**; 1.
2. Kochanek, C.S. 1992. ApJ **384**; 1.
3. Helbig, P. & R. Kayser. 1995. A&A submitted.
4. Kayser, R. & P. Helbig. 1995. A&A submitted.
5. Refsdal, S. & J. Surdej. 1994. Rep. Prog. Phys. **56**; 117.
6. Press, W.H., B.P. Flannery, S.A. Teukolsky & W.T. Vetterling. 1986. *Numerical Recipes.* Cambridge: Cambridge University Press, p. 469.

Gravitational Lensing with Polarization to Determine Galaxy Masses

C.C. DYER[a], P.P. KRONBERG[a], R.A. PERLEY[b] and H.-J. RÖSER[c]

Department of Astronomy[a], University of Toronto
60 St. George St., Toronto, Ontario, Canada

National Radio Astronomy Observatory[b]
P. O. Box 0, Socorro, NM 87801

Max-Planck-Institut für Astronomie[c]
Königstuhl 17, D69117, Heidelberg, Germany

INTRODUCTION

Radio jets frequently exhibit the property that the polarization angle is closely coupled with the local axial direction along the jet. This provides the link between the morphology and the polarization angle that allows our technique to function. Radio jets are also of particular interest because they usually have dimensions that allow their images to span the entire projected image, hence most or all of the mass scale, of an intervening galaxy, thus providing a *range* of impact parameters with the intervening lensing mass distribution.

Up to now, most gravitational lensing observations have consisted in searches for multiple imaging of single, "point" optical images of QSO's, multiple imaging of extended sources, such as the "double quasar" 0957 + 561 (Walsh *et al.*[1]), radio ring images (*e.g.* MG1131 + 0456 — Hewitt *et al.*[2]), optical arc images of galaxies due to foreground galaxy clusters (Soucail *et al.*[3]), or microlensing amplification of stellar images. All of these phenomena are the result of multiple imaging, rather than a milder distortion due to a mass intervenor.

Kronberg *et al.*[4] proposed, and applied (to the jet of 3C9) a technique in which an intervening mass can be measured *without multiple imaging.* Their method utilizes the fact indicated above – that a polarized radio jet represents a spatially contiguous array of background sources, whose overall dimension is of comparable size to the galaxy's mass scale, and in which there is a coherence between the local polarization orientation, and local jet direction.

The Optics of Linearly Polarized Rays

An important property of a single linearly polarized ray is that its polarization vector does not undergo a rotation under gravitational deflection if the deflecting, or lensing, gravitational field is generated by any bounded, non-relativistically rotating mass distribution. This approximation applies to all galaxy scale masses. Thus, whereas an intervening mass can bend a jet away from its intrinsic projected shape, the polarization orientation at each point along the jet remains *unchanged.* This results in a alignment-breaking between the polarization and the orientation of a gravitationally distorted jet segment. This alignment breaking can be detected even in the absence of multiple imaging.

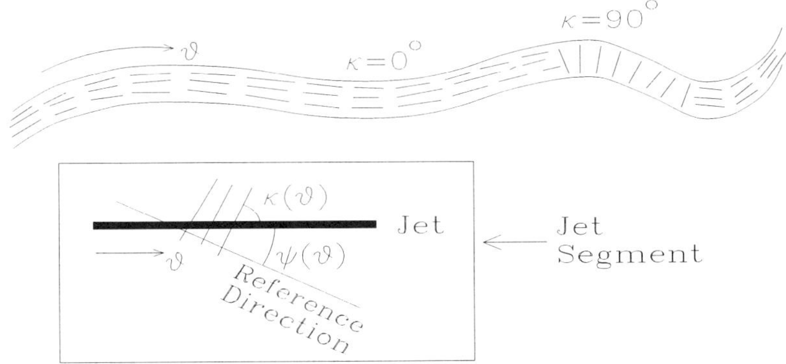

Figure 1: Schematic representation of a radio jet as a locus of linearly polarized emitters, in which the orientation of the plane of polarization is usually, although not necessarily always, coupled to the local jet direction. Intrinsic deviations from coupling are defined by $\kappa(\theta)$ as illustrated, in contrast to gravitationally-induced decoupling which is defined by $\eta_G(\theta)$.

More fundamentally, we distinguish here between two kinds of vectors; namely "single ray" vectors (*e.g.* polarization vectors) which characterise a single ray, and "connecting vectors" (sometimes called two-point vectors) – which connect two null geodesics by specifying their separation in the image plane. A point polarized source, described by a single-ray vector, is simply translated in the image plane by a lens. Its neighbour, elsewhere along the jet, is translated in a different direction in the image plane, but both of their polarization orientations remain unchanged, i.e. they are *not* rotated. Since the connecting vector between these two points on a lens-distorted jet will differ from what it would have been if there were no mass intervenor, any original alignment between the polarization angle (a ray vector) and the local jet orientation (a connecting vector) will be changed by the mass intervenor.

Thus it is appropriate to define a *gravitational alignment-breaking parameter*, which we define as

$$\eta_G(\theta) = \psi(\theta) - [\chi_o(\theta) + 90^\circ] + \kappa(\theta) \tag{1}$$

where $\psi(\theta)$ is the orientation of a jet segment at position θ along the jet, $\chi_o(\theta)$ is the intrinsic polarization direction, and $\kappa(\theta)$ represents any deviation, or "decoupling" of $\chi_o(\theta)$ from the local jet direction (see Figure 1). In an ideal jet, for our purposes, κ should be zero. In practice, the polarization angle in some jets occasionally does a sudden flip from being perpendicular to being parallel to the local jet direction. In addition, irregular deviations, and the effect of noise in the observations will also contribute to non-zero κ. $\kappa(\theta)$ variations can normally be "filtered out" of any truly gravitational effect, since gravitational intervenors cannot cause discontinuities (except possibly where there is multiple imaging). In general, variations of $\kappa(\theta)$ will have spatial frequency components which are unrelated to those which are largely predetermined by the observed lens galaxy position and redshift (see Figure 2). $\chi_o(\theta)$

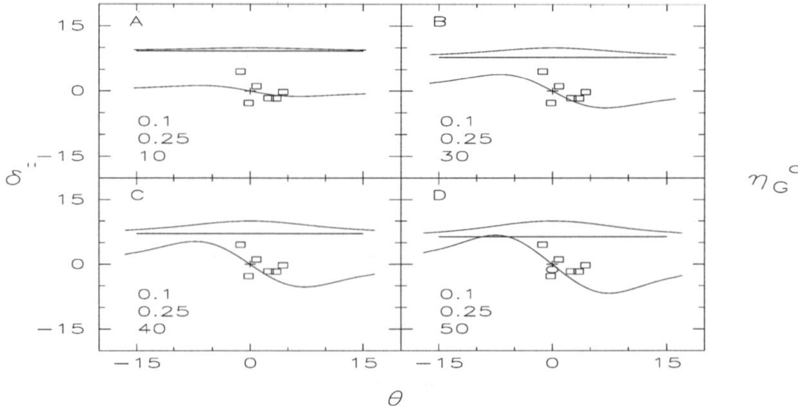

Figure 2: Shown at the top of each panel is the distortion of an originally straight jet due to an intervening King galaxy at the + position, which is also the origin of the S-shaped $\eta(\theta)$ curve, representing the alignment breaking induced by the lens. Model parameters shown in the lower left of each panel are q_o, z_{lens}, and m_G in units of 10^{11} solar masses.

requires that the measured polarization angles along the jet be corrected to "zero wavelength" (hence the "$_o$" subscript) so that any Faraday rotation is removed.

The Effect on Polarization of Non-Sphericity of the Lensing Mass

It is of interest to consider physical conditions in a mass intervenor which *could* cause a rotation of the intrinsic plane of polarization in the observer's frame. The first is the case of a rapidly rotating mass, which requires a Kerr metric to describe the distortion of its local space-time. The propagation of polarized electromagnetic waves through a Kerr space was investigated by Pineault and Roeder[5,6], who showed that a Kerr space-time can rotate the polarization of the lensed image. However, for galaxy-scale masses, Kerr metric terms are negligible, since the terms associated with angular acceleration are negligible compared with those associated with radially symmetric gravitational acceleration. Thus, measurable rotation will only occur for light rays that pass very close to, *e.g.*, a rapidly rotating black hole (*cf.* also Dyer and Shaver[7]).

Another more realistic example case of a non-rotating lens is a non-spherical mass distribution – *e.g.* of a highly flattened intervening galaxy. Dyer and Shaver[7] have shown that, also in this case any rotation by the lens is immeasureably small for galaxy-like lenses, indeed for almost any galaxy scale situation of astrophysical interest.

Multiple Lensing, Loops, and Mass Distributions

In the case where the lensing is strong enough to create multiple images of any portion of the jet, significantly more information can be obtained. Assuming that the lensing is strong enough to produce secondary/tertiary image pairs of parts of the jet, and that the ends of the jet are distant enough not to be multiply imaged, the locus of the secondary/tertiary images will form a closed loop opposite to the jet. The polarization and intensity properties of this loop can be determined from the original jet properties together with the postulated mass model properties. A form of closure is available in this situation, and furthermore, a given mass model can be studied *over significant ranges of radius*. In particular, information about the core structure of the lensing mass distribution is available, since photons from the loop have traveled through the inner regions of the lens. It is important to note that while there are many mass models available for the core and halo, this technique provides a direct consistency link between these two components of the mass distribution, since we have a definite relation between the polarization of the radiation from the long jet and the closed loop.

A useful parameter that will become available when a secondary/tertiary loop structure is observed is an estimate of the ratio of the core mass to the halo mass of a galaxy. The choice of mass distribution for the lens model crudely reflects this ratio, but a detailed choice among standard mass distributions, such as those of King, deVaucouleurs, or Bergeron and Gunn, is not required. Our approach is to allow an arbitrary mix of mass distributions and their parameters, with the aim of being able to determine such key parameters such as the core-halo mass ratio.

In particular, the technique provides mass distribution information at two distinct ranges of impact parameter. The primary image of the jet provides information most closely related to the total mass and halo mass of the galaxy. The secondary-tertiary loop, when it occurs, provides information on the mass distribution near the core of the galaxy. Our technique thus has the potential for providing mass information at both scales. The observation of such secondary/tertiary loops also have the potential of ensuring consistency in the interpretation of η_G since, given adequate sensitivity, the relationship between the polarization and morphology in two distinct image regimes can be compared.

Acknowledgements

C.C.D and P.P.K acknowledge research grant support from the Natural Sciences and Engineering Research Council of Canada (NSERC).

Bibliography

1. Walsh, D., Carswell, R.F., Weymann, R.J.: 1979, *Nat* **279**, 381
2. Hewitt, J.N., Turner, E.L., Schneider, D., Burke, B.F., Langston, G.L., and Lawrence, C.R.: 1988, *Nat* **333**, 537
3. Soucail, G., Fort, B., Mellier, Y., Picat, J.P.: 1987, *A&A* **243**, 23
4. Kronberg, P.P., Dyer, C.C., Burbidge, E.M., Junkkarinen, V.T.: 1991, *Ap J* **367**, L1
5. Pineault, S., Roeder, R.C.: 1977a, *Ap J* **212**, 541
6. Pineault, S., Roeder, R.C.: 1977b, *Ap J* **213**, 548
7. Dyer, C.C., and Shaver, E.G.: 1992, *Ap J* **390**, L5

AGAPE, an experiment to detect MACHO's in the direction of the Andromeda galaxy

R. Ansari[a], M. Aurière[b], P. Baillon[c], A. Bouquet[d], G. Coupinot[b], C. Coutures[e], C. Ghesquière[f], Y. Giraud-Héraud[f], P. Gondolo[d], J. Hecquet[b], J. Kaplan[d], A.L. Melchior[d], M. Moniez[a], J.P. Picat[b] and G. Soucail[b]

[a] *LAL, Université Paris Sud, Orsay, France,*
[b] *Observatoire Midi-Pyrénées Bagnères de Bigorre et Toulouse, France,*
[c] *CERN, Genève, Switzerland,* [d] *LPTHE, Universités Paris 6 et 7, France,*
[e] *DAPNIA, CEN Saclay, France,* [f] *LPC Collège de France, Paris, France.*

The M31 galaxy in Andromeda is the nearest large galaxy after the Small and Large Magellanic Clouds. It is a giant galaxy, roughly 2 times as large as our Milky Way, and has its own halo. As pointed by A. Crotts[1] and independantly by some of us[2] M31 provides a rich field of stars to search for MACHO's in galactic halos by gravitational microlensing[3]. M31 is a target complementary to the Large Magellanic Cloud and the galactic bulge wich are used by the three current experiments[4–6]. It is complementary in that it allows to probe the halo of our galaxy in a direction very different from that of the LMC. Moreover, the fact that M31 has its own halo and is tilted with respect to the line of sight provides a very interesting signature : assuming an approximately spherical halo for M31, the far side of the disk lies behind a larger amount of M31 dark matter, therefore more microlensing events are expected on the far side of the disk. Such an asymmetry could not be faked by variable stars[1].

In other words, M31 seems very appropriate to detect brown dwarfs through microlensing. However, as very few stars of M31 are resolved, we had to develop an approach to look for microlensing by monitoring the pixels of a CCD, rather than individual stars[2]. The AGAPE collaboration has set out to implement this idea.

MONITORING PIXELS

In the case of a crowded field such as M31, the light flux F_{pixel} on a pixel comes from the many stars in and around it, plus the sky background. The light flux of an individual star, F_{star}, is spread among all pixels of the seeing spot and only a fraction of this light, $F_{\text{pixel}} = \{\text{seeing fraction}\} \times F_{\text{star}}$, reaches the central pixel. If the star luminosity is amplified by a factor A, the pixel flux increases by :

$$\Delta F_{\text{pixel}} = (A-1)\ \{\text{seeing fraction}\}\ F_{\text{star}} \tag{1}$$

The amplification of the star luminosity allows an event to be detected if the flux on the brightest pixel rises sufficiently high above its rms fluctuation σ_{pixel} :

$$\Delta F_{\text{pixel}} > Q\ \sigma_{\text{pixel}} \tag{2}$$

Typically, in our simulations, we require Q to be larger than 3 during 3 consecutive exposures and larger than 5 for at least one of them.

EXPECTED STATISTICS.

We have performed numerical simulations using the above detection criterium. As anticipated, the number of events we expect to be able to detect depends strongly both on the stability of the pixel and on the average seeing. The numbers in Table 1 below assume a two meter telescope, a field of view of 60 by 20 arcminutes (planed for a second generation experiment) centered on the center of M31, 1 arcsecond pixels, 30 minutes exposures, and 120 consecutive nights. The brown dwarf mass is taken to be 0.08 $M_{\odot}$, and we assume a standard halo (see reference[2] for instance) with a local dark matter density of 0.3 GeV/cm^3 (0.0075 $M_{\odot}/pc^3$) and a core radius of 5 kpc.

Table 1: Expected number of events under various observing conditions

seeing	$\sigma_{\text{pixel}}/F_{\text{pixel}}$	number of galactic events	number of M31 events	total
1"	2%	9	15	24
	1%	15	29	44
	0.5%	27	50	77
2"	2%	3	4	7
	1%	7	10	17
	0.5%	11	21	32

These numbers have to be compared with the 2 events per year expected by the EROS collaboration for brown dwarfs with masses of order 0.1 $M_{\odot}$.

It is clear from Table 1 that the number of detectable events depends crucially on the relative flux fluctuation on the pixel, $\sigma_{\text{pixel}}/F_{\text{pixel}}$. To study the feasibility of the experiment, we have analyzed these pixel fluctuations in three series of real data (Table 2) :

• 1) 82 images of the Large Magellanic Cloud (LMC) taken by the EROS collaboration

• 2) 26 images of M31 taken with the one meter telescope at Pic du Midi, in collaboration with F. Colas (Bureau des longitudes, Paris) and J. Lecacheux (DESPA, Meudon)

Table 2: The mean relative fluctuation obtained for the three series of images listed above. The numbers in the first column refer to the numbers in the list.

Images	mirror size (meter)	pixel size (arcsec)	relative fluctuation
LMC EROS (• 1)	0.4	1.15	3%
M31 Pic (• 2)	1	0.7	1%
M31 Pic (• 3)	2	0.25	0.7%
M31 Pic (• 3)	2	1 ("superpixels")	0.23%

• 3) 4 images of M31 taken by E. Davoust (OMP, Toulouse) with the 2 m telescope at Pic du Midi. In this case, the angular size of the pixels is very small (0.23"), and we have also considered a rearrangement in 4×4 "superpixels".

The results given in Table 2 clearly show that the required photometric stability of pixels can be reached. Moreover, the analysis of the third series of data shows that pixels small compared to the seeing allow an efficient matching between images, whereas, once images are matched, superpixels are more appropriate for a stable photometry.

DISCRIMINATING AGAINST VARIABLE STARS AND OTHER VARIABILITIES

Variable stars should be the main background. The usual tools to discriminate against this background are available (symmetry, unicity, achromaticity of the light curve). Still, some points particular to our approach are discussed below.

Achromaticity. At first sight, one would think that there is no achromaticity as a star rising above a background of a different color will cause a color variation of the pixels involved. However, it is easy to show that when a star rises above the background in two color bands (say red and blue) then the ratio

$$\frac{(F_{pixel} - \langle F_{pixel} \rangle)_{red}}{(F_{pixel} - \langle F_{pixel} \rangle)_{blue}} = \frac{F_{star}|_{red}}{F_{star}|_{blue}} \quad (3)$$

is constant in time during a microlensing event.

High amplifications. To rise above the background an unresolved star needs a rather high amplification ($\langle A \rangle \sim 6$), which will exclude most variable stars.

On the other hand small secondary maxima indicating unstable stars with occasional strong flares will not be discriminated. Further study is required in this respect.

THE FIRST RUN OF AGAPE

We were given 57 half nights of observation on the 2 meter telescope "Bernard Lyot" at Observatoire du Pic du Midi in the French Pyrénées, from September 29 to November 24 1994. The field was $8' \times 8'$ only, covered by 4 exposures on a 800×800 thin Tektronix CCD camera with pixels 0.3" wide.

The data of this prototype run are currently under treatment. A key step of this treatment is the alignement of successive images both in position and in photometry. The photometric alignement is performed by linearly transforming the light flux of one image in such a way that the mean flux and the variance of the transformed image matches those of some reference image. The result of this alignement between images is illustrated in figure 1. After alignement the dispersion of the relative difference between the two images is 1.6%. This preliminary result is very encouraging as our alignement procedures are not yet optimized.

AKNOWLEDGMENTS We thank The EROS collaboration, and E. Davoust who allowed us to use their data, as well as F. Colas, and J. Lecacheux with whom we took data on the 1 meter telescope at Pic du Midi. The help of F. Colas during our first observation run has been particularly appreciated.

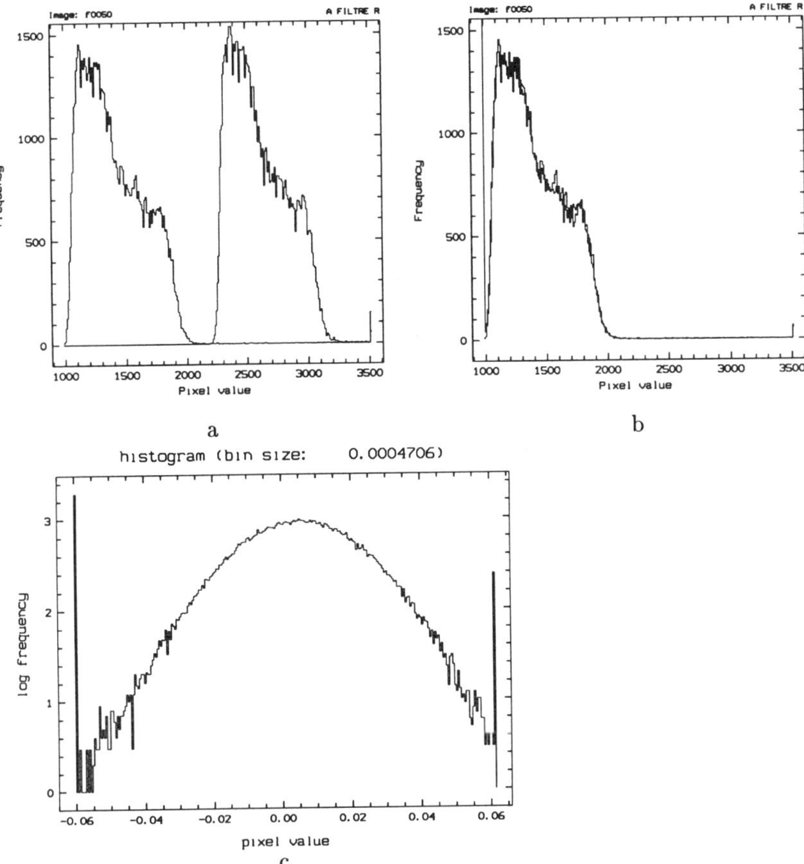

Figure 1: photometric alignement of two different images of the same field. Histogramms of pixel flux of two images, a: before treatment, b: after photometric aligement. c: Histogram of the difference of pixel fluxes between the two images, after photometric alignement.

References

[1] Crotts A. P. S. 1992. ApJ. 399; L43.

[2] Baillon P. Bouquet A. Giraud-Héraud Y. & Kaplan J. 1993. A&A 277; 1.

[3] Paczyński B. 1986. ApJ. 304; 1.

[4] Alcock et al., 1993. Nature 365; 621, and these proceedings.

[5] Aubourg E. et al. 1993. Nature 365; 623.

[6] Udalski A. et al. 1993. Acta Astronomica 43; 289.

THE CANADA-FRANCE REDSHIFT SURVEY

O. LE FÈVRE*
Dominion Astrophysical Observatory, National Research Council of Canada
D.A.E.C, Observatoire de Meudon, 92195 Meudon, France

S. J. LILLY*
Department of Astronomy, University of Toronto, Toronto, M5S 1A7, Canada

D. CRAMPTON*
Dominion Astrophysical Observatory, National Research Council of Canada
Victoria V8X 4M6, Canada

F. HAMMER*, and L. TRESSE*
D.A.E.C, Observatoire de Meudon, 92195 Meudon, France

ABSTRACT

The Canada-France Redshift Survey (CFRS), conducted with the CFHT MOS spectrograph, has provided secure redshifts for ~600 galaxies with $0\leq z \leq 1.3$ of which more than 350 are at $z\geq 0.5$. This unique data set allows to investigate the evolution of the luminosity function of galaxies over timescales ~50% the present age of the universe. We find that the luminosity function of red galaxies does not change much over the redshift range of the CFRS, while the luminosity function of blue galaxies is strongly evolving, consistent with a ~1 magnitude brightening at z~0.75 (but equivalently described by an increase in comoving density). Morphological information with HST will help to establish whether the evolution we observe is due to simple brightening of individual objects with look-back time.

1. Introduction

The study of the evolution of galaxies requires information about the population of galaxies at different epochs. Deep redshift surveys with efficient multi-slit spectrographs have recently allowed to build the statistical samples needed, and a number of redshift surveys have been carried out (Broadhurst et al., 1988, Colless et al., 1990, 1993, Lilly et al., 1991, Lilly, 1993, Tresse et al., 1993, Songaila et al., 1994). The initial survey galaxies were selected from the B band to understand why the galaxy number counts are steeper than predictions. However, at high redshift the B-band samples the rest frame UV and samples are then dominated by the most active objects. Moreover, k(z) corrections are more challenging to apply to B selected galaxies when redshifts become significantly higher than 0.3, as they are spanning 3 magnitudes at z~0.8. A deep survey selected from the I-band, which sample galaxies' older stellar populations for redshifts up to unity, allows to use smaller k(z) corrections. The CFRS is aimed at constructing an unbiased sample of 1000 objects selected solely from their apparent magnitude $17.5\leq I_{AB} \leq 22.5$, and to secure redshift of galaxies in the range $0\leq z \leq 1$.

---

* Visiting Observer with the Canada-France-Hawaii Telescope, operated by the NRC of Canada, the CNRS of France and the University of Hawaii

2. The CFRS sample

Deep imaging of 5 high galactic latitude fields was obtained with the FOCAM camera at CFHT. Careful photometry has allowed to construct catalogs complete down to I_{AB}=23 and central surface brightness $\mu_{AB}(I)$ ~24.5, sufficient to include normal surface brightness galaxies and extreme low surface brightness galaxies (Lilly et al., 1995, CFRS-I). Multi-object spectroscopy was conducted with the MOS spectrograph at CFHT (Le Fèvre et al., 1994) on more than 1000 objects selected with $17.5 \leq I_{AB} \leq 22.5$, with high multiplexing gains of up to ~80 slits per observations (Le Fèvre et al., 1995, CFRS-II). Very strict procedures have been followed to establish a reliable list of final spectroscopic measurements with fully independant processing of the data carried out by three members of the team for each spectroscopic data set. Final redshifts were assigned only after the careful comparison of the independant measurements and a confidence scheme was established.

The final spectroscopic catalog contain 200 stars, 591 galaxies with secure redshifts in the range $0 \leq z \leq 1.3$, 6 QSOs, and 146 objects with uncertain or unknown redshifts, leading to an overall success rate of identification of 85%. Data are presented in Le Fèvre et al. (CFRS-II), Lilly et al. (CFRS-III) and Hammer et al. (CFRS-IV). Examination of the overall properties of the sample showed that most unidentified objects are likely to be galaxies, and that half of them have redshifts similar to the redshifts of the identified galaxies, reducing the fraction of truly unknown objects to ~7% (Crampton et al., 1995, CFRS-V). The remaining unidentified galaxies have been demonstrated to be most probably at the high redshift end of our sample.

3. The evolution of the luminosity function

The 591 galaxies with secure redshifts constitute a unique sample with $< z >= 0.56$, increasing the number of known galaxies with z≥0.5 by an order of magnitude to more than 350 in the CFRS. This allows us to construct the luminosity function Φ(M,color,z), using a $1/V_{max}$ formalism (Lilly et al., 1995, CFRS-VI). No change is observed with epoch in the LF of the galaxies redder than a present day Sbc galaxy (Coleman et al., 1980). This probably implies that the population of ellipticals and bulge dominated galaxies is roughly constant with epoch, although one can imagine a combination of evolutionary processes that could produce the appearence of no-evolution.

The right panels in Figure 1 show the LF for galaxies bluer than the Sbc SED of Coleman et al. As redshift is increasing the evolution of the LF is clearly apparent and can be characterized by a brightening of about 1 magnitude between $0.2< z <0.5$ and $0.5< z <0.75$, with a saturation at bright magnitudes and a further increase in luminosity of ~ 1mag. at $M_{AB}(B)$ ~−20 for $0.75< z <1$. With the large redshift baseline of the CFRS sample, the evolution of the LF is evident within our sample and independant of uncertainties in the definition of the local population.

The evolution of the spectrophotometric properties with cosmic epoch is outlined in Figure 2, which shows an increase in the rest-frame equivalent width of [OII]3727A. A detailled analysis of the galaxies in the redshift range $0.05< z <0.3$, for which lines from

[OII] to Hα can be measured has been conducted (Tresse et al., 1994). Classical diagnostic diagrams, [OIII]5007/Hβ vs. [SII]6725/Hα and [OIII]5007/Hβ vs. [OII]3727/Hβ, have been used to show that ~20% of the galaxies in this redshift range have properties intermediate between SeyfertII and LINER galaxies, compared to 2% locally. This may indicate that nuclear activity is closely associated with the rapidly evolving blue population.

4. Morphology

To solve the ambiguity between luminosity evolution and number density evolution to explain the observed evolution of the LF, morphological information is of critical importance. With sufficient spatial resolution one hopes to measure the sizes of discs and bulges and the associated surface brightness, which would provide indication on the location and amount of brightening, and gain information on the frequency of recent or on-going mergers. To this effect, we have recently obtained deep HST images of ~30 galaxies in the CFRS sample. While the data is being processed (Schade et al., 1995, CFRS-IX), we are confident that we can measure from HST data Bulge/Total luminosity

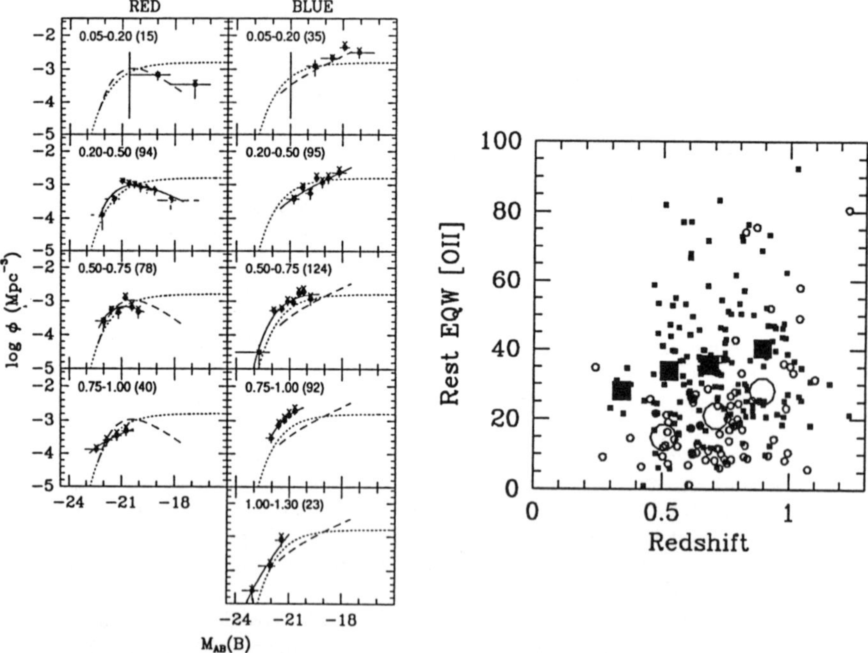

Fig. 1 (left). Luminosity functions for two color classes (redder or bluer than present-day Sbc), and 5 redshift ranges (Lilly et al., 1995, CFRS-VI). The Loveday et al. (1990) LF is drawn as a doted curve. The fit to the LF in the redshift range $0.2 \leq z \leq 0.5$ has been drawn for reference as a long-dash curve in the other redshift panels.

Fig. 2 (right). Increase of the [OII] equivalent width with redshift for galaxies redder (circles) and bluer (squares) than present day Sbc. The bigger symbols are the average EQW for these two populations.

ratio, disc and bulge scale length and central surface brightness up to redshifts close to unity. To allow for statistical analysis, the CFRS has joined forces with the group of Broadhurst, Colless, Ellis et al. to observe a sample of more than 200 galaxies during HST cycle5.

5. Summary

The CFRS sample provides for the first time a sample of 591 galaxies selected with $17.5 \leq I_{AB} \leq 22.5$, 350 of which are at z>0.5. This allows to explore the properties of normal galaxies when the Universe was 50% of its present age. The luminosity function of galaxies redder than present day Sbc does not seem to change with epoch. However, the LF of galaxies bluer than present day Sbc evolves dramatically with redshift, showing a brightening of 1 magnitude between z~0.3 and z~0.8, equivalent to an enhancement in the numbers of galaxies brighter than $M_B \sim -21$ by a factor ~3. Observation with the HST are on-going and will provide key parameters to evaluate the contribution of luminosity evolution and number evolution to the observed evolution of the galaxy luminosity function.

References

Broadhurst, T.J., Ellis, R.S., Shanks, T., 1988, MNRAS, 235, 827.

Coleman, G.D., Wu, C.C., Weedman, D.W., 1980, ApJ.Supp., 43, 393.

Colless, M.M., Ellis, R.S., Taylor, K., Hook, R.N., 1990, MNRAS, 244, 408.

Colless, M.. Ellis, R., Broadhurst, T., Taylor, K., Peterson, B., 1993, MNRAS, 261, 19.

Crampton. D., Le Fèvre, O., Lilly, S.J., Hammer, F., 1995, ApJ, submitted (CFRS-V).

Hammer, F., Crampton, D., Le Fèvre, O., Lilly, S.J., 1995, ApJ, submitted (CFRS-IV).

Koo, D.C.. Gronwall, C., Bruzual, G., 1993, ApJ.Lett., 415, L21.

Le Fèvre, O., Crampton, D., Felenbok, P., Monnet, G., 1994, A&A, 282, 325.

Le Fèvre, O., Crampton, D., Hammer, F., Lilly, S.J., Tresse, L., 1995, ApJ, submitted (CFRS-II).

Lilly, S.J., Cowie, L.L., Gardner, J.P., 1991, ApJ, 369, 79.

Lilly, S.J., 1993, ApJ, 411, 501.

Lilly, S.J., Le Fèvre, O., Crampton, D., Hammer, F., Tresse, L., 1995, ApJ, submitted (CFRS-I).

Lilly, S.J., Hammer, F., Le Fèvre, O., Crampton, D., 1995, ApJ, submitted (CFRS-III).

Lilly, S.J., Tresse, L., Hammer, F., Crampton, D., Le Fèvre, O., 1995, ApJ, submitted (CFRS-VI).

Loveday, J., Peterson, B.A., Efstathiou, G., Maddox, S.J., 1992, ApJ, 390, 338.

Schade, D., et al., 1995, ApJ, in preparation (CFRS-IX).

Songaila, A., Cowie, L.L., Hu, E.M., Gardner, J.P., 1995, ApJ, in press.

Tresse, L., Hammer, F., Le Fèvre, O., Proust, D., 1993, A&A, 277, 53.

Tresse, L., Rola, C., Hammer, F., Stasińska, G., Proc. 35rd Cambridge Conference, 1994, Maddox, Aragon-Salamanca Eds., World Publishing, Singapore.

Emission-line galaxies at $z \leq 0.3$ in the Canada-France Redshift Survey[†]

Cláudia Rola[a,b], Laurence Tresse[a], Grazyna Stasińska[a]

and

François Hammer[a]

[a]DAEC, Observatoire de Meudon, 92195 Meudon Cedex, France

[b]Centro de Astrofísica da Universidade do Porto, Portugal

1 Introduction

The I-band magnitude-limited ($17.50 \leq I_{AB} \leq 22.50$) Canada-France Redshift Survey (CFRS[†]) allowed to select field galaxies from their old stellar population, the selection of sources being independent of morphological appearance. The completeness of the CFRS $z \leq 1$ sample reaches about 85%, and the larger part of the incompleteness occurs at high redshifts (see O. Le Fèvre et al. contribution). The 4500 Å − 8500 Å spectral range allowed to measure the H_α line up to $z \approx 0.3$. Among the field galaxies in the $0 < z < 1$ redshift range, we have selected a sub-sample of 137 spectra in the $0 < z < 0.3$ range. Our goal was to identify the nature of the emission-line galaxies in this redshift range, by using their emission-line intensities.

2 The sample of Field Galaxies up to $z = 0.3$

The 137 spectra in the $0 < z \leq 0.3$ domain, were classified in three categories. Category I contains 36 sources (26 %) having a spectrum with H_α in emission and H_β in absorption; Category II contains 80 sources (58 %) with a spectrum presenting H_α and H_β in emission; Category III contains 21 sources (15 %) having a spectrum with only absorption lines, generally associated to a red continuum. The only spectra suitable for the analysis presented below are those from Category II. Their corresponding sources have low or moderate luminosities ($-20 < M_{B_{AB}} < -16.5$, $H_0 = 50\ km\ s^{-1}\ Mpc^{-1}$) and are bluer than Sbc galaxies ($0.3 < (V - I) < 0.7$).

2.1 Emission-line intensity measurements

The integrated intensities and errors at one sigma were computed for each emission-line with the package SPLOT (software IRAF) and the MEASURE tool, developed by Pelat[1]. When [N II]λ6583 was seen, we deblended H_α from [N II]λ6583 and [N II]λ6548 with the deblend utility under the SPLOT package. The reddening constant was computed using the Balmer decrement H_α/H_β equal to 2.86 (case B, T= 10 000 K, n= 100 cm^{-3})[2] and the Seaton[3] reddening law.

[†]The Canada-France Redshift Survey's team is composed by D. Crampton, F. Hammer, O. Le Fèvre, S. Lilly, and L. Tresse.

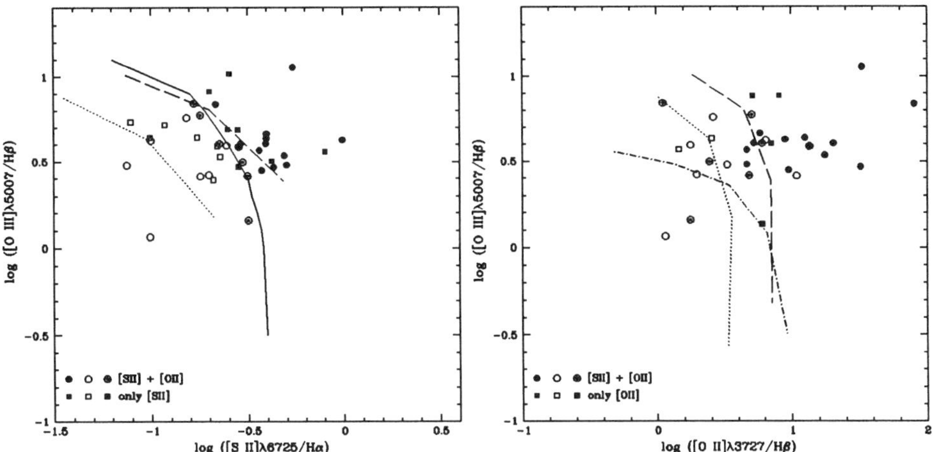

Figure 1: Position of our Category II objects in two diagnostic diagrams. Round symbols: sources with both [S II]λ6725 and [O II]λ3727 in observed spectral range. Square symbols: sources with only one of these lines inside the spectral range. Black symbols: sources classified as active galaxies. Open symbols: sources classified as H II region-like galaxies. White and black symbols: intermediate. The thick line represents the Veilleux & Osterbrock curve determined empirically from observational data[4], separating active galaxies from H II region-like objects. Upper limits for photoionisation by OB stars are represented by long dashed lines for $Z = 0.25\ Z_\odot$, by dot lines for $Z = 0.1\ Z_\odot$, and by dot-dashed lines for $Z = Z_\odot$.

3 Analysis of the emission-line intensities

3.1 Diagnostic diagrams

Several diagnostic diagrams relying on emission-line ratios[4,5] can be used to separate objects ionized by a stellar continuum (H II region-like galaxies) from objects ionized by a harder spectrum (active galaxies). With our data, we have been able to build two diagrams: [O III]λ5007/H_β versus [S II]λ6725/H_α and [O III]λ5007/H_β versus [O II]λ3727/H_β. The second one is more dependent on the redenning correction and on flux calibration, because of the large wavelength separation between [O II]λ3727 and H_β. Therefore, it has been used rather to check the consistency of the results obtained from the first one. All the objects from Category II for which we had the relevant data have been plotted in these diagrams (Fig. 1).

3.2 Photoionisation models

In order to determine more accurately the separation zone between the H II region-like galaxies and the active galaxies, we have done an extensive study covering a large domain of physical parameters for nebular regions photoionised by OB stars.

For this we used the photoionisation code PHOTO[6] and constructed a grid of steady-state spherically symmetric H II region models. This grid was build to determine the upper limits for photoionisation by hot main sequence stars (OB stars) in both diagnostic diagrams. We considered 4 abundance sets (2 $Z_\odot$, $Z_\odot$, 0.25 $Z_\odot$, 0.1 $Z_\odot$) and the models had a constant density (n_H =10 cm^{-3}). The ionisation parameter $U = Q_H/(4\pi R^2 n_H c)$ (where Q_H is the number of hydrogen ionizing photons, R is the Strömgren radius and c is the light speed) was varied roughly between 0.2 and 1×10^{-6}. For simplicity, we have considered the ionizing source as a single star with an effective temperature, T_{eff} ranging from 3×10^4 to 6×10^4 K. For the distribution of the ionizing radiation we have used the log $g = 5$ Kurucz model atmospheres[7] with an abundance consistent with the one in the nebula.

3.3 Results from diagnostic diagrams

Figure 1 shows the loci in the two diagnostic diagrams of the models which, for metallicities $Z_\odot$, 0.25 $Z_\odot$ and 0.1 $Z_\odot$, give the upper boundary for photoionisation by OB stars ($T_{eff} \leq$ 60 000 K). Among the objects shown in both diagrams, 27 have both [S II]λ6725 and [O II]λ3727 lines, and were considered in both diagrams and 45 have [S II]λ6725 or [O II]λ3727 out of the spectral range, and were considered only in one diagram. These plots show that a significant fraction of our objects lie above and to the right of the boundary defined by our models, and therefore cannot be ionized by main sequence stars. We find that 13 objects, out of the 27 present in both diagrams, are placed in the active galaxies region and seem to be intermediate between Seyfert 2 and LINER's.

3.4 Statistics

If we assume that the 27 spectra considered in both diagnostic diagrams are representative of all 80 category II spectra, we obtain that 28 % of all $I_{AB} \leq$ 22.5 field galaxies at $0 < z \leq 0.3$ are active galaxies. If the spectra which are only in one diagram are also considered, this percentage is practically unchanged. Furthermore, if we consider the less favorable case, where a small part of the incompleteness of the survey would be located at low redshift (see O. Le Fèvre et al. contribution), this would only diminish the previous percentage to 25% of active galaxies at $z \leq 0.3$. Regarding this possibility, we consider that about 25% of all $I_{AB} \leq$ 22.5 field galaxies at $0 < z \leq 0.3$ are active galaxies.

We found it instructive, for the 27 spectra represented in both diagnostic diagrams, to compare our classification with the observed [O II]λ3727 equivalent width, EW([O II]) (Fig. 2). The fact that we find no trend between our sources classification and EW([O II]) leads us to conclude that the separation between active galaxies and H II region-like galaxies is not possible from the EW([O II]) measurements.

4 Conclusions

The analysis of the nature of emission-line galaxies from the $0 < z < 0.3$ CFRS sample from diagnostic diagrams and photoionisation models shows that about 25 % of all $I_{AB} \leq$ 22.5 field galaxies at $z \leq 0.3$ are active galaxies, i. e. their emission-line

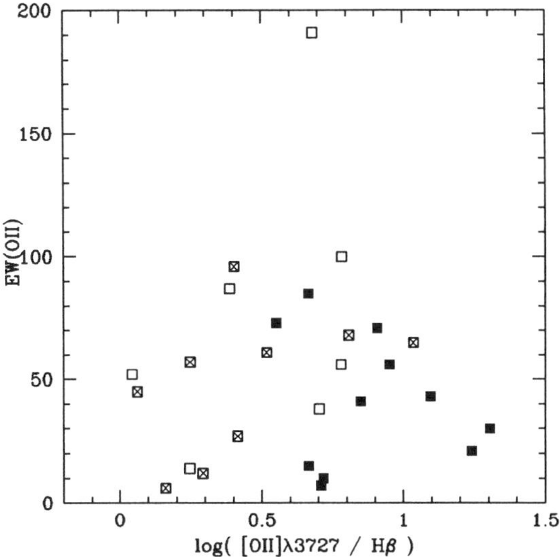

Figure 2: EW([O II]λ3727 vs [O II]λ3727/H_β intensities ratio for the 27 sources represented in both diagnostic diagrams. Active galaxies tend to have an higher [O II]λ3727/H_β intensity ratio, but no trend is seen relatively to their EW([O II]).

emitting gas is not ionized by OB stars. These numbers should be compared to the much smaller fraction (2 %) of Seyfert's found locally[8]. The CFRS images of these active galaxies show that some are spirals edge-on, others are compact or irregular sources.

A full version of this work will be submitted to MNRAS[9].

Acknowledgements

We thank C. Boisson, O. Le Fèvre, M. McCall, J.-R. Roy and R. Terlevich for helpful discussions. C. R. acknowledges the Symposium organisation for financial support.

References

[1]Rola, C. & Pelat, D., 1994, A & A 287, 676.

[2]Osterbrock, D., 1989, in *Astro. of Gas. Neb. & Active Gal. Nuclei*,Univ. Sci. Books.

[3]Seaton, M. J., 1979, MNRAS 187, 73.

[4]Veilleux, S. & Osterbrock, D. E., 1987, ApJS 63, 295.

[5] Baldwin, J.A., Phillips, M.M., Terlevich R., 1981, PASP 93, 5.

[6]Stasińska, G., 1990, A & A 83,501.

[7]Kurucz, R., 1992, in *The Stellar Pop. of galaxies*, Barbuy and Renzini eds., Kluwer Acad. Pub., 225.

[8]Huchra, J. & Burg, R., 1992, ApJ 393, 90.

[9]Tresse, L., Rola, C., Hammer, F. & Stasińska, G., 1995, MNRAS, in preparation.

REDSHIFT DISTRIBUTION & NATURE

OF μ-JY RADIO SOURCES

François Hammer[a], David Crampton[b], Olivier Le Fèvre[a], Simon Lilly[c]

[a]DAEC, Observatoire de Meudon, France

[b]DAO, University of Victoria, Canada

[b]University of Toronto, Canada

1 Introduction.

Deep 1.4GHz counts show an upturn below a few milliJanskys (mJy), corresponding to a rapid increase in the number of faint sources[1], with a decreasing fraction of sources which can be ascribed to elliptical galaxies and QSOs. Benn et al.[2], have identified the brightest optical counterparts of sub-mJy radiosources as z$\sim$0.2 blue galaxies which they interpreted as starburst galaxies. Towards the μJy levels ($\sim$ 16μJy at 4.86GHz), the projected number density of radiosources reaches the number density of B = 21.5 field galaxies, while the fraction of flat spectrum radiosources is increasing continuously (Fomalont et al.[3], hereafter FWKK). Because of this high surface density, the understanding of μJy radiosources may provide new and interesting constraints on the evolution of field galaxies.

During the preliminary deep imaging phase of our large spectroscopic survey of faint field galaxies (CFRS), one of our fields (10 arcmin $\times$ 10 arcmin) was chosen to coincide with the FWKK radiosource field[4], including 36 S$\sim$ 16μJy radiosources of the FWKK complete sample. All sources but two have been identified to V $<$ 25 and/or $I_{AB} \leq 24$, and/or $K_{AB} \leq 21$. Details on astrometry, photometry and spectroscopy are given elsewhere[5].

2 Spectroscopy of the 25 $I_{AB} \leq 22.5$ radio counterparts.

Among objects enough bright for spectroscopy we find:
-5 early type galaxies (26%)
-6 spirals (32%)
-5 emission line galaxies (26%)
- 1QSO, 1 M star and 3 spectroscopic failures.
The early type galaxies are rather luminous (L $\approx$ 2 L*), and have redshift ranging from 0.7 to 1, latter being the limit of our completness for our spectroscopy. Their spectra show a red continuum with a significant 4000Å break and very faint or no [OII] emission. The spirals present disk-like morphologies are rather luminous galaxies (L$\sim$1.5$\pm$0.4L*) and lie at moderately high redshift (0.37 $<$ z $<$ 0.81). They all show moderate [OII] emission (average W =12Å at rest), and relatively strong Balmer continuum and absorption lines (equivalent width ranging from 3 to 5Å), indicating the presence of A and F stars and suggesting that strong star formation occurred in these objects $\sim$1 Gyr ago. We classify them as 'S+A' galaxies by analogy with the E+A galaxies[6]. The emission line galaxies have relatively moderate luminosities (L$<$L*), while the intensity ratio of their emission lines classify all of them but one into the AGN ionisation region on the current diagnostic diagram[7].

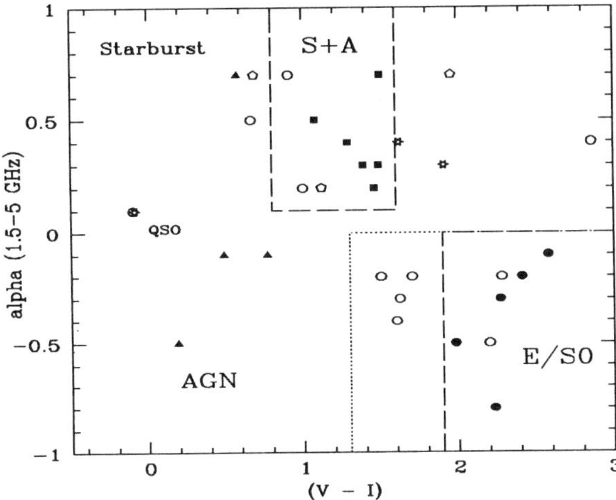

Figure 1: The radio spectral index versus V - I colour for all objects in the complete sample. Symbols are filled circles for early type galaxies (E/S0), filled squares for later type galaxies (~Sa to Sbc), filled triangles for emission-line galaxies, stars for other peculiar objects and open circles represent the faintest optical counterparts without spectral classification. Note the location of the early-type galaxies in the bottom right, the S+As (post starburst) in the middle top and the emission-line objects either in the top left (one starburst) or in the bottom left (AGN induced emission-line galaxies). Delimiting areas are indicated from the location of spectroscopically-identified objects, separating the early type galaxies (E/S0) from the post starburst galaxies (S+A). An extension of the E/S0 area is also indicated for large redshift where V-I is likely to decrease.

3 The radio spectral index-colour diagram and the faintest counterparts.

Each of the three μJy populations occupy distinct areas in the diagram (Figure 1): all red ellipticals have inverted radio spectra (α = -0.4±0.3), all the S+As have moderately steep spectra (α = 0.40±0.18), while the bluest emission-line galaxies have inverted spectra, with the notable exception of the most distant one (α = 0.7), which could be classified as a starburst. Inverted-spectrum radio emission from ellipticals has been observed in some nearby ellipticals[8]. This may indicate the presence of a low-power AGN[9], although other alternatives are possible. It is unlikely that the observed inverted radio spectra can be attributed to opacities at low frequencies since it only affects emissions below 1GHz at rest[10], i.e., below 0.5 GHz at z ~ 1. The post-starburst spirals have radio slopes noticeably flatter than the mJy starburst galaxies, which might indicate an increasing contribution of thermal radiation from remnant supernovae [10].Inverted radio spectra are also exhibited by the four very blue galaxies at low and moderate redshift, supporting the hypothesis that AGNs are present in their cores too, producing both the radio emission and the emission lines. Note also that the QSO lies in the area of blue emission-line sources.

Among the 11 I_{AB} >22.5 sources, two have no optical counterparts and one is only detectable in the combined B+V+I image, so it is impossible to speculate on the nature of these. There is considerable evidence that virtually all of the remaining sources are likely to be at z > 1. Indeed all the ones for which we have K photometry have color much redder than any object in the CFRS sample (I-K > 3.2) and are likely early type galaxies at z > 1. This is supported by the location of the faintest optical counterparts in Figure 1. Most of them lie in the area defined by early type galaxies with inverted radio spectra, and their optical color is slightly shifted to the blue V-I colour index, as one can expect for z>1 early type galaxies.

4 Discussion and conclusion.

An estimate can be made of the relative contribution of the different types of objects to the μJy population. We then find, among the 36 FWKK radio sources with S>16μJy in our MOS field:

5 definite and 6 probable z>1 early-type galaxies (~33 per cent of sample).
7 S+A's (21 per cent) including 1 possibly at z>1.
7 blue emission line objects (21 per cent), including 1 (or 2) QSOs.
1 starburst at z = 1.135.
3 sources with a very complex optical morphology.
3 candidates for very high redshift radiogalaxies.
1 M star and 3 undetected or barely detected objects.

The microJy population is hence mainly constituted of three distinct populations of galaxies with different redshift regimes: early-type galaxies at z > 0.75, S+A at intermediate redshifts (z = 0.375 to z = 0.8 or slightly > 1), and emission-line galaxies at z < 0.45 containing AGNs. The fraction of μJy sources with z > 1 is 38 - 42 per cent, depending on whether one assumes the unidentified sources are all at high z, or if they have the same distribution as the rest of the sample.

At bright flux levels near ~ 1 mJy, the radio source population is known to be composed largely of starburst galaxies at moderate redshifts. Our data indicate that the star formation activity, as well as the corresponding radio power, may decrease towards lower flux limits. Only one galaxy in our sample has a classical starburst spectrum, and it is at a much higher redshift than the Benn et al. sources. As noted above, most of the emission-line galaxies identified by us appear, from their emission line ratios and from their inverted or flat radio spectra, to be powered by AGN rather than starbursts. On the other hand, the S+A objects in our sample may well be the remnants of an active starburst population. Starburst activity is expected to last for only a relatively short period (typically 10^8 yrs), followed by a decrease in radio and emission line activity which in turn produces a less steep radio spectrum (increasing contribution from the thermal radiation by supernovae remnants) and optical spectra dominated by several Gyr old stars (A and F stars), with faint [OII] and no [OIII] emission. The surface density of the S+A post-starburst population found at μJy levels – assuming the count slope of 1.18 from FWKK – is indeed ten times larger than that of the mJy population at 1.4GHz, which is assumed to be mainly composed of starbursting galaxies. It is thus possible that the S+A galaxies are the remnants of generations of active starburst galaxies.

The presence of radio emission in a large fraction of nearby early-type galaxies was discovered by Wrobel and Heeschen[11] in a complete survey at 6cm. They found that ~30 per cent of nearby ellipticals have radio emission in their cores at powers ranging from 10^{19} to 10^{21} W Hz^{-1}, which are believed to be linked with low-power AGNs, since the core radio emission of a small subsample have flat or inverted radio spectra. The radio emission from the high redshift early-type galaxies in our sample probably has the same origin, since for all but one, the radio emission is concentrated in regions smaller than 0".2 to 2". However, we find that at z>0.75, more than a third of the early-type galaxies have $P > 10^{23}$ W Hz^{-1}, much larger than those in the Wrobel and Heeschen sample. The samples are not directly comparable, however, since only 4 of their galaxies have comparable optical luminosity to those in our sample, and their radio power is more than ten times lower. We thus conclude that we have detected a new, relatively numerous, population of radio flat-spectrum early-type galaxies at high redshift. The fact that the fraction of radio emitting early-type galaxies in the μJy population is increasing with the redshift (0 per cent at z<0.5, 25 per cent at z<1 and 40 per cent at z>1) also demonstrates that early-type galaxies are experiencing strong evolution of their radio properties beyond $z = 0.75$. Indeed, one third of them apparently exhibit AGN activity at $z \sim 1$.
The strong decrease of the radio spectral index from sub-mJy to μJy counts appears to be due to a combination of three factors: (1) the emergence of an elliptical population at high redshifts with moderate radio emission (2) an increasing fraction of narrow emission-line AGNs (Seyfert 2 and LINER); (3) a higher contribution of the thermal radiation to the radio emission from spirals, and the almost complete disappearance of starburst galaxies. The fact that radio sources have much flatter radio spectra at μJy levels compared to those above 0.1 mJy can thus be mostly attributed to the emergence of radio sources driven by low-power AGNs (> 50 per cent of the whole μJy population). Since the space density of AGN-driven sources apparently overtakes those powered by stellar emission, the contribution of AGN light to the faint source counts should be reevaluated.

References.

[1] Windhorst, R.A., 1984, PhD thesis, Univ. Leiden
[2] Benn, C.R.,*et al.* 1993, MNRAS, 263, 98
[3] Fomalont, E.B.,*et al.* 1991, AJ, 102, 1258(FWKK)
[4] Tresse, L., Hammer, F., Le Fèvre, O., Proust, D., 1993, A&A, 277, 53
[5] Hammer, F.,*et al.* 1995 MNRAS, submitted
[6] Dressler, A. & Gunn, J., 1983, ApJ, 270, 7
[7] Tresse, L., *et al.* 1995, in preparation
[8] Wrobel, J.M., Heeschen, D., 1984, ApJ, 335, 677
[9] Rees, M., 1984, ARA&A, 22, 471
[10] Condon, J.J, 1992, A&A Review, 30, 575
[11] Wrobel, J.M., Heeschen, D., 1991, AJ, 101, 148

From X-Ray Observations to Galaxy Formation

G. FABBIANO

Harvard-Smithsonian CfA, 60 Garden St., Cambridge, MA 02138

THE HOT ISM OF E AND S0 GALAXIES

X-ray observations of galaxies[9] have revealed the presence of hot ISM in E and S0 galaxies, sometimes giving rise to extended hot gaseous halos. In some galaxies, the hot gas can amount to as much as $\sim 10^{12} M_{\odot}$ and can totally dominate the X-ray emission, reaching luminosities of $L_X \sim 10^{42-43} ergs\ s^{-1}$, of the order of 100 times more than that expected from the stellar component. The presence of this hot ISM has given us a novel picture of early-type galaxies, which had been traditionally considered as old, gas-free systems.

While large hot gaseous halos have been found in E and S0 galaxies, not all of these galaxies retain their hot ISM. Figure 1 shows the distribution of E and S0 galaxies observed with *Einstein*[10,6] in the $L_X - L_B$ plane. There are two main features that can be extracted from a first look at this diagram: 1) the overall distribution of points follows a fairly steep power-law slope ($L_X \sim L_B^{1.7-2.2}$)[20,6]; 2) the scatter about the best-fit slope is very large, far in excess of what can be accounted for by measurement uncertainties: for a given L_B, one can find values of L_X differing by two orders of magnitude.

This diagram has been one of the main tools for understanding the properties and history of the hot ISM of early-type galaxies (see the review of Fabbiano 1989[9]). Based on both theoretical considerations and observational evidence[5,11,16], it has been suggested that the low L_X/L_B galaxies are mostly devoid of a hot ISM. Their X-ray emission would be due entirely to their stellar component. This view is also supported by the analysis of Eskridge, Fabbiano and Kim[6], which find significant differences in the correlation slopes, as a function of the X-ray luminosity. For galaxies with $L_X < 3\times10^{40} ergs\ s^{-1}$, $L_X \sim L_B^1$, while more luminous galaxies follow a correlation of the form $L_X \sim L_B^2$. The slope 1 correlation is consistent with that found in the *Einstein* sample of spiral galaxies (including bulge-dominated spirals), which covers the same luminosity range. By analogy it suggests that the X-ray emission of X-ray faint E and S0s is dominated by the stellar component.

Which are the important factors that favor or discourage the formation of hot gaseous halos?

– Environment may play a role, through tidal or ram pressure stripping, for galaxies with nearby companions and in a cluster/group environment. A positive correlation between environment and X-ray faintness was reported by White and Sarazin[21]. However, this cannot be the sole answer, because there are X-ray faint isolated galaxies.[17,13]

– SN rate is clearly a factor, because it regulates the energy input in the ISM. Models[5,18] are crucially dependent on the SN Ia rate.

– Recent galaxy merging (in the past ~1 Gy), appear to result in depletion of the hot ISM, either through the energy resulting from the merging, or via subsequent enhanced star formation.[13]

– Finally, the galaxy potential may play a crucial role.

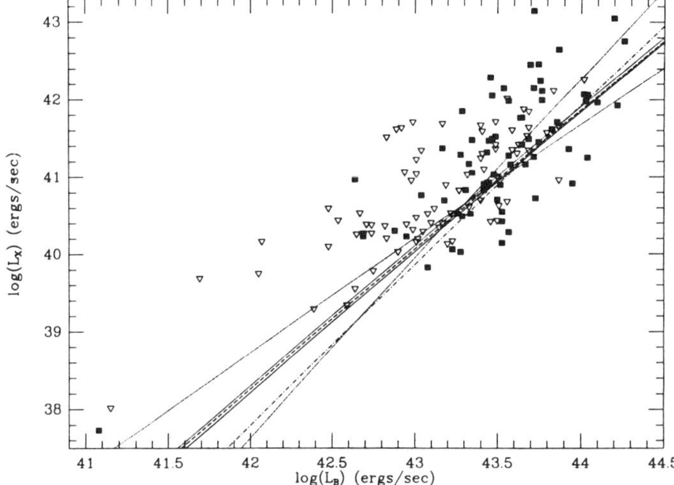

Figure 1. The distribution of the *Einstein* sample of E and S0 galaxies[6] in the $L_X - L_B$ plane.

In this talk I will report some recent results of statistical studies of the *Einstein Observatory* sample of early-type galaxies,[6,7,8] which address the questions of which parameters regulate the retention of the hot ISM, and consequently what this hot ISM can tell us about the history of the galaxies.

CORRELATION STUDIES

We explored the question of the important parameters affecting the presence of a hot ISM, and of the effect of the hot ISM on galaxy properties by studying the relations between *stellar* ($L_{12\mu m}, L_B, L_X$), *ISM* ($L_{HI}, L_{100\mu m}, L_X$), *nuclear* ($L_{6cm}$), and *structural* [potential shape: $a/b, a_4$,[1]; potential depth: σ_v, Mg_2] parameters.[6,7,8] We used censored analysis methods, including regression analysis and partial-rank tests.

The galaxy potential is clearly a major player. This may be reflected in the fact that S0s are systematically X-ray fainter than E galaxies, although in this case the increased fraction of rotational energy in the former may play a role.[6,17]

But there are stronger and more direct indications.[7] The *depth* of the potential is a key factor in promoting the retention of a hot ISM. There are strong correlations between L_X (and the L_X/L_B ratio) and both σ_v and Mg_2, suggesting that the deeper the potential well, the better the hot ISM retention ($L_X - \sigma_v$), and the better the retention of stellar ejecta from the early formation phase ($Mg_2 - \sigma_v$). Also, a certain depth appears to be critical for both retaining the hot ISM and fuelling an active nucleus: there are 'threshold' correlations between σ_v and both L_X/L_B and L_{6cm}.

The *shape* of the potential is also a key factor. Rounder (and boxier) galaxies have larger L_X, L_X/L_B, and L_{6cm}, with the strongest correlation being that between L_X/L_B and a/b. Also rounder galaxies tend to have boxier isophotes.

The potential shape and the depth parameters appear to be related, but not in a simple way. For instance, at a given value of a_4 the spread in σ_v appears directly related to the spread in Mg_2 such that the higher σ_v objects also tend to have higher Mg_2. Alternatively, for a given σ_v, boxy galaxies tend to have *lower* Mg_2 values (and thus lower stellar abundances) than disky galaxies. However, the galaxies with the lowest σ_v tend to have disky isophotes and low metallicity (Figure 2).

The coupling between the shape and depth parameters is clearly an important input for understanding the formation and evolution of early-type systems. A galaxy may have a deep potential either because of initial conditions, or because it is the product of mergers of a number of less massive objects.[2,19] In the case of relatively high σ_v galaxies, for a given central velocity dispersion (depth of the potential), galaxies with disky inner isophotes tend to be more metal rich. If disky isophotes signal 'primordial' ellipticals, then a disky elliptical of a given mass would have higher abundance than a boxy elliptical, if the Mass-Abundance relationship[14] is a relic of a fundamental contraint of the galaxy formation process. In this picture a boxy galaxy would be a merger product that was formed from lower mass (and thus lower abundance) precursors. Alternatively, disky isophotes may signal a merger system that has either formed a secondary nuclear disk,[15] or captured a self-gravitating spiral disk. If so, then the higher Mg_2 at a given σ_v may be due to high abundance in those disks.[3]

The effect of merging is not clear. Although it was suggested that this may be a key factor for the presence of a hot ISM, a comparison of galaxy emission properties with the κ parameters[8] suggest a conflicting possibility. While optically luminous galaxies tend to concentrate in the merging corner of the fundamental plane of elliptical galaxies (the $\kappa_1 - \kappa_2$[4] plane), the same is not true for high-L_X galaxies. The age of the merging episode may also be a complicating factor: there are indications that very recent merging may remove the hot ISM (e.g., NGC 4365[12]; NGC 3610 and NGC 4125[13]).

The link between L_X and potential depth is confirmed by the comparisons with the κ parameters[8], suggesting that L_X is more strongly correlated with inner mass to light ratio[4] than any other luminosity indicators. This suggests that galaxies with central excesses of dark matter also have more massive extended dark matter halos, providing a mechanism for retaining larger amounts of hot ISM. We also find that galaxies with higher inner M/L for their mass also tend to be more metal rich. This may be due to the enhanced ability of such galaxies to mantain ongoing or episodic central star formation for extended periods, or to retain the enriched ejecta of early epochs of star formation, or both. This trend becomes stronger when tested for constant a_4, suggesting that this effect is not connected to the presence of metal-rich nuclear disks.[3]

Acknowledgments. This work was partially supported by NASA contract NAS8-39073 (AXAF Science Center), and by NASA grant NAGW2681 (LTSA).

References

1. Bender, R., Surma, P., Doebereiner, S., Moellenhoft, C., and Madejsky, R. 1989, A&A, 217, 35.

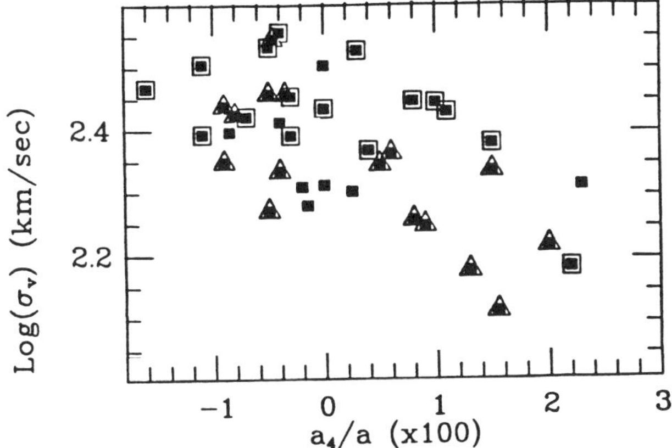

Figure 2. Central velocity dispersion σ_v versus boxiness parameter a_4. The points enclosed in squares are the systems with the highest 40% of Mg_2 values; those enclosed in triangles are the lowest 40%.[7]

2. Bender, R. et al 1992, in "Structure, Dynamics, and Chemical Evolution of Elliptical Galaxies", I. J. Danziger, W. W. Zeilinger, and K. Kjar, eds., ESO Proc. No. 45, p. 3.
3. Bender, R. and Surma, P. 1992, A&A, 258, 250.
4. Bender, R., Burstein, D., and Faber, S. M. 1992, ApJ, 399, 462.
5. Ciotti, L., D'Ercole, A., Pellegrini, S., and Renzini, A. 1991, ApJ, 376, 380.
6. Eskridge, P. B., Fabbiano, G., and Kim, D.-W. 1995a, ApJS, 97, 141.
7. Eskridge, P. B., Fabbiano, G., and Kim, D.-W. 1995b, ApJ, in press.
8. Eskridge, P. B., Fabbiano, G., and Kim, D.-W. 1995c, ApJ, in press.
9. Fabbiano, G. 1989, Ann. Rev. Ast. Ap., 27, 87.
10. Fabbiano, G., Kim, D.-W., and Trinchieri, G. 1992, ApJS, 80, 531.
11. Fabbiano, G., Gioia, I. M., and Trinchieri, G. 1989, ApJ, 347, 127.
12. Fabbiano, G., Kim, D.-W., and Trinchieri, G. 1994, ApJ, 429, 94.
13. Fabbiano, G., and Schweizer, F. 1995, ApJ, in press.
14. Faber, S. M. 1973, ApJ, 179, 731.
15. Hernquist, L., and Barnes, J. E. 1991, Nature, 354, 210.
16. Kim, D.-W., Fabbiano, G., and Trinchieri, G. 1992, ApJ, 393, 134.
17. Pellegrini, S. 1994, A&A, in press.
18. Pellegrini, S. and Fabbiano, G. 1994, ApJ, 429, 105.
19. Stiavelli, M. 1992 in "Structure, Dynamics, and Chemical Evolution of Elliptical Galaxies", I. J. Danziger, W. W. Zeilinger, and K. Kjar, eds., ESO Proc. No. 45, p. 303.
20. Trinchieri, G., and Fabbiano, G. 1985, ApJ, 296, 447.
21. White, R. E. III, and Sarazin, C. L. 1991, ApJ, 367, 476.

Photoionization Effects During Galaxy Formation

MATTHIAS STEINMETZ

Max–Planck–Institut für Astrophysik
Postfach 1523
85740 Garching, Germany

INTRODUCTION

In addition to gravitation, radiative cooling and feedback processes due to supernovae, radiative heating by photoionization might play a major role in the hydrodynamics of forming galaxies. In the first place, it provides an important feedback process on scales smaller than 1 kpc[1], which can be even more efficient than supernovae. Secondly, an external UV background field can significantly affect the collapse and cooling history of the gas: the photon background raises the gas temperature to a few 10^4 K increasing the Jeans mass to $\approx 10^8\,M_\odot$ in baryons[2]. On the other hand, the photoionization depletes the fraction of neutral hydrogen and helium and, therefore, reduces the cooling due to collisional excitation, the dominant cooling process for a gas of primordial composition. Compared to the case of no background radiation, the cooling time of the gas can be enlarged by 2 to 3 orders of magnitude[3].

It is widely believed, that quasars are responsible for the UV background. Hence, the UV spectra are usually assumed to be quasar like[3,4]:

$$I(\nu) = I_{-21}\left(\frac{\nu}{\nu_{\mathrm{H}}}\right)^{-\alpha}\ [10^{-21}\,\mathrm{erg/cm^2/Hz/sec/sterad}] \qquad (1)$$

where ν_{H} corresponds to the hydrogen ionization energy of 13.6 eV. For photoionization to be an important process, intensities I_{-21} of $1-10$ and rather hard spectra ($\alpha < 2$) are required. Both requirements are within the limits imposed on the UV background by the Gunn–Peterson effect and the proximity effect[3], but possibly in conflict with the observed X–ray background.

The photoionization rate is proportional to the density, whereas the recombination rate is proportional to ϱ^2. The UV background is therefore most important at low densities ($< 10^{-2}\,\mathrm{cm}^{-3}$). Indeed, even collisionless systems which collapses at redshift $z > 3$, which are typical for haloes less massive than $10^{11}\,M_\odot$, are too dense for photoionization to become important. Therefore, the importance of the UV background does not only critically depend on its spectrum and intensity, but also whether it becomes sufficiently strong at epochs when haloes leave the linear regime. Insufficient numerical resolution can lead to a completely different and even qualitatively wrong result: Due to the missing resolution the formation period of

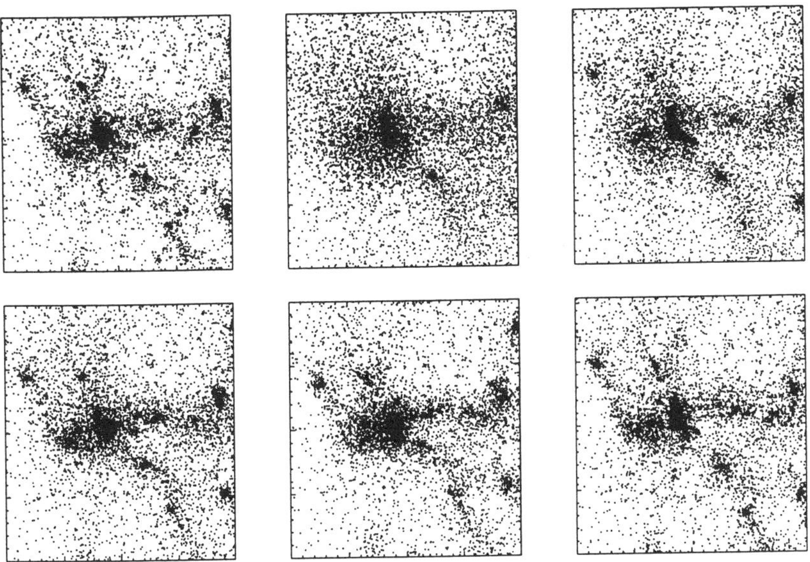

Figure 1: Distribution of gas particles at $z = 2.9$ for four models with identical initial conditions but different UV background fields. Every frame is 200 kpc across (in physical coordinates). Upper left: no UV background; upper middle: strong UV background ($\alpha = 1$, $I_{-21} = 10$) from the beginning; upper right: Strong UV background, exponentially switched on between $z = 8 - 5$; lower left: $\alpha = 1$, $I_{-21} = 1$ from the beginnning; lower middle: $\alpha = 5$, $I_{-21} = 10$ from the beginnning; lower right: $\alpha = 5$, $I_{-21} = 1$ from the beginnning. Absorption effects are included for the simulations shown in the lower three panels.

the smallest structures might be delayed to epochs when the photon field is strong enough to prevent collapse, whereas in a higher resolved simulation these objects are already collapsed and inert against photoionization. Most work so far was restricted to one–dimensional hydrodynamical simulations[5].

Numerical Simulations

We have performed three-dimensional, high–resolution hydrodynamical simulations with *Smoothed Particle Hydrodynamics.* The fully time dependent ionization network equations are solved simultaneously on the hydrodynamical or cooling time step whichever is smaller. Effects due to collisional excitation and ionization, recombination as well compton heating and cooling with the microwave and UV background are taken into account. At the relevant densities below $10^{-3} - 10^{-2}\,\mathrm{cm}^{-3}$, the cooling, ionization and recombination time scales are comparable or even longer than the dynamical time scale and a non equilibrium description is required.

The simulations start with a spherically symmetric volume at a redshift of $z = 25$ and contain a mass of $8\ 10^{11}\ M_\odot$ assuming vacuum boundary conditions[6,7]. Small scale power is added according to a $b = 2$ biased CDM spectrum. The influence of modes with larger wavelengths is approximated by raising the mass of the sphere by 27% corresponding to a $2.5\,\sigma$ peak of a biased CDM spectrum on a mass scale of $8\ 10^{11}\,M_\odot$. The tidal field is approximated by setting the sphere initially into rigid rotation with a spin parameter of $\lambda = 0.05$. Since the sphere reaches turn around at $z \approx 5$ and is fully collapsed only at $z = 2$, the growth of structures is fairly well described up to redshifts $z \approx 2.5$. The simulations were done with 18000 particles in the dark matter and also in the gas ($\Omega_{\rm bary} = 0.1$), corresponding to individual masses of $4\ 10^6\,M_\odot$ and $3.6\ 10^7\,M_\odot$, respectively. The gravitational softening is 1 kpc for the dark matter and 0.5 kpc for the gas.

For the UV background, we assume a power law as given by Eq. (**1**) with I_{-21} = 1or 10 and $\alpha = 1$ or 5. For some models, absorption effects were taken into account. In principle this requires a solution of the radiation transport equations and is still beyond the capabilities of the current generation of supercomputers. We, therefore, calculate the intensities modified by the spatially averaged opacities of the gas in the considered volume. The resulting spectrum is similar to those calculated by Madau[3] for photon energies between 13.6 and 100 eV, although it is noticeably stronger than those obtained by Cen and Ostriker[2]. Furthermore, in some models the UV background was not present from the beginning but was exponentially switched on or off over a redshift interval of about 3. A detailed description of the way the network equations are solved, and how absorption effects are taken into account are postponed to a forthcoming publication.

RESULTS

Only in the case of an intense ($I_{-21} \gtrsim 1$) and hard ($\alpha \lesssim 2$) quasar like UV background, is the formation of small structures ($M \lesssim 3{\cdot}10^8\,M_\odot$) significantly suppressed (see also Fig. 1,2). Also the mass of the most massive object ($2\ 10^{10}\,M_\odot$) can be reduced by 20%. However, such models are not compatible with the observed X–ray background, unless absorption effects are sufficiently strong[2] (which would, of course, again lower the photoionisation heating). In addition to small structures being depleted, cooling due to collisional excitation is completely suppressed. In contrast to simulations without a UV background, where the gas can cool efficiently and is kept at temperatures around 10^4 K, in the case of a strong UV background energy is gained by gravitational collapse and cannot be radiated away. The gas temperature can reach the virial temperature of the halo, i.e. a few 10^6 K for a Milky Way sized halo. As one might expect from the different density scaling of photoionisation and recombination described above, the effects of photoionisation are more prominent for a lower baryon fraction ($\Omega_{\rm bary} < 0.1$).

The influence of the UV background field is almost negligible in this simulations, if the spectrum becomes softer ($\alpha \gtrsim 2$) or less intense ($I_{-21} \lesssim 1$). The effect also becomes negligible, if the UV background is turned on too late when most of the structures with masses of $M \lesssim 3\ 10^8\,M_\odot$ have already collapsed (Fig. 1,2 lower left panel). For a biased CDM spectrum ($b = 2$) this corresponds to redshifts of $z \lesssim 7$. Switching off the radiation field at some point in the evolution, the gas cannot be held outside the dark haloes. It cools and collapses as soon as the background becomes weaker than $I_{-21} \approx 1$.

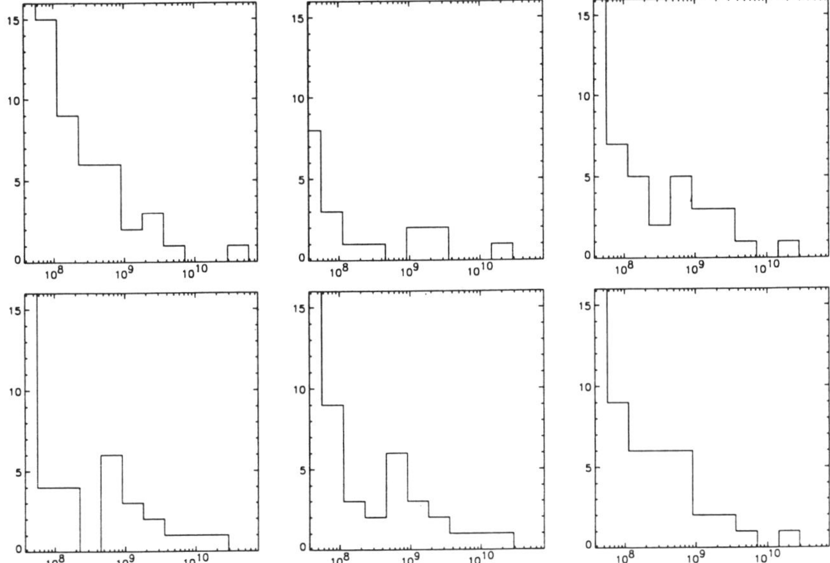

Figure 2: Number of groups N of mass M at $z = 2.9$ for the for models in Fig. 1.

If absorption is taken into account, the quasar-like external background field is reduced up to 2 orders of magnitude in the range between 13.6 eV and 1 keV. Only in the most extreme model ($I_{21} = 10$, $\alpha = 1$) small scale objects are suppressed, but only if the intensity of the background radiation is already high at $z > 10$.

In summary, extreme, if not unrealistic assumptions for the UV background must be made in order to significantly change the formation of galaxies more massive than a few $10^8\,M_\odot$.

ACKNOWLEDGMENTS

It is a pleasure to thank Martin Haehnelt, Ewald Müller, Jerry Ostriker and Simon White fore many helpful discussions.

REFERENCES

1. Haehnelt, M. 1994. Mon. Not. Roy. Atron. Soc.: in press.
2. Cen, R. & J. Ostriker. 1993. Astrophys. J. 417; 404.
3. Efstathiou, G. 1992. Mon. Not. Roy. Atron. Soc. 262; 43p.
4. Madau,P. 1992. Astrophys. J. 389; L1.
5. Thoul, A. & D. Weinberg. 1995. Astrophys. J.: in press.
6. Katz, N. 1991. Astrophys. J. 368; 325.
7. Steinmetz, M. & E. Müller. 1995. Mon. Not. Roy. Atron. Soc.: submitted.

On the origin of halo globular clusters and spheroid stars

By MARIO VIETRI[1] AND ENRICO PESCE[2]

[1]Osservatorio Astronomico di Roma, 00040 Monte Porzio Catone (Roma), Italy

[2]Istituto Astronomico, Università di Roma, Via Lancisi 29, 00100 Roma, Italy

A scenario is presented for the simultaneous formation of halo globular clusters and spheroid stars, in which shocks occuring in the gaseous, primordial ($Z = 0$) Galaxy play the central role. Denser-than-average subcondensations are shown to cool quickly and implode. Small condensations ($M < M_l = 5\times10^4 M_\odot$) are returned to the hot surrounding phase by the reflected shock, large ones ($M > M_u = 2\times10^7 M_\odot$) fragment into stars in the dense shell behind the implosion shock, thus generating the first spheroid stars. Intermediate clouds, suddenly cooling to $100K$ because of collisional H_2 excitation, are compressed and left in approximate pressure equilibrium with the hot medium: they give rise to halo GCs. The dependence of the proposed formation mechanism upon overall galactic metallicity and galaxy rotation velocity is weak, so that a universal upper mass limit is deduced. A mass range for halo GCs ($2\times10^3 - 6\times10^5 M_\odot$) and a velocity dispersion for spheroid stars in agreement with observations are predicted.

1. Introduction

The formation of globular clusters is a difficult problem: their masses vary only by two orders of magnitude in contrast to the five orders of magnitude spanned by the masses of their parent galaxies and their luminosity functions are quite similar in all well known galactic systems (Magellanic Clouds, M31, M87, M33 ...), so that a universal physical formation mechanism is required.

Previous attempts concentrated on locating a particular time in the history of protogalaxies at which the Jeans mass of the gaseous material equal the tipical GCs mass. Fall & Rees (1985) proposed that gas of primordial composition would undergo a thermal instability leading to a two phase medium, the cold phase of which, unable to cool below the Lyman barrier ($T \simeq 9000K$), would be imprinted by its Jeans mass ($\simeq 10^6 M_\odot$) and fragment into GCs. It was pointed out, however (Shapiro *et al.*, 1987), that H_2-formation via H^-, in conditions of non-equilibrium ionization, would lead to runaway cooling to temperatures below $10^3 K$, and a reduction of the Jeans mass by 4 orders of magnitude.

In order to rescue this model, Kang *et al.* (1990) postulate an intense UV-field from a pre-existing AGN or previous populations of stars to photodissociate H_2 and prevent the rapid cooling of the gas and the decrease of the Jeans mass to stellar values, but such modified scenario is not free from objections and at most marginally consistent with observations of metallicity. We propose a different solution (Vietri & Pesce, 1995).

2. The Model

We follow Fall & Rees (1985) in the description of the early Galaxy. The gravitational potential is due to a dark halo, with density distribution $\rho(R) = V_c^2/4\pi G R^2$, and $V_c = 220\, km\, s^{-1}$, which is adequate for $3\, kpc < R < 30\, kpc$. Gas falls freely into the potential well, in time

$$\tau_{ff} = \left(\frac{\pi}{2}\right)^{1/2}\left(\frac{R}{V_c}\right) = 5.6\times10^6 yr \frac{R}{1\, kpc}$$

and is heated to the virial temperature of the halo $T_h = \mu_h V_c^2/2k_B \simeq 1.7 \times 10^6 K$, and its density is determined by requiring that its free-fall time equals the cooling time: $\rho_h = 1.7 \times 10^{-24} \left(\frac{R}{1\,kpc}\right)^{-1}$ $gm\ cm^{-3}$.

3. Evolution of shocks and mass limits

Overdense regions in the protogalaxy cool more effectively and are compressed to higher densities by the hot surrounding gas. Let us consider one such region. The self-gravitational free-fall time-scale $\tau_{grav} = (3\pi/32G\rho)^{1/2} = 5.4 \times 10^7 yr(R/1\,kpc)^{1/2}$ is very long at such low densities, compared to all other relevant time-scales, so that the cloud shall initially evolve isochorically. We numerically followed the temperature history of one such subregion for $n = 1cm^{-3}$ and demonstrated that H_2-formation via H^- is efficient even at such low densities: it manages to cool the clouds in a time $\tau_{ic} \ll \tau_{ff}, \tau_{grav}$ down to very low temperatures $T < 10^3\,^oK$, so that the hot surrounding medium shall drive shocks into any such cloud.

Three different types of clouds can be distinguished according to the cloud crossing time by the shock R_c/V_S, where $V_S \simeq 100 kms^{-1}$ in the conditions under investigation. If $R_c/V_S < \tau_{cool}$, where τ_{cool} is the cooling time of the post-shock layer, the cloud material does not have time to cool before the shock reaches the cloud center and bounces back, and is then returned to the hot phase. In larger clouds, the shocked material has time to cool and form a cold, dense shell moving with the shock which in time τ_g (Vishniac 1983) shall collapse because of self-gravity. If $R_c/V_S > \tau_g$, fragments of roughly stellar size shall form, which collapse to make stars in time τ_g. These stars, moving at the shock speed, cannot be decelerated by the bounced shock, nor can be bound by the cloud self-gravity, and thus disperse as stars into the galaxy. The derived velocity dispersion for such fragments is close to the observed value $\sigma \simeq 100\ kms^{-1}$ for the spheroid stars. In the intermediate range, $\tau_{cool} < R_c/V_S < \tau_g$, the return shock decelerates the collapsing shell and leaves a much smaller, colder, denser cloud in approximate pressure equilibrium with the hot medium. These intermediate clouds are the progenitors of GCs. For a reasonable star formation efficiency $\eta = 0.03$, the predicted GC masses, M_{GC}, are bracketed by:

$$\eta M_l = 1.5 \times 10^3 M_\odot \left(\frac{R}{1\,kpc}\right)^2 < M_{GC} < \eta M_u = 6 \times 10^5 M_\odot \left(\frac{R}{1\,kpc}\right)^{1/2} ;$$

the details of this calculation are in Vietri & Pesce (1995). Figure 1 shows these curves, together with masses of known Galactic globular clusters: solid circles are halo GCs for which individual M/L_V have been determined (Pryor & Meylan, 1993), open triangles those for which a fiducial $M/L_V = 3$ was assumed, for lack of exact values. Most triangles lie outside $R = 20\ kpc$, where our model looses predictive power.

4. The dependence on galaxy size and metallicity

To the best of our knowledge, this is the first time an upper limit to the GC mass is found and it can also be shown that $\eta M_u \propto V_c^{1/2}$, so that the largest allowed mass is quite a universal limit, almost independent of the host galaxy. The argument goes as follows. The density of the hot phase is determined under the assumption that its cooling time be about equal to its collapse timescale τ_{ff}; since τ_{cool} is mostly determined by bremsstrahlung, for which radiative losses scale as $\propto \rho_h^2 T_h^{1/2}$, we have

$$\tau_{cool} \approx \frac{\rho_h T_h}{\rho_h^2 T_h^{1/2}} \simeq \tau_{ff} \approx \frac{R}{V_c} \Rightarrow \rho_h \approx V_c T_h^{1/2} \approx V_c^2$$

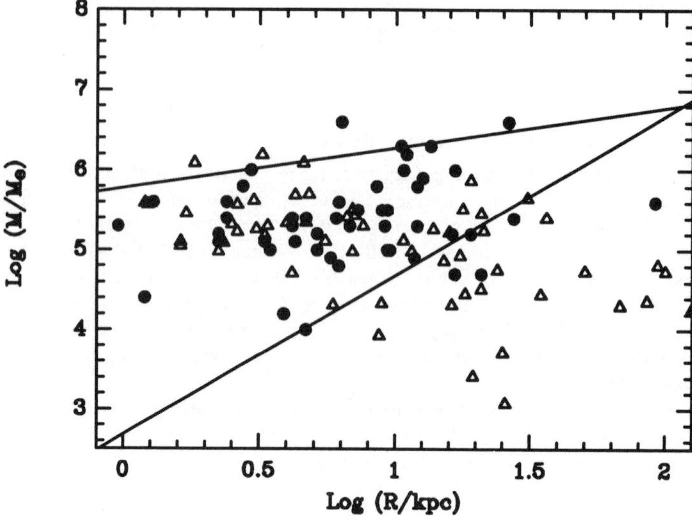

FIGURE 1. Limits obtained for GCs masses as function of the galactocentric radius.

because the virial temperature $T_h \propto V_c^2$. Furthermore, since obviously $V_S \propto V_c$, we obtain

$$\tau_g \approx \left(\frac{1}{\rho_h V_S}\right)^{1/2} \approx V_c^{-3/2}$$

for the self-gravitational fragmentation timescale of the cold post-shock shell. Finally we have

$$M_u \approx \rho_h (V_S \tau_g)^3 \approx V_c^{1/2} .$$

Stated otherwise, using the Faber-Jackson relation $L \propto V_c^4$, we have that $M_u \propto L^{1/8}$, so that a variation of L over four orders of magnitude (*i.e.*, 10 astronomical magnitudes, more than from Cen A to the LMC), corresponds to a range of M_u of just a factor of 3. On the contrary, in the theory of Fall and Rees, the mass limit decreases as $M_J \propto V_c^{-2}$, a much stronger and somewhat counterintuitive dependence.

In the scenario of Fall and Rees, the dependence of the value of the imprinted M_J on the cooling function makes the formation of GCs and stars a sort of threshold process: for $Z \leq 0.01 Z_\odot$ only GCs can be formed, whereas for higher Z only stars are expected, because gas can then cool below the Lyman barrier in a collapse timescale. This is in contrast with determinations of metallicity distributions for globulars and halo stars and with observations of galaxies still forming GCs. On the contrary, in our model, the presence of more metals in the initial gaseous medium simply implies a lower temperature ($T \simeq 10K$) of the post-shock material, which then fragments into stars more easily. The upper mass is thusly reduced to:

$$M_{GC} < \eta M_u \left(\frac{T_{ps}}{100\ K}\right)^{3/4} \simeq 10^5 M_\odot,$$

so that later GCs can form and are only a bit smaller than earlier ones.

We notice also that this formation mechanism, being single–cloud, avoids several problems characteristic of cloud–collision scenarios. In that case, in fact, the temperature to which gas is raised after the shock depends sensitively upon the clouds' relative velocity, thus making the formation of GCs strongly dependent upon the parent galaxy's mass. This is in stark contrast with the observations, which give a galaxy–independent upper mass limit to GCs. Lastly, we remark that we turned the Fall and Rees scenario upside down: far from steering away from H_2 cooling, we used it exactly to show why the GCs properties are remarkably similar in the whole observed Universe, independently of the cloud's metallicity.

REFERENCES

FALL, S. M. & REES, M. J. 1985 *Astrophys. J.* **298**, 18.

KANG, H., SHAPIRO, P. R., FALL, S. M. & REES, M. J. 1990 *Astrophys. J.* **363**, 488.

PRYOR, C.& MEYLAN, G. 1993 in "Structure and Dynamics of Globular Clusters" *Astron. Soc. of the Pacific Conf. Ser.* Vol.**50**, 357.

SHAPIRO, P. R. & KANG, H. 1987 *Astrophys. J.* **318**, 32.

SILK, J. & WYSE, R. G. F. 1989 in "The Epoch of Galaxy Formation", ed. by Frenk, C. S. *et al.* **Dordrecht**, 285.

VIETRI, M. & PESCE, E. 1995 *Astrophys. J.*, April 1st 1995 issue.

VISHNIAC, E. T. 1987 *Astrophys. J.* **274**, 152.

Clusters and Large-Scale Structure

NETA A. BAHCALL

Astrophysical Sciences, Princeton University
Princeton, NJ 08544-1001

INTRODUCTION

One of the fundamental questions in cosmology is what is the structure of the universe on large scales, and how did it form and evolve. A detailed understanding of this topic should shed light on the cosmology we live in, the amount, composition and distribution of dark matter in the universe, the initial spectrum of density fluctuations, and the formation and evolution of galaxies, clusters and other structure.

In this talk I review some of the observational studies of clusters and large-scale structure and discuss the implications of these results for constraining cosmological models and for estimating the mass-density of the universe

CLUSTERS OF GALAXIES: OPTICAL, X-RAYS, AND LENSING RESULTS

The binding gravitational potential of clusters of galaxies has traditionally been traced by the velocity dispersion of the cluster galaxies using the virial theorem.[1,2] The cluster potential can also be estimated by the temperature of the hot intracluster gas as determined from X-ray observations[3,4] and by the weak gravitational lens distortion of background galaxies caused by the intervening cluster mass.[5] Sufficient data are now available in both the optical (galaxy velocity dispersion in clusters) and X-ray (the temperature of the intracluster gas) so that these two methods can be compared. In addition, some current lensing data are becoming available. We compare these three independent methods for cluster mass determination.

Optical-X-Rays

The standard hydrostatic, isothermal model for clusters assumes that both the gas and the galaxies in the cluster are isothermal and in hydrostatic equilibrium within a common potential. The relation between the galaxy velocity dispersion (σ) and the temperature (T) of the intracluster medium (ICM) is then expected to be $\sigma \propto$

$T^{0.5}$. Lubin and Bahcall[6] analyzed the largest available sample of clusters for which both σ and T have been measured. They find a strong correlation between σ and T, $\sigma = 332\pm52(kT)^{0.6\pm0.1}$ km s^{-1}, where kT is in keV (Fig. 1); the observed relation is consistent with that expected from the isothermal, hydrostatic model. Previous results, obtained with smaller cluster samples,[7,8] are generally consistent with these findings.

The observed σ (T) relation suggests that the galaxies and the intracluster gas trace the same potential. To determine the average energy per unit mass in the galaxies and the gas, we use the β parameter,[4,7-9] $\beta = \sigma^2/(kT/\mu m_p)$, where μm_p is the average particle mass, σ is the one-dimensional rms velocity dispersion of the galaxies, and kT is the temperature of the ICM. In the self-gravitating isothermal model (assuming an isotropic velocity distribution), $\beta = 1$; this implies that the galaxies and the ICM trace the same potential and that the energy per unit mass in the gas and in the galaxies are equal. Lubin and Bahcall[6] used the observed σ(T) relation to determine the best-fit β parameter for clusters; they find $\beta = 0.94\pm0.08$, and a median of $\beta_{(med)} = 0.98$. These values suggest that the galaxies and gas trace each other with little or no velocity bias in the average cluster. It also suggests that both the galaxies and the gas are good tracers of the underlying cluster mass.

The comparable kinetic energies of the gas and galaxies in clusters suggest that the gas and galaxies should also have similar density profiles in the clusters. Bahcall and Lubin[10] find that this is indeed satisfied thus resolving the long-standing "β–discrepency" problem in clusters. The average gas and galaxy distributions in clusters are therefore consistent with an approximate hydrostatic, isothermal cluster model where both the galaxy velocities and the gas temperature reflect the underlying cluster potential.

Lensing

Recent observations of weak gravitational distortions of background galaxies by intervening cluster masses[5] yield positive detections, thus providing estimates of the lensing cluster mass within a given radius. Smail *et al.* [11] used this method to investigate two luminous X-ray clusters of galaxies, CL1455+22 at z=0.26 and CL0016+16 at z=0.55; Falham *et al.* [12] investigated the X-ray cluster MS1224+20 at z=0.33; and Kaiser[13] studied the cluster A2163 at z=0.17. Smail *et al.* [11] find that the mass distribution derived from the lensing signal in their two clusters is strikingly similar to that traced by the galaxies and by the hot X-ray gas in the clusters. Their inferred cluster mass profile is compared with the galaxy distribution profile in Fig. 2. These findings agree well with our conclusions that the galaxies, gas, and mass approximately trace each other. They find marginal evidence for a greater central concentration of dark matter with respect to the galaxies, but the uncertainties are large. Smail *et al.* also find that the cluster masses determined from lensing are consistent with the masses determined from the hot gas using the hydrostatic assumption. For CL1455+22 they find $M_{cl}(\leq 0.45$

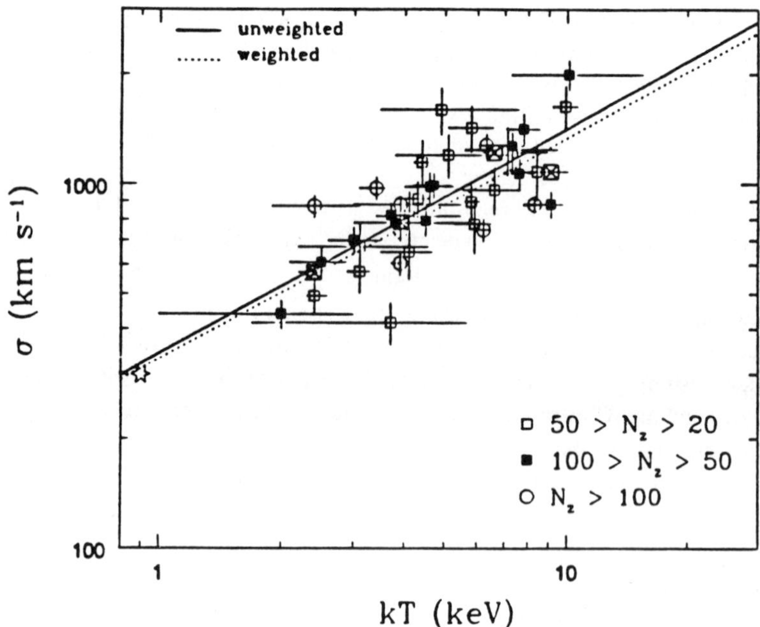

Figure 1. Cluster velocity dispersion versus intracluster gas temperature. N_Z is the number of measured velocities per cluster (Lubin & Bahcall 1993). The best-fit line is $\sigma=332\pm52\ (kT)^{0.6\pm0.1}$ km s^{-1}.

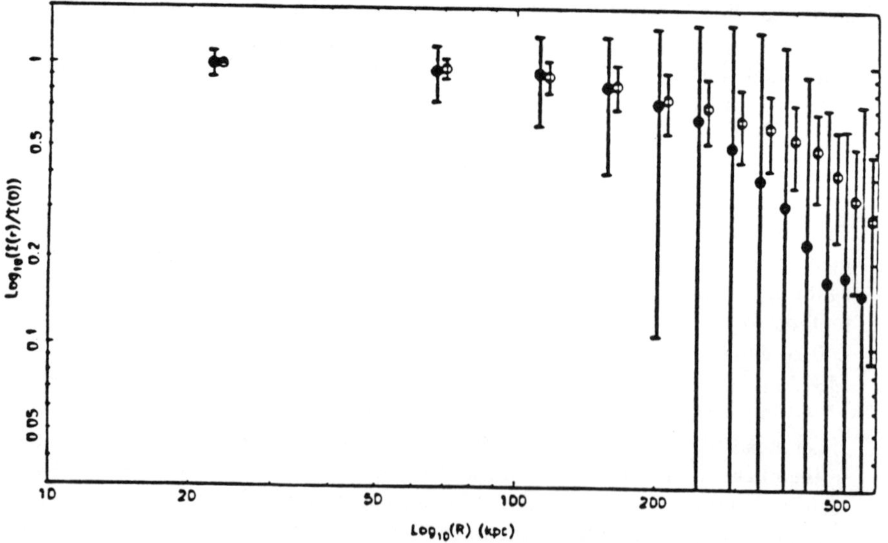

Figure 2. A comparison of the surface density profile of mass (•) (from lensing) and galaxies (O) in the cluster 14455+22. From Smail *et al.* (1994). (h=1/2 in this plot).

h_{50}^{-1}) $\simeq$ 3.2x$10^{14}M_\odot$ from lensing as compared with 3.6x$10^{14}M_\odot$ from X-rays; for CL0016+16 they find ~ 10x$10^{14}M_\odot$ from lensing and 7.5x$10^{14}M_\odot$ from X-rays, both within 0.6h_{50}^{-1} (all values are for h=0.5). Falham *et al.* [12] find a considerably larger mass (by a factor of ~ 3) for MS1224+20 from lensing as compared with the virial mass estimate. For A2163, the opposite is found: lensing yields a small or no signal compared with the virial X-ray mass. The lensing versus the optical or X-ray mass comparisons is summarized below:

– CL0016+16	z = 0.55	r ≤ 0.23h^{-1}Mpc	$M_{lens}/M_{opt/X\text{-}ray}$ = 0.9 ± (~50%)
– CL1455+22	0.26	r ≤ 0.30	1.3 ± (~50%)
– MS1224+20	0.33	r ≤ 0.45	2.9 ± (~50%)
– A2163	0.17		<1

The ratio of the gravitational lensing mass to the optical or X-ray mass within the radius r ranges from <<1 to ~3 for the four clusters studied. What can we learn? First, it is clear that a large cluster sample is needed before reliable conclusions can be reached. Individual clusters are *expected* to vary significantly in their mass ratio estimates due to various effects including cluster elongation (elongation along the line-of-sight will yield $M_{lens}/M_{opt/X\text{-}ray} > 1$ while flattening will yield a ratio < 1), anisotropic velocity orbits, etc. Therefore, a mean of a large sample is needed. In addition, effects such as boundary corrections in the lensing mass estimate can produce a lowered lensing mass, while substructure will overestimate the lensing mass.[14] Based on the optical and X-ray results and the preliminary lensing results shown above we assume that cluster masses, *on average*, are properly traced by galaxies and gas. More accurate lensing data should clarify this assumption.

THE MASS, M/L, AND BARYON CONTENT OF CLUSTERS

Cluster Masses and Mass-to-Light Ratio

Detailed optical and X-ray observations of rich clusters of galaxies yield virial cluster masses that range from ~10^{14} to ~ 10^{15} h^{-1} $M_\odot$ within 1.5 h^{-1} Mpc radius of the cluster center (above sections). When normalized by the cluster luminosity, a median value of $M/L_B \simeq 300h$ is observed for rich clusters. This mass-to-light ratio implies a dynamical mass density of Ω_{dyn} ~ 0.2. If, as desired by current theory, the universe has critical density (Ω=1), then most of the mass in the universe *can not* be concentrated in clusters, groups and galaxies; the mass distribution in this case would be strongly biased (i.e., does not follow the light).

A recent analysis of the mass-to-light ratio of galaxies, groups and clusters by Bahcall, Lubin and Dorman[15] suggests that while the M/L ratio of galaxies increases with scale up to radii of R ~ 0.1-0.2h^{-1} Mpc, due to the large dark-halos around galaxies, this ratio appears to flatten and remain approximately constant for groups and rich clusters, to scales of ~ Mpc. The flattening occurs at $M/L_B \simeq$ 200-300h, corresponding to Ω ~ 0.2. This observation may suggest that most of the dark matter may be associated with the dark halos of galaxies and that clusters do *not* contain a substantial amount of additional dark-matter, other than that associated with (or torn-off from) the galaxy halos. Unless the matter distribution is highly biased (with large amounts of dark matter in the "voids" or on very large scales), this suggests that the mass-density in the universe may be low, $\Omega \sim 0.15\text{-}0.2$.

Baryons in Clusters

Clusters of galaxies contain a large amount of baryons. Within 1.5 h^1 Mpc of a rich cluster, the X-ray emitting gas contributes ~ 5-10 $h^{-1.5}$% of the cluster virial mass (or ~ 15-30% for h=1/2).[16] Visible stars contribute only a small addition to this value. Standard Big-Bang nucleosynthesis limits the mean baryon density of the universe to $\Omega_b \sim 0.015h^{-2}$. This suggests that the baryon fraction in some rich clusters exceeds that of an Ω=1 universe by a large factor. Detailed hydrodynamic simulations[17,18] suggest that baryons are not preferentially segregated into rich clusters. The above results therefore imply that either the mean density of the universe is considerably smaller than the critical value, or that the baryon density of the universe is much larger than predicted by nucleosynthesis. The observed baryonic mass fraction in rich clusters, combined with the nucleosynthesis limit suggest $\Omega \sim 0.15\text{-}0.2$; this estimate is consistent with $\Omega_{dyn} \sim 0.2$ determined from cluster dynamics.

THE CLUSTER CORRELATION FUNCTION

The correlation function of clusters is one of the most efficient tools in tracing the large scale structure of the universe. Clusters are correlated more strongly than individual galaxies, by an order of magnitude, and their correlation extends to considerably larger scales (~ $50h^{-1}$ Mpc). The cluster correlation strength increases with richness ($\propto$ luminosity or mass) of the system from single galaxies to the richest clusters.[19,20] The correlation strength also increases with the mean spatial separation of the clusters.[21,22] This dependence results in a "universal" dimensionless cluster correlation function; the cluster dimensionless correlation scale is constant for all clusters when *normalized* by the mean cluster separation.

The two universal relations of the cluster correlation function are represented by:[23] (I) $A_i \propto N_i$, and (II) $A_i \simeq (0.4\, d_i)^{1.8}$, where A_i is the amplitude of the cluster correlation function, N_i is the richness of the galaxy cluster i, and d_i is the mean separation of the clusters. (Here $d_i = n_i^{-1/3}$, where n_i is the mean spatial number density of clusters i). The relation can be described as a universal dimensionless correlation function of a scale-invariant nature, with a correlation scale $r_{o,i} \simeq 0.4\, d_i$.

The $A_i(d_i)$ relation for groups and clusters of various richnesses is presented in Figure 3. The recent automated cluster surveys of APM[24] and EDCC,[25] shown in Fig. 3, are consistent with the predictions of the above relations. The correlation function of X-ray selected ROSAT cluster of galaxies[26] is also consistent with the relations presented above. Bahcall and Cen[27] show that the spatial correlation of a flux-limited sample of X-ray selected clusters will exhibit

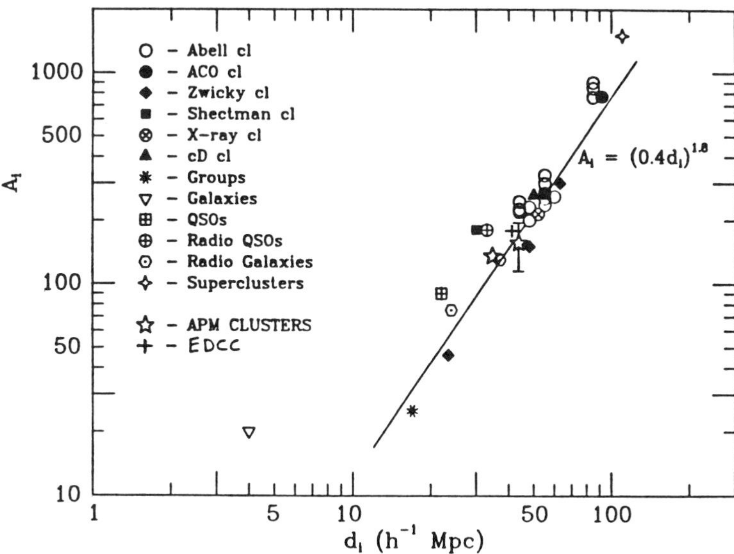

Figure 3. The dependence of the cluster correlation amplitude on mean separation (Bahcall and West 1992).

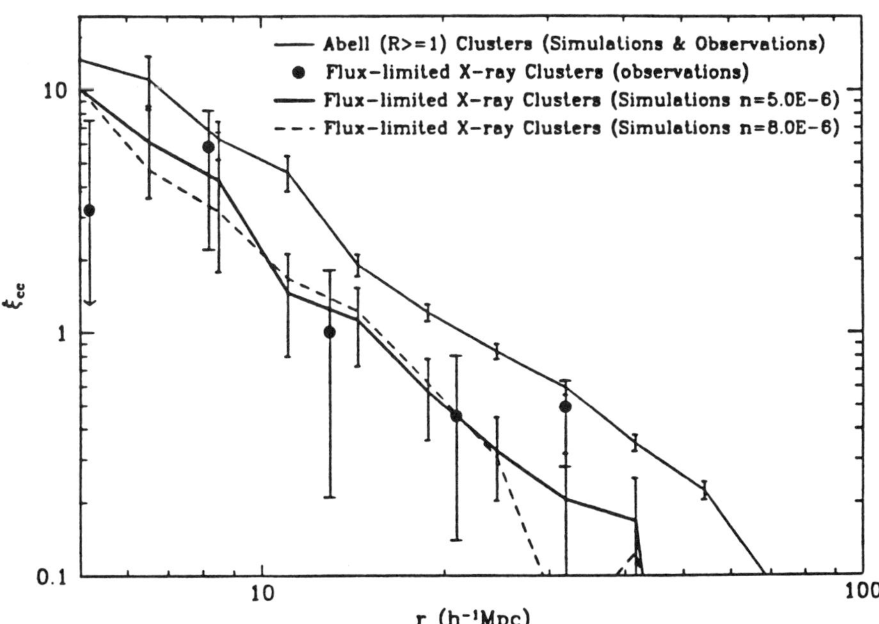

Figure 4. The cluster correlation function of a volume-limited R≥1 cluster sample (faint solid line) and the relevant flux-limited x-ray cluster correlation for that sample (dark and dashed lines). The X-ray cluster observations (Romer *et al.* 1994; dots) are consistent with expectations (Bahcall and Cen 1994).

a correlation scale that is smaller than that of a volume-limited, richness-limited sample of comparable apparent spatial density. The flux-limited sample contains poor groups nearby and only the richest clusters farther away. Using the richness-dependent cluster correlations discussed above, Bahcall and Cen[27] find excellent agreement with the flux-limited X-ray cluster correlations of Romer *et al* [26] (see Fig. 4).

THE MASS-FUNCTON OF CLUSTERS OF GALAXIES

The mass function (MF) of clusters of galaxies, representing the number density of clusters above a threshold mass M, n(>M), constitutes an important test of theories of structure formation in the universe. The richest, most massive clusters form from rare peaks in the initial mass density fluctuations; poorer clusters and groups form from smaller, more common fluctuations. Comparison of the observed abundance distribution of these clusters as a function of their mass with specific model calculations places strong constraints on the cosmological model and its parameters (see following section).

Bahcall and Cen[28] determined the MF of clusters of galaxies using both optical and X-ray observations of clusters. Their MF is presented in Fig. 5. The function can be described analytically by $n(>M) = 4 \times 10^{-5} (M/M^*)^{-1} \exp(-M/M^*)\ h^3\ Mpc^{-3}$, with $M^* = (1.8 \pm 0.3) \times 10^{14}\ h^{-1}\ M_\odot$, (where the mass M represents the cluster mass within $1.5h^{-1}$ Mpc radius). The use of the MF in constraining cosmological models is discussed below.

CLUSTERS AND COSMOLOGICAL MODELS

Two of the fundamental observed properties of clusters of galaxies – the cluster correlation function and the cluster mass function – can be used to place strong constraints on cosmological models and the density parameter Ω by comparison with model expectations. Bahcall and Cen[29] contrasted these cluster observations with standard and nonstandard CDM models using large N-body simulations ($400h^{-1}$ box, $10^{7.2}$ particles). They find that none of the standard $\Omega = 1$ CDM models, with any bias parameter, can fit consistently both of these cluster observations. A low-density, low-bias ($\Omega \sim 0.2$–0.3, $b \sim 1$–1.3) CDM-type model, however, provides a good fit to both sets of observations (see Figs 5-8). This is the first CDM-type model that is consistent with the high amplitude and large extent of the correlation function of the Abell, APM, and EDCC clusters. Such low-density, low-bias models are also consistent with other observables, e.g., the angular correlation function of galaxies,[30] and, with $\Lambda = 1 - \Omega$, the recent COBE results of the microwave background fluctuations.[31] The Ω constraints of this result are model dependent; a mixed hot + cold dark matter, for example, with $\Omega = 1$, is also consistent with these cluster data.

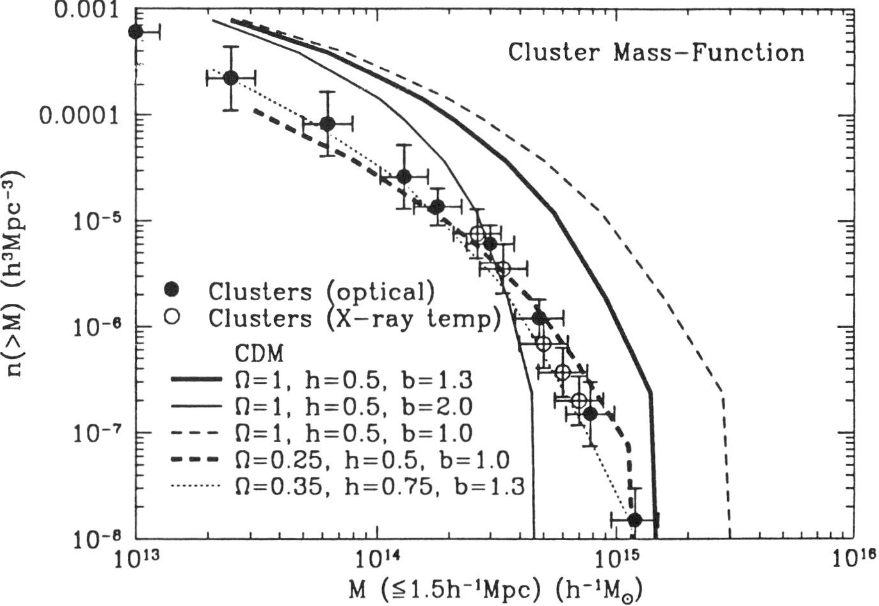

Figure 5. The mass-function of clusters of galaxies from observations and simulations (Bahcall and Cen 1992, 1993).

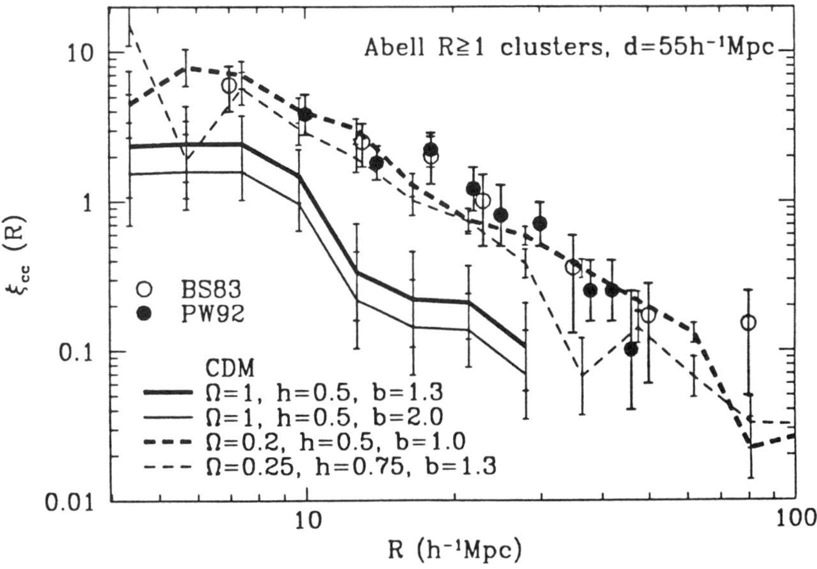

Figure 6. The correlation function of rich (R≥1) Abell clusters, with $d=55h^{-1}$ Mpc, from observations and simulations (Bahcall and Cen 1992).

The Cluster Mass Function

The observed cluster mass function is presented in Figure 5 together with the CDM simulations results. The standard CDM model ($\Omega = 1$) is presented for several bias parameters b; only the unbiased case, b ~ 1, is consistent with the COBE results. This unbiased model, however, is clearly excluded by the observed mass function: it predicts a much large number of rich clusters than is observed. The larger bias model[32] of b ~ 2, while providing a more appropriate abundance of rich clusters, is too steep for the observed mass function; it is also incompatible with the COBE results.

The low-density, low bias CDM models, with $\Omega \approx 0.25$ and $b \approx 1$, (with or without Λ) are consistent with the observed mass function (Fig. 5).

The Cluster Correlation Function

How does the CDM model agree with the observations of cluster correlations? Bahcall and Cen [29] used the large simulations above to answer this question. The model cluster correlation function was calculated for different cluster mean separations *d*, since the correlation function depends on d. The most massive clusters in the simulations that comprise a sample of the required mean separation d is used.

The CDM results for clusters corresponding to the rich Abell clusters ($R \geq 1$) with $d = 55\ h^{-1}$ Mpc are presented in Figure 6 together with the Bahcall & Soneira[19] and Peacock & West[33] observed correlations. The simulations provide ~ 400 $R \geq 1$ clusters – considerably larger than any previously run models. The results indicate that the standard $\Omega = 1$ CDM models are inconsistent with the observations; they cannot provide either the strong amplitude or the large scales ($\geq 50\ h^{-1}$ Mpc) to which the cluster correlations are observed. Similar results are found for the APM and EDCC clusters.

The low-density, low bias models are consistent with the observed cluster correlation function. They reproduce both the strong amplitude and the large scale to which the cluster correlations are detected. Such models are the only scale-invariant CDM models that are consistent with both cluster correlations and the cluster mass function. The models are also consistent with the APM and EDCC cluster correlations.

The Universal Dimensionless Cluster Correlation

The dependence of the observed cluster correlation on d was tested in the simulations.[29] The results are shown in Figure 7 for the low-density model. The dependence of correlation amplitude on mean separation is clearly seen in the simulations. To compare this result directly with observations, we plot in Figure 8 the dependence of the correlation scale, r_o, on d for both the simulations and the observations. The low-density model agrees well with the observations, yielding $r_o \approx 0.4d$, as observed. The $\Omega = 1$ model, while also showing an increase of r_o with

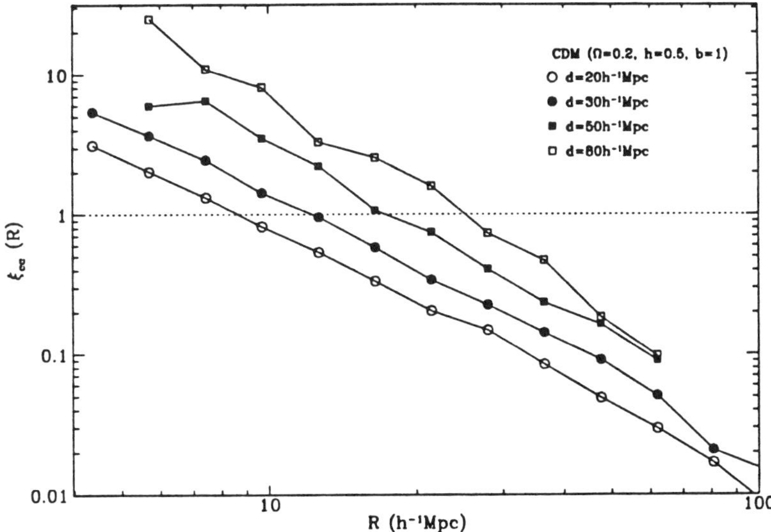

Figure 7. Dependence of the model cluster correlation function on mean separation d.

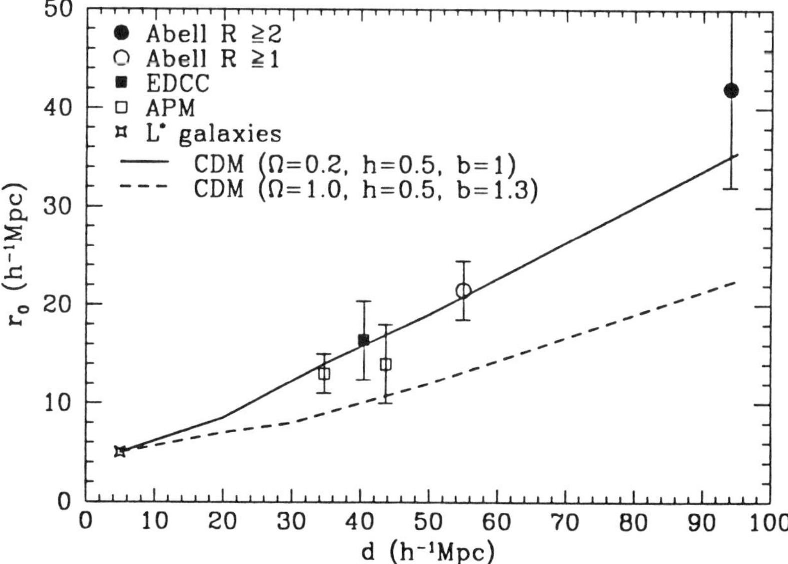

Figure 8. Correlation length as a function of cluster separation from observations and simulations (Bahcall and Cen 1992).

d, yields considerably smaller correlation scales and a much slower increase of $r_0(d)$ (see Fig. 8).

What causes this $r_0 \propto d$ dependence? The dependence, seen both in the observations and in the simulations, is most likely caused by the statistics of rare peak events, suggested by Kaiser[34] as a way of explaining the strong increase of correlation amplitude from galaxies to rich clusters. The correlation function of rare peaks in a Gaussian distribution increases with their selection threshold. Since more massive clusters correspond to a higher threshold, implying rarer events and thus larger mean separation, the above dependence results. A similar dependence $r_0 \propto d$ is also expected in a fractal distribution of galaxies and clusters.[21]

LARGE SCALE MOTIONS OF GALAXY CLUSTERS

Clusters of galaxies can also serve as efficient tracers of the large-scale peculiar velocity field in the universe.[35] Using large-scale cosmological simulations Bahcall *et al.*[35] find that clusters move reasonably fast in all models studied, tracing well the underlying matter velocity field on large scales. The clusters exhibit a Maxwellian distribution of peculiar velocities as expected from Gaussian initial density fluctuations. The cluster 3-D velocity distribution, presented in Fig. 9, typically peaks at $v \sim 600$ km s^{-1} and extends to high cluster velocities of ~ 2000 km s^{-1}. The low-density CDM model exhibits somewhat lower velocities (Fig. 9). Approximately 10% of all model rich clusters (1% for low-density CDM) move with $v \gtrsim 10^3$ km s^{-1}. A comparison of model expectation with the available data of cluster velocities is presented in Fig. 10 (Bahcall *et al.* 1994). The cluster velocity data is not sufficiently accurate at present to place constraints on the models; however, improved cluster velocities, expected in the near future, should help constrain the cosmology.

Cen, Bahcall and Gramann[36] have recently determined the velocity correlation function of clusters in different cosmologies. They find that close cluster pairs, with separations $r \lesssim 10h^{-1}$ Mpc, exhibit strong *attrractive* peculiar velocities; the pairwise velocities depend sensitively on the model. The mean pairwise attractive cluster velocities on $5h^{-1}$ Mpc scale ranges from ~ 1700 km s^{-1} for $\Omega = 1$ CDM to ~ 700 km s^{-1} for $\Omega = 0.3$ CDM.[36] The cluster velocity correlation function, presented in Fig. 11, is negative on small scales – indicating large attractive velocities, and is positive on large scales, to ~ $200h^{-1}$ Mpc –- indicating significant bulk motions in the models. None of the models can reproduce the very large bulk flow of clusters on $150h^{-1}$ Mpc scale, $v \simeq 689 \pm 178$ km s^{-1}, recently reported by Lauer and Postman.[37] The bulk flow expected on this large scale is generally $\lesssim 200$ km s^{-1} for all the models studied ($\Omega = 1$ and $\Omega \simeq 0.3$ CDM, and PBI).

SUMMARY

Clusters and large-scale structure provide powerful tools in constraining

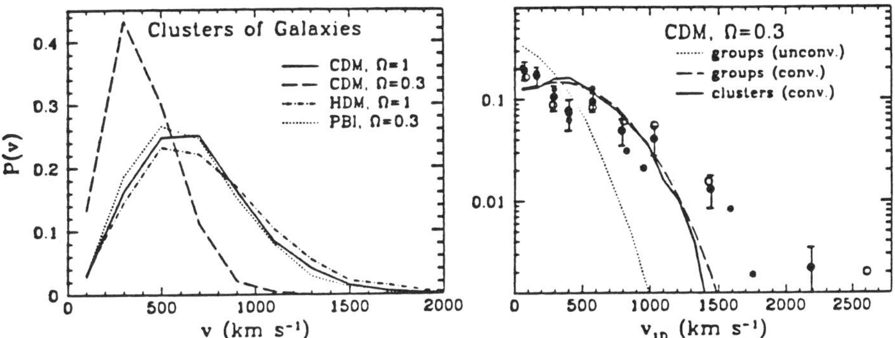

Figure 9. (left) The cluster 3-D velocity distribution in four cosmological models (Bahcall, Gramann and Cen 1994).

Figure 10. (right) Comparison of the currently observed 1-D velocity distribution of groups and clusters of galaxies with expectation from an Ω=0.3 CDM model. The dotted line is the model expectation; the dashed line is the model convolved with the observed velocity uncertainties (Bahcall *et al.* 1994).

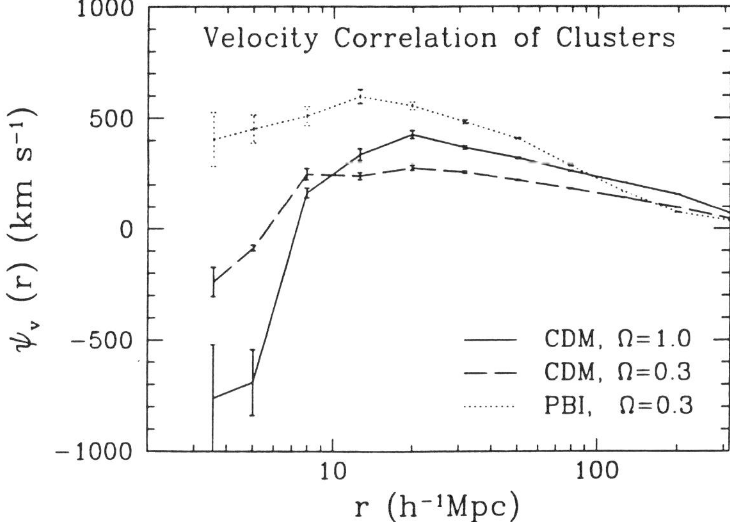

Figure 11. The velocity correlation function of rich (R≥1) clusters for three models (Cen, Bahcall and Gramann 1994).

cosmological parameters. On cluster scales of ~ Mpc, the dynamical estimates of cluster masses - using optical, X-ray, and some gravitational lensing data – suggest low-densities: $\Omega \sim 0.2$. A large fraction of the cluster virial mass is baryonic: ~ 10% for h = 1. This suggests that either the baryon density implied by nucleosynthesis, $\Omega_b \simeq 0.015h^{-2}$, is underestimated, or the total mass-density of the universe is low: $\Omega \sim 0.15 - 0.2$. This Ω estimate is consistent with the dynamical mass estimate from clusters of galaxies.

Comparing fundamental cluster and large-scale structure properties such as the observed mass-function and correlation-function of clusters of galaxies with expectations from various cosmological models also suggests that a low-density ($\Omega h \sim 0.2$), low-bias ($b \sim 1$) model (with a CDM or PBI-type spectrum) best fits the data. This Ω constraint, however, is model dependent. The models that best fit the cluster correlation and mass functions are low-density CDM (open or flat) and a mixed cold + hot dark matter (with Ω=1). However, when combined with *all* cluster and large-scale structure data, including the constraints on Ω suggested above, a low-density model (Ωh~0.2; open or flat) with a CDM-type spectrum appears to best fit the overall data.

This work is supported by NSF grant AST93-15368 and NASA grant NGT-51295

REFERENCES

1. Zwicky, F., 1957, *Morphological Astronomy*, Springer-Verlag, Berlin.
2. Peebles, P.J.E., 1980, "The Large Scale Structure of the Universe," Princeton University Press, Princeton, NJ.
3. Jones, C. and Forman, W., 1984, *ApJ*, **276**, 38.
4. Sarazin, C.L., 1986, *Rev. Mod. Phys.*, **58**, 1.
5. Kaiser, N. and Squires, G. 1993, *ApJ*, **404**, 441.
6. Lubin, L.M. and Bahcall, N.A., 1993, *ApJ*, **415**, L17.
7. Mushotsky, R., 1984, *Phys.Scripta.*, **T7**, L157.
8. Edge, A. and Stewart, G.C., 1991, *MNRAS*, **252**, 428.
9. Mushotzky, R., 1988, *Hot Astrophysical Plasma*, ed. R. Pallavicini, (Dordrecht: Kluwer), 114.
10. Bahcall, N.A. and Lubin, L.M., 1994, *ApJ.*, **426**, 513
11. Smail, J., Ellis, R., Fitchett, M., 1994, *MNRAS* (in press).
12. Fahlman, G.G., Kaiser, N., Squires, G., and Woods, D., 1994, preprint.
13. Kaiser, N., 1995, "Large Scale Structure of the Universe" 10th Potsdam Cosmology Workshop, World Scientific Publishing.
14. Narayan, R. 1994, in "Dark Matter," 5th Annual Astrophysics Conference in MD, AIP.
15. Bahcall, N.A., Lubin, L., Dorman, V., 1995, *ApJ* (submitted).
16. Briel, U.G., Henry, J.P., and Boringer, H., 1992, *A&A*, **259**, L31.
17. White, S.D.M., Navarro, J.F., Evrard, A., and Frenk, C.S., 1

993,*Nature*, **366**, 429.

18. Lubin, L., Cen, R., Bahcall, N.A., Ostriker, J.P., 1995, *ApJ* (submitted)
19. Bahcall, N.A. and Soneira, R.M., 1983, *ApJ,* **270**, 20.
20. Bahcall, N.A., 1988, *ARA&A*, **26**, 631.
21. Szalay, A., Schramm, D., 1985, *Nature*, **314**, 718.
22. Bahcall, N.A. and Burgett, W.S., 1986, *ApJ,* **300**, L35.
23. Bahcall, N.A. and West, M.L., 1992, *ApJ,* **392**, 419.
24. Dalton, G., Efstathiou, G., Maddox, S. and Sutherland, W. 1992, *ApJ*, **390**, L1.
25. Nichol, R., Collins, C., Guzxzo, L., and Lumsden, S. 1992, *MNRAS*, **255**, 21.
26. Romer, A., Collins, C., Boringer, H., Ebeling, H., Voges, W., Cruddace, R. and MacGillivray, H. 1994, *Nature*, (in press).
27. Bahcall, N.A. and Cen, R.Y. 1994, *ApJ*, **426**, L15.
28. Bahcall, N.A.and Cen, R.Y., 1993, *ApJ,* **407**, L49.
29. Bahcall, N.A. and Cen, R.Y., 1992, *ApJ,* **398**, L81.
30. Maddox, S., Efstathiou, G., Sutherland, W. and Loveday, J.,1990, *MNRAS*, **242**, 43.
31. Smoot, G.F., *et al.*, 1992, *ApJ*, **396**, L1.
32. White, S.D.M., Frenk, C.S., Davis, M., and Efstathiou, G., 1987, *ApJ*, **313**, 505.
33. Peacock, J. and West, M., 1992, *MNRAS*, **259**, 494.
34. Kaiser, N., 1984, *ApJ*, **284**, L9.
35. Bahcall, N.A., Gramann, M. and Cen, R. 1994, *ApJ*, **436**, 23.
36. Cen, R., Bahcall, N.A. and Gramann, M. 1994, *ApJ*, in press.
37. Lauer, T. and Postman, M. 1994, *ApJ.*, **425**, 418.

The Primordial Perturbation Spectrum and Large Scale Structure

STEFAN GOTTLÖBER

Astrophysikalisches Institut Potsdam
An der Sternwarte 16
D-14482 Potsdam
sgottloeber@aip.de

INTRODUCTION

The standard model of structure formation is based on two assumptions, namely, that the primordial perturbation spectrum generated during inflation is of Harrison-Zeldovich type and, that the dark matter, which gives the main contribution to the density of the Einstein-deSitter universe, is cold. Recent observations have shown that this model cannot explain all observational facts together[1–4]. In order to change the theoretical predictions of the standard model, one can modify either of these assumptions. Widely discussed candidates for closing the universe are mixed dark matter[5] or a cosmological term[6]. Due to the additional matter component the evolution of the perturbations is changed so that the resulting spectrum differs from the standard one. Improved inflationary scenarios lead to other than Harrison-Zeldovich spectra[7,8] or even to an open cosmological model[9]. Changing the primordial perturbation spectrum one can maintain both the successful description of structure formation over a wide range of scales by the CDM model and the Einstein-deSitter model predicted by almost all inflationary scenarios. Our underlying inflationary model has two consecutive stages of exponential expansion with a short intermediate stage of power law expansion[7]. In consequence of the intermediate stage the scale invariance of the resulting perturbation spectra is broken, i.e. they are of Harrison-Zeldovich type only in the limit of very small and very large scales. In the intermediate range the power is scale dependent. Besides the normalization the BSI (Broken Scale Invariance) spectra are characterized by the ratio Δ of the power on large scales to that on small scales, and a scale k_{br}^{-1} denoting the onset of the break in the perturbation spectrum at small scales.

OBSERVATIONAL CONSTRAINTS

At large scales we have to normalize the resulting power spectra by means of the COBE data. There are different approaches to this procedure. We are using the normalization proposed by Górski et al.[10] who calculated the multipole a_9 of the CMB temperature fluctuations $\Delta T/T$ from the COBE data. In this case the normalization of the spectra is independent of the spectral index which is for our spectra slightly different from 1 on COBE scales (cp. Fig. 1 and 2). We have calculated the multipole moment of the CMB background fluctuations for different primordial perturbation spectra[11]. Our approximation agrees within a few percent with the approximation proposed by Naoshi Sugiyama (see this proceedings), in particular it gives exactly the same result for low multipoles which are relevant for the normalization. Normalizing the perturbation spectra with the COBE data we find the biasing parameter $b = \sigma_8^{-1} \approx 1.7$. In Fig. 1 the BSI spectrum is shown in comparison with a Harrison-Zeldovich spectrum (both for $\Omega = 1, H = 50$ km/s/Mpc) and a spectrum for a model with cosmological term ($\Omega + \lambda = 1$, $\lambda = 0.8$).

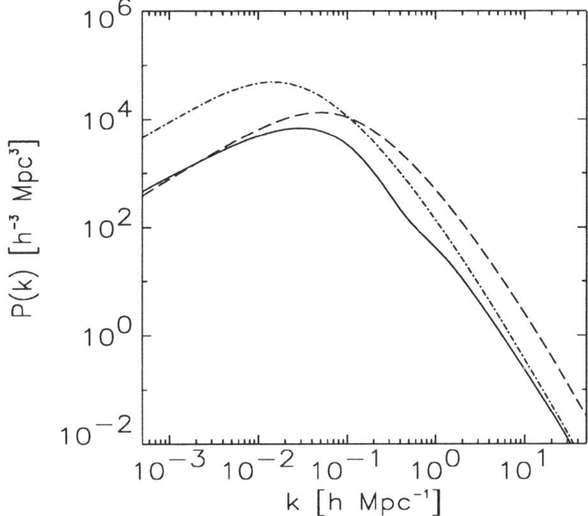

Figure 1: The linear perturbation spectra of the BSI model (solid line), the standard CDM model (dashed line), and a model with a cosmological term (dash-dotted line).

We have compared the theoretical predictions of different BSI spectra with such observational data as the variances calculated from the counts in cells, the angular correlation function, the Mach numbers, or the rms mass fluctuations necessary for quasar and galaxy formation[12]. These data appear to favour a spectrum with $\Delta = 3$ and an onset of extra power at about 2 h^{-1}Mpc. In that case the Hubble parameter is assumed to be 50 km/s/Mpc. On the other hand, recent observations with the HST suggest a big Hubble parameter of about 75 km/s/Mpc or even higher (see the contributions of W. Freedman and T. Shanks in these proceedings). Such a high Hubble parameter requires a nonzero cosmological term. This changes the predicted multipoles of the CMB fluctuations. In Fig. 2 the multipoles for the Harrison-Zeldovich and BSI spectra in a CDM model with $H = 50$ km/s/Mpc are shown in comparison with a model with $\lambda = 0.8$, $\Omega = 0.2$, $H = 75$ km/s/Mpc.

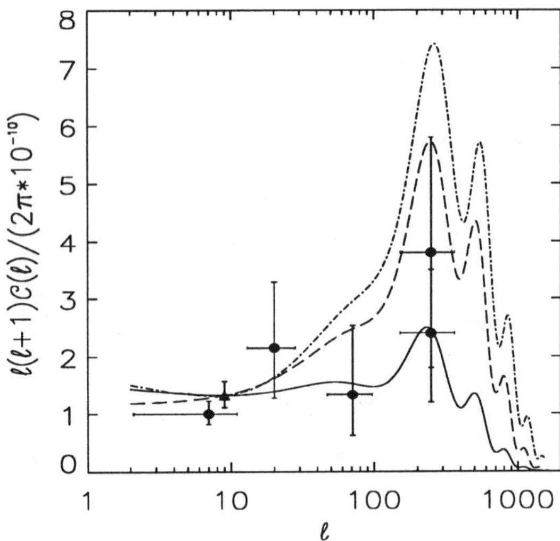

Figure 2: The multipole moments of the fluctuations of the CMB radiation for the BSI model (solid line), the standard CDM model (dashed line), and a model with a cosmological term (dash-dotted line).

The triangle in Fig. 2 denotes the Górski normalization. The left experiment is the slightly lower σ_{10} COBE measurement. It follow from left to right the Teneriffe, SK93 and MSAM3 (full and source free) data points (courtesy B. Ratra, from Bond[13]). The data points and error bars are an illustration. One has to calculate the predicted $\Delta T/T$ of each model by convolving the multipole moments with the filter function of the experiment in order to compare models correctly. As one can see easily from Fig. 2 models with a cosmological term predict a much higher first Doppler peak than the standard model. On the other side BSI models predict a lower Doppler peak. Thus better measurements of the multipoles around 100 will become a crucial test to discriminate between these models.

N-body simulations

Using the BSI spectrum shown in Fig. 1 and a Harrison-Zeldovich spectrum we have performed N-body simulations and calculated from these simulations galaxy correlation functions and higher moments, the cluster abundances, reconstructed power spectra for the density fields, the angular correlation function and further statistics[14,15]. The simulations show that the model with broken scale invariance fits the observational data quite well also in the nonlinear regime. In particular, the BSI model yields the right slope of the APM angular correlation function over all simulated angular scales, whereas the standard CDM model shows a too steep slope[14]. The observed cluster mass function is fitted very well by the BSI models.

The slope of the mass function in the standard CDM model is too steep on cluster scales[15]. In the BSI model the structure formation occurs later than in the CDM model with a Harrison-Zeldovich spectrum, i.e. the standard model shows to much structures at $z = 0$.

Obviously, one uses only a small part of the spectrum in the simulations. From Fig. 1 it is clear that for small boxes ($l_{Box} < 25h^{-1}$ Mpc) with decreasing box sizes the simulations become more and more equivalent to a CDM model with Harrison-Zeldovich spectrum and a normalization of $\sigma_8 \approx b_{lin}^{-1} = 0.6$. The simple reason is that simulations in these boxes feel only the reduced power produced during the second inflationary stage (the slope of the BSI spectrum approximately coincides with the Harrison-Zeldovich spectrum). Thus, the promising results obtained in simulations at low normalization (σ_8 between 0.6 and 0.8; note however, that the COBE normalized Harrison-Zeldovich spectrum with CDM yields $\sigma_8 > 1$.) of small scale structure within the CDM model are preserved in the BSI model.

CONCLUSION

Based on confrontation with observations and comparison with the COBE-normalized CDM model, we conclude that the perturbation spectra with broken scale invariance (BSI models) represent a promising modification of the CDM picture. Under the standard assumption of an Einstein-deSitter model ($\Omega = 1$) with $\Omega_{bar} = 0.06$ the BSI models allow to describe many details of large scale structure formation.

REFERENCES

1. Maddox, S.J., Efstathiou, G., Sutherland, W.J., & Loveday, J., 1990, MNRAS 242, 43P.
2. Loveday J., Efstathiou G., Peterson B. A., & Maddox S.J. 1992, ApJ 400, L43.
3. Fisher, K., Davis, M., Strauss, M., Yahil, & A. Huchra, J.P., 1993, ApJ 402, 42.
4. Baugh, C. M., & Efstathiou, G., 1993, MMNRAS 265, 145.
5. Klypin, A., Holtzman, J., Primack, J., & Regös, E. 1993, APJ 416, 1.
6. Kofman, L., Gnedin, N., & Bahcall, N. 1993, ApJ 413, 1.
7. Gottlöber, S., Müller, V., & Starobinsky, A.A., 1991, Phys. Rev. D43, 2510.
8. Cen, R. , Gnedin, N., Kofman, L., & Ostriker, J., 1992, ApJ 399, L11.
9. Ratra, B., & Peebles, P.J.E., 1995, Phys. Rev. D, submitted.
10. Górski, K.M., Hinshaw, G., Banday,A.J., Bennett, C.L., Wright, E.L., Kogut, A., Smoot, G.F. & Lubin, P. 1994, ApJ 430, L89.
11. Gottlöber, S., & Mücket, J.P., 1993, A & A. 272, 1.
12. Gottlöber, S., Mücket, J.P., & Starobinsky, A.A., 1994, ApJ 434, 417.
13. Bond, J. R., 1994, CITA preprint CITA 94-5.
14. Amendola, L., Gottlöber S., Mücket J.P., & Müller, V., 1995, submitted to ApJ.
15. Kates, R., Müller, V., Gottlöber, S., Mücket, J.P., & Retzlaff, J., 1995, submitted to MNRAS.

The Density of the Universe from the Peculiar Velocities of Sc Galaxies

Wolfram Freudling
Space Telescope–European Coordinating Facility
D-85748 Garching b. München, Germany
Luiz N. Da Costa
Institut d' Astrophysique
F75014 Paris, France
and
Observatorio Nacional
Rio de Janeiro, Brazil
Riccardo Giovanelli and Martha P. Haynes
Center for Radiophysics and Space Research and National Astronomy and Ionosphere Center
Cornell University, Ithaca, NY 14953
John J. Salzer
Dept. of Astronomy, Wesleyan University
Middletown, CT 06457

and

Gary Wegner
Dept. of Physics and Astronomy, Dartmouth College
Hanover, NH 03755
and
Astronomisches Institut, Ruhr-Universität Bochum
D-44780 Bochum, Germany

ABSTRACT

The all-sky survey of Sc galaxies with estimated redshift–independent distances presented by Haynes et al. (1993) is used to map the peculiar velocity field in the local universe. After suitable correction for Malmquist and selection biases, the resulting peculiar velocities are used to estimate the mass density in the local universe using the POTENT method (Dekel 1994). The resulting density field is compared with the distribution of galaxies.

1. The Sc Sample

We have obtained photometric I-band CCD images and high signal/noise HI spectra for a sample of about 1500 Sbc–Sc galaxies located at declinations $> -45°$ and of redshifts $cz < 7500$ km s^{-1} . The galaxies were selected by their blue angular diameters a in the following way: $5' > a > 2.5'$ for $cz < 3000$ km s^{-1} , $5' > a > 1.6'$ for 3000 km s^{-1} $< cz <$ 5000 km s^{-1} , $5' > a > 1.3'$ for 5000 km s^{-1} $< cz < 7500$ km s^{-1} . In addition, we obtained data for a cluster sample, including spiral galaxies in 20 clusters to $cz \simeq 12,000$ km s^{-1} . Velocity widths were derived from the 21cm observations made at Arecibo, Nançay, Green Bank and Effelsberg. Total magnitudes and inclination estimates were derived from the I band observations which were carried out at MDM, Kitt Peak, CTIO and ESO. Details of the selection criteria are given in Giovanelli et al. (1994, 1995).

In order to achieve full sky coverage with the Sc sample, our data were combined with those of Mathewson et al. (1992) in the South polar cap. Only the subsample of their data which met our set of criteria were used. The published data were reprocessed with the same procedures applied to ours. The combined all sky sample of Sbc and Sc galaxies contains a total of about 1600 objects. Hereafter, this sample is referred to as "Sc sample".

2. Magnitude and Width Corrections

The purpose of the Sc sample is to estimate distances from the Tully-Fisher (TF) relation (Tully & Fisher, 1977). Therefore, the total magnitude estimates have to be corrected for external extinction. We have taken advantage of our large and homogeneous set of data to derive a luminosity dependent correction for internal extinction (Giovanelli et al. 1995).

An important component of observational scatter in the TF relation are errors in the HI line widths. We have measured the line width for most of our galaxies from high-resolution spectra without previous smoothing. A large fraction of galaxies with small line widths was reobserved to reduce the uncertainties for that part of our sample. The line width were measured with an optimized algorithm which takes into account the line shape, S/N and resolution of the data.

3. Derivation of Tully-Fisher Relation

The first step in the derivation of redshift-independent distances is the derivation of the appropriate TF relation for the considered sample of galaxies. In order to derive this correct underlying relation, biases introduced by incompleteness of the cluster samples and Malmquist bias has to be taken into account (e.g. Teerikorpi 1987, Lynden-Bell et al. 1988, Willick 1994).

We used a sample of Sc galaxies from 15 clusters to derive a TF template relation. The magnitude zero–point of our TF relation was derived from the mean relation obtained from the subset of most distant clusters in the sample ($cz > 4000$ km s^{-1}) as seen by an observer at rest with respect to the CMB. The use of the distant clusters rather than the whole cluster sample for this purpose allows for the smallest margin of error in the approximation of a CMB rest frame by our TF template.

4. Malmquist and Selection Biases

The interpretation of maps of the peculiar velocity field must take into account biases introduced by the coupling of errors in the distance measurements and the spatial distribution of galaxies and/or selection effects in the samples considered. Biases originating from the spatial distribution are usually referred to as Malmquist bias and to correct for it requires a detailed knowledge of the underlying distribution of galaxies from which the sample is drawn. Biases introduced by the selection of the sample, which we will call selection biases throughout this paper, are caused by the correlation of the parameters used to select the sample with parameters used to estimate distances. In order to correct for such biases, the distribution function describing relevant galaxy properties is needed, as well as the detailed criteria adopted in selecting the sample being considered.

We evaluated such biases by creating simulated redshift – distance surveys with the same properties and thus the same biases as the Sc sample. To achieve this, we simulated properties like the spatial distribution, magnitude – diameter function, color relations, distance relations and their respective scatter properties. In addition, the selection criteria and completeness of the observed sample were taken into account.

For the spatial distribution of galaxies, we used the density field derived from redshift surveys corrected for peculiar velocities. The derived density field depends on the assumed $\beta = \Omega/b^{0.6}$, where Ω is the density parameter and b is the linear biasing of galaxies relative to matter. For the simulations, we assumed that $\beta = 1$. It should be emphasized that the density field and therefore the simulation is only slightly model dependent.

We used this density field to create a large number of artificial galaxies, to which we assigned properties similar to those of the observed galaxies. We then simulate the selection of catalogs from which observational samples are drawn. Subsequently, a observational sample of about 10^5 galaxies was drawn from the simulated catalog, following the same selection criteria used for the Sc sample, including any known incompleteness of the observational sample. Finally, distances are derived from the assumed distance relation.

The final sample, which contains both the true and the estimated distances, was used to compute the bias field in estimated distance space. For each artificial galaxy, the bias is computed by subtracting from the estimated distance the known "true" distance of galaxies. Since the artificial catalog of galaxies is large, the uncertainty in the determination of the biases can be reduced by binning them on a grid either in redshift or in estimated distance space. This average bias can then be applied to observed samples selected like the simulation. The raw estimated distance of the observed Sc sample was used to select the appropriate bias from the simulation and the estimated distance was corrected for it. Peculiar velocities were subsequently computed from bias corrected distances.

The main advantage of this approach is the flexibility it provides to evaluate a variety of effects and their impact on the estimated distances, including the scatter in the various relations used, the properties of the distance relations and the

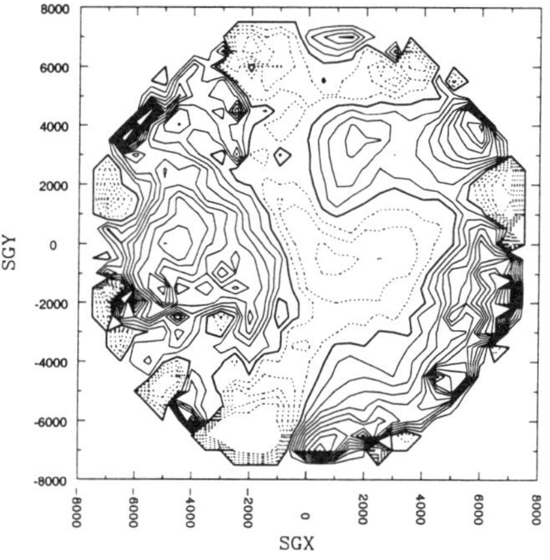

Figure 1:
Preliminary POTENT map of the Supergalactic Plane from the Sc data.

selection and incompleteness of observational samples.

5. Preliminary results

The list of cluster peculiar velocities is consistent with the picture that cluster peculiar velocities are smaller than 500 km s^{-1} , even though much larger peculiar velocities have been measured for some of the clusters in this list. This is in agreement with the expected distribution of cluster peculiar velocities in most scenarios (Bahcall et al. 1994).

A preliminary bias corrected map of peculiar velocities of the field galaxies was used to derive the distribution of matter in the local universe from the POTENT method (Dekel et al., 1990). The resulting density map of the supergalactic plane is shown in Figure 1. A comparison of such maps with the distribution of galaxies from redshift surveys allows the derivation of the β parameter. Preliminary results indicate that $\beta < 1$ but very small values of β are excluded.

6. References

1. Bahcall, N.A., Cen, R., Gramann, M. (1994), *Ap. J.* **430**, L13
2. Dekel A., Bertschinger E., Faber S.M., 1990, *Ap. J.* **364**, 349
3. Giovanelli, R., Haynes, M.P., Salzer, J.J., Wegner, G., da Costa, L.N. and Freudling, W. (1994), *Astron. J.* **107**, 2036
4. Giovanelli et al. (1995), *Astron. J.* submitted
5. Han, M. & Mould, J. (1992), *Ap. J.* **396**, 453
6. Lynden–Bell, D., Faber, S.M., Burstein, D., Davies, R.L., Dressler, A., Terlevich, R.J. and Wegner, G. (1988), *Ap. J.* **326**, 19
7. Mathewson, D.S., Ford, V.L. and Buchhorn, M. (1992) *Ap. J. Suppl.* **81**, 413
8. Teerikorpi (1987)*A&A***173**, 39
9. Tully, R.B. & Fisher, J.R. (1977), *Ap. J.* **54**, 661
10. Willick 1994, *Ap. J. Suppl.* **92**, 1

HST Cepheid Distance to Leo Group Galaxy M96.

T. SHANKS[a], N.R. TANVIR[b], H.C. FERGUSON[c] AND D.R.T. ROBINSON[b]

[a]*Department of Physics, University of Durham, South Road, Durham DH1 3LE, UK.*

[b]*Institute of Astronomy, University of Cambridge, Madingley Road, Cambridge CB3 0HA, UK.*

[c]*Space Telescope Science Institute, 3700 San Martin Drive, Baltimore MD21218, USA.*

INTRODUCTION

The simplest prediction of the theoretically favoured inflation model is that $\Omega_o=1$[1,2]. But if $H_o=100kms^{-1}Mpc^{-1}$ then this implies that the age of the Universe is only 6.5Gyr. This is in serious contradiction with the age of globular clusters at 16±2Gyr[3]. Even if $H_o=70kms^{-1}Mpc^{-1}$ then the age of the $\Omega_o=1$, Einstein-de Sitter Universe is only 9.3 Gyr and there is formally a 3σ rejection of the Einstein-de Sitter model. Therefore it is clearly important to obtain an accurate value for H_o to make a crucial test of the $\Omega_o=1$ model and thus of the inflation theory itself.

Fortunately the refurbished Hubble Space Telescope makes it now possible to resolve Cepheid variables in more distant galaxies than ever before. Already Freedman et al [4] have reported an HST Cepheid distance to the M100 Virgo spiral of 17.1±1.8Mpc, and a Hubble constant of $H_o=77\pm16kms^{-1}Mpc^{-1}$. However, the width of the Tully-Fisher relation in Virgo and the very flat velocity distribution of the Virgo spirals have led to suggestions that the Virgo spirals may not be at the same location as the Virgo ellipticals[5,6,7]; the biggest component of the error associated with the above H_o estimate indeed turns out to be the unknown Virgo cluster depth.

Here we describe an alternative route to H_o which bypasses this Virgo cluster problem. The Leo-I galaxy group with a velocity of $\approx700kms^{-1}$ is the nearest galaxy group which contains both spiral and elliptical galaxies. It is a compact group with a surrounding HI ring which strongly argues that this group has no significant depth along the line of sight. We have therefore used the Hubble Space Telescope to obtain a Cepheid distance to the Leo-I spiral M96 which then also provides a distance for the Leo-I elliptical galaxies such as NGC3377 and

NGC3379. In this way it will then be possible to obtain the first direct calibration of early-type galaxy distance indicators such as D_n:σ[8], Surface Brightness Fluctuations (SBF),[9] and the Planetary Nebula Luminosity Function (PNLF)[10] and thus a value of H_o based on early type galaxies in clusters as distant as Coma.

M96 CEPHEID DISTANCE VIA HST DATA

The Leo-I galaxy M96 has been observed by the Hubble Space Telescope over the 8 month period April-December 1994. Thirteen epochs of WFPC-2 observations have been obtained. At each epoch we obtained two 1200s V exposures. At 3 epochs we also obtained two 1200s I exposures.

Full details of the reduction procedure are given by Tanvir et al [11]. Only a brief description of the data reduction procedures are given here. Stellar aperture magnitudes were obtained to V=26.5 at each epoch using the DAOPHOT-2 PHOT routine[12]. The zeropoints and colour terms for the magnitudes are taken from Holtzmann et al [13]. The rms scatter at $V=26^m$ is $\pm0.^m18$. Variables were identified using a variety of automated algorithms. All candidate variables were tested for periodicity using the Lafler-Kinman string length statistic[14]. From these, 8 Cepheid variables were selected on the basis of visual inspection of the light curve.

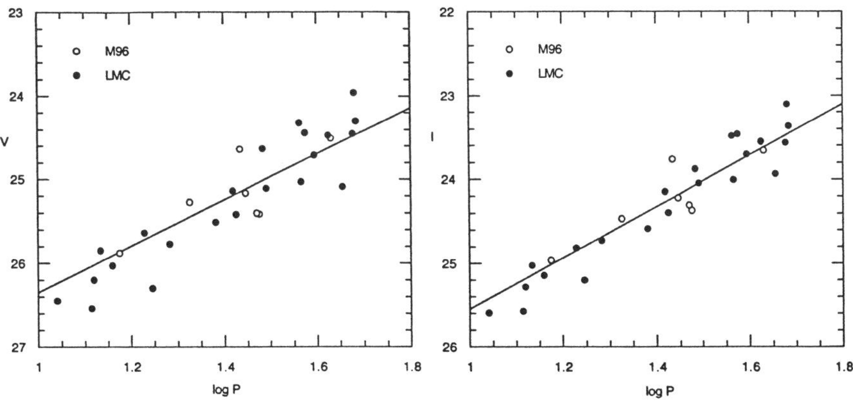

Fig. 1. The P-L relations for the M96 Cepheids in V and I compared to the P-L relations for the LMC Cepheids[15], and also to fits to the LMC P-L relations[16]. The LMC data is shifted to our observed apparent distance moduli for M96 in both V and I. The derived absolute distance modulus to M96 is $(m\text{-}M)_0=30.32\pm0.15$ or a distance of 11.6±0.8 Mpc.

The V and I period-luminosity relations for 7 Cepheids with periods between 10 and 60 days are shown in Fig. 1. Also shown are the LMC Cepheid P-L relations, as summarised by Madore[15], shifted to the derived apparent distance moduli in V and I. The M96 P-L relations are seen to be similar in both slope and scatter. The Madore & Freedman[16] fits to the LMC P-L relations are also shown. Assuming these fits, we obtain the apparent distance moduli for M96, $(m\text{-}M)_{AV}=30.51\pm0.12$, and $(m\text{-}M)_{AI}=30.42\pm0.09$. The derived absorption towards the M96 Cepheids is $A_V=0.19\pm0.11$. We thus obtain an absolute distance modulus to M96 of $(m\text{-}M)_o=30.32\pm0.15$ or a distance of 11.6±0.8Mpc.

THE DISTANCE TO VIRGO, COMA AND THE VALUE OF H_o

We now use relative Virgo-Leo and Coma-Leo distances from early-type galaxy distance indicators to obtain the distances to Virgo and Coma. The most direct route is via the D_n-σ relative distance between Coma and Leo. Faber et al[17] list the ratio of distances as 9.53±1.5 or a relative distance modulus of $\Delta(m\text{-}M)=4.90\pm0.34$. The resulting distance to Coma is $(m\text{-}M)=35.2\pm0.37$ or 111±19Mpc. Assuming a Coma velocity of $7200\pm300\text{kms}^{-1}$ this implies $H_o=65\pm12\text{kms}^{-1}\text{Mpc}^{-1}$.

A more accurate value for H_o can be obtained by using early-type distance indicators other than D_n-σ. A summary of these other distance indicators' estimates of the Virgo-Leo relative distance modulus is given by Table 1.

Table 1. Virgo-Leo Relative Distance Moduli

Surface Brightness Fluctuations	1.04±0.05	Ref.18,19,20
Planetary Nebula Luminosity Function	0.85±0.09	Ref. 21, 22
Globular Cluster Luminosity Function	1.42±0.41	Ref. 23
Colour-Magnitude Relation	0.77±0.50	Ref. 24
Weighted Mean and estimated error	0.99±0.15	

Thus the implied Virgo distance modulus is $(m\text{-}M)_o=31.3\pm0.2$ or $d_{Virgo}=18.3\pm1.8$Mpc. This value is intermediate between the traditional "short-scale" Virgo distance of 13-15Mpc and the "long scale" distance of 20-22Mpc. The Leo-Virgo modulus from D_n-σ is 1.16±0.35[17], consistent with the above value; however, we have not used it in Table 1 to maintain as much independence as possible between these two methods for estimating Hubble's constant. The value of the Coma-Virgo distance modulus is now quite well determined at $\Delta(m\text{-}M)=3.75\pm0.1$[25,26,27,28], resulting in a relative Coma-Leo modulus of $\Delta(m\text{-}M)=4.74\pm0.18$. This implies a Coma distance modulus of 35.1±0.23 ie 103±11Mpc. The resulting value of H_o, assuming a recession velocity for Coma of 7200kms^{-1}, is $H_o=70\pm8\text{kms}^{-1}\text{Mpc}^{-1}$. Taking the weighted average of the two nearly independent values of the Coma-Leo modulus gives $\Delta(m\text{-}M)=4.77\pm0.16$, which with our Leo result implies our final estimate of the distance to Coma of 104±11Mpc and our final estimate of H_o=69±8kms^{-1}Mpc^{-1}.

COMPARISON WITH OTHER CEPHEID RESULTS

We now compare our Cepheid calibration of the early-type route to H_o, with other recent ground- and space-based Cepheid results. First we consider the impact of recent ground-based[29,30] and HST[31] observations of M31 and M81 Cepheids, on the bulge D_n:σ distance to Virgo[32]. The distances to M31 and M81 have increased by an average of 13%, which increases d_{Virgo} from 19.8±2.4Mpc to 22.4±2.7Mpc. Using the same Coma-Virgo distance modulus and Coma recession velocity as above gives H_o=57±8kms^{-1}Mpc^{-1}.

When new Cepheid distances to M31, M33, NGC2403[29,30] are included in the calibration of the Tully-Fisher relation used by Pierce & Tully[33] a distance of 16.3±1.6Mpc to Virgo is obtained. Stepping from Virgo to Coma as we have done above, a value of H_o=79±8kms^{-1}Mpc^{-1} is obtained, similar to the value obtained when Tully-Fisher is used to step to Coma directly from the local calibrators[34].

So far there have been two ground-based searches and one HST-based search for Cepheids in Virgo spirals. Shanks et al[35] used the MARTINI adaptive optics instrument at the William Herschel 4.2m Telescope put a lower limit of $d_{Virgo} \gtrsim$ 20Mpc from the lack of detection of both Cepheids and Red Supergiants in 2 Virgo galaxies to R=24.^{m}5. From the same Virgo-Coma route as above this suggests $H_o \lesssim$ 64kms^{-1}Mpc^{-1}. Pierce et al[36] using HRCAM at CFHT claimed 3 Cepheids in NGC4571 at relatively bright magnitudes which results in a distance to Virgo of d_{Virgo}=14.9±1.2Mpc and a value of H_o=87±7kms^{-1}Mpc^{-1}. As discussed above, Freedman et al[4] used HST to obtain a Cepheid distance to the M100 Virgo galaxy of d_{Virgo}= 17.1±1.8Mpc for a value of H_o=77±16kms^{-1}Mpc^{-1}. It should be noted that none of these results are significantly different from the distance to the Virgo ellipticals of 18.3±1.8Mpc implied by our M96 observations. In any case, a real line-of-sight depth to the Virgo cluster as suggested by the wide Tully-Fisher relation for Virgo spirals[5,6,7], could also be contributing to the dispersion in distance measurements for Virgo spirals.

HST has also been used to observe Cepheids in galaxies IC4182 and NGC5253 where Type Ia supernovae have occurred[37,38]. When used to calibrate the Type Ia scale, a value of H_o=52±8kms^{-1}Mpc^{-1} is obtained. When a correction is made for the possible correlation between light decay rate and supernova maximum luminosity[39,40] the value of H_o may rise to between 58-67kms^{-1}Mpc^{-1}.

Thus from the HST M96 data we estimate H_o=69±8kms^{-1}Mpc^{-1}. Values of $H_o \gtrsim$ 75kms^{-1}Mpc^{-1} begin to be inconsistent with the bulge D_n:σ and the Type Ia methods; values of $H_o \lesssim$ 60kms^{-1}Mpc^{-1} begin to be inconsistent with the Tully-Fisher relation. Thus the intermediate range for H_o which our Leo observations suggest, is a range which is consistent with a large number of other H_o estimates.

CONCLUSIONS

We have described the HST detection of Cepheids in the Leo-I Group spiral galaxy M96. We obtain a distance of 11.6±0.8Mpc to the Leo Group and have used this to calibrate early-type galaxy distance indicators via the Leo-I ellipticals. Using the early-type distance scale, we then stepped to the Coma cluster via Virgo. We derive a distance of 18.3±1.8 Mpc to Virgo and a distance of 104±11Mpc to Coma. From the latter, we obtain a value for Hubble's Constant of H_o=69±8kms^{-1}Mpc^{-1}, implying an age for the theoretically favoured, Einstein-de Sitter, Ω_o=1 Universe of 9.3±1.1 Gyr, which is only marginally consistent with the globular cluster age of 16±2 Gyr. Significantly lower values of H_o would now seem to require either that some fundamental problem exists in the calibration of Cepheid variables or that the velocity field of clusters at the distance of Coma is much more complicated than has been so far suspected.

REFERENCES

1. Guth, A.H. 1981. Phys. Rev. D, **23:** 347-354
2. Linde, A.D. 1982. Phys. Lett., **108B**: 389-396.
3. vandenBerg, D.A. 1988. *In* The Extragalactic Distance Scale. S. van den Bergh & C.J. Pritchet. Eds.: 187-200. Astr. Soc. Pac. San Francisco.
4. Freedman, W.L., Madore, B.F., Mould, J.R., Hill, R., Ferrarese, L., Kennicutt, R.C., Saha, A., Stetson, P.B., Graham, J.A., Ford, H., Hoessel, J.G., Huchra, J., Hughes, S.M., & Illingworth, G.D. 1994. Nature, **371**: 757-762.
5. Pierce, M.J. 1991. *In* Observational Tests of Cosmological Inflation. T. Shanks et al. Eds.: 173-178, Kluwer:Dordrecht.
6. Shanks, T., Tanvir, N.R., Major, J.V., Doel, A.P., Dunlop, C.N., Myers, R.M. 1992. MNRAS, **256**: 29P-32P.
7. Fukugita, M., Okamura, S., Tarusawa, K., Rood, H.J. & Williams, B.A. 1991. ApJ, **376**: 8-22.
8. Dressler, A., Lynden-Bell, D., Burstein, D., Davies, R.L., Faber, S.M., Terlevich, R.J. & Wegner, G. 1987. ApJ, **313**: L37-L42.
9. Tonry, J.L. & Schneider, D.P. 1988. AJ, **96**: 807-815.
10. Jacoby, G.H., Ciardullo, R., Ford, H.C. & Booth, J. 1989. ApJ, **344**: 704-714.
11. Tanvir, N.R., Shanks, T., Ferguson, H.C. & Robinson, D.R.T. 1995 in prep.
12. Stetson, P.B. 1987. PASP, **99**: 191-222.

13. Holtzmann, J.A. et al. 1995. PASP in press.
14. Saha, A. & Hoessel, J.G. 1990. AJ, **99**: 97-148.
15. Madore, B.F. 1985. *In* Cepheids: Theory and Observations. B.F. Madore. Ed.:166-198. Cambridge Univ. Press. Cambridge.
16. Madore, B.F. & Freedman , W.L. 1991. PASP, **103**: 933-957.
17. Faber, S.M., Wegner, G., Burstein, D., Davies, R.L., Dresler, A., Lynden-Bell, D. & Terlevich, A. 1989. ApJS, **69**: 763-808.
18. Ciardullo, R., Jacoby, J.G. & Tonry, J.L. 1993, ApJ, **419**: 479-484.
19. Tonry, J.L. 1991. ApJ, **373**: L1-L4.
20. Tonry, J.L., Ajhar, E.A., Luppino, G.A. 1990. AJ, **100**: 1416-1423.
21. Ciardullo, R., Jacoby, G.H. & Ford, H.C. 1989. ApJ, **344**: 715-725.
22. Jacoby, G.H., Ciardullo, R. & Ford, H.C. 1990. ApJ, **356**: 332-349.
23. Harris, W.E. 1990. PASP, **102**: 966-971.
24. Visvanathan, N. & Sandage, A. 1977. ApJ, **216**: 214-226.
25. Sandage, A. & Tammann, G. 1990. ApJ, **365**: 1-12.
26. van den Bergh, S. 1992. PASP, **104**: 861-883.
27. Rowan-Robinson, M. 1988. Space Science Reviews, **48**: 1-71.
28. de Vaucouleurs, G. 1993. ApJ, **415**: 10-32.
29. Freedman, W.L. 1990. ApJ, **355**: L35-L38.
30. Metcalfe, N. & Shanks, T. 1991. MNRAS, **250**: 438-452.
31. Freedman, W.L., Hughes, S.M., Madore, B.F., Mould, J.R.,Lee, M.G., Stetson, P., Kennicutt, R.C., Turner, A., Ferrarese, L., Ford, H.C., Graham, J.A., Hill, R., Hoessel, J.G., Huchra, J., & Illingworth, G.D. 1994. ApJ, **427**: 628-655.
32. Dressler, A. 1987. ApJ, **317**: 1-10.
33. Pierce, M.J. & Tully, R.B. 1988. ApJ, **330**: 579-595.
34. Aaronson, M., Bothun, G., Mould, J., Huchra, J., Schommer, R.A. & Cornell, M.E. 1986. ApJ, **302**: 536-563.
35. Shanks, T., Tanvir, N.R., Major, J.V., Doel, A.P., Dunlop, C.N., Myers, R.M. & Ratcliffe, A. 1993. Gemini: RGO Newsletter, **42**: 1-5.
36. Pierce, M.J., Welch, D.L., McClure, R.D., van den Bergh, S., Racine, R. & Stetson, P.B. 1994. Nature, **371**: 385-389.
37. Saha, A., Labhardt, L., Schwengeler, H., Macchetto, F.D., Panagia, N., Sandage, A. & Tammann, G.A. 1994. ApJ, **425**: 14-34.
38. Saha, A., Sandage, A., Labhardt, L., Schwengeler, G.A., Tammann, G.A., Panagia, N. & Macchetto, F.D. 1995. ApJ, **438**: 8-26.
39. Hamuy, M., Phillips, M.M., Maza, J., Suntzeff, N.B., Schommer, R.A. & Aviles, R. 1995. Preprint.
40. Tammann, G.A. & Sandage, A. 1995. ApJ submitted.

E.R.O.S. SEARCH FOR MICROLENSING OF STARS BY LOW MASS OBJECTS IN THE GALACTIC HALO

Christophe MAGNEVILLE

CEA, Direction des Sciences de la Matière,
DAPNIA/SPP
CE-Saclay, F-91191 Gif-Sur-Yvette CEDEX, *France.*

MICROLENSING

As suggested by Paczyński[1] dark massive compact objects (MACHOs) could be detected in the halo of our Galaxy by monitoring the brightness of the individual stars in the Large Magellanic Cloud (LMC) using the gravitational microlensing effect. Gravitational lensing leads to an apparent temporary brightening of stars outside our Galaxy as the unseen object passes near the line of sight. EROS monitors stars in the LMC ($L=$ 55 kpc).

EXPERIMENTAL TECHNIQUE

The EROS collaboration designed two experiments. Both experiments take place in the European Southern Observatory in Chile and are observing the LMC.

The first experiment uses photographic plates on the wide field Schmidt $1m$ telescope. A plate covers about $5° \times 5°$ in the LMC and we monitor 8 10^6 stars. We are taking one plate with a red filter and one with a blue filter every few days, five months a year. The program is therefore sensitive to microlensing events with time scales larger than few days which correspond to MACHO above $10^{-3}M_{\odot}$.

The second experiment uses a CCD wide field camera[2] built for this purpose mounted on a 40 cm reflector (F/10). The covered angular area is $(1.1° \times 0.4°)$. The exposure time was typically 10 minutes with up to 46 alternating red and blue images taken per night. The program is therefore sensitive to microlensing events with time scales larger than 30 minutes which correspond to MACHO below $10^{-3}M_{\odot}$.

THE PHOTOGRAPHIC PLATES EXPERIMENT

The plates experiment consists of 304 plates taken at the Schmidt telescope. This represents 3 years of data taking: 56 plates in 1990-91, 198 plates in 1991-92 and 50 plates in 1992-93.

We begin with 8.5 10^6 stars on our reference catalog. After the elimination of images of poor quality, of stars too bright or too faint and of stars in a difficult or crowded environment, we remain with 3.8 10^6 stars detected with sufficient accuracy in both colours. Each light curve is then subjected to a series of cuts chosen to isolate microlensing-like events. The efficiency of these cuts to accept real microlensing events is determined by applying the same cuts to Monte Carlo microlensing events that are constructed by amplifying points on randomly selected experimental light curves.

The analysis of microlensing events first uses the unicity of the microlensing on one star. For this, we first reconstruct the largest variations and compute their significance for the red (P_R^1) and for the blue (P_B^1) curves. Then we reconstruct the second largest variation and compute its significance (P_R^2 and P_B^2).

We first require that the largest variation be significant. The great majority of stars exhibit only random fluctuations due to measurement errors. These stars are mostly eliminated by a loose requirement that the most significant variation in the blue be compatible in time with the most significant variation in the red. After these cuts 9100 stars remain. Intrinsically variable stars with a very significant second variation are eliminated by requiring that $P_R^2 + P_B^2$ for the second variation be small in each colour (remains 900 stars). At this point we ask that the size of the second variation be small relative to the size of the first variation ($\frac{P_R^2}{P_R^1} + \frac{P_B^2}{P_B^1}$) and remain with 60 stars. Each of the three years are studied separatly and we do not allow a microlensing amplification to be over 2 years of data. We then ask that the dispersion in the studied year be greater than the dispersion on the over two years. This selects light curves with a large dispersion where the microlensing is supposed to be and we are left with only 4 stars. We require that the amplification be achromatic and only keep 2 stars as microlensing candidates.[3]

The first candidate (EROS 1) is shown in figure 1. The amplification has been fitted to be 2.5 and the time scale 26 days. The amplified star has been studied through absolute photometry and spectroscopy. It was found to be a *Be* star with a visible magnitude $M_V = 19.11$, a color index $B - V = 0.35$ and emission line in H_α.

For the second candidate (EROS 2), the amplification has been fitted to be 3.3 and the time scale 30 days. It was found to be a $A0 - 2$ main sequence star with a visible magnitude $M_V = 19.38$, a color index $B - V = 0.04$.

Using the time of amplification distribution, a $10^{-1} M_\odot$ MACHO mass have been roughly estimated.

THE CCD EXPERIMENT

During two years of data taking, 8000 CCD frames were exposed. About 45000 useful stars were monitored between December 18, 1991 and March 31, 1992 while about 82000 stars were monitored between August 21, 1992 and March 31, 1993.

The CCD analysis is performed in the same spirit as the plate analysis. We just tune the algorithms to take into account the different time structure of the data. For example, when studying the 1992-93 data, we begin with 82000 stars, the cuts on the significance of the main variation and of the second variation together with the time compatibility of the main variation in the blue and red filters cut 85% of stars and 12000 stars remain. After the cut on the relative significance of the second and the main variation we remain with 88 stars. Their light curves are then examined in detail and fitted for the theoretical microlensing light curve. Most of the 88 stars show an "unphysical" discontinuous flux variation, generally due to inaccurate photometry due to bad atmospheric conditions or inaccurate telescope guiding. These stars are eliminated by requiring a good agreement between the time of maximum variation in the red and in the blue. After this cut 11 stars remain. All of them are located in the upper part in a region where we only have 5% of our total sample. This disagrees with a microlensing amplification which should not depend on the physical properties of the amplified stars. Six of the remaining stars have variations on long time scales ($\tau > 7$days) and are concentrated in regions of the colour-magnitude diagram known to contain many variable stars. For the purposes of short time-scale microlensing, we make a cut requiring $\tau < 7$ days leaving us with five stars. This significantly reduces our efficiency for microlensing events only if the lensing objects have $M_d > 10^{-3} M_\odot$.

The five remaining stars show very small flux variations of an amplitude comparable with the photometric resolution. All events have reconstructed amplifications less than 1.16 which, if they were indeed microlensing events, would correspond to impact parameters, $u > 1.4$. In contrast to the observed events, the expected distribution for microlensing events is concentrated at small impact parameters. We therefore make a final cut requiring impact parameters $u < 1.3$ leaving no candidates.

CONCLUSIONS

Using Montecarlo technique, we compute the number of events expected for our experiments as a function of the MACHO mass for a standard isothermal halo.

In the CCD experiment we do not see any microlensing event. The expected number of events is greater than 2.3 for $5\ 10^{-8} < M_d/M_\odot < 7\ 10^{-4}$ so we exclude this mass range at the 90% C.L. under the assumption that all objects in the Halo have the same mass. The expected number of events is greater than 6.9 for $3\ 10^{-7} < M_d/M_\odot < 1.5\ 10^{-5}$ so in this mass range we exclude the possibility that such objects could account for as much as one third of the halo.

The plate experiment shows two candidates. From the time of amplification we crudely compute a MACHO mass of the order of 10^{-1} solar mass (with a formal one

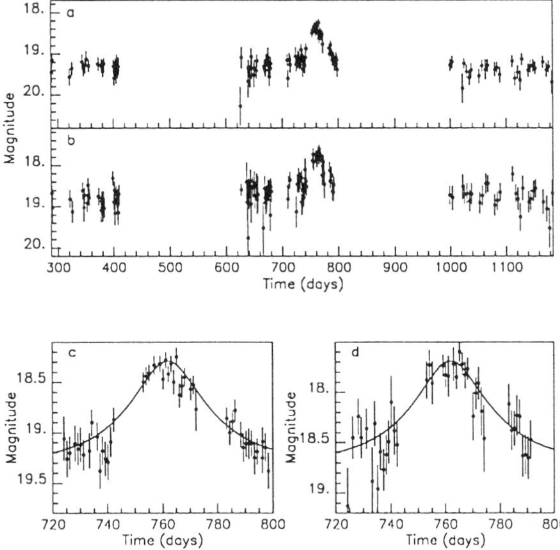

Figure 1: The first EROS candidate. Magnitude is plotted versus time in days (a) and b) are for the blue filter, c) and d) for the red filter).

standard deviation error of a factor 3). At this mass we are expecting around 8 events. Microlensed stars are from different spectral classes, one is a *Be* star, the other is an $A0 - 2$ main sequence star. This variety of lensed stars is in agreement with what is expected for microlensing amplifications. At this point, we may just say that our two candidates are compatible with microlensing but are not incompatible with pre or post-nova bursts or a new type of cataclysmic variable star.

References

[1] Paczyński B. 1986, ApJ 304, 1-5;

[2] Arnaud M. et al. 1994, Exp Ast 4, 265-278 and 279-296;

[3] Aubourg E. et al. 1993, Nat 365, 623-625.

[4] Aubourg et al., submitted to Astronomy and Astrophysics.

CMB Anisotropies: An Overview

DOUGLAS SCOTT

Department of Astronomy and Center for Particle Astrophysics,
University of California, Berkeley, CA 94720

The results from the *COBE* satellite have had an enormous impact on cosmology. The spectral results from FIRAS are impressive, to say the least, and provide strong support for a hot big bang model. However, it seems clear that the day belongs to the DMR measurement of anisotropy[1] and the subsequent flood of other experimental results[2]. Since the *COBE* announcement, there have been roughly 15 reported detections from around 10 separate experiments (see figure). So what progress has been made? How much better do we understand the microwave sky? Can we rule out specific models? What has happened over the last three years — in other words, where are we now and where are we going?

Progress There has been genuine progress in both experiment and theory. With the coming of data, theorists and experimenters have actually been talking with one another! One clear indication of progress is that both sets of people are now talking the language of ℓ-space (where $\ell \sim \theta^{-1}$). The squares of coefficients of spherical harmonics (the C_ℓ's) are the natural way to describe a power spectrum on a curved sky. And they are quantities whose expectation values directly come out of theoretical calculations. Perhaps we can finally lay the correlation function to rest.

Precision Theorists are now becoming much more precise in their analysis and predictions for experiments. There are many physical effects that should be correctly included. With the potential to measure anisotropies accurately over a range of scales, it is now worth concentrating on understanding C_ℓ calculations to at least better than 10%.[3] Among other things it is necessary to properly account for polarization, neutrinos and the physics of recombination

Normalization The CMB has also revolutionized another field, namely Large-Scale Structure. Anisotropies on the largest scales are now clearly *the* way to normalize theories. The *COBE* data have enough information about spectral shape that it is not sufficient to use just the rms amplitude, i.e. theories should be normalized to *COBE* in detail. An example of this is the accurate *COBE* normalization of variants of the Cold Dark Matter model[6], e.g. for 'vanilla-flavoured' CDM, $\sigma_8 = 1.34 \pm 0.10$. A point worth stressing, however, is that it is no longer enough to assume that $n \equiv 1$, since there are perfectly reasonable inflationary models with $n \simeq 0.9 - 0.95$, say. A tilted CDM model with some gravity wave component can have a respectable $\sigma_8 \simeq 0.7$. So before ruling out models, we have to be more careful about the freedom in the initial conditions.

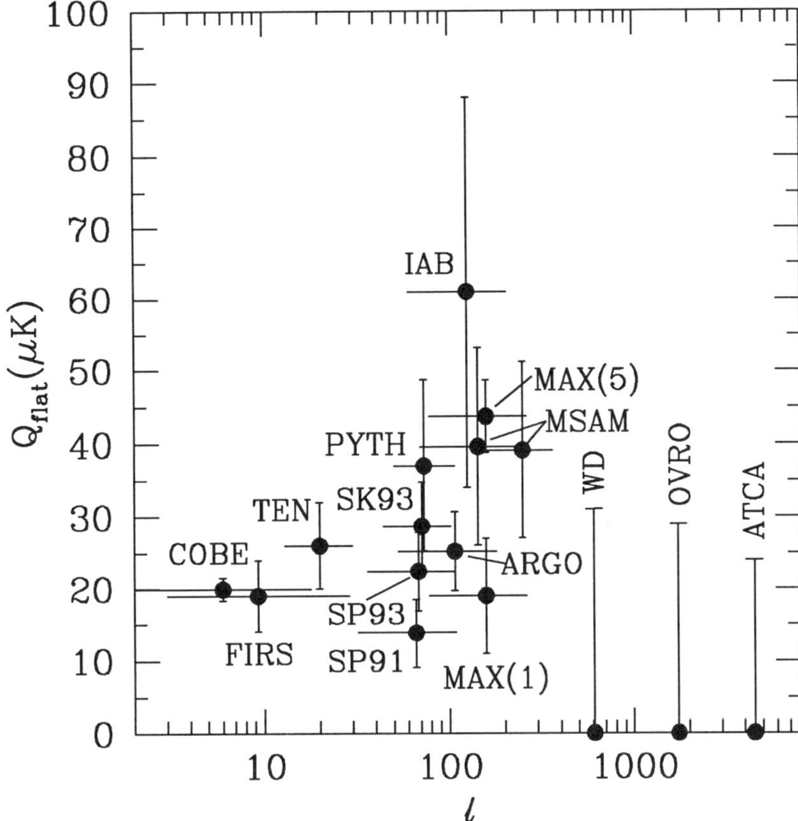

The 'power' in each experiment as a function of scale (multipole $\ell \sim \theta^{-1}$). $Q_{\rm flat}$ is the best-fitting amplitude of a flat power spectrum through the window function of the experiment, quoted at the quadrupole. The vertical error bars are $\pm 1\sigma$, while the horizontal lines represent the half-power ranges of the window functions. References and discussion of the data are presented elsewhere[4,5]. The results of 5 MAX scans have been combined into one point, with the discrepant μ Peg point plotted separately. The MSAM experiment has two independent modes. The three smaller-scale upper limits are plotted at the 95% confidence level. The general rise in the area around $\ell \simeq 200$ is evidence for a Doppler peak in the radiation power spectrum.

Consistency In order to compare experimental results, one crucial task is to translate all the experimental data into a common framework ('$\Delta T/T$' is simply not precise enough any more). This involves accurately calculating the window function for each experiment, then convolving a theoretically predicted curve with the window functions to normalize that theory for each experiment. Since the signal-to-noise for most experiments is still fairly modest, they are only really sensitive to the total 'power' seen through the window, and not to the detailed shape of the C_ℓ's. Power through the window is equivalent to the normalization for a theory that is a *flat* power spectrum. Because of this fact, we prefer to translate the results of an experiment into a value $Q_{\rm flat}$, which is the amplitude of a flat spectrum, quoted at the quadrupole scale for definiteness. This is plotted in the figure for all of the published anisotropy detections[4], together with the newer Saskatoon[7] and ACME/South Pole[8] results. A careful comparison of these $Q_{\rm flat}$ values shows that, in fact, the experimental results are not 'all over the place', as often stated. It is possible to find a smooth curve that is consistent with all of the results at about the 90% confidence level. Assuming that the data are fairly consistent then, can any conclusions be drawn about cosmology?

Doppler peaks? It seems that there is weak evidence for the presence of the so-called Doppler peak at around a degree, i.e. the degree-scale experiments tend to have higher values of $Q_{\rm flat}$ than the larger angular-scale experiments. No Doppler peak is a worse fit at about the 95% level (assuming Gaussian errors) than the best-fit Doppler peak model[4,5]. This conclusion means that we are starting to learn about the physics of the photon-baryon interactions at redshift ~ 1000. The existence of extra power at ℓ's of a few hundred is a direct and testable prediction of inflationary models like CDM and its variants.

Reionization Even if you are not prepared to take the current evidence for a Doppler peak seriously, it is clear that the Universe cannot have been ionized forever, unless you are prepared to disbelieve the results of several different experiments on degree-scales. This is because such fluctuations would be erased in a fully reionized scenario. Adopting a CDM model with maximal Ω_B gives an upper limit on the optical depth since reionization of $\tau \lesssim 0.7$. This, I believe, is something new that we have already learned about the Universe from CMB studies.

Ω_0 & Λ What about open or Λ models? The position of the main Doppler peak is a very clear indicator of the value of the density parameter Ω_0 — it moves to smaller angular scales for lower Ω_0. It seems that in the next year or two we will have data which will at least allow us to argue about the value of Ω_0 from the CMB. On the largest scales, it is still unclear whether the predictions of open inflationary models[9] will allow a test from *COBE* data. However, it is clear that a constraint on Λ, competitive with lensing constraints, could be obtained from the 4 years of DMR data[10]. There are also constraints that come from combinations of CMB and galaxy-scale measurements, since the matter amplitude at $z \simeq 0$ and the radiation amplitude at $z \simeq 1000$ have Ω_0 and Λ through the growth rate, and the power spectrum shape depends primarily on $\Omega_0 h$.

Defects Models with topological defects (e.g. cosmic strings or textures) provide an alternative to inflation for generating fluctuations. Such models may be constrained from *COBE* data alone and generally give non-Gaussian fluctuations at some level. However, it is crucial to obtain detailed calculations of degree-scale fluctuations in non-reionized defect models. Hand-waving arguments indicate that the Doppler peaks may have a testably different shape. There seems little point in discussing such models until the predictions can be confronted with the available data.

PIB: RIP? The baryon isocurvature scenario is difficult to rule out because it has a great deal of freedom. However, it seems fair to say that it has a hard time fitting the *COBE* DMR slope and the FIRAS y-constraint[11], not to mention the possible Doppler peak. For the CMB, the best-fit model tends to seem contrived to look as much like CDM as possible. But it is still sufficiently different that, as the degree-scale situation improves, it should provide the definitive test.

Satellites If there were to be a repeat of the efforts of the *COBE* satellite with today's technology and a significantly smaller angular scale, then we could gather a huge amount of information about the Universe. It would be possible to simultaneously measure Ω_0, Ω_B, H_0 and n, as well as constraining the cosmological constant, the contribution from gravity waves and the amount of reionization. This is the goal of a number of satellite projects currently being discussed (FIRE, PSI, MAP, COBRAS/SAMBA, ...). If such a project is successful and has enough angular resolution and frequency coverage (to subtract foreground contamination), then many of today's outstanding cosmological questions ought to be answerable.

Small scales Fluctuations at the smallest angular scales are best observed from the ground, and this is another area where there will be great expansion in the near future. Arc minute and arc second measurements will come into their own with interferometers, SCUBA, etc. looking for Sunyaev-Zel'dovich fluctuations and dust at high z. Theorists have still done little here[12], but there is great scope for learning about galaxy and cluster formation in the mm waveband.

The next few years will be very interesting for getting cosmological parameters from CMB anisotropy studies. Most experimental groups that have reported results have about the same amount of data again, awaiting full analysis. In only a year or two we are likely to be arguing about the question of Ω_0, as well as constraining isocurvature baryon models and defect models for structure formation. And the more distant future looks astonishingly promising. The CMB anisotropy field is currently in a phase of rapid progress (so much so that these comments are likely to be out of date by the time they see print!).

Acknowledgements Most of these comments are based on work or discussions with my collaborators Martin White, Wayne Hu, Ted Bunn, Naoshi Sugiyama and Joe Silk.

References

1. Smoot, G. *et al.*, 1992, *ApJ*, **396**, L1.
2. White, M., Scott, D. & Silk, J., 1994, *ARAA*, **32**, 319.
3. Hu, W., Scott, D., Sugiyama, N. & White, M., 1995, in preparation.
4. Scott, D. & White, M., 1994, in *CWRU CMB Workshop: 2 years after COBE*, ed. L. Krauss, World Scientific, in press.
5. Scott, D., Silk, J. & White, M., 1995, *Science*, in press.
6. Bunn, E. F., Scott, D. & White, M., 1995, *ApJ Lett.*, in press.
7. Netterfield, C. B., Jarosik, N., Page, L., Wilkinson, D. & Wollack, E., 1995, *ApJ Lett.*, submitted.
8. Gundersen, J. O., *et al.*, 1995, *ApJ Lett.*, submitted.
9. Bucher, M., Goldhaber, A. & Turok, N., 1995, *PRD*, in press.
10. Bunn, E. F. & Sugiyama, N., 1995, *ApJ*, in press.
11. Hu, W., Bunn, E. F. & Sugiyama, N., 1995, *ApJ*, submitted.
12. Bond, J. R., 1988, in *The Early Universe*, eds. W. G. Unruh & W. G. Semenoff, Reidel, Dordrecht, p. 283.

MAPPING WITH THE JODRELL BANK-TENERIFE RADIOMETERS

R.D. DAVIES[a], C.M. GUTIÉRREZ[a], R.REBOLO[b], R.A. WATSON[b], A.N. LASENBY[c] AND S. HANCOCK[c]

[a] *University of Manchester, NRAL, Jodrell Bank, Macclesfield SK11 9DL, UK*
[b] *Instituto de Astrofísica de Canarias, 38200 La Laguna, Tenerife, Spain*
[c] *MRAO, Cavendish Laboratory, Madingley Road, Cambridge CB3 OHE, UK*

INTRODUCTION

Following the first identification in 1967 of the cosmic microwave background[1] (CMB) and the detection of the dipole component[2,3] due to the Local Group motion, there has been a $\sim$ 25 year search for intrinsic structure in the CMB. The earliest predictions[4] suggested that *rms* values of $\Delta T/T$ for this structure might be of order 10^{-4} to 10^{-3} (where T is the mean temperature of 2.7 K). As experimental limits on $\Delta T/T$ were pushed successively lower, theoretical models for CMB structure were refined to match the observations. The recent unambiguous detection of CMB structure[5–8] confirmed by independent experiments has established the fluctuation amplitude to be $\Delta T/T \sim 10^{-5}$ on scales of several degrees and larger. The subject is now ready to move on to the mapping of individual structures in the CMB to provide the basic observational input to cosmogony.

On the angular scales investigated in the Tenerife experiments, the main contribution to the CMB structure is the Sachs-Wolfe effect[9] in which the CMB intensity is affected by the gravitational potential of large-scale mass concentrations in front of the last-scattering surface. In contrast to this scalar contribution, there is a possible tensor contribution[10] to the CMB structure arising from gravitational radiation originating in the inflationary era.

The weakness of the intrinsic signals (*rms* $\sim 30 - 50\ \mu$K) requires a careful assessment of the systematic effects. These can arise from the environment of the experiment, the atmosphere, solar system objects (Earth, Moon and Sun), the Galaxy and celestial point sources. The contribution of the effects related to the environment and the solar system objects can be established over long observing periods (ideally years) while the extra-solar system contribution is determined from observations covering a sufficiently wide range of frequencies. The ultimate confirmation comes from a comparison of independent experiments over a wide frequency range.

EXPERIMENTAL STRATEGY

The Tenerife experiments were designed to search for CMB structure in the region of the angular spectrum dominated by the Sachs-Wolfe effect in contrast to the Doppler peak at an angular scale of $\sim 1°$. We used a beamwidth of 5° and a switch of $\pm 8°$ on the sky to give a response of $(-0.5, +1.0, -0.5)$. Our experiments were accordingly sensitive to structure on scales of $5°-15°$. This triple-beam response was effective in removing a large fraction of the atmospheric water vapour emission and eliminated receiver drifts. Observations were made by having the radiometers fixed to the ground and making scans in RA at fixed declination using Earth rotation. Data were taken continuously over several years. Details of the radiometers are given elsewhere[11)].

Three frequencies were chosen for the observations to cover a range of $\sim 3:1$. The highest frequency was 33 GHz, compatible with obtaining a significant ($\sim$ 30 %) amount of observing time unaffected by water vapour at Teide Observatory which is located at an elevation of 2400 m. The other two frequencies were 10 and 15 GHz. The lowest of these was used to obtain a first order indication of the Galactic contribution. At high Galactic latitude, where our observations were concentrated, the lower frequency surveys at 408 and 1420 MHz[12,13)] are not suitable for determining the Galactic contribution to the higher frequency scans[14)] and accordingly the 10 GHz and 15 GHz observations are fundamental for this purpose.

RESULTS FOR A SINGLE DECLINATION (+40°)

The first deep integrations with the Tenerife radiometers were made at Dec=+40°, chosen because it contains the strong radio sources Cyg A, Cyg X and Per A (3C 84) which can be used to authenticate the daily scans. Data at 10 GHz have now been taken over ten years; the Dec=+40° observations have been interspersed with other declinations. At 15 GHz and 33 GHz the data cover four and two years respectively. After removal of data at times when water vapour increases the receiver *rms* by a factor of 3 and the times when the Sun and the Moon are within 50° and 30° respectively, the data were stacked at each frequency to obtain a high sensitivity scan. Long-term drifts in the baseline of individual days were removed by a maximum entropy technique. Independent data sets were obtained at each of the three frequencies covering entirely different epochs. Structure common to the independent data sets is therefore not expected to be due to the atmosphere, the Sun and Moon or receiver drifts.

Correlated astronomical signals of comparable amplitude were found in the high sensitivity data at frequencies of 15 and 33 GHz. These are considered to originate in the CMB and were accordingly added to provide a more sensitive display of the CMB structure. The equivalent amplitude of the structure can be expressed either as the amplitude $\sqrt{C_0}$ of the intrinsic auto-correlation function of the signal giving $\sqrt{C_0} = 54 \pm 14\ \mu$K, or as an equivalent quadrupole amplitude based on a Harrison-Zeldovich spectrum in which case one obtains $Q_{rms} = 26 \pm 6\ \mu$K[6)]. These observations gave the first unambiguous identification of hot and cold spots in the CMB.

We compared these results with those from the COBE DMR which has a similar, although somewhat larger, angular resolution. The DMR scans in this region are a

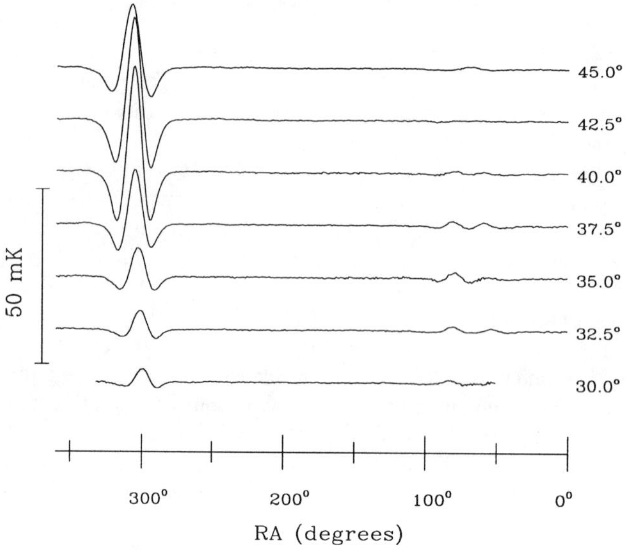

Figure 1: 15 GHz RA scans with the Tenerife beam-switching radiometer at fixed declination in the range +30° to +45°. The beamwidth is 5° and the beamswitch is ±8°.

factor ~ 2 more sensitive than in the rest of the sky by virtue of its high ecliptic latitude. The most prominent feature in the Tenerife data (~ 80 μK) is evident in both the 53 and 90 GHz DMR maps and has the Planckian spectrum expected for CMB structure[8)]. These results are described by Lineweaver *et al.* in a companion paper in this volume.

MAPPING AT 5° RESOLUTION

The objective of our mapping programme is to cover the declination range +30° to +45° at 10, 15 and 33 GHz. This area will be covered at the full sampling theorem separation of 2°.5 in declination for a beamwidth of 5°. The survey will therefore cover an area of 1.5 steradians and provide more than 100 beam-sized areas for study at high Galactic latitudes.

Figure 1 shows the data stacks at present available at 15 GHz for declinations +30° to +45°. The strong and weak crossings of the Galctic plane at 20^h30^m and $4^h - 6^h$ are evident in each stack. The high latitude regions are used for studies of structures in the CMB. Table 1 gives the sensitivity per beam area achieved up to the present in the high Galactic latitude RA range of 160° to 250° at each of the frequencies. The sensitivity per beam area is approaching our goal of 20 μK at 15 and 33 GHz where the structure is dominated by the CMB, and 50 μK at 10 GHz which is used to assess the Galactic contribution. We have identified a CMB feature at

Dec=+40°, RA=188° and a likely candidate at Dec=+35°, RA=218°. Both features have brightness temperatures of ~ 80 μK as measured in our 5° beamwidth.

Table 1: Rms sensitivities (in μK) per 5° beamwidth for the Tenerife radiometers operating at 10, 15 and 33 GHz in the declination range +30° to +45°, the RA range is 160° to 250°.

	Declination						
Experiment	30°0	32°5	35°0	37°5	40°0	42°5	45°0
10 GHz	–	69	97	62	57	75	77
15 GHz	22	27	30	19	30	25	24
33 GHz	–	38	54	47	21	–	–

The results from this larger area of the sky will allow us to determine more accurate values of the *rms* amplitude ($\sqrt{C_0}$ and Q_{rms}) of CMB structure with reduced contributions from receiver noise and from sample variance. Of particular interest will be a comparison with the 4-year data from COBE DMR to obtain the index n of the power law describing the amplitude of CMB structure as a function of angular wave number. The value of n is critical in establishing the structure formation scenario in the early Universe, and in particular the role of gravitational waves in producing such structure.

ACKNOWLEDGEMENTS

The Tenerife experiments are supported by the UK PPARC, the European Community "Science" programme SC1-CT92-830, and the Spanish DGICYT science programmes.

REFERENCES

1. Penzias, A.A. & Wilson, R.W. 1965, *Astrophys. J.*, **142**, 419-421.
2. Conklin, E.K. 1969, *Nature*, **222**, 971-972.
3. Corey, B.E. & Wilkinson, D.T. 1976, Bull. A.A.S., **8**, 351.
4. Silk, J. 1968, *Astrophys. J.*, **159**, 459-471.
5. Smoot, G.F. *et al.* 1992, *Astrophys. J.*, **396**, L1-L5.
6. Hancock, S. *et al.*, 1994, *Nature*, **367**, 333-338.
7. Ganga, K., Cheng, E., Meyer, S., & Page, L. 1993, *Astrophys. J.*, **410**, L57-L60.
8. Lineweaver, C.H. *et al.*, 1995, *Astrophys. J.*, (*in press*).
9. Sachs, R.K. & Wolfe, A.M. 1967, *Astrophys. J.*, **147**, 73-90.
10. Davis, R.L., Hodges, H.M., Smoot, G.F., Steinhardt, P.J. & Turner, M.S.
11. Davies, R.D. *et al.*, 1992, *Mon. Not. Roy. Astron. Soc.*, **258**, 605-615.
12. Haslam, C.G.T., Salter, C.J., Stoffel, H., & Wilson, W.E. 1982, *Astron. Astrophys. Suppl. Ser.*, **47**, 1-143.
13. Reich, P., & Reich, W. 1988, *Astron. Astrophys. Suppl. Ser.*, **74**, 7-28.
14. Davies, R.D. & Watson, R.A. 1995 (in preparation).

Comments on the Comparison of the COBE DMR and Tenerife Data

CHARLES. H. LINEWEAVER

Université Louis Pasteur
Observatoire Astronomique de Strasbourg
11 rue de l'Université, 67000 Strasbourg, France
charley@cdsxb6.u-strasbg.fr

INTRODUCTION

In the Spring of 1992, the COBE DMR team announced the discovery of anisotropies in the cosmic microwave background radiation[1] (CMB). The two year results[2] confirm the first year detection. The importance of this discovery motivates every possible effort to verify it. To confirm or falsify the DMR detection, independent measurements at large angular scales over a broad range of frequencies are needed, i.e., experiments at frequencies below and above the DMR frequencies with window functions which overlap with the DMR window function. Two experiments in a position to provide confirming or falsifying evidence are the FIRS balloon-born experiment at 170 GHz with a single beam FWHM of 3.8° and the Tenerife mountain-based triple-beam experiment (5.2° FWHM with a 8.1° chop). Ganga et al. (1993) performed a cross-correlation of the first year DMR map with the 170 GHz FIRS map covering $\sim$ 20% of the sky. Their results support the DMR anisotropy detection. Here we discuss the recent comparison by Lineweaver et al. (1995) of the DMR two year data with the Tenerife data. The main importance of this work is that it extends the range of frequencies over which validation of the DMR anisotropy detection has been done, thus substantially reducing the probability that a systematic error or Galactic foreground is responsible for the anisotropies.

GALACTIC FOREGROUND

Perhaps the largest difficulty in the detection of CMB anisotropy is the emission from our own Galaxy. There are three known sources of Galactic foreground: synchrotron radiation from relativistic electrons spiralling in the Galactic magnetic field, free-free emission (bremsstrahlung) from high density regions as electrons are deflected by electric fields of ions, and thermal emission from warm dust. Estimates of these three foregrounds as a function of frequency are displayed in Figure 1. The antenna temperatures of these foregrounds can be represented by $T \propto \nu^{\beta}$ where the exponent is either β_{syn}, β_{ff} or β_{dust} equal to -2.8, -2.1 and $+1.5$ respectively. The thickness of the grey bands in Figure 1 are estimates of the variation within the Galactic latitude range $15° \lesssim |b| \lesssim 70°$. Differentiation of these foregrounds

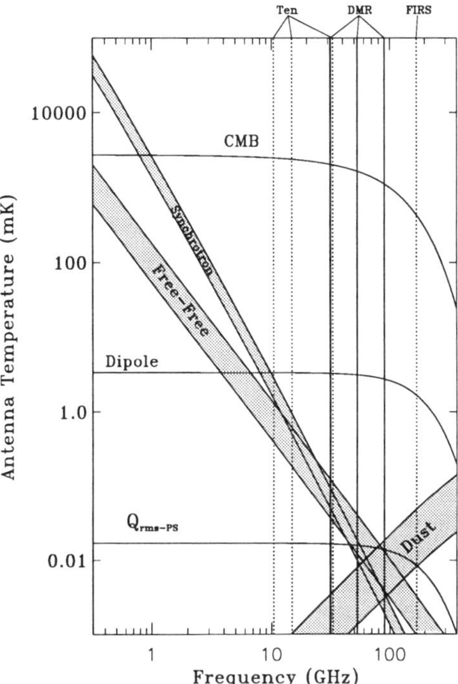

Figure 1. Galactic Foregrounds The main importance of the DMR-Tenerife comparison is in the extension of the frequency coverage over which the same structures have been see. The frequencies of the Tenerife, DMR and FIRS experiments are indicated at the top. The frequency dependences of the synchrotron, free-free and dust emission from our Galaxy are shown. The gray bands indicate the galactic latitude dependences within the range $15° \lesssim |b| \lesssim 70°$. The wide frequency range over which spectrally flat CMB anisotropies are detected is strong evidence for their non-Galactic nature. The CMB monopole, dipole and $Q_{rms-PS} = 17\mu$K are shown for reference. Figure adapted from Bennett et al. (1992).

is facilitated by the fact that the flattest spectrum foreground (dust) is also the most concentrated in the Galactic plane, while the least concentrated component (synchrotron) can be most easily separated by its strong spectral dependence. The frequencies of the DMR, Tenerife and FIRS experiments are indicated at the top. At the two most sensitive DMR frequencies (53 and 90 GHz), the free-free contribution is probably the most problematic. In the FIRS 170 GHz map, dust is the major Galactic foreground while synchrotron and free-free are the major Galactic foregrounds at the Tenerife frequencies. The comparison of the DMR and Tenerife data was in the high Galactic latitude, $|b| > 56°$ region. The synchrotron dominated 10 GHz channel of the Tenerife data was not used.

The Galactic foreground can be a useful calibrator. For example, Lineweaver et al. (1995) used the bright free-free source in Cygnus in the Galactic plane to show that the DMR and Tenerife experiments are cross-calibrated to better than 5%. Thus, in this comparison, the relative normalization is not as big a problem as it is for the DMR-FIRS comparison.

WINDOW FUNCTIONS

The DMR is most sensitive to fluctuations at large angular scales $\theta \gtrsim 7°$ while the Tenerife experiment is sensitive to fluctuations in the range $10° \gtrsim \theta \gtrsim 4°$. The comparison of the DMR and Tenerife data is made difficult by the limited overlap of the window functions (Figure 2). The window function G_ℓ^2 of a given experiment

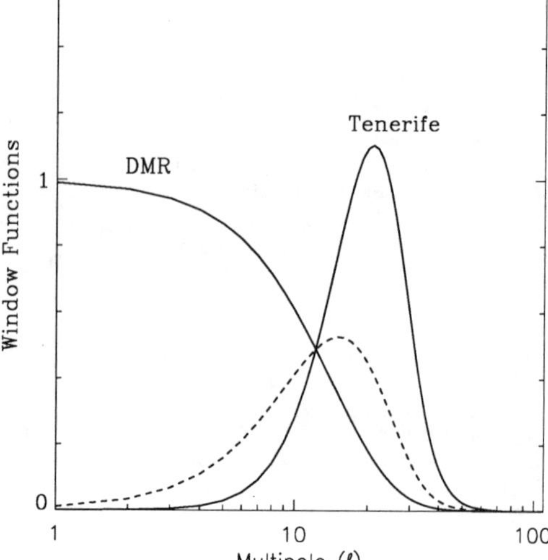

Figure 2. Window Function Comparison
Both experiments can see structure at a scale of $\ell \sim 15$ (FWHM$\sim 8°$). The dotted curve is the square root of the product of the two window functions.

can be defined by

$$\sigma^2 = \frac{1}{4\pi} \sum_{\ell} a_{\ell,m}^2 G_{\ell}^2, \tag{1}$$

where σ^2 is the temperature fluctuation observed by the experiment and the $a_{\ell m}$ are the coefficients of the spherical harmonic decomposition of the true CMB sky. Figure 2 compares the DMR window function[6] and the Tenerife window function. See White (1995) for a good discussion of window functions of CMB experiments.

RESULTS

One of the main results of the Lineweaver et al. (1995) comparison is shown in Figure 3 (their Figure 3h). The probability that the level of agreement seen between these two scans could be the result of instrument noise is less than 5%. The most

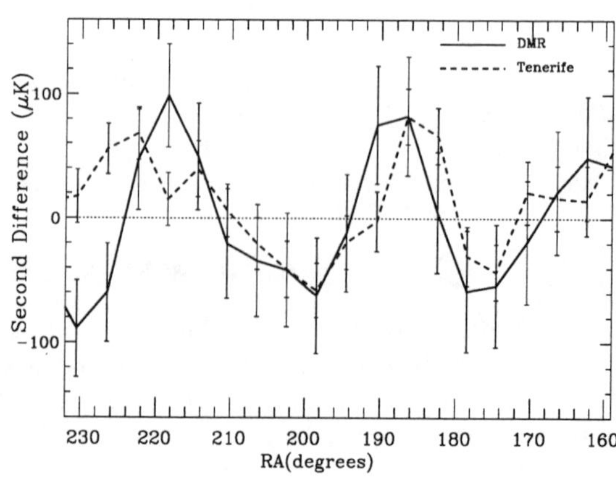

Figure 3. Scan Comparison
The DMR scan is the weighted addition of three scans resulting from passing the Tenerife triple-beam pattern over the DMR 31, 53 and 90 GHz maps. The Tenerife scan is the weighted addition of scans at 15 and 33 GHz. The probability that the level of agreement seen between these two scans could be the result of instrument noise is less than 5%. These scans are at declination +40°. The y-axis is in thermodynamic (not antenna) temperature. See Lineweaver et al. (1995) for further details.

prominent feature in the Tenerife data[7], ($\Delta T \sim 80\mu K$), is centered at $\sim 186°$ R.A. It appears to be present in the DMR maps and has the Planckian spectrum expected for CMB anisotropy. The combination of the spatial and spectral information from the two data sets is consistent with the presence of cosmic microwave background anisotropies common to both. Known sources of Galactic foreground do not seem to be a problem here because of their strong spectral dependences over the wide spectral range of the comparison. For example, since the frequencies in the two experiments differ by as much as six (= 90 GHz/15 GHz) the amplitudes of synchrotron, free-free or dust bumps at 15 GHz are expected to be respectively, 125, 43 or 1/15 times their amplitudes at 90 GHz.

DISCUSSION

So far, confirmation of the DMR discovery has come from

- The DMR two year results[2]
- The DMR-FIRS comparison[3].
- The DMR-Tenerife comparison[4].
- Claimed detections at smaller angular scales at the approximate level expected from the DMR results.

The consistency of the second year DMR data with the first year is a necessary but not sufficient condition for confirmation, (we could have made the same mistake in analyzing the second year as we did in analyzing the first). Similarly, statistical confirmations, i.e., comparable rms fluctuation levels at smaller angular scales, are useful but are model dependent and thus do not provide direct confirmation. The spot by spot comparison of the FIRS and now the Tenerife data sets seem to provide strong evidence that the anisotropies in the COBE DMR maps are real CMB fluctuations.

We thank all our colleagues involved in the Tenerife and DMR projects whose efforts have made this work possible.

References

1. Smoot, G. F., et al. 1992, Ap.J, **396**, L1
2. Bennett, C. L., et al. 1994, Ap.J , **436**, 423
3. Ganga, K., Cheng, E., Meyer, S., Page, L. 1993, Ap.J., **410**, L57
4. Lineweaver, et al. 1995, Ap.J., in press
5. Bennett, C. L., et al. 1992, Ap.J., **396**, L7
6. Wright, E. L., et al. 1994, Ap.J., **420**, 1
7. Hancock, S., et al. 1994, Nature, **367**, 333
8. White, M, 1995, Ap.J., in press

The Slope of Matter Density Perturbations from Tenerife and COBE/DMR

F. ATRIO–BARANDELA[a], L. CAYON[a,b] AND J. SILK[a,c]

[a] *Center for Particle Astrophysics, University of California, Berkeley, CA 94720*

[b] *Lawrence Berkeley Laboratory, University of California, Berkeley, CA 94720*

[c] *Department of Astronomy, University of California, Berkeley, CA 94720*

INTRODUCTION

Temperature fluctuations on the cosmic microwave background provide a unique test of models of galaxy formation. Comparison of theoretical predictions and observations requires a careful analysis of the statistical uncertainties associated with each measurement –systematic errors, cosmic and sampling variances– to conclude when a given model is ruled out by observations. Currently, the observational situation is far from clear. For example, on large angular scales comparison of COBE/DMR and Tenerife indicates that the slope of matter density perturbations is $n > 1$ at 2σ confidence level[1], the favourite value of inflation, when the COBE/DMR data alone indicates that n is close to unity[2].

In this paper we propose a new method to analyze observations. Our purpose is to use not only the variance of the temperature field measured in a particular region of the sky but also the full information provided by the observed temperature pattern. As an example we use this extra information to determine the level at which the observations of COBE/DMR and Tenerife are compatible.

CONSTRAINED SKY SIMULATIONS

In theoretical studies of the CBR it is conventional to express the temperature anisotropy in terms of spherical harmonics: $\Delta T/T(\mathbf{x}) = \sum_{lm} a_{lm} Y_{lm}(\Omega_{\mathbf{x}})$. In a Gaussian theory like inflation the coefficients a_{lm}

are gaussian random variables and their variance can be numerically calculated for each particular model of large scale structure. Given a realization of a GRF it is possible to construct a different one, with the same power and statistical properties, but restricted to have a particular value at some prespecified locations[3,4]. Let us assume that we want to generate a constrained realization of the sky $\Delta T/T_{const}$ that takes values $\{c_i\}(i=1,..,N)$ at N particular directions $\mathbf{x_i}$ on the sky. If we denote by $\xi_{ij} = \sum_{l,m} a_{lm}^2 P_l(\mathbf{x}_i \cdot \mathbf{x}_j)$ the correlation matrix between the directions $\mathbf{x}_i$ and $\mathbf{x}_j$ $(i,j=1,..,N)$, and $\xi_i(\mathbf{x})$ the correlation function between $\mathbf{x}_i$ and $\mathbf{x}$, the Hoffman–Ribak prescription states

$$\Delta T/T_{const}(\mathbf{x}) = \Delta T/T(\mathbf{x}) + \sum_{i,j=1}^{N} \xi_i(\mathbf{x})\xi_{ij}^{-1}(c_j - \tilde{c}_j) \quad (1)$$

is a realization derived from the same power spectrum than the original one. It can be easily verified that $\Delta T/T_{const}(\mathbf{x}_i) = c_i$ at the N previously specified locations $\mathbf{x}_i$. In the previous expression ξ_{ij}^{-1} represents the inverse of the correlation matrix ξ_{ij}.

COMPARISON OF TENERIFE AND COBE/DMR

Recently, comparison of Tenerife and COBE/DMR has shown that the structure seen on the Tenerife scan is also present in the COBE/DMR second year data[5]. To investigate whether this is the source of the aparent contradiction between the value of the slope n of matter density perturbations extracted from COBE/DMR or from Tenerife we estimate it by sampling a very active region of the microwave sky obtained from a $n = 1$ power spectrum.

First, we generate a whole sky map by drawing the amplitude of the spherical harmonic expansion from a model with variance $< |a_{lm}|^2 >= 4\pi Q_{rms-PS}/5l(l+1)$. Only multipoles up to $l = 100$ are used to generate the unconstrained map. This map is filtered with a gaussian window of FWHM $\beta = 7^o$ that corresponds to the angular resolution of COBE/DMR. We superimpose a hot spot of amplitude $\nu\sigma$ on the area sampled by Tenerife –σ represents the dispersion of the unconstrained sky map–. Next, we recover the multipoles from the constrained map up to $l = 20$ since, due to filtering, higher multipoles are not affected by the constraining process. To those multipoles we add the other 80 up to $l = 100$. Finally, with these a_{lm} coeficients weighted with Tenerife window function we construct simulated scans. We substract any baseline to guarantee that the mean temperature is zero.

Since the amplitude of the spot is unknown, in each map we superpose five different spots of different amplitude: $\nu = 3.5, 3, 2.5, 2, 1.5$ times the variance of COBE/DMR and we calculate one scan for each height. We generate a total 1200 maps and 6000 scans and for each one we compare the amplitude

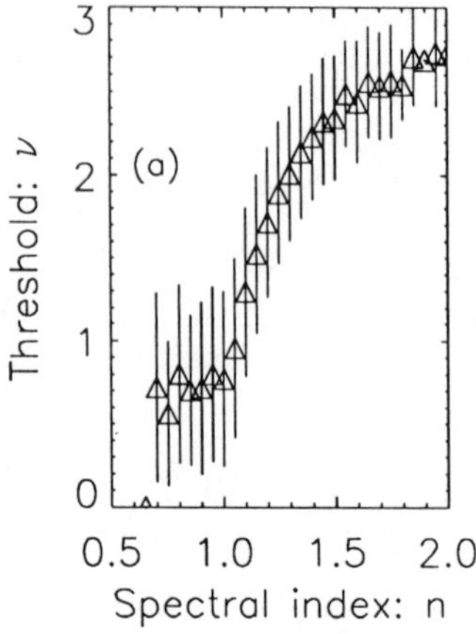

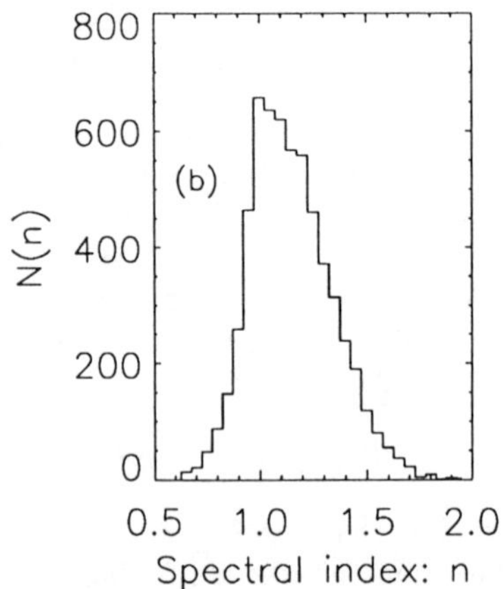

FIG 1.- (a) Amplitude of the spot in a Tenerife scan as a function of the slope computed from it. The error bars indicate the variance in the height of the spots that produce the same spectral index. (b) Indicates the total number of spots used to compute each error bars. Where there is no error bar, only one simulation fell into that interval.

of the correlation function at zero lag with the theoretical expected variance

$$C(0) = \frac{1}{4\pi} \sum_{l \geq 2} (2l+1) a_l^2 W_l \tag{2}$$

where W_l is the window function of Tenerife. In general, the multipole coefficients a_l^2 depend on the spectral index n and the relation above can be used to scale the amplitude measured in the COBE/DMR map to the *r.m.s.* temperature fluctuation on each scan as a function of n. Because our constrained realization sample a hot spot, the correlation function at the origin for those scans is efectively larger than it would be otherwise, so one would deduce a larger value of n when the COBE/DMR amplitude is scaled to Tenerife angular scales. In Fig.1a we show the estimated slope of the power spectrum as a function of the height of the maximum $\nu\tilde{\sigma}$, where $\tilde{\sigma}$ is the variance of the scan.

The error bars in Fig.1a represent the dispersion on the amplitude of the spot calculated from the fraction of scans that predict the same spectral index. The total number of those events is given in Fig.1b. From this study one can conclude that even though a Tenerife scan that crosses a very active region of the COBE sky would lead to a higher value of the spectral index n, only in very few cases -corresponding to spots of $\nu = 3, 3.5$ in the COBE/DMR map- one would obtain a value as high as $n = 1.7$.

Let us remark that if Tenerife is sampling a field on the sky that contains a significant spot on COBE/DMR scales the estimated value of the power spectrum can change significantly. For example, if the spot has amplitude of 1.5 σ we could overestimate the power spectrum by as much as 20% (see Fig.1a). However, it is very unlikely in any circunstance that a spectral index of $n = 1.7$ in Tenerife could come from a biased sample of an underline $n = 1$ temperature field. This statement is consequence of the well known fact that cosmic and sampling variances on Tenerife scales are rather small[6,7]. A full account of the likelihood that a given value of n is actually seen in one scan will be given elsewhere[8].

To summarize, we would like to remark that in this communication we have applied the formalism of constrained gaussian random fields to the analysis of CBR data. This technique could become useful when comparing data from different experiments. Since all experiments are bound to measure temperature fluctuation on the same sky, data observed by experiments with comparable angular resolution would be correlated. Therefore, prior information of an experiment should be used both to compare different data sets and to design optimal experiments.

REFERENCES

1. Hancock, S. et al. 1994, Nature, 367, 333.
2. Gorski, K. M. et al. 1994, Ap. J., 430, L89.
3. Bertschinger, E. 1987, Ap.J., 323, L103.
4. Hoffman, Y. & Ribak, E. 1991, Ap.J., 380, L5.
5. Linewaver, C. et al. 1994, astro-ph/9411097
6. Cayón, L., Martínez-González, E. & Sanz, J.L. 1991, M.N.R.A.S., 253, 599
7. Scott, D., Srednicki, M. & White, M. 1994, Ap. J., 421, L5
8. Atrio-Barandela, F., Cayón, L. & Silk, J. 1994, Ap. J. Lett. *submitted*

Estimating Microwave Power Spectra

MAX TEGMARK
Max-Planck-Institut für Physik
Föhringer Ring 6
D-80805 München, Germany
max@mppmu.mpg.de

INTRODUCTION

A new method for estimating the power spectrum C_ℓ from cosmic microwave background (CMB) maps was recently presented by the author[1] and applied to the 2 year COBE data, giving the results in Figure 1. It was found that the spectral resolution $\Delta\ell$ for COBE could be more than doubled at $\ell = 15$, thereby revealing previously unresolved features in the power spectrum. Whereas that paper was rather technical, the present paper shows that all qualitative features of this method can be understood from a simple analogy with quantum mechanics.

There has been a surge of interest in the cosmic microwave background radiation (CMB) since the first anisotropies of assumed cosmological origin were detected by the COBE DMR experiment[2]. On the experimental front, scores of new experiments have been carried out and many more are planned or proposed for the near future.

On the theoretical front, considerable progress has been made in understanding how the CMB power spectrum C_ℓ depends on various cosmological model parameters, both analytically[3] and quantitatively[4]. It is therefore quite timely to further strengthen the link between these two fronts, by better understanding how to extract more accurate power spectra from experimental data. Early work on this problem[5,6] has recently been extended and applied to the 2 year COBE data[7]. When estimating power spectra, it is customary to place both vertical and horizontal error bars on the data points, as in Figure 1. The former represent the uncertainty due to noise and cosmic variance, and the latter reflect the fact that

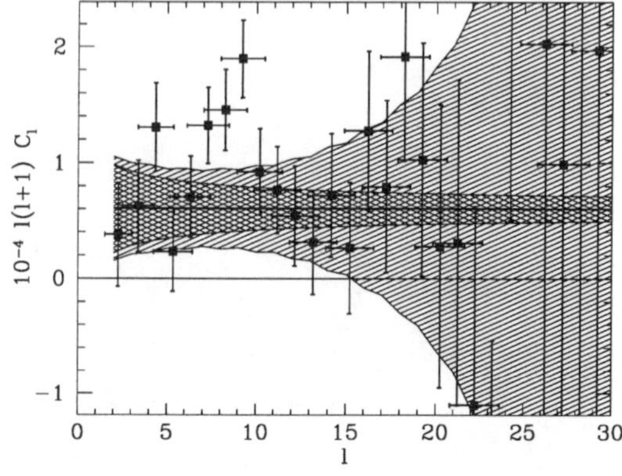

Figure 1. The power spectrum observed by COBE.
The observed multipoles are plotted with $1-\sigma$ vertical error bars, including both pixel noise and cosmic variance. The horizontal error bars show the r.m.s. window function widths. If the true power spectrum is given by $n = 1$ and $Q_{rms,ps} = 20\mu$K (heavy horizontal line), then the shaded region gives the $1-\sigma$ error bars and the double-shaded region shows the contribution from cosmic variance.

an estimate of C_ℓ inadvertently also receives contributions from other multipole moments. As is well-known, this unavoidable effect is caused by incomplete sky coverage, which destroys the orthogonality of the spherical harmonics. For typical ground- and balloon-based experiments probing degree scales, the spectral blurring $\Delta\ell/\ell$ tends to be of order unity, which makes it difficult to resolve details such as the number of Doppler peaks. By expanding the COBE sky map in spherical harmonics[7], the spectral resolution $\Delta\ell/\ell$ can be brought down to the order of 25%. The optimal method[1] reduces the horizontal error bars still further, down to their theoretical minimum, which is seen to be $\Delta\ell \approx 1$ for the COBE data in Figure 1. It involves solving a generalized eigenvalue problem numerically, were the matrices involved depend on the survey geometry and the pixel variances.

THE QUANTUM ANALOGY

Although all the elements of realism included in the numerical technique[1] are important when applying it to real data, one can in fact understand all of the qualitative features of what is going on by ignoring most of these complications.

First of all, let us ignore the fact that a real CMB sky map is pixelized. We let the function $x(\widehat{\mathbf{r}})$ denote the temperature fluctuation $\Delta T/T$ in the direction of the unit vector $\widehat{\mathbf{r}}$. In the discrete real-world case, each pixel has some r.m.s. noise variance σ^2 and effectively covers some solid angle Ω of the sky. For our continuous analogue, let us define the noise variance function $V(\widehat{\mathbf{r}})$ by $V \equiv \sigma^2\Omega$, where σ and Ω refer to a pixel in the direction $\widehat{\mathbf{r}}$. If we neglect contamination problems, it is easy to see that $V(\widehat{\mathbf{r}})$ is simply proportional to the inverse of the time spent observing the direction $\widehat{\mathbf{r}}$ (the small-scale details of the pointing of the antenna are irrelevant, as the beam smearing will ensure that the function V is smooth on angular scales below the beam width). Secondly, let us ignore the nuisance terms[8,1] from monopole, dipole, *etc.*, so that we can omit A from equation (1) of [1]. Finally, we let C_ℓ denote the power spectrum that results after the experimental beam smearing has been taken into account (this is simply the real power spectrum multiplied by some coefficients that approach zero for very large ℓ, corresponding to angular scales far below the beam width).

Let us focus on some fixed value $\ell = \ell^*$ for now, say $\ell^* = 17$. The analysis is then repeated from scratch for all other ℓ-values of interest. Since the power C_{ℓ^*} has units of $\mu\mathrm{K}^2$, we clearly want to estimate it by some quantity $\tilde{C}_{\ell^*}$ that is quadratic in the data. The simplest estimator of this type is

$$\tilde{C}_{\ell^*} \equiv \left|\int \psi(\widehat{\mathbf{r}})x(\widehat{\mathbf{r}})d\Omega\right|^2, \tag{1}$$

where ψ is some function on the sphere, and the most general estimate is readily shown to be just a weighted average of such estimates[1]. A straightforward calculation shows that

$$\langle\tilde{C}_{\ell^*}\rangle = \sum_{\ell=0}^{\infty}\sum_{m=-\ell}^{\ell}|\widehat{\psi}_{lm}|^2 C_\ell + \int|\psi(\widehat{\mathbf{r}})|^2 V(\widehat{\mathbf{r}})d\Omega, \tag{2}$$

where $\widehat{\psi}_{lm}$ denotes the spherical Fourier transform of ψ, *i.e.*, the coefficients in an expansion of ψ in spherical harmonics. In other words, we see that the expectation value of our estimator is the sum of two terms of quite different character. The first, the contribution from cosmology, is the power spectrum convolved with a window function $W_\ell \equiv \sum_m |\widehat{\psi}_{lm}|^2$. The second, the contribution from noise, is just an average value of V, the weights being $|\psi(\widehat{\mathbf{r}})|^2$. This is very similar to the result when estimating the power spectrum from a galaxy survey[9]. Just as in that paper, we will find it very convenient to use the standard Dirac quantum mechanics notation with kets, bras and linear operators. This allows us to write $\widehat{\psi}_{\ell m} = \langle \ell m|\psi\rangle$. A window function should always integrate to unity, so the correct normalization for ψ is just $\langle\psi|\psi\rangle = 1$. Defining the operator

$$L \equiv \sum_{\ell=0}^{\infty} \sum_{m=-\ell}^{\ell} \ell|\ell m\rangle\langle \ell m|, \tag{3}$$

equation (2) becomes simply

$$\langle \tilde{C}_{\ell^*}\rangle = \langle\psi|C_L + V(\widehat{\mathbf{r}})|\psi\rangle. \tag{4}$$

Note that L is a scalar operator satisfying $L|\ell m\rangle = \ell|\ell m\rangle$, and is related to the (vector) angular momentum operator $\mathbf{L} = -ir \times \nabla$ through $\mathbf{L}^2 = L(L+1)$.

Now what is the best choice of ψ? Equation (1) tells us that $\tilde{C}_{\ell^*}$ is the square modulus of a random variable whose real and imaginary parts are both Gaussian. Thus if ψ is real, the standard deviation of $\tilde{C}_{\ell^*}$ (the vertical error bar) is simply $\sqrt{2}$ times its expectation value. (If the real and imaginary parts contribute equally, this decreases by a factor of $\sqrt{2}$.) Basically, we minimize the vertical error bars by minimizing $\langle\tilde{C}_{\ell^*}\rangle$. But assuming that our window function is narrow enough that we are measuring mostly what we want to measure, C_ℓ, the first term in equation (4) satisfies $\langle\psi|C_L|\psi\rangle \approx C_\ell$, independent of ψ, so we minimize the vertical error bars by simply minimizing the second term, $\langle\psi|V(\widehat{\mathbf{r}})|\psi\rangle$. As a measure of the horizontal error bars, we will use $\Delta\ell$, the r.m.s. deviation of the window function from ℓ^*. With our quantum notation, we have simply $\Delta\ell^2 = \langle\psi|(L-\ell^*)^2|\psi\rangle$. As we will see, it is impossible to minimize both error bars at the same time, since there is a trade-off between them. It would be like asking for the best and cheapest car. Instead, the best we can do is minimize some linear combination $E \equiv \langle H\rangle$, where we have defined

$$H \equiv (L-\ell^*)^2 + \beta V(\widehat{\mathbf{r}}), \tag{5}$$

and the parameter β specifies how concerned we are about the vertical error bar relative to the horizontal one. Continuing our quantum analogy, we see that we want to find the ψ that minimizes the total "energy", where the "kinetic energy" $(L-\ell^*)^2$ corresponds to the horizontal error bar and the "potential energy" $\beta V(\widehat{\mathbf{r}})$ corresponds to the vertical error bar. If we for sake of illustration set $\ell^* = 1/2$, we simply want to minimize $\langle\psi|\mathbf{L}^2 + \beta V(\widehat{\mathbf{r}})|\psi\rangle$, given the constraint $\langle\psi|\psi\rangle = 1$. Introducing a Lagrange multiplier E, we arrive at the Schrödinger equation

$$[\mathbf{L}^2 + \beta V(\widehat{\mathbf{r}})]|\psi\rangle = E|\psi\rangle. \tag{6}$$

In other words, we want to find the ground state wavefunction for a particle confined to a sphere with some potential. From our knowledge of quantum mechanics, we can immediately draw a number of conclusions about the solution, all which turn out to agree well with the exact numerical results[1].

- $|\psi|^2$ will be small in regions where the noise variance is large, so regions that received little observation time will receive low weights in the analysis.
- Except for the case of complete sky coverage, we will have $\Delta\ell > 0$.
- If incomplete sky coverage confines ψ to a region of the sky whose angular diameter in the narrowest direction is of order $\Delta\theta$, then the uncertainty principle tells us that the minimum $\Delta\ell$ must be at least of order $1/\Delta\theta$.
- This limit on the spectral resolution is independent of ℓ.
- For a sky map of COBE type, where $\Delta\theta$ is of order a radian given a 20° galactic cut, the uncertainty principle thus gives $\Delta\ell \gtrsim 1$. This agrees well with the horizontal error bars actually attained in Figure 1.
- If the sky-coverage is incomplete, V is infinite outside of the region covered, and we recover the quantum-mechanical particle-in-a-box problem. From this we know that ψ will always go to zero smoothly as it approaches the survey boundary.
- This smoothness of ψ is really the gist of the method, as it radically reduces "ringing" in Fourier space, "kinetic energy", without increasing the "potential energy" $\langle\psi|V(\hat{\mathbf{r}})|\psi\rangle$ much at all.
- If the the survey volume consists of several disconnected parts, then $\Delta\ell$ is limited by the $\Delta\theta$ of the largest part. For the galaxy-cut COBE case, for instance, using only the northern half of the sky gives the same $\Delta\ell$ as using both the northern and southern skies combined. (However, including both of course helps reduce the *vertical* error bars.)

What typically happens when using the full optimization machinery[1] is that several solutions ψ are found to be almost equally good. For the COBE case, for instance, the ground state solution for a given ℓ^* comes out to be almost degenerate, with $(2\ell^*+1)$ orthogonal eigenfunctions all giving almost the same E. This is because the azimuthal symmetry of the galactic cut preserves the orthogonality of spherical harmonics corresponding to different m-values. When this happens, we obviously want to average the $(2\ell^*+1)$ different estimates of C_{ℓ^*}, as this reduces the vertical error bars by a factor of $\sqrt{2\ell^*+1}$ without hardly widening the horizontal ones at all. This is basically how Figure 1 was made.

The author wishes to thank Ted Bunn, George Efstathiou, Carlos Frenk and Joseph Silk for useful comments on the manuscript.

REFERENCES

1. Tegmark, M. 1994. Preprint astro-ph/9412064.
2. Smoot, G.F. *et al.* 1992. *Ap.J.* **396:** L1.
3. Hu, W. & Sugiyama, N. 1994. Preprint astro-ph/9411008.
4. Bond, J. R. *et al.* 1994. *Phys. Rev. Lett.* **72:** 13.
5. Peebles, P. J. E. 1973. *Ap.J.* **185:** 413.
6. Hauser, M. G. & Peebles, P. J. E. 1973. *Ap.J.* **185:** 757.
7. Wright, E.L. *et al.* 1994. *Ap.J.* **436:** 443.
8. Tegmark, M. & Bunn, E. F. 1994. Preprint astro-ph/941205.
9. Tegmark, M. 1995. Preprint astro-ph/9502012.

Structure Formation with Global Texture [1]

Ruth Durrer and Zhi–Hong Zhou
Universität Zürich, Institut für Theoretische Physik,
Winterthurerstrasse 190,
CH-8057 Zürich, Switzerland

INTRODUCTION

Observations show that the Cosmic Microwave Background (CMB) fluctuation spectrum on large scales as observed by COBE should be not very far from scale invariant[1]. This has been considered a great success for inflationary models of structure formation which predict a scale invariant fluctuation spectrum. Here we consider an alternative class of models which also yield scale invariant spectra: Models where perturbations are seeded by global topological defects which can form during symmetry breaking phase transitions in the early universe[2]. To be specific, we consider texture, π_3–defects which lead to event singularities in four dimensional spacetime[3]. The energy density of global topological defects scales like $1/t^2$ and thus always represents the same fraction of the total energy density of the universe.

$$\rho_T/\rho \sim 8\pi G\eta^2 \equiv 2\epsilon\ , \tag{1}$$

where η determines the symmetry breaking scale. The background spacetime is a Friedmann–Lemaître universe with $\Omega = 1$.

We present a fully gauge invariant and local system of perturbation equations. The (non–local) split into scalar, vector and tensor modes on hypersurfaces of constant time is not performed. We solve the equations numerically in a cold dark matter (CDM) universe with global texture. In this contribution, we present some of the main results, detailed derivations of the equations and a description of our numerical methods are given in [4] and partially in [5].

THE SYSTEM OF EQUATIONS

We calculate the CMB anisotropies on angular scales which are larger than the angle subtended by the horizon scale at decoupling of matter and radiation, $\theta > \theta_d$. For $\Omega = 1$ and $z_d \approx 1000$ we find $\theta_d = 1/\sqrt{z_d + 1} \approx 0.03 \approx 2^o$. It is therefore sufficient to study the generation and evolution of microwave background fluctuations after recombination. During this period, photons are influenced only by cosmic gravitational redshift and by perturbations in the gravitational field (if the medium is not reionized). This is described by Liouville's equation

$$X_g(f) = 0 \tag{2}$$

[1]This work has been supported by the Swiss National Science Foundation

for the photon distribution function which lives on a seven dimensional relativistic phase space $P_0\mathcal{M} = \{(x,p) \in T\mathcal{M} | g(x)(p,p) = 0\}$. (2) leads to a perturbation equation, which in terms of a gauge invariant variable

$$\chi = \nabla^2\langle\frac{\delta T}{T}\rangle + \text{monopole terms} + \text{dipole terms}$$

reads

$$(\partial_t + \gamma^i\partial_i)\chi = -3\gamma^i\partial^j E_{ij} - \gamma^k\gamma^j\epsilon_{kli}\partial_l B_{ij} \equiv S_T(t,\boldsymbol{x},\boldsymbol{\gamma}) \; , \qquad (3)$$

where ϵ_{kli} is the totally antisymmetric tensor in three dimensions, E_{ij} and B_{ij} are the electric and magnetic part of the Weyl tensor.

The divergence of the electric part of the Weyl tensor does not contain tensor perturbations. On the other hand, scalar perturbations do not induce a magnetic gravitational field. The second contribution to the source term in (3) represents a combination of vector and tensor perturbations. If vector perturbations are negligible, the two terms on the r.h.s. of (3) represent a split into scalar and tensor perturbations which is local. We find that the vector and tensor fluctuations represented by B contribute approximately 20 % to the total microwave background anisotropies[5].

Since the Weyl tensor of Friedmann–Lemaître universes vanishes, the r.h.s. of (3) is manifestly gauge invariant[6]. Therefore also the variable χ is gauge invariant.

The general solution to (3) is given by

$$\chi(t,\boldsymbol{x},\boldsymbol{\gamma}) = \int_{t_i}^{t} S_T(t',\boldsymbol{x}+(t'-t)\boldsymbol{\gamma},\boldsymbol{\gamma})dt' \; + \; \chi(t_i,\boldsymbol{x}+(t_i-t)\boldsymbol{\gamma},\boldsymbol{\gamma}) \; , \qquad (4)$$

where S_T is the source term given on the rhs of (3).

The electric and magnetic part of the Weyl tensor are determined by the perturbations in the energy momentum tensor via Einstein's equations[5]. In addition we need the evolution equation for the dark matter density perturbations and for the scalar field describing the textures[5].

RESULTS AND CONCLUSIONS

We have solved the above closed hyperbolic system numerically on a 200^3 grid for different random initial conditions on a NEC–SX3. The numerical methods employed and the different tests of our programs are described elsewhere[4]. Here we just want to present some main results.

We have determined the dark matter density and the temperature fluctuation spectra, $P(k) = |\delta(k)|^2$ and

$$\frac{\delta T}{T}(t_0,\boldsymbol{x},\boldsymbol{\gamma}) = \sum_{lm} a_{lm}(x)Y_{lm}(\boldsymbol{\gamma}) \; . \qquad (5)$$

As usual, we assume that the average over N_x different observer positions coincides with the ensemble average and determine

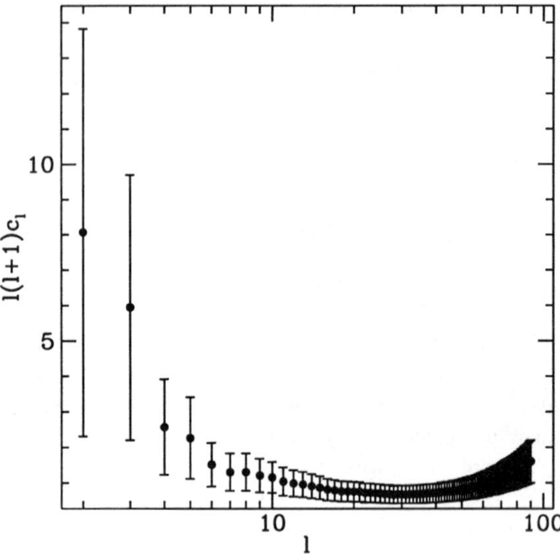

Figure 1: The harmonic amplitudes $l(l+1)c_l$ are shown with 1σ error bars. No smoothing is applied.

$$c_l = \frac{1}{(2l+1)N_x} \sum_{m,x} |a_{lm}(\boldsymbol{x})|^2 \ , \quad l \geq 2 \ . \tag{6}$$

We have performed several simulations on a 200^3 grid with 30 different observer positions for each simulation. The average harmonic amplitudes with 1σ variance are shown in Fig. 1. The low order multipoles strongly depend on the random initial conditions (cosmic variance), like in the spherically symmetric simulation[7]. From Fig. 1 it is clear that the texture scenario is compatible with a scale invariant spectrum (although a spectral index n slightly lower than 1 seems to be preferred). The main difference of the currently favored inflationary scenarios lies in the distribution of fluctuations which is non Gaussian in models with topological defects. The quadrupole amplitude is given by

$$Q = (0.65 \pm 0.16)\epsilon \ .$$

To reproduce the COBE amplitude[1] $Q_{COBE} = (0.6^{+0.37}_{-0.24})10^{-5}$, we have to normalize the spectrum by choosing the phase transition scale η according to

$$\epsilon = 4\pi G\eta^2 = (1.1 \pm 0.5)10^{-5} \ . \tag{7}$$

This value is comparable with other results[8, 9, 7].

The power spectrum of dark matter density fluctuations is shown in Fig. 2. It clearly fits very well a scale invariant spectrum. To be compatible with observations[10], a somewhat high bias factor of $b \approx 2 \pm 1$ is required on scales up to $20h^{-1}Mpc$.

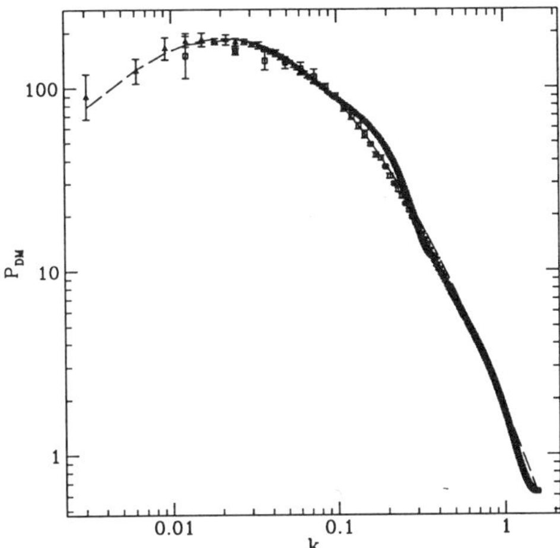

Figure 2: The power spectrum of the CDM fluctuations induced by texture. To enhance the dynamic range, three simulations with different physical grid size have been patched together. The vertical scale is arbitrary. The dashed line indicates the best fit scale invariant spectrum.

ACKNOWLEDGMENTS

We thank the staff at the Centro Svizzero di Calcolo Scientifico (CSCS) for valuable support. Especially we want to mention Andrea Bernasconi and Djiordic Maric.

References

[1] K.M. Gorski et al. *Ap. J.* **430**, L85 (1994).

[2] T.W.B. Kibble, *J. Phys.* **A9**, 1387 (1976).

[3] N. Turok, *Phys. Rev. Lett.* **63**, 2625 (1989).

[4] R. Durrer and Z.H. Zhou, in preparation.

[5] R. Durrer and Z.H. Zhou, *Phys. Rev. Lett.* in print (1995).

[6] J.M. Steward and M. Walker, *Proc. R. Soc. London* **A341**, 49 (1974).

[7] R. Durrer, A. Howard and Z.H. Zhou, *Phys. Rev.* **D49**, 681 (1994).

[8] D. Bennett and S.H. Rhie, *Astrophys. J.* **406**, L7 (1993).

[9] U.-L. Pen, D.N. Spergel and N. Turok, *Phys. Rev.* **D49**, 692 (1994).

[10] K.B. Fisher et al., *Ap. J.* **402**, 42 (1993).

Chasing CMB Photons Through the Nonlinear Universe[a]

ROBIN TULUIE AND PABLO LAGUNA

Department of Astronomy & Astrophysics
and
Center for Gravitational Physics & Geometry
The Pennsylvania State University
University Park, PA 16802

Detailed observations of the Cosmic Microwave Background (CMB) have renewed the interest in studying anisotropies of the CMB that are due to non-linear structures. Previous studies have considered the effect of a single, spherically symmetric density inhomogeneity on the CMB anisotropy using either a Swiss cheese model,[1,2,3,4] thin-shell approximation[5] or Tolman-Bondi solutions.[6,7,8,9] Other studies have parametrized the time dependence of the correlation function[10] or considered the effect of a single, non-symmetric structure.[11] The effect on the CMB of multiple, non-trivially distributed matter structures, as well as the complete and unambiguous continuous linkage between the linear, weakly clustered and fully non-linear epochs, has not been taken into consideration. Furthermore, even for single structures, the peculiar motion of matter structures has been ignored when considering late-time non-linear effects on the CMB.

Our study addresses the signatures in the microwave sky from time-varying gravitational potentials associated with non-linear density inhomogeneities in the universe without any of the above simplifications; in particular, we wish to find the correlation between these gravitationally induced anisotropies with the underlying density field and proper motion of the structures that photons encounter along their paths. Anisotropies in the CMB from time-varying potential are not only due to intrinsic changes in the density field of large scale structures, namely the Rees-Sciama effect,[1] but also due to the proper bulk motion of structures.[12] We consider two unbiased $\Omega = 1.0, h = 0.75$ Cold Dark Matter (CDM) models, $124h^{-1}$ Mpc and $248h^{-1}$ Mpc on a side. The typical void size in these models is $\sim 30h^{-1}$ Mpc and $\sim 60h^{-1}$ Mpc, respectively.

The temperature anisotropies of the CMB associated to gravitational potentials and proper motions are given by[13,14]

$$\frac{\Delta T}{T} = \phi_r - \phi_e + 2\int_r^e \frac{\partial \phi}{\partial t} dt \tag{1}$$

where each term represents the Sachs-Wolfe and Rees-Sciama, respectively. The numerical factors in equation (1) differ among various authors depending on the choice of gauge.[15] Our results were obtained using the longitudinal gauge.

[a]Work supported in part by NSF Young Investigator award PHY-9357219, NSF grant PHY-9309834 and NASA (at Los Alamos National Laboratory).

Our technique for calculating CMB temperature anisotropies proceeds in two stages.[14] First, starting at a redshift of $z = 1900$, just prior to recombination, a 64^3 particle-mesh model is used to simulate the dark matter collapse from an initial Harrisson-Zel'dovic spectrum modified by the appropriate damping term for CDM. The simulation evolves from the linear, through the weakly clustered, into the non-linear regime and terminates at z=0. No assumptions are made about the non-linear nature of the density perturbations; in particular, we allow large density inhomogeneities ($\delta\rho/\rho > 1$) as well as large peculiar bulk motions. We only require sub-horizon scales and non-relativistic proper velocities, so linear theory of cosmological perturbations holds.[16] The initial amplitude of the perturbations is the only free parameter in the model and chosen so that the r.m.s. overdensity in $8h^{-1}$Mpc scales, σ_8, reaches unity at z=0.

Next, an ensemble of CMB photons are propagated backwards in time through the evolving dark matter structures, starting at z=0 and finishing at z=1900. We propagate beams with up to 10^6 individual CMB photons, updating their individual temperature at every timestep by computing explicitly the integral in equation (**1**).

We have measured temperature anisotropies in a variety of directions and angular scales in our model, and the cosmic variance for each angular scale is typically less than 15 % for photon bundles with angular scales $> 2°$. Further simulations in different directions would increase the sky coverage and decrease the error due to cosmic variance.

We find that temperature fluctuations from time-varying potentials increase during the non-linear stage of the matter evolution and is almost entirely due to recent ($z \leq 3$) non-linear structures.[12] Furthermore, the accumulation of these temperature anisotropies is not exactly monotonic since CMB photons propagate through a series of clusters and voids in which the redshift and blueshift contributions do not exactly cancel, but accumulate in a non-trivial way.

Fig. 1

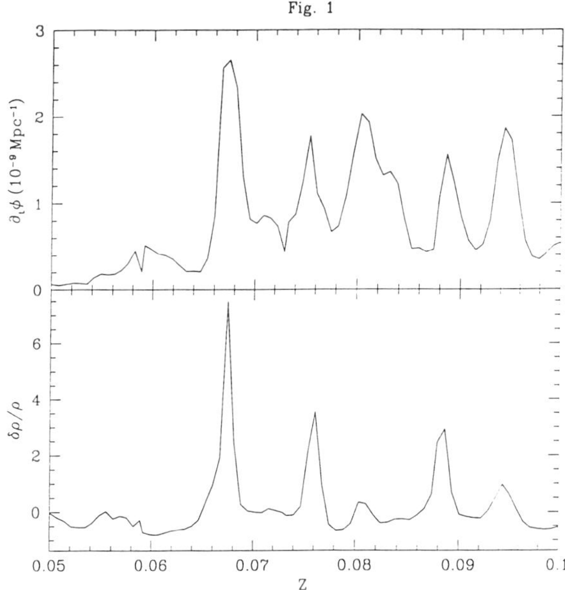

Fig. 1 shows the comparison between the average density contrast $\delta\rho/\rho$ and the r.m.s. change of the gravitational potential $\frac{\partial\phi}{\partial t}$ within the photon bundle for $z < 0.1$. The correlation between these quantities is very clear: the voids and overdensities that the photon beam traverses are directly mirrored by the temperature fluctuations through $\frac{\partial\phi}{\partial t}$. We also see from fig.1 that overdense structures provide the major contribution to the temperature fluctuations from time-varying potentials.

Fig. 2 shows a $8.0^\circ \times 8.0^\circ$ map of temperature fluctuations arising from the second term in equation (2), Rees-Sciama plus proper motion effects, for the $124h^{-1}$ Mpc CDM model. The anisotropies in this map are produced by several isolated structures $\sim 10h^{-1}$ Mpc across and located at $0.065 < z < 0.095$. The region was chosen for its prominent feature in fig. 1 and contains several voids and clusters. We have examined other regions as well, with similar results. The temperature fluctuations on $\geq 1^\circ$ angular scales range from 3×10^{-7} to 1×10^{-6} for the two CDM models we considered.

The correlation between temperature fluctuations and density inhomogeneities are performed via the column density contrast $\sigma \equiv \int_r^e \delta\rho/\rho\, dt$ of dark matter along the photon trajectories. Fig. 3 shows a map of σ, i.e. the galaxy clusters and voids "seen" by the CMB photons in fig. 2. There is a clear correlation between the temperature anisotropies (Fig. 2) and the column density contrast (Fig. 3), showing that indeed these temperature fluctuations from non-static potentials are caused by the matter structures.

The map in fig. 2 shows several dipole-like patterns; these dipoles are centered at the large clusters in the $0.065 < z < 0.095$ slab. The dipoles are due to the peculiar bulk motion of the large clusters or superclusters. Along the photon path, the motion of these structures effectively creates a change in the gravitational potential.[12] Hence even an intrinsically static potential can cause a significant temperature anisotropy as long as the *transverse* bulk velocity of the cluster is large enough. Indeed, we find that the effect is proportional to both the transverse cluster peculiar velocity and the maximum amplitude of the potential perturbation, and the separation between dipole peaks is indicative of the length scale of the gravitational potential perturbation from the cluster.[12,17]

In conclusion, our results show the gravitational influence of the dark matter dynamics on the CMB sky. The underlying matter structures used exhibit the sheet-filament-knot structure of voids, clusters and superclusters, providing a more realistic representation of the shape of density inhomogeneities in the universe. Since we propagate CMB photons through matter that evolves from the beginning of decoupling up to present time, our results naturally take into consideration not only the gravitational lensing of the CMB caused by the late, non-linear density perturbations, but also the cumulative effects on the CMB photons as they propagate through multiple and different regions of voids and clusters. Moreover, our numerical evolution avoids artificial splitting between linear, weakly clustered and nonlinear epochs. We find that the Rees-Sciama effect as well as the effect due to the proper

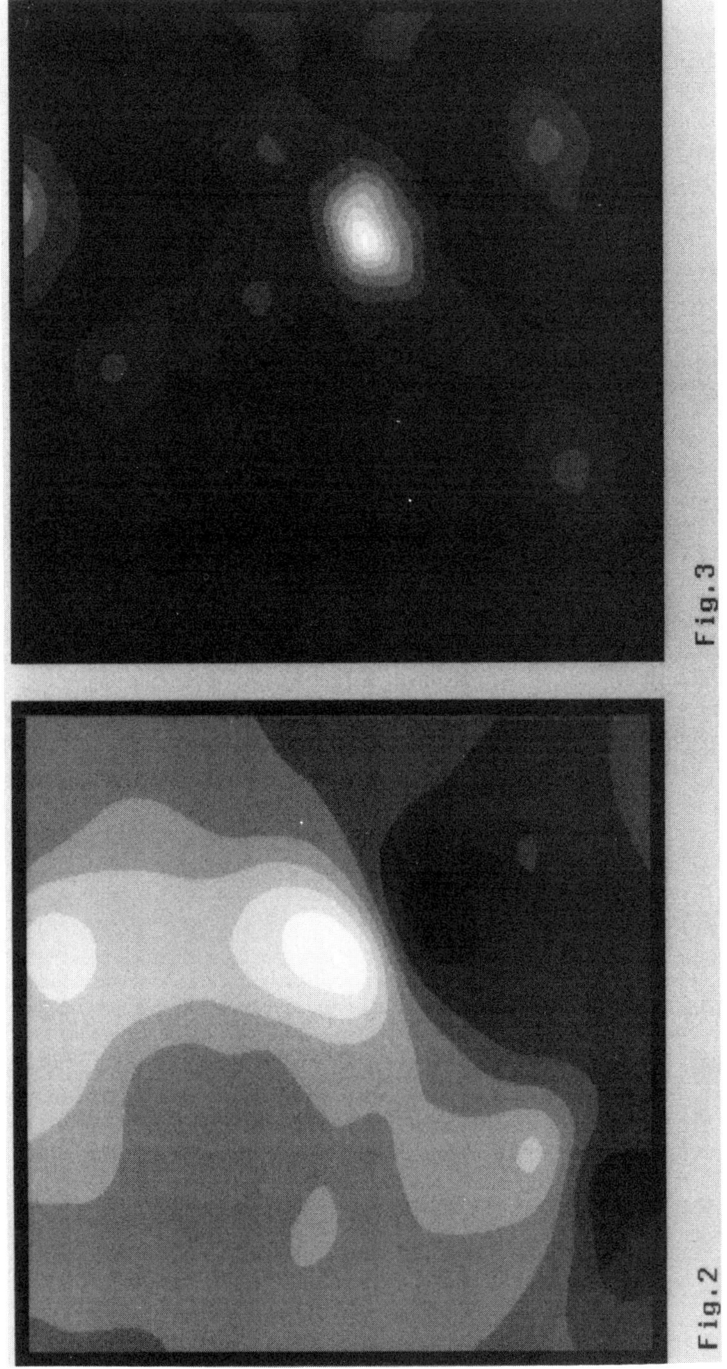

Fig.3

Fig.2

motion of the clusters and superclusters are small, with rms values on the order of a few $\times 10^{-7}$ – 10^{-6} for void scales $\sim 30 - 60h^{-1}$ Mpc and superclusters no larger than roughly $20h^{-1}$ Mpc. We also found that the temperature anisotropy created by the change of the potential as the cluster moves across the microwave sky is as important as that from the intrinsic evolution of the density field. The temperature anisotropies induced by peculiar velocities leave a dipole-like signature that in principle could yield information on the transverse velocity of structures.

REFERENCES

1. Rees, M.J. & Sciama, D.W. 1968, *Nature*, **517**, 611
2. Kaiser, N. 1982, *Mon. Not. R. Astr. Soc.*, **198**, 1033
3. Nottale, L. 1984, *Mon. Not. R. Astr. Soc.*, **206**, 713
4. Dyer, C.C., & Ip, P.S.S. 1988, *Mon. Not. R. Astr. Soc.*, **235**, 895
5. Thompson, K.L., & Vishniac, E.T. 1987, *Astrophys. J.*, **313**, 517
6. Mészarós, A. 1994, *Astrophys. J.*, **423**, 19
7. Fang, L. & Wu, X. 1993, *Astrophys. J.*, **408**, 25
8. Panek, M. 1992, *Astrophys. J.*, **388**, 225
9. Chodorowski, M. 1994, *Mon. Not. R. Astr. Soc.*, **266**, 897
10. Martínez-González, E., Sanz, J.L. & Silk, J. 1994, *Astrophys. J.*, **436**, L1
11. Van Kampen, E. & Martínez-Gonzá.lez, E. 1991 *in* Blanchard et al., eds., Second 'Recontres de Blois': Physical Cosmology, p. 522. Edition Frontièrs, Gif-sur-Yvette
12. Tuluie, R., & Laguna, P. 1995, submitted to *Astrophys. J. Lett.*
13. Sachs, R. K. & Wolfe, A. M. 1967, *Astrophys. J.*, **147**, 73
14. Anninos, P., Matzner, R. A., Tuluie, R. & Centrella, J. 1991, *Astrophys. J.*, **382**, 71
15. Padmanabhan, T. 1993, in *Structure Formation in the Universe* (Cambride), p227
16. Mukhanov, V.F., Feldman, H.A. & Brandenberger,R.H. 1992, *Phys. Rep.*, **215**, 203
17. Laguna, P. & Tuluie, R. 1995, to be submitted to *Astrophys. J.*

CMB Anisotropies Numerically vs. Analytically

NAOSHI SUGIYAMA[1,2] and WAYNE HU[1]

[1] *Department of Astronomy and Physics University of California, Berkeley, CA94720*

[2] *Department of Physics, University of Tokyo 116 Tokyo, Japan*

Because of rapidly increasing the number of observational data, Cosmic Microwave Background (CMB) anisotropies are becoming one of the most important tool to determine cosmological models and parameters, i.e., density parameter Ω_0, hubble constant h, cosmological constant Λ, baryon density Ω_B and so on. In order to get expected CMB anisotropies for each models with different cosmological parameters, we have to solve the Boltzmann equation for the distribution function of photons. There are several efforts to solve the equation numerically[1]. It is possible to get the power spectrum of CMB anisotropies in high accuracy from the numerical calculation. However, the final power spectrum contains a lot of different physical processes and it is very hard to separate each effects numerically. Moreover it takes long CPU time to calculate a specific model. Here we present an analytic treatment of CMB anisotropies[2] which makes us possible to separate each contributions. We can calculate CMB power spectrum by very short CPU time within about 10% accuracy in temperature. This is important for generating spectrum for many models.

There are several different contributions on CMB anisotropies[2,3].

(a)adiabatic oscillations: Before the recombination, photons and baryons are tightly coupled each other. Inside the sound horizon, fluctuations of this tight coupled fluid oscillate as acoustic waves. These acoustic oscillations create the peaks and wriggles on the power spectrum.

(b)diffusion Damping: The random walk of photons causes the exponential damping of fluctuations. The damping length is growing to infinity during the recombination although it is always smaller than the sound horizon before the recombination. On the other hand, the scattering probability between photons and baryons decreases and becomes negligibly small right after the recombination. Hence the total damping rate is smaller on larger scales.

(c)gravitational redshift: Climbing up the gravitational wall, photons loose energy and are redshifted. This redshift effect caused by the gravitational potential at the recombination (last scattering surface) is the famous Sachs-Wolfe (SW) effect. Gravitational potential stays constant during pure matter

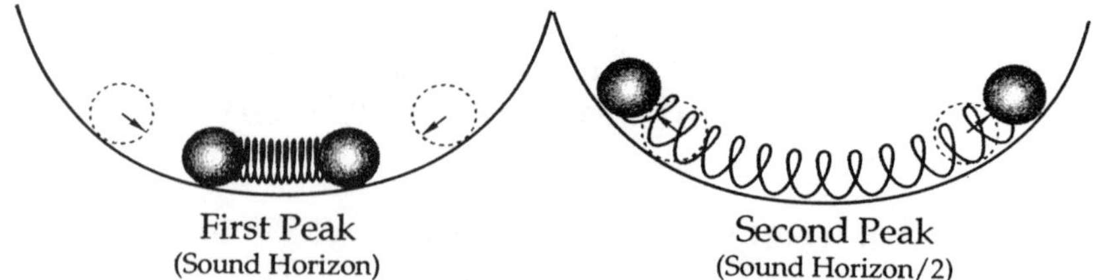

Figure 1: Analogy of accoustic oscillations. The gravitational potential wall is created by dark matter. The first compression mode at the last scattering surface corresponds to the first peak (left) and the first decompression mode to the second peak.

or radiation (for adiabatic perturbations) dominated regime in the flat $\Omega_0 = 1$ universe. However, it decays if one of these assumptions are broken. The decay of the gravitational potential also causes the redshift of photons. This effect is referred to here as the Integrated Sachs-Wolfe (ISW) effect. The ISW contribution is separated into two parts. Right after the recombination, if the universe is still not purely matter dominated, the potential decays. We call this as the early ISW effect. In case of the low density universe, however, the potential starts to decay very close to present epoch when the curvature or the cosmological constant starts to become the dominant component. We refer to this as the late ISW effect.
(d)doppler effect: If the baryon velocity is different from photon's, the baryon velocity induces temperature fluctuations. During the tight coupling regime, baryon and photon velocity are identical. If the universe is transparent, there is no interaction between photons and baryons. Therefore this effect only appears in late time reionized models.

We are taking into account all above physics in our analytic treatment. We solve the acoustic oscillations inside the gravitational wall which is generated by collisionless particles (dark matter) on small and intermediate scales. We can use the analogy of harmonic oscillator as shown in figure 1. A first compression mode at the last scattering surface corresponds to the first peak and a first depression mode corresponds to the second peak. On larger scales, we solve SW and ISW effects exactly. In order to get correct evolution of CMB fluctuations, we need to put recombination process, time evolution of the gravitational potential and even the anisotropic stress of massless neutrino as precisely as possible.

Although our treatment is very general, we show the application for CDM models as an example. Figure 2 is the comparison of angular power spectrum of CMB anisotropies C_ℓ between our analytic result and a numerical calculation. Very good agreement (within roughly 10% in power) is shown.

Using our simple analytic picture, we can easily explain the dependence

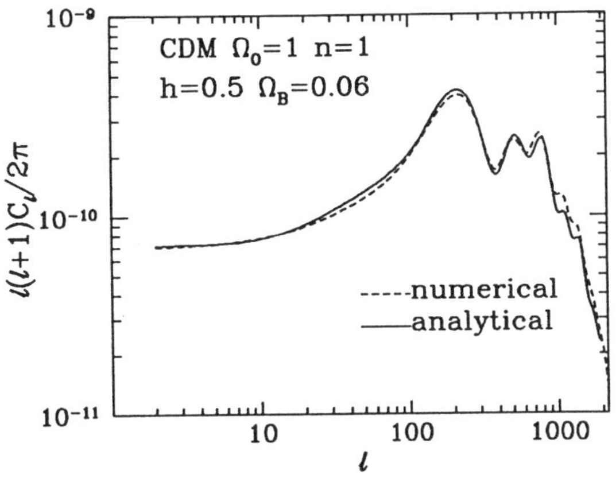

Figure 2: Angular power spectrum C_ℓ's of the CDM model. ℓ is the multipole moment which roughly corresponds to inverse of observed angle. A dashed line is the numerical calculation and a solid line is the analytical calculation.

of CMB spectrum on various cosmological parameters. First, we investigate the height of the peaks. General statements are followings. Pressure of photon-baryon fluid prevents fluctuations from growth. Hence more pressure (which corresponds to the stiffer string in our analogy) means lower peaks. On the other hand, Gravitational potential induces adiabatic growth. This causes blue-shift. Deeper potential seems to generate higher peaks. However, deeper potential wall means bigger SW effect which causes red-shift. Therefore deeper potential doesn't necessarily produce higher peaks. In case of small Ω_B which is consistent with current primordial big bang nucleosynthesis (BBN) limits, the period of adiabatic growth by the gravitational potential is very short since pressure is staying almost always $c/\sqrt{3}$ before the recombination. Therefore the deeper potential makes lower peaks because of the SW effect.

How about the dependence of each cosmological parameters? Increasing $\Omega_B h^2$ decreases sound speed. Hence we can expect higher peaks. Increasing $\Omega_0 h^2$ pushes the matter-radiation equality earlier. This causes potential to be deeper and peaks to be lower. We can explain most of parameter dependence from above arguments. For example, if we increase Ω_B with fixing Ω_0 and h, sound speed becomes smaller. Therefore we get larger peaks as shown in figure 3 (left panel). Because above two effects compensate each other, h dependence with fixing Ω_B and Ω_0 is complicated. However, if we fix $\Omega_B h^2$ which is determined by BBN, increasing h simply means lower peaks.

Secondly, we show dependence of the location of peaks on cosmological parameters. The peak location is determined by the projection of the sound horizon[3]. Because the sound horizon is weak function of $\Omega_B h^2$, the peak location is almost independent of Ω_B and h. However, it depends on Ω_0. The low density universe has longer age. Therefore the last scattering surface is further than the high density universe. This effect makes the sound horizon correspond to the smaller angular scale. Moreover, in the open universe,

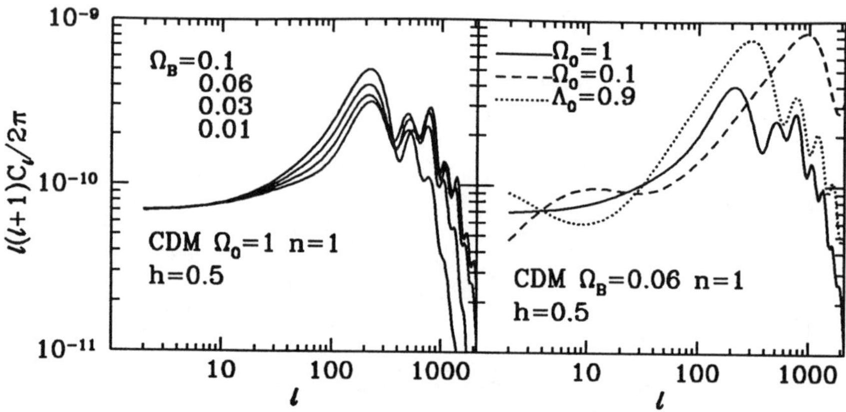

Figure 3: Numerical C_ℓ's of CDM models. Left panel: changing Ω_B for $\Omega_0 = 1$ models. Right panel: $\Omega_0 = 0.1$ open (dashed) and flat cosmological constant dominated (dotted) models with the $\Omega_0 = 1$ model (solid).

there is a geodesic effect which causes a same but bigger effect. As shown in figure 3 (right panel), peak locations are placed in order of $\Omega_0 = 1$, Λ and open models from large to small scales.

Finally, we have to mention the peculiar behaviour of CMB power spectrum on large scales (small ℓ) for low density models which is shown in figure 3. On large scales, the SW effect which predicts the flat tail makes the dominant contribution for the $\Omega_0 = 1$ model. On the other hand, there is the late ISW contribution for low density models. As for the Λ model, the ISW effect dominates on large scales. Because this effect has the damping caused by the finite tickness of *gravitational last scattering*[4], it is only significant on very large scales (small ℓ). For the open model, there is another effect, i.e., the curvature cutoff. Because the fluctuations outside the curvature scale do not contribute on C_ℓ, there is a cutoff on very large scales.

The analytic approach enables us the complete understanding of each physical processes which are working on CMB anisotropies. Practically, we can build a very fast but enough accurate Boltzmann code. Moreover, if once we know the power spectrum of CMB anisotropies precisely on various scales by observations, this analytic technique is very powerful to reconstruct the theoretical models from these observations.

1. Peebles, P.J.E. & J.T. Yu. 1970. ApJ. **162**: 815; Wilson, M.L. & J. Silk, 1981. ApJ. **243**: 14; Bond, J.R. & G. Efstathiou, 1984. ApJ. **285**: L45.
2. Hu, W. & N. Sugiyama, 1994. ApJ. : in press.
3. Hu, W. & N. Sugiyama, 1994. Phys.Rev.D : in press.
4. Hu, W. & N. Sugiyama, 1994. Phys.Rev.D **50**: 627.

THE GRISHCHUK-ZELDOVICH EFFECT IN THE OPEN UNIVERSE

David H. Lyth

School of Physics and Materials,
University of Lancaster, Lancaster LA1 4YB. U. K.

Introduction

When considering perturbations in an open universe, cosmologists retain only sub-curvature modes (defined as eigenfunctions of the Laplacian whose eigenvalue is less than -1 in units of the curvature scale, in contrast with the super-curvature modes whose eigenvalue is between -1 and 0). Mathematicians have known for almost half a century that all modes must be included to generate the most general *homogeneous Gaussian random field*, despite the fact that any square integrable *function* can be generated using only the sub-curvature modes. The former mathematical object, not the latter, is the relevant one for physical applications. This article summarizes recent work with A. Woszczyna. The mathematics is briefly explained in a language accessible to physicists. Then the effect on the cmb of any super-curvature contribution is considered, which generalizes to $\Omega_0 < 1$ the analysis given by Grishchuk and Zeldovich in 1978.

The mode expansion

According to the Einstein field equation, the energy density Ω of the universe is given by

$$1 - \Omega = -\frac{K}{(aH)^2} \tag{1}$$

Here K is a constant, $H = \dot{a}/a$ is the Hubble parameter, and Ω is the energy density measured in units of the critical density $3H^2/8\pi G$. The spatial curvature scalar is $R^{(3)} = 6K/a^2$ and we set $K = -1$ so that a is the curvature scale. Then the case $\Omega = 1$ corresponds to the limit $a \to \infty$, with physical distances like H^{-1} remaining constant.

We are interested in the stochastic properties of the perturbations, at fixed time. To define them we will take the approach of considering an ensemble of universes of which ours is supposed to be one. If, in some region of space, a perturbation f can be written as a sum of terms, with the coefficient of each term having an independent Gaussian probability distribution, it is a *Gaussian random field*, and its stochastic properties are completely determined by its correlation function. There is no requirement that the region of space be infinite, or that the terms be linearly independent. If the correlation function depends only on the geodesic distance between the points, the field is said to be *homogeneous* with respect to the group of transformations that preserve this distance. We assume that each cosmological perturbation in the observable universe is a typical realisation of some homogeneous Gaussian random field.

Spherical coordinates are defined by the line element

$$dl^2 = a^2[dr^2 + \sinh^2 r(d\theta^2 + \sin^2\theta d\phi^2)] \tag{2}$$

Any homogeneous Gaussian random field can be generated[1,2,3] by expanding it in eigenfunctions of the Laplacian with eigenvalue $(k/a)^2 < 0$,

$$f(r,\theta,\phi,t) = \int_0^\infty dk \sum_{lm} \tilde{f}_{klm}(t) Z_{klm}(r,\theta,\phi) \tag{3}$$

Here $q^2 = k^2 - 1$, and the mode functions are $Z_{klm} = \Pi_{kl}(r)Y_{lm}(\theta,\phi)$. For $q^2 > 0$ the radial functions are

$$\Pi_{kl} \equiv N_{kl}\tilde{\Pi}_{kl} \tag{4}$$

$$\tilde{\Pi}_{kl} \equiv |q|^{-2}(\sinh r)^l \left(\frac{-1}{\sinh r}\frac{d}{dr}\right)^{l+1} \cos(qr) \tag{5}$$

$$N_{kl} \equiv \sqrt{\frac{2}{\pi}}|q| \left[\prod_{n=1}^{l}(q^2+n^2)\right]^{-1/2} \qquad (l>0) \tag{6}$$

with $N_{k0} \equiv \sqrt{2/\pi}|q|$. For $-1 < q^2 < 0$, $\cos(qr)$ is replaced by $\cosh(|q|r)$. The spectrum $\mathcal{P}_f$ is defined by

$$\langle \tilde{f}^*_{klm}\tilde{f}_{k'l'm'}\rangle = \frac{2\pi^2}{k|q^2|}\mathcal{P}_f(k)\delta(k-k')\delta_{ll'}\delta_{mm'} \tag{7}$$

and the correlation function is

$$\xi_f(r) = \int_0^\infty \frac{dk}{k}\mathcal{P}_f(k)\frac{\sin(qr)}{q\sinh r} \tag{8}$$

For $r \gg 1$, the contribution to the correlation function from a mode with $k^2 \ll 1$ is $\xi_f(r) \propto \exp(-k^2 r)$ Thus the correlation length, in units of the curvature scale a, is of order k^{-2}. This is in contrast with the flat-space case, where the contribution from a mode with $k \ll 1$ gives a correlation length of order $1/k$.

The Grishchuk-Zeldovich effect

The cmb anisotropy is generally ascribed to a primeval curvature perturbation (measured by comoving observers), conveniently taken to be $\mathcal{R}$ defined by

$$4(k^2+3)\mathcal{R}_{klm}/a^2 = \delta R^{(3)}_{klm} \tag{9}$$

In the limit $\Omega \to 1$,

$$4k^2\mathcal{R}_{klm}/a^2 = \delta R^{(3)}_{klm} \tag{10}$$

During matter domination $\mathcal{R}$ is constant until Ω breaks away from 1. For $l \lesssim 30$ the mean square multipole C_l of the cmb anisotropy is given by the Sachs-Wolfe approximation

$$C_l = 2\pi^2 \int_0^\infty \frac{dk}{k} \mathcal{P}_{\mathcal{R}}(k) I_{kl}^2 \tag{11}$$

$$|q| I_{kl} = \frac{1}{5}\Pi_{kl}(\eta_0) + \frac{6}{5}\int_0^{\eta_0} dr \Pi_{kl}(r) F'(\eta_0 - r) \tag{12}$$

$$F(\eta) = 5\frac{\sinh^2\eta - 3\eta\sinh\eta + 4\cosh\eta - 4}{(\cosh\eta - 1)^3} \tag{13}$$

with $\eta = 2(aH)^{-1}$ and $r = \eta_0$ the last scattering surface. In this regime the COBE measurements give $l^2 C_l \simeq 8 \times 10^{-10}$. Within the observational uncertainties this is consistent[4] with a flat spectrum for all $0.1 \leq \Omega_0 \leq 1.0$, of magnitude $\mathcal{P}_{\mathcal{R}} \sim 10^{-9}$ to 10^{-10}. The corresponding mean square curvature perturbation $\langle \mathcal{R}^2 \rangle$ is of the same order.

Now suppose that the spectrum $\mathcal{P}_{\mathcal{R}}$ rises sharply on some very large scale $k_{\rm VL}$ *but that the perturbation is still a typical realisation of a Gaussian random field* so that it can be discussed using the above formalism. What is the effect on the cmb anisotropy? For $\Omega_0 = 1$ this question was asked and essentially answered by Grishchuk and Zeldovich[5]. The large scale contribution can be taken to be

$$\mathcal{P}_{\mathcal{R}}^{\rm VL} \simeq \delta(\ln k - \ln k_{\rm VL}) \langle \mathcal{R}^2 \rangle \tag{14}$$

In a sphere whose radius is of order the correlation lenght $d_{\rm VL} = a_0/k_{\rm VL}$, $\mathcal{R}$ is roughly constant with typical value $\langle \mathcal{R}^2 \rangle^{1/2}$. From Eq. (10) it is of order $d_{\rm VL}^2 \delta R^{(3)}$, which is a dimensionless measure of the geometry distortion due to the perturbation (recall that the background curvature $R^{(3)}$ vanishes for $\Omega_0 = 1$). For instance it is roughly equal to the fractional departure from $4\pi d_{\rm VL}^2$ of the sphere's area. The maximal distortion, corresponding to regions of positive curvature closing on themselves, is $\langle \mathcal{R}^2 \rangle \lesssim 1$. The quadrupole dominates for $d_{\rm VL} \gg H_0^{-1}$, and is given by

$$C_2^{\rm VL} \sim (d_{\rm VL} H_0)^{-4} \langle \mathcal{R}^2 \rangle \tag{15}$$

If the geometry distortion is maximal, then $d_{\rm VL} H_0 \sim (C_2^{\rm VL})^{1/4} \gtrsim 10^2$. In words, the correlation length is then more than two orders of magnitude bigger than the size of the observable universe.

To generalize this analysis to $\Omega_0 < 1$ one needs to take the background spatial curvature into account, and to note that the limit of large scales corresponds to $k \to 0$, not $q \to 0$. This has not been done to date. The only relevant publications of which we are aware either ignore the spatial curvature[6] or take use the $q \to 0$ limit[7,8]. Consider therefore Eq. (14) with $k_{\rm VL} \ll 1$, and suppose that Ω_0 is significantly below 1 so that $a_0 H_0 \sim 1$. In the absence of perturbations the geometry distortion of the observable universe is $a_0^2 R^{(3)} \sim 1$, and the addition distortion caused by the perturbation is $\sim a_0^2 \delta R^{(3)} \sim \mathcal{R}$. Thus, $\mathcal{R}$ meaures the fractional change in the geometry distortion, and since the relation between $\delta R^{(3)}$ and $\mathcal{R}$ is now scale independent this remains true on larger scales. The maximal distortion, corresponding to regions of space closing on themselves, is still $\langle \mathcal{R}^2 \rangle \lesssim 1$.

As $k \to 0$, $\Pi_{k0} \to 1$, but the other radial functions are proportional to k. Defining $I_l \equiv \lim_{k\to 0} I_{kl}/k$,

$$C_l^{\rm VL} = I_l^2 k_{\rm VL}^2 \langle \mathcal{R}^2 \rangle \tag{16}$$

Since I_l is roughly of order 1 for low multipoles, and also $a_0 \sim H_0^{-1} \sim 1$ we can write this

$$C_l^{\rm VL} \sim (d_{\rm VL} H_0) \langle \mathcal{R}^2 \rangle \tag{17}$$

The prefactor is not now raised to the fourth power as it is for $\Omega_0 = 1$, so that for maximal distortion $d_{\rm VL}$ must now be ten orders of maginitude bigger than the size of the observable universe! There are two physical reasons for the difference. One is that the correlation length is $a_0 k_{\rm VL}^{-2}$ instead of $a_0 k_{\rm VL}^{-1}$. The other is that the presence of background curvature allows the geometry distortion to be of order 1 in the observable universe, whereas before it was at most of order $k_{\rm VL}^2 \ll 1$.

In the case $\Omega_0 = 1$, the Grishchuk-Zeldovich effect contributes only to the quadrupole, and is not seen in the data (ie., the quadrupole is not anomalously high). In the case $\Omega_0 < 1$ it contributes to all multipoles up to some maximum, which is probably within the regime of validity of the Sachs-Wolfe approximation. It would be worth evaluating the l dependence to see whether it is the same as the observed $C_l \propto l^{-2}$ for some range of Ω_0. If so the effect might be present, and one could see whether this was so by looking at higher multipoles. (A more bizarre possibility would be that the effect persists even beyond the range of the Sachs-Wolfe approximation, in which case a full calculation would be necessary. The formalism is already in place, and has already been used for the sub-curvature modes[9,10,11].)

For ease of visualisation we have used the concept of the correlation length $d_{\rm VL}$, which presupposes that the perturbation continues to be a typical realization of a Gaussian random field in a region around us whose size is bigger than $d_{\rm VL}$, and therefore much bigger than the observable universe. The effect is really calculated on the hypothesis that the perturbation is a typical realization of a Gaussian random field *within the observable universe*, and can be written in terms of $k_{\rm VL}$ without reference to a correlation length. However, if the hypothesis is valid for k down to some minimum value, it is reasonable to suppose that it can indeed be extended out to a region bigger than the corresponding correlation length. Thus a positive detection of the Grishchuk-Zeldovich effect would suggest that this is the case. On the other hand a failure to detect the effect, which seems more likely, will tell us essentially nothing!

Finally, let us ask whether one should expect the effect to be present even below the level of detectability. For the case $\Omega_0 = 1$ the usual hypothesis is that the curvature perturbation comes from a vacuum fluctuation of the inflaton field, and in 1990 this was extended[13] to the case $\Omega_0 < 1$. To the extent that this is true there are no super-curvature modes, which means that for Ω_0 appreciably less than 1 there is no Grishchuk-Zeldovich effect. Like any hypothesis in physics this will be at best an approximation, and it will fail above some large scale. (As we just discussed, 'scale' strictly means simply some large value of k^{-1}, but one can probably think of it a also a large correlation length.) However, the hypothesis that the curvature perturbation in the observable universe is a typical realisation of a homogeneous Gaussian random field will also fail above some large scale, and this might well be the same as the scale on which the vacuum fluctuation hypothesis fails. If so, there will be no Grishchuk-Zeldovich effect.

It would be instructive to see how all this works in bubble model[13] of the $\Omega < 1$ universe. According to this model we inhabit the interior of the bubble extending far beyond the observable universe. Within the bubble the perturbation is well approximated by a typical realization of a random Gaussian field, which has only sub-curvature modes because it originates as a vacuum fluctuation. As the boundary is approached the nature of the perturbation changes and it no longer corresponds to a typical realization of the random field. Thus one expects in this model the coincidence of scales mentioned in the last paragraph, and no Grishchuk-Zeldovich effect.

ACKNOWLEDGEMENTS

This work was started with the help of EU research grant ERB3519PL920782(10835). One of us (DHL) thanks the Isaac Newton Institute for a visiting Fellowship while the work was being completed, and Bruce Allen, Robert Caldwell, Misao Sasaki and Neil Turok for useful discussions.

REFERENCES

1. YAGLOM, M. 1961 *in* Proceedings of the Fourth Berkeley Symposium Volume II, J. Neyman, Ed. University of California Press, Berkeley.
2. KREIN, M. G. 1949. Ukrain. Mat. Z. **1**, No. 1, 64; *ibid* 1950 **2**, No. 1, 10.
3. LYTH, D. H. & A. WOSZCZYNA. preprint astro-ph/9501044, submitted to Phys. Rev. D.
4. KAMIONKOWSKI, M., D. N. SPERGEL & N. SUGIYAMA. 1994. Astrophys. J. **426**, L57; GORSKI, K. M., H. RATRA, N. SUGIYAMA & A. J. BANDAY, preprint.
5. GRISHCHUK, L. P. & Ya. B. ZELDOVICH. 1978. Astron. Zh. **55**, 209 [Sov. Astron. **22**, 125 (1978)].
6. TURNER, M. S.. 1991. Phys Rev D **44**, 12.
7. KAMIONKOWSKY, M. & D. N SPERGEL. 1994. Astrophys. J. **432**, 7.
8. KASHLINSKY, A., I. TKACHEV & J. FRIEDMAN. 1994. Phys. Rev. Lett., **73**, 1582.
9. SUGIYAMA, N. & J. Silk. 1994. Phys. Rev. Lett. **73**, 509.
10. KAMIONKOWSKI, M., D. N. SPERGEL & N. SUGIYAMA. 1994. Astrophys. J. **426**, L57.
11. GORSKI, K. M. , H. RATRA, N. SUGIYAMA & A. J. BANDAY, preprint.
12. LYTH, D. H.& E. D. STEWART. 1990. Phys Lett. B **252**, 336.
13. COLEMAN, S. & F. DE LUCCIA. 1980. Phys. Rev. D **21**, 3305; GOTT, J. R.. 1982. Nature **295** , 304; GUTH, A. H. & E. J. WEINBERG. 1983. Nucl. Phys. **B212**, 321; GOTT, J. R. & T. S. STATLER. 1984. Phys. Lett. **B136**, 157; SASAKI, M., T. TANAKA, K. YAMAMOTO & J. YOKOYAMA. 1993. Phys. Lett. B **317**, 510; SASAKI, M., T. TANAKA, K. YAMAMOTO & J. YOKOYAMA. 1993. Prog. Theor. Phys. **90**, 1019; BUCHER, M., A. GOLDHABER & N. TUROK. 1994. preprint; TANAKA, T. & M. Sasaki. 1994. two preprints; YAMAMOTO, K., T. TANAKA, & M. SASKAKI. 1994. preprint.

HOW ANISOTROPIC CAN A UNIVERSE BE?

JOHN D BARROW

Astronomy Centre, University of Sussex
Brighton BN1 9QH, UK

Providing an explanation for the impressive large–scale isotropy of the Universe is a classic problem of cosmology. Inflationary universes[1,2] offer to explain it as a consequence of a period of accelerated expansion in the early universe. Unfortunately, there is no decisive evidence that inflation did occur in the past. Efforts to confirm the its predictions using recent observations of microwave–background anisotropies are not clearcut: different inflationary models predict different density fluctuation spectra[3] and the COBE data are only marginally consistent with the most "natural" prediction of a Zeldovich–Harrison spectrum[4,5]. In such ambiguous circumstances it is important to explore alternative cosmic histories which might explain some, or all, of the large scale features of the Universe.

Although one could simply appeal to isotropic initial conditions to 'explain' the present isotropy of the Universe[6] inflation aims to explain the present structure of the Universe without recourse to special initial conditions. If successful, this strategy ensures that observations of the large–scale structure of the Universe will be unable to tell us anything about its pre–inflationary stage and possible initial state. In this respect inflation is the latest manifestation of the chaotic cosmology programme[7], but it gives rise to local isotropy without dissipating initial anisotropies.

The chaotic cosmology programme investigated the range of cosmological initial conditions that would tend to isotropy with time[8] and studies were made of the effects of anisotropy upon physical processes like primordial nucleosynthesis[9]. It was widely concluded that, in general, anisotropic universes need not become as isotropic as our Universe by the present day[10]. However, it was not appreciated that the initial conditions required for the counter–examples are entirely unphysical. Consider a radiation–dominated anisotropic universe with isotropic spatial curvature. If the radiation pressure is isotropic then the shear anisotropy energy density, σ^2, will fall as the sixth power of the mean expansion scale factor a(t), with $\sigma^2 \propto a^{-6}$. The radiation density falls as $\rho_\gamma \propto a^{-4/3}$, so the shear anisotropy will dominate ρ_γ at early times. This is shown by the Friedmann–like equation governing a(t),

$$3\dot{a}^2/a^2 = \rho_\gamma + \sigma^2 \qquad (1)$$

where we neglect the curvature term and set $8\pi G \equiv 1$. Hence, if $\sigma^2 \sim \rho_\gamma$ at any time (eg $t \sim 1$s for nucleosynthesis or $t \sim 10^{13}$s for recombination) then at earlier times we will have $\sigma^2 >> \rho_\gamma$, culminating in a physically absurd initial condition at $t_{p\ell} \sim 10^{-43}$s because the anisotropy energy density residing in the anisotropic gravitational–wave modes will vastly exceed the Planck energy density of equlibrium radiation and relativistic particles at that time. Even if this simple form of anisotropy ceased to be significant when the temperature was as high as 1 GeV, it would have started with an energy density $\sim 10^{38}\rho_{p\ell}$ at $t_{p\ell}$.

At the Planck epoch, gravitational interactions should keep all energy densities of the same order as the Planck density. We call this the *Planck Equipartition Proposal* (PEP). If the gravitational wave modes supporting anisotropic stresses were to carry far more energy than the equilibrium radiation, then quantum gravitational interactions would re–establish equilibrium through graviton exchanges, particle production and momentum transport.

One immediate consequence of the PEP is that the simple modes of anisotropy allowed when the 3–curvature is isotropic, $\sigma^2 \propto a^{-6}$, are never significant after the Planck epoch since $\sigma^2(t_{p\ell}) \simeq \rho_\gamma(t_{p\ell})$ and $\sigma^2/\rho_\gamma \propto a^{-2}$. But we must ask what happens when the PEP is applied to the most general forms of cosmological anisotropy. To answer this we need to consider the effects of the dominant forms of relativistic matter and anisotropy in the very early universe. Immediately following the Planck epoch, the Universe will be assumed to contain a mixture of collisional and collisionless radiation fields (each with a trace-free energy-momentum tensor). The collisionless radiation consists of gravitons and possibly other relativistic species since asymptotically-free interactions are slower than the Hubble expansion rate at energies between $\sim 10^{16}$ and 10^{18} GeV.

When collisionless and collisional radiation stresses co-exist with similar initial densities the evolution of any shear anisotropy is dictated by the pressure anisotropy of the collisionless particles. In general, if $\sigma^2 = 2\sigma_{ij}\sigma^{ij}$ defines the shear scalar and $H = \dot{a}/a$ defines the mean Hubble rate, then, when small, the shear tensor evolves as[11],

$$\dot{\sigma}_{ij} + 3H\sigma_{ij} = -\{R_{ij}^* - (1/3)\delta_{ij}R^*\} + \{T_{ij} - (1/3)\delta_{ij}T^a{}_a\} \qquad (2)$$

where R_{ij}^* is the Ricci 3-curvature tensor and T_{ij} is the energy-momentum tensor of the radiation fields. In the simplest possible case, where R_{ij}^* and T_{ij} are both isotropic, the right-hand side of equation (2) vanishes and we obtain the well-known $\sigma \propto a^{-3}$ evolution used in the discussion of (1) above. We need to consider the cosmological effects of including the two anisotropic stress terms on the right of (2). First, assume the 3-curvature anisotropy term is still zero and consider the effects of anisotropy in T_{ij}. The collisional radiation sea will not contribute to this term and the anisotropy will be determined by the pressure anisotropy of the collisionless particles. Following the PEP initial condition the model will have mean expansion dynamics determined to leading order by the isotropic (collisional) radiation density; thus $a(t) \propto t^{\frac{1}{2}}$ and $\rho_\gamma \propto a^{-4} \propto t^{-2}$. The energy density of the collisionless stresses evolves proportional to ρ_γ, up to logarithmic corrections. Thus, from (2), the shear to mean Hubble rate evolution in the radiation-dominated era, is given by

$$t^{-3/2}(\sigma_{ij}t^{3/2})^{\cdot} \simeq M_{ij}t^{-2} \;;\; M_{ij} \text{ constant,} \qquad (3)$$

and with $H = 1/(2t)$ to leading order, we have

$$\sigma/H \propto \Sigma_1 + \Sigma_2 a^{-1} \;;\; \Sigma_1, \Sigma_2 \text{ constants.} \qquad (4)$$

If one includes the next order of approximation then the Σ_1 term decays logarithmically, as found by Doroskevich *et al*[8]:

$$\sigma/H \propto \Sigma_1\{1 + \ell n(t/t_*)\}^{-1} + \Sigma_2 a^{-1} \;;\; t_* \text{ constant} \qquad (5)$$

The Σ_1 term is driven by the pressure anisotropy of the collisionless particles (gravitons) and vanishes when T_{ij} is isotropic, in which case $\sigma \propto a^{-3} \propto t^{-3/2}$ and $\sigma/H = \Sigma_2 t^{-\frac{1}{2}}$, during the radiation era. When T_{ij} is a sum of isotropic and anisotropic trace-free stresses then, to first order, the isotropising effect of the collisional radiation is balanced by the average energy density of the collisionless radiation, while the second-order pressure anisotropy maintains a small shear distortion proportional to the ratio of the densities of isotropic to anisotropic radiation[12,13].

When the radiation era ends at $t = t_{eq}$, the isotropic material becomes dominated by the gravitational effect of pressureless 'dust' with $\rho_m \propto a^{-3}$, but the decay of the shear anisotropy remains dominated by the pressure anisotropy. Since $a(t) \propto t^{2/3}$ for $t \geq t_{eq}$, (3) is replaced by

$$t^{-2}(\sigma_{ij}t^2)^{\cdot} \simeq M_{ij}a^{-4} = M_{ij}t^{-8/3} \;;\; M_{ij} \text{ constant,} \qquad (6)$$

hence, with $H = 2/(3t)$ to leading order, we have

$$\sigma/H \propto \Sigma_1 a^{-1} + \Sigma_2 a^{-3/2} \;;\; \Sigma_1, \Sigma_2 \text{ constants.} \qquad (7)$$

Again, the pressure anisotropy slows the decay of the shear distortion with respect to the $\Sigma_2 a^{-3/2}$ decay found when the pressure is isotropic.

The influence of a 3-curvature anisotropy term in equation (2) can be understood because curvature anisotropies are created by standing or travelling homogeneous gravitational waves on a simpler anisotropic space of isotropic curvature [8]. The evolution of general anisotropic cosmological models following imposition of the PEP boundary condition at $t_{p\ell}$ is therefore a further example of the behaviour leading to the solutions (4)–(7) because their 3-curvature anisotropies behave like collisionless gravitational radiation with anisotropic pressure and trace-free energy-momentum tensor. The anisotropic curvature terms will therefore also remain of order $H^2(t_{p\ell})$ at $t_{p\ell}$ because of quantum gravitational interactions,

In open universes there is a further curvature-dominated phase at late times (for redshifts $1 + z < \Omega_0^{-1}$). The right-hand side of (2) is driven by the spatial curvature $\propto a^{-2} \propto t^{-2}$ and hence

$$t^{-3}(\sigma_{ij}t^3)^{\dot{}} \simeq M_{ij}t^{-2} \ ; \ M_{ij} \text{ constant,} \qquad (8)$$

so, with $H = 1/t$ to leading order, we have

$$\sigma/H \propto \Sigma_1 + \Sigma_2 a^{-2} \ ; \ \Sigma_1, \Sigma_2 \text{ constants.} \qquad (9)$$

The observed temperature anisotropy of the microwave background radiation, $\Delta T/T$, on large angular scales is determined by the value of (σ/H) at the last-scattering redshift of the radiation, $z_{\ell s}$. Hence, the PEP enables us to predict the present microwave background anisotropy precisely. The result depends upon the redshift of last scattering for the radiation and the number of massive spin states annihilating into the equilibrium interacting radiation sea after $t_{p\ell}$.

If the present total density of the Universe in units of the critical density is Ω_0 and $H(t_0) = 100h_0$ Kms^{-1}Mpc^{-1} defines the Hubble constant today, so $h_0 \sim 0.5$–1, then the radiation era ends at a redshift z_{eq}, where $1 + z_{eq} = 2.4 \times 10^4 \Omega_0 h_0^2$ and last scattering occurs at a redshift $z_{\ell s}$ where $1 + z_{\ell s} = 1100$. If there is reionisation of the cosmic medium, last scattering may be delayed until the lowest redshift at which there is unit optical depth. This is [14],

$$1 + z_{\ell s} = 15\Omega_0^{1/3} (\Omega_{b0}h_0)^{-2/3} \text{ for } \Omega_0 z_{\ell s} >> 1 \qquad (10a)$$

$$1 + z_{\ell s} = 39(\Omega_{b0}h_0)^{-1} \qquad \text{for } \Omega_0 z_{\ell s} << 1 \qquad (10b)$$

where Ω_{b0} is the present baryon density parameter.

The PEP requires that $(\sigma/H) = 1$ at $t_{p\ell}$. Thereafter, (σ/H) evolves in accord with (5) for $t_{p\ell} < t \le t_{eq}$, and with (7) for $t_{eq} \le t \le t_{\ell s}$. If $z_{\ell s} < z_* \equiv (\Omega_0^{-1} - 1)$ this phase of evolution will end at t_* and be followed by (9). The annihilation of g_i relativistic spin states in between $t_{p\ell}$ and t_{eq}, taking the gravitons to be the only free particles, diminishes ρ_{free}/ρ_γ by a factor $(2/g_i)^{4/3} \equiv \epsilon$. This reduces σ/H, and hence also $\Delta T/T$, by ϵ. The microwave background anisotropy observed today will be

$$\Delta T/T = \epsilon(\sigma/H)_{p\ell} (1 + z_{\ell s})(1 + z_{eq})^{-1}\{1 + \ln(t_{eq}/t_{p\ell})\}^{-1} \qquad (11)$$

where $t_{eq}/t_{p\ell} = \{(1 + z_{eq})/(1 + z_{p\ell})\}^{-2}$. Using (10)–(11) we can compute the observed $\Delta T/T$. The result is proportional to ϵ since the anisotropic pressure of the free particles sustains the shear distortion ($\sigma/H \propto \rho_{free}/\rho_\gamma$). We choose $g_i \sim 80$, typical of a small GUT, like minimal SU(5), so $\epsilon = 7.3 \times 10^{-3}$. The predicted $(\Delta T/T)$ values for three choices of $\Omega_0 h_0^2 = 1, 10^{-1}, 10^{-2}$ and two ionisation histories (with $z_{\ell s} = 1100$ or $z_{\ell s}$ given by (12)) are displayed in the table below

Remarkably, we see that the generic microwave background anisotropy is predicted with little uncertainty. Even in the absence of long-lasting inflation, this anisotropy must be small ($\Delta T/T \sim 10^{-4} - 10^{-6}$) and its generic level is tantalizingly close to the quadrupole anisotropy, $\Delta T/T \sim 10^{-5}$, measured by the COBE satellite [4]. Not only does this leave no 'isotropy problem' for inflation or quantum cosmology to solve, but gravitational equilibrium at the Planck epoch ensures that realistic universe have levels of anisotropy that are both small yet measurable. Indeed, in our Universe that level may have recently been measured.

Last scattering redshift $z_{\ell s}$ =:	1100	$15\,\frac{(\Omega_0 h_0{}^2)^{1/3}}{(\Omega_{b0} h_0{}^2)^{2/3}}$, $\Omega_0 z_{\ell s} \gg 1$ $39(\Omega_{b0} h_0)^{-1}$, $\Omega_0 z_{\ell s} \ll 1$
	$\Delta T/T$ =:	$\Delta T/T$ =:
$\Omega_0 h_0{}^2 = 1$	2.7×10^{-6}	$7.8 \times 10^{-7}\ (0.01/\Omega_b h_0{}^2)^{2/3}$
$\Omega_0 h_0{}^2 = 0.1$	2.8×10^{-5}	$3.7 \times 10^{-6}\ (0.01/\Omega_b h_0{}^2)^{2/3}$
$\Omega_0 h_0{}^2 = 0.01$	2.9×10^{-4}	$1.0 \times 10^{-4}\ (0.01/\Omega_b h_0)$

Acknowledgement

The author was supported by a PPARC Senior Fellowship.

References

1. A.H. Guth, Phys. Rev. D **23**, 347 (1981).

2. A. Linde, *Particle Physics and Inflationary Cosmology*, (Gordon and Breach, New York, 1990).

3. A.R. Liddle and D.H. Lyth, Phys. Rep. **231**, 1 (1993); J.D. Barrow and A.R. Liddle, Phys. Rev. D. **47**, R5219 (1993).

4. G. Smoot & COBE team, Astrophy. J. **396**, 1 (1992); C.L. Bennett & COBE project team, Astrophys. J. (submitted) (1994).

5. S. Hancock *et al*, Nature **367**, 333 (1994).

6. R. Penrose, in *300 Years of Gravitation*, eds S.W. Hawking and W. Israel, (Cambridge UP, Cambridge, 1987). J. Hartle and S.W. Hawking, Phys. Rev. D **28**, 2960 (1983). A. Vilenkin, Phys. Rev. D **33**, 3560 (1982).

7. C.W. Misner, Nature **214**, 40 (1967);

8. C.B. Collins and S.W. Hawking, Astrophys. J. **181, 317** (1972); A.G. Doroshkevich, V.N Lukash and I.D Novikov, Sov. Phys. JETP **37**, 739 (1973).

9. J.D. Barrow, Mon. Not R. astron. Soc. **175**, 359 (1976).

10. C.B. Collins and J.M. Stewart, Mon. Not. R. astron. Soc. **153**, 419 (1971).

11. G.F.R. Ellis, in *General Relativity and Cosmology*, ed R.K. Sachs, (Academic Press, New York, 1971).

12. I.S. Shikin, Sov. Phys. JETP **36**, 811 (1972); V.N. Lukash and A.A. Starobinskii, *ibid.* **39**, 742 (1974).

13. An identical effect occurs if the collisionless radiation is replaced by a homogeneous cosmological magnetic field. Linearisation of the Einstein equations about the isotropic solution with zero shear and zero pressure anisotropy produces zero eigenvalues and the stability of the isotropy is not determined by the linear approximation -- hence the logarithmic decay of the shear distortion and also of the ratio of the collisionless and collisional radiation densities since it is proportional to (σ/H).

14. T. Padmanabhan, *Structure Formation in the Universe*, (Cambridge UP, Cambridge, 1993).

The Signatures of Voids and the CMBR

Sharon L. Vadas
Center for Particle Astrophysics, 301 Le Conte Hall
University of California; Berkeley, CA 94720-7304

We explore the signature of a cold dark matter compensated void in the quasi non-linear regime. We find that this void is entirely a cold spot on the microwave background, in contrast to a non-linear void. On the last scattering surface(LSS), it appears as either a hot or cold spot depending on where this surface cuts the void. In addition, because the usual cancellations do not occur, the void's LSS signature can be very large as it is proportional to R/H^{-1} rather than $(R/H^{-1})^3$. This implies strict limits for voids on the LSS.

I. Introduction

A void is a region with underdensity $\delta \equiv 1 - \rho_{in}/\rho_{out}$ and radius R, where ρ_{in} and ρ_{out} are the energy densities inside and outside the void, respectively. A cold dark matter (CDM) void has three evolutionary phases: linear($\delta \ll 1$), quasi non-linear($\delta \simeq .1$ to $.9$) and non-linear($\delta \simeq 1$). In the quasi non-linear regime the void deepens quickly and its wall thins. A void's signature (e.g. the Rees-Sciama effect[1]) is obtained by tracing geodesics as the void evolves; for a compensated, non-linear void[2], it is

$$\Delta T/T = .4(c^{-1}R_e/H_e^{-1})^3[1-(5/3)\cos^2\psi_0]\cos\psi_0, \tag{1.1}$$

where H is Hubble's constant and "e" is when the photon exits the void. Also, ψ_0 is the angle between the photon's direction and a line from the void's center to the photon at t_e. Note that a non-linear void is a cold spot ($\psi_0 < 39°$) surrounded by a hot ring.

II. Fluid and Geodesic Equations

The spherically symmetric, comoving metric we use here is

$$ds^2 = -c^2\Phi^2(t,r)dt^2 + \Lambda^2(t,r)dr^2 + R^2(t,r)(d\theta^2 + \sin^2\theta d\phi^2). \tag{2.1}$$

We refer the reader to Ref. [3] for the hydrodynamic equations and an explanation of how the code works. We define the comoving-frame radius $R_{\rm CF} = R(t_i, r)$, so that at t, the sphere labeled by r has physical radius $R(t)$ and comoving-frame radius $R_{\rm CF}(t)$. In addition, $Z_{\rm CF} = R_{\rm CF}\sin\theta$ and $X_{\rm CF} = R_{\rm CF}\cos\theta\cos\phi$.

The geodesic equations for a photon can be shown to be[4], $z = dr/dt$, $w = d\theta/dt$,

$$\frac{dw}{dt} = w\left[-\frac{2}{R}(\Phi U + R'z) + \frac{\dot\Phi}{\Phi} + \frac{R'U'}{\Gamma^2\Phi}z^2 + \frac{2\Phi'}{\Phi}z + \frac{RU}{\Phi}w^2\right], \tag{2.2}$$

$$\frac{dz}{dt} = z\left(\frac{\dot\Phi}{\Phi} + \frac{R'U'}{\Gamma^2\Phi}z^2 + \frac{2\Phi'}{\Phi}z + \frac{RU}{\Phi}w^2\right) - \frac{2\Phi U'}{R'}z - \frac{\Phi\Phi'\Gamma^2}{(R')^2} - \left(\frac{R''}{R'} - \frac{\Gamma'}{\Gamma}\right)z^2 + \Gamma^2 R\, w^2/R', \tag{2.3}$$

where $U = \dot R/\Phi$ and $\Gamma = R'/\Lambda$. In addition, the energy of a photon as measured by comoving observers at each photon's location is

$$E(t) = E(t_i)\frac{\Phi(t)}{\Phi(t_i)}\frac{w(t_i)}{w(t)}\left(\frac{R(t_i)}{R(t)}\right)^2, \tag{2.4}$$

where $E(t_i)$ is the photon's initial energy and the propagation is non-radial ($w \neq 0$).

III. Signature of a Void

Consider a compensated, quasi non-linear void with initially unperturbed velocity in a flat, Fridemann-Robertson-Walker (FRW) universe. The void and void wall expand outward faster than the expansion rate of the universe. There are three major effects contributing to the temperature distortion of a photon as it enters, crosses and exits the void. As it enters and exits, the distortion is blueshifted because the photon jumps into frames moving towards it. And as it crosses, it is redshifted because the photon moves into frames expanding away from it faster than the expansion rate of the universe.

Suppose a photon with energy $E(t)$ moves through any part of this void region from t_1 to t_2. At the same time, another photon travels outside the void region. The temperature distortion of the former photon at time t_2 is

$$\frac{\Delta T}{T}(t_1, t_2) = \frac{E(t_2)\, a(t_2)}{E(t_1)\, a(t_1)} - 1, \tag{3.1}$$

where $a(t)$ is the scale factor of the universe. (These two photons do not move the same distance in general). We apply this to the inner region of a void which expands outward as t^α with $\alpha = 2/3 + 2\delta/9$ for $\delta < 1$ [4]. Let t_i and t_f be the initial and final times, and let t_1 and t_2 be the times the photon leaves the wall and enters the void, and leaves the void and enters the wall, respectively. The total temperature distortion is

$$\frac{\Delta T}{T}(t_i, t_f) \simeq \frac{\Delta T}{T}(t_i, t_1) - \frac{2}{3}\delta(t_2)\frac{R(t_2)}{cH(t_2)^{-1}} + \frac{\Delta T}{T}(t_2, t_f). \tag{3.2}$$

The first (third) term is the contribution from the first (second) void wall. The second term shows that the redshifting acquired after crossing the void is first-order in R/H^{-1}, and depends only on δ, not on details of the void wall. It also increases linearly with distance across the void. Because the sum of the terms in Eq. (3.2) is third order for a non-linear void (see Eq. (1.1)), the temperature distortions acquired entering and leaving the void must also be first-order. We verify these results numerically.

Place a void with radius $R_{\text{void}}(t_i)$ at the origin, and an observer far outside the void on the $+Z_{\text{CF}}$ axis. Initially all photons have $\phi = 0$, start at the same value of Z_{CF}, and propagate parallel to the Z_{CF}-axis. They always remain in the X-Z plane, by symmetry.

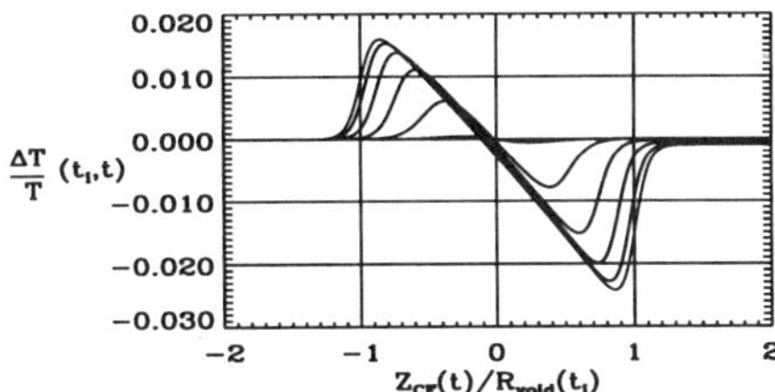

Figure 1: Temperature distortions of photons crossing an evolving void.

In Figure 1, we plot the temperature distortion, $\Delta T(t_i, t)/T$, as a function of $Z_{\text{CF}}(t)/R_{\text{void}}(t_i)$ for photons which pass through the void with $X_{\text{CF}}(t_i)/R_{\text{void}}(t_i) = .1, .3, .5, ..., 1.1$.

Table 1: Compensated Void with $\delta_e \simeq .4$ and $c^{-1}R_e/H_e^{-1} \simeq .2$

$X_{\rm CF}(t_i)/R_{\rm void}(t_i)$	$(c^{-1}R_e/H_e^{-1})^{-3}\Delta T(t_i,t_f)/T$	
	Quasi non-linear	Non-linear
.1	$-(.106 \pm .001)$	$-.259$
.3	$-(.087 \pm .001)$	$-.197$
.5	$-(.057 \pm .001)$	$-.0866$
.7	$-(.0226 \pm .0005)$	.0428
.9	$(.216 \pm .005)E-2$	.119
1.1	$-(.23 \pm .01)E-3$	0.0

(See Ref. [4] for details). Upon exiting the void, $c^{-1}R_e/H_e^{-1} \simeq .21$ and $\delta_e = .44$. (The exit location is where ρ/ρ_{out} is a maximum in the second void wall). As predicted, a photon's energy is blueshifted upon entering and leaving the void, and is redshifted while crossing the void. The curve with the largest temperature distortions has $X_{\rm CF}(t_i)/R_{\rm void}(t_i) = .1$. Using Eq. (3.2), $\Delta T(t_1,t_2)/T \simeq .06$, similar to Figure 1's more complicated "value" of .04. As $X_{\rm CF}(t_i)$ increases, the temperature distortions decrease because the parallel component of the wall's velocity decreases and the distance across the void decreases.

The final temperature distortions are much smaller than those obtained en-route (see Figure 1), and are listed in Table I along with the non-linear results from Eq. (1.1). The distortion is largest near the void's center, and is smaller than in the non-linear case because δ_e is not large enough[4]. In addition, the entire void appears as a cold spot. The large, hot ring is missing because the wall is thick, preventing the blueshifting from dominating over the the redshifting that occurs while crossing the fuzzy edge of the void.

IV. Signatures on the Last Scattering Surface

The optical depth of the LSS is $\tau(t_1,t_0) = \int_{t_1}^{t_0} \sigma_T n_{elec} c dt$, where n_{elec} is the electron density, σ_T is Thompson's cross section, and t_0 is the observer's time. We assume that the LSS is instantaneous; once $\tau(t_1,t_0)$ drops below one, the number of interactions between photons and electrons is negligible. Then the LSS has an optical depth of one.

Define t_f to be the time all LSS photons completely leave the void region. The optical depth between t_1 and t_f is $\tau(t_1,t_f) = cg\sigma_T \int_{t_1}^{t_f} n(t, X_{\rm CF}, Z_{\rm CF})dt$, where we assume that the light traces the mass: $n_{elec} = gn$. But because $\tau(t_1,t_f) + \tau(t_f,t_0) = 1$, we can move the void relative to the LSS (and therefore sample any part of the void) by changing $\tau(t_f,t_0)$ (i.e. moving the observer toward or away from the void). The last scattering surfaces then, are surfaces of constant optical depth, $\tau(t_1,t_f)$.

We can deduce the spatial geometry of the LSS when it crosses a void. Because the void is underdense, photons must travel farther inside than outside the void in order to have the same optical depth. Thus, photons that last scatter in the void are emitted earlier in time and therefore travel farther. If we plot the LSS in comoving-frame coordinates, the LSS curves away from the observer in the direction of the void.

To a given LSS, we apply the following transformation. Using Eq. (3.1) with t_2 replaced by t_f, the temperature distortion of a photon which last scattered at time t_1 is

$$\frac{\Delta T}{T}(t_1,t_f) = \frac{E(t_f)a(t_f)}{E(t_i)a(t_i)}\,\frac{E(t_i)a(t_i)}{E(t_1)a(t_1)} - 1 = \frac{\frac{\Delta T}{T}(t_i,t_f) - \frac{\Delta T}{T}(t_i,t_1)}{1+\frac{\Delta T}{T}(t_i,t_1)}. \tag{4.1}$$

For most LSS surfaces, $|\Delta T(t_i,t_f)/T| \ll |\Delta T(t_i,t_1)/T| \ll 1$ so that $\Delta T(t_1,t_f)/T \simeq -\Delta T(t_i,t_1)/T$; whatever temperature distortion is acquired getting into the void region at t_1 is permanently frozen into the LSS with an accompanying negative sign. In general then, the first-order effects shown in Figure 1 are frozen into the LSS, resulting in relatively large signatures. In Figure 2, we show the signature of the LSS, $\Delta T(t_1,t_f)/T$, as a function of $Z_{\rm CF}{}^{out}/R_{\rm void}(t_i)$ and $X_{\rm CF}(t_i)/R_{\rm void}(t_i)$, where $Z_{\rm CF}{}^{out}$ is the value of $Z_{\rm CF}(t)$ for a photon outside the void region on the same last scattering surface (i.e. constant optical depth). (A LSS can be represented uniquely by its value of $Z_{\rm CF}{}^{out}$). It is clear that the sign and magnitude of the signature depends sensitively on where the LSS slice the void. If the void is sliced near the back (front), the void is a cold (hot) spot on the microwave background. At the peaks of the distribution, the enhancement over $\Delta T(t_i,t_f)/T$ is $25 \simeq 2.5\delta_e(c^{-1}R_e/H_e^{-1})^2$. Only in one area is the signature relatively small.

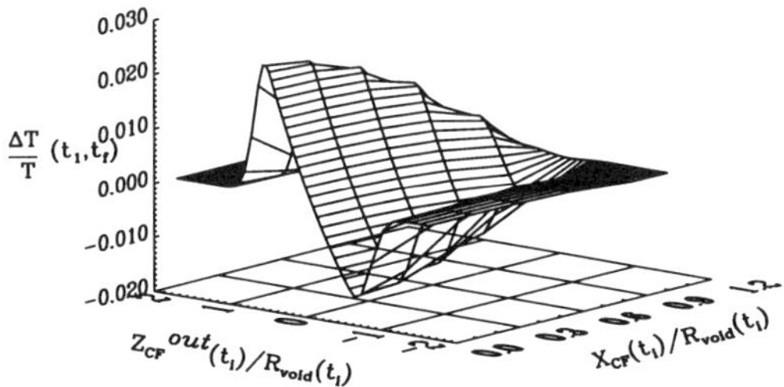

Figure 2: Signature of a void on the LSS—depends on where the LSS slices the void.

V. Discussion

In conclusion, when a quasi non-linear void lies in front of the LSS, it appears entirely as a cold spot on the microwave background. However, when this same void lies on the LSS, its signature depends on where the LSS cuts the void; the void can appear to be hot or cold. In addition, the void's LSS signature is much larger than its signature in front of the LSS because it is first-order in R/H^{-1}. Over the scales that $\Delta T/T \lesssim 3\times 10^{-5}$ experimentally, quasi non-linear voids with $\delta \simeq .3$ can be roughly limited to $c^{-1}R/H^{-1} \lesssim 10^{-4}$; they must be very small.

S.L.Vadas was supported by the President's Postdoctoral Fellowship Program at the University of California, and NSF Grant AST-9120005.

[1] M.J. Rees and D.W. Sciama, *Nature*, **217**, 511 (1968).

[2] K.L. Thompson and E.T. Vishniac, *Astro. J.* **313**, 517 (1987); E. Martinez-Gonzalez et al, *Astro. J.*, **355**, L5 (1990); E. Martinez-Gonzalez and J.L. Sanz, *Mon. Not. R. astr. Soc.*, **247**, 473 (1990).

[3] S.L. Vadas, *Phys. Rev. D* **48**, 4562 (1993); Sharon L. Vadas in *Numerical Simulations in Astrophysics*, Ed. by J. Franco, S. Lizano, L. Aguilar and E. Daltabuit, Cambridge University Press, 1994; S.L. Vadas, *Phys. Rev. D*, **50**, 7179 (1994).

[4] S.L. Vadas and J. Silk, "The Signatures of Voids and the CMBR", work in progress.

[5] E.W. Kolb and M.S. Turner, *The Early Universe*, (Addison-Wesley, 1990).

Ω FROM THE *COBE*-DMR ANISOTROPY MAPS

J.L. SANZ[1], L. CAYÓN[2,3], E. MARTINEZ-GONZÁLEZ[1],
N. SUGIYAMA[4] and S. TORRES[5]

[1]Dpto. Física Moderna, Universidad de la Cantabria and Instituto Mixto de Física de Cantabria, CSIC-U. de Cantabria, Santander (Cantabria), Spain.
[2]Lawrence Berkeley Laboratory, Berkeley, CA, USA.
[3]Center for Particle Astrophysics, Berkeley, CA, USA.
[4]The University of Tokyo, 113 Japan.
[5]Observatorio Astronómico Universidad Nacional and Centro Internacional de Física, Bogotá, Colombia.

INTRODUCTION

The angular correlation of the Cosmic Microwave Background maps can be used to constrain galaxy formation scenarios. Low-density (Ω) models can generate inflationary epoch adiabatic density perturbations in open universes (Lyth & Stewart 1990[1], Ratra & Peebles 1994[2]). Other low-Ω models have power spectra as given by simple prescriptions. The pioniering calculation of Wilson (1983)[3] assumed a power law in wavenumber while Kamionkowski and Spergel (1994)[4] considered power laws in spatial volume and scalar spatial Laplacian eigenvalue. Here we consider a low-Ω model where the cosmological constant is zero. We assume a cold dark matter cosmogony with baryon density $\Omega_B = 0.03$ and $h = 0.5$ and as primordial spectrum, we will use the Harrison-Zeldovich one where $P(k) \propto k$.

At angular scales greater than $(2\Omega^{1/2})^\circ$ a generalized formula for non-flat universes can be obtained (Anile and Motta 1976[5], Abbott and Schaefer 1986[6], Traschen and Eardley 1986[7], Gouda et al. 1991[8]) which gives the CMB anisotropy in terms of gravitational potential fluctuations at recombination with an integrated time-varying gravitational potential along the photon trajectory that depends on curvature. Recently low-Ω models and CMB anisotropies have used a multipole expansion for the temperature anisotropies $\Delta T/T = \Sigma_{l,m}\, a_{l,m}\, Y_{lm}$. To study the *COBE* cross-correlation data only multipoles up to 30 are considered (Sugiyama & Silk 1994[9], Kamionkowski et al. 1994[10]) .

We are interested to test with the most simple model (a CDM cosmogony with a HZ primordial spectrum) to what extend the *COBE*-DMR two-year data imply or not the flatness of the universe, i.e. $\Omega = 1$. To do this, we used two-year data from the most sensitive radiometers (53 and 90 GHz) and we obtained the limits on Q_{rms-PS} and Ω imposed by the cross-correlation of the two maps.

MONTE CARLO SIMULATIONS

Taking into account instrument noise, non-uniform sky coverage, galactic cut, smearing, pixelization scheme, and the DMR beam characteristics we simulated a set of cosmological models defined by Ω, Q_{rms-PS}, and the HZ primordial spectrum. We generated 1600 simulated CMB sky maps using a harmonic expansion of the temperature (Smoot et al. 1991[11]) for each of the 6144 DMR pixels:

$$T(\theta,\phi)=\sum_{l=2}\sum_{m=0}^{l}k[b_{l,m}\cos(m\phi)+b_{l,-m}\sin(m\phi)]N_l^m W_l P_l^m(\cos\theta), \qquad (1)$$

where

$$N_l^m=\left[\frac{(2l+1)(l-m)!}{4\pi(l+m)!}\right]^{1/2}.$$

$k=\sqrt{2}$ for $m\neq 0$ and $k=1$ for $m=0$. $P_l^m(\cos\theta)$ are the associated Legendre polynomials. $b_{l,m}$ are real stochastic, Gaussianly distributed variables with zero mean and model dependent variance $\langle b_{l,m}^2\rangle$ that are easily obtained from the multipole coefficients C_l ($\langle b_{l,m}^2\rangle = C_l/4\pi$) given by Sugiyama & Silk (1994)[9] in their Figure 1a. The weights, W_l, for DMR given by Wright et al.(1994)[12] are used. For each realization two maps are generated by adding to the cosmic signal the noise corresponding to the combination of the channels $\frac{1}{2}$(A+B) of *COBE*-DMR 53 GHz and 90 GHz. The noise is determined by instrument sensitivity and the number of observations per pixel. A small beam smearing correction is applied which accounts for the spacecraft motion during the 1/2 second integration time. The angular cross-correlation for $|b|>20°$ is calculated and binned in 36, 5° bins in the manner of the COBE data.

The standard χ^2 statistic associated to the cross-corelation is

$$\chi^2=\sum_{i=1}^{36}\sum_{j=1}^{36}(\langle C_i\rangle - C_i^{COBE})M_{ij}^{-1}(\langle C_j\rangle - C_j^{COBE}). \qquad (2)$$

$\langle C_i\rangle$ is the average of the cross-correlation for the 800 realizations at bin i. C_i^{COBE} is the cross-correlation for the COBE map at angular scale i. M_{ij} is the covariance matrix calculated with the Monte Carlo realizations:

$$M_{ij}=\frac{1}{N_{realiz.}}\sum_{k=1}^{N_{realiz.}}(C_i^k-\langle C_i\rangle)(C_j^k-\langle C_j\rangle). \qquad (3)$$

ANALYSIS OF THE *COBE*-DMR DATA

We used two-year data from the most sensitive radiometers (53 and 90 GHz). To determine the limits on Q_{rms-PS} and Ω imposed by the *COBE*-DMR cross-correlation, a grid of Monte Carlo data sets was generated for the HZ primordial spectrum, the seven selected Ω values ($\Omega = 0.1, 0.2, 0.3, 0.4, 0.6, 0.8, 1$) and Q_{rms-PS} between 7 and 30 μK in 2 μK steps. For each simulated $\frac{1}{2}$(A+B)53 and 90 GHz map, the cross-correlation was obtained and the χ^2 statistics was calculated. The realizations were 1600, a sufficient number as found after testing for the convergence of the relevant quantities.

For each model (Q_{rms-PS}, Ω) we searched for the model that minimized χ^2 using the Monte Carlo data set. A first result is that the minimum χ^2 leads in the plane (Ω, Q_{rms-PS}) to a cubic relation between these two variables. A qualitative origin of such a relation can be understood in terms of partial cancellation of the two terms which appear associated to anisotropies (potential fluctuations and integrated effect due to curvature). A second result is that the absolute minimum, $\chi^2 = 69.8$, is obtained for ($\Omega = 1$, $Q_{rms-PS} = 23$), although the χ^2 is not significantly worse at lower values of Ω (e.g. $\chi^2 = 70.8$ for $\Omega = 0.4$, $Q_{rms-PS} = 23$). In fact, all (Ω, Q)-values satisfying the previous mentioned relationship are statistically similar (see Cayon et al. 1995[13] for more details).

Finally, considering a likelihood statistics, instead of the χ^2 one, the result is different: in this case the most favoured value is at $\Omega = 0.1$, $Q_{rms-PS} = 15$ but other models are statistically similar. This result, as compared to the one obtained with the χ^2 , can be understood with the relation: $-2 \ln L = \chi^2 + \ln(det M) + n \ln(2\pi)$, where $n = 36$ and M is the correlation matrix, and the second term is (Ω, Q_{rms-PS})-dependent. On the other hand, for $\Omega = 1.0$ the most favoured normalizations for the χ^2 and likelihood cases are $Q_{rms-PS} =$ 23 and 19, respectively. The last number is in agreement with Bennett et al.(1994)[14] likelihood analysis of the $53 \times 90\, GHz$ cross-correlation including the quadrupole (although our multipole coefficients C_l are slightly diferents because we are including CDM). The difference of 4 units in Q, using the two methods, comes from the fact that the function to be minimized in the likelihood case differs from the χ^2 in the term $det(M)$, which is $\propto Q^{4n}$ if the elements of M are dominated by cosmic variance.

CONCLUSIONS

We have made a χ^2 statistical analysis of the angular correlations in the *COBE*-DMR two-year sky maps by Monte Carlo simulation of the temperature fluctuations. We assume an open universe with low-Ω and an early inflationary era, and consider a CDM cosmogony with a Harrison-Zeldovich primordial power spectrum.

We find that the flatness ($\Omega = 1$) of the universe is favoured by the data (i.e. minimum χ^2) although other values of Ω lead to similar χ^2. Therefore, our conclusion is that from the *COBE*-DMR two-year data is not possible to draw any definitive conclusion about the flatness of the universe. A second result is that there exists a relationship between Ω and the quadrupole normalization amplitude, Q_{rms-PS}, obtained for the minimum χ^2.

Another interesting result is that two different statistics, χ^2 and likelihood, lead to different results regarding the absolute minimum. Also for the likelihood technique all models are statistically similar.

ACKNOWLEDGEMENTS

LC, EMG and JLS were supported in part by the European Union, Human Capital and Mobility programme of the European Union, and the Spanish DGICYT; NS would thank for support from JSPS; ST by the European Union contract CI1-CT92-0013; and LC by a Fulbright fellowship. The COBE datasets, developed by NASA Goddard Space flight Center under the guidance of the COBE Science Working Group, were provided by the NSSDC.

REFERENCES

1. Lyth, D. H. & Stewart, E. D., 1990, Phys. Lett. **B252**, 336
2. Ratra, B. & Peebles, P. J. E., 1994, Astrophys.J. **432**, L5
3. Wilson, M. L. 1983, Astrophys.J. **273**, 2
4. Kamionkowski, M. and Spergel, D. N., 1994, Astrophys.J. **432**, 7
5. Anile, A. M. & Motta, S. 1976, Astrophys.J. **207**, 685
6. Abbott, L. F. & Schaefer, R. K. 1986, Astrophys.J. **308**, 546
7. Traschen, J. & Eardley, D. M. 1986, Phys. Rev.**34**, 1665
8. Gouda, N., Sugiyama, N. & Sasaki, M. 1991, Prog. Theo. Phys. **85**, 1023
9. Sugiyama, N. & Silk, J., 1994, Phys. Rev. Lett. **73**, 509
10. Kamionkowski, M., Ratra, B., Spergel, D. N. & Sugiyama, N., 1994, Astrophys.J. **434**, L1
11. Smoot, G.F. et al. 1991, Astrophys.J. **371**, L1
12. Wright, E. L., et al. 1994, Astrophys.J. **420**, 1
13. Cayón, L., Martínez-González, E., Sanz, J. L., Sugiyama, N. & Torres, S. 1995, in preparation
14. Bennett, C. L., Kogut, A., Hinsaw, G. et al. 1994, Astrophys.J. **436**, 423

Topology of the microwave background on scales $1^o - 2^o$ and reionization

P. D. NASELSKY[1] AND D. I. NOVIKOV[2]

1) Rostov State University, Rostov-Don, Russia
2)Astro Space Center of P.N.Lebedev Physical Institute, Moscow, Russia

During the last several years the angular anisotropy of Cosmic Microwave Background (CMB) has been widely investigated.

We discuss the following problems:

First, we propose a new test to determine whether the $\frac{\Delta T}{T}$ field is Gaussian or not. Second, we propose a new method of filtering the Gaussian $\frac{\Delta T}{T}$ field with the subsequent cluster analysis of the filtered field to determine the existence of Sakharov's oscillations[1,2,3]. This allows us to investigate the ionization history of the Universe.

We have calculated the correlation function of $\frac{\Delta T}{T}$ fluctuations for the CDM models with the following parameters $\Omega_b h^2 = 0.001$ (model I) and $\Omega_b h^2 = 0.1$ (model 2). In both cases the total mass density in the Universe is $\Omega = 1$, where $R = R_{LS}$is the scale of the acoustic horizon at the moment of recombination. Model II has the peak-like anomaly (Sakharov's oscillations)[1,2,3] at the angular scale $\theta_{ls} \approx R/\xi_n$. Model I has no such anomaly. The models with the secondary ionization of plasma are examples of the models without the peak-like anomaly (they are analogous to the model I).

We generated two maps for the models I and II (see Fig.1) using the algorithm of Bond and Efstathiou[4].

We will use the following notation $\nu = \alpha\sigma$, where $\sigma = \sqrt{\langle\left(\frac{\Delta T}{T}\right)^2\rangle_{L^2}}$ and α is the crossing parameter. We define an area limited by the line $\frac{\Delta T}{T} = \nu$ inside which $\frac{\Delta T}{T}(x,y) > \nu$ as a "hot" zone, and the "cold" zone has the area with $\frac{\Delta T}{T}(x,y) < \nu$. For each cross level we calculate the value of area of the "cold" - S_c and "hot" - S_h zones. For the highest ($\nu \sim \sigma$) cross levels there is a percolation over "cold" zones (see Fig.1a,b). Here we have $S_h/L^2 \ll 1$ and $S_c/L^2 \sim 1$. L is the angular size of the area ($L \approx 10^o$). Under decreasing of the cross level the topology of light and dark zones begins to change, followed by increasing of S_h/L^2 and decreasing of S_c/L^2. Finally, at the level $\frac{\Delta T}{T}(x,y) = 0$ there arises a point of the phase transition when any small fall of the cross level of the map leads to the percolation over "hot" zones. The critical point of the phase transition corresponds to the condition $\frac{S_h}{L^2} = \frac{S_c}{L^2} = \frac{1}{2}$, $\nu = 0$. This property is determined by the Gaussian character of the distribution of the initial fluctuations only.

In a separate realization of the Gaussian process the condition $\frac{S_h}{L^2} = \frac{S_c}{L^2} = \frac{1}{2}$ arises when ν is slightly different from zero because of the statistical character

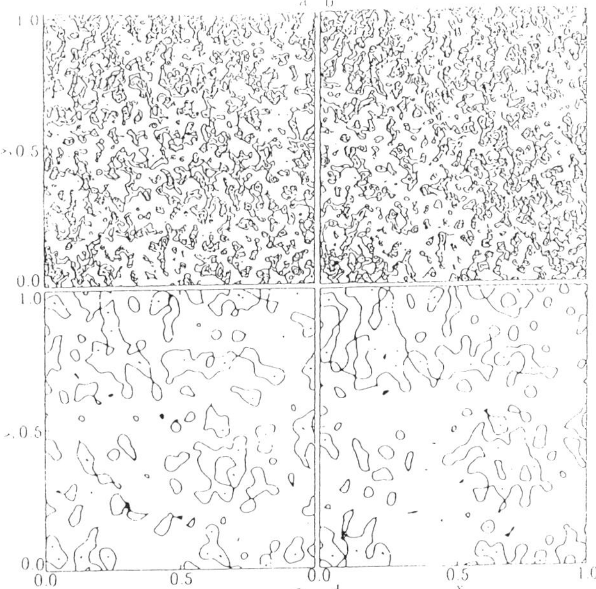

Fig.1

a, b - maps of CMB for oscillating and smoothed spectrum respectively $\theta_c \ll \theta_{ls}$;

c, d - the same as **a, b** but after filtering $\sigma \sim \theta_{ls}$. Asterisks are maxima with height $> 2\sigma$. Dots are maxima with height $> 1.5\sigma$. Lines correspond to $\frac{\Delta T}{T}(x, y) = \nu = 1\sigma$.

of the process. The possible difference of ν from zero can be evaluated as $\frac{\theta_c}{L}$.

Let us consider how the situation changes in the case of the existence of arbitrary noise connected with, for example, unresolved discrete sources. This problem will be described in detail in a separate paper. Here we describe the simplest model of such noise. Thus let the $\frac{\Delta T}{T}(10^o \times 10^o)$ map contain a primordial signal $\frac{\Delta T}{T}(pmdl)$ and a noise component $\frac{\Delta T}{T}(ns)$, then:

$$\frac{\Delta T}{T} = \frac{\Delta T}{T}(pmdl) + \frac{\Delta T}{T}(ns). \tag{1}$$

We introduce the new fluctuation field $\frac{\Delta \tilde{T}}{T}$ in such a way that $\langle \frac{\Delta \tilde{T}}{T}(x, y)\rangle_{L^2} = 0$. Then:

$$\frac{\Delta \tilde{T}}{T}(x, y) = \frac{\Delta T}{T}(pmdl) + \frac{\Delta T}{T}(ns) - \langle \frac{\Delta T}{T}(ns)\rangle_{L^2}. \tag{2}$$

It is clear that for the field $\frac{\Delta \tilde{T}}{T}(x, y)$ the cross section $\frac{\Delta \tilde{T}}{T}(x, y) = 0$ corresponds to the following condition

$$\frac{\Delta T}{T}(pmdl) = \langle \frac{\Delta T}{T}(ns)\rangle - \frac{\Delta T}{T}(ns). \tag{3}$$

Let us assume that the noise "signal" has the form of a narrow peak with height $h \sim \sigma$ and the halfwidth $\theta_n \sim \theta_A$. In this case the averaging (3) over the map leads to the following expression

$$\langle \left[\frac{\Delta T}{T}(pmdl)\right]^2 \rangle_{L^2} = \langle \left[\frac{\Delta T}{T}(ns)\right]^2 \rangle - \langle \frac{\Delta T}{T}(ns)\rangle^2 = M^2. \tag{4}$$

As one can see from (4) the map of $\frac{\tilde{\Delta T}}{T}(x,y) = 0$ is approximately equivalent to the map $\langle\frac{\Delta T}{T}(pmdl)\rangle = M$, and if $M > \frac{\theta_c}{L}$ we can say that additional noise is most probable present in this map.

Our discussion demonstrates the sensitivity of the percolation method for the determination of the Gaussian character of the signal under consideration.

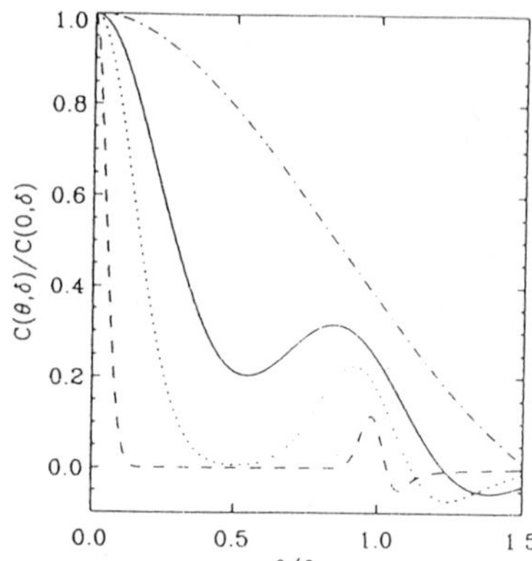

Fig.2 Correlation function for different filters δ: Dashed line for $\delta = 0$, dot line for $\delta = 0.2\theta_{ls}$, solid line for $\delta = 0.3\theta_{ls}$, dashed-dot line for $\delta = \theta_{ls}$.

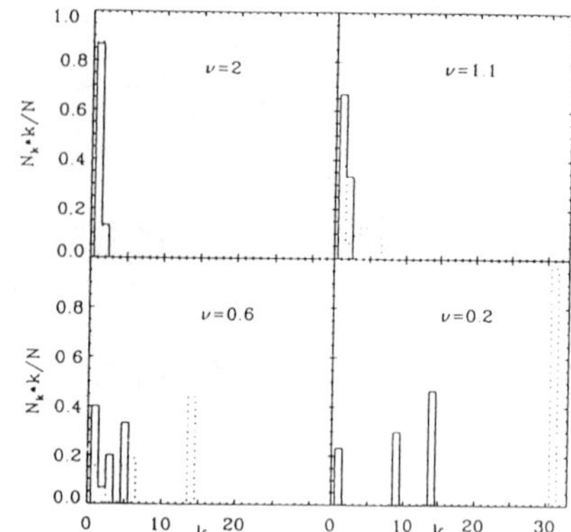

Fig.3 Number of cluster containing k maxima with height equal to 2σ for the different value of ν. Dot line for smoothed spectrum, solid line for oscillating one. N is the number of maxima, N_k the number of clusters with the length equal to k

Now we will investigate some special peculiarities of $\frac{\Delta T}{T}$ maps which are connected with Sakharov's oscillations of the spectrum. Using standard methods of filtering we introduce the smoothed field $\overline{\frac{\Delta T}{T}}(x,y)$;

$$\overline{\frac{\Delta T}{T}}(x,y) = \int_{-\infty}^{\infty}\int_{-\infty}^{\infty} \frac{d\mu d\xi}{8\pi^2\delta^2}\frac{\Delta T}{T}(\mu,\xi)e^{\left(\frac{-[(x-\xi)^2+(y-\mu)^2]}{2\delta^2}\right)} \tag{5}$$

where δ is the half-width of the filter. Using (5) we have

$$\overline{\frac{\Delta T}{T}}(x,y) = \sum_{ij} a_{ij}C^{1/2}(m_{ij})e^{-\frac{m_{ij}^2\delta^2}{2}}\cos\left[\frac{2\pi(ix+jy)}{L} + \varphi_{ij}\right], \tag{6}$$

where $C(m_{ij})$ is the power spectrum; a_{ij} are the independent Gaussian values normalized to 1; $m_{ij} = \frac{2\pi}{L}\sqrt{i^2+j^2}$; φ_{ij} are random values uniformly distributed in $(0, 2\pi)$.

Now we increase the halfwidth of the filter δ from $\delta \simeq \theta_A$ up to $\delta \gg \theta_A$ and plot the $\overline{\frac{\Delta T}{T}}$ maps for different parameters δ. On Fig.2 we plotted the dynamics of varying of the correlation function $C(\theta)$ for different values of δ. As one can see on Fig.2 the second maximum connected with Sakharov's oscillations, disappears when the filter δ becomes equal to $\delta = \theta_{ls}$, where $\theta_{ls} = \frac{1}{2}\frac{R}{\xi_n}$

In order to investigate the influence of the specific shape of the spectrum of $\frac{\Delta T}{T}$ with Sakharov's oscillations on the topology of the distribution of $\frac{\Delta T}{T}$, we generated the filtered maps using Eq.(6) for models I and II. We recall that Sakharov's oscillations are absent in model I and they are present in model II. We use the same Gaussian process for the construction of maps for different kinds of spectra. When δ is much less than all characteristic scales of correlations, the peak of the correlation function corresponding to the oscillating spectrum is much greater than the correlation radius (see Fig.2). Practically, this peak does not influence the topology of $\frac{\Delta T}{T}$ which is determined by the behavior of the correlation function in the vicinity of the correlation radius (see Fig.1). When δ increases to the critical value $\delta = \theta_{ls}$ then the difference of the behaviors of the two correlation functions becomes perceptible on scales comparable with the correlation radius (see Fig.2). After filtering, the topological properties of models I and II become different.

In order to give the qualitative characteristics of the different topologies, we made two maps cross at identical levels ν (maps 1c,d on Fig.1). On these maps, solid curves are the line of level $\overline{\frac{\Delta T}{T}}(x,y) = \nu$, and the asterisks mark all the maxima which are higher than a threshold value of $\nu_{th} > 2\sigma$.

Let us define a cluster of length k as the totality of k maxima within the closed line of level. After performing a statistic analysis of these clusters for different values of ν we see that for the same values of ν and ν_{th} the number of big clusters is essentially greater for the model II than for the model I. We plotted the histograms interpreting these results on Fig.3.

Conclusions

We have shown that percolation and cluster analysis are useful for the determination of the Gaussian nature of $\frac{\Delta T}{T}$ fluctuations and for the detection large scale correlation. This method can be used for processing the real signal for the determination of amount of baryons in the Universe and its ionization history. Of course, one needs for that to analyze a signal together with noise as we mentioned and discussed above and will discussed in some details in a separate paper. The questions connected with the deformation of the topology of primordial $\frac{\Delta T}{T}$ distribution will be discussed in a separate paper also.

References

1. Sakharov, A.D., 1965, JETP, **49**, 345
2. Naselsky, P. & I. Novikov, 1993, ApJ. **413**,14.
3. Jørgensen, H., E. Kotok, P. Naselsky, I. Novikov, 1993, NORDITA preprint n 93/74 (Astr & Astroph. in press)
4. Bond, J.R. & G. Efstathiou, 1987, MNRAS **226**, 655.

CMB ANISOTROPY DUE TO COMPTON SCATTERING IN CLUSTERS OF GALAXIES

S. COLAFRANCESCO[1], P. MAZZOTTA[2], Y. REPHAELI[3] & N. VITTORIO[2]

[1] *Osservatorio Astronomico di Roma*

[2] *Dip. di Fisica, Università di Roma "Tor Vergata"*

[3] *Center for Particle Astrophysics, University of California, Berkeley*

1. BACKGROUND

Compton scattering of the cosmic microwave background (CMB) photons by the hot gas in clusters of galaxies results in a systematic shift of the CMB photons from the Rayleigh-Jeans to the Wien side of the spectrum, a process known as the Sunyaev-Zel'dovich (SZ) effect[1]. We consider here the effect that this scattering produces on the angular distribution of the CMB. The main input for the calculation of the SZ effect is the cluster Comptonization parameter, $y_c = \int (kT/mc^2) n\sigma_T dl$, where T is the gas temperature and n the electron density; here m is the electron mass, c is the speed of light, σ_T is the Thomson cross section, and the integral is over a line of sight through the cluster.

Predictions of the CMB spectral distorsions and anisotropies due to galaxy clusters involve a detailed description of the basic properties of intracluster (IC) gas. The superposed effect due to all clusters is then determined by the cosmological model, the cluster abundance, and their dynamical and gas evolutions.

The cluster mass distribution (MD) is fitted (see CV[2] and references therein) by the PS[3] multiplicity function: $N(M,z) \propto \mathcal{I}(b\delta_v) M^{-2+a} exp[-0.5b^2\delta_v^2(M/M_*)^{2a}]$, where $a = (n+3)/6$, n is the index of the fluctuation spectrum, $P(k) = Ak^n$, and $M_* \sim 10^{15}h^{-1}\,M_\odot$. We assume $\delta_v = 2.2$ to select those fluctuations which have already virialized at redshift z. We fix the amplitude A of the primordial power spectrum by requiring $\sigma_\rho = 1/b$, where b is the biasing factor. Then, $N(M,z)$ depends, in a given model, mainly upon the combination $\delta_v b$, and we use here the values derived[2] from the observed abundance of X-ray clusters.

Galaxy clusters are extended X-ray sources with gas density profiles fitted by $n(r) = n_o[1+(r/r_c)^2]^{-3\beta/2}$, where n_o is the central electron density and r_c is a core radius. Observations[4] yield $\beta \sim 0.6 \div 0.8$; the value $\beta = 2/3$ is used for its analytic convenience. Gas core radii in local rich clusters are observed[4] to be in the range $r_c \sim 0.1 \div 0.2h^{-1}$ Mpc. In hierarchical clustering scenarios the following scaling for spherical systems holds:

$$r_c = r_{c,15}\frac{M_{15}^{1/3}}{(1+z)}\,, \qquad (1)$$

where $r_{c,15} = 0.15h^{-1}$ Mpc is the core radius of a local cluster with $M_{15} = 1$ ($M_{15} = M/10^{15}h^{-1}M_\odot$). The temperature of an isothermal gas in hydrostatic equilibrium within a virialized cluster extending out to a radius $10r_c$ ($\simeq$ the Abell radius) is:

$$T \approx (1.8\cdot 10^8 K)\, M_{15}\left(\frac{r_{c,15}}{r_c}\right) \qquad (2)$$

The central mass density can be expressed in terms of n_o and the fraction $f \equiv M_g/M$ of mass in IC gas: $\rho_o = n_o(1+X)m_p/(2f)$, so that[5]:

$$n_o \approx (1.2\cdot 10^{-3}cm^{-3})\left(\frac{f}{0.1}\right)(1+z)^3 \tag{3}$$

(In eq.2 and below we assume the mean molecular weight for solar abundances, $\mu \equiv 4/(5X+3) = 0.62$, for a hydrogen mass fraction of $X = 0.69$). Assuming f =const, may be unrealistic. In fact, the IC gas is not entirely primordial, as shown by its enriched chemical composition[6]. Moreover, the infall of external gas onto the cluster may further increase the gas fraction[7]. Thus, given uncertainties in the gas evolution we use the simple parameterization (following CCM and CV) $f = f_o M_{15}^{\eta}(t/t_o)^{\xi}$, where t is the age of the universe at redshift z and $f_o \simeq 0.1$, in local rich clusters. Comparison of theoretical X-ray luminosity functions (XRLF) with the EMSS data[8,9]. yield $\xi \simeq 1.2$. Also, $\eta \approx 0.1$ reproduces reasonably well the distribution of the gas masses in local clusters[2].

2. TEMPERATURE ANISOTROPIES

The scattering of the CMB in clusters results in an intensity change $\Delta I = 2(kT_r)^3 y_C g(x)/(hc)^2$, where T_r is the CMB temperature, $x = h\nu/kT_r$, and $g(x) = \frac{x^4e^x}{(e^x-1)^2}s(x)$ is a spectral function, with $s(x) = [x\coth(x/2) - 4]$. In the RJ limit ($x \ll 1$), $\Delta I/I = \Delta T_r/T_r = -2y_C$. The CMB temperature change across a single cluster is

$$\Delta(\hat{\gamma}) \equiv \frac{\Delta T_r}{T_r}(\hat{\gamma}) = \Delta_o \cdot \zeta(|\hat{\gamma}-\hat{\theta}|, M, z) \tag{4}$$

where $\Delta_o = (k_B\sigma_T/m_ec^2)T_o n_o r_c s(x)$, is the central value and the function $\zeta = \frac{1}{n_o r_c}\int_{l_{min}}^{l_{max}} n(l)dl$ describes the variation in Δ when observed along a line of sight $\hat{\gamma}$ different from $\hat{\theta}$, the line of sight intersecting the cluster center[5]. For a fiducial rich cluster at the current epoch, with $M = 10^{15}h^{-1}M_{\odot}$, $r_c = 0.15h^{-1}$ Mpc, $f_o = 0.1$, $\beta = 2/3$, and $h = 0.5$, we compute $\Delta_o \approx 4.5\cdot 10^{-5}$ in the RJ limit.

We derive the predicted CMB anisotropy induced by the cluster population averaging the single cluster effect in eq.(4) over the sky and over the statistical ensemble of the possible realizations of the cluster distribution in the (M, z) space:

$$\Delta_{rms} = \int r^2\frac{dr}{dz}dz\int dMN(M,z)\Delta_o(M,z)\bar{\zeta}(M,z) \tag{5}$$

where $\bar{\zeta}(M,z)$ is the sky averaged brightness profile. The variance of a single or double subtracted CMB temperature difference is given by $2[C(0,\sigma_b) - C(\alpha,\sigma_b)]$, or $1.5C(0,\sigma_b) - 2C(\alpha,\sigma_b) - 0.5C(2\alpha,\sigma_b)$, respectively. Here $C(\alpha,\sigma_b)$ is the correlation function of the power spectrum of the process $\Delta(\hat{\gamma})$, α is the angle between two lines of sight and σ_b is the dispersion of a Gaussian approximating the angular response of the receiver.

We evaluated Δ_{rms} (see Fig.1) in the following flat LSS models: standard CDM ($n = 1$, continuoue curves), a CDM model with a tilted ($n = 0.8$, dotted curves) power spectrum, and flat hybrid models ($n = 1$) with $\Omega_\nu = 0.1$ (dot-dashed curves) and $\Omega_\nu = 0.3$ (dashed curves). In Fig.1 we plot Δ_{rms} as a function of the beam-switching angle α, for different values of the antenna beam σ_b. The level of Δ_{rms} is

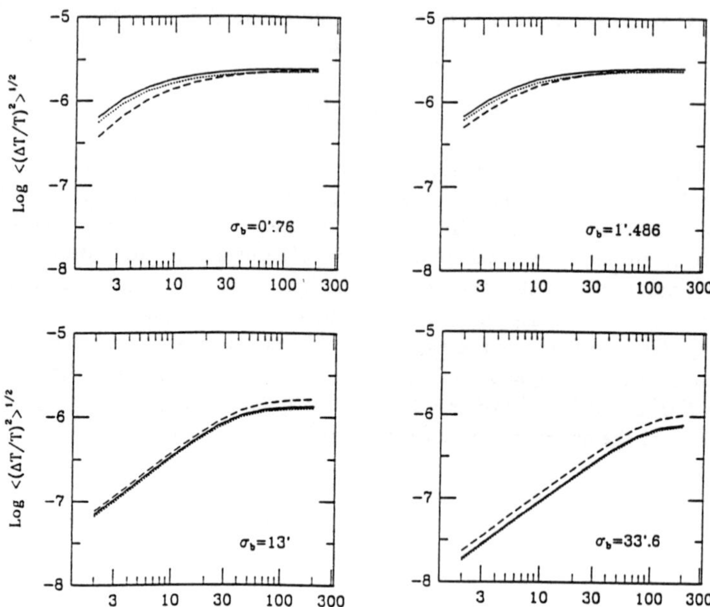

Figure 1: The predictions for Δ_{rms} in different scenarios of structure formation

not greatly different from model to model: this is mainly due to our normalization of the parameters $\delta_v b$ and ξ to the observed distribution of X-ray clusters. For small σ_b the standard CDM model produces more anisotropy than hybrid models, the reverse is true for large σ_b. This is expected, because the CDM model has more power on small scales than hybrid models. The intrinsic coherence angle of the cluster induced anisotropy is found to be [5] $\alpha_c \gtrsim 5'$; this suggests that most of the anisotropy is determined by massive ($M \approx 10^{14}h^{-1}M_{\odot}$), nearby ($z \lesssim 0.3$) clusters, as the characteristic scale of the anisotropy is related to the typical size of these, more dominant clusters. This behaviour is due to two main reasons. From one hand, we are considering LSS models where structure formation is by itself a late event. On the other hand, we are modeling the IC gas evolution by the X-ray cluster data, which seem to indicate[8,9] a sharp decline in the number density of bright clusters already at $z \approx 0.3$.

The present results were obtained with a particular choice of the basic parameters which determine cluster dynamics and gas evolution. While this choice is reasonable, it is by no means unique because of the inherent uncertainty in the description of the cluster evolution. The level of anisotropy depends linearly on f_o. Thus, any change in f_o yields a similar change in Δ_{rms}. Moreover, our parametrization of the gas evolution is based on fits to the X-ray cluster data, which indicate a sharp decline of bright clusters at $z \gtrsim 0.4$. Since this result is preliminary, we also consider the extreme case f = const Consequently, removing gas evolution increases the contribution of farther groups and clusters at $z \gtrsim 0.4$ and this results in an anisotropy higher by a factor ~ 2. We also considered uncertainties in modelling the cluster structure. Changing the scaling and the normalization in eq.(1) introduces variations by less than a factor ~ 1.5 in the final results. Note also that Comptonization of the CMB occurs in various classes of objects, from galactic halos to superclusters of galaxies. The gas pressure (see eqs. 2 and 3) decreases as the system gets smaller, while $N(M,z)$ increases. Because of the dominance of the pressure, the contribution to Δ_{rms} of galactic halos can be ignored[5]. Due to the inherent uncertainties in the prediction of the cluster structure and distribution, we have calculated[5] the anisotropy in the most optimal (but still within the allowed range in the context of our theoretical approach) combination of parameters which maximizes the anisotropy. Specifically, we took $M_{min} = 5 \cdot 10^{12} h^{-1} M_\odot$, $r_{c,15} = 0.1 h^{-1}$ Mpc, and we assumed $\eta = 0$ and the minimum possible level of gas evolution that is still consistent, given the uncertainties in the current data on the XRLF, *i.e.* $\xi = 1$. With this choice of parameters, $\Delta_{rms} \sim 5 \times 10^{-6}$, if measured with a Gaussian beam $\sigma_b = 0', 76$ (as in the OVRO experiment).

3. CONCLUDING REMARKS

Our calculated low-level of anisotropy, $\Delta_{rms} \approx$ few $\cdot 10^{-6}$, (quite independent of the considered model, see Fig.1) is due primarily to the assumed level of IC gas evolution. Higher levels of anisotropy were obtained in previous works[10,11,12], $\Delta_{rms} \sim 10^{-5}$; this is mainly due to the fact that no gas evolution was assumed, although uncertainties in the normalizations of the MD and of the value of f_o also account for some of the differences. In any case, secondary anisotropies are $\sim 10\%$ of the primary ones ($\lesssim 30\%$ maximing the effect, see §2). However, making a closer comparison between primary and secondary anisotropy requires a full Monte-Carlo simulations of the SZ effect, because of its intrinsic non-Gaussian nature. Also, considering a late reheating of the intergalactic medium, the primary anisotropy at scales smaller than few degrees is strongly suppressed[5]. So, our results can be interpreted as the minimal anisotropy at sub-degree angular scales.

REFERENCES

1. Sunyaev, R.A. & Zeldovich, Y.B., 1972, Comm. Astrophys. Sp. Phys., 4, 173
2. Colafrancesco, S. & Vittorio, N.: 1994, ApJ, 422, 443 (CV)
3. Press, W.H., & Schechter, P.: 1974, ApJ, 187, 425 (PS)
4. Jones, C. & Forman, W.: 1991, in *Clusters and Superclusters of Galaxies*, A.C. Fabian *et al.* eds., (Cambridge: Cambridge University Press), p. 49
5. Colafrancesco, Mazzotta, Rephaeli & Vittorio 1994, ApJ, 433, 454
6. Edge, A.C., & Stewart, G.C. 1991, MNRAS, 252, 414
7. Cavaliere, A., Colafrancesco, S., & Menci, N. 1993, ApJ, 415, 50
8. Gioia, I.M. et al. 1990, ApJ, 356, L35
9. Henry, J.P. *et al.* 1992, ApJ, 386, 408
10. Cole, S., & Kaiser, N. 1988, MNRAS, 233, 637
11. Markevitch, M. *et al.* 1991, ApJ, 378, L33
12. Makino, N. & Suto, Y. 1993, ApJ, 405, 1

Index of Contributors

Ansari, R., 608-611
Aschenbach, B., 196-199
Asseo, E., 261-264
Atrio-Barandela, F., 680-683
Aurière, M., 608-611

Bahcall, N. A., 636-649
Bailes, M., 279-282
Baillon, P., 608-611
Ballet, J., 312-315
Barrow, J. D., 706-709
Becker, W., 250-256
Bell, J., 279-282
Bennett, K., 226-231, 397-400
Bertsch, D. L., 275-278, 421-424
Bicknell, G. V., 530-533
Bisnovatyi-Kogan, G. S., 340-343
Blanchet, L., 507-511
Blandford, R., 534-537
Blinnikov, S., 352-355
Bloemen, H., 226-231, 397-400
Bocko, M., 485-488
Böhringer, H., 67-86
Boratav, M., 450-453
Bouchet, L., 312-315
Bouquet, A., 608-611
Bowyer, S., 241-245
Briggs, M. S., 416-420, 434-439
Burrows, A., 375-381
Bykov, A. M., 406-409

Caraveo, P. A., 246-249
Cayón, L., 680-683, 714-717
Cen, R., 563-567
Centrella, J., 503-506
Chakrabarti, S. K., 546-549
Churazov, E., 312-315
Colafrancesco, S., 722-725
Collmar, W., 226-231, 425-428
Connors, A., 425-428
Cordes, J. M., 279-282
Coupinot, G., 608-611
Courvoisier, T. J-L., 517-520
Coutures, C., 608-611
Crampton, D., 612-615, 620-623
Czerny, B., 538-541

Da Costa, L. N., 654-657
Danzmann, K., 481-484
Davies, R. D., 672-675
Dawson, B. R., 460-463
Deal, K. J. 308-311
Dean, T., 401-405
Diehl, R., 226-231, 384-387
Dingus, B. L., 421-424
Di Stefano R., 328-331
Dominik, C., 348-351
Durrer, R., 688-691
Dyachkov, A., 312-315
Dyer, C. C., 583-586, 587-590, 604-607

Eckart, A., 38-55
Ellis, J., 170-187

Fabbiano, G., 624-627
Fabian, A. C., 521-525
Ferguson, H. C., 658-663
Ferlet, R., 56-66
Fichtel, C. E., 221, 222-225, 275-278, 421-424
Finn, L. S., 489-492
Finogenov, A., 312-315
Fishman, G. J., 232-235, 308-311, 410
Fomin, Yu. A., 477-480
Freedman, W. L., 192-195
Freudling, W., 654-657

Garcia-Senz, D., 352-355
Genzel, R., 38-55
Geppert, U., 287-290
Ghesquière, C., 608-611
Gilfanov, M., 312-315
Gioia, I., 568-572
Giovanelli, R., 654-657
Giraud-Héraud, Y., 608-611
Glendenning, N. K., 303-307
Goldwurm, A., 312-315
Gondolo, P., 608-611
Gottlöber, S., 650-653
Greiner, J., 425-428, 429-433
Gutiérrez, C. M., 672-675

Hammer, F., 568-572, 612-615, 616-619, 620-623
Hancock, S., 672-675
Hanlon, L. O., 425-428
Harmon, B. A., 308-311
Hartman, R. C., 421-424
Hartmann, D. H., 388-391, 434-439
Hasinger, G., 200-205
Hawkins, M. R. S., 596-599
Hayes, J., 375-381
Haynes, M. P., 654-657
Hecquet, J., 608-611
Helbig, P., 600-603
Hermsen, W., 226-231, 397-400, 425-428
Hillebrandt, W., 352-355
Hirano, S., 295-298
Hirotani, K., 542-545
Hoffman, R. D., 388-391
Höflich, P., 348-351, 360-367
Hu, W., 697-700
Hunter, S. D., 421-424

Iwamoto, K., 360-367

Janka, H-T., 269-274, 368-374

Kahabka, P., 324-327
Kanbach, G., 275-278, 421-424
Kaplan, J., 608-611
Karim, M., 485-488
Kawai, N., 316-319
Khavenson, N., 312-315
Khokhlov, A., 348-351
Khristiansen, G. B., 477-480
Kippen, R. M., 425-428
Kirsten, T. A., 1-20
Kniffen, D. A., 275-278, 421-424
Kokkotas, K. D., 493-497
Kouveliotou, C., 411-415
Kovtunenko, V., 312-315
Kozlovsky, B. 392-396
Krabbe, A., 38-55

Królak, A., 493-497
Kronberg, P. P., 604-607
Kuiper, L., 226-231, 425-428
Kulikov, G. V., 477-480
Kundt, W., 265-268
Kurfess, J. D., 236-240
Kuulkers, E., 344-347

Laguna, P., 692-696
Landry, S., 583-586
Lasenby, A. N., 672-675
Le Fèvre, O., 568-572, 612-615, 620-623
Leibundgut, B., 356-359
Levinson, A., 534-537
Lichti, G. G., 226-231, 397-400
Lilly, S. J., 612-615, 620-623
Lin, Y. C., 421-424
Lineweaver, C. H., 676-679
Lingenfelter, R. E., 392-396
Loeb, A., 558-562
Lorenz, E., 472-476
Loska, Z., 538-541
Lovelace, R. V. E., 554-557
Luppino, G., 568-572
Lyth, D. H., 701-705

Magneville, C., 664-667
Mandrou, P., 312-315
Marchese, L. E., 485-488
Marleau, F., 587-590
Martín, E. L., 332-335
Martinez-González, E., 714-717
Mayer-Hasselwander, H. A., 275-278, 421-424
Mazzotta, P., 722-725
McConnell, M. L., 226-231, 397-400, 425-428
McMillan, S., 503-506
Melchior, A. L., 608-611
Mészáros, P., 440-445
Michelson, P. F., 421-424
Mirabel, I. F., 21-37
Mitsuda, K., 213-216
Moniez, M., 608-611
Morfill, G. E., xv-xvi
Morris, D. J., 397-400
Müller, D., 454-459
Müller, E., 269-274, 348-351, 368-374, 498-502
Mushotzky, R. F., 521-525

Nandra, K., 521-525
Naselsky, P. D., 718-721
Niemeyer, J., 352-355
Nolan, P. L., 275-278, 421-424
Nomoto, K., 360-367
Novikov, D. I., 718-721

O'Flaherty, K. S., 425-428
Ohashi, T., 217-220
Ostriker, J. P., 563-567

Paciesas, W. S., 308-311
Palmer, J. H., 587-590
Paul, J., 312-315
Pavlov, G. G., 291-294
Pendleton, G. N., 434-439
Perley, R. A., 604-607
Pesce, E., 632-635
Petre, R., 208-212
Picat, J. P., 608-611
Pinto, P. A., 388-391
Podsiadlowski, P., 283-286
Pols, O. R., 360-367
Ptuskin, V. S., 464-467

Qin, L., 291-294, 299-302

Rachen, J. P., 468-471
Ramanamurthy, P. V., 275-278
Ramaty, R., 392-396
Rappaport, S., 328-331
Rebolo, R., 672-675
Rees, M. J., 283-286
Rephaeli, Y., 722-725
Reynolds, C. S., 521-525
Robinson, D. R. T., 658-663
Rodriguez, L. F., 21-37
Rola, C., 616-619
Romanova, M. M., 554-557
Roques, J-P., 312-315
Röser, H-J., 604-607
Ruderman, M., 283-286
Ryan, J. M., 226-231, 397-400, 425-428

Salzer, J. J., 654-657
Sanz, J. L., 714-717
Sasaki, M., 512-516
Sasorov, P., 352-355
Schäfer, G., 493-497
Schneid, E. J., 421-424
Schönfelder, V., 226-231, 382-383, 397-400, 425-428
Schroeder, P. C., 279-282
Scott, D., 668-671
Seitz, C., 578-582
Seitz, S., 591-595
Setti, G., 110 126
Shaham, J., 336-339
Shanks, T., 658-663
Shaver, P. A., 87-109
Shibanov, Yu. A., 291-294
Shibata, M., 512-516
Shibazaki, N., 295-298
Silk, J., 680-683
Soucail, G., 573-577, 608-611
Spyromilio, J., 356-359
Sreekumar, P., 275-278, 421-424
Stasińska, G., 616-619
Steinmetz, M., 628-631
Stockman, H. S., 188-191
Strong, A., 226-231
Sugiyama, N., 697-700, 714-717
Sunyaev, R., 312-315
Supper, R., 320-323
Suzuki, T., 360-367
Swordy, S., 454-459

Tagoshi, H., 512-516
Tanaka, T., 512-516
Tanaka, Y., 206-207
Tanvir, N. R., 658-663
Tegmark, M., 684-687
Thielheim, K. O., 257-260
Thompson, D. J., 275-278, 421-424
Thorne, K. S., 127-152
Timmes, F. X., 388-391
Tomimatsu, A., 542-545
Torres, S., 714-717
Tresse, L., 612-615, 616-619
Trümper, J. E., xv-xvi
Tsuruta, S., 291-294, 299-302

Tuluie, R., 692-696
Turner, E. L., 563-567
Turner, M. S., 153-169

Ulmer, M. P., 279-282
Urpin, V., 287-290

Vadas, S. L., 710-713
van den Heuvel, E. P. J., 360-367
van der Klis, M., 344-347
Varendorff, M., 425-428
Vargas, M., 312-315
Vedrenne, G., 312-315
Vietri, M., 632-635
Vikhlinin, A., 312-315
Vittorio, N., 722-725
von Ballmoos, P., 401-405
von Montigny, C., 421-424

Wagner, S., 526-529
Wambsganss, J., 563-567
Watson, R. A., 672-675
Weaver, T. A., 388-391
Weber, F., 303-307
Wegner, G., 654-657
Wheeler, J. C., 348-351
Wilson, C. A., 308-311
Winkler, C., 226-231, 401-405, 425-428
Woosley, S. E., 352-355, 388-391, 446-449

Yamaoka, H., 360-367

Zakharov, A., 550-553
Zavlin, V. E., 291-294
Zhang, G., 485-488
Zhang, S. N., 308-311
Zhou, Z-H., 688-691
Zhuge, X., 503-506
Zwerger, T., 498-502
Życki, P. T., 538-541